EXPLORING THE UNKNOWN

For sale by the Superintendent of Documents, U.S. Government Printing Office
Internet: bookstore.gpo.gov Phone: (202) 512-1800 Fax: (202) 512-2250
Mail: Stop SSOP, Washington, DC 20402-0001

ISBN 0-16-061774-X

NASA SP-2001-4407

EXPLORING THE UNKNOWN

Selected Documents in the History of the U.S. Civil Space Program

Volume V: Exploring the Cosmos

John M. Logsdon, Editor
with Amy Paige Snyder, Roger D. Launius,
Stephen J. Garber, and Regan Anne Newport

The NASA History Series

National Aeronautics and Space Administration
NASA History Office
Office of Policy and Plans
Washington, D.C. 2001

Library of Congress Cataloguing-in-Publication Data

Exploring the Unknown: Selected Documents in the History of the U.S. Civil Space Program/ John M. Logsdon, editor ... [et al.]
p. cm.—(The NASA history series) (NASA SP: 4407)

Includes bibliographical references and indexes.
Contents: v. 1. Organizing for exploration
1. Astronautics—United States—History. I. Logsdon, John M., 1937–
II. Series III. Series V. Series: NASA SP: 4407.
TL789.8.U5E87 1999 96-9066
387.8'0973-dc20 CIP

Dedicated to
Hugh L. Dryden
and
Homer E. Newell

Fathers of the NASA Space Science Program

Contents

Chapter Three

Acknowledgments

This volume is the fifth in a series that had its origins more than a decade ago. The individuals involved in initiating the series and producing the initial four volumes have been acknowledged in those volumes—Volume I: *Organizing for Exploration* (1995), Volume II: *External Relationships* (1996), Volume III: *Using Space* (1998), and Volume IV: *Accessing Space* (1999). Those acknowledgments will not be repeated here.

We owe thanks to the individuals and organizations that have searched their files for potentially useful materials, and for the staffs at various archives and collections who have helped us locate documents. Among them are Alan Bunner, George Carruthers, Riccardo Giacconi, Robert Jastrow, William Kraushaar, John Mather, Michael Minovitch, Robert Mitchell, Charles Pellerin, Nancy Grace Roman, David H. Smith, the late Gerald Soffen, Harvey Tananbaum, Harley Thronson, and Gerald Wasserburg. The staffs of the National Archives and Record Administration, the Smithsonian Institution Archives/National Air and Space Museum Garber Facility, the National Academy of Sciences, and the American Institute of Physics library were particularly helpful. At the Space Policy Institute, research assistant Amy Paige Snyder made so many contributions to this volume that she deservedly has been fisted as co-editor. Graduate students Holly Carter, Brian Dewhurst, Erin Hatch, and Becky Ramsey also helped in the preparation of the volume. Michelle Treistman and Kim Holiday of the Institute staff supported the effort throughout.

My thanks go to all those mentioned above, and again to those who helped get this effort started almost a decade ago and who have been involved along the way.

John M. Logsdon, George Washington University

****

There are numerous people at NASA associated with historical study, technical information, and the mechanics of publishing who helped in myriad ways in the preparation of this documentary history. Regan Newport prepared the biographical appendix. Stephen J. Garber helped in the overall editing and production. M. Louise Alstork and Nadine J. Andreassen of the NASA History Division performed editorial and proofreading work on the project and prepared the index; and the staffs of the NASA Headquarters Library, the Scientific and Technical Information Program, and the NASA Document Services Center provided assistance in locating and preparing for publication the documentary materials in this work. The NASA Headquarters Printing and Design Office developed the layout and handled printing. Specifically, we wish to acknowledge the work of Jonathan L. Friedman, Bryan Elrod, Joel Vendette, and Timothy E. Lara for their editorial and design work. In addition Stanley Artis, Michael Crnkovic, and Warren Owens saw the book through the publication process. Thanks are due to them all.

Roger D. Launius
NASA Chief Historian

Introduction

One of the most important developments of the twentieth century has been the movement of humanity into space with machines and people. The underpinnings of that movement—why it took the shape it did; which individuals and organizations were involved; what factors drove a particular choice of scientific objectives and technologies to be used; and the political, economic, managerial, and international contexts in which the events of the space age unfolded—are all important ingredients of this epoch transition from an Earthbound to a spacefaring people. This desire to understand the development of spaceflight in the United States sparked this documentary history series.

The extension of human activity into outer space has been accompanied by a high degree of self-awareness of its historical significance. Few large-scale activities have been as extensively chronicled so closely to the time they actually occurred. Many of those who were directly involved were quite conscious that they were making history, and they kept full records of their activities. Because most of the activity in outer space was carried out under government sponsorship, it was accompanied by the documentary record required of public institutions, and there has been a spate of official and privately written histories of most major aspects of space achievement to date. When top leaders considered what course of action to pursue in space, their deliberations and decisions often were carefully put on the record. There is, accordingly, no lack of material for those who aspire to understand the origins and evolution of U.S. space policies and programs.

This reality forms the rationale for this series. Precisely because there is so much historical material available on space matters, the National Aeronautics and Space Administration (NASA) decided in 1988 that it would be extremely useful to have available to scholars and the interested public a selective collection of many of the seminal documents related to the evolution of the U.S. civilian space program. While recognizing that much space activity has taken place under the sponsorship of the Department of Defense and other national security organizations, within the U.S. private sector, and in other countries around the world, NASA felt that there would be lasting value in a collection of documentary material primarily focused on the evolution of the U.S. government's civilian space program, most of which has been carried out since 1958 under the Agency's auspices. As a result, the NASA History Office contracted with the Space Policy Institute of George Washington University's Elliott School of International Affairs to prepare such a collection. This is the fifth volume in the documentary history series; three additional ones detailing programmatic developments with respect to aspects of space science not covered in the current volume, and to human spaceflight, will follow.

The documents collected during this research project were assembled from a diverse number of both public and private sources. A major repository of primary source materials relative to the history of the civil space program is the NASA Historical Reference Collection of the NASA History Office located at the Agency's Washington, D.C., headquarters. Project assistants combed this collection for the "cream" of the wealth of material housed there. Indeed, one purpose of this series from the start was to capture some of the highlights of the holdings at headquarters. Historical materials housed at the other NASA installations, at institutions of higher learning, and at presidential libraries were other sources of documents considered for inclusion, as were papers in the archives of individuals and firms involved in opening up space for exploitation.

Copies of the documents included in this volume in their original form will be deposited in the NASA Historical Reference Collection. Another complete set of project materials is located at the Space Policy Institute at George Washington University. These materials in their original form are available for use by researchers seeking additional information about the evolution of the U.S. civil space program or wishing to consult the documents reprinted herein in their original form.

The documents selected for inclusion in this volume are presented in three major sections, each covering a particular aspect of the origins, evolution, and execution of the U. S. space science program. Chapter 1 deals with the origins, evolution, and organization of the space science program. Chapter 2 deals with solar system exploration. Chapter 3 deals with NASA's astronomy and astrophysics efforts. Volume I in this series covered the antecedents to the U. S. space program, as well as the origins and evolution of U.S. space policy and of NASA as an institution. Volume II dealt with the relations between the civilian space program of the United States and the space activities of other countries, the relationship between the U. S. civilian and national security space and military efforts, and NASA's relations with industry and academic institutions. Volume III provided documents on satellite communications, remote sensing, and the economics of space applications. Volume IV covered various forms of space transportation. Future volumes will cover solar and space physics, earth science, and life and microgravity science (Volume VI), and human spaceflight (Volumes VII and VIII).

Each chapter in the present volume is introduced by an overview essay. In the main, these essays are intended to introduce and complement the documents in the chapter and to place them in a chronological and substantive context. Each essay contains references to the documents in the chapter it introduces, and may also contain references to documents in other chapters of the collection. These introductory essay s are the responsibility of their individual authors, and the views and conclusions contained therein do not necessarily represent the opinions of either George Washington University or NASA.

The documents included in each section were chosen by the project team in concert with the essay writer from those assembled by the research staff for the overall project. The contents of this volume emphasize primary documents or long-out-of-print essays or articles and material from the private recollections of important actors in shaping space affairs. The contents of this volume thus do not comprise in themselves a comprehensive historical account; they must be supplemented by other sources, those both already available and to become available in the future.

The documents included in each section are arranged chronologically, with the exception that closely related documents are grouped together. Each document is assigned its own number in terms of the chapter in which it is placed. As a result, the first document in the third chapter of this volume is designated "Document III-1." Each document or group of related documents is accompanied by a headnote setting out its context and providing a background narrative. These headnotes also provide specific information about people and events discussed. We have avoided the inclusion of explanatory notes in the documents themselves and have confined such material to the headnotes.

The editorial method we adopted for dealing with these documents seeks to preserve spelling, grammar, paragraphing, and use of language as in the original. We have sometimes changed punctuation where it enhances readability. We have used the designation [not included, or omitted] to note where sections of a document have not been included in this publication, and we have avoided including words and phrases that had been deleted in the original document unless they contribute to an understanding of what was going on in the mind of the writer in making the record. Marginal notations on the original documents are inserted into the text of the documents in brackets, each clearly marked as a marginal comment. Except insofar as illustrations and figures are necessary to understanding the text, those items have been omitted from this printed version. Page numbers in the original document are noted in brackets internal to the document text. Copies of all documents in their original form, however, are available for research by any interested person at the NASA History Office or the Space Policy Institute of George Washington University.

We recognize that there are certain to be quite significant documents left out of this compilation. No two individuals would totally agree on all documents to be included from the many we collected, and surely we have not been totally successful in locating al relevant records. As a result, this documentary history can raise an immediate question from its users: why were some documents included while others of seemingly equal importance were omitted? There can never be a fully satisfactory answer to this question. Our own criteria for choosing particular documents and omitting others rested on three interrelated factors:

- Is the document the best available, most expressive, most representative reflection of a particular event development important to the evolution of the space program?

- Is the document not easily accessible except in one or a few locations, or is it included (for example, in published compilations of presidential statements) in reference sources that are widely available and thus not a candidate for inclusion in this collection?

- Is the document protected by copyright, security classification, or some other form of proprietary right and thus unavailable for publication?

As general editor of this volume, I was ultimately responsible for the decisions about which documents to include and for the accuracy of the headnotes accompanying them.

It has been an occasionally frustrating but consistently exciting experience to be involved with this undertaking. My associates and I hope that those who consult it in the future find our efforts worthwhile.

John M. Logsdon
Director
Space Policy Institute
Elliott School of International Affairs
George Washington University

Biographies of Volume V Contributors

Stephen J. Garber is a policy analyst in the NASA History Office, Washington, D.C. He is the author of numerous articles on aerospace history and space policy.

Roger D. Launius is Chief Historian of the National Aeronautics and Space Administration, Washington, D.C. He has produced several books and articles on aerospace history, including *Innovation and the Development of Flight* (Texas A&M University Press, 1999); *NASA & the Exploration of Space* (Stewart, Tabori, & Chang, 1998); *Frontiers of Space Exploration* (Greenwood Press, 1998); *Organizing for the Use of Space: Historical Perspectives on a Persistent Issue* (Univelt, Ina, AAS History Series, Volume 18, 1995), editor; *NASA: A History of the U.S. Civil Space Program* (Krieger Publishing Co., 1994); *History of Rocketry and Astronautics: Proceedings of the Fifteenth and Sixteenth History Symposia of the International Academy of Astronautics* (Univelt, Inc., AAS History Series, Volume 11, 1994), editor; *Apollo: A Retrospective Analysis* (Monographs in Aerospace History, No. 3, 1994); and *Apollo II at Twenty-Five*, electronic picture book issued on computer disk by the Space Telescope Science Institute, Baltimore, Maryland, 1994.

John M. Logsdon is Director of the Space Policy Institute of George Washington University's Elliott School of International Affairs, where he is also Professor of Political Science and International Affairs and Director of the Center for International Science and Technology Policy. He holds a bachelor of science degree in physics from Xavier University and a Ph.D. in political science from New York University. He has been at George Washington University since 1970, and previously taught at The Catholic University of America. He is also a faculty member of the International Space University. He is an elected member of the International Academy of Astronautics and the Board of Trustees of the International Space University and Chair of the Advisory Council of the Planetary Society. Dr. Logsdon has lectured and spoken to a wide variety of audiences at professional meetings, colleges and universities, international conferences, and other settings, and he has testified before Congress on numerous occasions. He is frequently consulted by the electronic and print media for his views on various space issues. Dr. Logsdon has been a Fellow at the Woodrow Wilson International Center for Scholars and was the first holder of the Chair in Space History of the National Air and Space Museum. He is a Fellow of the American Association for the Advancement of Science and the American Institute of Aeronautics and Astronautics.

John E. Naugle devoted many years to aerospace research and administration. He began his career by studying cosmic rays, using nuclear emulsions exposed during high-altitude balloon flights. Later, the same emulsions were carried on sounding rockets. In 1959, Mr. Naugle joined the Goddard Space Flight Center, where he continued to use sounding rockets to study the protons in the magnetosphere. In 1960, he took charge of the Fields and Particles Program at NASA Headquarters, where he remained until his retirement in 1981. During his years at NASA Headquarters, Mr Naugle served as a principal investigator, director of the Physics and Astronomy Program, associate administrator for the Office of Space Science, associate administrator of the Agency, and finally as chief scientist. He is the author of *First Among Equals: The Selection of Space Science Experiments* (NASA SP-4215, 1991). Dr. Naugle is now enjoying his retirement on Cape Cod, where he continues to write about the early history of space science.

Regan Anne Newport is an undergraduate in mathematics at the American University in Washington, DC. She worked on this documentary history during an internship in the NASA History Division in the autumn of 2000.

Nancy Grace Roman had the scientific responsibility for NASA's astronomy program from 1959 through 1979, although various portions of the program were split off as it grew. Her final title was Chief of Astronomy and Relativity Programs. Since that time, she has remained involved with the NASA science program in various capacities. Before joining NASA, she taught graduate courses and did research in stellar astronomy at the University of Chicago and conducted research in radio and radar astronomy at the U.S. Naval Research Laboratory. She earned a B.A. from Swarthmore College and a Ph.D. from the University of Chicago. She is a member of the International Astronomical Union, and a fellow of the American Astronautical Society and the American Association for the Advancement of Science.

Amy Paige Snyder is a space transportation analyst in the Federal Aviation Administration's Office of Commercial Space Transportation in Washington, DC. Ms. Snyder graduated with an M.A. in Science, Technology, and Public Policy from George Washington University's Elliott School of International Affairs in May 2000. She has authored several magazine and encyclopedia articles on current and historical aspects of the space program.

Glossary

AACS . . . . .Attitude and Articulation Control Subsystem
AAP . . . . .Advanced Apollo Program
AAS . . . . .American Academy of Sciences
ABRL . . . . .Army Ballistics Research Laboratory
ACE . . . . .Advanced Composition Explorer
ADSP . . . . .Advanced Digital SAR Processor
AEP . . . . .Apollo Extension Program
AFCRC . . . . .Air Force Cambridge Research Laboratory
AMB . . . . .Astronomy Missions Board
AMBA . . . . .Army Ballistic Missile Agency
ANS . . . . .The Netherlands Astronomical Satellite
AOSO . . . . .Advanced Orbiting Solar Observatory
APG . . . . .Aberdeen Proving Ground
APL . . . . .Applied Physics Laboratory
ARPA . . . . .Advanced Research Projects Agency
ASCA . . . . .Advanced Satellite for Cosmology and Astrophysics
ASI . . . . .Italian Space Agency
AXAF . . . . .Advanced X-ray Astrophysics Facility
BATSE . . . . .Burst and Transient Source Experiment
BBXRT . . . . .Broad Band X-ray Telescope
BMD . . . . .Ballistic Missile Division
CC&S . . . . .Central Computer and Sequencer
CCAS . . . . .Cape Canaveral Air Station
CDS . . . . .Command and Data System (Galileo)
CGRO . . . . .Compton Gamma Ray Observatory
CIDEX . . . . .Comet Ice and Dust Experiment
CIT . . . . .California Institute of Technology
CMB . . . . .Cosmological Microwave Background
COBE . . . . .Cosmological Background Explorer
COMA . . . . .Comet and Matter Analyzer
COMPLEX . . . . .Committee on Planetary and Lunar Exploration (Space Science Board)
CONTOUR . . . . . . .Comet Nucleus Tour
COSPAR . . . . .Committee on Space Research (International Geophysical Union)
CRAF . . . . .Comet Rendezvous Asteroid Flyby
DCS . . . . .Data Conditioning System
DIRBE . . . . .Diffuse Infrared Background Experiment
DMR . . . . .Differential Microwave Radiometer
EGRET . . . . .Energetic Gamma-Ray Telescope
ESA . . . . .European Space Agency
ESOC . . . . .European Space Operations Center
ESRO . . . . .European Space Research Organization
EUVE . . . . .Extreme Ultraviolet Explorer
FBC . . . . .Faster, Better, Cheaper

FEIS . . . . .Final Environmental Impact Statement
FIRAS . . . . .Far-Infrared Absolute Spectrometer
FOC . . . . .Faint Object Camera
FOPS . . . . .Faint Object Spectrograph
FPSO . . . . .Flight Projects Support Office
FUSE . . . . .Far-Ultraviolet Spectroscopic Explorer
GGS . . . . .Global Geosciences
GPHS . . . . .General Purpose Heat Source
GRO . . . . .Gamma Ray Observatory
HEAO . . . . .High Energy Astronomical Observatories
HGA . . . . .High Gain Antenna
HRS . . . . .High-Resolution Spectrograph
HSP . . . . .High-Speed Photometer
HST . . . . .Hubble Space Telescope
ICBM . . . . .Intercontinental Ballistic Missile
ICSU . . . . .International Council of Scientific Unions
IGY . . . . .International Geophysical Year
IPS . . . . .Instrument Pointing System
IR . . . . .Infrared
IRAC . . . . .Infrared Array Camera
IRAS . . . . .Infrared Astronomical Satellite
IRBM . . . . .Intermediate Range Ballistic Missile
IRP . . . . .Infrared Photometer
IRS . . . . .Infrared Spectrometer
ISAS . . . . .Japanese Institute of Space and Astronautical Sciences
ISPM . . . . .International Solar Polar Mission
IT . . . . .Information Technology
IUE . . . . .International Ultraviolet Explorer
KAO . . . . .Kuiper Astronomical Observatory
LAT . . . . .Large Airborne Telescope
LDR . . . . .Large Deployable Reflector
LESA . . . . .Lunar Exploration System for Apollo
LISA . . . . .Laser Interferometer Space Antenna
LOT . . . . .Large Orbital Telescope
MAPPER . . . . .Molecular Astrophysics Pencil-Beam Explorer
MCCC . . . . .Mission Control and Computing Center
MESSENGER . . . . .Mercury: Surface Space Environmental, Geochemistry, and Ranging
MESUR . . . . .Mars Environmental Survey
MeV . . . . .Million Electron Volts
MGCO . . . . .Mars Geoscience/Climatology Orbiter
MGS . . . . .Mars Global Surveyor
MIPS . . . . .Multiband Imaging Photometer
MIT . . . . .Massachusetts Institute of Technology
MORL . . . . .Manned Orbiting Research Laboratory
MOT . . . . .Manned Orbiting Telescope

MOWGSA . . . . .Management Operations Working Group for Space Astronomy
MSA . . . . .Mission Support Area
MSFEB . . . . .Manned Space Flight Experiments Board
NACA . . . . .National Advisory Committee for Aeronautics
NAR . . . . .Non-Advocate Review
NAS . . . . .National Academy of Sciences
NEAR . . . . .Near-Earth Asteroid Rendezvous
NEPA . . . . .National Environmental Policy Act of 1969
NGST . . . . .Next Generation Space Telescope
NRL . . . . .Naval Research Laboratory
NSCAT . . . . .NASA Scatterometer
NSE . . . . .National Space Establishment
OAO . . . . .Orbiting Astronomical Observatories
OART . . . . .Office of Advanced Research and Technology
OMB . . . . .Office of Management and Budget
OMSF . . . . .Office of Manned Space Flight
ORL . . . . .Orbiting Research Laboratory
OSAT . . . . .Office of Space Access and Technology
OSO . . . . .Orbiting Solar Observatories
OSS . . . . .Office of Space Sciences (NASA)
OSSA . . . . .Office of Space Science and Applications (NASA)
OSTP . . . . .Office of Science and Technology Policy
PC . . . . .Planetary Camera
QPOs . . . . .Quasi-Periodic Objects
REX . . . . .Relativity Explorer
RHUs . . . . .Radioisotope Heater Units
ROSAT . . . . .Roentgen Satellite
RTGs . . . . .Radioisotope Thermoelectric Generators
RXO . . . . .Redundant Crystal Oscillator
RXTE . . . . .Rossi X-ray Timing Explorer
SAM . . . . .Simultaneous Astrophysics Mission
SAMPEX . . . . .Solar, Anomalous, and Magnetospheric Particle Explorer
SAR . . . . .Synthetic Aperture Radar
SAS . . . . .Small Astronomical Satellites
SCEL . . . . .Signal Corps Engineering Laboratory
SELV . . . . .Small Expendable Launch Vehicle
SEMPA . . . . .Scanning Electron Microscope and Particle Analyzer
SEPS . . . . .Solar Electric Propulsion System
SESAC . . . . .Space and Earth Science Advisory Committee (NASA)
SETI . . . . .Search for Extraterrestrial Life
SIPS . . . . .Small Instrument Pointing System
SIRTF . . . . .Space Infrared Telescope Facility
SOA . . . . .Smithsonian Astrophysical Observatory
SOFIA . . . . .Stratospheric Observatory for Infrared Astronomy
SPAC . . . . .Space Program Advisory Council

SR&T . . . . .Supporting Research and Technology
SRM . . . . .Solid Rocket Motor
SRMU . . . . .Solid Rocket Motor Upgrade
SSB . . . . .Space Science Board
SSED . . . . .Solar System Exploration Division
SSM . . . . .Support System Module
SUOT . . . . .Spacelab Ultraviolet Optical Telescope
SWAS . . . . .Submillimeter Wave Astronomy Satellite
TDRSS . . . . .Tracking and Data Relay Satellite System
TOMS . . . . .Total Ozone Mapping Spectrometer
TOPS . . . . .Toward Other Planetary Missions
TPESP . . . . .Technical Panel for the Earth Satellite Program
TRMM . . . . .Tropical Rainfall Measuring Mission
UV . . . . .Ultraviolet
VLBI . . . . .Very-Long Baseline Interferometer
VLST . . . . .Very-Long Space Telescope
VOIR . . . . .Venus Orbiting Imaging Radar
VRM . . . . .Venus Radar Mapper
VSOP . . . . .Very-Long Baseline Interferometry Space Observatory Program
VVEJGA . . . . .Venus-Venus-Earth-Jupiter Gravity Assist
VVVGA . . . . .Venus-Venus-Venus Gravity Assist
WFC . . . . .Wide-Field Camera
WIRE . . . . .Wide-field Infrared Explorer
WSPG . . . . .White Sands Proving Ground
XMM . . . . .X-ray Multi-Mirror Satellite
XTE . . . . .X-ray Timing Explorer

Chapter 1

Space Science: Origins, Evolution, and Organization

by John E. Naugle and John M. Logsdon

Modern space science really began in 1946 when scientists first started to use balloons and sounding rockets to carry instruments to the outer fringes of Earth's upper atmosphere. With the latest technological advances, balloons could float at an altitude of 100,000 feet for several hours, enabling scientists to study cosmic rays and other atmospheric and stellar phenomena. Soon after, sounding rockets soaring to 400,000 feet gave scientists a fleeting glimpse of the ultraviolet and x-ray radiation from the Sun and stars. Almost ten years of upper atmosphere science using these new tools created a community of scientists eager to extend their observations. These efforts were followed by the International Geophysical Year in 1957–1958, when scientists planned to orbit satellites for their research. They recognized that satellites could provide months of observing time hundreds of miles above Earth's atmosphere, something neither balloons nor sounding rockets could do.[1]

When the Soviet Union launched Sputnik I on October 4, 1957, public reaction fostered greater efforts in space science as an attempt to atone for the Cold War humiliation. As a result, the United States began to pour previously undreamed of resources into space science. Hundreds of scientists shifted their research arena from Earth-bound research laboratories to Earth orbit and the remote reaches of the solar system. Some were driven by the opportunities to discover new phenomena; others were enticed by the resources available in a growing, exciting, and dynamic field.

To manage these efforts, the United States created a new agency in July 1958, the National Aeronautics and Space Administration (NASA). Americans expected NASA to organize a coherent national space science program that would regain U.S. leadership in space science and technology. It took NASA and space scientists nearly six years to achieve a coordinated, mutually agreeable program, but from 1964 to the present NASA has conducted a sophisticated, productive space science program, though not without continuing tension between the space agency and space scientists. During this period several thousand astronomers, physicists, chemists, and life scientists conducted experiments on NASA missions, and the results of NASA's programs have revolutionized human understanding of Earth's place in the cosmos.

---

1. On the IGY see, Constance McL. Green and Milton Lomask, *Vanguard: A History* (Washington, DC: NASA SP-4201, 1971), pp. 6–39; Rip Bulkeley, *The Sputniks Crisis and Early United States Space Policy: A Critique of the Historiography of Space* (Bloomington, IN: Indiana University Press, 1991), pp. 89–122; R. Cargill Hall, "Origins and Early Development of the Vanguard and Explorer Satellite Programs," *Airpower Historian 9* (October 1964): 102–108; Robert A. Divine, *The Sputnik Challenge: Eisenhower's Response to the Soviet Satellite* (New York, NY: Oxford University Press, 1993); Walter A. McDougall, . . . *The Heavens and the Earth: A Political History of the Space Age* (New York, NY: Basic Books, 1985).

The Origins of Space Science (1946–1958)

In the years immediately following World War II, the U.S. scientific community turned its attention from support of the war effort to the scientific questions that had been the focus of attention before the war. As they did so, new techniques for obtaining data became available; that development marked the beginning of the U.S. space science effort.

Balloons, Cosmic Rays, and Mesons

After World War II, the Office of Naval Research started the Skyhook balloon program. Although the Navy justified support of the program because of its eventual value to military systems, civilian scientists established the objectives of the program and conducted the research. Large plastic balloons carried cosmic ray instruments to altitudes above 100,000 feet. Because the cost of the balloon program was relatively low and because graduate students could assemble a payload in a few months in a university laboratory, the Skyhook program enabled many academic scientists and graduate students to study cosmic rays.[2]

V. F. Hess, an Austrian physicist, discovered cosmic rays in 1911 while searching for the source of a highly penetrating radiation. He personally carried a Wulf electrometer to a height of 5,000 meters, in an open gondola, where he found the ionization to be sixteen times that at the surface. Hess correctly interpreted his observations as demonstrating that the highly penetrating radiation came from outside the atmosphere rather than from the surface of Earth. Physicists could not immediately determine the nature of the radiation. Since it came from outside Earth—from the cosmos—they named the phenomenon "cosmic rays." By 1940, when World War II stopped cosmic ray research, physicists knew that most cosmic rays were positively charged particles of great energy that, upon entering the atmosphere, generated cascading showers of electrons, positrons, gamma rays, and some kind of unknown, highly penetrating charged particles.

In 1947, scientists at the University of Bristol exposed thick, very sensitive photographic "nuclear emulsions" to cosmic rays on a mountaintop in the Alps. In the emulsions they found the tracks of two new particles, heavier than a proton and lighter than an electron. They named these new particles the *pi* and *mu* mesons.[3] The *pi* meson proved to be the glue that held a nucleus together and the *mu* meson, the mysterious highly penetrating particle in cosmic rays. In 1948, using a Skyhook balloon to expose nuclear emulsions to cosmic rays at high altitudes, scientists at the University of Minnesota and the University of Rochester discovered that, in addition to protons and electrons, cosmic rays also included high-energy atomic nuclei stripped of their electrons.[4] These two scientific discoveries sparked intense interest in cosmic rays. For about a decade, until proton accelerators replaced cosmic rays as a source of mesons, cosmic rays and their nuclear interactions occupied center stage in theoretical physics. Many future space scientists took up the study of cosmic rays in this period.

---

2. David H. DeVorkin, *Race to the Stratosphere* (New York, NY: Springer-Verlag, 1989), pp. 296–304.

3. B. Occhialini and C. F. Powell, "Nuclear Disintegrations Produced by Slow Charged Particles of Small Mass," *Nature* 159 (February 8, 1947): 186–90.

4. P. Freier, E. J. Lofgren, E. P. Ney, F. Oppenheimer, H. L. Bradt, and B. Peters, "Evidence for Heavy Nuclei in the Primary Cosmic Radiation," *Physical Review* 74 (July 15, 1948): 213–17.

In the mid-1950s, just as proton beams from accelerators were replacing cosmic rays as sources of mesons, scientists discovered that during a solar flare the Sun emitted large numbers of cosmic rays and modulated the flux of cosmic rays coming from outside the solar system. These two discoveries, coupled with anticipation of the onset of a period of high solar activity in 1957, stimulated a renewed interest in cosmic rays. Unfortunately, the time of onset of a solar flare is unpredictable and it reaches its peak intensity in a few minutes. Sounding rockets and balloons were not good enough platforms to study solar flares because it was difficult to launch them on such short notice. On the other hand, a satellite would be an ideal platform. From such a platform outside Earth's atmosphere, a scientist could continuously monitor cosmic rays. After 1958, large numbers of cosmic ray physicists entered the fray determined to be the first to get a cosmic ray detector on a satellite. They were young, eager, and full of ideas for experiments that required satellites and space probes. Meanwhile, other groups of scientists had been using rockets to observe the Sun and stars. They were equally interested in getting their telescopes onto satellites.

V-2 Upper Atmosphere Panel

At the end of World War II in May 1945, the U.S. Army acquired a number of German V-2 rockets, together with many of the engineers who had developed them (including Wernher von Braun), and brought the rockets back to the United States for examination and testing. In late 1945, the Army offered scientists the opportunity to put experiments aboard these rockets as they were launched for engineering tests. This offer led to the formation of an ad hoc "V-2 Upper Atmosphere Panel" in February 1946 to "develop a scientific program, assign priorities for experiments to fly on the V-2s, and to advise the Army Ordnance Department on matters essential to the success of the program."[5]

Working through the panel, astronomers and geophysicists used these sounding rockets to study the properties of the upper atmosphere, solar and stellar ultraviolet radiation, and the aurora. Scientists used all the V-2s, then used new sounding rockets developed to replace them, and continued to control the nation's sounding rocket program until NASA Headquarters took over this function in 1958.[6] The minutes of the meetings of the V-2 Upper Atmosphere Panel provide a vivid history of the many failures and occasional triumphs of these first space scientists. [I-1, I-2, I-3] After the formation of NASA, several members of the V-2 Panel joined the space agency and applied the experience they had gained to the organization and management of NASA's space science program. Together with those scientists who had been conducting balloon experiments, these "rocket scientists" formed the nucleus of the initial U.S. space science community.[7]

---

5. The Panel started life as the V-2 Upper Atmosphere Panel; in 1948 it became the Upper Atmosphere Rocket Research Panel, and in 1957, the Rocket and Satellite Research Panel. On this Panel see, David H. DeVorkin, "Organizing for Space Research: The V-2 Panel," *Historical Studies in the Physical and Biological Sciences* 18 (1987): 1–24.

6. See James A. Van Allen, *Origins of Magnetospheric Physics* (Washington, DC: Smithsonian Institution Press, 1983), and Homer E. Newell, *Beyond the Atmosphere: Early Years of Space Science* (Washington, DC: NASA SP-4211, 1980), for details of the Panel. Newell's book is an outstanding summary of the development of the NASA space science program by its chief architect.

7. See R. Cargill Hall, *Lunar Impact: the History of Project Ranger* (Washington, DC: NASA SP-4210, 1977).

International Geophysical Year

The members of the V-2 Panel and the cosmic ray physicists were a small minority of the many other astronomers and geophysicists interested in the intense solar activity predicted for 1957. In 1952, scientists Lloyd Berkner, Sidney Chapman, and Marcel Nicolet persuaded the International Council of Scientific Unions (ICSU) to organize an International Geophysical Year (IGY), a cooperative scientific endeavor to study solar-terrestrial relations during the period of maximum solar activity. Some sixty-seven nations, including the Soviet Union, agreed to conduct cooperative experiments to study solar-terrestrial relations during the IGY.

In October 1954, ICSU challenged the United States and the USSR to use their missiles, which were being developed for war, to launch scientific satellites as part of the IGY program. In July 1955, the United States responded by announcing that it would develop a new rocket, the Vanguard, to launch scientific satellites. A year later, the Soviets announced that they too would launch scientific satellites as a part of the IGY. Thus began a race to see who would be first to launch an Earth satellite.

A National Security Council white paper approved by President Dwight D. Eisenhower, "U.S. Scientific Satellite Program," provided the rationale behind the satellite program. [Volume I, Document I-10] This paper, discussed on May 20, 1955, by the White House National Security Council (NSC), encouraged the Department of Defense to develop and launch a small scientific satellite "under international auspices, such as the International Geophysical Year, in order to emphasize its peaceful purposes . . . considerable prestige and psychological benefits will accrue to the nation which first is successful in launching a satellite . . . especially if the USSR were to be the first to establish a satellite." This document summarizes many of the forces that shaped space science over the coming years. The paper justified space science because of its contribution to national security, not because it was an activity worthy of support on its own merits. President Eisenhower and his associates were primarily interested in establishing the international legal principle that national sovereignty did not extend to the altitudes at which a satellite would orbit, and thus that there was no obstacle in international law to the overflight of a reconnaissance satellite over Soviet territory. To them, the scientific purposes of the satellite were of secondary importance. [Volume I, Document II-12]

Within two months, the Naval Research Laboratory's (NRL) proposal to develop a new Vanguard rocket to launch initial U.S. scientific satellites was chosen over the Army's competing Project Orbiter proposal by an "Ad Hoc Advisory Group on Special Capabilities." [Volume IV, Document I-4] The organization, conduct, and initial failure of the Vanguard Program stimulated the space science effort and helped shape its organization.

Although NRL managed the Vanguard program, the overall scientific and technical direction came from a Technical Panel for the Earth Satellite Program (TPESP). The National Academy of Sciences and its operating arm, the National Research Council, organized TPESP, which consisted mostly of scientists. [I-4] Richard Porter, an engineer from General Electric who had been in charge of the U.S. V-2 program, chaired TPESP. The Panel directed the work, set policies, selected experiments, and formulated scientific objectives for the Vanguard project. The National Science Foundation and the U.S. Army, Navy, and Air Force participated in the work and provided funds to pay for it.[8] However, it was the TPESP, which met about once a month, that controlled the pace and scientific content of Vanguard. Although NRL was responsible for building Vanguard, it could not start work on the payload for a mission until the Panel had established the objectives of the mission and selected the experiments.[9] [I-5]

The Vanguard program proved much more difficult to accomplish, and therefore much more costly, than had been anticipated. The original cost estimate for the program was $15–20 million; by the spring of 1957, the estimate had grown to $110 million, with possible growth to $150–200 million. President Eisenhower and the National Security Council considered canceling the program in May 1956, but decided to let it continue. The Eisenhower administration in mid-1956 also considered, but rejected, the possibility of authorizing the Army to attempt a satellite launch in advance of the first scheduled launch in the Vanguard program. [Volume IV, Document I-7, I-8] By 1957, neither the Department of Defense nor the National Science Foundation was eager to provide the additional funds to complete Vanguard; it took White House intervention to force them to do so. [I-8]

Sputnik

In the fall of 1957, as TPESP's cumbersome machinery slowed progress and Vanguard continued to fall behind schedule, the governing body of the IGY met in Washington, DC. On Friday, October 4, 1957, the Soviet Union hosted a party for the group at its Embassy. Midway through the party, Lloyd V. Berkner, ICSU president and a prime mover behind the IGY, announced that the Soviets had just launched a satellite.[10] The dramatic launch of the first satellite, Sputnik I, surprised the world. The Soviets had opened the age of space exploration.

The successful launch of Sputnik set off an accelerated U.S. effort to launch a satellite, despite attempts by President Eisenhower and his associates to minimize the significance of the Soviet accomplishment. [I-9 and Volume IV, Document I-9 and I-10] At its November 6 meeting, TPESP agreed that if there were a decision by the Department of Defense to provide one or more of Wernher von Braun's Jupiter C rockets as a backup to the Vanguard, one of the experiment packages it had approved would be shifted from a Vanguard launch attempt to the initial Jupiter C attempt. [I-10] On November 8, 1957, Secretary of Defense Neil H. McElroy indeed did direct the Army to use its Jupiter C launch vehicle to launch two satellites.

In early November 1957, the Soviets launched Sputnik II with a dog named Laika aboard. In December, the United States attempted to launch the first Vanguard. The rocket burst into flames, crumpled, and dumped its satellite back onto the launch pad. So far, the score was 2–0 in favor of the Soviets.

On January 31, 1958, Wernher von Braun, leader of the German engineers who developed the original V-2s, the Redstone, and the Jupiter C, used the Jupiter C to place

8. When Vanguard developed budget problems, even the Central Intelligence Agency (because of its interest in establishing the principle of free satellite overflight) provided some of the project's funding. See Dwayne A. Day, John M. Logsdon, and Brian Latell, *Eye in the Sky: the Story of the CORONA Spy Satellite* (Washington: Smithsonian Institution Press, 1998).

9. See James A. Van Allen, *Origins of Magnetospheric Physics*, and John E. Naugle, *First Among Equals: The Selection of Space Scientists* (Washington, DC: NASA SP-4215, 1991), for more information about the role of TPESP.

10. "Soviet Embassy Guests Hear of Satellite from an American as Russians Beam," *New York Times*, October 5, 1957, p. A3.

Explorer 1, the first American satellite, into orbit. The satellite was developed by the Army's Jet Propulsion Laboratory and carried an experiment designed by James Van Allen of the University of Iowa, one of the individuals involved in shaping initial U.S. involvement in space science research. Even after the 1955 selection of Vanguard as the sole U.S. scientific satellite project, Van Allen had remained in touch with the von Braun team in Alabama, and thus was quickly able to switch his payload from the Vanguard launcher to the Jupiter launcher after that opportunity became available. [I-6, I-7]

The Van Allen Belts

Riding on Explorer 1 were Geiger counters built by Van Allen. On May 1, 1958, at a joint session of the American Physical Society and the National Academy of Sciences, Van Allen announced the most significant discovery from Explorer 1 and the subsequent Explorer 3 mission, then in orbit: that there were high energy radiation belts surrounding the Earth.[11] [I-12] These "Van Allen Belts" consisted of doughnut shaped regions of space centered on the geomagnetic equator and filled with high energy (40 MEV) protons orbiting around magnetic lines of force and oscillating back and forth between the northern and southern hemispheres. The belts proved to be more than just an exciting scientific discovery; the radiation level in the belts was so intense that, if a human or a satellite were orbiting within them, he or she would receive a lethal dose of radiation in a few hours, solar cells would rapidly deteriorate, and electronic equipment would malfunction. Because of the belts, almost all Earth satellites have been placed in orbits either below or beyond them. Whereas the Sputniks had not produced any exciting or significant scientific results, Van Allen's discovery electrified the scientific community. Van Allen and his hard-working graduate students had demonstrated that a team of academic scientists could design and build instruments that worked in space. Later, when engineers argued that academic scientists were not qualified to build instruments for spacecraft, someone was sure to remind them of Explorer 1. Van Allen's unexpected discovery and the worldwide acclaim he received attracted many young people to space science.

Creating a National Space Science Organization

In the near hysteria that prevailed after the success of the first two Sputniks and the failure of the first Vanguard, the United States began an intense effort to create a space program that would restore American pride and prestige. The Speaker of the House, Joe Martin, and the Senate Majority Leader, Lyndon Johnson, each chaired hearings to learn why the United States had fallen behind and how best to organize the U.S. response. Many organizations fought to gain control of the nation's space effort.[12] After Sputnik I, the Rocket and Satellite Research Panel (the successor to the V-2 Upper Atmosphere Panel, which had become the Upper Atmosphere Rocket Research Panel in 1948 and had

---

11. J. A. Van Allen, G. H. Ludwig, E. C. Ray, and C. E. McIlwain, "Observation of High Intensity Radiation by Satellites 1958 Alpha and Gamma," *Jet Propulsion 28* (September 1958): 588–92.

12. Newell, *Beyond the Atmosphere*; Robert L. Rosholt, *An Administrative History of NASA, 1958–1963* (Washington, DC: NASA SP-4101, 1966); and McDougall, . . . *The Heavens and the Earth*, provide more details and many references on the forces at work in the 1957–1958 period.

changed its name again in 1957) doubled its membership. Its members prepared a plan for a civilian agency to take over the exploration of space and then testified before Congress in favor of their plan. [I-11] That plan best represented the views of the nascent U.S. space science community as the nation organized its space response to Sputnik.

The Space Act

In March 1958, President Eisenhower, under attack by the media and a Democratically controlled Congress, selected the National Advisory Committee for Aeronautics (NACA) to become the core of the new space agency. He sent a bill to Congress, which when revised became the National Aeronautics and Space Act of 1958. The President signed this Act into law on July 29, 1958, creating the National Aeronautics and Space Administration (NASA). [Volume I, Document II-17]

The Space Act was, and still is, significant to space science, not only because of what it says about space science, but also by what it left unsaid. The Act stated that the general welfare and security of the United States required space activities and listed eight objectives for those activities. The first objective: "the expansion of human knowledge of phenomena in the atmosphere and space" made space science a high priority for NASA. The fifth objective set a goal for space science: "the preservation of the role of the United States as a leader in aeronautical and space science and technology . . . ." This statement set a relative rather than an absolute goal for space science. In 1958, there were only two countries with space programs; therefore it tied federal support for space science directly to the relative status of the United States and the Soviet space science programs.

The Act directed the administrator of NASA to arrange for scientists to help plan the scientific measurements and observations to be made, to conduct itself or arrange for another party to make those measurements and observations, and to provide for the widest possible dissemination of their results. The Act did not state how to involve scientists in planning, but directed that the administrator be responsible for planning and conducting space science. If a spacecraft failed, or the Soviets scored a first, Congress wanted one individual held accountable, not a committee or two or three cooperating agencies.

The Space Science Board

As the Administration and Congress moved to create NASA and the staff of the NACA worked to make that organization the core of the new space agency, scientists organized themselves to participate in the planning and execution of the program. On June 4, 1958, the president of the National Academy of Sciences, Detlev Bronk, created a Space Science Board. The members of the Board, mostly senior academic scientists, were asked to draft a space science program, identify institutions and scientists to conduct the program, and provide their recommendations to the Administrator of the new NASA, once it began operations. Bronk appointed Lloyd V. Berkner, a dynamic, hard-driving scientist, to be Chairman of the Space Science Board. [I-13]

Between June and October, Berkner organized the Board and sent a telegram to scientists and scientific institutions that invited them to propose space science experiments. [I-14] He created committees to evaluate the two hundred proposals the Board received.

In December 1958, the Board recommended an initial scientific program of over thirty missions to the NASA Administrator, and issued a primer on space science to allow other scientists to propose additional experiments. [I-15, I-16] These missions ranged in size from sounding rockets to solar and astronomical observatories. The members of the Board thought they were recommending a program for the next two or three years. It took NASA the better part of the next decade to complete that program.

NASA Establishes Its Space Science Program

On October 1, 1958, when T. Keith Glennan, the first NASA Administrator, opened the doors of the new agency, he had no space scientists on his staff and no space science program at any of the NASA centers. Under the Space Act, Glennan had the option of either having NASA conduct the space science program or arranging for other agencies, such as the National Science Foundation (NSF), to conduct it. He decided that NASA should be responsible for space science and created the Office of Space Flight Programs at NASA Headquarters. He appointed Abe Silverstein, a propulsion engineer from the NACA Lewis Research Laboratory, as its director. In turn, Silverstein appointed Homer E. Newell[13] to be his Assistant Director for Space Sciences at NASA Headquarters. To conduct the program, the Eisenhower administration transferred from the Army to NASA control over the Jet Propulsion Laboratory (JPL) in Pasadena, California, which was operated by the California Institute of Technology, and created a new "Field Center," the Goddard Space Flight Center (GSFC) in suburban Maryland near Washington.[14] Besides Newell, an additional fifty NRL scientists transferred to NASA. Most went into a Space Science Division at GSFC. Two came to NASA Headquarters to help Newell administer the space science program.

In December 1958, Administrator T. Keith Glennan issued a document that specified how he intended to plan and conduct the space science program. This document outlined the objectives for NASA's space flight experiments, and stated that the research program would be national in scope and would be based on recommendations from, among other groups, the Space Science Board. NASA would ask educational and research institutions, industry, and federal laboratories to participate in the program. NASA, not the Space Science Board, would establish the priorities for experiments and projects.[15]

By the beginning of 1959, Newell had a clear and unambiguous mandate to organize and manage a comprehensive space science program. At NASA Headquarters he had only a three-man staff, hardly adequate to administer a large and complex program that involved NASA Centers, universities, and industry. [I-18] In addition, he was engaged in a

---

13. At the time, Newell was Superintendent of the Atmosphere and Astrophysics Division, Naval Research Laboratory (NRL). NRL was the organization in charge of the Vanguard Program. Newell and Silverstein had been discussing the transfer of NRL scientists for sometime before NASA opened its doors.

14. On the Goddard Space Flight Center and Jet Propulsion Laboratory see, Alfred Rosenthal, *Venture into Space: Early Years of Goddard Space Flight Center* (Washington, DC: NASA SP-4301, 1985); Lane E. Wallace, *Dreams, Hopes, Realities: NASA's Goddard Space Flight Center, The First Forty Years* (Washington, DC: NASA SP-4312, 1999); Clayton R. Koppes, *JPL and the American Space Program: A History of the Jet Propulsion Laboratory* (New Haven, CT: Yale University Press, 1982).

15. NASA, *Policy on Space Flight Experiments* (Washington, DC: NASA, December 12, 1958); T. Keith Glennan, *The Birth of NASA: The Diary of T. Keith Glennan*, edited by J.D. Hunley (Washington, DC: NASA SP-4105, 1993), pp. 6–15.

tug-of-war with the Space Science Board for control of the space science program. He also had a battle going with William Pickering, the director of JPL, as to whether NASA Headquarters or JPL would formulate the lunar and planetary program that JPL had chosen as its desired share of the space science effort.

The Space Science Board expected to function with NASA somewhat as TPESP had functioned with respect to the Navy in the Vanguard program. During 1959, as Newell increased the size of his staff and moved to take charge of the program, the Board continued in its self-appointed role. Finally, on October 29, 1959, NASA used the power of the purse to take control. In a letter that provided funds for the operation of the Board for 1960, NASA directed the Board to focus on long-range strategy for space science and leave the detailed planning and conduct of the program to NASA. [I-17]

The Reorganization of 1959

In addition to the external problems with the Space Science Board, there were internal problems with NASA's organization. In December 1959, to clarify the roles and missions of the Centers (and for a variety of other reasons), Glennan reorganized NASA. He assigned the responsibility for all automated lunar and planetary missions to JPL. All Earth satellite and sounding rocket missions went to Goddard. [I-19]

Following Glennan's reorganization, Silverstein reorganized the Office of Space Flight Programs; he abolished Newell's Office of Space Sciences (OSS) and created two new program offices—Lunar and Planetary Programs and Satellite and Sounding Rocket Programs—to replace it. He appointed Newell, a scientist, as his deputy. He appointed former NACA engineers to head each of the program offices and appointed scientists to be their deputies. This pairing of scientists and engineers at each administrative level proved to be a good technique for assuring that each program office identified and, where possible, resolved its own scientific and technical issues.[16]

Newell chaired a Space Science Steering Committee; the director and deputy director of each of the two new program offices were its other members. To provide technical support to the Steering Committee, he created several scientific subcommittees. An "administrative scientist" from Newell's staff chaired each of these subcommittees. The membership consisted of academic scientists and scientists from Goddard and JPL; each had a mixture of wise-old-heads and "young Turks." These subcommittees planned the program in their discipline, reviewed the proposed experiments for a specific scientific mission, and established priorities for their flight on that mission. Through the Steering Committee and its subcommittees, Newell brought together the scientific and engineering talent needed to assure him that a mission was ready for development and that the program office had chosen the best possible experiments. The external scientific community was ambivalent about these changes, which further reduced its influence over NASA's space science program. [I-18]

Newell decreed that all proposals for scientific experiments would come to NASA Headquarters, and administrative scientists at Headquarters would manage the process of soliciting, receiving, and evaluating all scientific proposals and selecting the scientists for

16. Glennan, *The Birth of NASA*, pp. 21–30; Naugle, *First Among Equals*, pp. 58–78.

all space science missions. In April 1960, NASA issued Technical Management Instruction 37–1–1, which specified exactly how NASA would select space scientists to participate in missions and outlined what their role would be during a mission. This has proved to be a durable procedure. Four decades later, NASA continues to use the same basic approach established in this document. [I-20]

James Webb Takes Charge

By the end of 1960, NASA had created a space science organization and established the broad policy and procedures for planning and conducting a space science program, and the program was beginning to produce scientifically valuable results. [I-21] Space scientists, however, still had a fundamental problem with the organization at NASA Headquarters; while there were scientists at each level of the organization, those in charge were always ex-NACA engineers, while a scientist served as the deputy at each level. In addition, as new projects began to mature and budgets tightened, a host of serious technical issues emerged.

Overshadowing all of these issues was the future of NASA. Would the new President, John F. Kennedy, support a vigorous space program? Would the new Administrator of NASA continue the policies laid down by Glennan or would he make major changes? What would happen to the fledgling space science program?

On February 14, 1961, James E. Webb took charge as the new NASA Administrator.[17] Although he continued the basic policies established by Glennan, Webb took several steps to strengthen the space science program. As part of the Apollo buildup, in 1962 he added a Sustaining University Program with a $40 million annual budget to provide funding for new facilities and graduate student fellowships. (See Volume II, Chapter 3, for a discussion of this program.) In November 1961, he reorganized NASA, abolishing the Office of Space Flight Programs and appointing Homer Newell as associate administrator of a new Office of Space Science (OSS). Thereafter, Newell reported directly to the Administrator of NASA.[18]

Newell continued, but in reverse, the policy of pairing a scientist and engineer. He selected Edgar M. Cortright, an ex-NACA engineer, to be his deputy. He changed the name of the Sounding Rocket and Satellite Program to Geophysics and Astronomy and appointed a scientist as director. He created a Bioscience Program office and appointed a scientist to head it. Each of these new directors selected engineers to be their deputies. This arrangement established the tradition that the associate administrator of the Office of Space Science would always be a scientist,[19] and that he would always have an engineer as his deputy. The director of a program office could be either a scientist or engineer, but he and his deputy must constitute a scientist-engineer pair.

Webb placed Newell in charge of NASA's launch vehicles for robotic missions. He also placed JPL and Goddard under Newell's direction. He assigned the Sustaining University Program to Newell. Thus, with the exception of the tracking and data acquisition system,

17. On Webb see, W. Henry Lambright, *Powering Apollo: James E. Webb of NASA* (Baltimore, MD: Johns Hopkins University Press, 1995).

18. Rosholt, *An Administrative History of NASA, 1958–1963*, pp. 217–26.

19. This tradition was violated from 1982–1987, when an engineer, Burton I. Edelson, served as NASA's Associate Administrator for Space Science and Applications.

Newell had under his direct control all the programs and all the institutions, capabilities, and facilities needed to conduct the space science program. He still, however, had to fight for his share of the NASA budget. He was also directed to provide data on the lunar surface and the radiation levels in cislunar space to the Apollo Program. Otherwise, Newell controlled an independent, self-sufficient space science organization.

The space science organization that Webb and Newell established in 1961 continued approximately unchanged until 1974, when NASA Administrator James C. Fletcher reorganized NASA.[20] Fletcher shifted control of JPL and Goddard from OSS to a new Office of Institutional Affairs to assure even-handed treatment of all NASA Centers. In anticipation of the Shuttle Era with a single transportation system, he transferred control of all launch vehicles from OSS to the Office of Manned Space Flight (OMSF).[21] These two changes substantially weakened the control of the Office of Space Science over its destiny. No longer was the associate administrator of the Office of Space Science head of an independent, self-sufficient office. If the associate administrator had a problem at a Center they had to work through another associate administrator, rather than directly with the Center director, to resolve it. Instead of controlling the performance and schedule of their own fleet of launch vehicles, the associate administrator now had to compete with the other users of the Shuttle.

In the early 1980s, another NASA Administrator, James M. Beggs, restored control of JPL and Goddard to OSS, but not control of launch vehicles. Otherwise, the basic space science organization established in 1960, and modified in 1961, continues to administer the program.

Learning to Conduct a Sustainable Space Science Program

By the mid-1960s, the OSS organization, its two Field Centers, and associated space scientists were seasoned veterans, able to plan and conduct a successful space science program. Budgets were increasing, annual launch rates were increasing, and scientists were making discoveries. Space scientists sensed no limitations, other than their own energies and imaginations, to their desire to explore and understand the universe. [I-22]

In June 1967, concerned by the management deficiencies found during the investigation of the Apollo 1 fire and looking for ways to improve the management of human space flight programs, NASA Administrator Webb asked Homer Newell to review the philosophy and techniques that he was using so successfully to manage the space science program. The resulting review summarized the hard lessons learned during the earlier troubled times. [I-24]

In the summer of 1967, shortly after the Webb review, Newell and his staff suffered a major setback. Though they had mastered the technical skills required to conduct a space science program, they had not learned the political skills required to maintain scientific support for a mission, or to accurately judge the support they could expect from Congress.

Planetary exploration had begun, like lunar exploration, in a race between the United States and the USSR to see who would be the first to get some sort of spacecraft near Mars or Venus. The Soviets tried for Venus first, launching on February 12, 1961. Unlike the lunar

---

20. On Fletcher see, Roger D. Launius, "A Western Mormon in Washington, D.C.: James C. Fletcher, NASA, and the Final Frontier," Pacific Historical Review 64 (May 1995): 217–41.

21. The formal name of this organization at NASA Headquarters was the Office of Manned Space Flight from the beginning of NASA until August 2, 1982, when it became the Office of Space Flight.

contest, however, the Soviets did not win the race to Venus; their spacecraft failed before reaching Venus. The U.S. Mariner 2 flew by Venus on December 14, 1962. In June 1963, the Soviets got to Mars first, but with little scientific return. The United States did not get to Mars until July 15, 1965, when Mariner 4 took twenty-two pictures as it flew past the planet.[22]

In the mid-1960s, as OSS planned its future Mars programs, two problems confronted NASA senior managers. They needed to find missions that required the big Saturn V launch vehicles developed for Apollo, and they had to decide whether to focus the entire planetary program on the exploration of Mars, or to have a more modest Mars program and explore other planets such as Venus and Jupiter. They turned to the Space Science Board for help.

In the summer of 1965, the Space Science Board conducted a summer study that recommended that NASA focus its space science program on exploring Mars, a recommendation that had first emerged in 1964. [I-23] NASA used this recommendation in an attempt to solve both its problems. It formulated a $2 billion program, Voyager, to search for life on Mars, and it canceled plans for missions to other planets. Voyager consisted of a pair of orbiter-landers to be launched on one Saturn V. Despite the positive recommendation of the Space Science Board, Voyager was controversial. Few scientists supported the mission; most opposed it as too risky and too expensive. In the summer of 1967, because of the conflicting testimony from scientists and because of the general shortage of funds due to the cost of the Vietnam War and the needs of the Great Society, Congress killed the project.[23] Voyager was the first major space science project to be killed by Congress. (See Chapter 3 of this volume for further discussion of this controversy over the future of the planetary program.)

OSS and its space scientists learned some hard lessons in practical politics from the Voyager fiasco and the highly constrained budgets of the late 1960s and early 1970s. They learned to resolve their differences in internal scientific discussions, not in complaints to the media or in testimony before Congress. Massive scientific backing could not guarantee Congressional support for a mission, but massive scientific opposition could certainly kill it. They also learned that there was an ill-defined limit to the size of a space science mission and the annual space science budget that Congress would support.

In the fall of 1967, Webb reorganized the Office of Space Science and renamed it the Office of Space Science and Applications (OSSA). He promoted Homer Newell to be NASA Associate Administrator, the agency's number three job. He replaced Newell with John E. Naugle, a scientist, appointed Newell's deputy, Edgar Cortright, to be deputy director of OMSF, and replaced him with Oran Nicks, an engineer.

To avoid future problems, OSSA formed a Lunar and Planetary Mission Board and an Astronomy Mission Board to assist in planning future missions and to provide a forum to identify and resolve differences among scientists and between the scientists and OSSA.[24]

Between November 1967 and November 1968, NASA's Office of Space Science, the Lunar and Planetary Mission Board, the Space Science Board, and the scientific community hammered out a mutually acceptable planetary program for the 1970s. Although the program continued to emphasize the exploration of Mars by recommending the Viking orbiters and soft

22. Roger D. Launius, *Frontiers of Space Exploration* (Westport, CT: Greenwood Press, 1998), pp. 35–36.

23. For a more complete treatment of Voyager, see also Edward C. Ezell and Linda N. Ezell, *On Mars: Exploration of the Red Planet, 1958–1978* (Washington, DC: NASA SP-4212, 1984).

24. NASA Office of Space Science and Applications, *Planetary Program Review* (Washington, DC: NASA, July 11, 1969).

landers and two other Mars orbiters, it also included a Venus-Mercury flyby, two Pioneer missions to Jupiter and Saturn, and a "Grand Tour" of all the outer planets except Pluto. But for a two-year delay in Viking, loss of one of the other Mars orbiters, and a downsizing of the Grand Tour spacecraft to become Voyager, this program was carried out exactly as planned.[25]

Relations between NASA's Office of Space Science and Office of Manned Space Flight were strained throughout the 1960s. Space scientists resented the priorities and media attention enjoyed by the Apollo program. They complained about the lack of plans or funding in the Apollo program for lunar research. When NASA decided to include lunar research in Apollo, questions arose as to whether OSS or OMSF should be responsible for it. In September 1966, Robert C. Seamans, NASA Deputy Administrator, assigned responsibility for all space science, including that to be performed on crewed spacecraft, to the Office of Space Science, but decreed that the funding be carried in the OMSF budget and then transferred to OSS after congressional approval.[26] This arrangement further exacerbated the tension between OSS and OMSF. The scientific staff of OSS complained that OMSF would not adequately fund scientific work; OMSF engineers complained that OSS scientists neglected lunar research in favor of other areas of space science. To solve the problem, Newell created a Manned Space Science Division, staffed it with OSS scientists and OMSF engineers, and required that the head of the division report to him on scientific issues and to the head of OMSF on technical and funding issues.

Even so, tensions between the "manned" and "unmanned" elements of NASA and the relevant external communities persisted as the Apollo program reached its end. [I-25] They were made worse when NASA appeared to ignore the advice of the scientific community as it planned its "post-Apollo" program. [I-26, I-27] These tensions have continued until the current time, and appear to be an unavoidable feature of a U.S. civil space program that combines the drama of human space flight activities with a commitment to obtaining top-quality scientific results.

The Emerging Crisis in Space Science

The 1970s were to all appearances a "golden age" for space science. In 1976, two Viking spacecraft landed on the surface of Mars, and in 1977, two Voyager spacecraft began their journeys to Jupiter and Saturn, and on to Uranus and Neptune. Also in 1976, President Gerald Ford approved "new starts" for two large science missions for launch in the 1980s—a Galileo spacecraft to do in-depth exploration of Jupiter and its moons, and a large space telescope (later named Hubble) that had been a high priority for space scientists for almost three decades. But there were also troubling longer-range trends. The administration of President Jimmy Carter did not give high priority to the space program, and the budget demands of Space Shuttle development made approval of additional large space science missions difficult. [I-28] Both the Carter administration and, in 1981, the new administration of Ronald Reagan refused to approve a U.S. mission to Halley's Comet.[27] In addition, the Reagan administration directed NASA to cancel one

25. See, Robert S. Kraemer, *Beyond the Moon: A Golden Era of Planetary Exploration, 1971–1978* (Washington, DC: Smithsonian Institution Press, 2000).
26. On Seamans see, Robert C. Seamans, Jr., *Aiming at Targets: The Autobiography of Robert C. Seamans, Jr.* (Washington, DC: NASA SP-4106, 1996).
27. John M. Logsdon, "Missing Halley's Comet: the Politics of Big Science," *Isis* 80 (1989): 254–80.

of its ongoing space science missions, and seriously considered terminating the Solar System Exploration Program and transferring JPL to some other government agency. (See Volume II, Documents I-8 and I-9, on the cancellation of the International Solar Polar mission that resulted from the 1981 Reagan administration directive, and Chapter 3 of this volume on the threat to terminate the Solar System Exploration Program.)

By 1986, the space science community perceived itself to be in a crisis situation. NASA's Space and Earth Science Advisory Committee concluded that the space science program was "facing grave difficulties" leading to "a growing sense of unease and frustration over the program's diminishing pace." The Committee noted that "more and more missions were being identified as candidates for 'new starts' at a time when prospects for new starts were becoming uncertain," and that as a result "the competition among prospective missions had escalated to a counterproductive level."[28] [I-29]

New Approaches to Managing Space Science

In 1988, a new head of NASA's Office of Space Science and Applications, Lennard Fisk, took a new approach to dealing with this competition. Rather than having potential missions compete with each other annually to determine which of them NASA would recommend as its next new start, Fisk created a strategic planning process. The first version of this plan, issued in April 1988, noted the "trend toward large, complex, long-duration missions" that had become characteristic of the space science enterprise in the 1980s.[29] The plan set scientific priorities for prospective missions and programs, and thereby determined the order in which various proposed missions would be put forth by NASA for White House and congressional approval. The plan assumed continuing growth in NASA's budget. It thus proposed that NASA would initiate one major or moderate new mission each year. This was clearly a bullish outlook, given the difficulties of obtaining new start approvals during the preceding decade. [I-30]

The strategic planning approach was initially successful. NASA received new start approvals for three major space science missions between 1989 and 1991—the Advanced X-Ray Astronomical Facility, a Comet Rendezvous-Asteroid Flyby mission, and a mission to do in-depth exploration of Saturn, called Cassini.

A new NASA Administrator, Daniel S. Goldin, came to the space agency on April 1, 1992, with a very different approach to future space science missions and different expectations for the future of NASA. He directed his associates to plan for a level NASA budget in the future, rather than continued growth. He indicated that Cassini would be the last major space science mission that NASA would propose for some time, and emphasized a "faster, better, cheaper" approach to future mission planning. The rationales behind such an approach rejected the trend toward large, complex, long duration missions that had been stressed in the 1988 and subsequent strategic plans, suggesting that undertaking more missions, each at lower cost and with shorter times between approval and launch, would produce better scientific returns, allow more scientists an opportunity to get involved with NASA missions, and better accommodate an occasional mission failure. [I-31, I-32] Goldin also reorganized the Office

---

28. Space and Earth Science Committee, NASA Advisory Council, *The Crisis in Earth and Space Science* (Washington, DC: NASA, November 1986), p. iii.

29. Office of Space Science and Applications, NASA, *Strategic Plan 1988* (Washington, DC: NASA, 1988), p. 2.

of Space Science and Applications, dividing its programs among three new offices—Space Science, Mission to Planet Earth (later renamed Earth Science), and Life and Microgravity Science and Applications.

The continuing tension between space scientists and the human space flight program resulted in a congressional suggestion that all of NASA's science programs be gathered into a quasi-autonomous "National Institute for Space Science," modeled on the organization of the National Institutes of Health. The thought behind this suggestion was that by making space science somewhat independent of the rest of NASA, its budget could be stabilized rather than be traded off against the budget needs of the human space flight program (in this case, the International Space Station and Space Shuttle operations). A panel of the Space Studies Board (the new name for the Space Science Board, adopted in the late 1980s to signal a broader mandate for the Board) examined this and other suggestions for changing the way that NASA managed space science. While rejecting the idea of a National Institute for Space Science, the Board made other, less far-reaching, suggestions for improving the management of the space science effort. [I-33]

Space Science in the Twenty-First Century

Although he rejected the specific content of earlier space science strategic plans, NASA Administrator Goldin was a strong advocate of the strategic planning process in general. Between 1992 and 1996, he and his associates considered various strategic visions to guide future space science efforts. These efforts were accelerated by President Bill Clinton's call for rethinking NASA's space science program following the August 1996 announcement that a Martian meteorite contained possible evidence of ancient fossilized organisms. (See Chapter 3 for additional discussion of this announcement.) NASA and the National Research Council in October 1996 convened a workshop attended by leading space scientists to consider a reformulation of the rationale for NASA's scientific efforts; the results of that workshop were presented to Vice President Al Gore in December 1996 and formed the basis for White House approval on a new space science initiative organized around the theme "Origins." [I-34] A new space science strategic plan, issued in November 1997, spelled out the initiative in detail and identified the future missions needed to accomplish it. [I-35]

The Origins initiative is addressing a set of "fundamental questions." They include:

- How did the Universe begin and what is its ultimate fate?
- How do galaxies, stars, and planetary systems form and evolve?
- What physical processes take place in extreme environments such as black holes?
- How and where did life begin?
- How is the evolution of life linked to planetary evolution and to cosmic phenomena?
- How and why does the Sun vary and how does Earth and other planets respond?
- How might humans inhabit other worlds?[30]

In addressing questions such as these, the space science enterprise has in just over a half-century evolved from modest attempts to put a few scientific instruments aboard captured weapons of war to a comprehensive attack on questions that have puzzled humans for millennia. Whatever else happens in space in the twenty-first century, space science is poised to thrive.

---

30. NASA, *The Space Science Enterprise Strategic Plan: Origins, Evolution, and Destiny of the Cosmos and Life* (Washington, DC: NASA, November 1997), p. 4.

Document I-1

Document title: G. K. Megerian, General Electric Company, "V-2 Report #2, Minutes of Meeting," February 27, 1946.

Source: National Air and Space Museum, Smithsonian Institution, Washington, D.C.

Document I-2

Document title: G. K. Megerian, General Electric Company, "V-2 Report #13, Minutes of the V-2 Upper Atmosphere Research Panel Meeting," December 29, 1947.

Source: National Air and Space Museum, Smithsonian Institution, Washington, D.C.

Document I-3

Document title: G. K. Megerian, General Electric Company, "Panel Report #35, Minutes of the Upper Atmosphere Rocket Research Panel," April 29, 1953.

Source: National Air and Space Museum, Smithsonian Institution, Washington, D.C.

At the end of 1945, the U.S. Army brought to the United States a number of German V-2 rockets and key members of the team, led by Wernher von Braun, that had developed them. At the end of 1945, the Army notified U.S. scientists that it was willing to allow scientific instruments to be placed atop the V-2s during the series of test firings scheduled for the White Sands range in New Mexico. These would not be attempts to enter orbit, but rather vertical flights to heights of as much as 160 kilometers, which provided heretofore unavailable opportunities to gather data about the upper atmosphere and beyond. This came as welcome news to scientists already planning experiments in the upper atmosphere. To "develop a scientific program, assign priorities for experiments to fly on the V-2s, and to advise the Army Ordnance Department on matters essential to the success of the program," which was managed for Army Ordnance by the General Electric Company, a "V-2 Panel" was formed. It included from its inception individuals such as James Van Allen who was to become a pioneering space scientist; other members added subsequently included William Pickering, future head of the Jet Propulsion Laboratory, and Homer Newell, first head of NASA's space science program after 1958.

These minutes of several of the Panel's meetings capture the intense character of its activities. The last V-2 was fired in 1952; thereafter most launches used an Aerobee rocket developed by the Applied Physics Laboratory. By 1948, the group had dropped the "V-2" from its name and operated as the Upper Atmosphere Research Panel until early 1957, when its name was changed to the Rocket and Satellite Research Panel.

Document I-1

[no page number]

RESTRICTED

V-2 REPORT #2

SUBJECT: Minutes of Meeting
PLACE: Princeton University, Princeton, N.J.
DATE: Wednesday, Feb. 27, 1946

PRESENT: Capt. G. D. Bagley - Signal Corps
Dr. E.C. Buckley - NACA
Dr. W. G. Dow - U. of Michigan
Mr. R.G. DuBois - Wright Field
Dr. C. F. Green - General Electric Co.
Dr. K. H. Kingdon - General Electric Co.
Capt. W.W. Kellogg - AAF
Maj. E. Kotcher - Wright Field ATCS
Dr. E. H. Krause - Naval Research Lab.
Dr. M. H. Nichols - Princeton U.
Dr. E. O. Salant - Applied Physics Lab.
Dr. C. N. Warfield - NACA
Dr. F. L. Whipple - Harvard
Lt. Col. H. A. Zahl - SECL
Mr. G. K. Megerian - General Electric Co.

PURPOSE OF MEETING

1. To elect a chairman
2. To decide on matters of organization
3. To discuss generally the construction of the warhead
4. To establish a schedule for rocket firing
5. To outline the problems of those present at the meeting
6. To discuss telemetering and recording
7. To make recommendations to Army Ordnance Dept.

SUMMARY OF RESULTS

1. CHAIRMANSHIP

It was the general feeling of those present that inasmuch as Dr. Krause is devoting 100% of his time to problems in physics of the upper atmosphere, he was the logical candidate. Dr. Krause was elected chairman of the Panel. There were no dissenting votes. It was agreed that the chairman's functions should be both technical and administrative.

[2]

2. ORGANIZATION and REPRESENTATION

The original panel was to have consisted of one representative from each of the following groups –

a. Naval Research Laboratory
b. General Electric Co.
c. Princeton University
d. Harvard University
e. Vacancy

All members were to be working members.

The University of Michigan subsequently nominated Dr. W.G. Dow to fill the vacancy. Dr. Dow is also acting in an advisory capacity for the Pilotless Aircraft Branch of the Air Tech Service Command at Wright Field.

It was stated that Col. [J.G.] Bain had suggested that Dr. M.J.E. Golay also be a member representing the Signal Corps on this Panel.

At the beginning of this meeting therefore, the Panel consisted of representatives from –

a. Naval Research Laboratory
b. General Electric Co.
c. Princeton University
d. Harvard University
e. University of Michigan
e. U.S. Army Signal Corps

Dr. E. O. Salant of the Applied Physics Laboratory, John [sic] Hopkins University, asked at this meeting that his groups also be considered for representation. The Panel agreed to accept a representative from APL.

The Panel now consists of a representative from each of the following groups –

Organization	
Naval Research Laboratory	Dr. E. H. Krause
General Electric Co.	Dr. C. F. Green
	Dr. K. H. Kingdon
Princeton University	Dr. M. H. Nichols
Harvard University	Dr. F. L. Whipple
University of Michigan	Dr. W. G. Dow

U.S. Army Signal Corps	Dr. M. J. E. Golay *
John [sic] Hopkins University	Dr. J. A. Van Allen *
G.E. Co. Technical Aide	Mr. G. K. Megerian

*Not present at this meeting

[3] This group is primarily interested in the physics of the upper atmosphere. The functions of the Panel will be –

a. To advise the super-advisory panel on matters relating to technical phases of the tests, and –
b. To supervise the design and construction of the necessary scientific equipment.

The super-advisory panel mentioned above is a group to be organized in the near future. It will not be directly associated with this Panel and will report to Gen. Barnes on all firings.

3. CONSTRUCTION of the WARHEAD

There are 2 warheads in the U.S. at White Sands, N.M. – 50 warheads have been lost en route from Germany to the U.S. NRL is now proceeding with the design and construction of 25 additional warheads – approximate cost is $1000 each.

The general construction is as follows –

a. Overall length 7 ft.
b. Dia. at base 4 ft. 5 1/2 in.
c. Empty wt. = 250 kg or 560 lbs.
d. Tot. cap. = 16.7 cu.ft. (This is not usable cap.)
e. Contour is not a true cone, but an ogive.
f. Warheads to be cast in accordance with German dimensions.
g. Warhead consists of 4 sections: A - B - C - D
h. Nose section (A, B) may be aluminum casting.
i. Nose section (A, B) to be bolted to base section (C, D).
j. Base section (C, D) to be steel cstg.1/4 in. wall thk.
k. Base section (C, D) to be pressurized.
l. Nose section (A, B) 50 in. long – vol. 1151 cu. in.
m. Base section (C) 24 in long – vol. 1634 cu. in.
n. Base section (D) 30 in. long – vol. 25000 cu. in.
o. Tube (dia. approx. 4 in.) thru sections C, D for passing air to alcohol tank
p. C.G. to be kept as far forward as possible.
q. Lens of collective type to be mounted in nose.
r. Flush mounted windows for cosmic ray studies. Maximum material NRL wants to go thru for cosmic ray measurements is 1.4 cm of Hg or its equivalent.

[4] Generally speaking, the reason for the above type of construction is that pressure, spectroscopic and cosmic ray measurements are all contending for the nose position because of roll and stability considerations. Complete working drawings of the warhead which is to be built will be available within the next few weeks and will be forwarded to all Panel members.

The ogive results in what is considered to be a better aerodynamic curve than a true cone, but aerodynamic characteristics are not available for an ogive of this form. The group agreed to accept the warhead with an ogive contour as designed. A cone may be tried later for Sect. A, B.

In a phone conversation between Dr. Whipple of Harvard and Dr. Johnson at Aberdeen on this matter of ogive vs. cone, Dr. Johnson stated that he saw no difficulty in use of a right circular cone contour for either a part of or all of the warhead. He felt this would not disturb the aerodynamic of the rocket, but added that he had no authority to make a decision. In a subsequent discussion of this matter with Dr. R. W. Porter of the G.E. Co. at Schenectady, Dr. Porter stated he does not believe there will be any difficulty in substituting a tangent cone for the section A, B of the warhead.

The possibility of ejecting instruments from the warhead at the top of its trajectory was discussed, but the remaining time before firing does not permit the design and construction of a satisfactory ejecting mechanism.

Balance will be achieved by means of lead poured or secured into containers welded in the warhead.

Accessibility will be provided by means of 2 large flush-mounted doors in Section D and thru the ends of Sect. C and D. Sect. A, B will be detachable from Sect. C and D.

German calculations indicate that skin temperatures will read 750°F on the rise and 1250°F on the fall of the missile. The former temperature occurs after 75 seconds and the latter occurs after 315 sec.

There were several questions raised regarding the possible use of aluminum casting for nose Sect. A, B of the warhead. This matter is to be explored further, but in the meantime steel casting will be used.

The warhead will be secured to the instrument chamber by means of 20 bolts. The instrument chamber will house (among other equipments) –

[5] a. Telemetering equipment (10 channels)
b. Amplifiers for temperature measurements
c. Receivers
d. Transmitters
e. Batteries

It was stated that deviations of 10° to 17° or more may be expected from the trajectory of the missiles.

4. SCHEDULE of ROCKET LAUNCHINGS

Dr. Green stated that static tests on No.1 missile had been postponed from March 8 to March 15. The first 6 missiles fired will be primarily for General Electric Company's Project Hermes. The Panel agreed to adopt a "hands off" policy on missiles No. 1, 2 and 3. Various members of the Panel indicated interest, however, in being able to introduce certain equipment in missiles No. 4, 5 and 6 provided this would be agreeable to the G.E. Co. In a subsequent discussion between Dr. Green and Dr. Porter at Schenectady on March 1, Dr. Porter agreed to wreckage tests on missiles No. 4, 5 and 6.

The minimum number of missiles required is as follows –

a. General Electric Co.	(6)
b. NRL	(5)
c. Princeton	(5)
d. University of Michigan	(5)
e. APL	(5)
Total	26

Minimum requirements -	26 missiles
Number available -	25 missiles

The schedule of PROBABLE launching dates for the first 25 missiles is –

Missile	Assigned to		Probable LAUNCHING Date 1946
#1)	G.E. Co.		April
#2)			
#3)			
#4)	G.E. Co.		May
#5)			
#6)			

[6]

#7	NRL	)	
#8	Princeton University	)	
#9	APL	)	June
#10	University of Michigan	)	
#11	NRL	)	
#12	Princeton University	)	
#13	APL	)	

#14	University of Michigan	)	July
#15	NRL	)	
#16	Princeton University	)	
#17	APL	)	
#18	University of Michigan	)	
#19	NRL	)	August
#20	Princeton University	)	
#21	APL	)	
#22	University of Michigan	)	
#23	NRL	)	September
#24	Princeton University	)	
#25	APL	)	

Because of the present shortage of one missile, Princeton and University of Michigan will work out between them a plan as to which of them will use missile No. 24.

The Naval Research Laboratory is planning to furnish all the warheads and telemetering equipment for the first 25 missiles. Dr. Krause agreed to check whether any financial arrangements need to be made between NRL and each of the other groups to whom missiles have been assigned. Dr. Krause estimated that deliveries on warheads would begin on April 15, 1946. One telemetering set will be assigned with each warhead.

Dr. Whipple suggested that the biological sciences be considered when obtaining data in the upper atmosphere. This suggestion was placed in the form of a motion and passed by the Panel. Dr. Whipple will follow the matter and report his findings at a later date.

The possibility was discussed of using WAC corporals in conjunction with the above firing tests. The Panel agreed, however, that while the WAC corporal is a device for measuring temperatures and pressures, it is not in the same class with a V-2.

The Panel passed a motion recommending that 25 additional V-2 missiles be built on an extended time schedule of 5 months (or longer) beyond the present schedule.

[7] Princeton would like to install a cosmic ray transmitter in missiles No. 4 and 5.

5. OUTLINE OF INVESTIGATIONAL PROGRAMS

General Electric Co.

Currently, General Electric's primary interest is in Project Hermes. Dr. Green stated that G.E. will be busy with V-2 launchings until June 1, 1946. From that date on G.E. will be in a position to assist other groups in the panel with their work.

Princeton University

Dr. Nichols outlined Princeton's program as follows –

a. Cosmic rays
   Measurements of total intensity
b. Propagation
   Ground meas. of e layer
   Ionization effects, concentration, etc.
   Propose to use Doppler-method
c. Optics
   Photocell measurements
d. P, T measurements
   Ratio of undisturbed pressure (ρ) to stagnation pressure (ρ_0)

University of Michigan

Dr. Dow submitted a written report by the Air Technical Services Command which contained an extensive list of desired data. This report is attached as ENCLOSURE 'A,' but should not be construed to represent University of Michigan's program. U. of M.'s present responsibility has to do only with the propagation properties of the upper atmosphere. The immediate program is part D (below) which is to be accomplished by microwave radar on the ground and a low frequency transmitter with 2 beacons installed in the missile. Dr. Dow expressed a desire to have as many missiles as possible be equipped with these beacons.

[8] The report of the Army Air Forces – Air Technical Service Command – (enclosure 'A' Attached) outlines a list of desired data. The groups who will provide the answers have been indicated on the enclosure.

John [sic] Hopkins University

Dr. Salant outlined the Applied Physics Laboratory's program as follows –

a. Cosmic rays
   Total counting rate
   Counter telescope
   Film for stars
   Cloud chamber
b. Spectroscopy
   Photoelectric cells
   Vacuum (and quarts) spectrograph
   Interior of warhead
c. Sampling of gases in warhead
d. Temperature measurements in warhead
e. Magnetic tape recording of data.

Dr. Salant expressed a desire to include film and magnetic tape in the first 5 missiles merely to see what happens to the film and tape. This will be suggested to Dr. Porter of G.E. for his comments.

Harvard University

Dr. Whipple of the Harvard College Observatory stated that he was also acting as a representative for M.I.T. The Harvard-MIT program involves –

a. Meteor Studies
   Densities in the upper atmosphere
   Pressures in the upper atmosphere
b. Desire to receive German data requiring interpretation

Harvard's program involves –

a. Solar phenomena
   [9] Design of spectro equipment for study of the sun (if any branch of the armed services desires work of this nature). Feel that ascents should be made to near the 100-mi. level in order to clear the ozone bands.

   Measuring upper atmosphere winds. No financial arrangements made to date. No equipment facilities available at this time.

U.S. Army Signal Corps.

Lt. Col. Zahl stated that all data the Signal Corps is interested in will be recovered from the Panel. The Signal Corps is all interested in the phases following the V-2 firings, but only from a passive research viewpoint.

Dr. Golay will attend future Panel meetings as the representative for the Signal Corps.

Lt. Col. Zahl will supply the Panel with copies of a report on Detection of Tracking on V-2 Rocket Firings at White Sands.

Naval Research Laboratory

Dr. Krause outlined NRL's program as follows –

a. Cosmic rays
   Total counting ratio and relation to –
      Hard shower components
      Soft shower components
   Cloud chamber (long range plan)
b. Spectroscopy
   Vacuum (and quartz) spectrograph

c. Sampling of gases in warhead (long range)
d. Sampling temperature on skin of warhead at several points, inside of warhead at several points, and air temperature at some distance (few mm) from skin by means of a resistance thermometer.
e. Stagnation pressure at nose
f. Propagation measurements

Method involving harmonic relationship between 3 frequencies, one of which is the critical [10] frequency. This will involve direct determination of phase delay.

Thru jet exhaust streams by means of x-band transmitter and x-band receiver located in fins of V-2 at one point.

Dr. Krause stated that there was considerable duplication in the objectives of NRL, Princeton and APL, but that this duplication was desirable.

Dr. Krause indicated that NRL's long range program (8 to 10 months after firing V-2s) involved missiles reaching an altitude of 300,000 ft. NRL is also planning smaller rockets and will try to interest some manufacturer in manufacturing rockets to specifications.

On the immediately forthcoming tests, the accuracy of radar track is expected to be approximately 15 yds. for range and 1 mil for angularity. The missile will be above 100,000 ft. between 60 sec. and 360 sec. after takeoff.

6. TELEMETERING – RECORDING – POWER SUPPLY

NRL is planning to furnish all telemetering equipment. The telemetering will consist of 10 channels and cost approximately $3000 each. The telemetering will operate with

a. 1000 Mc frequency band
b. 1 to 3 micro-sec. pulses
c. repetition rate of 1/500 sec.

Delivery on the first 5 sets is expected approximately April 16, 1946, plus 30 additional sets by May 15, 1946. One telemetering set will be assigned with each warhead.

Antennas will be installed in the warhead either in horizontal or vertical slots (Sect. D of warhead).

Recording - will use string recording oscilloscopes. Movement of film will be about 4 in./sec. Propose the use of absolute time bases of 440 c from WWV. Ten meters will be set up on the ground for calibrating the system.

The Panel recommended that the time base and record-tracking data tie in on the same film. Also that take-off (t_0) be plotted. Slant range and altitude are independent.

[11] Power Supply –

a. Input voltage to system 0^v to 5^v
b. Number of channels = 10 (tentatively established)
c. Peak power output = 500 watts
d. Power requirements

Power requirements	1500^v	10 ma.
	105^v	40 ma.
	250^v	7 ma.
	-150^v	1 ma.
	6.3^v	7 amp.

7. RECOMMENDATIONS

The Panel makes the following recommendations to the Army Ordnance Department.

a. Approve immediate construction of 25 additional warheads on an extended time schedule of 5 months or longer.
b. When firing missiles, that fuel be burned completely and not cut off thereby avoiding danger of fire and subsequent destruction of equipment when missiles crash.
c. Tie in time base and record-tracking data on same film – also plotting take-off (to)
d. That the matter of coordinating frequency with other groups be brought to a head as soon as possible.
e. That the Signal Corps will supply 3 more Communication Stations with personnel to operate them.

8. GENERAL

The Panel discussed briefly the German drawing showing the tail section of the warhead with a view to installing equipment in the tail fins. It was suggested that all available German data and all information on roll and stability be gathered together and sent to the writer for distribution to all members of the Panel.

The matter of power supply and frequency allocation and interference was discussed, but due to lack of time it was agreed to take this up for further discussion at the next meeting.

[12] It was definitely agreed that matters of frequency and time coordination are all-important and should be effected as soon as possible with groups working on other phases of the problem.

Enclosure 'B' is a map of the White Sands NM Proving Grounds. This is being distributed only to members of the Panel.

9. NEXT MEETING

The next meeting of the Panel will be held –

Date – Wednesday, March 27, 1946

Time – 10:00 A.M.

Place – Washington, D.C.
Naval Research Laboratory
Bldg. 42, Room 318
Phone Trinidad 2424 Ext. 323, 325, 326

Prepared by: [signature]
G.K. Megerian
General Electric Co.
Aeronautics & Marine Eng. Div.
Schenectady, N.Y.

DISTRIBUTION:

(Please see attached) [omitted]

Document I-2

[no page number]

V-2 REPORT #13

SUBJECT: Minutes of V-2 Upper Atmosphere Research Panel Meeting

PLACE: Hotel Sherman
Chicago, Illinois

DATE: Monday, December 29, 1947

- PRESENT -

Lt. Col. J.G. Bain	OCO	Washington, D.C.
Mr. E.W. Beth	AMC/CFS	Cambridge, Mass.
Mr. T.R. Burnight	NRL	Washington, D.C.
Dr. W.G. Dow	U. of Michigan	Ann Arbor, Mich.

Mr. H.C. Early	U. of Michigan	Ann Arbor, Mich.
Mr. J.B. Edson	BRL	Aberdeen, Md.
Maj. J.O. Fletcher	AMC	Wright Field, Ohio
Dr. G.W. Gardiner	N. Mex. A&M	State College, N.M.
Dr. M.J.E. Golay	SCEL	Belmar, N.J.
Mr. W.F. Gould	Franklin Inst.	Philadelphia, Pa.
Dr. C.F. Green	G.E. Co.	Schenectady, N.Y.
Dr. D. Hagelbarger	U. of Michigan	Ann Arbor, Mich.
Mr. R.P. Haviland	G.E. Co.	Schenectady, N.Y.
Mr. J.J. Hopfield	APL/JHU	Silver Spring, Md.
Mr. H.L. Karsch	WSPG	WSPG, N.M.
Dr. E.H. Krause	NRL	Washington, D.C.
Mr. D.P. LeGalley	Bu. Ord.	Washington, D.C.
Mr. H.A. Martens	U. of Michigan	Ann Arbor, Mich.
Mr. G.K. Megerian	G.E. Co.	Schenectady, N.Y.
Dr. H.A. Miley	AMC/CFS	Cambridge, Mass.
Mr. H.W. Neill	U. of Michigan	Ann Arbor, Mich.
Dr. H.E. Newell, Jr.	NRL	Washington, D.C.
Mr. H.E. Norton	Army Chem. Center	Edgewood Arsenal, Md.
Dr. M.D. O'Day	AMC/CFS	Cambridge, Mass.
Dr. W.H. Pickering	CIT	Pasadena, Calif.
Mr. F.V. Schultz	U. of Michigan	Ann Arbor, Mich.
Capt. J.P. Smith	Air Weather D.	WSPG, N.M.
Mr. N.W. Spencer	U. of Michigan	Ann Arbor, Mich.
Mr. W.J. O'Sullivan	NACA	Langley Field, Va.
Mr. J.A. Van Allen	APL/JHU	Silver Spring, Md.
Mrs. F.L. Whedon	OC Sig. O.	Washington, D.C.
Dr. F.L. Whipple	Harvard & Navy Ord.	Cambridge, Mass.
Col. M.C. Young	Hq. USAF	Washington, D.C.

* * * *

[2] AGENDA OF MEETING

1. Resignation of Dr. E. H. Krause from the V-2 Upper Atmosphere Research Panel.

2. Selection of chairman to preside at this meeting.

3. New members - Dr. Newell (NRL), Dr. Pickering (CIT).

4. Election of new chairman.

5. Preparation and distribution of data reports on completed firings.

6. V-2 Firing schedule.

7. Reports on completed firings.

8. Reports on future firings - experiments of special interest.

9. Letter from RDB re budget support for V-2 Panel agencies.

10. Letter from OCO re classification of V-2 information.

11. Consideration of OCO letter re change in scope of V-2 Upper Atmosphere Research Panel.

12. New NRL telemetering system.

13. Recovery.

14. Next meeting.

SUMMARY OF RESULTS

1. RESIGNATION OF DR. E. H. KRAUSE FROM THE V-2 UPPER ATMOSPHERE RESEARCH PANEL

In a letter dated 21 November, 1947 addressed to the Secretary of the V-2 Panel via Col. Toftoy, Office Chief of Ordnance, Dr. E. H. Krause of the Naval Research Laboratory submitted his resignation both as member and chairman of subject Panel because of his new work at NRL. This letter was read to the Panel members present at this meeting.

Dr. Krause, who was also present, personally expressed regret that it had become necessary for him to resign from the Panel. He urged that basic research (as compared with applied research) be pursued as much as possible on all V-2 rockets.

[3]

2. SELECTION OF CHAIRMAN TO PRESIDE AT THIS MEETING

A motion that Dr. Krause be chairman pro temp of this meeting was voted on and carried unanimously.

3. NEW MEMBERS – DR. NEWELL (NRL), DR. PICKERING (CIT)

The Secretary of the Panel received a letter dated 28 November 1947 from the Naval Research laboratory via Col. Toftoy, Office Chief of Ordnance, nominating Dr. H.E. Newell, Jr. to membership on the V-2 Panel. Dr. Newell has taken over Dr. Krause's former duties as Head of the Rocket Sonde Research Section at NRL.

Attached to the above mentioned letter was one dated 4 December 1947 from Col. Toftoy, Office Chief of Ordnance, addressed to the Secretary of the Panel and requesting that the nomination of Dr. Newell to replace Dr. Krause as Naval Research Laboratory Representative be acted on favorably by the Panel.

A motion was made and seconded that Dr. Newell be a member of the V-2 Panel. The vote on this motion was unanimously favorable.

In another letter dated 21 November 1947 from Col. Toftoy to the Chairman of the V-2 Upper Atmosphere Research Panel, the Panel was advised that in accordance with conversations between Dr. Krause, Dr. Newell and Col. Toftoy with reference to expansion of the scope and activities of the V-2 Panel, the Office of Chief of Ordnance had invited the California Institute of Technology to membership on the Panel. This letter requested that Dr. W.H. Pickering (and Dr. H.S. Seifert, alternate) nominated by the California Institute of Technology be recognized as the representatives of CIT on the Panel.

A motion was made and seconded that Dr. Pickering be a member of the V-2 Panel. The vote on this motion was unanimously favorable.

For purposes of record, the V-2 Upper Atmosphere Research Panel as now constituted is indicated on the following page.

[4]

Name of Member	Agency Represented
Dr. W.G. Dow	University of Michigan
Dr. M.J.E. Golay	Signal Corps Engr. Laboratory
Dr. C.F. Green	General Electric Company
Dr. H.E. Newell, Jr.	Naval Research Laboratory
Dr. M.D. O'Day	Air Material Command
Dr. W.H. Pickering	California Institute of Technology
Dr. N. Smith	National Bureau of Standards
Dr. J.A. Van Allen*	Applied Physics Laboratory/JHU
Dr. F.L. Whipple	Harvard College Observatory
Mr. G.K. Megerian	Executive Secretary (G.E. Co.)

*Dr. Van Allen is now the new chairman of the V-2 Upper Atmosphere Research Panel. Please see section 4 below.

4. ELECTION OF NEW CHAIRMAN

In this matter of the election of a new chairman of the V-2 Panel, Col. Bain stated that from the Office of the Chief of Ordnance's point of view, it would be desirable to have the new chairman located in the D.C. area. As in the past with Dr. Krause as chairman, a comparable arrangement would facilitate and expedite the disposition of business matters between OCO and the V-2 Panel. Close proximity of the new chairman to OCO was felt to be an important consideration.

Concurring with the above feeling, the V-2 Panel nominated Dr. Van Allen despite his declination and Dr. Newell as candidates for the Panel chairmanship. In a secret ballot vote taken immediately after the nominations, Dr. Van Allen was elected as the new chairman.

Following this election, a motion was made by Dr. O'Day that Mr. Megerian be made an executive secretary to act as vice–chairman of the Panel in the possible absence of the chairman at future meetings. This motion was seconded and the subsequent vote of the Panel was favorable.

5. PREPARATION AND DISTRIBUTION OF DATA REPORTS ON COMPLETED FIRINGS

There was considerable discussion concerning the preparation and distribution of data reports on completed firings. Dissatisfaction still prevails among active Panel agencies because of reports now issued failing to reach these agencies and particularly because of continued lack of ballistic and trajectory data from the Ballistics Research Laboratory at Aberdeen.

[5] Mr. Karsch stated that WSPG issues 2 reports on all firings. The first of these is a preliminary report describing what is in the missile, experiments to be performed, etc. This is issued prior to the firing. Within a week after the firing, a second report is issued containing all available results at that date. These reports apparently have not been reaching all actively interested agencies. It was agreed that the Secretary of the Panel will advise Mr. Karsch at WSPG of the desired distribution of these reports for Panel agencies. WSPG in turn will forward sufficient copies as indicated.

In this connection, Mr. Karsch stated that if WSPG could obtain more information from all active agencies on V-2 rocket experiments, a report more satisfactory to all agencies could be issued. A letter on this subject has been written by Gen. Blackmore to the V-2 Panel Chairman requesting such information. The attention of all active agencies is invited to this request. A copy of Gen. Blackmore's letter is appended to these minutes as Enclosure A [omitted].

The problem of obtaining ballistic and trajectory data from Aberdeen has been a troublesome one since the start of the V-2 program at WSPG. Mr. Edson of BRL reported that their problem is not an easy one because of a critical lack of computing and reducing personnel. There are no indications of any immediate relief in this respect. During discussions of this entire problem, it appeared that Dr. Gardiner of State College NM can furnish askania trajectory data very shortly after each firing. Interested agencies may request this information directly. Additional data will be requested from Aberdeen through a single BRL representative such as Mr. Clancy who is located at WSPG and to whom the official requests will be directed.

The Aberdeen problem is one that causes real concern to all active agencies. Significantly, the problem will become worse because as reported by Col. Young, AAF firings at [6] Alamagordo will increase after the first of the year (1948) and BRL's analytical load will become correspondingly heavier.

In discussion of Dr. Pickering's remark that full reduction of data is frequently not necessary, it was suggested that each Panel agency state in its basic firing requests for each missile the extent of tracking data reduction required. Such statements will in many cases alleviate the burden of data reduction.

6. V-2 FIRING SCHEDULE

A copy of the V-2 firing schedule is appended to these minutes as Enclosure B [omitted]. There are no significant revisions other than a request by Dr. O'Day to move up AMC's firing date in the ninth cycle from 19 August to 22 July. The affected agency in this case, the General Electric Company, is agreeable to the change. Consequently, the General Electric Company missile will be fired on 19 August.

To assist WSPG in launching V-2 rockets on a regular schedule, all active agencies will please refer to Enclosure C [omitted] appended to these minutes. Enclosure C is a normal rocket assembly and test schedule as prepared by Mr. L. D. White of the General Electric Company at WSPG.

It is intended to reduce the excessive last-minute load of work on WSPG personnel in the last few days before each firing and should be adhered to by all agencies to the full extent practical.

[7]
7. REPORTS ON COMPLETED FIRINGS

MISSILE #24 – Pressure measurements made by the General Electric Co. on this missile are classified. Mr. Haviland reported that reports have been prepared and these are obtainable through Office Chief of Ordnance.

MISSILE #27 – G.E. Co. – Mr. Haviland stated that the primary experiment in this missile involved heat transfer data. The data has been reduced for the period from 0 to 64 sec. at which time telemetering stopped functioning. A complete report is being prepared which will include skin and boundary temperatures. The max. skin T on the nose was 610°F.

TEST ROUND – NOV. 20, 1947 – Mr. Haviland reported that the overall appraisal of the new American made components is that -they definitely worked satisfactorily up to the time of motor failure. The trouble was attributed to a mechanical failure and not control failure of the new components. This was borne out by the fact that the new components worked perfectly in the Dec. 8 missile.

The new components consisted of the following –

(a) Gyros - German design with modifications.
(b) Mixer computer.
(c) Rebuilt main distributor.
(d) Changed all wiring to stranded wire.

The instrumentation included 3 auxiliary displacement gyros in roll pitch and yaw plus 1 rate gyro. There were 3 pick-off units which isolated the main gyros and 4 pick-off units which isolated the output of the mixer computer.

Mr. Haviland reported that the cause of the turbine shutting down is still unknown.

Mr. Karsch stated that the new components in this test round were very extensively tested at WSPG and found to be fully as good as, if not better than, the German units. The feeling at WSPG is that there will be less [sic] failures in the future than in the past.

The test round missile performance was as follows:

(a) Max. ht. 17 mi.
(b) Range 1 mi. E and 1 1/2 mi. N of blockhouse.
(c) Max. vel. 1700 ft/sec.
(d) Side drift to E believed caused by wind.

[8] MISSILE #28 – AMC – Dr. O'Day reported that the experimental equipment worked up to 100 sec. with the exception of the blossom experiment. At 100 sec. about one-half of the experiments ceased to function because of an explosion in the warhead due to some unknown cause. The blossom experiment was not ejected as planned.

Mr. Gould stated that to date it has not been definitely determined as to what happened to the blossom experiment. Blossom was supposed to go off at Zenith. When the missile failed an actual attempt was made to eject the parachute after Zenith was reached, but the ejection equipment did not function - cause unknown. The parachute was destroyed, but some film was recovered from the containers. The films are now being processed

All missile control equipment functioned perfectly.

This missile remained on the launching platform for some period of time after loading with oxygen. Both Mr. Karsch and Mr. Norton reported on measurements made on the mid-body of the missile during this period. The temperature went down 1°F every 20 minutes. This was felt to be insignificant since the temperature differential in total would be only 6°F in 2 hrs. Although the parachute-camera equipment was close to the O_2 tank, the problem of temperature was not considered to be critical.

On this flight, 2 of the 3 Douglas aspect cameras installed in the missile were recovered. The films are now being analyzed. This is the first successful recovery of the Douglas aspect cameras in a V-2 missile.

In connection with the above blossom experiment there was some feeling that the trouble may have been attributable to the use of 2 cut-off receivers in the missile – this practice is still in use at WSPG. It is believed that the connection of 2 receivers in parallel can interact and cause trouble instead of doubling the safety of the flight, which was the original intent of the decision to use 2 receivers in each V-2. WSPG has recommended to OCO that the number of radio fuel cut-off receivers in a V-2 be reduced from 2 to 1.

8. <u>REPORTS ON FUTURE FIRINGS – EXPERIMENTS OF SPECIAL INTEREST</u>

<u>MISSILE #25</u> – SC – Dr. Golay reported that missile #25 will again contain the sulphuric acid smoke experiment. Launching is scheduled for 8 Jan. at 20 to 25 minutes before sunrise. The missile will also carry a grenade experiment to determine the velocity of sound at various altitudes.

[9] In response to Dr. Golay's request for space on other missiles to be used by the Signal Corps for installation of heliographs (total weight per unit 55 lbs. – desirable complete set comprises 3 units) and sampling bottles, the following agreements were reached.

(a) Missile #36 – 5 Feb. – G.E. Co.
OK for SC to include 3 heliographs
(b) Missile #35 – 19 Feb. – APL
OK for SC to install 3 heliographs plus sampling bottles

<u>MISSILE #34</u> – NRL – Dr. Newell reported that this missile scheduled for launching on 22 Jan. will include the following major experiments.

(a) Cloud chamber experiment.
(b) Pressure measurements with refined gages [sic] to measure pressures up to 135 Km within few % accuracy.
(c) Ionosphere experiment – relative electron and ion densities.

<u>MISSILE #36</u> – G.E. Co. – Mr. Haviland reported that this will be a specially controlled missile.

<u>MISSLE #35</u> – APL – Dr. Van Allen reported that APL will repeat its cosmic ray experiment using a single counter. There will also be included a new spectrograph with a grating of 30,000 lines/in. with two axis sun seeker to hold the light on the slit during roll and tumble. Other experiments to be included –

(a) Heavily lead shielded cosmic ray ionization chambers (in warhead)
(b) Sampling bottles (cleared with Dr. Nichols)

(c) K-25 camera
(d) 3 Heliographs (if OK with WSPG)

This missile is scheduled for launching a half-hour after dawn on 19 Feb.

MISSILE #37 – AMC – Dr. O'Day reported that plans for this missile have been changed. The missile is scheduled for launching on 4 March and was to have contained blossom 3. This is to be a night firing containing the following experiments.

[10] (a) Temperature measurements – especially in the ozone layer.
(b) Sky brightness.
(c) Experimenting with light sources for emission and absorption spectrum.

* * * * * * * *

Dr. Whipple informed active agencies that there are 6 Pohl crystals on hand available for future V-2 tests. These crystals were developed by the Germans and have the property of reacting to ultra-violet light and measuring the intensity between 2000 A° and 2260 A° and below 2000 A°. The crystals do not react to ordinary visual light. Interested agencies may contact Dr. Kuiper of the Yerkes Observatory.

During the discussion of these crystals, Dr. Krause stated that NRL had considered the use of such crystals. Two significant problems are believed to be present; one is that of containing the crystals, the other is the effect of temperature on the crystals. If the crystals are exposed, they may get hot and also on impact the crystals may get hot.

In this connection, Dr. Van Allen stated that APL is equipped to measure the sensitivity of such crystals to temperature changes. APL feels that the temperatures attained on impact are not too significant because to date there has been no evidence of fogging on films recovered after impact.

* * * * * * * *

Mr. Edson informed the Panel that Aberdeen Proving Ground has shipped its new tracking telescope to WSPG for installation. It is expected to be ready for use approximately by March 1. The telescope has a 16" aperture, is mounted on a 90 mm gun mount and has a special recording head. It will be located on Mule Peak, altitude 8000 ft., a few miles south of Alamagordo.

Mr. Edson is also arranging spectrographic attachments to be used in connection with the telescope for studying the spectroscopy of jet flames on missiles in flight.

[11]
9. LETTER FROM RDB RE BUDGET SUPPORT FOR V-2 PANEL AGENCIES

It will be recalled that at the previous V-2 Panel meeting held on 1 Oct. 1947, the Panel voted that in connection with the problem of budgets and expenditures, a letter be written to the Upper Air Panel of RDB stating the way operations have been carried out, justifying the basic principle of this procedure, pointing out the budgetary problem and recommending that RDB support and defend the budgets as submitted by all the active V-2 Panel agencies. On 13 Oct. 1947, Dr. Krause wrote such a letter.

In a letter dated 20 Nov. to Dr. Krause as chairman of the V-2 Panel, C.S. Piggott, Executive Director of RDB's Committee on Geophysical Sciences, wrote that the RDB Panel had regarded favorably the matter of supporting budgets of V-2 Panel agencies and had passed a resolution, to this effect. A copy of Mr. Piggott's letter is appended to these minutes as Enclosure D [omitted].

Since the matter of this year's budgets has been disposed of, it was suggested at this V-2 Panel meeting that the time for action on next year's budgets should be before Sept. 1948.

10. <u>LETTER FROM OCO RE CLASSIFICATION OF V-2 INFORMATION</u>

Appended to these minutes as Enclosure E [omitted] is a copy of a letter from Col. H.N. Toftoy to the Commanding General, WSPG dated 21 October 1947 stating the policy of the Office of the Chief of Ordnance in regard to classification of V-2 information.

11. <u>CONSIDERATION OF OCO LETTER RE CHANGE IN SCOPE OF V-2 UPPER ATMOSPHERE RESEARCH PANEL</u>

Prior to the date of this V-2 Panel meeting, all Panel members were mailed a copy of a letter from Office Chief of Ordnance dated 6 Nov. to the Chairman of the V-2 Panel proposing a plan for increasing the scope and activities at the working level (of the V-2 Panel) to cover upper atmosphere research other than the V-2.

The general feeling was that the Panel members would like to have more time to think about this proposal change. Consequently, the Panel voted to carry over this item until the next meeting.

[12]

12. <u>NEW NRL TELEMETERING SYSTEM</u>

Dr. Newell reported briefly that the first V-2 flight with the new system will be on April 15. A new ground station is being set up. The new system will have the same range as the old system.

13. <u>RECOVERY</u>

Mr. Karsch stated that equipment from V-2 flights is sometimes recovered after the organized search has been discontinued. Identification and disposition of this material presents a problem at WSPG. It is requested therefore that all agencies participating in V-2 rocket firings fill out a form describing the items desired to be recovered. A copy of WSPG's suggested form is appended hereto as Enclosure F [omitted].

All active agencies are urged to cooperate with WSPG in this matter.

14. NEXT MEETING

The next meeting of the V-2 Panel will be held on Wednesday, January 28, 9:30 A.M. (EST) at the Naval Research Laboratory, Bldg. #9, Washington, D.C. Persons planning to attend this meeting will kindly advise the Secretary in order that clearances may be established at NRL.

Prepared by: [signature]
G. K. Megerian
General Electric Co.
Aeronautics & Ord. Systems Div.
Schenectady, New York

Document I-3

[no page number]

PANEL REPORT #35

SUBJECT: Minutes of Meeting of Upper Atmosphere Rocket Research Panel

PLACE: Navy Department
T-Building #3, Room 1803
Washington, D.C.

DATE: Wednesday, April 29, 1953

*** PRESENT ***

Mr. W. W. Berning	APG/BRL	Aberdeen, Md.
Cdr. K. W. Cramp	OCNO Op-51	Washington, D.C.
Capt. G. D. Dean	OC Sig. O	Washington, D.C.
Lt. E. W. Diehl, USN	NOMTF	WSPG, New Mexico
Dr. W. G. Dow	U. of Mich.	Ann Arbor, Mich.
Mr. M. Dubin	AFCRC	Cambridge, Mass.
Dr. H. D. Edwards	AFCRC	Cambridge, Mass.
Dr. M. Ference, Jr.	SCEL	Belmar, N.J.

Dr. C. F. Green	G.E. Co.	Schenectady, N.Y.
Lt. Cdr. G. Halvorsen	OCNO Op-51	Washington, D.C.
Mr. E. E. Harriman	BU. ORD.	Washington, D.C.
Mr. L. M. Jones	U. of Mich.	Ann Arbor, Mich.
Lt. M. S. Jones, Jr. USN	ONR	Washington, D.C.
Dr. L. D. Kaplan	USWB	Washington, D.C.
Maj. T. S. Lewis, Jr.	USAF AWS	Andrews AFB
Mr. J. R. Lein	AFCRC	Cambridge, Mass.
Mr. K. R. Medrow	NRL	Washington, D.C.
Mr. G. K. Megerian	G.E. Co.	Schenectady, N.Y.
Dr. H. E. Newell, Jr.	NRL	Washington, D.C.
Dr. M. D. O'Day	AFCRC	Cambridge, Mass.
Mr. W. J. O'Sullivan	NACA	Langley Field, Va.
Dr. W. H. Pickering	CIT	Pasadena, Calif.
Mr. N. W. Spencer	U. of Mich.	Ann Arbor, Mich.
Mr. J. W. Townsend, Jr.	NRL	Washington, D.C.
Dr. J. A. Van Allen	U. of Iowa	Iowa City, Ia.
Mr . T. B. Walker	Aerojet-General Corp.	Azusa, Calif.
Mr. J. R. Walsh	SCEL	Belmar, N.J.
Dr. F. L. Whipple	Harvard (AF & ONR)	Cambridge, Mass.
Mr. P. H. Wyckoff	AFCRC	Cambridge, Mass.

[2]

*** AGENDA ***

1. Discussion on Oxford meeting.

2. Reports on completed firings

(a) Aerobee AF-34 (GRD-9)	18 Feb. 1953	USAF
(b) Aerobee AF-35 (GRD-10)	14 Apr. 1953	USAF
(c) Aerobee NRL-12	10 Feb. 1953	NRL
(d) Aerobee NRL-13	12 Feb. 1953	NRL
(e) Aerobee SC-27	17 Feb. 1953	SCEL
(f) Aerobee SC-28	24 Apr. 1953	SCEL
(g) Aerobee SC-30	23 Apr. 1953	SCEL

3. Report on results of previous firings

4. Future experiments of special interest

5. Coordinated Panel program for Northern latitude firings

6. Next meeting

*** SUMMARY OF DISCUSSION ***

1. DISCUSSION ON OXFORD MEETING

The Panel was informed by Dr. Van Allen that although there was no recent correspondence from Professor Massey immediately prior to this meeting, he has had substantial assurances that plans are well in hand for the meetings in Oxford the week of August 24.

A firm Panel program has been transmitted to the Oxford group by the Chairman on the basis of the diverse titles and abstracts of papers to be presented by individual Panel members and others. It was suggested that about 150 copies of these papers be sent to the Secretary by mid-July to be forwarded to Oxford for the advance information of participants. This would allow for the preparation of questions to be asked during the symposium, thereby assuring fruitful discussion. All papers, of course, should be cleared through the proper military channels as necessary.

In view of the substantial nature of the contributions from the U.S., England and European countries to the Oxford conference, Dr. Van Allen proposed that durable publications of the papers in collected form should be provided for. He had inquired of the plans of the Gassiot Committee, but had not yet received a reply. It was felt that any British publication which is planned should provide suitable credit to the Panel in the title page, etc. Publication in a special issue of the Reviews of Modern Physics was discussed as a possibility for obtaining automatic distribution of some 5,000 copies in a reasonably prompt way. The possibility that one of the university presses (Princeton, Chicago, etc.) might be interested in publishing the compilation as a bound book was also considered.

It was agreed that the Chairman would ascertain further information on various modes of publication and proceed with the matter as he saw proper.

[3] There was some discussion concerning attendance at the Oxford meeting by persons outside of the Panel. Such persons are thoroughly welcome, although the problem of transportation will be an individual one for them.

Dr. Dow and Dr. O'Day both expressed the thought that contractors' representatives with field experience should be present to discuss actual techniques on instrumentation and rocket handling. It was brought out that Dr. C. F. Green will present a paper on this topic.

It was Dr. Whipple's feeling that the symposium group will be interested more in the scientific aspects of instrumentation rather than techniques and actual operational details.

With regard to the problem of transportation to England, Lt. Jones (USN) of ONR informed Panel members that arrangements can be made to fly with MATS, either from Westover Field, Mass. or from Patuxent, Md. The USAF has daily flights from Westover on all types of aircraft while the Navy flies DC-6's from Patuxent on Tuesdays to London and on Fridays to Paris.

Requirements due to the Coronation make it advisable to submit reservations early and also to obtain priority "One" if possible. A priority "One" may be necessary because it is likely that there will be a sizeable waiting list of military personnel for these flights. Civilian passports and medical injections will be necessary.

The Panel Chairman was advised by Capt. Sanders of ARDC that the Oxford symposium is regarded as a meeting sponsored by the Federal Government. Consequently, USAF personnel are confronted with no travel problem. Naval contractors can become temporary Naval technicians and if the need arises, invitational travel orders can be issued to facilitate travel.

Because of differences in individual plans there will be no effort by the Panel to travel as a group.

* * * * * *

2. REPORTS ON COMPLETED FIRINGS

AEROBEE AF-34 (GRD-9) 18 Feb. 1953 USAF - This flight was a test vehicle reported by Dr. O'Day. Launching took place at 1042 hours after encountering a delay of 2 hours due to trouble with the HADC monitor equipment. The actual versus predicted performance was as follows:

	ACTUAL	PREDICTED
Peak altitude (miles)	77	71
Time to peak (sec.)	178	176
Burnout time (sec.)	32.2	34
Payload (lbs.)	163	–
Dry wt. missile CG	130 in. from tip	
Time to impact (sec.)	545	540

The rocket had a very high roll rate of 7 rps immediately after burnout. Fibreglass, which had been used on the nose cone and fins, did not adhere properly and was believed to have contributed to the high roll rate due to partial peeling.

[4] This flight was instrumented for the purpose of localizing causes of failure experienced previously. Mr. Walker reported that a lateral accelerometer was located in the nose for measuring vibration or violent yawing motion. Temperature-indicating varnishes were painted inside the nose and fins, thermistors were located at various points, a pressure gauge was connected to the thrust chamber, and contacts of the cut-off valve were monitored. In addition an aspect camera was provided by the New Mexico College. Lateral vibration in the nose did not exceed 0 ± 1/4 g. Vibrations, temperature and aspect data will be available later.

Mr. O'Sullivan raised a question whether some structural failure had occurred and called attention of the Panel to research studies by NACA showing the effects of aerodynamic heating on structures flying through the air. These studies have shown

(a) Weakening of structural members due to high temperature.
(b) Temperature gradients resulting in unequal expansion, even for structures of homogeneous material.
(c) Augmentation of the effects (b) in structures of heterogeneous material due to different thermal coefficients of expansion.
(d) Distortion and strain result. Common symptoms are loosening of the skin on aerodynamic structures and subsequent skin flutter.

It was suggested by Dr. Dow that detailed reports be issued to the Panel on these test flights.

AEROBEE AF-35 (GRD-10) 0845 hours 14 April 1953 USAF - This second test vehicle was also a type RTV-A-la Aerobee similar to AF-34 except that there were thin stainless steel "cuffs" (16 mm wide) on the leading edges of the fins to protect the welds from excessive heating. There was no abnormal roll in this flight. Performance exceeded expectations.

Actual versus predicted performance of this flight was reported by Dr. O'Day as follows:

	ACTUAL	PREDICTED
Peak altitude (miles)	77 ±2	72
Time to peak (sec.)	200	175
Burnout time (sec.)	-	34
Payload (lbs.)	159	
Dry wt. missile CG	130 in. from tip	
Time to impact (sec.)	535	540

The burnout time on this flight is not known because of some confusion between two recording stations. Dry weight refers to missile weight less fuel.

There was a single camera inside the rocket looking at one fin and one cable type antenna having 1/16-inch diameter. The fin showed no evidence of flutter at all. The cable type antenna (U. of Utah) was one of three extending from the tip of a fin forward to the nose cone.

Dr. Van Allen mentioned that APL had used with considerable success similar type "outrigger" stainless steel antennas for telemetering and radio cut-off on the first Aerobee flight in 1947 and 1948. A full description of such [5] antenna systems for 85 mc/sec and 34 mc/sec use is contained in the APL Aerobee report.

An aspect camera again was included in AF-35 by the State College of New Mexico.

On both AF-34 and AF-35 no parachute was flown, but the tail cone was separated approximately 20 seconds after zenith time.

It was reported by Mr. Walker that the fins of AF-35 were not located after impact, but the nose section was recovered in good condition. The fibreglass on the nose section indicated an excellent bond. Mr. Walker mentioned also that UCLA under an Air Force contract has developed a "temp-tab" device for determining temperature in a rocket. This device consists of 17 small triangles of different alloy materials which define the maximum temperature reached by fusing in the range 105°F to 550°F. The device has a short time constant.

On the basis of these two test vehicle firings, Dr. O'Day expressed the belief that future Aerobee firings under similar conditions would have a high probability of success. The AFCRL now regards this new 4000 lb. thrust Aerobee as proved in and available for use by upper air research groups.

<u>AEROBEE NRL-12 10 Feb. 1953 NRL</u> - With a payload of 120 pounds, this flight reached an altitude of about 86 miles (138 km) above mean sea level at 195.2 seconds. Other performance data reported by Dr. Newell were as follows:

(a) Burnout - 44.2 seconds.
(b) Velocity at burnout - 4700 ft/sec.
(c) Altitude at burnout – 95,730 ft. (above WSPG)
(d) No recovery of tail section
(e) Nose section recovered

The rocket showed severe heating and discoloration from the nose tip to a slot section 25 inches back from the nose tip. Below this point there was evidence of paint discoloration.

A modified Bennett RF mass spectrometer failed to operate in this flight. The most probable cause is believed to be leakage of pressure seals.

The altitude was also verified by rocket borne pressure gauges. The Aerobee was one of the original 2600 lb. thrust type. The payload was especially light.

<u>AEROBEE NRL-13 12 Feb. 1953 NRL</u> - Results of this flight were reported by Dr. Newell as follows:

(a) Launched at 0008 MST
(b) Peak altitude 84.5 ± 0.5 miles (136 km) above sea level
(c) Burnout 46.2 sec.
(d) Velocity at burnout 4640 ft/sec
(e) Altitude at burnout 98,460 ft (above WSPG)
(f) Time to peak 196.2 sec.
(g) Impact NE corner White Sands region
(h) Recovery of nose section successful
(i) No tail parts found.

[6] The rocket showed very severe heating from the tip to the slot section 25 inches back from the nose tip, severe heating from this slot to the end of the ogive and discoloration of paint below this point.

The r.f. mass spectrometer operated successfully; 97 samplings of composition were taken between 85 km and 106 km on the ascent of the rocket and 103 samplings between 136 km and 89 km on the descent of the rocket at a sample rate of 0.94 per second. Of these samples, those between 85 km and 106 km on the ascent must be discarded because the mean free path of the gas within the spectrometer was too short for proper operation. The range covered was 54 AMU to 6 AMU. Analysis of these data is almost complete and a report will be issued in late spring.

Mr. Townsend of NRL observed that there was no change in the A/N2 ratio during flight although the gas pressure within the tube did vary—decreased—until peak and then increased on the way down in a reasonable manner. Water vapor was recorded as well as several peaks identified as coming from higher hydrocarbons. There is a strong possibility that the air in the spectrometer was contaminated with gas carried along by the rocket, gas arising from the burned paint on the nose cone, and residual rocket fuel vapor.

Dr. Whipple reiterated his suggestion of a previous meeting that the deliberate inclusion of a contaminant of distinctive molecular weight in the pressurizing gas of the nose cone will provide a good check of validity of the high altitude results.

Mr. Townsend reported that Freon had been introduced for this purpose in NRL-12 but unfortunately not in NRL-13. The seal of the spectrometer was opened at 93 km on NRL-13. Emission was off for the first 10 or 12 sweeps, then settled down. The spectrometer appeared to be operating properly from 106 km over the peak of flight and back down to 106 km on the decent [sic].

Apart from the contaminants noted above, the composition of the samples closely resembled sea level air in #13. In later flights it is planned to expose the complete spectrometer to the atmosphere without the usual entrance canal and glass envelope. It may be possible in this way to learn something of the degree of dissociation of O_2 and N_2.

Flights 12 and 13 also carried pressure and density instrumentation which operated satisfactorily. A comparison of day-night data will thus be possible.

In connection with mass spectrometry, Dr. O'Day cited the work of Mr. Friedman of the University of Connecticut. He has a small mass spectrometer for the analysis of Geiger counter gases. Such an instrument might be adaptable to rocket use. Dr. Nier at Minnesota has recently developed a compact permanent magnet type mass spectrometer.

In June on the NRL Viking rocket flight, the r.f. mass spectrometer experiment will be repeated. The nose cone will not be opened up, but an intake tube will be utilized having a very high conductance, an open area diameter of 2 1/2 cm and a length of 6-7 cm.

[7] AEROBEE SC-27 17 Feb. 1953 SCEL - This rocket flew successfully to an altitude of 66 miles with a payload of 155 1/2 lbs. Additional performance data reported by Mr. J. R. Walsh was as follows:

(a) Time of firing 2350 hrs.
(b) Predicted peak altitude 67 1/2 miles
(c) Actual peak altitude 66 miles
(d) Time to peak alt. (Pred.) 176 sec.
(e) Time to peak alt. (Act.) 174 sec.
(f) Max. velocity at burnout (Pred.) 4180 ft. /see.
(g) Max. velocity at burnout (Act.) 4111 ft. /sec.
(h) Burnout time 45.1 sec.

Aerobee SC-27 with a 2600 lbs. thrust motor was instrumented with seven grenades to measure upper air temperatures in the winter season. Of these 7 grenades, 5 exploded in space at estimated altitudes from 127,200 ft. to 258,000 ft. Grenade rounds #2 and #7 failed.

Weather at the time of the firing was optimum - clear, calm and ideal for sound ranging measurements. The ballistic cameras and sound ranging equipment worked satisfactorily. Telemetering failed at about 75 seconds, but ground flash detectors observed the grenade detonations, so its loss was a non-essential one.

Data from this flight had not been reduced at the time of this meeting, but are believed to be of exceptionally good quality.

Both the temperature and the horizontal wind structures in the above mentioned altitude range will be obtained, assuming zero vertical wind velocity.

Dr. Ference reported that the analysis of data of this nature from previous flights had shown a maximum wind velocity as high as 200 mi./hr. at 50 km altitude. The altitude increments between grenade bursts are usually about 8-10 km.

There was considerable discussion on winds and on visible meteor trains in the atmosphere. The determination of the actual wind fields up to 80 km altitude has much scientific interest as does the observation of various types of trains.

Dr. Whipple stated that wind shears of 100 mph in less than one kilometer have been observed in rare cases. Accumulating evidence suggests that there may be as much structure in the 50-100 km altitude range as in the troposphere.

There was extended discussion of possible rocket methods for making persistent smoke trains or other type trains at high altitudes, which could be photographed for at least several minutes in order to directly observe wind velocities. In spite of earlier failures by various groups to make satisfactory trails, Dr. Whipple stated his belief that it should be possible to do. This belief was based on the observed persistence of meteor trains (a few minutes up to an hour) in spite of the low energy delivered to the atmosphere. Meteor trains are self luminous but it is not known whether it is the amount of material or amount of energy which is of significance in determining the visibility.

[8] Dr. Ference commented on the potential meteorological significance of a knowledge of the wind structure in the 45-70 mile altitude region for understanding large scale movements of the atmosphere. At altitudes above 100 km, ionospheric methods are now yielding a considerable body of wind information.

The Air Force Cambridge Research Laboratory is sponsoring the development of a sodium vaporizer by the Arthur D. Little Company. Dr. O'Day described it as a simple device using thermite burning at 3000°F as a source of heat.

Mr. Walker and Dr. Van Allen suggested the use of a rocket itself as a possible generator of a visible train. It would probably be a simple matter to add sodium in the propellant of a Jato and arrange for it to be fired at high altitude.

Vikings burn to an altitude of about 65 km but no night flights have been made yet. The WAC Corporal in successful Bumper burned into the E layer, but this was a day flight. Most rockets—Aerobees, V-2, Corporals, etc.—burn out at much lower altitudes.

The temperature of a Jato flame is 3000 - 4000° F.

AEROBEE SC-28 24 April 1953 SCEL - Only a very preliminary report was available on this firing. Mr. Walsh reported that the flight was instrumented with 7 grenades to measure temperature and winds. The last 2 grenades failed to detonate. Launching took place at 0319 hours immediately following the U. of Michigan "falling sphere" flight, in an effort to correlate temperature data. Velocity at burnout on SC-28 appeared close to normal. Data from the five successful grenades are probably good.

It was observed by Dr. Ference that the Signal Corps appears to have experienced the lowest altitudes on Aerobee flights taking payload into account. Burning times and velocities at burnout appear to be normal, but peak altitudes have fallen short when compared with APL and NRL Aerobee flights.

Fabrication at Aerojet of these Aerobee rockets takes place separately for the NRL and SCEL contracts. Handling procedures at WSPG, however are identical for all Aerobees.

Lt. Diehl, Officer in Charge of Aerobee firings at the Naval Ordnance Missile Test Facility at White Sands, was present at the meeting. He reported that they are equally puzzled by the subnormal performance of the SCEL units. SC 27 and NRL 12 and NRL 13 were fired within a few days of each other. Acid from the same batch with the same storage was used for all three. Fuel for the three was all mixed at the same time. Handling procedure was as nearly identical as possible. All three were pressurized with helium to 2250 psi.

The quality of the acid has been the subject of considerable investigation. Mr. Walker stated that it has been conclusively established that for the old thrust chambers (APL-NRL type) acid containing about 1% of solids (Fe_2O_3 etc.) gives much superior performance. The catalytic action of these traces of impurities is apparently the reason for the mysteriously superior performance of acid stored in stainless steel drums over that stored in aluminum drums.

[9] AERORBEE SC-30 24 April, 1953 SCEL - This flight was instrumented by the U. of Mich. for SCEL with a "falling sphere" experiment to determine upper air densities. Weather conditions at the time of launching were poor. Time of launching was 1230 hours. Performance and instrumentation reported by Mr. Jones were as follows:

(a) Peak altitude 365,000 ft.
(b) Time to peak 185 sec.
(c) Payload 110-120 lbs.
(d) Extended blunt ogive nose section and very long rocket
(e) Sphere ejected from front end by Doppler Command
(f) Single antenna used for receiving and transmitting

The inflated sphere is 4 ft. in diameter and weighs about 50 lbs. It is perfect to about ± 1/2 inch. The use of a single antenna for receiving and transmitting makes less ambiguous the corrections for spin in reduction of the Doppler data. The thirteen Doppler receiving stations were each provided with crossed dipole antenna.

Data were received by the 13 Doppler receiving stations, but between x + 250 sec. and x + 370 sec. there was a power failure at one of the ground sources supplying power to 8 of these 13 recorders. Despite this failure, there were adequate data from the remaining stations.

An 80°F rise in temperature was noted on the sphere. The inflated diameter of the sphere is 4 ft. and the weight is 50 lbs. There are no projections on the sphere and the Doppler unit is inside the sphere. Measurements can be made beginning at 260,000 ft. on the way down.

No leaks were experienced in this flight. The pressure was 3.7 psi absolute. In the range of 365,000 to 300,000 ft. an acceleration accurately equal to g was observed. At 270,000 ft. there was about 0.1 g drag. The Doppler positional error is ± 0.1 cycle or about 1.3 ft. (random). The ambient temperature can be determined to about ± 8°C with present technique. It is hoped to ultimately reduce the error to ± 4°C. The present battery and Doppler unit weigh 9 pounds. The internal pressurizing gas bottle weighs 20 lbs.

The Michigan group would prefer to work with a rigid sphere 15" diameter—the same diameter as the Aerobee. This will be aerodynamically equal to the present sphere if the weight can be reduced to five pounds.

Dr. Whipple inquired if a light in the sphere might not make a much lighter body (with subsequent greater drag acceleration). Optical tracking can give an accuracy of ± 10 feet at 50 miles using the star background in a night firing. Dr. Whipple's meteor cameras give 60 measures/sec as presently set up near Las Cruces.

Mr. Jones replied that this possibility has been under consideration

Vertical winds do not cause an appreciable error in temperature as long as they are negligible in velocity compared to the velocity of the sphere (Mach No. = 2).

****

[10] 3. <u>REPORTS ON RESULTS OF PREVIOUS FIRINGS</u>

There were no further data to report at this time.

4. <u>FUTURE EXPERIMENTS OF SPECIAL INTEREST</u>

Dr. Van Allen reported that the State University of Iowa is preparing another summer expedition to Thule, Greenland, and then back South. Fourteen balloon-launched rocket flights are planned to measure total primary cosmic ray intensity and to measure separately the primary intensities of heavy nuclei. Seven of these flights will be instrumented with single Geiger counters while the remaining seven will be equipped with ionization chambers.

In addition, Dr. Van Allen's group will collaborate with NRL in another 7 rounds to obtain pressure and temperature measurements as far north as possible.

It is planned in these firings to employ the same techniques used last summer. The highest successful flight at that time with a Deacon rocket was 295,000 ft. from a balloon launching altitude of 57,000 ft. It is hoped that a Deacon summit altitude of 325,000 ft. will be achieved from a balloon launching altitude of 70,000 ft. When this is accomplished, an attempt will be made to reach 360,000 ft. from a launching altitude of 95,000 ft. This is believed to be the limit of the techniques using the Deacon rocket.

****

Mr. Lein of AFCRC revealed that the Air Force plans to instrument several Aerobee rockets for sky brightness and electron density experiments. The first of these firings will take place in June and July. The June rocket will contain the propagation experiment, while the July rocket will have a bi-axial pointing control to measure the solar constant. Other firings will follow in September and October.

****

At NRL, plans are being made to instrument Viking #10 for a group of experiments on propagation, temperature, pressure and high altitude densities. The flight is scheduled for 23 June and represents the last of initial upper air series Viking rockets. One of the camera installations will represent an effort to evaluate the light of the day sky.

Following Viking #10, there will be an Aerobee flight with a sun-follower and set of photon counters in the autumn of 1953.

****

In view of anticipated reductions in Government appropriations, Dr. Dow raised the issue of how the Panel might seek to obtain more information per dollar. It was his observation that the State University of Iowa group under [11] Dr. Van Allen was accomplishing this objective by the technique of launching a small rocket from a balloon.

Dr. Van Allen stated that except for two men, his group carries on this activity as thesis work with graduate students. A large rocket requires an expensive outlay, but if the scale of operations can be reduced to the level of usual university research activities, fruitful and thoughtful high altitude work can be done with modest expenditures.

****

5. <u>COORDINATED PANEL PROGRAM FOR NORTHERN LATITUDE FIRINGS</u>

This subject encompasses the general idea of upper air studies by means of rockets fired around the world and the International Geophysical Year 1957-58.

Because of the Panel's broad interest in this subject and with specific reference to a recent letter by the Signal Corps to CNO indicating interest in Aerobee rocket-firings from the *U.S.S. NORTON SOUND* at northern latitude, Dr. Van Allen invited Lt. Cdr. Halvorsen of OCNO to discuss the matter informally.

Cdr. Halvorsen was in charge of Aerobee firings at White Sands for several years and subsequently was missile officer on the *U.S.S. NORTON SOUND* during Aerobee and Viking firing expeditions.

He pointed out that the *NORTON SOUND* is presently assigned to high priority military missile evaluation programs. The first reaction in C.N.O is that it would be very difficult to arrange time for upper atmosphere research expeditions. However, there are occasional lulls in the missile activities and he stated his belief that with a sufficiently flexible program on the part of the upper atmosphere groups involved, advantage might be taken of these lulls to conduct Aerobee firings in the Alaskan region. It will also be necessary that preparations for the work be as complete as possible in order to reduce the time required of the ship.

The Signal Corps at present is looking for an expression from the Navy before deciding on a site and preceding [sic] with any definite plans. If the Navy Dept.'s response is favorable, the Signal Corps will prepare a detailed study of all requirements. Serious consideration already has been given to instrumentation requirements (land based) for use in conjunction with shipboard firings.

Instrumentation requirements in the launching area present a major problem unless military installations are already there.

A launching crew would have to be supplied by the Signal Corps, although these men can be supplemented to some extent by the ship's crew in a distant land-based operation.

It is recognized that the Signal Corps' request to CNO is of a type that will recur periodically from other agencies and, in Cdr. Halvorsen's opinion, CNO will be able to accommodate such requests only on the basis of free periods between the *NORTON SOUND* missions.

There was some discussion concerning an alternate plan involving a mobile [12] Aerobee tower platform and utilizing a railroad flat car for transportation to other latitudes. A highway caravan also was mentioned.

Mr. Jones of the Univ. of Michigan has expressed his thoughts on this subject in a recent letter to the Signal Corps appended hereto as enclosure A [omitted].

In the course of discussion, Dr. Whipple expressed the thought that a mobile carrier not only has the advantage of low maintenance, but also should be of vital interest to Army Ordnance who may wish to undertake a program separate from that of the Panel.

Mr. Townsend suggested that it might be advantageous to mount the Aerobee tower, associated tracking and telemetering facilities and work shops on automotive trailers of the sort available in the Army Transportation Corps. Such trailers could then be moved over roads or could be transported on railroad flat cars as appropriate.

It became evident that a new Aerobee tower would have to be built; split perhaps in 3 sections for convenience in handling, transportation aboard a flatcar and erection. This would appear to be cheaper than attempting to rebuild the existing Norton Sound Aerobee tower.

The Panel is agreed that the basic idea of mobile Aerobee firing facility is an excellent one, but that one agency should take the initiative with the Panel's support to get the job done. Other agencies would also use the facilities.

As a consequence, the Signal Corps will undertake to contact Army Ordnance in this matter. The Ordnance Department will be furnished with a technical plan covering mobile launching equipment for Aerobees, an estimate of the cost involved and the urgency for such project.

It was felt there may be a possibility that some equipment of this type exists in the Ordnance Department. This will be investigated, but in any event one of the largest items of expense appears to be the new tower.

In this connection, Mr. Berning stated that missiles fired from a mobile caravan—either railroad or highway—pose a major problem in obtaining accurate tracking data. The estimated cost would be $150,000 for a Doppler station to give triangulation on a 200 ft. base line. Two right angle 200 ft. base lines would be needed. Increasing the base line involves increased costs and additional Doppler stations.

Dr. Dow suggested emphasis on experiments which are not dependent on the accuracy of tracking data; i.e., sampling, pressure measurements, etc.

****

With reference to the experiments to be performed during the international geophysical year 1957-1958, Dr. Newell submitted a preliminary list of experiments for the consideration of Panel members. This list is appended hereto as enclosure B [omitted].

Prior to this meeting and, in response to Dr. Newell's letter, Mr. Berning suggested complete Doppler instrumentation would be required for the sphere [13] experiment. Also, Dr. Ference had suggested that photography of the earth's surface be added to the list to determine cloud structure, cover, albedo and other geophysical data.

Dr. Dow indicated that he would like to see a globe of the world with a master plan outlining what is to be done in 1957 so that the Panel could work in the direction of filling in the vacancies. It was emphasized that the Panel must think in terms of priorities and dollar costs.

In this connection, Dr. Van Allen felt that there is much to be said for an inexpensive procedure which can be extended to numerous sites on a high quantity basis. Balloon-launched Deacon rockets cost only a total of $1,000 each and fill this category for a number of experiments. He expressed the opinion that only in some such inexpensive manner can the Panel look forward to continuing high altitude research with rockets over a considerable period of time.

A distinctively different view was expressed by Dr. O'Day and Dr. Pickering. They both commented upon the increasing tempo of development and production of large military missiles and of proving ground and missile ship facilities. If military need for these missiles is reduced in the coming period of years, they may be available in large quantities for use by scientific groups.

Dr. Newell pointed out the role of large rockets for exploratory measurements at ever higher altitudes whereas the smaller, inexpensive rockets are suitable for a wide range of synoptic observations.

In further discussion of plans for the International Geophysical Year, Dr. O'Day suggested that the Oxford Conference in August would be an appropriate time to discuss the matter with our foreign colleagues who will be present there. Of particular interest will be the coordination of ground-based experiments with rocket firings of the Panel agencies. He believes that the Panel should agree on a few important areas of investigation; that collective effort should be concentrated on these in the interest of conclusiveness and full comprehension.

It was agreed to defer further discussion until the Oxford Conference.

****

Before the close of the meeting, Mr. O'Sullivan informed the Panel of a new rocket unit designated as the T-40 which is available now to all the Services. The T-40 has half the thrust and twice the burning time of the Deacon, but the specific impulse is considerably better than that of the Deacon.

****

6. NEXT MEETING

The next meeting of the Panel will be held at the Air Force Cambridge Research Center during the first week in October. Further details will be issued at a later date.

Members are reminded to keep in touch with the Secretary concerning the development of transportation and housing arrangements for the Oxford Conference.

Prepared by G. K. Megerian [signature]
General Electric Company
Schenectady, New York

Document I-4

Document title: Technical Panel on Earth Satellite Program, United States National Committee for the International Geophysical Year 1957–1958, "Minutes of the First Meeting," October 20, 1955.

Source: Archives, National Academy of Sciences, Washington, D.C.

Once the proposal that the United States would attempt to launch a scientific satellite during the International Geophysical Year (IGY) had been approved by the Eisenhower administration, and the Vanguard proposal of the Naval Research Laboratory had been selected for the satellite program, the National Research Council, the operating arm of the National Academy of Sciences that was in overall charge of U.S. preparations for the IGY, organized a Technical Panel on Earth Satellite Program (TPESP). This panel set the scientific objectives for the Vanguard program, selected experiments for various launch attempts, and oversaw the implementation of the program.

[no page number]

National Academy of Sciences
National Research Council
Washington 25, D.C.

ADMINISTRATIVELY RESTRICTED
FOR OFFICIAL USE ONLY

UNITED STATES NATIONAL COMMITTEE
for the
INTERNATIONAL GEOPHYSICAL YEAR 1957-58

Minutes
of the
First Meeting
Technical Panel on Earth Satellite Program
October 20, 1955
Room 716, 1145 19th Street, N.W.
Washington 6, D.C.

1. Attendance

1.1 Members: R. W. Porter (Chairman), Hugh Odishaw (Secretary), Joseph Kaplan, H. E. Newell, Jr., W. H. Pickering, A. F. Spilhaus, Lyman Spitzer, J. A. Van Allen, F. L. Whipple.

1.2 USNC Secretariat: G. F. Schilling, Marian McCray.

1.3 Invited Participants and Observers: S. E. Clements, J. P. Hagen, J. W. Joyce, T. J. Killian, P. A. Smith, C. S. Weaver.

2. Introductory Session

2.1. Dr. Porter as Chairman of the USNC-IGY Technical Panel on the Earth Satellite Program, convened the meeting and welcomed the participants. He proceeded to discuss the tasks before this Panel, making reference to letters of October 2, 1955, from Dr. Joseph Kaplan, Chairman of the U.S. National Committee for the IGY, appointing the members of this Panel (Attachment 1 to these Minutes). The Chairman summarized the principal tasks for the Panel:

a) To formulate the scientific program to be carried out by means of artificial satellites as part of the U.S. program for the International Geophysical Year.
b) To delegate and direct the execution of this program.
c) To establish policies and formulate procedures related to the program in

the fields of (i) budget, (ii) information policy, and (iii) institutional relationships.

[2] 2.2. The Chairman outlined the relationships of the Panel to other groups as follows:

2.2.1 Because the Panel is an instrument of the USNC, all Panel actions must be approved by the USNC or the USNC Executive Committee. In its advisory function to the USNC, the Panel is expected to take executive actions within the scope of responsibilities outlined in the letter of Dr. Kaplan, October 2, 1955.

2.2.2 The relation of this Panel to other USNC Technical Panels and Committees is one of independent equality, and the interchange of information, with appropriate discipline consultation as deemed necessary, represents the principal responsibility of this Panel to other Panels.

2.2.3 The relationship of this Panel to the Department of Defense will be carried out in the form of liaison on the appropriate level. The following discussion brought out that sufficient over-lap was in existence between this Panel and appropriate DOD Committees through joint-membership to ensure satisfactory initial working relations at this time.

2.2.4 The discussion did not reveal the necessity of immediate direct contact of this Panel with other Government agencies, since these relationships appear to be properly covered by existing USNC relationships.

2.2.5. In view of the critical time-table of the Earth Satellite Program, the Panel stressed the desirability of direct relationship to Project Vanguard. It was agreed in the following discussion that there would be complete information interchange with Project Vanguard and with the Office of the Assistant Secretary of Defense for Research and Development.

2.3 The Chairman made reference to the AGENDA (Attachment 2 to these Minutes) and the AGENDA DOCUMENT before the participants, and the Panel proceeded to discuss the various agenda items.

3. <u>Budget Problems</u>

3.1 Mr. Odishaw proceeded to review the historical developments pertaining to the USNC-IGY budget and detailed the present status. He explained the necessity of developing a budget for the Earth Satellite Program before November 7, 1955, for presentation to the USNC at this date; this stringent deadline had been made obligatory by present commitments of the USNC to submit a complete IGY supplemental budget to the NSF, the Bureau of the Budget, and the Congress for the next fiscal year. He alluded to the difficulty of preparing such a budget without accurate knowledge of financial requirements of Project Vanguard.

[3] 3.2. A discussion ensued as to a budgetary interpretation of DOD logistics support of the IGY Satellite Program as stated in the Presidential announcement of July 29, 1955. It was understood that the Department of Defense agreed, in principle, to furnish logistic support for the NAS-USNC Satellite Program within reasonable limits, similar to logistic support provided by DOD for USNC-IGY projects such as the Antarctic Program, the Fort Churchill Rocket operation, and other undertakings. The Panel felt that it was necessary to get a better understanding of the scope and limitations of Project Vanguard before detailing a budget for the USNC to be presented before the Congress.

3.3 It appeared that at the present time the objectives of Project Vanguard were to put one satellite into orbit with six vehicles being scheduled to try this. Since it obviously cannot be predicted what number of trials will have to be attempted before successful achievement of the objective, thus ending Phase I of Project Vanguard, the Panel faced the difficulty of budgeting for ten instrumented satellites as called for in the USNC-IGY LPR program and budget document of May 6, 1955, without being able to define the unknown number of Vanguard trys [sic] which may not be completely successful. A compromise solution was later found by the Panel (see Item 3.8 of these Minutes).

3.4 Dr. Spilhaus suggested that the Panel could budget regardless of n £ 6 unsuccessful Vanguard trys [sic], if broken down into the following categories:

a) Basic Instrumentation.
b) Number of scientific instrumentation units up to a maximum of six.
c) Operation costs depending on estimate of length of time.
d) Propulsion units plus instrument units needed above the number of six.

Tentative time estimates indicated that Project Vanguard would attempt to start earnest trys [sic] by the beginning of the IGY, i.e., in July 1957.

3.5 The Panel set up a Working Group on Budget, consisting of

Homer E. Newell, Jr., Chairman
J. A. Van Allen
F. L. Whipple

with the following assignment: to prepare a satellite budget, at least in preliminary form, for presentation to the USNC on November 7, 1955; this budget is to be prepared in cooperation with the USNC Secretariat. It was understood that this group would take proper cognizance of the LPR Program and budget document of May 6, 1955, as approved by the USNC, but would not be limited by this document because of further developments since that date.

[4] 3.6 On request, Mr. Odishaw presented an outline of general budget policies and procedures of the USNC as follows: The USNC-IGY has been set up by the National Academy of Sciences-National Research Council and charged with responsibility for planning, direction, and execution of a scientific program which constitutes the effort to be extended during the International Geophysical Year by the United States of America. To this effect, the USNC, on behalf of NAS-NRC, has set up technical panels and subcommittees and has developed appropriate scientific programs and budgets.

A source of support for the program was needed and the Academy decided that the National Science Foundation was the appropriate Federal agency through which the program could be presented to the Congress. The Foundation has been responsible for the Government's fiscal sponsorship of the Academy's IGY Program since then, and a procedure has been developed and successfully utilized whereby NSF cooperates with the USNC in submitting the Academy's IGY Program budget to the Bureau of the Budget and the Congress. Individual projects, after having been accepted, reviewed, and endorsed by USNC Technical Panels, and approved by the USNC, receive their funding in the form of individual grants from the USF upon request by the USNC.

3.7 It became apparent that no details were available to DOD with regard to a preliminary USNC-IGY budget of $5,300,000 for Project Vanguard. The Panel finally resolved that it would be advisable to follow the same pattern which has been set as a precedent by other IGY projects such as the conventional Rocket Program.

3.8 The Panel discussed and drew up the following tentative operational procedures:

	Item	Agency Budget	Responsibility
1.	Primary Costs, Missiles, Launchings, Logistics (6 trys [sic])	DOD	DOD
2.	Costs of Basic Instrumentation and Observation and Computations to place Missile in orbit, track it, and Telemetering Air to Ground. Possibly to include Orbit Computing Center.	Suppl. USNC – IGY	DOD (NAS advisory)
3.	Basic Scientific Instrumentation, Optical Observations, Communication, Data Reduction, and Publication.	Supply. USNC – IGY	NAS (DOD advisory)

3.9 A discussion developed on the present plans of NRL and the envisioned scope of Project Vanguard. It appeared that Project Vanguard has made tentative plans to include scientific instrumentation in addition to the setting up of observation stations, the cost for putting the experiments into satellites, and telemetering. An official statement was made on behalf of DOD that Project Vanguard plans were not yet approved by DOD, that DOD would secure this information and share it promptly with the USNC, and that DOD was in full accord with general NAS-USNC policies and attitudes vis-a-vis the NAS-USNC satellite program.

[5] 4. Scientific Program

4.1 A detailed discussion developed on technical and scientific aspects of the Earth Satellite Program for the IGY.

4.2 This discussion introduced the topic of security classifications. Mr. Odishaw stated that the USNC Secretariat had not made an official check on the security clearances of all participants, but informally had reason to believe that all participants had active security clearances.

It was realized that scientific instrumentation would be unclassified, but that certain technical information is to be considered, at least at present, as SECRET security information. The Panel passed a resolution that for the present everything pertaining to this meeting would be considered ADMINISTRATIVELY RESTRICTED–FOR OFFICIAL USE ONLY until such time when the Panel decides to release either portions of it or all of it. It was further resolved that at the present time the technical information contained in the discussion under item 4.1 above and item 4.3 below would be considered SECRET information with access limited to participants of this meeting. This portion of the Minutes is contained in Attachment 3 [omitted] to these Minutes and this Attachment carries a SECRET security classification as of 20 October 1955.

4.3 The Panel reviewed preliminary requirements and drew up specific requirements relative to Project Vanguard (see Attachment 3 to these Minutes), and unanimous agreement was reached that NRL will submit to this Panel specific formal proposals for proposed projects and related instrumentation of specific designs for consideration at the next Panel meeting.

5. Information Problems

5.1 The Panel discussed the topic of a Symposium on Scientific Merit of a Satellite Vehicle (refer to Tabs 6.1 of the AGENDA DOCUMENT). The Panel finally set up a Working Group on a Symposium consisting of Van Allen (Chairman) and Odishaw with the following assignment: evaluation of problems of a Symposium and Briefing Session as to timing, and preparation of a Symposium outline for consideration by the Panel at its next meeting.

5.2 The Panel discussed operational procedures of the scientific aspects of the Earth Satellite Program, including ways and means of inviting project proposals, reviewing them, and execution of same. Attachment number 4 is a copy of the general outline proposed by Dr. Van Allen. The Panel suggested that the Van Allen process be related to the Symposium and Briefing Session.

5.3 The Panel set up a Working Group on Optical Observations and Tracking, consisting of Dr. F. L. Whipple (Chairman) and Dr. Lyman Spitzer with the following assignments: study and planning of the optical tracking portion of the satellite program and submission of a report to the Panel at its next meeting. The Working Group was advised to work closely with NRL.

5.4 The Panel agreed to release a public announcement (probably at or shortly following the next meeting) relative to the existence and membership of this Panel.

5.5 The date of the next meeting of this Panel was set for 9:30 a.m., Monday, November 21, 1955, in Room 716, 1145 19th Street, N.W. Washington 6, D.C.

****

[Attachment 1]

[no page number]

COPY

ADMINISTRATIVELY RESTRICTED
FOR OFFICIAL USE ONLY

National Academy of Sciences
2101 Constitution Avenue, N.W.
Washington 25, D.C.

UNITED STATES NATIONAL COMMITTEE
for the
INTERNATIONAL GEOPHYSICAL YEAR 1957-58
October 2, 1955

Dear Dr. . . . :

On behalf of the USNC Executive Committee, I am writing to ask you to serve as a member of the USNC Technical Panel on the Earth Satellite Program. The enclosed document provides some recent information on the U.S. - IGY program, and some reference to the satellite program appears in the section on rocketry. To this brief description, I should like to add the following comments:

The Committee's thinking about an instrumented satellite program began last October during the international meeting on IGY at Rome. Prior to this meeting, the IUGG and

URSI had adopted resolutions recommending such an effort. The ICSU Special Committee for IGY, concerned with international coordination of the IGY, considered these resolutions and adopted a similar one. It was this which initiated our planning at Rome.

Upon the return of our group to the United States, the USNC carefully considered the resolution and adopted a favorable position by March 10. In the meantime, a special study group was established, drawn largely from our panel on rocketry and including certain specialists, to consider aspects of technical feasibility and the scientific experiments that might be undertaken. These studies (which, of course, will be made available to the Panel) permitted the Executive Committee to propose a preliminary program and budget document designed to serve as the basis for securing our Government's position. As you know, the President announced the Nation's backing of the program on July 29, 1955.

The document alluded to above recognized that a satellite program would require the close collaboration and assistance of the Department of Defense. Accordingly, the satellite program was formulated in terms analogous to those already established for our "conventional" rocket exploration program, in which the Committee assumed responsibility for the planning and direction of the scientific program and provided funds, through the National Science Foundation, for the rockets and instruments while Defense agreed to provide facilities, launching equipment, and various logistic support. A somewhat similar pattern is envisioned for [2] the satellite program and I am pleased to report that preliminary work is already proceeding along these lines.

The USNC Technical Panel on the Earth Satellite Program provides, in our opinion, a challenging opportunity for contributions in a pioneering area: I am sure that I need not elaborate on this point. The problems confronting the Academy and the Committee are important ones. The Panel must consider, for example, a variety of scientific and technical problems, must establish various policies concerning the projected program, evaluation of suggested research projects, and direction of the effort. The activities of the Panel will require meetings of the group and may well entail a fair amount of time—how much it is difficult to say. The Panel will undoubtedly add working groups and consultants, and such moves will, perhaps, spread the burdens facing us while simultaneously drawing in specialists that the group may desire on various topics, e.g., orbit problems, vehicle instrumentation observing station instrumentation, etc. The proposed Panel group is listed on the attached sheet.

I have discussed this subject and your nomination with Dr. Bronk, President of the Academy, and he joins me in the hope that you can serve on this Panel, whose activities may well be destined to play a historical role in the research of outer atmosphere. It is urgent that the Panel begin its operations soon: therefore, may I hear from you promptly by airmail or by collect telegram.

Sincerely yours,

[signature]
Joseph Kaplan
Chairman

Enclosure:
Proposed United States Program for the International Geophysical Year 1957-58
August 1955

****

Attachment 2
ADMINISTRATIVELY RESTRICTED
FOR OFFICIAL USE ONLY

National Academy of Sciences
2101 Constitution Avenue, N.W.
Washington 25, D.C.

UNITED STATES NATIONAL COMMITTEE
for the
INTERNATIONAL GEOPHYSICAL YEAR 1957-58

First Meeting of the USNC-IGY Technical Panel
on the Earth Satellite Program to be held on
October 20, 1955, at 9:30 a.m. in Room 716
1145 19th Street, N.W. Washington 6, D.C.

AGENDA

1. Terms of Reference: (Tab l).

2. Review of USNC-IGY Satellite Program Background (Tab 2).

3. Review of DOD Developments.

4. Scientific Program:
   4.1 Choice and Design of Scientific Experiments (Tab 2.3).
      (i) The Instrumentation Problem.
      (ii) The Ground Station Observational Problem.
   4.2 Handling of Program Proposals (Tab 4).

5. Budget Problems:
   5.1 The USNC-IGY Program Budget.
   5.2 The USNC-IGY Supplemental Budget.
   5.3 The ESP Budget Document of 6 May 1955 (Tab 2.4).

6. Information Problems:
   6.1 Security Classifications.
   6.2 Scientific Information: The Proposed Symposium (Tab 6.1).
   6.3 Public Relations:
      (i) Background: The International Context and the NAS Position (Tab 2.1); White House Announcement (Tab 2.5); Vanguard Announcements (Tab 6.2); DOD Meeting on 12 October 1955.
      (ii) USNC Recommendations: The Basic TPESP Public Relations Policy. The implementation of this Policy.

(iii) Public Relations vis-a-vis the Panel: Public Announcement. Panel Member Positions on Inquiries and in Addresses.

7. Next Meeting: November 18, 1955.

****

[Attachment 4]

COPY

[handwritten: "Van Allen Outline of Operational Procedures"]

19 October 1955

DEPARTMENT OF PHYSICS
State University of Iowa

Scientific Program with Satellite

A. Technical Aspects Common to Nearly All Experiments

1. The vehicle: acceleration; flight path; payload volume, weight, and configuration.

2. Source of Circuit Power.

3. Telemetering Transmitter.

4. Telemetering Receiving Stations (ground based or shipboard)

5. Tracking
   (a) Initial tracking and guidance (Missile point of view).
   (b) Rudimentary tracking adequate for many experiments; e.g., solar U.V., meteoric impacts.
   (c) Intermediate gravity tracking for other experiments; e.g., cosmic ray surveys, geomagnetic field plotting.
   (d) High quality tracking; e.g., geodetic and ray measurements.

6. Aspect Control, if any.

B. Choice of Scientific Experiments

1. Scientific Symposium.

2. Invitation for Proposals.

3. Panel Judgment on Relative Merits of Proposals Submitted and on Competence of Persons Making Proposals (Written Proposals and Personal Presentation).

4. Allocation of Funds, if necessary.

5. Assignment of Vehicles.

6. In view of [the] relatively small number of persons engaged in this type of research and in view of close mutual familiarity, a considerable telescoping of the above may be possible.

Document I-5

Document title: Working Group on Internal Instrumentation of the I.G.Y. Technical Panel on Earth Satellite Program, "Minutes of First Meeting," March 6, 1956.

Source: Archives, National Academy of Sciences, Washington, D.C.

Once TPESP began operations, it established working groups on the various aspects of its responsibilities. Deciding what scientific instruments and other devices would be carried inside the Vanguard satellite was the principal responsibility of the Working Group on Internal Instrumentation. At its first meeting, the group reviewed the various candidate proposals and set forth selection criteria for choosing among them.

[no page number]

Administratively Restricted
For Official Use Only

Iowa City, Iowa
6 March 1956

Minutes of First Meeting of
Working Group on Internal Instrumentation of the I.G.Y.
Technical Panel on Earth Satellite Program

Place and Time of Meeting: 9:00 A.M., 2 March 1956, Room 4807 Enrico Fermi Institute for Nuclear Studies of University of Chicago.

Present: L. R. Alldredge, M. Ference, Jr., H. Friedman, W. W. Kellogg, L. Spitzer, Jr., J. A. Van Allen.

Absent: H. Odishaw, R. W. Porter.

1. The present technical, administrative and fiscal status of the Earth Satellite program was briefly reviewed.

2. Dr. Van Allen reported on the results of a meeting on the 24th of February with Drs. Porter, Hagen, Newell and Rosen at N.R.L.:

(a) The agreed objectives of the E.S.P. are, in order of decreasing priority:
(1) Place an object in orbit and prove by observation that it is there.
(2) Obtain a precision optical track for geodetic and high altitude atmospheric drag purposes and
(3) Perform experiments with internal instrumentation (After achievement of objective (2) in one or two flights, objective (3) will take precedence over (2)).

[2] (b) It was emphasized by Mr. Rosen that the necessary performance specifications for each of the three stages are very stringent indeed. Every effort will be made to deliver the nominal payload of 21.5 lbs. into a useful orbit. But the achievement of this objective within the established time scale is by no means certain. Reduction of third stage inert weight gives an increase of final velocity at the rate of 80 (ft/sec)/lb.

(c) If necessary to buy improved performance by reduction of payload it was agreed tentatively that this should be done by working up the following list in approximately the order 6, 5, 4, 3, 2, 1 as necessary:

(1) Empty third stage bottle (18"diameter by 50" length) - 0 lbs. payload
(2) #(l) plus minitrack, mounted in a minimum size and weight capsule - 6 lbs. payload
(3) Same as #(2), except mounted in a 20" sphere - 8.5 lbs. payload
(4) #(l) plus minitrack, telemeter and 2 lbs. experiment, mounted in a minimum size and weight capsule - 14 lbs. payloads
(5A) (Porter) #(4) except mounted in 20"sphere - 18.5 lbs. payload
(5B) (Van Allen and Newell) Same gross payload and general contents as #(5A) except mounted in a minimum size and weight capsule to allow a 5.5 lb experiment - 18.5 lbs. payload
(6) Greater total payloads and sizes as feasible

The tentative and controversial nature of this listing was realized. But the discussion served to develop the firm view that the payload must be regarded as a flexible element of the system and that a variety of payloads must be developed to take advantage of all reasonable degrees of overall vehicle performance.

[3] (d) It was agreed that the head end of the 3rd stage bottle will be designed with standard attachments so that various payloads can be freely interchanged.

(e) Project Vanguard will develop a set of "black-box" specifications for payload capsules for the guidance of groups which are developing instrumentations. Such specifications will include axial and radial g's, vibrational frequencies and accelerations, temperature limits, surface finish, data on center of gravity, moments of inertia, dynamic balancing, etc. It is anticipated that the vibration during third stage burning may provide the most rigorous feature of the specifications. A tentative set of specifications will be provided to the T.P.E.S.P on the 8th of March. Dr. Hagen stated that they expect to establish a testing laboratory for the assistance of groups who are developing instrumentation.

(f) Dr. Hagen agreed to excerpt pertinent characteristics of the minitrack system from reports of restricted distribution so that these characteristics can be made available to instrumentation groups.

(g) The outline of a proposed telemetering system was sketched by Mr. Townsend. Details will be presented to the T.P.E.S.P. on 8 March. Present planning on telemetering is still quite preliminary and it appears that several competing possibilities should be considered.

3. The principal business of the meeting was the detailed discussion of the merits of the various proposals for internal instrumentation which have been received to date. It was agreed that they should be assessed on the following four aspects:

[4] (a) Scientific Importance. This aspect was taken to be measured by the extent to which the proposed observations, if successful, would contribute to the clarification and understanding of large bodies of phenomena and/or by the extent to which the proposed observations would be likely to lead to the discovery of new phenomena.

(b) Technical Feasibility. This criterion emcompassed [sic] evidence for previous successful use of the proposed technique in rockets (or otherwise), apparent adaptability of the instrumentation to the physical conditions, and data transmission potentialities of presently planned satellites, nature of data to be expected, and feasibility of interpretation of observations into fundamental data.

(c) Competence. An assessment of competence of persons and agencies making proposals was attempted. The principal foundation for such assessment was previous record of achievement in work of the general nature proposed.

(d) Importance of a Satellite Vehicle to Proposed Work. The nature of each proposal was analyzed with respect to the questions: Is a satellite essential or very strongly desirable as a vehicle for the observing equipment proposed? Or could the observations be made nearly as well or better with balloons or conventional rockets as vehicles?

4. Discussion developed the point-of-view that the development of internal instrumentation was lagging far behind other aspects of the Earth Satellite Program. The Working Group agreed that it was its duty and the duty of the parent Panel to actuate such developments at the earliest feasible date in order that a comprehensive variety of internal instruments be proved-in [sic] in rocket flights and otherwise and [5] be available on a time scale consistent with the intended flight schedule. The most important aspects of the actuation of any such project are:

   (a) An assurance to the persons or agency in question that their apparatus is on the tentative "flight priority list." This assurance might be in the form of a "letter of intent" stating that—subject to successful development of the apparatus, subject to vehicle capabilities, and further subject to judgment of the Panel as developments proceed—their apparatus will be installed and flown on one or more of the I.G.Y. satellites.

   (b) Funding of the effort by contract or transfer of funds.

   (c) Supplying of technical information necessary for proceeding intelligently (Cf. para. 2(e), 2(f), 2(g)). This may be done in part by documents and in part by conference of persons concerned.

5. Part A. Proposals Formally Received to Date.
(Summarized by assigned serial number, Title, Principal Investigator, Agency and funds requested).

ESP-1 "A Proposal for Meteorological Observations from an Earth Satellite," W. G. Stroud, Signal Corps Engineering Laboratories, $93,000 for 2 years.

ESP-2 "Proposal for IGY Satellite Experiment to Detect Extreme Ultraviolet Solar Radiation by Photoelectric Techniques," H. E. Hinteregger, Air Force Cambridge Research Center, $5,000 for travel.

ESP-3 "Proposal for Thermal Measurements," P. R. Gast, Air Force Cambridge Research Center, $38,400 for 2 years.

ESP-4 "Proposal for the Measurement of Interplanetary Matter from the Earth Satellite," Maurice Dubin, Air Force Cambridge Research Center, $90,000 for 2 years.

[6]
ESP-5 "Detection of the Far Infrared Band Emission of the Planet Earth," Jean I. F. King, Air Force Cambridge Research Center (and J. Strong, Johns Hopkins University), $25,000 for 2 yrs.

ESP-6 "Ionospheric Structure as Determined by a Minimal Artificial Satellite," Warren W. Berning, Ballistic Research Laboratories, Aberdeen Proving Ground, $15,000 to $50,000 for various alternatives.

ESP-7 "Proposal for Measurement of Meteoric Dust Erosion of the Satellite Skin," S. F. Singer, University of Maryland, $47,150 for 2 1/2 yrs.

ESP-8 "Satellite Environmental Measurements," H. E. La Gow, Naval Research Laboratory. No funds requested.

ESP-9 "Solar Lyman-Alpha Intensity," H. Friedman, Naval Research Laboratory. No funds requested.

ESP-10 "Cosmic Ray Experiment," L. H. Meredith, Naval Research Laboratory. No funds requested.

ESP-11 "Proposal for Cosmic Ray Observations in Earth Satellites," J. A. Van Allen, State University of Iowa, $106,375 for 3 yrs.

Part B. Proposals of Doubtful Status

ESP-12 "A Fundamental Cosmological Experiment for the Artificial Satellite," William A. Baum, Mount Wilson and Palomar Observatories. (Author's intention not clear.)

ESP-13 "Solar Ultraviolet Measurements," Fred L. Whipple, Smithsonian Astrophysical Observatory. (Informally submitted only.)

ESP-14 "Release of Balloons for Air Density Near Apogee," Fred L. Whipple, Smithsonian Astrophysical Observatory. (Informally submitted only.)

[7]
ESP-15 "Observations of the Earth's Magnetic Field in Artificial Satellites," S. F. Singer, University of Maryland, $61,230 for 2 yrs. (Temporarily withdrawn due to payload limitations announced in Ann Arbor.)

6. The Working Group tentatively recommends the following "Flight Priority Listing." This listing is based on detailed discussion of individual proposals per paragraph 3 of these minutes. The proposals are arranged in decreasing order of overall priority:

ESP-8* [handwritten: 1/4 in [illegible] with ESP 9]
9 [handwritten: 1 3/4]
11 [handwritten: 2 lb.]
4 [handwritten: 3 lb. +]
6 [handwritten: 0 lb. or 2 lb., [illegible handwriting]]
10 [handwritten: 1 lb.]

5 [circled; handwritten: out - too horoz.]
1 [handwritten: 1 lb.]
7 [handwritten: very light, 2 1/2 oz.]
2 [handwritten: not proven [illegible]]
12 [circled; illegible handwriting]
3 [circled; illegible handwriting]

(*Environmental measurements, though not of fundamental scientific interest, are recommended, for obvious practical reasons, to be done in the first one or two flights.)

It further recommends that at least a selected group of these programs be actuated at the earliest possible time.

7. The Working Group notes the absence of proposals in the following important fields:
   (a) Magnetic field in the vicinity of the earth.

   (b) Radio noise measurements at frequencies below the ionospheric cut off.

   (c) Atmospheric density by internal methods (e.g., by such methods as described by Dr. Spitzer and by Mr. Jones at the Ann Arbor symposium).

[8] (d) Worldwide ionospheric transmission measurements (e.g., as described by Drs. Hartman and Haviland at the Ann Arbor symposium).

8. The Working Group recommends that sympathetic consideration be given to proposals which will serve to develop a reservoir of broad scientific interest and competence in a continuing program of national scientific satellite flights even though such work may not yield practical apparatus for the short range I.G.Y. program.

9. The Working Group proposes to give further consideration to the establishment of a world-wide set of telemetering receiving stations for the continuous or nearly continuous reception of observed data.

10. The Working Group proposes to consider concerted action on development of solar batteries, telemetering systems of more general applicability, data storage and read out devices, etc.

Respectfully submitted,

[signature]
J. A. Van Allen,
Chairman

Document I-6

Document title: Ernst Stuhlinger, Army Ballistic Missile Agency, to Dr. James A. Van Allen, State University of Iowa, November 23, 1956.

Source: Special Collections, University of Iowa Library, Iowa City, Iowa.

Document I-7

Document title: James A. Van Allen to Ernst Stuhlinger, February 13, 1957.

Source: Special Collections, University of Iowa Library, Iowa City, Iowa.

Ernst Stuhlinger was one of Wernher von Braun's closest associates and served as a link between the von Braun team and the scientific community. Even after the Army's proposal to launch the first U.S. scientific satellite was rejected in mid-1955, Stuhlinger kept in touch with scientists such as Van Allen who might be interested in putting their experiments aboard an Army-launched satellite, should the Army receive authorization to attempt a satellite launch (as it did in November 1957). Van Allen, perhaps at the time the most influential of the scientists hoping to get space-based data, was indeed interested in that possibility, and forwarded a copy of his reply to William Pickering, head of the Army's Jet Propulsion Laboratory, who presumably would be responsible for satellite design for an Army launch. Interestingly, none of the possible experiments Van Allen listed in his letter flew aboard the first U.S. satellite, Explorer 1, which was launched by von Braun's Jupiter C and designed by JPL. That first experiment was a geiger counter originally intended to fly on the second Vanguard scientific satellite. However, the existing link between Van Allen and the von Braun team facilitated their collaboration, once the authorization to attempt a satellite launch was granted.

Document I-6

[no pagination]

U. S. ARMY ORDNANCE CORPS
ARMY BALLISTIC MISSILE AGENCY
HUNTSVILLE, ALABAMA

IN REPLY
REFER TO
ORDAB-DV

23 November 1956

Dr. Van Allen
Department of Physics
Iowa State University [sic]
Iowa City, Iowa

Dear Jim:

I wish to thank you again very sincerely for the most pleasant and stimulating day in your Physics Department and with your family. In particular, I was very happy to realize that our thoughts regarding satellites were so much alike. Wernher von Braun was no less intrigued when I told him of our discussions. We would be very glad to have you here on a visit to show you what we have and can do. We even think that we could send one of our own planes to Iowa City to pick you up and take you home again later.

We are now looking forward to receiving some of the design drawings of your instruments. I am enclosing some papers which we mentioned in our talks for your retention.

Please give my sincerest regards to Mrs. Van Allen.

Yours very sincerely,

[signature]
ERNST STUHLINGER
Director,
Research Projects Office
Development Opns Division

Incls:
a/s

Document I-7

[no page number]

STATE UNIVERSITY OF IOWA
IOWA CITY
DEPARTMENT OF PHYSICS

13 February 1957

Dr. Ernest [handwritten line drawn through second "e" in correction] Stuhlinger
Army Ballistic Missile Agency
Huntsville, Alabama

Dear Ernest [handwritten line through second "e"]:

1. We are delighted to know that there is a possibility of flying some scientific apparatus on one or more of your orbiters.

It is my understanding that a total payload of 15 pounds is now regarded as feasible. In considering what types of scientific apparatus may be appropriate, I have taken two pounds as a reasonable weight. And, of course, I have depended rather heavily on the considerations in which our I.G.Y. Working Group on Internal Instrumentation has been engaged for over a year.

I have assumed no data storage of the type which requires command read-out and have also assumed that the I.G.Y. 108 mc/sec telemetering stations will be available, or that substantial Microlock array will be available.

2. The following are of very great interest and appear feasible:

(a) Comic Ray Observations (See attached more detailed description and "Scientific Uses of Earth Satellites" Chapter 20).
(b) Time Fluctuations of Solar Ultraviolet and X-Radiation (S.U.E.S. Chapters 16 and 19).
(c) Meteoric Erosion and Penetration (S.U.E.S. Chapters 8, 32, 33).
(d) Air Density (20" expandable sphere of very light weight 0.7 pound total installation weight and optical tracking).
(e) Radiative Energy Balance of the Earth.
(f) Cloud Cover and Weather Patrol.
(g) Ionospheric Measurements Using Two On-Board Transmitters of Different Frequency (S.U.E.S. Chapter 30).

[2] 3. In addition there may be mentioned the ionospheric measurements which can come from the comparison of optical and radio tracking data. The apparent radio position of the satellite transmitter is influenced by ionospheric refraction in a quite different way and to a much larger extent than is the optical position. Also the (much smaller) optical refraction is almost perfectly calibrated-out by photography against the star background. These ionospheric measurements require no on-board instrumentation other than the beacon transmitter itself. But they do require radio tracking of the highest feasible accuracy and establishment of a precise orbit over a long period or nearly simultaneous optical observations if the elements of the orbit are rapidly changing.

The air density can be determined simply from ground based tracking data, without on-board apparatus, as can also all of the geodetic data of interest. The air drag object of paragraph 2(d) above is a device of good optical visibility and of greater drag sensitivity, though the latter property will be desirable only if the altitude of perigee is substantially above 200 miles.

4. Needless to say, our group here at the State University of Iowa is very eager to participate in your program. We now have all the appropriate elements of a suitable cosmic ray apparatus well developed, as well as the foundations for interpretation of the observed data. We can make several sets of flight gear (See enclosure) [omitted] within about a month after receipt of definite packaging details. The only other significant factors which are not presently known to us are the impedance, voltage and pulse width of our signal for modulating the transmitter.

We shall await further information with great interest.

Sincerely yours,

[signature]
J. A. Van Allen
Head, Department of Physics

Enclosure: Cosmic Ray Apparatus for Satellite Observations

[handwritten: "Copy sent to W.H. Pickering on 4/16/57"]

Document I-8

Document title: Percival Brundage, Director, Bureau of the Budget, Memorandum for the President, "Project VANGUARD," April 30, 1957.

Source: Dwight D. Eisenhower Presidential Library, Abilene, Kansas.

As the Vanguard program continued, so did its cost growth. President Eisenhower and the National Security Council reviewed the program's cost growth in May 1956, and decided to let the program continue. Even the Central Intelligence Agency, presumably because of Vanguard's importance in establishing the right of satellite overflight, contributed emergency funding to keep the program going. The question of canceling Vanguard arose again in April 1957, as the Department of Defense and the National Science Foundation resisted providing the additional funds needed to keep up with cost growth. Once again, the Eisenhower administration decided to continue the program. The General Cutler mentioned in this memorandum was Lloyd Cutler, Special Assistant to the President.

[stamped: UNCLASSIFIED]

[no page number]

EXECUTIVE OFFICE OF THE PRESIDENT
BUREAU OF THE BUDGET
WASHINGTON 25, D.C.

MEMORANDUM FOR THE PRESIDENT

[stamped: APR 30 1957]

Subject: Project VANGUARD

The Department of Defense advises that developmental difficulties requiring additional time and effort have resulted in further upward revision of the estimated total cost of Project VANGUARD and that it will not be possible to complete the presently authorized six vehicle project within the January estimate of $83.6 million for the total cost. Arrangements have been made to fund approximately $70 million to date. Of this amount, some $50 million is being provided by the Department of Defense for the launching vehicles and related activities of which $25 million was advanced from the fiscal year 1957 Department of Defense emergency fund and has not been replaced. A fiscal year 1956 appropriation for the National Science Foundation has provided funds for the satellites themselves and the scientific instrumentation and ground observations.

We have been advised that it is currently estimated that if no further major developmental problems are encountered, the project may be completed within a total of $110 million. With respect to the probability of success of the project within this level of funding, the Department of Defense has reviewed and reconfirmed its statement to the National Security Council at the meeting of January 24, 1957, that in the technical judgment of Defense scientists and their consultants at least one successful satellite should result from six launchings of the presently planned Project VANGUARD launching vehicle. Since arrangements have been made to fund approximately $70 million, an additional amount of $40 million would be required to complete the project on present assumptions.

While no further major technical difficulties are now anticipated, it must be recognized that flight tests have not yet been completed. We have been advised that in the event unforeseeable developments should make it necessary to incorporate fundamental changes in the present approach or to employ an alternative approach, substantial additional funds beyond the $110 million estimate might be required.

When continuation of the policy established under NSC 5520 was considered at the NBC meeting of May 8, 1956, it was decided that this policy should be continued "with the understanding that the program developed thereunder will not be allowed to interfere with the ICBM and IRBM programs but will be given sufficient priority by the Department of Defense in relation to other weapon systems to achieve the objectives of NSC 5520."

The use of Department of Defense emergency funds in late fiscal year 1956 as well as during fiscal year 1957 was necessary because costs of [2] development and procurement of the launching vehicles increased much higher than the original estimate. The Central Intelligence Agency had made $2.5 million available to the Department of Defense, and the National Science Foundation was able to transfer $5.8 million when the decision was made to plan for no more than six launchings. It is the position of the Department that use of its funds was not based on any understanding by the Department that it had a continuing responsibility for funding this project but rather that the Department has used its funds thus far because no other clear-cut assignment of responsibility for funding the launching vehicles has been made and because it was assured that funds advanced to this project would be replaced, at least insofar as advances were made from fiscal year 1957 funds.

The Secretary of Defense has now concluded that it is not advisable for the Department to provide further support of the project in fiscal year 1957 or future years from the emergency fund. In addition to the fact that the Department does not consider that it has a continuing responsibility for funding the project, the Secretary's position is understood to result from the fact that the Department has not been reimbursed for fiscal year 1957 emergency funds already provided as well as from congressional criticism of the use of emergency funds for this purpose. In this connection it is noted that in view of established fiscal policies limiting supplemental appropriations to the most urgent cases, the Bureau of the Budget recently disapproved a request of the Department of Defense to reimburse the emergency fund.

The Bureau of the Budget has reviewed this problem with staff of the Department of Defense and the National Science Foundation. From the evidence at hand, the Bureau of the Budget believes that the project cannot go forward without additional funding. Taking into consideration the fact that this project has all the elements of a guided missile development program together with additional problems of a novel and difficult character, it is not surprising that substantial cost increases have occurred. However, inasmuch as the Department is now well into the project and states that it has already resolved a number of the technical problems, the present estimate of $110 million may be more reliable than previous estimates.

On the other hand, in the light of past experience with this project and in the absence of flight test results confirming the soundness of the present approach, I believe that it should be recognized that the final cost of the project may be as high as $150 to $200 million. In weighing the benefits deemed to be derived from the project and its priority in comparison with all the other current projects, it was initially approved in the expectation that the cost would be between $15 and $20 million. I question very much whether it would have been authorized, at least on a crash basis, if the actual cost had been know at that time. [3] It is hoped that in the future more careful estimates will be made as to the total cost or range in possible costs before such projects are initially approved. Furthermore, this seems to offer an opportunity to give up a desirable project for something else which is considered to be of higher priority in relation to cost and benefits to be derived. We are presently developing nine inter-continental and intermediate missiles with a range of over 1,000 miles, some of which involve comparable techniques and which will require difficult priority decisions as to programming and funding. Some eliminations will have to be made.

The Department of Defense has indicated interest in this program to about the same degree it has shown on some other basic research projects, but has stated that its interest is not sufficient to justify the project's continuance with Department of Defense financing. Therefore the Department believes that the program must be justified on the basis of the several national objectives stated in NSC 5520 rather than on the Department's interest.

The Department of Defense believes that to prosecute the balance of the program successfully, adequate financing should be arranged by supplemental requests submitted for appropriation to the National Science Foundation, which the Department considers to be the sponsor of the program. The Department would assist in justifying the supplemental requests of the National Science Foundation by assuming the burden of justification as to the technical difficulties encountered and the cost elements involved.

It should be noted that one of the important considerations has been and is the completion of the project during the period of the International Geophysical Year. If you desire the project to be continued in accordance with the existing policy under NSC 5520, it is suggested that the following actions could resolve the current financing problem:

1. The Department of Defense should be directed to provide immediately $5.8 million from the emergency fund to continue the project from May 1 through approximately August 1. The Department feels it must clear this use of the emer-

gency fund with the Appropriations Committees who have questioned the propriety of its use for this purpose. It should be recognized that the Department would prefer that these funds be replaced.

2. A fiscal year 1958 budget amendment should be submitted requesting an additional $34.2 million for appropriation to the National Science Foundation to cover costs to completion of the project, assuming that current cost estimates are valid, that no further major difficulties are encountered in the course of completing the development, and that the [4] Department of Defense would continue to provide general support for which no special funding has been considered necessary. Upon availability to the National Science Foundation these funds would be transferred to the Department of the Navy to complete the program.

The National Science Foundation believes that in view of the national interests involved the program cannot be permitted to fail at this stage. If it were the only possible alternative to cancellation of the project, the National Science Foundation would consider it necessary in the total national interest to request a supplemental appropriation to cover the costs required to complete the responsibilities undertaken by the Department of Defense under NSC 5520. However, the National Science Foundation recommends that the Department of Defense provide the necessary funds to complete the project for the following reasons: (1) the Department of Defense is responsible under the present terms of NSC 5520 for the portion of the program requiring additional funds; (2) the Department of Defense is best qualified to justify to the Congress the reasons for present cost increases.

Apparently, both the Department of Defense and the National Science Foundation are very reluctant to continue to finance this project to completion. But each is quite prepared to have the other do so.

General Cutler believes the following considerations are particularly relevant to a decision in this matter:

"1. The substantive scientific information concerning upper atmospheres which might be acquired by the launching of a successful satellite. Included in this information would be data as to the content of the upper atmosphere (such as invisible heavenly bodies) through which the very costly intercontinental ballistic missiles, if perfected, must pass.

"2. The world reaction to an abandonment by the U.S. in mid-stage of the satellite program. A conclusion that the richest nation in the world could not afford to complete this scientific undertaking would be unfortunate. Even more unfortunate would be an inevitable inference that American scientists were not up to bringing the project to a successful conclusion.

"3. The reaction of the scientific community to the abandonment by the U.S. in mid-stage of the satellite program. A time when the Free World is coming more and more to depend on advanced technology and scientific accomplishment is not a time to alienate the scientific community at home and lead it to believe that the Government has lost faith in scientific accomplishment.

[5] From what I hear and read, the scientific community and those in highly technical industry who work with them are already sensitive in this regard.

"4. A final decision on the satellite program should be made by the President on an integrated presentation of the views of all concerned in this matter. The integrated process of presentation, such as is illustrated in the National Security Council, is a primary achievement of this Administration. Where so much, beyond financial considerations alone, is at stake, the President should have the benefit of an integrated presentation and discussion. This point of view is important, irrespective of what the President's decision might ultimately be."

It should be noted that the Air Force has already started its own project for a much larger reconnaissance satellite vehicle and is spending approximately $10 million in fiscal year 1957 and is currently planning additional funding of at least $10 million for fiscal year 1958. Therefore, whether or not the International Geophysical Year satellite project is completed, research in this area will not be dropped.

[signature: Percival Brundage]
Director

Document I-9

Document title: Memorandum for the President, "Earth Satellite," unsigned but written by Deputy Secretary of Defense Donald Quarles, October 7, 1957.

Source: Dwight D. Eisenhower Presidential Library, Abilene, Kansas.

President Eisenhower called a meeting with his top advisers on the morning of October 8, 1957, to discuss the implications of the Soviet launch of Sputnik I and how the United States should respond to it. This memorandum was given to the President at the start of the meeting and formed the background for the discussion. Since 1955, Donald Quarles had been the highest-level Department of Defense official most closely following the attempt to develop a U.S. scientific satellite. He had been the primary architect of the strategy to use that satellite to establish the right of free satellite overflight over all areas of the globe, a necessary precondition for the development of U.S. reconnaissance satellites. This link between space science and national security objectives was one of the major realities of the early U.S. space program.

["TOP SECRET" stamped on each page but crossed out and stamped "UNCLASSIFIED"]

[no pagination]

THE SECRETARY OF DEFENSE
WASHINGTON

7 October 1957

MEMORANDUM FOR THE PRESIDENT

SUBJECT: Earth Satellite

The first serious discussion of an earth satellite as a scientific experiment to be incorporated in the program for the International Geophysical Year took place at a meeting of the International Council of Scientific Unions in Rome, Italy, in October 1954. At this meeting which Russian scientists attended, a resolution was adopted recommending—"In view of the advanced state of present rocket techniques, ... that thought be given to the launching of small satellite vehicles . . . . ." We assumed at that time that the Russian scientists were innocently concurring in this resolution. It now seems likely that it was part of a deliberate plan.

We in Defense were concerned at that time about international reactions to a reconnaissance satellite that the Air Force was giving serious study to. It was felt that scientific satellites which would be clearly non-military and clearly inoffensive might help to establish the principle that outer space is international space. Thus, reconnaissance satellites travelling in it could not be objected to by the countries over-flown because the space is free and the satellite itself is inoffensive in character.

Pursuing this line of thought with our own scientists led to the adoption by the U.S. National Committee for the International Geophysical Year of a resolution recommending that the U.S. institute a scientific satellite program. This matter was considered at the highest government level with the conclusion that such a program should be pursued as part of IGY. Within the Government, responsibility for scientific aspects was assigned to the National Science Foundation. Defense participation would be to supply the rocketry needed to place such a satellite in orbit. Since such rocketry would follow the same general lines as our long-range ballistic missile developments, it was part of the stated policy at that time that the scientific satellite should not interfere with the top priority ballistic missile program. In line with the recommendations of the Special Scientific Advisory Committee that studied the matter at that time, the Navy's proposals were accepted, and the scientific satellite project was assigned to the Naval Research Laboratory as Project VANGUARD.

In order to meet the requirement that it be non-interfering with top priority ballistic projects, and for other technical reasons, the Navy pursued an independent course of rocketry involving improvements in earlier Navy high-powered rockets, as well as the development of new equipment.

The National Science Foundation worked with the U.S. Committee of IGY to formulate plans for the satellite as such and its instrumentation, as well as for the preparation and deployment of the ground observer equipment required for the program. To guide public relations in this area, a special sub-committee of OCB was established with representation from the White House, State, Defense, CIA, USIA, National Science Foundation and the National Academy of Sciences. On July 29, 1955, the White House announced that plans "are going forward for the launching of small, unmanned earth-circling satellites as part of the U.S. participation in the International Geophysical year." The military participation in rocketry was de-emphasized as being incidental to the scientific program. All subsequent public releases have followed this same line.

The VANGUARD program was last presented to the NSC on May 10, 1957. The program outlined at the time contemplated the launching of certain test vehicles during the rest of this year and the launching of the first fully instrumented satellite vehicle about the end of March 1958. It was planned that certain of the earlier launchings would carry a smaller satellite sphere which would be placed in orbit and tracked as a check on the rocketry, instrumentation and ground stations. The earliest of these experimental part-size satellites is now scheduled to be launched about December 1, 1957. A current review of the Navy's programs indicates that from where we are today there would be little to be gained by attempting to accelerate or substantially modify the VANGUARD program in an attempt to launch a satellite at an earlier date than now planned. At best, the changes that could be made would produce either marginal accelerations of schedules or high risk programs that would have a fair chance of failure. Somewhat similar remarks apply to the possibility of paralleling the Navy program with an Army program based on the REDSTONE missile. Since, in any event, the U.S. satellite would be second rather than first, it appears sound to adhere to our program as presently planned.

The satellite which the Soviets launched on October 4th was generally in line with the planning of the International Scientific Committee, but deviated in certain respects, the most important of which was the change in the radio signal frequency which had been agreed upon internationally as 108 megacycles and which the Soviets abandoned for reasons of convenience and, no doubt, speed, substituting signals at much lower frequencies (around 20 and 40 megacycles) where their techniques were more readily available.

There is considerable intelligence to indicate that the Russian satellite work has been closely integrated with and has drawn heavily on their ballistic missile developments, including the range facilities. In fact, it seems quite likely that the ICBM test which they announced on August 27th and follow-on experimental work since that time was either related to or even an integral part of their satellite program. Their scientists who came to Washington to participate in the IGY conferences on the satellite programs must have known when they left Moscow that the first Soviet launchings were scheduled for approximately October 4th. In retrospect one sees that their whole behavior has been carefully planned to fit either with a successful launching or with a failure, depending on information that they would later receive from Moscow. The fact that what they claim was their first attempt was successful, and that it was timed perfectly in relation to the IGY conferences in Washington supports the thesis that this was all a very carefully laid plan to make maximum cold war capital out of their satellite program.

The satellite they have actually launched is said to carry only radio signaling instrumentation and, of course, this is all we have observed. They describe the satellite as being a 22 inch diameter sphere, about the same as our own, but claim that it has a weight of about 185 pounds as compared with our 21 pounds. This leaves some uncertainty as to whether what they are calling their satellite is not a combination of the 22 inch sphere and the last stage of rocketry required to give this sphere its orbital velocity. In our planning, we will separate the sphere from the last stage rocket. Another difference between the two plans is that the

Russian satellite has been placed in an orbit averaging about 370 miles above the surface of the earth, whereas the U.S. VANGUARD plan involves orbit in the range of 300 miles above the surface. Still another interesting point is that the Soviets launched their satellite on an orbit inclined about 65° to the equator (which is probably consistent with their ballistic missile range in Siberia); whereas our plans are to launch VANGUARD in an orbit only about 25° from the equator (which is consistent with our Long-Range Proving Ground layout southeast from Cape Canaveral, Florida). The orbit of the Russian satellite derives less advantage from the rotational velocity of the earth and is, therefore, a more difficult orbit requiring more powerful rocketry, other things being equal. If the 185 pounds is in fact the weight of the sphere, placing it in the Soviet satellite orbit 370 miles above the surface would require substantially more powerful rocketry than that planned for VANGUARD. This again is consistent with the thesis that the Soviets have used their ballistic missile rocketry which we know to be powerful enough to launch a satellite of the general character now being observed.

As we see it, two main cold war points are involved: (1) the impact on public imagination of the first successful invasion and conquest of outer space, and (2) the inferences, if any, that can be drawn about the status of their development of military rocketry. As to the former, we are face to face with the basic unfavorable fact that the Russians have been first. We can take the position, however, that our satellite program has been coordinated throughout with the International Scientific Community and that it has been programmed as a part of the International Geophysical Year (July 1, 1957 through December 31, 1958). It can be pointed out, that as to instrumentation and as to availability of ground observation points and other appropriate scientific arrangements, our launching schedule follows a carefully prepared plan. The question of the first country to launch is minor compared with the question of the success of the program in achieving scientific objectives. Moreover, the Russians agree that their first satellite falls short of scientific objectives and for this reason they propose to launch additional ones from time to time. Consistent with our international planning, we propose to continue with our plans and on our schedule.

As to the military implications of the Soviet satellite, the facts as indicated above appear to be that the satellite success does indicate competence in long-range ballistic missiles and does tend to corroborate their ICBM claim of August 27. Parenthetically, one might observe that the Russians, if they were sure that we would not start a war, could properly conclude that their speed in developing ICBM was not so important as to require non-interference from their satellite program. In other words, both their objectives would be cold war objectives, and they would therefore logically follow the program that would yield the maximum cold war results. In our own case, this logic would not apply, and our decision to make the VANGUARD program non-interfering with the high-priority ballistic missiles was certainly reasonable at the time, even though it may appear questionable in retrospect. On this second point, our public position might well be that our own VANGUARD program was divorced from military rocketry as much as possible, and the fact that our schedules have not, produced a satellite at as early a date as the Russians have succeeded in doing is without military significance. The rocketry we are using is completely separate from ICBM and IRBM rocketry. Other technical requirements than the mere production of high-powered rockets have controlled our schedules.

A proposed public announcement stressing these two key points is attached. [omitted]

Attachment

Document I-10

Document title: Technical Panel for the Earth Satellite Program, United States National Committee for the International Geophysical Year 1957–1958, "Minutes of the Fourteenth Meeting," November 6, 1957.

Source: Archives, National Academy of Sciences, Washington, D.C.

Meeting one month after the launch of Sputnik 1, TPESP considered what to recommend if, as was possible (and indeed happened two days later), the Department of Defense authorized the use of two Jupiter-C rockets as backups to Vanguard for launching a U.S. scientific satellite. The Panel decided that if a launch opportunity on a Jupiter-C became available, it would withdraw the experiment prepared by James Van Allen from the Vanguard program and assign it to Jupiter-C. The Panel was aware that Van Allen had kept in touch with Wernher von Braun and his associates about the possibility of launching a Geiger counter on the Army rocket even after it had been eliminated in 1955 from the competition to launch the first U.S. scientific satellite (see Documents I-6 and I-7).

[no page number]

FOR OFFICIAL USE ONLY

National Academy of Sciences
2101 Constitution Avenue, N.W.
Washington 25, D.C.

UNITED STATES NATIONAL COMMITTEE
for the
INTERNATIONAL GEOPHYSICAL YEAR 1957-58

MINUTES
of the
Fourteenth Meeting
Technical Panel for the Earth Satellite Program
November 6, 1957
Associations Building, 1145 19th Street, N.W.
Washington 6, D.C.

Call to Order

The fourteenth meeting of the USNC-IGY Technical Panel for the Earth Satellite Program convened at 9:30 a.m., November 6, 1957, with Chairman R.W. Porter presiding.

Members Present:	R.W. Porter (Chairman); J.G. Reid, Jr. (Secretary); Michael Ference, Jr.; J.A. Hynek (for F.L. Whipple); W.W. Kellogg (for J.A. Van Allen); H.E. Newell, Jr; Hugh Odishaw; W.H. Pickering.
Members Absent:	G.M. Clemence; J. Kaplan; A.F. Spilhaus; J.A. Van Allen; F.L. Whipple.
USNC Secretariat:	L.N. Cormier.
Invited Participants:	W. Berning (BRL); J.E. Froehlich (JPL); A.M. Gerlach (AFCRC); G. Grimminger (USAF); T.J. Killian (ONR); P.H. Kratz (NSF); W.J. O'Sullivan, Jr. (NACA); E. Rechtin (JPL); R.W. Stroup (NRL); J.W. Townsend, Jr. (NRL); H.A. Zahl (USASEL); H.K. Ziegler (USASEL).

*** Denotes formal action taken by the TPESP.

[2] Chairman Porter announced that this meeting had been called for the consideration of a number of urgent items. He suggested that the following agenda be followed and there was assent by all present:

1. Review of the USSR satellite data.
2. Backup Vehicle Program.
3. Internal Instrumentation Program.
4. Tracking Program.
5. Panel Organization.
6. Long-Range Program.

1. <u>Review of the USSR Satellite Data</u>. A comprehensive discussion was held on Russian satellites 1957-alpha 1, 1957-alpha 2, and 1957-beta. Many of the technical details in connection with these satellites were based on conjecture; there was no Panel action taken on this. At the end of this discussion, Porter emphasized the importance of channeling all data obtained on Russian satellites to the NRL Control Center.

2. <u>Backup Vehicle Program</u>. Porter announced that information discussed on this subject is for official use only and should not be given dissemination until the Department of Defense gives permission. He then informed the Panel that DOD had been considering the addition to the satellite program of two backup rounds of the Jupiter-C type. He announced that approval of these plans had gone through the lower echelons of authority, and that the plans were then at the highest level for approval.

Configuration of the payload carried by the Jupiter-C would be cylindrical rather than spherical, with a total weight amounting to 18 or 20 pounds and a diameter of six inches. The spin rate of this vehicle would be about two to three times as great as that of the Vanguard, and the acceleration would be about twice as great. The instrumentation section would not be separated from the last stage of the launching vehicle.

Porter announced that JPL will be doing most of the work on the configuration for the Jupiter-C launchings, and that JPL's position in the Jupiter–C series is analogous with that of NRL in the Vanguard series. He also stated that he had discussed the instrumentation with Kellogg quite thoroughly. No action had been taken by the Working Group on Internal Instrumentation, but members available for comment appeared to have no obvious dissent to this program. The following is a cursory review of the status of the various experimental packages in relation to their possible adaptation to the Jupiter–C series:

Package I - Is in advanced stages of development, design, and testing. Could not be easily modified to the Jupiter-C configuration; although this could be done, possible wasted effort might indicate that some other experiment would be a better choice.

[3] Package II - Is in advanced stages of development, but has not had the time and effort for NRL testing invested in it that Package I has had. It is consistent with the required configuration weight-wise. This experiment would be left attached to the final stage of the rocket vehicle; the resultant possibility of increased visibility would be an advantage. Presence of the last stage should not affect the cosmic ray instrumentation.

Package III - The magnetometer in this experiment might be adversely affected by the steel contained in the last stage of the Jupiter-C.

Package IV-a - This experiment might also be affected by the presence of the last stage, and a greater correction factor might have to be taken into account.

Package IV-b - This experiment has had considerable design work in the Jupiter-C configuration and would probably be a better candidate for this series than IV-a. However, it contains no environmental instrument while Package II does.

*** In view of the above considerations it appeared more practical for Package II to be withdrawn from Vanguard and placed in the Jupiter-C series rather than any of the other experimental packages. It was also pointed out that Van Allen and others working with this experiment had given consideration and had perhaps applied actual design work to the adaptation of the experiment to the Jupiter-C configuration. Further, it appeared that there was general agreement on this by JPL and Army people as well as by those working on the experiment. Final decision would be contingent upon the definite availability of Jupiter-C vehicles to the satellite program, and upon Van Allen's approval of the plan.

Porter stated that in connection with the backup launching there was a question of whether a backup for the cosmic ray experiment should be prepared. This is based upon (a) the possibility that for any reason the cosmic ray experiment may not be ready in time; or (b) the more realistic possibility that Package II would have a successful ride on the first attempt and another experiment would be able to ride in the second firing. While it is possible that DOD would not let a second firing be made if the first were successful, it would be well to be prepared in case this could be done. Porter expressed the thought that NRL might be asked to work up a package of environmental equipment suitable for this purpose to be interchangeable with the cosmic ray package, or that provision might be made for a backup which is a different kind in that it would provide a highly visible experiment of some sort and would be useful in getting air density at very high altitudes.

[4] Porter stated that the one high visibility experiment that had been found to have any real feasibility or any practical scientific utility appeared to be the large inflatable balloons that have been developed by NACA. It also appeared that a 12-foot sphere would be preferable to the 12-foot corner reflector, since the reflector is too small to be a good radar reflector at the frequencies of interest to ionospheric investigations. The sphere would have greater visibility and would be more susceptible to interpretation of air drag and radio reflection data. The very high drag sensitivity of the large inflatable sphere is at once its principal disadvantage and advantage. In order to have a lifetime long enough to be at all usable, the large inflatable sphere must be in an orbit with a perigee well above 300 miles. On the other hand, this technique appeared to be the only practical way of getting air drag data at such altitudes. It appeared that a 300-mile or better perigee can be obtained by using an apogee impulse technique. Although untried, there appeared to be no reason why such a technique should not work. Froehlich explained that neither of the two Jupiter-C vehicles under discussion is appropriate for modification for the apogee impulse technique. He also stated that if another Jupiter-C were made available, the technique could be incorporated by, but not before, June 1958. Porter noted that a Vanguard vehicle might also be adapted to use the apogee impulse technique.

*** The Panel agreed that a large inflatable sphere orbiting at a high altitude would constitute a valuable scientific experiment which should be conducted eventually. Although immediately available vehicles are not suitable, the Panel desired to urge NACA to continue the development and testing of a 12-foot inflatable sphere with the hope that a suitable vehicle would become available, possibly as soon as June 1958.

*** Kellogg moved, and it was unanimously agreed, that the Secretary write letters insuring that a) NACA would be encouraged to continue work on inflatable spheres, and b) that DOD be urged to develop a suitable vehicle.

Pickering announced at this point that he would like to explain in some detail the things to be included in the first two Jupiter-C vehicles. He also indicated concurrence with Porter's announcement of JPL's position with Jupiter-C, and pointed out that JPL would have complete responsibility for the high speed stages and for the instrumentation payloads as far as the Army is concerned.

Pickering explained that there would not be sufficient time to adapt the SUI cosmic ray experiment to the Jupiter-C configuration if the command receiver and tape recorder storage readout system are to be included. However, unless the storage feature is included in at least one flight, SUI would probably be unwilling to transfer its experiment from the Vanguard series. Therefore JPL proposes to prepare two types of instrumentation, the first of which for the sake of simplicity and reliability would use continuous telemetry only; and the second of which would more closely resemble the original experiment and employ both continuous and command readout telemetry.

Both types of instrumentation would measure the cosmic ray count, four temperatures, and the effects on two erosion gauges. For added reliability each type would employ two independent transmitters. The first transmitter would operate at a power of about 100 milliwatts for about two or three weeks; scientific data from the Geiger counter and from one erosion and two [5] temperature gauges would be contained in the frequency variation of an amplitude modulated 108.03 mc carrier. The second transmitter would operate at a power of about 10 milliwatts for about two or three months; redundant cosmic ray data, and data from the other erosion and the other two temperature gauges would be contained in the frequency variation of a small–deviation phase modulated 108.00 mc carrier. Each transmitter would telemeter four measurements on standard telemetering bands 2, 3, 4, and 5. The low-power transmitter, which uses only small amplitude phase modulation, would be a highly stable transmitter for tracking.

There was some discussion on the number and type of erosion gauges to be used. It was agreed that this choice should remain flexible and with JPL. The AFCRC gauge developed for use in conjunction with the SUI experiment would probably be used, but additional and/or substitute gauges were offered by NRL.

*** The Panel agreed to accept the instrumentation proposal for the backup program as outlined above contingent upon availability of two Jupiter-C vehicles and upon Van Allen's approval.

Porter noted that implicit in this agreement was the decision to drop the proposed NRL environmental package from the backup program.

Pickering stated that it would be highly desirable to have additional telemetry coverage, particularly for the cosmic ray package containing no storage and command readout system. Three sets of Microlock receiving Systems are available. Pickering proposed that one be placed at the launching site, and that the remaining two be placed near the geomagnetic equator at separated longitudes. Singapore and Nigeria would be ideal but there would probably be insufficient time to complete all of the necessary negotiations and preparations. Hawaii and the Philippines would probably be more practical sites if Army support could be obtained for the operation of the stations. However, it was agreed that the possibility of locating stations at Singapore and Ibadan, Nigeria should be investigated. These two stations would be primarily for telemetering rather than for tracking and would provide valuable additional coverage for any satellite experiments using continuous telemetry on 108 mc.

3. <u>Reminder of Internal Instrumentation Program</u>. Newell and Townsend pointed out the necessity of having at the earliest possible date an absolute decision on the disposition of the SUI experiment, since it would be necessary for NRL to drop this experiment almost immediately from its development and test program and work instead on its substitute. Newell stated that it is probably already too late to move any of the other experiments into position for the second earnest try, should Lyman-alpha be launched successfully on the first earnest try. As a backup for this possibility NRL can prepare an X-ray/environmental package from spare Lyman-alpha/environmental packages. Essentially, this can be done by changing the ionization chamber. Although the change is somewhat complicated by the need for either more sensitive instrumentation or a larger ionization chamber and consequent modification of the pressure zones, the X-ray experiment could be prepared on relatively short notice.

[6] The decision to transfer the cosmic ray experiment from the Vanguard to the Jupiter-C series automatically eliminated the necessity of choosing between the radiation balance and cloud cover experiments. After some discussion it was agreed that of the remaining experiments, the cloud cover experiment could be brought into the number two position with the least disruption of the present satellite program. Further, it was believed that the instrumentation for this experiment had progressed sufficiently to be able to meet the launching dates originally intended for the cosmic ray experiment.

Zahl and Ziegler explained that if the cloud cover experiment is to meet the early launching date, the project would have to be given top priority consideration by the Army and full cooperation would have to be received from NRL. They further explained that while there were no technical risks, procurement difficulties caused by strikes, for example, might prevent SEL from meeting the schedule. Finally, they warned that although rapid data analysis and presentation is especially desirable for this experiment, the data evaluation equipment may not be ready until three months after the satellite instrumentation. They stated it may be possible to obtain sample pictures of a crude type fairly soon after the data is obtained.

Porter stated that he would work closely with the IGY Secretariat, and that he would endeavor to obtain the necessary high-level Army endorsement of the project.

Townsend noted that NRL support for the cloud cover experiment would be extremely difficult to fulfill under the present launching schedule. He referred to the following chart which had been prepared and distributed earlier by Newell and Stroup:

Launching Vehicle:	TV-5	SLV-1	SLV-2	SLV-3	SLV-4	SLV-5	SLV-6	TV-4bu
Package I, Lyman-alpha Env. (NRL)	X	X	X	X	X	X	X	X
Package II, Cosmic Ray-Meteorite Detection (SUI-AFCRC)			X	X	X	X	X	X
Package III, Magnetometer/ Sub-Satellite (NRL-NACA)					X	X	X	X
Package IV, Radiation Balance or Cloud Cover (Univ. of Wisc. or USASEL)							X	X
Package Ia, X-ray-Environmental (NRL)				X	X	X	X	X
Package IVa, Second of Group IV experiments	This experiment could not be readied in time to be included in the present launch schedule.							

[7]*** Ference moved and the Panel agreed to authorize Porter, together with the IGY Secretariat, to take such steps as seem necessary and possible to place the cloud cover experiment in the position vacated by the cosmic ray experiment in light of information which may develop as a result of further inquiries about the speed-up program and the capabilities of SEL and NRL.

Newell noted that additional funds would be necessary to bring both meteorological experiments to tested-for-flight configuration. The Secretariat was requested to work with NRL on this matter.

*** After some discussion Porter requested that the minutes show that it is the desire of the Panel that if it is not possible, either by extending the program or by some means for meeting the speeded up program with the cloud cover package, then the next best alternative in the minds of the Panel members is to substitute the X-ray/environmental package.

The Panel discussed the implications of the speeded up launching schedule to the scientific program not only with reference to the increased difficulty of preparing experiments but in particular with reference to possible damage which closely spaced launchings might inflict upon scientific observations. To this end the Panel resolved:

*** "In view of information presented today by the Naval Research Laboratory concerning the proposed speed-up of IGY satellite firings, the Panel desires to point out that any such schedule as is now proposed would adversely affect the possibility of completing the orderly series of scientific experiments now contemplated.

"It should be noted that it now appears that it will be necessary to change the experimental program somewhat as a result of the probable addition of a different type of back-up vehicle and the consequent need to derive the maximum of meaningful scientific results from these satellites as well as from those originally planned. The Naval Research Laboratory has stated that it will not be able to provide the required support in the way of engineering coordination and environmental testing for this modified program. The Panel believes that even if some way could be found to provide the necessary NRL support, the scientific program would still be jeopardized by vehicular difficulties as well as by problems associated with tracking and telemetering which would inevitably result from the proposed speed-up.

"It is the opinion of the Panel that the primary value of the IGY satellite program, after a first successful satellite has been achieved, will reside in the scientific accomplishments resulting from an orderly series of experiments and the Panel strongly recommends that no action be taken that would jeopardize these accomplishments."

The Panel drafted a second resolution to allow for the possibility that the request contained in the first resolution cannot be realized:

[8] "On the assumption that there exist reasons of such overpowering importance that the previous recommendation relating to a speed-up in the firing of the IGY satellite vehicles cannot be considered, it is requested that the Department of Defense make adequate provisions for supporting work on the part of NRL for this series of experiments and possible ways and means by which the currently contemplated series of experiments, which includes the Lyman-alpha, magnetometer, radiation balance, and cloud cover experiments, can be satisfactorily handled by the Naval Research Laboratory without unduly sacrificing any work elements which are vital to the obtaining of maximum reliability in flight."

*** As a result of a discussion regarding the advisability of requesting additional vehicles for additional experiments, the Panel agreed that the USNC-IGY Secretariat should make an urgent inquiry to the Department of Defense requesting official information on the status of the backup program. However, at this time there would be no request made for additional vehicles to be added to either the Vanguard or Jupiter-C series of firings.
*** Porter then announced that ONR had recommended an experiment having to do with life processes which, it is believed, could be available for inclusion in the first test sphere scheduled for December. He stressed that this experiment was not for publication at this time. It is lightweight and extremely simple. It is proposed to put a colony of simple yeast organisms in a nutrient medium in a capsule, and measure the pressure in a void which will be proportional to the generation of carbon dioxide. The rate of generation of carbon dioxide from a known sample with a given start of its life history will show the reproduction of successive generations, and the rate of life processes in each generation. A comparison of this experiment with the control experiment which would be run on the ground, reproducing as nearly as possible the same stresses which would pertain, tends to indicate very well the stresses of flight in the satellite. This is a statistical experiment, and is in many ways more meaningful and fundamental than the much-publicized USSR experiment with the dog. The Panel expressed general approval of this experiment and its inclusion in the test spheres. However, Porter suggested that the TPESP go on record as recommending that the first flight of this experiment be made as nearly at zero G as it is technically possible. Hynek so moved and Newell seconded the motion, proposing that the following be added to the statement: " ... since it is felt very strongly that the zero G feature of the experiment is its primary justification." There was no dissent, and it was agreed that Kellogg would take the necessary action in this connection.

4. Tracking Program. Hynek reviewed the history of Project MOONWATCH budgetwise, pointing out that it has developed into a more important operation than was intended or even thought of at the time funding was obtained, and with the increase in number of teams wishing to participate, plus the demand for information relating to satellite observations, he desired a statement from the TPESP as to what the Smithsonian Observatory's responsibilities in connection with the Project should be. He also noted that the SAO had received criticism for not providing sufficient information to observatories and others, and that the question had been raised about extending visual observations to more northern latitudes. Hynek pointed out that he was not at this time [9] recommending any course of action, but was requesting advice. He said that they had outlined three plans for consideration in this matter of MOONWATCH operations: 1) Maintain status quo—i.e., no further expansion beyond present commitments insofar as registering teams is concerned. It was also pointed out that by maintaining status quo, the Project still could not operate on the available $15,000 for the rest of the year. 2) Expand into the higher latitudes organizationally and consequently operationally. 3) The third alternative would be to end the project. Hynek then presented a detailed summary relating to these MOONWATCH proposals.
*** After discussion, the Panel approved the immediate allocation of $25,000 to Project MOONWATCH subject to the satisfaction of Odishaw on details of the expenditure of the funds. Hynek then requested, and Porter approved, that the record show that responsi-

bility for curtailing MOONWATCH operations, should this be necessary, would fall directly on the USNC-IGY Secretariat.

Porter announced that he had made no arrangements for locating two meteor cameras at interim sites since it did not appear that Satellite 1957-beta would have a long life.

A brief discussion was held on radio tracking, but no report on recent operations of the prime Minitrack stations was available. When it was learned that the Minitrack station in South Africa did not include telemetering facilities, the TPESP urged that telemetry equipment be added to the station if this were at all possible.

Newell reported that work was being done in connection with the Panel's earlier request that several organizations which had gathered substantial radio tracking data be thanked for their assistance to NRL and be encouraged to continue this assistance. The specific organizations involved here are: NBS, Stanford, University of Alaska, RCA, ARRL, and the University of Illinois.

The Panel noted the possible radio interference problem which might result from several satellites operating on the same frequency. In addition to adequate spacing of U.S. satellites, the members of the Panel believed coordination to avoid such radio interference might be an appropriate item for international consideration.

5. Panel Organization. Porter announced that at the last meeting of the USNC Executive Committee held the previous week, there had been considerable discussion about the fact that orbit data on USSR satellites was being received, but that it was not being scientifically exploited. Of primary concern here was the obtaining of ionospheric data from the 20-40 mc transmission of the satellites. He stated that his position at the Executive Committee meeting had been that the TPESP charter had never clearly defined that Panel's responsibility in connection with ionospheric experiments using artificial satellites, and that reliance had been placed on the Technical Panel for Ionospheric Physics to outline what is desired of the TPESP in this connection. As a result of the discussions, the Executive Committee had reached a decision and stated that satellite ionospheric research matters would be the responsibility of the TPESP.

Document I-11

Document title: J. A. Van Allen, Chairman, "National Space Establishment: A Proposal of the Rocket and Satellite Research Panel," December 27, 1957.

Source: National Air and Space Museum, Smithsonian Institution, Washington, D.C.

The Rocket and Satellite Research Panel was the expanded successor to the panel of scientists that had guided the U.S. upper atmosphere research effort using balloons and sounding rockets since 1946. The Panel recognized that the Eisenhower administration and the Congress were, in the aftermath of the first two Sputnik launches, thinking about how best to organize for an expanded U.S. space program. The Panel wanted to make sure that the views of the directly interested scientific community were considered in these deliberations; this statement represented their recommendations. After Sputnik 1, the

Rocket and Satellite Research Panel doubled its membership. Its members prepared a plan for a civilian agency to take over the exploration of space and then testified before Congress in favor of their plan.

[no page number]

27 December 1957

NATIONAL SPACE ESTABLISHMENT

A Proposal of the Rocket and Satellite Research Panel

Summary of Proposal

It is proposed that there be created a unified National Space Establishment for the purpose of carrying out the scientific exploration and eventual habitation of outer space.

It is imperative that the United States establish and maintain scientific and technological leadership in outer space research in the interests of long-term human progress and national survival.

1. Role

The role of the National Space Establishment shall be to unify and to greatly expand the national effort in outer space research, specifically excluding areas of immediate military urgency (e.g., the development, production and fielding of intercontinental and intermediate-range ballistic missiles).

2. Mission

The broad mission of the National Space Establishment shall be to establish United States leadership in space research by 1960 and to maintain it thereafter.

Accomplishment of this mission requires the following specific achievements:

(a) An intensified program of scientific soundings with high-altitude rockets, immediately.

(b) An intensified program of scientific and technical developments with small instrumented satellites of the earth, immediately.

[2] (c) Impact on the moon with non-survival of apparatus, by 1959.

(d) Placing an instrumented satellite in an orbit about the moon, by 1960.

(e) Impact on the moon with survival of scientific instruments, by 1960.

(f) Returnable, manned satellites in flight around the earth, by 1962.

(g) Manned circumnavigation of the moon with return to the earth, by 1965.

(h) Manned permanent satellite, by 1965.

(i) Manned expedition to the moon by one or two men, by 1968.

(j) Manned expedition to the moon by a sizeable party of men, by 1971.

A thorough analysis of existing capabilities shows that all of these objectives are within reach of a unified, vigorous national effort.

3. Funds Required

A detailed analysis shows that the accomplishment of the basic mission will require a national expenditure of ten billion dollars over the next decade.

4. Administrative Status of National Space Establishment

(a) It is strongly desirable that the N.S.E. be given statutory status as an independent agency in order that its work can be freely directed toward broad cultural, scientific and commercial objectives. Such objectives far transcend the short term, though vitally important, military rocket missions of the Department of Defense.

(b) If the proper creation of an independent agency is judged to require an intolerable delay, then it is believed that statutory existence under [3] the Secretary of Defense (but not within the jurisdiction of any one of the military services) will be a workable arrangement for the immediate future. But in this event, it is urged that the "charter" of the agency explicitly provide for its independence as soon as its stature and achievements make this advisable.

(c) It is explicitly advised that the National Space Establishment not be placed within the jurisdiction of any one of the three military services. There are many reasons, growing out of extensive professional experience, for this view. The military services are basically operating agencies, not research ones. The research talent of any branch of the military services is almost inevitably turned toward helping meet short-term, limited objectives. Such a point of view would assure the failure of a National Space Establishment in its broad mission—which is truly a national one, far beyond the mission of any one of the services or of the Department of Defense taken as a whole. During the early phases of space research, it is evident that existing facilities and existing missile technology of the Department of Defense can make enormous contributions. The National Space Establishment must be set up in such a way that it enjoys the unqualified support of all three services, and not merely one of them. Such a situation is believed to be possible only if the N.S.E. is an independent agency from the outset or if it is directly responsible only to the Secretary of Defense during its early years—with the clear prospect of independence at the earliest possible date.

(d) There must be clear channels for mutual cooperation between the proposed N.S.E. and all levels of the Department of Defense, in order to assure no jeopardy of short term, vital military need on the one hand and in order to assure maximum rate of advance of space research on the other.

[4] 5. Remarks on the Long Range Importance of Space Research

It is already clear that international leadership hinges, to a very great extent, on preeminence in scientific and technological matters.

Space research will contribute enormously to the educational, cultural, and intellectual character of the people of the United States and of the world. Indeed, the exploration and eventual habitation of outer space are the finest examples of the "Endless Frontier." It is for such bold endeavors that the highest motives of men should be invoked.

There will be a rich and continuing harvest of important practical applications as the work proceeds. Some of these can already be foreseen—reliable short-term and long-term meteorological forecasts, with all the agricultural and commercial advantages that these imply; rapid, long-range radio communications of great capacity and reliability; aids to navigation and to long-range surveying; television relays; new medical and biological knowledge, etc. And these will be only the beginning. Many of these applications will be of military value; but their greater value will be to the civilian community at large. (To use a homely example, the telephone is certainly a valuable military device, but its importance to the civilian population is vastly greater.)

6. Availability of the Rocket and Satellite Research Panel for Consultation and Participation

The Rocket and Satellite Research Panel comprises a broad membership of persons of extensive experience in all aspects of the proposed program of outer space research. Its members are professionally dedicated to national leadership in this field. They offer their services, individually and collectively, in the conduct of the broad mission of the National Space Establishment.

[5] The Rocket and Satellite Research Panel

Berning, W. W.	Army Ballistics Research Laboratory
Delsasso, L. A.	Army Ballistics Research Laboratory
Dow, W. G.	University of Michigan
Ehricke, K.	Convair Corp.
Ference, M.	Ford Research Laboratory
Green, C. F.	General Electric Company
Greenberg, M.	Air Force Cambridge Research Center

Jones, L. M.	University of Michigan
Kaplan, J.	University of California
Kellogg, W. W.	Rand Corporation
Newell, H. E.	Naval Research Laboratory
Nichols, M. H.	University of Michigan
O'Day, M. D.	Air Force Cambridge Research Center
Pickering, W. H.	Jet Propulsion Laboratory
Spencer, N. W.	University of Michigan
Stehling, K.	Naval Research Laboratory
Stewart, H. J.	Jet Propulsion Laboratory
Stroud, W. G.	Army Signal Engineering Laboratory
Strughold, H.	Randolph Air Force Base
Stuhlinger, E.	Army Ballistic Missile Agency
Townsend, J. W.	Naval Research Laboratory
Van Allen, J. A. Chairman	University of Iowa
Von Braun, W.	Army Ballistic Missile Agency
Whipple, F. L.	Smithsonian Astrophysical Observatory
Wyckoff, P. H.	Air Force Cambridge Research Center
Zelikoff, M.	Air Force Cambridge Research Center
Megerian, G. H. Secretary	General Electric Company

Document I-12

Document title: J. A. Van Allen, G. H. Ludwig, E. C. Ray, and C. E. McIlwain, "Observation of High Intensity Radiation by Satellites 1958 Alpha and Gamma, *Jet Propulsion* 28 (September 1958): 588–92.

Source: Copyright American Rocket Society (now American Institute of Aeronautics and Astronautics), 1958. Used with Permission.

These are excerpts from the first published paper reporting the findings of James Van Allen and his colleagues at the State University of Iowa from the experiment they placed aboard the first U.S. satellite, Explorer I, launched January 31, 1958, and re-flew aboard Explorer III, launched March 26, 1958. (These satellites were also designated 1958-alpha and 1958-gamma.) Van Allen's results were first announced at a May 1, 1958, meeting of the American Physical Society and the National Academy of Sciences. This paper reflects a presentation made six weeks later, with additional data analysis, at a meeting of the American Rocket Society.

---

[588]

Observation of High Intensity Radiation by Satellites 1958 Alpha and Gamma[1]

J. A. VAN ALLEN,[2] G. H. LUDWIG,[3] E. C. RAY[4] and C. E. McILWAIN[5]
State University of Iowa, Iowa City, Iowa

Introduction

This is a preliminary report of results obtained concerning radiation intensities measured with a single geiger tube carried by the artificial earth satellites 1958 a and 1958 γ.[6]

The counting rate of the counter in 1958 α was transmitted continuously, and the data were recorded only when the satellite was quite near one of the 16 receiving stations distributed over the earth.

The data collected by 1958 γ were also telemetered continuously. In addition, a small magnetic tape recorder stored the data obtained during each entire orbit. Then, as the satellite passed near one of the receiving stations, a radio command from the ground caused these data to be read out.

A preliminary study of the data obtained from 1958 α and several interrogations of 1958 γ has been carried out, with the following results.

Reasonable cosmic ray counting rates have been obtained for altitudes below about 1000 km. In particular, we have obtained a plot of omnidirectional intensity vs. height in the vicinity of California for the first two weeks in February. This curve, extrapolated down to altitudes previously reached by rockets, agrees with earlier data.

At altitudes greater than about 1100 km, very high counting rates were obtained. This conclusion is the result of a somewhat lengthy analysis. Geiger tube output rates up to about 140/sec have actually been observed. In addition, periods have been found during which the geiger tube put out less than 128 pulses in 15 min. (We have a scaling factor of 128.) The considerations detailed in section 3 cause us to conclude that this is not due to equipment malfunction, but is caused by a blanking of the geiger tube by an intense radiation field. We estimate that if the geiger tube had had zero dead time, it would on these occasions have been producing at least 35,000 counts/sec.

We surmise that the radiation we have found is closely related to the soft radiation previously detected during rocket flights in the auroral zone.[7]

The radiation intensity necessary just to blank the geiger tube is equivalent to 60 mr/hr. In this connection the recommended permissible dose for human beings is 0.3 r/week.[8] The present radiation is 0.3 r in 5 hr or less.

Several geophysical effects of this radiation seem possible. It is very likely closely related to aurorae and geomagnetic storms. In addition, a rough calculation suggests that the

---

1. Assisted by U.S./IGY Project 32.1 of the National Academy of Sciences and the National Science Foundation.
2. Head, Department of Physics.
3. Research Assistant, Department of Physics.
4 Assistant Professor, Department of Physics.
5. Research Assistant, Department of Physics.
6. These satellite are sometimes called Explorer 1 and Explorer 3, respectively.
7. Meredith, Gottlieb and Van Allen, J.A., *Physics Reviews,* Vol. 97, 1955, p. 201
8. Kinsman, S. "Radiological Health Handbook," U.S. Dept. of Health, Education and Welfare, 1955, p. 292.

radiation may be sufficiently intense to contribute important heating to the upper atmosphere. It will be important to investigate the amount of atmospheric ionization, light and radio noise which would be produced, under various assumptions as to the nature of the radiation.

1. Instrumentation for 1958 α and 1958 γ

The instrumentation for 1958 α consisted essentially of a single Geiger Mueller tube, a scaling circuit for reducing the number of pulses to be worked with, and telemetry systems for transmitting the scaler output to the ground receiving stations. The system contained in 1958 γ was identical, with the addition of a miniature tape recorder for storing the data for the duration of each orbit and a command system to cause the telemetry of the stored information over a ground receiving station (Fig. 1) [all figures omitted].

Identical G.M. counters, scaler input circuits and scaling circuitry were used in the two cases. The G.M. counters were Anton halogen quenched counters having approximately 0.050 in. thick stainless steel walls. In addition, the counters were surrounded by the stainless steel cases of the payload, which were 0.023 in. thick. Thus the total absorption was approximately 1.5 gm cm-2 of stainless steel (approximately [589] 15 per cent iron, 25 per cent chromium). The G.M. tubes had essentially infinite lives, small variation in counting efficiency for the range -55 to 175 C, approximately 85 per cent counting efficiency for cosmic rays, and about 0.3 per cent counting efficiency for photons of energy 660 kev. The dead time of the counters was approximately 100 microsec. The length of the counter wire was 4 in.; the inside diameter of the counter was 0.781 in.

Following the counters were current amplifiers, which directly fed the first scaler stages. The scalers were bistable transistor multivibrators, which operated over a wide range of supply voltage and over a temperature range of -15 to 85 C. This limitation was caused by the supply batteries. The scaler resolving time was 250 microsec. If input pulses at higher rates than 4000 per sec periodic were received, the scaler simply indicated a constant rate of 4000 per sec. That is to say, the scaler would not go out of operation if this rate was exceeded. It did, however, have an input pulse amplitude discrimination level, so that counter pulses of less than approximately one eighth normal were not counted

In each of the satellites, the output of a scale of 32 was telemetered directly by the low power transmitter. In addition, it was transmitted by the high power transmitter in 1958 α. In all cases, the shift of state of the output scaler stages caused a discontinuous shift in the frequency of the subcarrier oscillators, of which the outputs were transmitted by the appropriate transmitters. The data telemetered in this manner have been readable when the rates of input pulses to the scalers were between 0.14 pulses per sec (16 pulses or the change of state per 2 min pass) and 80 pulses per sec by the bandwidth of the receiving and data reduction systems.

In 1958 γ additional scaling circuits were included to provide a total scaling factor of 128 for the data to be stored. It was also necessary to include a time base, in order that a proper correlation could be established between the data and the satellite position. These two bits of information were combined in such a way that they could both be stored and telemetered on a single channel. Fig. 2 indicates the manner in which an inhibitor circuit effected [sic] this combination. The time base input was a train of pulses at the rate of one each sec. These pulses appeared at the output of the inhibitor, and were recorded, unless one was preceded by an output from the scale of 128, in which case it was suppressed.

The tape recorder was advanced in a discontinuous manner at the rate of one step per sec. As the tape advanced, it wound a spring for the eventual return of the tape to the starting point.

Upon receipt of a properly coded interrogation signal by the command receiver in the satellite, a relay system was activated which caused the higher power transmitter to be turned on and the tape to be released, so that the spring was free to return it to zero. The return tape speed was controlled by an eddy current damping system, so that the playback time was approximately 5 sec. As the tape returned, the information was read off the tape, telemetered, and the tape was erased. Upon completion of the cycle, the relays were reset, the transmitter turned off, and the next recording begun.

The information thus telemetered to the ground was the train of pulses emanating from the inhibitor circuit, except that it was much compressed in time. It can be seen then that scaler input pulse rates between 0 and 128 per sec were properly passed on, and that all rates above 128 per sec appeared as a rate of 128 per sec, that is, all pulses missing.

2. Summary of Preliminary Observations

Table I is a list of the stations receiving data and reporting them to us. The stations labeled JPL are operated under the auspices of the Jet Propulsion Laboratory at Pasadena, Calif. Those labeled NRL are operated by the Naval Research Laboratory in Washington, D.C. Data were obtained from 1958 α only when it was reasonably near one of these stations, since it had no provision for storing data for a later readout. We have already analyzed most of the data from the JPL stations, and some of that from the NRL stations as well. This work is continuing.

A small magnetic tape recorder in 1958 γ stored the cosmic ray information for an entire orbit, and then played it into a transmitter on command from the ground. Data from nine of these orbits have been reduced in a preliminary way. We already have on hand many more of these passes, and are reducing the data from them in a routine way.

It is evident from the above summary that the present report is a very preliminary one. The nine cases from 1958 γ occur during the last four days of March, and we expect ultimately to have data obtained during several weeks after those days. In addition, we have so far reduced the data from 1958 γ only in a rather rough way, as explained in the following paragraphs. Finally, we do not yet have highly accurate data on the satellites' orbits. We do have the position of 1958 α as a function of time tabulated in 1 min intervals as supplied by the Vanguard computing center for the month of February. These data seem to be in error by several minutes in time, but apparently are sufficiently accurate for the purposes of the present report. So far for 1958 γ, we have only a set of orbital elements for March 26 and position vs. time for one orbit on April 1, together with estimates of the various perturbations. This information, supplied to us by the Vanguard Computing Center, has made it possible for us to estimate the orbit during the last days of March with reasonable accuracies. In particular, we estimate that our error in determining the time of passage through perigee is not more than about 5 min on March 31, and is less on earlier dates. Our errors in estimating latitude and longitude may amount to 10 deg in some cases.

Accurate orbital data will ultimately be supplied to us by the Vanguard Computing Center.

Table I Receiving stations

Blossom Point, Md.	NRL
Fort Stewart, Ga.	NRL
Antigua, Br. W. Ind.	NRL
Havana, Cuba	NRL
San Diego, Calif.	NRL
Quito, Ecua.	NRL
Lima, Peru	NRL
Antofagasta, Chile	NRL
Santiago, Chile	NRL
Woomera, Aus.	NRL
Patrick Air Force Base, Fla.	JPL
Earthquake Valley, Calif.	JPL
Singapore	JPL
Ibadan, Nigeria	JPL
Temple City, Calif.	JPL
Pasadena, Calif.	JPL

[590] We discuss first the data obtained from 1958 α.

Fig. 3 is a plot of height against counting rate near the California coast. All of the passes recorded by JPL stations in California are included in this graph. There is some variation in latitude, which presumably accounts for some of the scatter of the points. In addition, as explained above, the orbital data are not yet known with good accuracy, and this presumably contributes significantly to the scatter. A linear extrapolation down to a height of 100 km yields a value of omnidirectional intensity of 1.22 $(cm^2\text{-}sec)^{-1}$, in adequate agreement with values we have previously obtained from rocket flights, considering the crudity of the extrapolation. The data shown figure were nearly all taken before Feb. 11.

The data obtained by the NRL stations in South America during the first two weeks of February are altogether different from those just discussed. The passes fall into two classes. In the first case, one obtains a counting rate of about 30/sec, a roughly reasonable value. In the second case, the telemetered signal fails to show a single scaler output pulse during the approximately 2 min of clean signal. This represents an input rate to the scaler of less than about 0.1 sec. There are, in addition, a few cases showing a strong change in counting rate during the pass.

For reasons discussed in section 3, we believe that the extremely low output rate of the scaler is caused by very intense radiation which "jams" the geiger tube so that it puts out pulses of such low height that they are below the threshold of the counting circuits. Laboratory tests show that this first happens for the present equipment when the radiation reaches such an intensity that a counter of the same effective dimensions and efficiency as the present geiger counter but with a zero dead time would produce 35,000 counts/sec.

Fig. 4 is a plot of height vs. geographic latitude in the vicinity of 75 W longitude. The positions of 1958 α during reception of its telemetering signal by various of the NRL stations are marked. A code designates the kind of information received. It is at once evident that the extremely low counting rates observed all occur at a high altitude, while the more or less normal rates occur at a low altitude. Transitional cases occur at intermediate altitudes.

Quite similar behavior is observed near Singapore, and probably also Ibadan. In these two cases no thorough study has been made, mostly because of the lack of trajectory data for the dates on which extremely low telemetered counting rates occur. In the one case at Singapore where such a rate occurred on a date for which orbital data were available, the extremely low counting rate observed occurred at an altitude of about 2000 km.

Fig. 5 is a plot of geographic latitude vs. geographic longitude for various orbits. Only the high altitude cases are plotted on this figure. The fact that the segments of data do not correspond to positions of closest approach to the interrogating stations is due to our so far inaccurate knowledge of the trajectory.

These data already suggest a picture of the geophysical phenomenon being measured. The data from 1958 γ are much more explicit. Fig. 6 is a plot of the scaler output as a function of time as given by the tape recorder readout for the pass ending near San Diego on March 28, 1748 UT. Since the tape recorder can only record one scaler output pulse each second (see section 1) the maximum indication on the tape recorder output corresponds to 128 counts/sec for the geiger tube output rate. (Our scaling factor is 128 in this case.) It is evident from the figure that reasonable counting rates occur near the two ends of the pass. These ends correspond to the most northern latitudes and the lowest heights above the earth. The section where the counting rate indication is zero corresponds to a portion of the magnetic tape where no tuning fork pulses were missing, and hence no scaler output pulses occurred. This condition lasted 15 min, and 128 pulses were fed to the scaler during this time. This is an [591] average counting rate for the interval of 0.14/sec, to be compared the usual cosmic ray rate for a geiger tube of this sort of about 50/sec. The counter goes through the transition from putting out essentially no counts to putting out a great many very quickly, and we presume that most of the 128 counts observed during this 15 min interval occurred near the ends of the interval. There is, of course, no real evidence for this.

As discussed in detail in the next section, we believe that if we had had a detector with zero dead time, and a storage mechanism of unlimited capacity, Fig. 6 would begin where it does now, and at about 13 min would have begun rising rapidly to a peak near 25 min at which point the counting rate would have been greater than 35,000 counts/sec. After this time, the rate would gradually have subsided, returning finally to about the value actually recorded near the end of the pass.

Fig. 7 is a plot of geographic latitude vs. geographic longitude of those orbits for which the tape recorder readout data have so far been analyzed. We have simply identified the transition points between portions of the record where no tuning fork pulses are missing, all tuning fork pulses are missing, or some tuning fork pulses are missing. These three different kinds of regions are identified on the graph as >15,000/sec, 128 to 15,000, and < 128, respectively. The dashed portions of the various curves represent regions where the identification as to counting rate range is uncertain. Since these passes all occurred during March 28 through March 31, the orbit did not have time to precess appreciably. Since perigee was near the most northern latitude, a given latitude corresponds closely to a given altitude. It is evident that at high altitudes and low latitudes, mostly in a certain range of longitude, the counting rate is very high. Near perigee the counting rate is low. Elsewhere intermediate counting rates occur. Possible interpretations of this result will be discussed in section 3.

3. Interpretation of Observed Data

We now propose to justify our claim that when essentially no scaler output pulses occur, the apparatus is, in fact, exposed to very intense radiation.

Three possibilities are immediately evident. The apparatus may have some simple malfunction. This possibility can immediately be rejected except for the scalers, geiger tubes, and geiger tube voltage supplies, since the subsequent treatment information is completely different in the 1958 α and 1958 γ. Some effect of temperature seems the only reasonable possibility here. The temperature of the geiger tube was measured in 1958 γ and telemetered on the continuously operating transmitter. The observed temperatures range from zero to about 15 C. As discussed in section 1, the operating range of the circuitry is -15 to 85 C. In addition, the frequencies of the continuously telemetering channels which carried the cosmic ray information are significantly temperature sensitive. These showed that no extreme temperatures occurred at the location of the corresponding sub-carrier generators.

Another possibility might be that the satellite passed through regions which very few cosmic rays could reach. This is extremely unlikely. A magnetic field of the order of one gauss extending over thousands of kilometers and remaining unbelievably free of local irregularities would be required to exclude a sufficient fraction of the cosmic radiation.

The possibility that we firmly believe is correct is that the geiger tube encountered such intense radiation that dead time effects reduced the counting rate essentially to zero. In order explore this possibility, we have carried out the following experiments.

A spare flight unit for 1958 a was placed in an X-ray beam which was hardened by a 3/8 in. thick brass absorber. The voltage on the X-ray tube was varied between 50 and 90 [592] kev to vary the flux over a wide range. The counting rate was measured with and without lead shields which permitted only part of the beam to reach the geiger tube. In this manner the counting rates with and without the dead time effects were determined. As shown in Fig. 8, the dead time effects are negligible up to highest rates which can be handled by the telemetering systems. At high fluxes few of the pulses from the geiger tube have sufficient amplitude to operate the scaling circuit, and the counting rate returns to the range which can be telemetered. At very high fluxes no pulses have sufficient amplitude, and the counting rate is zero.

An ion chamber placed in the position of the satellite apparatus measured an intensity of 60 milliroentgens per hr at the minimum flux required to reduce the counting rate to zero. The ionization produced by different energy X-rays or by charged particles producing this effect would of course be different from this measurement. The X-rays used for this measurement had energies in the range 50 to 90 kev.

We have little concrete evidence concerning the nature of this radiation. Apparently, however, it is not electromagnetic. It makes its effects felt through the 1.5 g/cm^2 of absorber which constitute the hull of the satellite and the walls of the counter. Photons with such energy should then be seen down to the lowest altitudes our equipment reaches. The radiation can presumably be either protons or electrons. If it is electrons, we then are probably detecting bremsstralung formed in the satellite shell.

4. Implications

Any reasonable identification of this radiation strongly suggests several geophysical consequences. It is unlikely that the particles have several Bev of energy each. Then in order to reach such low heights through the geomagnetic field they must at least initially be associated with plasmas which seriously perturb the magnetic field at an earth radius or so. We presume that this plasma is closely related to geomagnetic storms and aurorae.

Secondly, at heights only a little above 1000 km, there is still some atmosphere. Crude quantitative estimates suggest that the energy loss in this residual atmosphere of the radiation we detect may contribute significantly, if not dominantly, to the heating of the high atmosphere. In addition to considering this heating effect, it will be important to calculate, on various assumptions as to the nature of the radiation, the amount of visible light, radio noise, and ionization produced.

Finally, there are obvious biological implications of these results. As discussed in section 3, if photons are being detected directly by the geiger tube, and if these photons are in the energy range 50 to 90 kev, then the radiation field inside the satellite corresponds to about 0.06 r/hr. The maximum permissible dose for human beings is 0.3 r/week. Other assumptions as to the nature of the radiation would obviously lead to different results.

Acknowledgments

We owe a large debt of gratitude to many individuals and agencies. We are indebted to the Jet Propulsion Laboratory at Pasadena, Calif., for the high speed rocket cluster and for assembly of the satellite payload. The Army Ballistic Missile Agency at Huntsville, Ala., supplied the booster stage and conducted the launching. Project Vanguard of the Naval Research Laboratory assisted in the early design of the instrumentation. They also set up and operated the minitrack tracking and telemetering stations, with cooperation and assistance from the countries in which the stations are located. They supplied us with orbital information for both satellites. The Jet Propulsion Laboratory set up the microlock stations for telemetry reception and operated all of them except those at Ibadan and Singapore. These last two were operated by students at University College, Ibadan, and the University of Maylaya, Singapore, as a part of the British IGY effort.

Document I-13

Document title: Space Science Board, National Academy of Sciences, "Minutes of the First Meeting," June 27, 1958.

Source: Archives, National Academy of Sciences, Washington, D.C.

Document I-14

Document title: L. V. Berkner, Chairman, Space Science Board, National Academy of Sciences, July 3, 1958.

Source: Archives, National Academy of Sciences, Washington, D.C.

Document I-15

Document title: Space Science Board, National Academy of Sciences, *Scientific Research in Space: Procedures for Conducting Experiments in Satellites and Space Probes* (Draft II – For Review), 1958.

Source: Archives, National Academy of Sciences, Washington, D.C.

Document I-16

Document title: Hugh Odishaw, Executive Director, Space Science Board, National Academy of Sciences, to Administrator, NASA, Director, NSF, and Director, ARPA, "Recommendation of the Space Science Board for Space Experiments," December 1, 1958.

Source: Archives, National Academy of Sciences, Washington, D.C.

Document I-17

Document title: Hugh Dryden, Deputy Administrator, NASA, to S. Douglas Cornell, Executive Officer, National Academy of Sciences, October 20, 1959.

Source: Archives, National Academy of Sciences, Washington, D.C.

Even before President Eisenhower signed the Space Act on July 29, 1958, that established the new National Aeronautics and Space Administration, the National Academy of Sciences moved to assert its influence in shaping the U.S. space science program. Academy President Detlev Bronk established a Space Science Board on June 4, 1958, and named the strong-willed Lloyd V. Berkner to be its chair. The Board held its first meeting on June 27, at which it reviewed existing plans for space activities and began the process of identifying scientifically valuable experiments to be carried out in space. On July 3, Berkner sent a telegram to a number of scientists soliciting ideas for such experiments. The Board also prepared a primer on space science with the goal of broadening the base of scientists interested in putting experiments and instruments into orbit or beyond. On December 1, 1958, the Board submitted its recommendations for an initial space science program to NASA, NSF, and ARPA, the latter of which had been set up to manage Department of Defense space efforts.

This quick-moving effort did not receive a totally positive reception from those in the new NASA who believed that it was their responsibility, not that of the Space Science Board, to plan the nation's space science program. In the work request to the Space Studies Board for fiscal year 1960, NASA asked the Space Studies Board for "guiding principles" for the space science effort "rather than a detailed program formulation." The tension between the external space science community and NASA managers has been a constant feature of the U.S. space science effort.

Document I-13

[no page number]

National Academy of Sciences
2101 Constitution Avenue, N. W.
Washington 25, D.C.

PRIVATE

MINUTES
of the
First Meeting
Space Science Board
June 27, 1958
Rockefeller Institute
New York, New York

Members

Dr. Lloyd V. Berkner, Chairman
Dr. Harrison S. Brown
Dr. Leo Goldberg
Dr. H. Keffer Hartline
Dr. Donald F. Hornig
Dr. W. A. Noyes, Jr.
Dr. R. W. Porter
Dr. Bruno B. Rossi
Dr. Alan H. Shapley
Dr. John A. Simpson
Dr. S. S. Stevens
Dr. Harold C. Urey
Dr. James A. Van Allen
Dr. O. G. Villard, Jr.
Dr. Harry Wexler

Invited Participants

Dr. Detlev W. Bronk, President
Dr. S. D. Cornell
Dr. Hugh L. Dryden
Dr. Robert O. Piland
Dr. Alan T. Waterman
Dr. Herbert F. York

Members Secretariat

Dr. Hugh Odishaw, Executive Director
Mr. L. N. Cormier
Mr. G. A. Derbyshire
Mr. J. G. Reid, Jr.

[2] Dr. Lloyd V. Berkner, Chairman, presiding.

Chairman Berkner announced the charter of the Space Board is contained in President Bronk's letter of June 26, 1958, which states:

> "We have talked of the main task of the Board in three parts – the immediate program, the long-range program, and the international aspects of both. In all three we shall look to the Board to be the focus of the interests and responsibilities of the Academy-Research Council in space science; to establish necessary relationships with civilian science and with governmental scientific activities, particularly the proposed new Space Agency, the National Science Foundation, and the Advanced Research Projects Agency; to represent the Academy–Research Council in our international relations in this field on behalf of American science and scientists; to seek ways to stimulate needed research; to promote necessary coordination of scientific effort; and to provide such advice and recommendations to appropriate individuals and agencies with regard to space science as may in the Board's judgment be desirable.
>
> As we have already agreed, the Board is intended to be an advisory, consultative, correlating, evaluating body and not an operating agency in the field of space science. It should avoid responsibility as a Board for the conduct of any programs of space research and for the formulation of budgets relative thereto. Advice to agencies properly responsible for these matters, on the other hand, would be within its purview to provide."

The work of the Board will cover three phases:

1. Immediate program
2. Long-range program
3. International program

Responsibilities of the Board – Detlev W. Bronk, President

President Bronk described the need for the Board and the National Academy's hopes for accomplishment. He pointed out his wish that the Board provide for an orderly extension and continuation of the rocket and satellite work of the USNC/IGY. With rockets and satellites there are opportunities for many fields of science through the application and development of new techniques. He called the Board's attention to its membership which did represent many fields of science as well as different geographical locations. President Bronk stated that the Board was being formed [3] because of the urgent need to provide help and advice to a possible new civilian space agency (the National Aeronautics and Space Agency), the National Science Foundation and the Advance Research Projects Agency. Continuity of Program would be insured through common membership of R. W. Porter, Chairman of the Technical Panel for the Earth Satellite Program, and, he hoped, through members of the Secretariat.

<u>Introduction to Tasks of the Board – L. V. Berkner, Chairman</u>

Chairman Berkner introduced the Board to some of its tasks:

1. Encourage participation of scientists from universities and institutions outside of government to ensure U.S. space science development on a broad base. While government participation was essential, it would be unwise for space science to develop entirely within the bounds of government activity. Consequently, the Board must encourage initiative outside the bounds of government laboratories.

2. Provide guidance to scientific endeavor in the field of space science, encouraging the participation from all fields of science, guiding integration of similar proposals, and eliminating that which is inappropriate. He noted that these functions would be best provided by a board broadly representative of U.S. science outside direct government channels.

3. Be aware of the military and commercial aspects of space science as well as the purely scientific.

    He listed as primarily military applications: reconnaissance, intelligence and communications – jamming activities; and as an example of commercial application, the use of satellites as communication and TV links. He pointed out the effect that would be produced on the other two by pure science use. Chairman Berkner made clear that the launching of a space vehicle has become an international symbol of scientific success and strongly influences a desire for cooperation. As an example, he mentioned the IGY which had a tremendous effect on international relations.

4. Work to prevent contamination of moon and planets.

    Through ICSU and other international bodies, obtain recognition of the problem, and prevent irresponsible or unnecessary contamination of moon and planet surfaces and atmosphere.

5. Work with government space agencies.

Provide advice, guidance and assistance to all government space agencies to aid in development of effective space science programs and experiments.

[4] Chairman Berkner presented his plans for the meeting as follows:

1. Review the government plans and programs

2. Discuss the work of the Board

   a. Immediate program
   b. International activities (joint or shared experiments)
   c. Long-range program plans

3. Organize the Board on an ad hoc basis

   a. Ask each member to analyze the problems in his own scientific field, perhaps with a small subject committee
   b. Work out the means of international co-operation; i.e., consider the proposal for the ICSU Special Committee on Space Science
   c. Establish the secretariat

Remaining Space Program of the IGY – R. W. Porter

There are in the Vanguard program five vehicles remaining for three flight–ready experiments. The experiments are: (1) cloud-cover: cloud mapping for weather forecasting; (2) the earth's magnetic field: satellite-borne magnetometer and several ground-based magnetometers for synchronized comparison measurements; (3) earth's energy balance: energy radiation and re-radiation. (He also stated that there is considerable pressure to reschedule the solar x-ray and ultraviolet experiments which have not been successfully launched.)

In the Explorer program, there are three vehicles and two experiments being readied for flight: (1) a repeat of the radiation experiments with improved instrumentation to provide adequate range of counting rates; (2) the 12-foot inflatable sphere to obtain air-drag data together with suitable radio beacons.

In the Lunar Probe program, there are five vehicles (three Air Force and two Army).

Experiments:

(1) Photoelectric scanners to show where it has gone.
(2) Magnetometer with a sensitivity of 10 gamma.
(3) Cosmic ray intensity experiment (Simpson, University of Chicago) – Air Force probe.
(4) Soft radiation experiment (Van Allen, State University of Iowa) – Army probe.

Final Satellite for the IGY Program.

Juno II type vehicle – 120-pound payload containing:

(1) A cosmic ray experiment
(2) Solar x-ray and ultraviolet experiments
[5] (3) Energy balance experiment
(4) Meteorite measurement (size distribution)
(5) Possible corner reflectors (geodetic experiment)
(6) Stabilized harmonically-related transmitters for ionospheric measurements.

This program concludes the IGY satellites and lunar probes with the last launching to take place in about March 1959.

Space Payloads in 1959 – Dr. Herbert F. York

Dr. York stated that ARPA has no charter for planning satellite programs in pure science beyond February 1959; that in the military program a reconnaissance satellite with attitude control is planned. As far as vehicular development is concerned, the Advanced Research Projects Agency program is as follows:

1. Composite Thor

Authorized for development in 1959, a vehicle to put 1800 pounds total weight in orbit, including 400-pounds useful payload; available for scientific use roughly one year later.

2. Composite Atlas

Target date for development July 1959; to fly six months later. Total weight in orbit 4000 pounds, useful payload 3000 pounds. Perigee altitude 300 miles. Available for science possibly late in 1959. It was noted that experimental payloads might be carried "piggyback" in several Atlas vehicles in the military series of nose cone tests.

3. Juno IV

The Juno IV, would orbit 1600 pound payload including 1200 pound instrument package (this might provide soft landing on moon for 100 pound-payload); not now established as a military project but might be ready by mid-1959 if funded soon.

ARPA has no funds to proceed with scientific vehicle construction. In summary, Dr. York stated:

(1) No additional vehicles are available for space science beyond IGY.
(2) Space vehicles for scientific use will require on the order of one year after request.
(3) Capability for space probes (with limited instrument payloads) to nearer planets should be available in the 1960 period.

(4) Orbit control of suitable precision will be available within one year with a sacrifice in payload for control use.

[6] (5) If ARPA's proposed "Man in Space" program is approved, recovery of payload from any vehicle can be expected in 1959-60.

At this point, Chairman Berkner advised the Board that it was his hope that Board meetings could be conducted with free and open discussion without danger of publication or release of information concerning operations of other agencies by Board members.

Upon question of Dr. Urey, it was agreed that Board members should be free to discuss launching potentialities with members of the scientific community to ensure adequate development of plans; but the Board should endeavor to avoid advance public release of the tentative plans of launching agencies.

He stated that Dr. Cornell would prepare an announcement of the activation of the Board for publication in Science, Physics Today, and perhaps the newspapers, but that no press conferences would be held.

He advised the Board that:

(1) Two indoctrination works would be provided by the Academy

a. Exploration in Space by Clarke

b. Scientific Uses of Earth Satellites edited by Van Allen

c. Space Research and Exploration by Bates (distributed at the meeting)

(2) Insofar as security clearance was concerned, the Board agreed that it should not be cleared for security as a Board, but it could adequately follow classified projects through cleared Board members strategically placed on the ad hoc committees.

Future Vehicles and Science Payloads – Dr. Herbert F. York

Future developments were summarized by Dr. York as follows:

1960 – Up to 3000 pounds available for satellites with 1/3 to 1/5 of the 3000 pounds for moon or planet exploration.

Post 1960 – (1961-1962) Using Atlas or Titan, up to 6000 to 8000 pounds for satellites, 1/3 to 1/5 of the 6000 to 8000 pounds for moon or planet exploration.

Mid 1960's – (1965) Up to 25,000 to 50,000 pounds for satellites using a single rocket with 1,000,000 to 2,000,000 pounds thrust. (Payload in orbit can be calculated at 2% to 3% of the rocket thrust in pounds).

Post
1965 – 50 tons into orbit by multiplexing of rockets.

[7] York stated that nuclear-powered rockets would not be available under ten years and that, in his opinion, chemical engines are adequate for satellite use and that nuclear engines should be good primarily for producing large velocity changes. With regard to space stations, Chairman Berkner suggested that the Board must eventually study this problem thoroughly before taking a position.

Present and Future Organizations

Advanced Research Projects Agency – Dr. Herbert F. York

ARPA is established to carry out assignments for the Secretary of Defense in all research fields and to study and recommend solutions to the Secretary of Defense where cross-service problems are involved. Currently, its mission includes the study of ballistic missile defense where it is both an operating and a staff agency; develop-ment of solar propellants where its function is staff; the study and development of what needs to be done – staff responsibility.

He pointed out that temporarily, ARPA is the only agency operative in the space program and that it provides coordination with NACA, NSF, and NAS.

National Aeronautics and Space Agency – Dr. Hugh L. Dryden, Director, NACA

Dr. Dryden stated that since 1926 NACA has been an operating research rather than an advisory agency. He pointed out to the Board some general characteristics and some major differences between the Senate and House versions of the Space Agency bills. He expressed confidence that these differences would soon be resolved satisfactorily. In both versions of the bills, Congress has tried to give the Space Agency very broad powers. For example:

(1) To accept services, to contract or to cooperate with individuals, agencies, corporations, universities or others.

(2) To permit reorganization across departmental lines thus consolidating space programs in one agency.

(3) Probably to provide for the arrangement of international scientific experiments.

National Science Foundation – Dr. Alan T. Waterman, Director

Dr. Waterman stated that the Science Foundation's activity is concerned primarily with basic research as contrasted to development activities. Currently, the Foundation provides financial administration and support for the scientific portion of the IGY satellite program. In the future, it will be most anxious to provide similar support to experimenters and will, through

its staff, support (a) experiments that must be worked out in the laboratory to assess their feasibility in space science; (b) the invention of hardware to carry out experiments in space; (c) the reduction and analysis of data representing scientific results of experiments. The National Science Foundation will provide for a review of proposed projects.
[8] Dr. Waterman defined the government's position in research as being that each agency should have funds for the support of basic research in areas related to its particular problems. The National Science Foundation's interest is not limited but is much more general: It assists all agencies, groups, and individuals where the research does not relate obviously to their missions. Dr. Waterman stated that the National Science Foundation would be able to provide financial support for the Space Science Board if it is required. (Dr. S. D. Cornell has the responsibility for developing support and will discuss this further with Dr. Waterman.)

National Academy of Sciences – Dr. Hugh Odishaw, Executive Director, USNC-IGY

Dr. Odishaw outlined the operation of the National Academy of Sciences-IGY relationships. International coordination is under the general cognizance of the International Council of Scientific Unions, through its Annee Geophysique Internationale (1957-58) Comite Special, which is really the general assembly of IGY National Committees. The National Academy of Sciences set up the U.S. National Committee for the IGY, which formulated over-all plans and policy and now meets on the order of two times per year. The operating agency is the Executive Committee of the U.S. National Committee which now meets about once per month. There are technical panels drawn from each scientific area having general responsibility for the conduct of experiments in each of the IGY disciplines. The panels which are of particular interest to the Board are those for rockets and for satellites. These panels specifically provide stimulus, direction, and project review. The Committee further provides for international co-operation and directs and plans the U.S. efforts in the IGY. The staff currently consists of a total of about fifty, including secretarial and clerical. Dr. Odishaw pointed out that close relationship had been maintained with the Department of Defense, the National Science Foundation and other government agencies. He added that the Committee was anxious to see an orderly continuation of both the rocket and satellite programs and that it had strongly urged for some time the establishment of a Space Science Board. He warned that rocket and balloon potentialities for scientific research should not be neglected in the face of the prospective satellite program.

President Bronk remarked that there were many points of similarity between the work of the Space Science Board and the IGY. However, he felt that since government agencies would play a major part in launching of Space Science Board experiments, the work of the Space Science Board would be less operational than the IGY and involve more co-ordination of planning.

Existing and Planned Funding – 1959

ARPA – Dr. York advised that funds are available only for the existing programs already outlined.

[9] NASA – Dr. Dryden stated that the NASA budget is currently unclear. The 1959 budget will contain $72,000,000 specifically for space science with about $250,000,000 included for related and complementary areas. Dr. Dryden stated that some funds (perhaps $10,000,000) may be transferred from ARPA but that this situation is also not clear.

NSF – Dr. Waterman said that NSF has its standard budget for basic research. It currently plans only through the continuation of IGY in the space field. Moreover, the Bureau of the Budget will not consider specific requests until the space agency is established. NSF's 1960 budget discussions have included $20,000,000 for instrumenting 12 satellites (one per month). However, again the Bureau of the Budget will not accept a formal request for these funds. In discussion of these reports, Dr. Dryden stated that NASA will want to pick up where the IGY program ends in such fields as optical and radio tracking, computing and so forth. There was some discussion in the Board on the possible sources of funds for non-governmental space science proposals, if Dr. Waterman's request was not approved.

Dr. Porter informed the Board of the TPESP and USNC recommendations for immediate fundings for future space experiments to be administered by the National Science Foundation. President Bronk emphasized the need for arousing Congress and the public to the continuing and expanding needs of scientific research. Dr. Odishaw pointed out that no IGY money would be left on the conclusion of the IGY programs. He restated the USNC resolutions concerning immediate needs for $6,000,000 to insure [sic] that experimental packages will continue to be available in the period immediately following IGY. In accordance with Dr. Waterman's request, specific experiments for immediate activation will be provided in connection with this resolution.

<u>Proposed Outline of Work of Board - L. V. Berkner, Chairman</u>

Chairman Berkner proposed that the work of the Board be apportioned among the members on an ad hoc basis. The permanent organization would be determined by experience.

After discussion of tentative ad hoc committees, the Board agreed on the following assignments.

(1) Geo-Chemistry of Space and Exploration of Moon and Planet – Chairman, H. C. Urey; Vice-Chairman, Harrison S. Brown

(2) Astronomy and Radio Astronomy – Leo Goldberg, Chairman

(3) Future Vehicular Development (Beyond Military Requirements and Including Space Stations) – Donald F. Hornig, Chairman

(4) International Relations Field – W. A. Noyes, Chairman
Co-ordination with ICSU. ICSU will decide whether an inter-national Space Board is necessary and the Board should be able to provide a U.S. position to them by September. Other problems in the international field will include the international sharing of payloads, environmental testings, advice on regulation, and ICSU representation.

[10] (5) Immediate Problems – R. W. Porter, Chairman
Space laboratories, orbits, immediate research projects, reorganization of TPESP, continuity and immediate support of NSF and ARPA.

(6) Space Projects – Bruno B. Rossi, Chairman
Long-range planning and general guidance.

(7) Ionosphere – A. H. Shapley, Chairman
Experiments pertaining to ionospheric studies including ionization, whistlers, and special propagation effects.

(8) Physics of Fields and Particles in Space – Chairman, J. A. Simpson;
Vice-Chairman, J. A. Van Allen

(9) General Engineering Service and Co-ordination – O. G. Villard, Chairman
Telecommunications, telemetry, guidance, environmental conditions, components and functions of the central laboratory for NASA.

(10) Meteorological Aspects of Satellites – Harry Wexler, Chairman

(11) Psychological and Biological Research – Chairman, H. K. Hartline;
Vice-Chairman, S. S. Stevens

(12) Geodesy –

The Board discussed the need for a member skilled in the field of geodesy. It was agreed that the Chairman should discuss this matter further with President Bronk.

Chairman Berkner outlined the scope of activity of the Committees as follows:

(1) In general, as designated in the Space Board charter provided by President Bronk. More specifically, for each committee within its field to investigate all aspects of problems such as payload compositions relative importance of experiments, expectancy, timing, environmental effects, orbital requirements and so forth, in relation to the effort and cost involved.

(2) To develop knowledge through symposia, publications, committee member-ship and so forth. (The need for such activities was emphasized by Dr. Urey.)

(3) To make reports to the Space Science Board. In turn, based on the work of the committees, the Space Science Board must issue studies of:

[11] a. Scientific programs and timing.
b. Vehicle requirements and timing.
c. Extent and character of support.

d. Long-range national plan.

Chairman Berkner asked the Board to consider this proposed outline of assignments and committee organizations and be prepared for discussion later in the day. This discussion produced the following:

On the selection of committee memberships, chairmen were urged to keep their committee small although it was recognized that all committees will not be the same size. Recommendations for membership and organization are to be made to Chairman Berkner and President Bronk.

The Secretariat will assist Committee Chairmen and the Board when it is organized. Hopefully this will be soon.

Secretariat

Dr. Odishaw reported on his study of Secretariat organization required for Board support.

With a number of ad hoc committees to support, as well as the Board itself in meetings, (international as well as Board and committees), symposia, adequate documentation possibly Congressional testimoney [sic] and the like, a minimum of 6-7 professional staff members with adequate secretarial assistance is required. Funds in the order of $250,000-$300,000 are required for the first year of operation.

Subsequent discussion established this level of operation to be necessary and reasonable. President Bronk agreed and thought that the Secretariat, when appointed, working with Dr. Cornell, could arrange side support from government agencies that will be associated by the work of the Board. If necessary, further support of private found will be sought.

Drs. Van Allen and Shapley pointed out that the sunspot cycle which has been obtained throughout IGY is very high but is decreasing rapidly and to be exploited requires the immediate availability of satellite experiments to follow the IGY series. Dr. Berkner emphasized the importance of Dr. Porter's committee, and stated that he would like to see from Dr. Porter a proposal for an immediate program, within one month or six weeks. This program is to include recommendations for specific experiment packages and satellites over the next two years. The other committees must develop experimental programs geared to achieve results in three to five years; but, inevitably, they must also assist Dr. Porter in unravelling [sic] the immediate problems in their respective fields.

[12] Information on the current efforts was provided by Dr. Van Allen as follows:

(1) The TPESP continuing program document (copies provided).

(2) The NACA committee under Guyford Stever which includes representation of all interests (military, scientific, commercial).

(3) Pressing needs are for an immediate program
   a. Publication of the needs through symposia.
   b. Determination of where and to whom to submit proposals.

c. Sifting and consideration of proposals.
d. Establishment of adequate funds.
e. Flight assignment to a vehicle.

(4) Experience shows that flight engineering consumes a long time and, therefore, planning for experiments is urgently needed.

(5) It must be recognized that space science is expensive.

Discussion then proceeded to the work of Dr. Porter's committee.

Dr. Porter stated that he did not feel a satisfactory study could be conducted and a report and recommendation made to the Board within this six weeks time. He recommended that each Board member (as a committee chairman) assume responsibility, as indicated, for evaluation of the following experiments or areas which the TPESP had recommended as requiring immediate activation.

Experiments	Committee No.
1. Solar Corpuscular Radiation	
2. Mass Spectrometers & Pressure Gauge	
3. Magnetometer – light pumping	
4. Cosmic ray package	
5. Color of extragalactic light	
6. Relativity experiments	
7. Astronomical problems	
8. Geodetics (perigee motion)	
9. Nocturnal U.V.	
10. Life Sciences – Growth of Living Tissue	
11. Psychological Experiments	
12. 100-Ft. Inflatable Sphere	
13. Ionospheric Experiments	
14. Meteorological Package (Advanced)	
15. Gamma Ray Astronomy	

Dr. Porter stated that these questions need to be answered:

(1) What can be done that can be instrumented by June 1959 to January 1960?
a. The weight of the instruments (less than 100 pounds per experiment includes power supplies and data storage and transmission).
[13] b. Orbit required (any, lower than 1000 miles).
c. Stabilization (1∞ to 2∞ for short periods to four weeks).

(2) What agency is best prepared to develop the experiment?

(3) When can instruments be ready?

(4) What is the cost?

(5) What is the weight?

(6) What are the vehicular requirements?

For establishing an immediate program Dr. Porter requested that each committee chairman address himself to these questions for each pertinent experiment within the scope of his committee. In the meantime, he would endeavor to collect all current proposals for the next meeting. Out of these, he hoped to select a few of immediate significance. He would ask for help from members of the Board in their respective fields.

Discussion by Committee Chairmen

Urey-Brown – Geo-Chemistry of Space and Exploration of Moon and Planet(s)

1. Backside of the moon
2. Chemistry of moon surface
3. Seismology of the moon
4. Magnetic properties of the moon
5. Soft landings on the moon
6. Study of use of environmental materials as propellants.

Goldberg – Astronomy and Radio Astronomy

Four months of work and study will be required for a good experimental program; this would require some funds which Dr. Waterman indicated NSF can supply.

Hornig – Future Vehicular Development

Primary objectives are:

1. Stay ahead of activities and keep Board informed of developments.
2. Recommend on future vehicular development and needs.

Noyes – International Relations Field

1. Provide recommendation to ICSU on proposed establishment of International Space Board by September 1958.
2. Develop a feel for the character of the regulations problem; while this is a governmental or inter-governmental problem, the Board must understand it.

[14] 3. Develop long-range plans for co-operative efforts in experiments, i.e., shared payloads. (It was recommended that Noyes attend the CSAGI R&S meetings in Moscow early August.)

Rossi – Special Projects

Gamma ray astronomy – gravitational red-shift.
Long-range visionary experiments and/or programs.

Shapley – Ionospheric Projects

Experiment search and evaluation is already underway through an ad hoc working group of the TPESP.

Simpson-Van Allen – Physics of Fields and Particles in Space

Suggested consolidation of committees appears workable and justified.

Hartline-Stevens – Psychological and Biological Research

Suggested consolidation to study the problem is acceptable and appears workable. W. R. Lovelace and Orr Reynolds were mentioned as being excellent sources of information. The recent Satellite-Life Sciences Symposium was also indicated to provide a basis for study.

Wexler – Meteorology

It is possible that space science will revolutionize meteorology. It was noted that IGY experimenters will run out of money soon. Suomi of the University of Wisconsin and the Signal Corps team of Stroud were cited as examples. It was agreed that every effort should be made to provide for continuation of their work. Funds should be provided for data analysis by meteorologists.

A meteorological committee exists in ARPA and Wexler will use it in Board activity.

Villard – General Engineering Service and Co-ordination

Responsibilities of this group include information theory, telemetry, environmental conditions; liaison aspects and components specifications for all other committees. In addition, evaluation of need for, and characteristics of, a NASA Space Laboratory are [sic] required.

In summary Chairman Berkner summarized the Tasks of the Board thus:

1. To collect information.

2. To broaden the base of Space Science.

3. To develop a national Space Science program that is effective scientifically.

[15] Action Porter is to present a basic document evaluating the fifteen experiments and with the answers to the questions raised to the next Board meeting.

Second Meeting Space Science Board
Casperi Hall, Rockefeller Institute, 9:30 A.M., July 19, 1958

First meeting of the Board adjourned at 4:50 P.M.

Document I-14

[handwritten "night letter 7/3/58 – USNC/IGY ESO"]
[handwritten "cc: Berkner
Porter
Joyce
Odishaw"]
[no pagination]

WESTERN UNION TELEGRAM
CONFIRMATION COPY

Academy has been asked by government to assess possible experiments that might be designed and constructed for satellite flights during next two years. This is admittedly a preliminary study but could lead to support of some experiments in near future. Approximate payload per flight perhaps as high as hundred pounds and within this limit several smaller non-conflicting experiments might be accommodate. Should appreciate your assistance and following information airmail within one week. Do you or your colleagues have experiment(s) that could be developed to point of final environmental test by mid-1959 or earlier. If so, please provide following information on each proposed experiment. First, several paragraphs describing each experiment, its scientific value, and the proposed instrumentation. Include estimated weights. Second, provide best possible estimate of total cost for design and construction of four complete hardware units, flight liaison personnel, data reduction and analysis. Third, provide estimate of months required between granting of funds and completion of hardware. Regret need to ask for such information on so short notice but cannot avoid.

L.V. Berkner Chairman NAS Space Science Board
C/O Associated Universities Inc. 10 Columbus Circle, New York 19, New York

---

Document I-15

[cover page]

DRAFT II – FOR REVIEW
FOR SPACE SCIENCE BOARD USE ONLY

SCIENTIFIC RESEARCH IN SPACE:
PROCEDURES FOR CONDUCTING EXPERIMENTS IN
SATELLITES AND SPACE PROBES

Space Science Board
National Academy of Sciences
Washington, D.C.

[1]

Contents

7. Flight Package Considerations
    a. Weight Availability
    b. Space Availability
    c. Payload Power Availability
    d. Payload Temperature Control
    e. Shock and Vibration Requirements
    f. Laboratories Equipped to Make Environmental Tests
    g. Telemetry
    h. Tracking Facilities Available
    i. Data Handling

Illustrations
[list of all figures and tables omitted]

[4] 1. The Challenge of Scientific Research in Space

Recent achievements in rocket technology have cleared the way for a revolution in physical science: experimentation in space. Measurements performed above the earth's atmosphere can be expected to have an impact on astronomy comparable to that of the invention of the telescope. New information thus gained on chemical reactions occurring on other planets and stars cannot fail to influence our understanding of the basic physics and chemistry of the universe. This understanding can then be tested with the aid of lunar, planetary, and solar probes. The possibility of studying life forms on other planets should be of the utmost significance to biology and the several life sciences.

It can be expected that the next few years will bring to hand a flood of new information, of the greatest consequence to mankind.

To make the most of this challenging opportunity, it is essential that the United States have a vigorous and effective program of research in space. This can only be achieved by ensuring the fullest possible participation by U.S. scientists and scientific institutions. At the present time, it appears likely that the production of rocket and satellite vehicles may far outstrip the supply of scientific experiments ready to be flown.

One purpose of this booklet is to interest U.S. scientists in participating in this country's space science program. Another is to summarize the essential information which will be needed by anyone deciding to take part.

[5] At the present time, it is likely that every qualified person who is interested in conducting scientific experiments in space can find an opportunity to do so. There is no shortage of funds for supporting really worthwhile projects. The initiative clearly rests with the individual investigator.

Research in space is not simple. It calls for unfamiliar techniques; it involves uncertainties and it requires coordination with other experimenters. Many aspects of the work are not under the immediate control of the participating scientist.

However, all of these difficulties can be minimized or overcome. In this connection, the National Academy of Sciences has set up a Space Science Board whose purpose is to assist those wishing to participate in research with the aid of rockets or satellites. This booklet was prepared with the same objective. Inquiries and correspondence will be welcome.

Lloyd V. Berkner,
Chairman

[6] 2. Purpose of This Booklet

This booklet is addressed to the many scientists of the United States who have not yet given serious consideration to the possibility of conducting experiments with the aid of satellites and space probes. Its purpose is to point out the opportunity for worthwhile research utilizing this new technique; to suggest that the obvious practical difficulties are by no means so formidable as they might seem; and to encourage new investigators to try their hands in this field. The general aim is to make the United States space research program more effective by ensuring participation by the largest possible number of competent investigators.

Later sections of this booklet will discuss reasons why experimentations in space are worthwhile from the standpoint of an individual scientist—for example, a member of the faculty of a university. There will then be presented a brief account of how research in rockets and satellites is done, with notes on the principal facilities available, the organizations that might be in a position to provide support, and other similar information.

3. Space Experimentation as a Research Field

To an individual investigator, such as a member of the faculty of a university, research in satellites and space probes may seem at first glance almost hopelessly formidable. There are a number of reasons, however, why this point of view is unjustified. In fact, it is possible to say that at the present time, the general area of scientific research in space is one which offers unprecedented opportunity. The reasons for this point of view are summarized below.

[7] a. Money to Support Space Research is Available. A number of agencies, ranging from the National Science Foundation through the Advanced Research Projects Agency, are interested in supporting studies and experiments having to do with space science. At the present time, no really worthwhile proposal can fail to find support, and this situation is likely to persist for a long time.

b. Available Vehicles Are Outstripping Available Experiments. Both the number of rockets in production, and their reliability, are rapidly increasing; the cost of placing a pound of material in orbit is rapidly falling. Vehicles can be placed on a production-line basis; but for the most part scientific experiments, being individual and sequential, cannot. Thus the number of tested experiments ready for flight may fall well behind the number of payloads which, for one reason or another, will be flown anyway.

c. Larger Payloads Coming. Payload weight restrictions lengthen the time requested to engineer the hardware of a given experiment. Payloads for typical satellite orbits will soon rise to values of the order of 100 pounds, thus increasing the range of possible experiments and easing the engineering problem.

d. Proposer of Experiment Need Not Build Hardware. It is possible for the proposer of an experiment to have as much or as little to do with the hardware as he desires. He can turn the entire proposal over to an agency, such as NASA, for implementation. Those who wish to are encouraged to develop their own "breadboard" models, to see them through the flight testing phase, and to supervise preflight testing.

[8] e. In Many Research Areas, Space Experiments Will Supersede Conventional Techniques. In many fields, the great advances made possible by research in space will render research by conventional techniques obsolete. Many workers will sooner or later wish to avail themselves of the opportunity for conducting research in space.

4. Special Aspects of Space Research: Role of the Space Science Board

Conducting experiments in satellites bears little resemblance to the traditional physics research conducted in the basement of a lecture hall on Saturday afternoons with the aid of a few graduate students. Yet the historical record shows that the contributions of individual scientists and their student collaborators have been enormously productive. It is accordingly a challenge to see how research in space can be organized so as to be as appealing as the traditional informal variety.

Realizing the importance of the individual investigator to a national program of scientific research in space, the National Academy of Sciences has set up a Space Science Board whose assignment is to encourage the fullest possible participation in research done with the aid of satellites and space probes, and to assist and represent individual scientists in every step of the necessarily lengthy procedure between conception and conclusion of an experiment.

A very brief outline of this procedure now follows. First, an experiment must be proposed and reviewed. After approval is obtained, a sponsor is found, and a place for the experiment [9] on the test schedule determined. This may call for some compromise in the details of the experiment, since economy usually requires that several be flown at once, and different experiments may interfere with one another. Furthermore the orbit or path selected for a given vehicle may not be optimum for all the experiments. After necessary compromises have been made, the task of engineering begins. A suitable physical embodiment must be capable of withstanding the enormous static and dynamic forces to which the apparatus is subjected during the launching phase, as well as the variations in temperature encountered during every phase of flight. At the present time, only a few institutions have the necessary facilities for conducting environmental tests. The model, once completed, must perform in accordance with the experimenter's specifications. Launchings for the most part take place at military installations; data are received and sent back to the investigator from tracking stations in various parts of the world.

At the present time, the scheduling of payloads available for unclassified research is being handled by the National Aeronautics and Space Agency. The N.A.S.A. will also sponsor the preparation of experiments, and will ready them for flight. Experiments of primarily military interest are coordinated by the Advance Research Projects Agency of the Defense Department. Contracts for experiment preparation can be let by A.R.P.A., or by other military agencies. Preparation of equipment for flight will be assigned by A.R.P.A. to the appropriate service laboratory.

[10]The flow chart of Fig. 1 [all figures omitted] illustrates these procedures graphically for the case of an unclassified proposal.

It is appreciated that this technique for doing research has formidable aspects. In an effort to answer questions and possible objections to space research which might be raised by individual experimenters, Table 1 has been prepared. It is the aim of the Space Science Board to see that as many as possible of these objections are minimized or removed.

5. Function and Policies of the Space Science Board

The Space Science Board of the National Academy of Sciences has been requested by N.A.S.A., N.S.F., and A.R.P.A. to review and assess the relative scientific importance of proposals for space research which have been submitted to those organizations. The Board operates in a purely advisory capacity; it makes recommendations only, and concerns itself with scientific matters only. (The Board may endorse a particular experiment, as being scientifically important, but the choice of a contractor is up to the agency which sponsors the work.)

Chairman of the Board is Dr. Lloyd V. Berkner; its executive officer is Dr. Hugh Odishaw, and its secretary is Mr. Ross Peavey. The Board maintains offices at the National Academy of Sciences, 2101 Constitution Avenue, Washington 25, D.C. The Board consists of 15 members representing the various fields of scientific endeavor. These members, in turn, are heads of committees consisting of competent scientists and experts in the Board member's special field.

[11]How the Board Can Help Individual Investigators

Proposals for scientific research in satellites and space probes should be sent to the Board for consideration. To save time, copies can simultaneously be sent to possible sponsoring agencies (N.A.S.A., A.B.P.A., etc.). If doubt exists as to the most suitable sponsoring agency, the Board will be glad to advise.

The Board will endeavor to keep itself informed of all U.S. and foreign research in the space field, so that it can inform the proposer of a particular experiment of the relationship of that experiment to other work being performed elsewhere.

The Board will consider proposals, and will provide its recommendations concerning the suitability and relative priority of experiments to potential sponsors and to the originator of the proposal. When, in the Board's opinion, a particular branch of science or class of experiments is not receiving sufficient attention, the Board may take an active role in promoting interest in that particular area.

Specific Policies

The broad aim of the Board is to do everything possible to further and to strengthen the United States space research effort. With this aim in mind, the Board has adopted certain policies, which are outlined below.

a. Unclassified Research
The Board will be concerned with unclassified scientific research only, and will encourage early publication of essential scientific results.

[12] b. Largest Possible Participation
In general, the Space Science Board wishes to encourage participation in space research by as many qualified investigators as possible in the largest possible number of institutions. The Board will be grateful for any suggestions that will help it achieve this goal.

c. Encouragement of the Individual Investigator
The Board believes that the overall effectiveness of a national program of research on space science is related directly to the number of qualified scientists who participate in it. Hence it desires to encourage participation by as many individual investigators as possible, even though they may be at relatively isolated institutions.

d. Encouragement of Individual Initiative
Recognizing that many important discoveries in science have stemmed from speculative, ad hoc, or unprogrammed research, the Board wishes to encourage individual initiative and will give fair consideration to research proposals that may seem to run counter to the prevailing scientific thought of the day.

e. Encouragement of Individual Control of Experiments
It is the Board's policy that the individual experimenter should have as much control over his own experiment as possible. Thus, to the extent that practical circumstances permit, the individual experimenter should be able to:

1. have adequate liaison with those responsible for model construction and environmental testing.
2. [13] test his own equipment before firing.
3. determine the actual time of firing.
4. have full and complete access to the resulting data.

f. Encouragement of Experimenter's Participation in Hardware Preparation
To the extent that circumstances permit, the Board feels it desirable for individual experimenters to take an active part in the preparation and testing of experimental equipment. At the present time, model construction and environmental testing can be done only in a very few large centers for such work. It is felt that there should be a reasonable number of university laboratories having the facilities to determine whether a given experiment is flight feasible or not. The main engineering, however, will have to be done at the large centers.

g. Encouragement of Contracting Procedures that Serve the Above Objectives

Since it takes roughly a year to prepare a space experiment a year to collect data, and then another year to analyze the results, the traditional system of one-year contracts with yearly renewals is felt to be cumbersome. Three to five year contracts should be established whenever possible.

The chief investigator's own institution should have the contract for final equipment testing and data reduction, except when the investigator himself should desire otherwise.

In the event that an industrial firm prepares some portion of the experimental equipment, the contract for this work should, [14] whenever possible, be a subcontract from the investigator's institution, rather than a direct contract from the sponsoring agency.

h. Protection of Individual Authorship and Priority of Conception

The Space Science Board will keep a careful record of the date of receipt of each suggestion or proposed experiment. From time to time the Board will publish a list consisting of a one-sentence description of the general field of each such proposal, the author's name, and the date of receipt.

i. Disclosure of Proposal Contents

The Board will treat all proposals and correspondence as private material, unless the author desires otherwise. In the event that it would be desirable for author A to know the contents of a proposal by author B, A's permission will always be obtained before releasing anything to B.

j. Handling of Simultaneous Proposals

Often the next step forward in a given field of experimental science will become apparent to several individuals or groups of individuals at once, thus resulting in proposals which compete or conflict. Although the Board cannot make recommendations concerning the choice of contractors, it can serve as an intermediary in the event that two or more individuals or agencies wish to join forces in performing an experiment which both have proposed.

k. Proprietary Rights

In view of the large number of its Committees, [15] subcommittees, and consultants, the Board cannot accept responsibility for maintaining the privacy of proprietary or "company confidential" material. Every effort will be made to respect the wishes of those who submit proposals, but it must be understood that the submission, and the later conduct of the experiment, is at their own risk.

l. Meetings, Symposia, and Publications

The Board will endeavor to develop a national program for research in space science. It plans to hold meetings and symposia, and to support publication whenever feasible.

m. Publicity

In view of the current public interest in space research, the Board feels that the following rules concerning release of publicity should be followed to the fullest extent possible.

1. In general, releases should be cleared in advance through the National Academy of Sciences.
2. There should be no advance description of specific experiments. Only general classes or kinds of experiments may be mentioned.
3. In general, there should be no general publicity release concerning an experiment until after the launching.
4. News stories concerning the results of experiments should be timed to coincide with publication of a scientific paper, or presentation of a full discussion at a scientific meeting.
5. The chief investigator himself has the right to release the results of his own experiment. This must be made [16] clear to subcontractors and others in a privileged position who may be consulted by the press.

n. Preparation of Proposals

No fixed form need be followed in submitting proposals to the Space Science Board. However, the following pieces of information are of great value in assessing the relative merit of proposals.

1. It should be made clear why a given experiment must be performed in a satellite or space probe, rather than in a rocket or sounding balloon.
2. The qualifications and previous experience of the chief investigator should be listed.
3. The percent time to be devoted by the chief investigator to the proposed project should be given. It will be helpful to list other commitments and responsibilities of the chief investigator.
4. The significance to science of the proposed experiment should be pointed out.

6. Sources of Financial Support for Space Research

a. Study and Research Leading to the Design of Experiment

The value and feasibility of an experiment usually develops rather directly out of the scientific familiarity and activity of a scientist in a particular field. The conception of the experiment and its design have usually already been worked out in some detail by the time the first proposal is contemplated.

However, there are undoubtedly situations where the complexity of an experiment, feasibility in terms of state of the art, [17] or other considerations, may justify an extensive preliminary analysis or feasibility study. In such cases, a request for support may be addressed to the National Science Foundation if the experiment is of basic scientific interest. Studies dealing with engineering or environmental aspects of space research might find support from NASA. Those of military significance may be supported by ARPA.

b. Development of Experiment and Instrumentation

Further development of the experiment and instrumentation prior to the establishment of a flight package design constitutes a major area of responsibility for the scientist. Funds for supporting the extensive work in this phase of the program may be obtainable from NASA or, for military applications, from ARPA. Where special zones of technical interest are involved, supporting funds may be obtainable directly from research offices of the respective military services, e.g., ONR, ARDC, ARO, AFCRC, WADC, USAESL.

c. Development of the Flight Package

The program for adapting acceptable instrumentation to its place in a flight package is largely the responsibility of NASA or of ARPA (for military type experiments). In any case, an experiment which has progressed to this point will have already established support from one of these agencies for collaborative work required of the author of the experiment, or of his staff.

d. Liaison With Launching, Data Handling and Data Analysis

The cognizant government agency, NASA or ARPA, maintains responsibility for the launching and tracking of a space [18] experiment as well as for the recording and reduction of telemetry data. The experimenter has the responsibility for maintaining all requisite liaison and final reduction and analysis of all telemetry data. Support for these phases would undoubtedly derive from NASA or ARPA. However, there is the possibility that support for analysis of data of basic scientific interest may also be obtainable from NSF.

7. Flight Package Considerations

The launching of an experiment in a satellite or space probe involves a most significant expenditure of time, effort, and engineering capabilities. Preparation of a suitable scientific flight package accordingly requires careful collaboration between authors of experiments and the engineers and scientists conversant with flight package design. The following guidelines are offered for the use of scientists in evaluating the possibilities of placing a particular experiment in space.

a. Weight Availability

Launching capabilities have thus far limited our scientific payloads for satellites to about 20 pounds in orbits of some 300 miles mean altitude. During 1959 the availability of improved booster and upper-stage rockets should increase payload capabilities to more than 100 pounds. (With the present state of the art, about 25 pounds is required for a five-watt power supply using solar cells; about 35 pounds for the satellite body, hardware, and radio transmitter; with approximately 40 pounds remaining for [19] instrumentation.) By late 1959 or early 1960 a still larger booster rocket should make it possible to place in orbit payloads in the range of one to two tons. At the same time, improved guidance should facilitate the precise attainment of higher altitude orbits with appropriate reduction in payloads.

By 1962 improvement in upper-stage rockets should make possible the launching of several tons into a 300-mile orbit or more than a ton into an orbit of co-rotation with the earth at 22,300 miles altitude. In the period beyond this, very powerful launching rockets should be available. It is then likely that scientific justification rather than launching capability may play the determining role in our science program.

b. Space Availability

For the immediate future, weight rather than space appears to impose the greater restriction on scientific satellite packages. Both Explorer and Vanguard satellites have carried most of their scientific equipment in a centrally located cylindrical instrument compartment about 6" in diameter. Therefore, instrument assemblies have taken the form of disk-shaped modules or decks of various heights, arranged in a stack. The Pioneer, on the other hand, disposed its instrument packages around the equator of its top-shaped shell.

Components and fabrication techniques similar to those for miniaturized airborne and missile electronics have proved satisfactory for space use. Instruments requiring exterior surface mounting or bulky units of non-modular dimensions can probably [20] be accepted in prospective launching vehicles, which should permit satellite volumes as much as several cubic feet.

c. Payload Power Availability

The electrical power supply carried by the space vehicle ordinarily determines the total amount of scientific data which can be detected, amplified, and telemetered to the earth. Reasonable flexibility exists for the band width of communication and observation in comparison to the operating lifetime of the system. Thus far, satellites and space probes have depended almost entirely upon chemical batteries. These have in general encountered no serious difficulties in meeting the outer space environment. In comparison with optimum figures of about 80 or more watt hours per pound of batteries, space vehicles have designed for and obtained performance of about 50 watt hours per pound. Mercury cells have been used, and silver-zinc cells have been projected for use in a later satellite. On the average, our satellites have utilized about twenty-five per cent of available weight for the battery pack.

In 1958 Beta, the 6" Vanguard test sphere, sufficient power was developed by banks of solar cells to operate a transistor transmitter at about twelve milliwatts. This operation, which has continued for sunlight periods since March 17, 1958, indicates that silicon solar cells are not rapidly damaged by the space environment. Their more extensive use in future vehicles is expected. For the range of variation of aspect in an uncontrolled satellite, for the storage of energy during dark periods, for regulation and [21] voltage conversion about five pounds are presently required for the supply of one watt. This is equivalent to more than 1,500 watt hours per pound, if a one-year life is assumed. A 5-watt power system of this type will be used in a late IGY satellite.

Nuclear powered devices are expected to provide efficiencies up to perhaps 2,000 watt hours per pound for high power, long life applications in space vehicles. Although such systems are well advanced in development, the problems of their application, particularly shielding and dissipation of the generated heat, may present difficulties. In summary, U.S. satellites to date have operated at a fractional watt level with lifetimes up to a few months. Prospective space vehicles should operate at a few watts for periods up to a year. It may be expected that the 1960-62 period will see the operation of space science vehicles at power levels of 50 to 100 or more watts.

d. Payload Temperature Control

The approximate solution of the problem of the temperature of a satellite or space vehicle was worked out prior to launching, both for the Explorers and for Vanguard I. In both cases, temperature range reported from the satellites agreed with predictions within design limit. The space vehicle temperature represents a balance between radiation absorbed by or heat dissipated within the satellite and the heat lost by radiation, or latent heat exchanges within the satellite. By suitable adjustment of the absorption and emission characteristics of the various portions of satellite surfaces over the full spectral [22] range, and with due consideration of the satellite orbit, the shell temperature for Explorer I was restricted to a range of 25°C. to a probable 90°C. cyclic range. This may increase to a variation of from 0°C. to about 40°C. within the shell.

In the case of Vanguard I, temperature stabilization at about 40°C. appears to have been reached within one day after launching.

In summary, it is indicated that the present state of the art makes possible the design of instrument compartments adequate for the maintenance of temperature limits and within the requirements of instrument components of a type suitable for airborne or other mobile applications. The possibility of maintaining a close temperature regulation for a portion of the instrument compartment to any reasonable degree is attainable at the sacrifice of some payload weight and operating power.

e. Shock and Vibration Requirements

Once an instrument package has achieved orbit or a condition of coasting flight in space, it is essentially in a force-free condition except for the effects of residual spin, attitude control, or a possible meteoritic impact. In this environment, large-size light-weight structures may be developed for scientific purposes, e.g., by the inflation of plastic balloons. However, prior to the attainment of this free flight condition the instrument payload package must withstand appreciable shock and vibration during an initial period of shipment and handling leading to its being placed on the launching pad; as well as the subsequent [23] launching period where acceleration forces of both setback and spin are encountered along with random vibration over a wide spectrum along all three axes.

For establishing dependability of the payload package to meet its launching environment, an extensive program of shock and vibration testing has been evolved in the satellite program. Test limits are dictated in large measure by the shock and vibration characteristics of the launching rocket system and the mounting characteristics of the payload itself. Specific test routines have been worked out for the Vanguard launching system, the Jupiter-C, the Pioneer, and the Juno-2. These have included: dynamic balancing; acceleration with spin tests; vibration (random noise, band limited 20 - 1500 cps) along all three axes.

Although the test limits differ in some degree for the various launching systems, the following, taken from the type approval tests for the Juno-2 flight prototype payloads, may be taken as typical:

(1) Shock

Complete payload subjected to about four 100G shocks parallel to axis of launching thrust. Tests by means of ballistic hammer.

(2) Vibration

Random noise, 15G rms parallel to thrust axis for two minutes. Random noise, 12G rms along two planes mutually orthogonal and perpendicular to thrust axis. Two minutes for each plane. A test on electrodynamic [24] shaper with white noise drive.

(3) Static Acceleration
Payload to be held at 75G for two minutes, by means of centrifuge.
(4) Spin
After dynamic balancing, the payload is spun at 900 rpm for ten minutes.

The foregoing tests are applied to the flight prototype sample. Somewhat less rigorous flight acceptance tests are then applicable to identical payloads scheduled for actual flight.

The development of the payload instrument package for withstanding these mechanical requirements, as well as thermal and low pressure requirements, is the responsibility of an expert space package design group having cognizance of the launch in question. Ideally, this group should begin to work cooperatively with the scientist carrying out the experiment at an early phase of his instrument development, even prior to the completion of a laboratory bench model of the instrumentation. Following the successful testing of individual components, there is the qualification of the complete instrument configuration for flight readiness through a flight approval test. At this point, the flight package engineers assume responsibility for adapting the approved instrumentation into a flight prototype package which will meet test requirements and, at the same time, be functionally acceptable to the responsible scientists.

[25] f. Laboratories Equipped to Make Environmental Tests

Massive test equipment is required for carrying out the full range of environmental tests for payload instrument packages, particularly for payload capabilities of 100 pounds or more. Adequate installations of test equipment are located at laboratories engaged in the development and design of airborne or space vehicle instrumentation. These include NASA facilities at NRL, Washington, D.C.; the Jet Propulsion Laboratories at the California Institute of Technology, Pasadena, California; The Army Ballistic Missile Agency in Huntsville, Alabama; and the Air Force Ballistic Missile Division, Inglewood, California. Others having such test equipment are a number of major commercial contractors active in the Defense Program, such as, the Space Technology Laboratories, Lockheed Aircraft, Douglas, Convair, General Electric, and others. These facilities would not ordinarily be available for other than military program use. Less extensive test equipment is adequate for the test of individual components and sub-assemblies leading to the flight approval test. This class of shock and vibration equipment and thermal-vacuum test chambers is doubtless in use at the State University of Iowa and is probably established in great part in a number of other university laboratories that have been active in programs of airborne electronics. It is likely that a number of such test facilities will soon be established throughout the country for assisting scientists in the initial development of flightworthy space instrumentation.

[26] g. Telemetry

Scientific satellites have used two general methods for the telemetry of scientific data to the earth. Most of these have continuously impressed one or more channels of data on radio frequency carriers of about 108 mc, which also served as beacons for radio tracking. Both phase and amplitude modulation have been used. Reception of a complete record for a satellite employing continuous telemetry requires an extensive worldwide network of receiving stations placed so that one is always within line of sight of the satellite. Despite the severity of this requirement, excellent, though not complete, records have been obtained for the U.S. satellites using continuous telemetry.

An alternate scheme planned for the Vanguard satellites and so far used successfully in Explorer III utilizes the readout upon command of satellite data. With the satellite continuing a data storage system sufficient for one orbit, it is possible for a complete data record to be obtained from readouts made once each orbit as the satellite passes over the "picket fence" array of Minitrack stations.

Our satellite experiments have been basically simple and their results have involved a communication rate of only a few cycles per second. Even when the synoptic data for one entire orbit has been compressed for readout transmission during passage over a tracking station, the communications bandwidth has not exceeded 15 kc. Design of experiments for narrow band signals was considered preferable not only for its reduced communications power [27] requirement but also for the attendant increased reliability from instrumental simplification and the lessened requirement for data reduction and analysis.

As space science experiments become more complex, improvements of telemetry capabilities may be expected. With the present state-of-the-art, communications capabilities approach video bandwidths for satellite altitudes of several hundred miles. Limitations are of an engineering rather than basic nature. Current capabilities also exist for maintaining signal bandwidths of a few tens of cycles out to distances of several hundred thousand miles. It has been estimated by the Jet Propulsion Laboratory that improvements should make it possible by 1962 to communicate at 30 cps bandwidth to a distance of about 5 billion miles or, alternately, to five hundred million with a voice channel of 3 kc.

h. Tracking Facilities Available

Facilities for tracking satellites by radio and optical means and for the computation of orbital position as a function of time have been established as a part of the U.S. IGY program. These facilities are now being continued and expanded as part of the U.S. Space Program under NASA. The radio tracking network consists of Minitrack (interferometer) stations at the following locations:

Blossom Point, Maryland
Savannah, Georgia
Havana, Cuba
Mt. Cotopaxi, near Quito, Ecuador
Lima, Peru
Antofagasta, Chile
Santiago, Chile
San Diego, California
Woomera, Australia

[28] These stations are also equipped for command readout of telemetry. Some are also equipped for tracking at 40 mc. A network of Micro-lock stations for increased longitudinal coverage includes stations at the following locations: San Gabriel and Earthquake Valley, California; Cape Canaveral, Florida; Ibadan, Nigeria; and Singapore. A number of

additional radar and space vehicle reception stations have also been set up through ARPA and the military services.

Precise observations leading to the computation of definitive orbits for satellites are carried out by an optical tracking network which is operated by the Smithsonian Astrophysical Observatory. This includes the following stations, each equipped with an F-1, 20" photo telescope: White Sands, New Mexico; Florida, near Palm Beach; Curacao, Netherlands West Indies; Arequipa, Peru; Villa Dolores, Argentina; Olifantsfontein, South Africa; Cadiz, Spain; Shiraz, Iran; Naini Tal, India; Woomera, Australia; Mitaka, Japan; Haleakala, Maui, Territory of Hawaii.

i. Data Handling

Centralized headquarters at NASA are now being organized for reduction and compilation of both telemetry and orbital data. Thus it may be expected that scientists who will engage in satellite or space experiments may expect to receive reduced and compiled data, ready for study.

Telemetry data in the form of the original magnetic tapes with time base, upon receipt from the telemetry recording stations, are broken down into individual tape records of the respective telemetry channels. Data can be furnished to the experimenter in this [29] form or in the form of a continuous strip oscillographic record, suitable for direct application of the channel calibration.

Orbital data provide a correlation of scientific data with the position of the instrument package in space. These data can be furnished to the scientist as a tabulation of coordinates of orbital subpoints and vehicle altitudes, given at specified instants of time. Presently, the uncertainty of the positions in space so defined is probably about 5 miles. For experiments requiring higher precision of position, some improvement is undoubtedly possible.

Document I-16

[no page number]

NATIONAL ACADEMY OF SCIENCES
SPACE SCIENCE BOARD
2101 Constitution Avenue, N.W.
Washington, D.C.

Memorandum Report, December 1, 1958

TO: Administrator, National Aeronautics & Space Administration
Director, National Science Foundation
Director, Advanced Research Projects Agency

FROM: Hugh Odishaw, Executive Director, Space Science Board

SUBJECT: Recommendation of the Space Science Board for Space Experiments

This document presents, in outline form, current recommendations of the Space Science Board. These recommendations are based on studies conducted by the appropriate committees of the Board during the summer and early fall of 1958. They were formally adopted by the Board at its meeting on October 24-25. The content of the recommendations is known to the principal federal agencies having responsibilities in space work (National Aeronautics & Space Administration, National Science Foundation, Advanced Research Projects Agency), by virtue of their participation in the meeting of the Board, while one of these agencies (NASA) received upon request a complete set of the proposals on November 4, 1958. The Board is continuing its studies and will submit further recommendations of its findings.

Insofar as possible, the Board has considered five implementation phases with respect to the experimental proposals:

<u>Phase 1.</u> Feasibility study.

<u>Phase 2.</u> Development of a bench model. (This includes the experimental demonstration of the principles using transistor circuitry and other critical components required in satellites and rockets. It may also include preliminary balloon flights to test the apparatus.)

Phase 3. Development of a flight prototype model. (This model uses flight type components or the physical equivalents and is ready for a design test program although it may not have the final configuration required for a specific satellite.)

[2] <u>Phase 4.</u> Checkout, launching and coordination. (This includes flight package design and fabrication, test of flight packages, preparation for a launching, and participation by the experimenter in a launching to the extent required.)

<u>Phase 5.</u> Data reduction and analysis.

Wherever possible, the recommended agency and recommended experimenter are indicated. Supporting documents received from interested proposers are attached to provide more details of the proposed experiment where this material has been available to the Board. With respect to such proposals, the Board believes that the proprietary interests of the submitting scientists, whether with respect to unique concepts or instrumental apparatus, must be kept in mind as a matter of principle and statute.

I. ASTRONOMY

A. Solar Physics

1. <u>Title</u>: Solar Lyman-alpha Radiation Measurements
   <u>Institution</u>: U.S. Naval Research Laboratory
   <u>Project Director</u>: Herbert Friedman

Remarks: This experiment is essentially the same as that supported under IGY Project 32.8 and which is still unflown. Herbert Friedman of the Naval Research Laboratory and William A. Rense of Upper Air Laboratories, University of Colorado, are considered to be the two investigators most ready for immediate work in the investigation of solar ultraviolet and X-ray radiation.

Board Action: The Board recommends support of this experiment for phases 1 through 5. (See also item 2 below.)

Supporting Document: Paper, "The Lyman-alpha Experiment."

2. Title: A Solar Lyman-alpha Intensity Monitor
Institution: Upper Air Laboratories, University of Colorado
Project Director: William A. Rense

Board Action: The Board recommends support for phases 1 through 5. (See also "Remarks," item 1 above.)

Supporting Document: Letter dated July 9, 1958, from William A. Rense to Chairman, Space Science Board.

[3] 3. Title: An Experiment for Mapping the Sun in the X-ray and Far UV Regions by Means of a Satellite
Institution: U.S. Army Signal Research and Development Laboratory
Project Director: W. G. Stroud

Remarks: The Board notes that Stroud has indicated a willingness to collaborate with Herbert Friedman of the U.S. Naval Research Laboratory in this experiment and that to a certain extent Stroud's qualifications are complementary to Friedman's.

Board Action: The Board recommends that Stroud be encouraged to collaborate with Friedman on this experiment.

Supporting Document: Proposal, titled as above.

B. Stellar Astronomy

1. Title: Proposed Study for a Satellite Telescope
Institution: Princeton University Observatory
Project Director: Lyman Spitzer

Board Action: The Board recommends support for phase 1 of this work and suggests that the National Science Foundation may be an appropriate source of support for the study phase of the program.

Supporting Document: Proposal, titled as above.

2. Title: Feasibility Studies of the Development and Operation of an Astronomical Telescope in a Satellite Orbit
   Institution: Smithsonian Astrophysical Observatory
   Project Director: Fred L. Whipple

   Remarks: The Board recognizes the scientific importance of an orbiting astronomical telescope, as proposed by the Smithsonian-Harvard Observatories, for observing radiation in the far ultraviolet. There is further agreement on the general feasibility of this proposal and the preliminary plans for executing it. However, since the formal proposal on the project was not available for Board consideration, further action is deferred.

   Board Action: The Board recommends phase 1 support.

   Supporting Documents: Proposal, titled as above and supporting documentation entitled "Notes on the Development and Operation of an Astronomical Telescope in a Satellite Orbit."

[4] C. General

1. The Board reaffirms its recommendation that consideration be given to providing support as soon as possible, for the development of a flashing light system, suitably packaged, for incorporation in a geodetic or astronomical satellite and that a study of ground tracking facilities be concurrently made to determine their adequacy. (See letter of July 24, 1958, from Executive Director, Space Science Board for prior recommendation.)

II. THEORY OF RELATIVITY

1. Title: Relativistic Clock Experiment
   Institutions: (1) National Bureau of Standards – Rubidium gas cell clock
   (2) Massachusetts Institute of Technology – Cesium beam clock
   Project Directors: (1) Peter Bender, National Bureau of Standards
   (2) Jerrold R. Zacharias, Massachusetts Institute of Technology

   Remarks: Because of the fundamental scientific value of an experimental validation of the general theory of relativity, the Board recommends that a satellite clock experiment be carried out as soon as possible. The Board is advised that suitable launching vehicles will probably not be available before 1960, and therefore recommends the following program schedule:

   a. Completion of feasibility study (phase 1) on both the rubidium gas cell clock and the cesium beam clock (by the respective institutions) with definitive technical reports by May 1, 1959.

b. Completion of three (3) flight prototype models of each type (phase 3) by December 31, 1959.

Board Action: The Board recommends immediate support to:

(1) National Bureau of Standards, for phases 1 through 3.
(2) Massachusetts Institute of Technology, for phases 1 through 3.

In addition to the two experiments above, the Board recommends that support also be provided to Peter Bender, National Bureau of Standards, for instituting a program with a qualified lamp manufacturer for the improvement of alkali vapor lamps for light pumping applications.

[5] Supporting Documents:
a. Proposal from National Bureau of Standards
b. Proposal from the Massachusetts Institute of Technology and the National Radio Company
c. Minutes of the Special Ad Hoc Committee for the Consideration of the Relativistic Clock, October 7, 1958

III. IONOSPHERIC PHYSICS

1. Title: Satellite Ground-Based Ionospheric Measurements

Remarks: The Board recommends that the following IGY satellite ground-based ionospheric measurement programs be continued for an additional year in view of the observational opportunities that will probably be available during the period July 1, 1959 to June 30, 1960.

IGY Project Number	Short Title	Institution	Project Director	Estimated Amount
32.40	Radio Interferometry and Data Analysis	U. of Illinois	G. W. Swenson	$ 66,000
32.41	Interferometer/Doppler Recording & Analysis	National Bureau of Standards	Ralph J. Slutzl	21,500
32.42	Auroral Ionosphere Studies	U. of Alaska	C. Elvey	66,000

IGY Project Number	Short Title	Institution	Project Director	Estimated Amount
32.43	Electron Density & Propagation Locations	Stanford U.	O. G. Villard	70,400
32.44	Doppler Measurements from Spaced Locations	Penn State	A. H. Waynick	82,200
32.46	Absolute Signal Strength & Frequency Measurements	Linfield Research Institute	W. P. Dyke	31,600
32.47	True-Height Electron Density Profiles	National Bureau of Standards	Ralph J. Slutz	75,000
32.48	Polar Satellite Propagation Measurements	Geophysics Research Directorate, AFCRC	W. Pfister	25,000
		TOTAL		$537,700

Supporting Documents: These projects are described, respectively, in the attached IGY Earth Satellite Program documents.

[6] 2. Title: Multifrequency Ionospheric Beacon Transmitter

Remarks: The Board strongly recommends that a multifrequency ionospheric beacon transmitter be included in an early flight package, preferably for a launching not later than June 1959, and endorses the following optimum characteristics as developed by the Working Group on Satellite Ionospheric Measurements of the IGY Technical Panel for the Earth Satellite Program:

Frequencies: Approximately 20, 40, 108, 400, 1000 Mcs, harmonically related.

Transmitter stability: $1{:}10^7$

Minimum power: 100 milliwatts to 1 watt with a stability for field strength measurements of less than 1%.

Antenna: Linearly polarized along the spin axis.

Modulation: Amplitude modulation or pulse modulation keyed off for one second period every thirty seconds on 20 and 40 Mc.

Lifetime: One year (self-destruction feature to be included if possible).

Orbit: 75° preferred but 51° would be acceptable; a 63° orbit should be avoided.

Perigee: 150 to 200 miles.

Apogee: Consistent with one-year lifetime.

It is also recommended that consideration be given to the addition of a frequency close to 40 Mc, i.e., 38 Mc, to permit better studies of Faraday rotation effects.

3. Title: Direct Atmospheric Electric Measurements from Satellites
Institution: U.S. Naval Research Laboratory
Project Directors: R. E. Bourdeau and J. F. Clark

Remarks: This group has instrumented such an experiment in rockets and could draw on the experience of the NRL miniaturization group.

Board Action: The Board recommends immediate funding, phases 1 through 5, of this proposal.

Supporting Document: Proposal, titled as above.

[7] 4. Title: Development of Modified Langmuir Probe
Institution: U.S. Naval Research Laboratory
Project Directors: C. A. Pearse and Willard H. Bennett

Board Action: The Board recommends support of this experiment through phase 2.

Supporting Document: Proposal, "Charge Density and Ionic Composition."

5. Title: Ion Density Probe
Institution: Geophysics Research Directorate
Project Director: Mrs. R. C. Sagalyn

Remarks: This experiment is based on a modification of the ion density probe in Sputnik III. The experimenter requires three or four months to complete her study of improvements on the USSR experiment, but this could be accelerated with additional funds.

Board Action: The Board recommends support through phase 2.

Supporting Document: Minutes of Second Meeting, Committee on Ionospheres of Earth and Planets, Space Science Board, October 20, 1958.

6. Title: VLF Pulse Experiment
   Institution: Stanford University/Stanford Research Institute
   Project Director: R. A. Helliwell

   Remarks: This experiment involves development of a pulsed transmitter by which pulse signals are received from the ground and monitored in the satellite. The study would require two to three months.

   Board Action: The Board recommends support for feasibility studies (phase 1).

   Supporting Document: "Proposal for a Very Low Frequency Satellite Experiment."

[8] 7. Title: VLF Continuous Wave Experiment
   Institution: Stanford Research Institute
   Project Director:

   Remarks: This experiment embodies a CW transmitter experiment in which signal strength from a CW vlf station is measured on mutually perpendicular magnetic loop and electrical dipole antennas.

   Board Action: The Board recommends support for a review in detail (phase 1) of scientific and engineering design problems.

   Supporting Document: Preliminary Proposal from Stanford Research Institute (par. 4).

8. Title: Topside Ionospheric Sounder
   Institution: National Bureau of Standards, Central Radio Propagation Laboratory, Boulder, Colorado
   Project Director:

   Board Action: The Board recommends support for a study (phase 1) to determine feasibility and design requirements.

   Supporting Document: Minutes of Second Meeting, Committee on Ionospheres of Earth and Planets, Space Science Board, October 20, 1958.

IV. PHYSICS OF FIELDS AND PARTICLES IN SPACE

A. Magnetic Fields

1. Title: Magnetic Field Studies from an Earth to Moon Package and a Polar Orbit Earth Satellite
   Institution: U.S. Naval Research Laboratory

Project Director: J. P. Heppner and L. E. Meredith (with Varian Associates)

Remarks: Instrumental development includes light-pumping alkali vapor magnetometer (scalar) having approximately 10-5 gauss sensitivity. This is for use in studies within 2 earth radii, fringe or transition fields as functions of time and position (3-20 earth radii), interplanetary fields (solar connective field) as functions of time, and lunar field.

Board Action: The Board recommends support for phases 1 through 5.

Supporting Document: Proposal, titled as above.

[9] 2. Title: Search for Hydromagnetic Waves above the Ionosphere
Institution: Lockheed Aircraft Corp., Missile Systems
Project Directors: Francis S. Johnson and A. J. Dessler

Remarks: Project involves no significant instrument development.

Board Action: The Board recommends support for phases 1 through 5.

Supporting Document: Lockheed document LMSD-5134, Section 1.

3. Title: Mapping of the External Geomagnetic Field from Satellites
Institution: University of New Mexico
Project Director: V. H. Regener

Remarks: This proposal involves the development of instrumentation and its use in studies of fringe or transition fields as function of time and position.

Board Action: The Board recommends support through phase 2.

Supporting Document: Proposal, titled as above.

4. Title: High Altitude Studies of the Earth's Magnetic Field
Institution: State University of Iowa
Project Director: L. J. Cahill, Jr.

Remarks: Involved here are studies within 2 earth radii and studies of fringe or transition fields as function of time and position at greater altitudes.

Board Action: The Board recommends support for phases 1 through 5 with the advice that the simplest possible equipment should be used for immediate exploration.

Supporting Document: Memorandum for Space Science Board from the State University of Iowa dated July 12, 1958, p. 14.

[10]5. Title: Development of Minimum Weight Prototype Instrumentation for Geomagnetic Measurements with Earth Satellites
Institution: Smithsonian Astrophysical Observatory collaborating with Batelle Memorial Institute and Harvard College Observatory
Project Director: Fred L. Whipple, Smithsonian Astrophysical Observatory

Remarks: This proposal involves a vector magnetometer utilizing the Hall effect in a semi-conductor, with a vector sensitivity approaching 10^{-4} gauss. This development appears desirable because of light weight, simplicity, and possibility of future improvement.

Board Action: The Board recommends support through phase 2.

Supporting Document: Proposal, titled as above.

General

1. With regard to basic instrumentation requirements for the measurement of magnetic fields, the Board recommends immediate support for the development of magnetometers to the following specifications:

(a) scalar (light pumping), of sensitivity 10^{-5} gauss and sampling time a few seconds.

(b) vector, of sensitivity 10^{-4} gauss, angular precision 2°-5° at 10^{-4} gauss level. Here, improved semi-conductor magnetometers offer promise. The possibility is suggested that MAD unit or Navy 3-component pendulum unit may be improved to approach these specifications.

(c) There is need for an extremely simple instrument suitable for routine use in rockets and satellites for exploratory purposes. Here, small size and weight, simplicity of operation, and sensitivity are most important.

2. An extended period of geomagnetic field observations should be initiated as soon as possible, 1959 at the latest, in order to take advantage of observational possibilities during the post maximum period of the solar cycle. These observations should continue for a year or more.

[11] B. Low Energy Particle Radiation (< 100 Mev)

1. Title: Low Energy Particle Studies
Institution: State University of Iowa
Project Director: J. A. Van Allen

Remarks: These studies are aimed at the mapping of low-energy particle density and spectrum with pole-to-pole orbits; and for particle identification, which is considered to be of greatest immediate importance (search for protons, electrons, He^{++}, etc.).

Board Action: The Board recommends support for phases 1 through 5.

Supporting Document: Memorandum for Space Science Board from State University of Iowa dated July 12, 1958.

2. Title: Satellite Auroral Particle Measurements
Institution: U.S. Naval Research Laboratory
Project Director: L. R. Davis and L. H. Meredith

Board Action: The Board recommends support for phases 1 through 5.

Supporting Document: Proposal, titled as above.

C. Cosmic Rays

1. Title: Cosmic Ray Investigations
Institutions: Bartol Research Foundation and Rias, Inc., Division of The Martin Company
Project Directors: Martin A. Pomerantz – Bartol
Gerhart Groetzinger – Rias, Inc.

Board Action: The Board recommends support through phase 2 for the following experiments:

(a) composition, intensity and variations with time of relativistic particle energies
(b) exploratory studies of particles with atomic number greater than 6.
(c) searches for ± electrons.

Supporting Documents: Proposal from Bartol dated July 11, 1958; Letter from Gerhart Groetzinger to Chairman, Space Science Board, dated July 11, 1958.

[12] 2. Title: Studies Concerning Relativistic and Nonrelativistic Particle Energies
Institution: State University of Iowa
Project Director: J. A. Van Allen

Remarks: Involved here are studies in composition, intensity, and variations with time of relativistic and nonrelativistic particle energies.

Board Action: The Board recommends support through phase 3.

Supporting Document: Memorandum for Space Science Board from State University of Iowa dated July 12, 1958.

3. Title: Satellite-Borne Cosmic-Ray Experiments
Institution: The University of Chicago
Project Director: J. A. Simpson

Board Action:

1. The Board recommends support through phase 3 for those portions of the proposal dealing with:
   (a) studies in composition, intensity, and variations with time of relativistic and nonrelativistic particle energies.
   (b) experiments in the use of cosmic rays as probes of geomagnetic field, solar and interplanetary magnetic fields.
2. The Board recommends support through phase 2 for that portion of the proposal on the search for ± electrons.

Supporting Document: University of Chicago documents CML-PR-E-1150 and 127.

4. Title: High-Energy Gamma-Ray Satellite-Borne Experiment
Institution: Massachusetts Institute of Technology, Laboratory for Nuclear Science
Project Director: W. Kraushaar

Board Action: The Board recommends support through phase 2.

Supporting Document: Proposal, titled as above.

[13] 5. Title: A Proposal for the Measurement of Cosmic Light and Radiation from an Earth Satellite
Institution: California Institute of Technology, Jet Propulsion Laboratory
Project Director: H. Victor Neher, for cosmic-ray portion

Board Action: The Board recommends support for phases 1 through 5 for the cosmic-ray portion of the experiment.

Supporting Document: JPL Publication No. 70 attached.

6. Title: Proposal to Explore the Properties of High Energy Radiation at Rocket Altitudes
Institution: University of Chicago
Project Director: Marcel Schein

Remarks: This proposal has already been endorsed by the Space Science Board and presumably has been included in the Department of Defense Program by the Geophysics Research Directorate.

Board Action:
1. The Board recommends support through phase 3 for experiments with recoverable emulsion blocks carried in rocket nose cones.
2. The Board recommends support through phase 2 for experimental searches for magnetic monopoles.

Supporting Document: Proposal, titled as above.

D. General

1. Title: Observation of X-rays ($\geq$0.5 Mev)
   a. Map of celestial sphere in the light of x-rays
   b. Image of sun and solar flares
   c. X-rays of terrestrial origin
   d. Spectral distribution

   Remarks: Adequate instrumentation appears to be available. Attitude control and registration appear to be required.

   Board Action: The Board recommends that proposals be solicited.

[14] 2. Title: Special Aurora and Airglow Observations
   a. Pictures of Earth in the λ 5577A and λ 3914A Bands
   b. Associated low energy detectors on vehicles with experiment
   c. Associated magnetometers
   d. Far ultraviolet exploration of auroral distributions

   Remarks: To meet the instrumentation requirements for image registration and transmission, improvements may be needed in attitude control and registration, photoelectronic devices, image storage, and wider band telemetry.

   Board Action: The Board recommends that proposals be solicited.

3. Title: Detection of Interplanetary Particles: 0 to 50 kev Ions, Electrons, Neutral Atoms, and Molecules, etc. as follows:
   a. Energy spectrum
   b. Ratio of neutral to charged
   c. Composition
   d. Directional detection
   e. Time dependence
   f. Associated magnetic fields

g. Plasma experiments (<100 ev particles)
h. Search for free radicals in space

Remarks: These measurements will require development of new kinds of detectors supported by improvements in attitude registration, electron-multiplier development, and radio probe methods.

Board Action: Proposals should be solicited and interest stimulated in measurements of this type.

[15]

V. METEOROLOGY

1. Title: Investigation of Thermal Radiation Budget of the Earth and Survey of World-Wide Thunderstorm Activity Using an Earth Satellite
Institution: University of Wisconsin
Project Director: V. E. Suomi

Remarks: The Board recommended to the Government on July 24, 1958, that support be given to the development of equipment for a directional bolometer experiment designed to make measurements of global radiant energy. Details for this bolometer are given in Section IV of the enclosed proposal.

Board Action: The Board recommends immediate support for the following experiments:

Improved Radiation Balance Experiment, Phases 1 through 5 (Section II)

Improved Meteorological Experiments, Phases 1 through 5 (Section III)

Experiment Using Directional Bolometers, Phases 1 through 5 (Section IV)

Radiation Cloud Cover (combination experiment), Phases 1 through 5 (Section V)

World-Wide Thunderstorm Survey, Phase 1 (Section VI)

Supporting Document: Proposal, titled as above.

VI. INTERPLANETARY PROBES AND SPACE STATIONS

In order to develop a Board position with regard to interplanetary probes, considerable study was given to the general consideration of the problem and to two proposals, one for a Venus probe and one for a Mars probe. The Board did not consider these last proposals per se but used, as the basis for formulating its recommendation, the general recommendations which were transmitted to the Government by our letter of October 3, 1958, summarized herewith as follows:

1. It is urgently necessary to begin the exploration of space within the solar system with any means at our disposal if a continuing U.S. program of space science and exploration is to proceed at an optimum rate. To this end, a comprehensive program of deep space probes should be initiated.

[16] 2. With vehicles of the Thor-Able it appears possible to get a payload of the order of 50 lbs. out to interplanetary distances in the fairly near future, while maintaining communication and control. As far as the Thor is concerned, it is recommended that it be used without a control retro rocket as part of the payload and that the additional weight thereby made available be used to increase the reliability of communications, and perhaps for additional experiments.

3. It is recommended that a program aimed at launching a Mars probe during the 1961 conjunction be immediately initiated.

4. With a combination such as Atlas and a high-performance second stage, a payload in excess of 1,000 lbs. seems feasible. It is therefore recommended that immediate steps be taken to begin the development of a space vehicle based on the Atlas plus a high-performance second stage, together with suitable communications and controls, in order to provide a payload sufficient to carry out a more scientifically satisfying set of experiments on the planets Venus and Mars. In addition, because of the long lead time involved in such a program, development of vehicle telemetry and experimental equipment should be started soon.

5. A study of appropriate scientific packages for different classes of space probes is now in progress. Recommendations resulting from this study will be provided to the Government shortly after the first of the year.

6. With regard to manned space stations, the Board feels that further study is required before specific recommendations can be provided.

Supporting Document: Minutes of the Ad Hoc Committee on Interplanetary Probes and Space Stations, September 13, 1958.

Document I-17

NATIONAL AERONAUTICS AND SPACE ADMINISTRATION
1520 H Street Northwest
Washington 25, D.C.

[handwritten: Oct. 20, 1959]

Dr. S. Douglas Cornell
Executive Officer
National Academy of Sciences
2101 Constitution Avenue, N.W.
Washington 25, D.C.

Dear Dr. Cornell:

The enclosed work request for the Space Science Board has been sent to the National Science Foundation to be used as a basis for discussion in negotiating the renewal of the Space Science Board contract for Fiscal Year 1960. If you desire, we would be pleased to discuss this with you in detail.

Sincerely yours,

[signature]
Hugh L. Dryden
Deputy Administrator

Enclosure:
Work Req. to SSB
fm NASA, dtd 12 Oct 59

cc: Dr. H. Odishaw, NAS

****

[no page number]

WORK REQUEST TO THE SPACE SCIENCE BOARD
FROM THE
NATIONAL AERONAUTICS AND SPACE ADMINISTRATION

1. Long Range Planning

The National Aeronautics and Space Administration would like to have from the Space Science Board a continuing input of thoughts, ideas and recommendations on the broad overall objectives, and the course that the space science activities in the United States should take. A prime question is: What are the basic philosophical objectives that should underlie the space sciences activities and program? Guiding principles are needed, rather than a detailed program formulation, which must be worked up in the NASA in consideration of a variety of factors, such as budget, availability of rockets, testing facilities, the balanced program emphasis between space sciences and other NASA activities, and so forth.

The following several paragraphs include some of the NASA thinking on the overall problem and question.

Any program is naturally composed of individual tasks that stem from the ideas and activities of the individual workers. In the case of the space sciences program these tasks are generally in the form of a rocket, satellite, or space probe experiment, and occasionally may be a related or supporting theoretical or laboratory investigation. These individual tasks are themselves best described in terms of the scientific disciplines in which they fall. Sometimes groups of tasks are gathered together into convenient packages for management or budgeting purposes.

But underlying the science program there should be a philosophical pattern that ties the various tasks together into a coherent and unified program, and which provides a compelling motivation that in itself can be accepted as adequate justification for the program. The underlying philosophy and basic motivation should be such that (in addition to the scientific specialists themselves who are working in the field, and who would naturally approve) the scientific community in general would feel the necessity of supporting the program in principle, that the Government recognize the desirability and necessity of supporting the program, and that the public accept the value of the program and support it.

One such basic philosophical objective might be to learn as much as possible about the earth, its atmosphere, and its environs. The idea here would be to put man in a position of understanding thoroughly the planet on which he lives. This is a worthy objective, one that may be expected to lead to both scientific and practical benefits. One may in all conscience ask the people of the United States to support such an objective. It is, in fact, the very motivation that underlay [sic] the International Geophysical Year. In the area of space science such an objective would call for a broad and substantial program.

[2] A somewhat broader philosophical basis for a space science program might be to learn as much as possible about the solar system, with particular emphasis upon solar terrestrial relationships. Since the sun is the primary source of energy for activity on the earth, in fact the very basis of man's ability to exist on the earth, a vigorous program directed at obtaining a thorough understanding of solar terrestrial relationships closely concerns the daily interests of mankind.

Another, and very exciting, philosophical basis for a space science program would be to learn as much as possible about the behavior of terrestrial life forms in space and under the conditions of space flight, and to seek out extraterrestrial life. The philosophical implications of a discovery that life does indeed exist elsewhere than on earth are tremendous, and surely of interest to the entire world, as well as to the scientist.

Finally, one might set as one objective of a space science program, a concerted search for the fundamental nature of the universe, of its origins, and of the bodies within it, including the sun and earth. Included here would be the search for experimental and observational evidence that could be used to seek out the fundamental nature of gravitational forces, or to determine the relationships between electromagnetic and gravitational fields, for example.

The NASA would appreciate having from the Space Science Board a continuing input on what should be the philosophical guidelines to use in building up the NASA space science program. Are those briefly stated above appropriate? Are there better ones? Has anything been left out? Where should the initial program emphasis lie? Should all of the above philosophical objectives be pursued vigorously simultaneously, or should there be some time phasing of the pursuit of the different objectives? What should be the broad lines of attack (a) to start, (b) after 5 years, (c) even later?

2. Discipline Planning

The strength of a scientific research program rests on good ideas and properly conducted experiments. The individual scientists are the source of both of these. The various discipline committees of the Space Science Board can serve a valuable function as a forum for discussion, and as a stimulus to the scientific community in their respective disciplines. To NASA the most valuable product of the Space Science Board committees would be a continuing outpouring of ideas for individual experiments, broad lines of attack, and relative emphases, all properly related to broad philosophical objectives as discussed above. NASA would also appreciate being informed of the names of scientists who would be interested in participating in the program. (In this connection NASA would, of course, undertake to honor and protect the rights of the individual scientists submitting original ideas for research.)
[3] NASA would find such an input from the different committees of great value in the detailed formulation of the NASA national space sciences program. In this connection, it may be of value, from time to time, to call upon individual Space Science Board committees to meet at NASA for a working session to consider with the NASA space sciences staff specific problems of program planning.

3. International Programs

The NASA wishes to establish both the fact and the posture of a sound and substantive program of international cooperation in space research. To this end, NASA proposes to utilize all appropriate media. The ICSU Committee on Space Research (COSPAR) is regarded as a particularly appropriate medium for this purpose. In keeping with the U.S. tradition of maintaining contact with international scientific bodies through the U.S. National Academy of Sciences, NASA would like to maintain contact with COSPAR through the NASA and its Space Science Board.

NASA anticipates that COSPAR will serve as a focal point and means of suitable endorsement for cooperative activities in space research, as a forum for scientific discussion, and as a means of stimulating scientific interest and participation in space research. It is understood that COSPAR will not be an operating group.

NASA will undertake, through the Space Science Board, to keep COSPAR informed of the U.S. space science program and its scientific results. NASA will look forward to having the thoughts, ideas and suggestions of COSPAR in the area of space research. It is hoped that the Space Science Board will act to stimulate and transmit such contributions.

NASA presently contemplates two types of cooperation in developing its international program. The first involves bilateral arrangements in which each participant meets the costs of its own contributions, so that there is no interchange of funds; however, there is no requirement that the contributions of the different participants be equal. Where such cooperative projects involve sizeable efforts and sums of money, the agreements between technical agencies must ultimately be formulated in government-to-government agreements. COSPAR's contribution to such cooperation would come in the form of stimulus, comment, general aegis, and assistance in disseminating information as desirable.

The second type of cooperation involves the participation of scientists abroad in NASA experiments where such participation is possible merely by exercise of their own efforts, as in ordinary ground base applications. Such cooperation will, however, require adequate information. It is to be hoped that the Space Science Board will devote considerable thought and effort to the establishment, through COSPAR, of effective and rapid channels of communication to facilitate the types of cooperation described above.

[4] 4. Data and Results

The Space Science Board could provide a most useful service by arranging to continue the operation and functioning of the World Data Center A for Rockets and Satellites after the close of the International Geophysical Cooperation—1959. NASA would be happy to discuss the possibility of defraying the costs of this Center. NASA would also undertake to forward to the Center the results and data obtained from the space sciences basic research program.

As part of the Center activity, it would be of great value to have a continuing literature search and abstracting activity in the field of space research, coordinating and supplementing other similar activities. Timely reports of current activities and results from rockets, satellites, and space probes would be of value to the scientific community, and also to NASA operations. Such an effort should cover not only U.S. activities but also those of other countries.

Document I-18

Document Title: John A. Simpson, University of Chicago, to Lloyd Berkner, Associated Universities, July 30, 1959.

Source: Archives, National Academy of Sciences, Washington, D.C.

Together with James Van Allen and Herbert Friedman of the Naval Research Laboratory, John Simpson was one of the most prestigious of the pioneering space scientists. In this letter, Simpson expresses his concern regarding the organization of the U.S. space science program in its first months of NASA's operation.

[no page number]

FOR OFFICIAL BOARD USE ONLY

THE UNIVERSITY OF CHICAGO
Chicago 37, Illinois
The Enrico Fermi Institute
for Nuclear Studies

July 30, 1959

Dr. Lloyd Berkner
Pres. Associated Universities
10 Columbus Circle
New York 19, New York

Dear Lloyd:

On my way back from Europe I had time to reflect on the events of the past year related to the participation of scientists in experiments on space vehicles and to the question of stimulation of research requiring these vehicles.

It is clear that the U.S. scientific community must rely heavily upon NASA in these matters, and, therefore, my question is: Are the policies of NASA, as they are now being evolved, directed to the achievement of the goals we all feel are so necessary for the strengthening of U.S. science?

Although my knowledge of their plans is quite limited, as are my contacts with their current programs, my personal opinions are mixed regarding the developments in NASA. On the credit side, they have done an excellent job of getting under way rapidly and have acquired adequate funds to carry out a strong program in pure and applied research. They have brought some excellent people into their organization and there is a tremendous amount of good will and enthusiasm for the success of the organization.

However, I write you because of a growing realization that NASA's policies do not lead to the most effective use of the talents of U.S. scientists deeply interested in this subject. I give three examples which illustrate my point.

1. Most acutely needed are large payload engineering facilities capable of integrating the scientific requirements of various scientific investigators into unified payloads for satellites and space probes. University and other research groups do not have these facilities and in general do not want to undertake this part of the job. Subcontracting (for example as in the case of Space Technology Laboratories) has proven defects which I do not need to restate here. All this was clear by January 1959. In my opinion a bold effort is needed on the part of NASA to establish one or more payload engineering centers as part of [2] NASA capable of serving the scientific community and its expanding interests. Instead of this, however, it appears that NASA is [in] an ambivalent position. It has a very modest payload engineering laboratory derived from the Vanguard program and has directed its main effort to the acquisition of scientific staff to increase its capability in under-

taking scientific experiments within its own organization. Today we are wondering where we shall find a payload engineering group oriented primarily toward the interests of the scientists rather than the military organizations.

2. The scientific community needs to be assured that adequate backup vehicles and payloads are available for their experiments and that not too much time will elapse before experiments are repeated following vehicle failures. No reasonable guarantees have come forth as of this writing. This inevitably eats into the morale of the participating scientists. In addition, it has been very difficult to get information on what will be accepted on board the different payloads for 1960.

3. The concept of the NASA "working groups" of participating scientists is not clear. It is not yet certain whether the working group will decide which experiments go on a given payload of [sic: or] whether the working group simply carries out decisions already made within the NASA organization. For example, a meeting for a Lunar Probe Working Group was called early in May but since that time we have had no communication whatever regarding further plans even though we hope there will be a shot in the first three months of 1960. The "working group" concept at the present moment leaves us with real uncertainties as to whether we are doing our own experiments as we conceive them, or are part of a technician team supplying instrument payloads for a NASA project. It is not obvious which approach is the most practical and I do not try to judge this here. However, indecision, in my opinion, is hurting the progress in this field.

I believe they must have a scientific group within the organization so as to be able to judge the over-all effectiveness of their programs and to undertake research of importance. However, this can be done without neglecting the national service aspect of the NASA program. In my opinion, we can make immediate progress by NASA's taking a strong lead in developing its own large diversified payload engineering facility to serve the scientists.

In spite of these criticisms, I have considerable admiration for the way Glennan and his organization have tried to face up to a very difficult situation in our government. I write these opinions in confidence to you and would be delighted to have them discussed at our next Board meeting.

Sincerely,

J. A. Simpson

Document I-19

Document title: Richard A. Horner, Associate Administrator, NASA, to William H. Pickering, Director, Jet Propulsion Laboratory, December 16, 1959.

Source: NASA Historical Reference Collection, History Office, NASA Headquarters, Washington, D.C.

As the capabilities needed to carry out a comprehensive space science program were either created by NASA or transferred to it during 1958 and 1959, it became imperative for NASA's top managers to assign the agency's "role and missions" to the various elements of the organization. The Jet Propulsion Laboratory had hoped not only to be the lead NASA element for deep space exploration, but also to have a role in developing at least the upper stages of the launch vehicles needed to undertake deep space missions. However, with the November 1959 decision to transfer the Wernher von Braun "rocket team" from Army to NASA control, NASA Headquarters, through this letter, told JPL that its mission was to be limited to deep space exploration, and that NASA Headquarters in Washington, not JPL, would be responsible for overall planning of lunar and planetary missions in the context of NASA's space science program, and that JPL would then be responsible for detailed mission planning and implementation, the latter in collaboration with universities and the aerospace industries.

It is also worth noting that NASA by December 1959 had decided, as indicated in this letter, to focus its initial deep space robotic missions on the moon rather than on planetary exploration.

---

[1]

COPY

December 16, 1959

Dr. William H. Pickering, Director
Jet Propulsion Laboratory
California Institute of Technology
4800 Oak Grove Drive
Pasadena 3, California

Dear Bill:

It is my purpose in writing this letter to restate some of the program policies of the Administration that we discussed during your recent visit to Headquarters and relate them to the current circumstances.

There have been two recent decisions of major importance insofar as the implementation of our space exploration program is concerned. The first was the announcement by the President of his intention to transfer the space oriented component of the Army Ballistic Missile Agency at Huntsville, Alabama, to the NASA, and the second was the more recent determination to cancel the development of the Vega vehicle. I know this second decision must be disturbing in many respects to you and your staff and will certainly necessitate a major reorinetation [sic] of the Laboratory work program. It will, however, afford us an opportunity to advance toward our long-term objective of having each of the Centers directly involved in the space experimentation program assigned a major functional area of responsibility. Briefly, this arrangement can now be described as follows:

The NASA Huntsville facility under the direction of Dr. von Braun will have development responsibility for the launch vehicle systems. It will also carry out vehicle launching operations to the point of injection. The Goddard Space Flight Center will be responsible for earth satellite space craft, and sounding rocket payloads in both their development and operation. The development and operation of space craft for lunar and interplanetary exploration will be the responsibility of the Jet Propulsion Laboratory. It is pertinent to note here that the Administrator has decided that our efforts for the present planning period should be concentrated on lunar exploration as opposed to exploration of the planets.

In consonance with these assigned responsibilities, the Headquarters staff is being reorganized such that there will be two components sharing the responsibility for the space experimentation program. Dr. Abe Silverstein will direct the staff elements responsible for space craft development and operation. The Jet Propulsion Laboratory and the Goddard Space Flight Center will report to him in this capacity. Staff responsibility for the launch vehicle activity will be directed by Major General Don Ostrander with the Huntsville Center reporting to him. It is apparent that each of the three [2] laboratory Centers will have direct program interest at each of the launch sites.

These functional areas of responsibility have been assigned with full recognition that it may be necessary to change their boundaries in the coming years as the program develops, and it will undoubtedly be necessary to cross functional lines in specific work assignments where individual Center competence or facility capability indicates the desirability. For example, it might be decided at some future time that manned space flight to the moon will be the responsibility of Goddard while JPL would concentrate its attention on exploration of the planets. The primary purpose of these assignments is to set responsibility for program planning and initiation and permit the Center Directors to formulate their supporting research and advance development efforts in accordance with the needs of these program responsibilities. It should also be noted that there will be a transition period of many months before current work assignments can be realigned to fit the pattern of functional responsibilities described herein.

In carrying out its responsibility for planning and execution of [the] lunar and interplanetary space exploration program, the Jet Propulsion Laboratory will accomplish detailed mission planning, develop space craft to carry out these missions, integrate the experiment instrumentation into the space craft, acquire and analyze the necessary data during the mission flight and record final results. It is apparent that at presently foreseen levels of program activity it will not be possible for the Laboratory to accomplish all of the space craft development solely with the use of its own staff. It is, therefore, expected that a part of the developments will be contracted with industry and the Laboratory will assume the responsibility of monitoring such contracts.

The proper development and integration of the entire NASA program to assure the optimum use of resources, a full use of research and development results by all of the Centers, and compatible timing in all of the program elements is the responsibility of the Headquarters staff. To this end, an annual program guidance document will be provided to the Centers in the early months of each calendar year. It will reflect in general terms the accomplishments to be sought in each functional area and the approximate level of resources upon which the Centers may base their planning. Additional information concerning the results of related research and development undertaken elsewhere in the NASA program or in that of the Department of Defense will be furnished as appropriate. Such additional information would include extracts from the long-range plan of the Administration or any ad hoc studies undertaken for the purpose of guiding our over-all space experimentation program.

[3] In response to this program guidance, the field Centers will be expected to formulate a detailed plan of work which sets forth their resource requirements and schedule of accomplishments. These submissions from each of the Centers will be in turn reviewed by the Headquarters staff. After this review it is my intention to convene a program council made up of the Directors of each of the field Centers and staff components concerned, to consider the staff and Center recommendations for program content. The deliberations of this program council will establish final guidelines for the program that will be submitted to the Administrator for his consideration in the annual budget review. Incidentally, it is my intent that the program council will be convened periodically through the year to consider program implementation problems as appropriate.

The procedure I have described will be initiated with the program guidance for the Fiscal Year 1962 program which you may expect at the Laboratory in February 1960. Immediate planning attention is necessary, however, to accommodate the changes incident to the Vega cancellation. I am sure you will proceed energetically and in consonance with the guidance from the Headquarters staff in planning a program of experiments using the substitute vehicle. You may be assured of our understanding in the difficulties caused by this dislocation in your work schedule. I am confident that with mutual effort a satisfying and meaningful program can be implemented with the resources available.

Sincerely yours,

[signature]
Richard E. Horner
Associate Administrator

Document I-20

Document title: Office of Space Flight Programs, NASA, "Establishment and Conduct of Space Sciences Program – Selection of Scientific Experiments," NASA Management Instruction 37-1-1, April 15, 1960.

Source: NASA Historical Reference Collection, History Office, NASA Headquarters, Washington, D.C.

By this management instruction, NASA set forth its basic process for the conduct of space science missions. This process, which specified NASA's responsibilities and participating scientists' responsibilities with respect to the conduct of a space science mission, remained in place for over thirty years, until it was replaced by the "faster, better, cheaper" approach to space science missions.

[no page number]

MANAGEMENT MANUAL
OFFICE OF SPACEFLIGHT PROGRAMS
TECHNICAL MANAGEMENT INSTRUCTIONS

NUMBER
37-1-1
EFFECTIVE DATE
April 15, 1960

SUBJECT: ESTABLISHMENT AND CONDUCT OF SPACE SCIENCES PROGRAM—SELECTION OF SCIENTIFIC EXPERIMENTS

1. PURPOSE
This Instruction defines responsibilities and establishes procedures for the conduct of the NASA Space Sciences Program.

2. BACKGROUND
Under the provisions of the National Aeronautics and Space Act of 1958 (42 U.S.C. 2451 et seq.), the NASA is responsible for developing and executing a program in space sciences which is scientifically sound and in which the scientific community has broad participation. Success of the program rests in large measure on the ideas and technical abilities of participating scientists, both within and outside NASA. It is essential, therefore, that such competence be utilized in developing and carrying out scientific space missions and experiments, in analyzing research and development requirements, and in recommending efforts to further national program goals.

3. PROGRAM RESPONSIBILITIES

a. Director of Space Flight Programs. The Director of Space Flight Programs is responsible for overall direction of the NASA space sciences program, including:
(1) Establishment of the short and long range scientific program;
(2) Selection of experiments, experimenters, and specific flight missions;
(3) Determining research and development needs to meet overall scientific objectives; and
(4) Appraising results of research efforts.

b. Space Sciences Steering Committee. The Space Sciences Steering Committee, appointed by the Director of Space Flight Programs, serves as the focal point for space sciences activities and is responsible for the review and approval for submission to the Director of Space Flight Programs of:
(1) Proposed short and long range space sciences programs;
(2) Proposed experiments, experimenters and contractors;
(3) Program and budgetary breakdowns and supporting research recommendations; and
(4) Scientific space science assignments for the Goddard Space Flight Center and the Jet Propulsion Laboratory.

[2] c. Space Sciences Steering Committee Subcommittees. Subcommittees are appointed by the Director of Space Flight Programs for various space science disciplines or groups of disciplines, including Aeronomy, Ionospheric Physics, Energetic Particles, Astronomy and Solar Physics, Lunar Sciences, and Planetary and Interplanetary Sciences. Such subcommittees serve in an advisory capacity to the Steering Committee and the Assistant Directors of Space Flight Programs and are responsible in their own areas of interest and competence for providing advice and assistance in:
(1) Formulating short and long range plans;
(2) Analyzing, evaluating, and recommending proposed flight experiments and supporting research; and
(3) Reviewing programs for weaknesses, gaps, and imbalances, and recommending necessary actions to correct such inadequacies.

d. Assistant Directors of Space Flight Programs. The Assistant Directors of Space Flight Programs are responsible for:
(1) Working with the subcommittees to organize "state-of-the-art" information in pertinent scientific disciplines;
(2) Making tentative selections of experiments and experimenters based on recommendations of subcommittees and field centers;
(3) Working directly with the appropriate field centers to secure the necessary budgetary backup and supporting documentation; and
(4) Supporting and coordinating the research and development work of the field centers in executing approved programs and missions.

e. Field Centers. The Goddard Space Flight Center is responsible for conducting missions involving earth satellites and sounding rockets. The Jet Propulsion Laboratory is responsible for conducting unmanned missions involving lunar and deep space probes. In carrying out these responsibilities, such installations will:
   (1) Initiate proposals for and participate in the performance of space science experiments and projects.
   (2) Technically evaluate proposals submitted by Headquarters for recommendations.
   (3) Analyze supporting requirements and recommend scheduling of space sciences programs.
[3] (4) Prepare and operate, either in-house or by contracts consistent with established policy, the necessary spacecraft to carry out approved scientific missions.
   (5) Conduct or contract for supporting research on advanced technology and instrumentation.
   (6) Monitoring of selected Headquarters research and development contracts.

4. SELECTION PROCEDURES

a. Proposed experiments submitted by scientists within and without NASA will be forwarded to the appropriate Assistant Director of Space Flight Programs. The Assistant Directors will submit such proposals to the appropriate advisory subcommittee and to Centers for review and advice. In selecting experiments, proposals from research scientists will be considered on the following basis:
   (1) Desirability within the discipline to which it pertains;
   (2) Probability of acquiring positive scientific results;
   (3) Worth and timeliness in comparison with other competing proposals; and
   (4) Competence and experience of its proposer.

b. With advice and assistance of the Center and appropriate subcommittees, the Assistant Directors of Space Flight Programs will make tentative selections of experiments and experimenters, and will submit such recommendations to the Space Sciences Steering Committee.

c. The Space Sciences Steering Committee will review the detailed plans and forward its recommendation, including the designation of the Center to be assigned the technical management responsibility, to the Director of Space Flight Programs for approval.

d. The Director of Space Flight Programs will approve the mission and will assign the responsibility for program execution.

5. RELATIONSHIP BETWEEN CENTERS AND EXPERIMENTING SCIENTISTS

a. After selection of flight experiments, funding for prototype models or design concepts of scientific instruments for the selected experiments will be provided either by NASA Headquarters or by the Center with the approval of the Office of Space Flight Programs. At this time, a Center may be assigned the responsibility for technical monitoring of selected Headquarters contracts. The schedule for completion of prototype models will be established by the Center, consistent with the spacecraft development program.

[4] b. Completed prototypes or design concepts will be delivered to the Center and evaluated by the Center personnel in collaboration with the experimenters. Additional development of the selected instruments will be made under the technical direction of the Center in collaboration with the experimenting scientists. If the Center personnel determine during the course of fabrication of the flight instruments that modification of the functional specifications are required in order that the instruments operate reliably in the overall system, such modifications will be made on the basis of agreement between the Center and the experimenters. For modifications which imply major changes in the scientific objectives of the experiment, concurrence of the Office of Space Flight Programs will be obtained by the Center.

c. Based on the functional specifications determined by the responsible experimeters, the following functions will be performed by or under the direction of the Center with the assistance of the experimenters:
(1) Fabrication,
(2) Testing,
(3) Calibration,
(4) Checkout of flight instruments,
(5) Integration of experiments into payload and/or spacecraft,
(6) Participation, as necessary, in field operations,
(7) Acquisition and reduction of data from measurements taken in flight.

d. Each selected experimenting scientist will be responsible for:
(1) Preparing the prototype instruments and associated equipment for his experiments,
(2) Cooperating in the preparation of flight instrumentation, its environmental testing and calibration for flight,
(3) Participating, as necessary, in field operations,
(4) Analyzing and reporting the data from his experiment.

6. PAYLOAD DESIGN AND FABRICATION

The responsible Center, with the concurrence of the Office of Space Flight Programs, will determine whether the Center or an outside contractor will design and construct the scientific instrument payload and/or spacecraft.

7. RELATIONSHIPS WITH UNIVERSITIES AND NONPROFIT ORGANIZATIONS

a. The Office of Space Flight Programs and Centers will inform each other concerning their plans involving universities and nonprofit organizations and of all concepts and dealings with the scientific community.

[5] b. With prior approval by the Office of Space Flight Programs, Centers may invite proposals for experiments, including the supplying of flight hardware, from universities and other nonprofit organizations in accordance with overall NASA program objectives. Centers are not authorized to proceed with negotiations for research and development effort with universities and nonprofit organizations without prior approval of the Office of Space Flight Programs.

c. Proposals received by Centers from universities and nonprofit organizations will be forwarded to the Office of Space Flight Programs for preliminary appraisal and, where appropriate, for assignment of detailed tech [sic] technical evaluation. The Director of Space Flight Programs will make the determination whether or not to proceed.

d. For those proposals which the Office of Space Flight Programs supports, copies of such proposals will be forwarded to the Director, Division of Research Grants and Contracts, Office of Business Administration, NASA Headquarters. The Division of Research Grants and Contracts, NASA Headquarters, will:

(1) Determine the form of the contractual arrangement to be used, that is contract or grant;
(2) Make the preliminary contact with the business management of the university or nonprofit organization leading to a contractual arrangement; and
(3) When requested, proceed to negotiate and consummate the contract.

Where the contract is a field assignment, the Office of Space Flight Programs, after obtaining the above clearances, will inform the Center that it is authorized to negotiate and consummate the contract.

e. Centers may be requested to monitor and administer contracts led by Headquarters in addition to monitoring and administering their own contracts. The field centers are responsible for keeping the Division of Research Grants and Contracts, NASA Headquarters, currently informed of the financial and management status of each assigned project.

[6] f. The procedure set forth herein does not apply to grants for basic research financed from non-project research and development funds. Such proposals will be handled by Headquarters in accordance with instructions to be issued.

[signature: "D. D. Wyatt"]
[handwritten: "for"] Director of Space
Flight Programs

Approved:

[signature]
T. Keith Glennan
Administrator

Document I-21

Document title: NASA, "Evaluation of USSR vs. US Output in Space Science, Release No. 61-34," February 27, 1961.

Source: NASA Historical Reference Collection, History Office, NASA Headquarters, Washington, D.C.

The January 1960 statement of National Space Policy [Volume I, Document II-21] declared that NASA should "select from among those current or projected U.S. space activities of intrinsic military, scientific or technological value, one or more projects which offer promise of demonstrably effective advantage over the Soviets and, so far as it is consistent with solid achievements in the overall space program, stress these projects in present and future programming." This policy allowed for a modest space race with the Soviet Union, as long as a particular mission also had "intrinsic" value. By early 1961, there was broad curiosity regarding comparative U.S. and U.S.S.R. space science achievements. This paper, developed at Goddard Space Flight Center, addressed this issue.

NATIONAL AERONAUTICS AND SPACE ADMINISTRATION
WASHINGTON 25, D.C.

RELEASE No. 61-34 February 27, 1961

NOTE TO EDITORS:

The attached paper entitled "Evaluation of USSR vs. US Output in Space Science" was prepared by NASA upon the request of the Committee on Science and Astronautics, U.S. House of Representatives.

O. B. Lloyd, Jr.
Director
Public Information

****

February 25, 1961

EVALUATION OF USSR VS. US OUTPUT IN SPACE SCIENCE

GENERAL IMPRESSIONS

The average quality of Soviet scientific research is the same as that of the United States. This conclusion is based on perusal of their literature and on personal contacts between scientists of both countries in conferences held between 1956 and 1960, both in nuclear physics and in areas related to space research.

The range of ability of Soviet scientists is also approximately the same as that of US scientists. A few are brilliant, as good as this nation's best physicists, and the majority do conventional but necessary research.

It is a striking fact that in spite of equality of talent in US and USSR science, nearly all the highly original work in space research has come out of the US program. The first two Sputniks had little or no scientific apparatus, apart from a biological experiment; and while the third Sputnik had a great deal of interesting geophysical apparatus, this flight was never followed up by the second generation of experiments with which the Russians could have capitalized on their experience with Sputnik III. However, these references relate only to basic scientific investigations and not to technology. The USSR has achieved a number of successes, such as the Sputnik III moon shot and the recent space cabin launchings, which were great achievements in space technology.

The US, on the other hand, has been responsible for:

1. first detection of trapped energetic particles (Van Allen belts);
2. launch of Explorer VI and Explorer VII energetic particle satellites; measurement of energy distribution and time variation of radiation in the Van Allen belts; coordinated observations of radiation belt and red auroral arc over Colorado;

[2] 3. launch of Pioneer V space probe; communication with earth out to distance of 23 million miles; study of properties of interplanetary space; detection of cloud of energetic particles sweeping over Pioneer V en route from sun to earth at the time of a solar storm; correlation between ground-based atmospheric data and data received simultaneously from Pioneer V in deep space and from Explorer VII near to earth;
4. correlation between solar weather activity and atmospheric density, via satellite drag measurements; first measurement of air density at an altitude of 1000 miles,

using drag data from the ECHO satellite;

5. launch of the Ionosphere Satellite; measurement of density and temperature of electrons and ions in the upper atmosphere; measurement of ionospheric disturbances during the solar storm of November 1960;
6. satellite contributions to geodesy and celestial mechanics: precise measurement of variations in sea level, gravitational perturbations by sun and moon, effect of solar radiation pressure;
7. TIROS cloud-cover photographs; and
8. ECHO passive communications experiment.

Another significant fact appears in the comparison of US and USSR papers published in the periodical literature or presented at international conferences: The US and USSR papers are of comparable quality but the number of US papers greatly exceeds the USSR contribution. This very sizeable discrepancy in the level of effort constitutes the most significant difference between the US and USSR space research programs. The USSR has capitalized skillfully on the advantage in payload capability which it acquired from its early successes in missile [3] development. It has concentrated on a small number of flights, and has chosen its missions for these flights with close attention to their impact on world opinion and their effectiveness in reinforcing the public image of USSR strength in science and technology. Yet the USSR has done relatively little in space science, considering the resources at its command in payload capability. It is very difficult, for example, to understand why the USSR failed to follow up the remarkable flight of Sputnik III with further geophysical satellite experiments.

Perhaps the explanation is that Soviet scientists are not in close contact with their program planning authorities, and have not been able to participate effectively in the formulation of their space program. The gathering momentum in the US space science program, and the remarkable variety of fields in which highly original results are being obtained, must be credited, in part, to the insistence of US authorities on the development of a sound and broadly-based program in space science, and to their continuing efforts to enlist in the program an increasing fraction of the country's scientific community.

There is another circumstance which has probably had a major effect in determining the relative levels of effort in US and USSR space research. This is the fact that the US has had a tradition of strong support for research over the last several decades, and has developed a powerful base for scientific operations in this country, including many large laboratories with highly trained staffs. This is true in nuclear physics and solid state physics, as well as in the earth sciences and other disciplines from which space research draws its problems and techniques. The base of research in the USSR, on the other hand, is not as deep as in the US. It is new and relatively thin. For this reason the USSR scientific effort is spotty, with excellent work in certain fields on which Soviet interest has been concentrated for one reason or another, whereas work in other fields of equal scientific interest is poor or entirely missing. Because the US has tremendous scientific resources at its command it has been able to develop a vigorous program in atmospheric physics, auroral phenomena, geophysics, sun-earth relationships, and trapped particle research without appreciably reducing its level of effort in other fields of research. We have in fact developed an entirely new field to a very high [4] level of activity in the short space of two years. The Russians may not have the reserve strength in laboratory facilities and trained talent to do this without disrupting established areas of research, and perhaps this is the primary reason for the paucity of their achievements in space science thus far.

It is important to note that this reflects the situation at the present moment. In making an assessment of the Soviet position it is also important to look at the rate of change of their effort, in addition to the current level. The USSR is believed to be training its scientists at a rate several times greater than the rate of training in the US, and it is entirely possible that the USSR may strengthen the weak spots in its scientific structure and move ahead in an effort to overtake us in every important area of research, in the course of the next several years or perhaps the next decade. It may be noted again that the intrinsic ability of the Russian scientific community is in no way inferior to that of our own; and that, therefore, there is no reason why they cannot overtake us in this period if we do not continue to develop and to strengthen our program.

Figures on the number of papers presented at international meetings suggest that the Russian level of effort in space science has not increased appreciably between the Moscow CSAGI meeting in 1958 and the Nice COSPAR meeting in 1960 in spite of the greater rate of training of scientists in the USSR. It is possible that the talent being trained in this area is still at the graduate student level and has not yet reached the level of responsibility at which this effect can be felt in Soviet contributions to conferences. It is the impression of a US authority on upper atmosphere theory that this is the case in atmospheric physics and in sun-earth relationships. In any case, the contrast between the output of the USSR space research program and our own cannot have failed to impress the Soviet scientific community itself, and some elements of the USSR government. USSR scientists may succeed soon in persuading their government to initiate a more vigorous program in space physics under the stimulus of our own successes in this field.

In summary, the USSR space science program has dissipated some of its momentum after the initial successes of the Sputnik launchings; while the US program has picked up momentum from a [5] standing start, and now surpasses the USSR effort in its breadth of interest, originality of concept, and volume of research. However, we should anticipate that the Russians will respond to this challenge with a more vigorous space science program of their own in the future.

SPECIFIC IMPRESSIONS

QUALITY OF PRESENTATIONS

USSR papers cover the same range as those in the US. The top stratum of individual talents in the USSR scientific community is also comparable to ours, according to impressions gained in meetings and by an examination of the translated literature. The Russians have very capable people working in some fields, such as cosmic radiation and energetic particle measurements, and are also strong in mathematics, celestial mechanics, astrophysics, seismology, and observational aspects of oceanography. In certain other areas of the earth sciences and astronomy, either directly or indirectly related to the space program, the USSR makes a less favorable showing in comparison with the work in the United States.

For example, the USSR has published little on the geodetic applications of satellites, a most fruitful field of endeavor in the United States, and a major source of important developments in geophysics. The analysis of satellite orbits to obtain density data also has been carried out at a high level of activity by several groups in the US. In particular, the discovery of a correlation between satellite drag and solar activity, one of the most significant developments in this field, was made by L.G. Jacchia at the Smithsonian Astrophysical Observatory, and his work has since been refined and extended to reveal important diurnal variations. There has been relatively little USSR work published on this important and interesting problem.

It is possible that this particular field of geodesy is considered sensitive by the Russians, and that for this reason a substantial amount of USSR material has been held back from publication.

[6] QUANTITY OF PRESENTATIONS

The United States stands strongly to the fore in the quantity of work, in its volume, and in the number and variety of fields in which we have been doing original work. The contrast between the level of our effort and that of the USSR appears both in the comparison of presentations at international meetings, and in the survey of US and USSR scientific periodicals.

Presentations at International Meetings. Two major international conferences in space research have been held in the last few years. The first of these was the CSAGI conference in Moscow, in July 1958. The US sent a large delegation to this conference. US contributions to the program were varied and extensive, and gave an impression of strength in our incipient space science effort, in spite of the weakness of our vehicle capability at that time. The next international meeting in the field was arranged by COSPAR, and convened in Nice in January 1960. The number of US and USSR contributions to these meetings is listed below:

	US	USSR
CSAGI (1958)	39	15
COSPAR (1960)	40	10

The US and USSR numbers are about the same for the two meetings. It might be noted that the Goddard Theoretical Division alone read five papers to the COSPAR meeting, half as many as the entire USSR contribution, and ranging over the fields of celestial mechanics, geodesy, the moon and planets, meteorites, and trapped particles. A perusal of the table of contents of the recently published Proceedings of the COSPAR Conference drives home the point that the US program contrasts very favorably with the USSR effort in its breadth and in the fullness of participation of the American scientific community.

Volume of Publication. A bibliography has been compiled of USSR periodical literature in space physics, appearing in reputable Soviet publications (PROCEEDINGS of the Soviet Academy, GEOPHYSICS BULLETIN of the Soviet Academy, SOVIET ASTRONOMY, SOVIET JETP, SOVIET PHYSICS EXPRESS) and also the (British) JOURNAL OF PLANETARY AND SPACE SCIENCE.

[7] For comparison, NASA has examined the contents of the JOURNAL OF GEOPHYSICAL RESEARCH for the years 1958-1960. The JOURNAL OF GEOPHYSICAL RESEARCH is the principal medium for the publication of space physics in the US, and contains approximately half of all papers on this subject and related topics in the earth sciences.

The Soviet series, ARTIFICIAL EARTH SATELLITES, has not been included in the USSR lists because it does not constitute a part of the periodical literature on current research to be compared with the JOURNAL OF GEOPHYSICAL RESEARCH.

The table below lists US and USSR contributions gathered from these sources for the years 1958-1960. The 1960 figures are extrapolated from the volume of publications for the first six months of 1960.

	1958	1959	1960
USSR	5	8	17
US (JGR)	5	64	111

We see that the US and USSR efforts started at comparable levels, that the USSR effort has shown a moderate increase in the last three years, and that the US effort has shown a greater increase. It appears from these figures that the US has reacted vigorously to the challenge of the first Russian successes in space rocket technology. Our space program has been able to draw on the resources of US research, and has greatly exceeded the modest increase in the Russian effort over this period.

The contrast in the level of US and USSR efforts in space physics is in fact so great that we may expect a counterreaction from the USSR, in the form of a greater emphasis on scientific effort in their space program. There may in fact be a hint of this development in a recent unusual display of initiative on the part of the USSR, through its proposal for the organization of a conference on problems of lunar research to be held in Leningrad in December of this year under the joint auspices of the USSR Academy and the IAU.

Detail of Publication; Openness at Meetings. In the first post-Sputnik period, USSR reports on satellite experiments and calculations were sketchy, both at meetings and in the periodical literature. The papers lacked the details needed for the [8] formation of an independent judgment regarding the validity of results presented. It should be noted that this secretiveness was confined to the rocket and-satellite area; in other fields of physics and the earth sciences USSR publications and presentations have been relatively open and detailed. Another factor may contribute to the sketchiness of some of the Russian presentations. It appears to be the practice of the USSR to send a relatively small delegation of leading Soviet scientists to these meetings, who then read review papers on the work of a large number of their colleagues. For this reason some of the areas of research covered in their papers may lack the intimate understanding of the subject which is required to answer pertinent questions.

Since 1958 USSR publication policy seems to have relaxed to some degree, although USSR publications on orbit data and tracking systems are still virtually non-existent.

Timeliness. Originally the USSR appeared to favor newspaper releases over publication in the professional literature, for early reports on their satellite results. After the Moscow CSAGI meeting this situation improved considerably. Preliminary notes on USSR results have been appearing as promptly as in the US, about six to eight weeks after launch for the most interesting results. For example, Sputnik I was launched on January 2, 1959, and the Vernov note describing cosmic ray and trapped particle measurements was submitted for publication in the USSR ACADEMY PROCEEDINGS on February 25, 1959. Sputnik III was launched early in September of 1959, and the paper describing the photographs of the hidden face of the moon was received for publication in the USSR ACADEMY PROCEEDINGS on November 14, 1959.

These examples refer to the publication of preliminary notes, equivalent to our letters to the editor. The detailed papers appear somewhat more slowly than in the US, typically after a delay of 12 months vs. six months in the US.

Channels of Communication. As noted above, in the first period of the USSR space program the Russians appeared to prefer qualitative newspaper accounts for the layman to quantitative publication in serious scientific journals. An examination of the literature suggests that this is no longer the case. As in the US, newspaper articles on achievements of exceptional interest [9] appear in the Russian press or public media at approximately the same time their preliminary scientific reports are published. The impression of this earlier but no longer valid circumstance has persisted because the newspaper stories are picked up immediately and translated quickly by the Department of Commerce and other agencies for distribution to government personnel; whereas the Academy proceedings and scientific literature do not appear in translated form until a much later time. For this reason US scientists often obtain their first reports on interesting results through a PRAVDA translation.

It appears that the Moscow CSAGI meeting was responsible for the change towards normal channels of scientific communication and away from the public press. The detail presented in the US papers at the Moscow meeting, and the strongly critical attitude of the US delegation regarding Soviet suppression of detail, seemed to have had a beneficial effect. The dates of publication quoted above for the Sputnik I and Sputnik III experiments demonstrate this clearly. It seems reasonable to say that Soviet officials and scientists have shown a positive response to US criticism in these matters and can no longer be censured as severely on this basis as in 1958. An exception is the area of tracking systems and orbit information, in which USSR disclosures are still inadequate.

Document I-22

Document title: Space Science Board, National Academy of Sciences-National Research Council, "A Review of Space Research," 1962.

Source: NASA Historical Reference Collection, History Office, NASA Headquarters, Washington, D.C.

With the mobilization to carry out Project Apollo in full swing and after almost four years of NASA carrying out the U.S. space science program, the non-governmental science community agreed with NASA that it would be timely to take a comprehensive look at the current and future space science pro-

gram. Accordingly, a Space Science Board "summer study" took place from June 17–August 10, 1962, at the State University of Iowa, with James Van Allen as its host. This report summarizes the results of that study.

[cover page]

A REVIEW
OF
SPACE RESEARCH

The Report of the Summer Study
conducted under the auspices of the
Space Science Board
of the
National Academy of Sciences

at the

State University of Iowa
Iowa City, Iowa
June 17-August 10, 1962

Publication 1079
National Academy of Sciences-National Research Council
Washington, D.C.
1962

[1-1] Chapter One

INTRODUCTION AND SUMMARY

I. Introduction

Toward the end of 1961, both the National Aeronautics and Space Administration and the Space Science Board recognized the timeliness of an evaluation of the national research program in space and its future objectives. Some five years of expanding space research activity provided a background of experience and discovery for an inquiry into the problems and opportunities before the scientific community. The same period had also seen an appreciable development in the nation's technological capabilities for space research: new vehicles had been brought to operational status and others were rapidly being developed, techniques of spacecraft maneuverability and orientation had advanced substantially, allowable payload weights no longer imposed such severe restrictions on the design of scientific instruments, and a world-wide net of tracking and telemetry stations had been established to assure ready acquisition of scientific data. In November 1961, NASA had established the Office of Space Sciences as one of its four primary divisions. And during this time NASA had acquired a useful body of administrative and management experience in the conduct and support of research in this complex new field.

Considering these factors, the Board and NASA concluded that the time was appropriate for a review of the primary problems in space research in each major scientific area and of the policies and procedures which would shape the national effort for the years ahead. It was recognized that such a review required an extended period of deliberation by specialists in various fields. Accordingly, plans were made to conduct an eight-week study of these matters under Board direction during the summer of 1962.

Boundaries of the subject matter for discussion by the Space Science Summer Study were set early in the planning. The Study would direct its attention to the objectives of basic research in space: the status of present achievement in each scientific field, current NASA programs in these disciplines, the goals toward which each scientific program should be directed during coming years, and related administrative and policy questions. On the other hand, the Study was not to be concerned with certain other aspects of the NASA program, such as the development of new propulsion systems, applied technology (e.g., communications and navigation satellites), etc., except as developments in these programs had direct relevance to the program of basic research in space. The man-in-space program, culminating with the Apollo manned mission to the Moon, while primarily a program of technological development in its present stages, was considered in terms of its scientific potentials, because it will certainly eventually lead to a greater capacity for science in space.

[1-2] The objectives of the Space Science Summer Study were outlined by Dr. Lloyd V. Berkner, then Chairman of the Space Science Board, to Summer Study participants at the opening session:

> "The first task is carefully to consider the future course of our nation's scientific program in space, and to help the government's planners to chart the way. This is a grave responsibility, and you have been chosen because you as a group represent a broad coverage of the disciplines involved in space research, and we have enough time this summer to think carefully about all facets. In particular, the backup research that underlies a comprehensive program must be fully elaborated.
>
> "The second task is similar, and involves aiding the government in its conduct of the space research program in such a way that maximum benefit will come from it. We all recognize the many opportunities opened by the space age for education, stimulation of industry and the nation's economy, research in many allied fields, collaboration and exchange of ideas with scientists in other countries, etc. These many extra benefits from our space activities can only be fostered if the program is wisely administered, and here, again, your advice has been sought and—we are assured—will be carefully heeded. If I may identify one of these aspects that deserves especial attention, it is education for space research and engineering. The burden of carrying out the education and training of new scientists and engineers rests with our universities, and it is not entirely clear how NASA, the National Science Foundation, the Department of Defense, Congress,

the President, and the university community can best work together in our changing world to do this job. The training of young people is a major national responsibility. We will consider this matter carefully.

"The third task is not so much a task as an inevitable consequence of our Summer Study. You, as spokesmen for the scientific and industrial community, will be privy to the problems being faced by our government administrators and scientists, and they in turn, will hear your views. The strength of our program will depend on mutual respect and understanding between the various interests, and it is most important for the university and industry people to comprehend the many broad problems and decisions that must be faced by our government people. The two-way exchange of ideas may perhaps, in the long run, be one of the most enduring benefits to come from our efforts. It is with this in mind that many of the key people from the government have agreed to spend time with us, and to be a part of our Study. Let us be careful at all times to listen to each other."

More than one hundred scientists participated, both full and part time, in the Summer Study. (For a list of participants, see Appendix II [omitted].) Many of these scientists received, for the first time, an opportunity for close association with scientific and administrative personnel of NASA and of other government agencies with space interests. This association was valuable for a number of reasons.

[1-3] First, it permitted the exploration and clarification of many of NASA's policies and procedures which of necessity deal with the complex nature of space missions: e.g., the technological framework necessarily surrounding the contributions of experimenters, the scheduling and launching problems associated with expensive and complicated space rocket systems, the many factors involved in tracking and data acquisition by telemetry that call for extensive networks of stations in many parts of the world. Just as a better appreciation of these problems was attained by the scientific community, so the NASA staff became more aware of the interests and problems of experimenters throughout the nation, particularly those involved in the conduct of research at universities.

Second, this association led to a growing appreciation of the quality of NASA's scientific staff and of the general excellence of its scientific program planning and execution.

Third, and very probably most important of all, this association, because it was of sufficient duration to afford ample time for exploring attitudes and views, and because it was characterized by candor and openness, provided a basis for a satisfactory examination of space science—achievements so far, current status, and plans for the years immediately ahead. The validity of the findings of the Summer Study rests significantly upon the qualities of this association.

The material from which this report is compiled comes from the reports of more than twenty different working groups and subgroups. It thus reflects varied approaches. Preliminary reports from each group were submitted to the entire Summer Study for review and comment before being prepared in final form for inclusion in this report. Opinions, even within each working group, were not always unanimous. The report thus does not pretend to reflect every shade of opinion: it is intended to be a consensus of the Summer Study participants. At the end of each chapter will be found an appendix listing the persons who participated in the discussion of that subject; not all those whose names appear necessarily subscribe personally to every opinion recorded in that chapter.

The scientific content of the space program has been treated very differently by the different groups of specialists participating in the Summer Study. For instance, the astronomers have, on the whole rightly, assumed that every astronomer would know what scientific data of value can be deduced from a given line of experimental development, and so have not elucidated the science in detail. On the other hand, the chapter on particles and fields summarizes what we know today and what we need to know; in this case the specialists have assumed that the experimental lines of attack will be obvious.

Some of the science reviewed in this report is similar to, or in elaboration of, that in earlier reports of the Space Science Board or of NASA committees and consultants. In particular, one can refer to the following three documents as supplementary reading, in which the scientific goals of space research are described in greater detail: two reports by the Space Science Board—Science in Space (L. V. Berkner and H. Odishaw, Eds., McGraw-Hill Book Company, New York, 1961) and The Atmospheres of Mars and Venus (W. W. Kellogg and C. Sagan, Eds., NAS-NRC Publication 944, 1961)—and NASA's Long-Range Thinking Document.

[1-4] Although some of the scientific discussion in this report necessarily represents a review and elaboration of earlier thinking, some new topics are given consideration. The chapter on biological researches is an example. Progress made in arriving at a policy on sterilization is reported, and also a plan for deriving the maximum scientific return from the manned exploration of space. In sterilization, enough experience has now been accumulated for setting up procedures realistic enough to be incorporated into the preparation of lunar and planetary probes, which will minimize the risk of contaminating extraterrestrial bodies. A series of recommendations for the man-in-space program sketches the scientific tasks that a man might carry out, especially in the exploration of the Moon, and outlines what we believe is a workable plan to select and train scientists as astronauts, and vice versa.

NASA's relationships with the outside academic and research world were reviewed. These relationships can be grouped under three broad headings: (1) the real significance of the long-range scientific potential of the man-in-space program, and the necessity for making this clear to the scientific public without extravagant claims; (2) NASA's responsibility to support academic institutions in the development and replenishment of the supply of educated manpower, on which NASA and other scientific and technological activities will impose a heavy drain during the foreseeable future; (3) planning the national space science program, as far as compatible with engineering and scheduling constraints, so as to encourage maximum freedom and flexibility for the individual working scientist and thus to produce an atmosphere conducive to scientific originality and initiative. (Participants referred to this last area of concern as "scientific elbowroom.") All three of these are large and complicated subjects. Although the Study reached very definite positions, these topics will require continuous review in the future.

With regard to the second of these areas—NASA's support of academic institutions as sources of scientific manpower—NASA's program of fellowships, training and facilities grants, and research contracts is a most encouraging sign that this set of problems is on its way to being solved. With regard to the other two areas—science in the man-in-space program, and maximum attainable freedom for scientific experimenters—the situation is not so clear-cut. For example, in its efforts to promote freedom and flexibility, the Study strongly recommends the allocation of blocks of payload space in a satellite or a series of satellites to scientists of demonstrated competence. The requirement for an exact description of the experiments as a condition for the assignment of payload space would be waived to allow a choice of experiments flexible enough to meet the change in circumstances that can take place between the first granting of payload space and the launch, perhaps two years later. The Study also strongly advocated the use of "small satellites, defined in this context not so much as having less than a certain size or weight, but as being suitable to carry a single experiment or single integrated set of related experiments under the direction of a single experimenter or group of experimenters. The Study also emphasized that the launching schedule should be flexible enough to allow the launching of such satellites on short notice, to take advantage of particular circumstances. It was pointed out, for example, that if such an arrangement had been in force during the summer and fall of 1962, much more could have been learned about the artificial radiation belts of Project Starfish.

Proposals like those just discussed—the block allocation of payload space and the use of satellites under the control of single experimenters—obviously presuppose [1-5] a very high degree of skill in space technology on the part of the experimenter. Not many scientists can at present meet these conditions, but the Study felt that many have the potential to do so, and would develop that potential if the scientific atmosphere were more nearly in accord with their desires for freedom of action, perhaps approaching what they have in their laboratories. The Study realizes that implementation of these proposals can result in a certain amount of dislocation of the present system of budgeting, payload space assignment, and flight scheduling, and that it will take considerable time, effort, and perhaps additional funds, to accommodate the system to these ideas.

A number of other important questions were on the agenda which need further study, because the participants did not feel themselves to be sufficiently expert, or because their views were too divergent to formulate a consensus, or because they were unable to find the time to treat the subject thoroughly, or because they did not or could not secure the precise information needed. Subjects that have not been adequately treated and that require further attention are noted in the following paragraphs.

<u>The assignment of relative priorities</u> to scientific programs and experiments was not adequately discussed for the space science program as a whole: their relative scientific importance, the proper time sequence for optimizing the scientific return, or the equitable and efficient distribution of funds from a limited budget. Thus, it is almost impossible to decide on rational grounds many questions that cut across the entire program as long as a variety of special interests are represented. A number of recommendations were made, however, dealing with priorities in specific, narrower fields of activity. For example, astronomers would postpone putting instruments on satellites to observe extraterrestrial objects in the infrared until the potentialities of observation from the ground or from balloons are more fully exploited. Lunar specialists emphasized that certain kinds of scientific data about the Moon must be obtained relatively early, not because they are of greater

scientific importance but because they are required for the proper execution of the Apollo mission. A comprehensive recommendation cutting across all other interests is that, in the early exploration of Mars, biological and biochemical studies must have the right-of-way until we find either that there is no life on Mars or that the risk of irrevocably destroying it by the introduction of terrestrial organisms is negligibly small. These examples do not by any means exhaust the list of specific priorities; nevertheless the subject as a whole is still largely open.

Questions connected with processing, distributing, and storing data were discussed. The Study is on the whole well pleased with NASA's current policy concerning the exclusive rights of an experimenter to his data for a time specifically agreed upon in each case, with flexible provisions for their later release. The mathematical problems of extracting the essential information from a vastly redundant quantity of data, technical and managerial problems of getting the data from the satellite to the experimenter, the filing and storage of data in a compact and easily accessible form, and the necessity for greater attention to these problems were all remarked; no definitive conclusions were drawn on these topics and they are recommended for further study.

Another question on which NASA asked the Summer Study for guidance concerns the scope of the ground-based researches that NASA should support; that is, is there some way to distinguish categorically between researches that are closely [1-6] enough related to NASA's mission to justify its support, and those that are not? Although this question was discussed by many of the working groups in many different contexts, no one succeeded in drawing a clear dividing line. In fact, the Study concluded that no such sharp line can or should be drawn. Some of the findings are rather obvious. For instance, certain types of ground-based research, such as that relating to the development of techniques and instrumentation or the acquisition of physical data about the space environment, are indispensable just to make the equipment function properly; other ground-based researches are necessary in order to derive the maximum scientific return on a given investment (for instance, the acquisition of astronomical and physical data that are complementary to the space results or that assist in the interpretation of space results, or the theoretical and interpretative studies themselves). Several lines of research in optical, radio, and radar astronomy or radar physics of the circumterrestrial medium (which could be considered classic examples of ground-based activities) were specifically recommended for NASA support. The Study was careful to make clear, however, that these recommendations were not meant to imply an obligation by NASA to support all of classical astronomy or aeronomy, including those programs which would more normally be within the purview of the National Science Foundation, for instance. On the other hand, it was suggested that NASA should be liberal, rather than strict, in its definition of what is relevant, when considering researches commonly thought of as part of a classical ground-based discipline—that NASA should feel free to support anything of interest to NASA itself, especially researches of an exploratory nature, without relying on some strict rule for justification. In these fringe areas, it is assumed that NASA will coordinate its activities and interests with those of the other federal funding agencies.

This subject has been incompletely treated, then, chiefly because the border line between space research (defined primarily in terms of the technique used) and the classical disciplines (defined in terms of subject matter, which happen to be identical with the scientific aims of space research) must, by its very nature, be ill-defined. The ground-based ramifications of space research can be circumscribed only by drawing artificial and arbitrary distinctions.

One working group considered some of the social implications of the national space program. This group consisted of specialists representing several disciplines within the social sciences. They met for four days in consultation with several physical scientists participating in the Summer Study. Our culture, economy and society obviously already have been modified in a number of ways by the impact of the space program. Even greater changes may be foreseen as the latent effects of the growing space program become more evident. A number of provocative questions were raised and a number of significant problems were identified which are suggested for appropriate study. The report of this working group makes clear, however, that the four days available for their sessions were not sufficient for a satisfactory review of the complex and diverse social implications of the national space program. This report is transmitted with full recognition that the majority of the Summer Study participants are not experts in the disciplines relevant for a competent evaluation of its findings.

Clearly this report should be regarded neither as a comprehensive master plan for the guidance of the space science program nor as a complete encyclopedia of all the science that is being performed or that should be performed. It is at [1-7] best a compendium of ideas which we believe are sound and will be useful to NASA in the conduct of its scientific program over the next few years.

II. Summary

This section summarizes the findings and recommendations of the various working groups, which were later presented to the entire Study for their comment. For the full text and discussion accompanying each finding or recommendation, see the chapter and page number in parentheses inserted in the text at appropriate points. The reader is cautioned to consult the full text and discussion, and to regard this summary only as an index with explanatory comment. The findings range from very general to quite specific; in this summary they have been grouped under several major headings:

Flight Program: Satellites and Space Probes

Flight Program: Rockets, Balloons, and High-Flying Aircraft

Ground-Based Research Activities

Science in the Manned Space Flight Program

Administrative and Policy Matters

International Cooperation

Social Implications of Space Activities

Document I-23

Document title: H. H. [Harry] Hess, Chairman, Space Science Board, National Research Council, to Dr. H. E. Newell, Associate Administrator, Space Science and Applications, NASA, "Future Goals of the Space Science Program," August 11, 1964.

Source: Archives, National Academy of Sciences, Washington, D.C.

With the implementation of Project Apollo in full swing by early 1964, new President Lyndon Johnson asked NASA to indicate its post-Apollo plans and priorities. [Volume I, Document III-17] As part of the process of preparing its response, NASA in turn asked the Space Science Board for its views on future scientific priorities. This memorandum provided the Board's initial response, which emphasized the scientific exploration of Mars and planetary exploration overall as NASA's top post-Apollo priorities. The Space Studies Board in 1965 conducted a summer study that reinforced this recommendation.

[no page number]

2101 CONSTITUTION AVENUE
WASHINGTON 25, D.C.

CABLE ADDRESS: NARECO
WASHINGTON, D.C.

NATIONAL ACADEMY OF SCIENCES
NATIONAL RESEARCH COUNCIL
OF THE UNITED STATES OF AMERICA

SPACE SCIENCE BOARD

11 August 1964

MEMORANDUM for Dr. H. E. Newell
Associate Administrator
Space Science and Applications
NASA

SUBJECT: Future Goals of the Space Science Program

This memorandum represents the consensus of the Space Science Board on a primary goal and a NASA program following Apollo, the manned lunar landing, and results from a review of NASA programs and earlier Board positions and was stimulated by a NASA invitation for the Board's views on its response to President Johnson's request.

Progress in the Apollo program, suggesting that its objective (manned lunar landing) may be achieved within less than a decade, coupled with President Johnson's request to NASA for a statement of its long-range goals, has led the Space Science Board to review the program recommendations made to the Government in March 1961. In those early days of space research a little over 3 years ago, the Board recommended:

> "that scientific exploration of the Moon and planets should be clearly stated as the ultimate objective of the U.S. space program for the foreseeable future. This objective should be promptly adopted as the official goal of the United States space program and clearly announced, discussed and supported. In addition, it should be stressed that the United States will continue to press toward a thorough scientific understanding of space, of solving problems of manned space exploration, and of development of applications of space science for man's welfare."

Planetary Exploration. The new goal should be scientific exploration of Mars and general planetary exploration primarily by unmanned probes, landers, and orbiters for scientific investigations. Mars is of great scientific interest: first because it offers the best possibility in our solar system for shedding light [2] on extraterrestrial life and, second, because as a planet it is dimensionally quite comparable to our own. One of the most exciting questions, and in the view of many scientists the outstanding problem of our times, is whether or not living forms have developed on Mars. It may be that organic compounds of inorganic origin may be found on or near its surface; such compounds, the progenitors of life systems, could lead to an understanding of the origin of terrestrial life. It may be that forms of life radically different from our own may be discovered, different in their chemistry, different in their cell structure, and different in their metabolism. Or perhaps we may find fossil evidence of earlier Martian life when perhaps Mars had a denser atmosphere and conditions more favorable to biological processes. The discovery of any of these situations would be of enormous scientific importance and perhaps the most important possible discovery of space research in our generation.

Mars is also an object of great physical and geological interest. For example, how does it compare with the Earth? Is it differentiated, as the Earth is? Does it have a core and is the core molten iron so that a magnetic field is present? Has it a crust differentiated from a mantle? Scientists are in the tantalizing position of trying to discover the general laws of planetary formation and evolution on the basis of one example—the Earth—plus deductions from meteorites. Other examples are essential—examples roughly similar to the Earth, like Mars or Venus, and ones entirely different, like Jupiter and Saturn or comets and asteroids. The anticipation of gathering and analyzing data from them is as exciting to scientists—and in a large and hitherto underestimated measure, we believe, to the general population—as if they had just been presented with a fresh and largely unknown Earth to explore. For a long time this exploration must rely on unmanned fly-bys, orbiters, and landers.

Biological Research. Biomedical research and development must be pressed. In the view of the Space Science Board the scientific exploration of Mars will require that man be present when it becomes technologically feasible to include him. To see that this is true, it is only necessary to imagine how difficult exploring the Earth by remote instruments would be, in comparison to the results of manned exploration. But this phase must be deferred until the biomedical problems of long journeys in space are solved.

Experimentation with man cannot be rushed; it must proceed at a measured pace. Moreover, there are a few significant types of basic biological investigations to be undertaken in the space environment, and these are relevant to manned operations of long duration. Such work may call for manned orbiting laboratories, as well as work in unmanned vehicles and on the ground.

[3] Astrophysical Research. New results in fundamental astrophysics must be closely watched. Ground-based observations have discovered massive objects radiating enormous amounts of energy—the so-called quasi-stellar radio sources. These observations in the visual and radio wavelengths suggest a fundamental connection between the physics of the very large (relativity) and the physics of the very small (elementary particles). The discovery of localized x-ray sources by rocket observations further underlines the basic connection between astrophysics and laboratory physics. As in the case of the quasi-stellar radio sources, a further investigation may bring about a new view of cosmology. These new astronomical discoveries again emphasize the importance of looking out into the universe for insight into the fundamental nature of matter and energy.

Beyond the classical electromagnetic radiation there is a very real possibility of the detection of gravitational radiation. It is not yet possible to define a specific program in this field; however, the results from the planned ground-based, orbiting observatory, and rocket observations may well compel major emphasis in the future. These results could have the most profound influence on the future philosophical and perhaps practical developments of science.

Continuation of the Present Science and Applications Program. We urge that these varied programs must continue and suggest that our 1962 Review of Space Research is for the most part still pertinent. In recommending Martian exploration as the primary objective of the space program, the Board has taken into account the present NASA program in manned and unmanned scientific research in space, with its many important investigations, and the anticipation of its success and growth. For the scientist, the most important thing is that well organized and coordinated geophysical, astronomical, and biological researches must go on—for example, satellite, rocket and ground-based investigations of the upper atmosphere, the magnetic fields and particle fluxes near the Earth and in interplanetary space, astronomical observations in a variety of spectral ranges of the planets, and Sun, none of which should be neglected. As a matter of fact, they continue to be critical scientific objectives in themselves and some are essential back-ups to lunar and Martian missions.

We invite your attention to a number of suggestions for the scientific program as the vehicle capability and performance improve. A concomitant improvement in the performance of orbiting observatories and probes—e.g., in precision of pointing stabilization, and guidance, all of which entail larger weight in orbit—opens up a wider range of scientific opportunity.

[4] Improvement in the capability and standardization of our vehicles should also permit us to consider probes to greater distances from the Earth; for example: penetrating far into the solar corona; out of the plane of the ecliptic; to a distance of 40-60 AU from the Sun; and to Mercury and Jupiter for a closer examination of these planets.

The space science program, while directed toward the ultimate goal of Martian exploration, must be designed so that it can be modified to take advantage of new discoveries as they occur. It would be unwise to sacrifice flexibility which must be available if the changing needs of science are to be accommodated. The Board expects to examine these long-range goals in more detail in the very near future.

It is also clear that the next generations of geodetic and navigation satellites should go forward. The Board believes that research meteorological and communications satellites (as distinct from operational units) will also continue to be of great importance. These responsibilities cannot be neglected, for it may well be that man will be soon able accurately to predict and ultimately to modify his weather at will, through an improved understanding of the mechanisms of the radiation received from the Sun and emitted or absorbed by the Earth.

Alternatives to Planetary Exploration. The Board recommends that the goal of planetary exploration be prosecuted consistent with the decreasing demands of the manned lunar landing program assuming that the space program which the nation should support will remain at or above the present level. It is aware of two possible major alternatives to planetary exploration which could be supported by these funds: (i) extensive manned lunar exploration, including the construction of a lunar base, and (ii) major manned orbiting space stations or laboratories. It is the Board's view that both of these choices have scientific merit and should be developed on a modest scale as our knowledge of their scientific value grows, but as adjuncts to the program of scientific exploration of the planets, rather than primary goals in themselves. In fact, it is conceivable that at some intermediate stage these auxiliary programs may require a large fraction of the total effort, and the total program should be flexible enough to accommodate such a contingency.

Spacecraft and Vehicle Development. A suitable vehicle and standardized communications and instrumentation must be developed. If the vehicle were capable of reaching all space objectives with suitable payloads and of covering the needs over a ten- to twenty-year span, it could be "mass-produced" with eventual great savings per unit, much increased reliability and amortization of development costs over its comparatively long useful life. We would like to see a vehicle capable of twice the estimated payload instead of a minus ten percent margin.

[5]

Major Recommendations

1. The establishment of a goal to focus attention and energies on a readily identifiable target is desirable just as President Kennedy's designation of a manned lunar landing in this decade was. This goal we believe should be the most important objective within our capabilities during the 1971-1986 period. The Board would designate as this goal the exploration of planets with particular emphasis on Mars leading toward eventual manned exploration. This objective includes the search for extraterrestrial life; it may provide clues on the origin of life itself and provides the opportunity to explore another planet comparable to the Earth in size.

2. The experimentation should be largely carried out by unmanned vehicles while the solution of difficult biomedical and bioengineering problems proceeds at a measured pace so that toward the end of this epoch we shall be ready for manned planetary exploration.

3. Alternatives to the Mars and planetary exploration goal, (i) extensive manned lunar exploration including lunar base construction, and (ii) major manned orbiting space station and laboratory program, are rejected as the primary goal because they have far less scientific significance though both have sufficient merit to warrant smaller programs.

4. The Board recommends development of a large rocket vehicle and spacecraft capable of serving the whole planetary exploration program over the 15-year interval so that it could be "mass-produced" with eventual great savings per unit, increased reliability and amortization of development cost over a long period of useful life.

H. H. Hess
Chairman
Space Science Board

Document I-24

Document title: Office of Space Science and Applications, NASA, "Program Review: Science and Applications Management," June 22, 1967.

Source: NASA Historical Reference Collection, History Office, NASA Headquarters, Washington, D.C.

NASA's space science program during the 1960s established a reputation for management excellence. This contrasted with the management problems in NASA's Office of Manned Space Flight that had been dramatized by the January 27, 1967, Apollo 1 fire that killed three astronauts. This document contains a review of space science program management philosophy and practice presented to all top NASA managers in mid-1967.

[title page]

PROGRAM REVIEW
SCIENCE AND APPLICATIONS MANAGEMENT

June 22, 1967

Presented by: Office of Space Science and Applications
Published by: Program and Special Reports Division, Executive Secretariat

****

CONTENTS

****

[1] SCIENCE AND APPLICATIONS MANAGEMENT

INTRODUCTION

By: Dr. Homer E. Newell

Today we would like to review management philosophy and practice in NASA's Space Science and Applications Programs. We plan to review the management scheme in totality rather than to emphasize detailed problems.

I will discuss management objectives and philosophy. Mr. Cortright will discuss organization and, with the assistance of Dr. Townsend, program and project management in the Headquarters and Field Centers. Dr. Naugle will discuss how we handle the problems of working with scientific and academic communities. He will also review the management aspects of planning future science and applications missions. Mr. Taylor will discuss the staff role of providing program support to the line operation.

PURPOSE OF REVIEW

These Program Reviews in the past have delved deeply into program planning and program results, with less attention on program management. Yet management is our main business in NASA Headquarters. We felt, therefore, that it would be of value to devote one session of these reviews to the subject of management.

We felt that such a review would be timely for several reasons. First, NASA has, in its first decade, been through an era of considerable learning in both technical and management areas. Secondly, NASA is entering upon a new era. As a result of our rapidly growing space capability, the large, complex missions will form an ever growing fraction of our program. Thirdly, the number of users of space techniques is growing rapidly, so that, especially in the applications area, we will have increasingly close and involved association with many other agencies. Finally, the creation of the Office of Organization and Management gives us all renewed opportunity to fit our management thinking and actions to the total needs of the NASA organization.

At the very outset, we wish to make perfectly clear that we in OSSA do not view our management arrangements as anything magic, or profound, or even greatly inspired. We certainly do not view what we shall present as the answer to all questions, even all OSSA questions, or as answers for all times. We present simply what we are using to meet current needs. It will be clear to you, as it is to us, that there are problems still to be solved, especially those associated with the changing character of the NASA program. You will see in the discussions to follow, some of the approaches we are taking in trying to work out solutions to those problems.

We especially hope that the members of the Office of Organization and Management will listen critically and give us the benefit of their thoughts, not only now but in the future, as we continue to hammer away at the problems. We in OSSA expect to learn from this afternoon's discussion. We hope that you and the rest of the audience will also find the review and discussion useful.

[2] MANAGEMENT PERSPECTIVE

The objectives of OSSA management are to achieve in an efficient, effective, and acceptable manner various objectives of the national Space Program. The objects of management include: technical problems, schedules, resources, facilities, people, institutions. The tools of management include: funding and other resources, personnel, organization, and a variety of processes and procedures.

Management is efficient if it conserves resources—including time, handles problems in an orderly fashion rather than on a crisis-to-crisis basis, exhibits a reasonable durability of management solutions, and evidences proper foresight of future developments. Management is effective if it accomplishes established objectives within applied resource constraints. Management is acceptable if it strengthens rather than weakens those with whom it deals—including ourselves, our sister offices within NASA, and agencies and institutions outside of NASA.

A management pattern should be an appropriate function of time. While a good management arrangement should exhibit substantial durability, there should also be an orderly mechanism for making changes to meet changing needs. Caution should be exercised to avoid becoming too enamored of a chosen scheme, or too comfortable with the status quo. The science and applications management pattern that we will discuss did not "just happen," but is the product of careful and serious thought on the part of many, including

Abe Silverstein, Al Siepert, Jack Young, Ed Cortright, John Clark, and myself. The OSSA management pattern is reviewed continually, and periodically is subjected to detailed scrutiny in order to maintain the proper continuing evolution in the context of changing needs. For example, a year-and-a-half ago we introduced the Management Information and Control System, which Mr. Cortright will describe to you, to improve both the efficiency and effectiveness of our management process. At the present moment, we are engaged in a reassessment of how we work with the scientific community on projects like Voyager and large orbiting astronomical facilities.

OSSA MANAGEMENT ARENA

We have intentionally narrowed our discussion today to selected topics of immediate management concern, as illustrated on Figure 1 [all figures omitted]. This is necessary in order to fit our presentation into a single afternoon's review. We wish, however, to make perfectly clear at the outset, our awareness of the total context of the OSSA management effort.

As we focus on OSSA, we must keep in mind the total arena in which we operate. Within NASA, that arena is illustrated by the Headquarters organization chart, Figure 2, with which we are all familiar. In addition to the normal relationships implied by the chart, [4] there are special assignments and arrangements designed to increase the effectiveness of NASA management. Some of these are indicated on the chart.

Our working relations with your office, the other Program Offices, and the functional and support offices, are necessarily extensive and continuing. You are well aware of the various interrelationships, so I shall not elaborate. It suffices to point out that the management arrangements in OSSA are established with the total context in mind.

Outside of NASA the arena is broad and varied, as illustrated in Figure 3. We deal with many agencies in a variety of contexts and arrangements—in some cases acting as the principal NASA contact, as with the Space Science Board or the National Environmental Satellite Center.

While in no way attempting to treat totally the subject of the OSSA management environment, I would like to speak further on a few of the more important aspects of that environment.

The Scientific Community

The viability of our Space Science Program depends on the competence and quality of the scientists working with us. It is our conviction that NASA must have highly competent scientific groups within its Centers. At the same time, the majority of our space science effort is, and we believe should be, carried out by scientists outside of NASA, primarily in [5] the university community. We draw the scientific community into our program through the opportunities to do research on space missions. We have also drawn leading scientists into our planning effort through the subcommittees to our Space Science Steering Committee—and through association with the National Academy of Sciences (especially the Space Science Board), the President's Science Advisory Committee, and other similar organizations. Special ad hoc studies, like the two Space Science Summer Studies conducted in the past, and the Applications Summer Study in prospect, add to the scope of our planning and expand our contact with the outside community.

The emergence of large-scale scientific projects, like Voyager and the creation of a large astronomical facility in orbit, appear to call for new arrangements for continuing our association with the scientific community. We are, as you know, working on this problem. As one step in the direction of organizing for the future, we have recently created the Lunar and Planetary Missions Board under the chairmanship of Dr. John Findlay. At the same time, we are discussing with leading members of the astronomical community appropriate arrangements for the future in the astronomy area.

The Universities

A large part of the outside scientific community resides in our universities and colleges. In carrying out our Space Science and Applications Programs we have, therefore, developed an extensive association with the university community. Past associations have been very productive and rewarding, and we will want to continue many of them in much the same form as in the past. On the other hand, just as the changing character of our program calls for changes in our relations with individual scientists, so also we will probably need to develop new types of association with the universities as institutions. At the present time there is under discussion the possibility of creating special institutes in areas such as lunar exploration and space astronomy. We are examining possibilities such as these with the help of the National Academy of Sciences. These matters are not simple. It would be very easy to create something undesirable to our long-lasting regret. The problem before us is to work with the Academy and the universities to strengthen all of us as we work together in carrying out the national space program. We in OSSA welcome the opportunity to work with the new University Affairs Office on problems such as these.

Relations with Other Government Agencies

OSSA has numerous relations with other Government agencies, such as the Departments of Defense, Commerce, Transportation, Interior, and Agriculture, and the Atomic Energy Commission. There is no need to elaborate upon these relationships in this brief review. I would, however, like to single out one important point for attention.

Many of the users of the space technology and capabilities that we are developing within NASA are to be found in our sister government agencies. The Department of Commerce has responsibilities and interests in meteorology, the space environment, and oceanography. The Departments of Interior and Agriculture have special interests in the earth resources area. The Federal Aviation Agency of the Department of Transportation is concerned with navigation and air traffic control and safety. The Federal Communications Commission has special interest in our communications developments. The Department of Defense has a wide ranging interest in those technologies that can contribute to military and national defense applications. [6] Our Applications Office must, therefore, develop a close rapport with these various agencies both in the planning and in the conduct of space programs designed ultimately to support the agencies in meeting important national needs. The molding of such relationships can become very complex, as in the case of the Communications Satellite Corporation, where both government and commercial interests become mixed in an arena where national policy has yet to be defined completely.

The Aerospace Industry

Over 82% of the Science and Applications R&D dollar is expended with industry in the development of hardware. The majority (58%) of this total is concentrated in 21 major contractors. Problems associated with schedule, cost and performance are primarily evident during the development phase, and we have experienced most of these problems in one or more of the several developments which have been successfully completed. We have worked closely with our industrial partners in developing science and applications projects, and through this relationship, seek to contribute to the national capability for accomplishing complex technological undertakings.

As a source of applied research, the aerospace industry bears to our Applications Program the sort of relationship that the scientific and university basic research communities bear to our Space Science Program. In the Applications Program, we already have a strong relationship with industry through the contract work on applications missions, and through special contracted studies. We are endeavoring to draw applications-oriented and engineering communities more into the total planning activity than has been true in the past. One of the means for this will be the Applications Summer Study, to be carried out by the National Academies of Sciences and Engineering during the next two summers.

The Congress

We share in the defense of the NASA program before various Congressional Committees. The major part of this workload stems from the Space Science and Applications Subcommittee of the House Authorization Committee, which always conducts lengthy and detailed hearings. In the defense of the FY 1968 budget, we prepared 79 separate statements amounting to an excess of 2,000 pages of prepared testimony. This was supplemented by approximately 800 graphics, a 30-minute film report on program progress, and about 150 written responses to direct queries by members of these Committees. This Congressional workload does not terminate with defense of the budget but, as you know, continues throughout the year in various forms.

Diversity

There are 30 sizable projects in the OSSA program, covering a large number of different disciplines. This leads to a sizable number of divisions. This circumstance requires top OSSA management to work continually to insure [sic] that the individual parts together make a sensible, balanced program, and that the different divisions pull together instead of separately.

[7] Communication

The diversity of the OSSA program just discussed poses unique problems of communication. This topic, which involves the pairing of scientists and engineers, our Management Information and Control System, the annual OSSA Prospectus, the Space Science Steering Committee and its Subcommittees, Missions Boards, Summer Studies, the OSSA Management Committee and staff meetings, and the OSSA Senior Council, will be discussed at length by Mr. Cortright.

Long-range Planning

Our environment is also one of continual planning on the part of many groups. There is a need for the varied planning activities to fit together in a sensible and effective way. Dr. Naugle will describe OSSA planning, including the key role played by our Prospectus, and how we try to make it fit with total NASA planning.

MANAGEMENT PHILOSOPHY

Management patterns reflect the basic philosophies of those who design and use them. In establishing the OSSA management pattern, we have had in mind several criteria which we think are fundamental to achieving efficiency, effectiveness, and acceptability. A few of these criteria are listed in Figure 4.

Personnel

People are the most important single ingredient in an organization. We have placed great importance on high quality in staffing OSSA. We have been able to attract and retain a highly qualified staff because of the opportunity to work at the center of things on the challenging problems of the Science and Applications programs. Having obtained good people, we must work to retain them by sustaining their interest and motivation. Management must provide for reasonable working conditions, professional recognition, organizational presitge [sic], personal career development, including the opportunity for promotions, and the various other factors of importance within an organization. We think that the remarkable stability of the key OSSA staff is evidence of some success in this direction. An extremely high percentage of the professionals at the Program Chief, Program Manager, Division Director, and OSSA top management level were with the Office of Space Flight Programs under Dr. Silverstein and have remained with the program as it evolved into the Office of Space Science and Applications.

[8] Integration

The management structure should provide for proper cohesiveness and unity within programs and projects. In the OSSA program, science, engineering, and administration must all contribute to the achievement of our ultimate objectives. As you will see in the following discussions, we have chosen an arrangement in which science, engineering, and administration are integrated in all phases of our planning and execution, and in our staffing and organizational arrangements. This permits us to maintain a constant communication among the scientists, engineers, and administrators, and generates a process of trade-offs across the total organization rather than only within the individual disciplines.

We have, from time to time, seriously considered other approaches to our organization which would, in effect, have strictly separate discipline-oriented science organizations, supported by separate project oriented engineering organizations, all in turn supported by separate functionally-oriented administrative organizations. There are advantages to this type of arrangement, and it may be that in very large size operations, this becomes the most manageable type of setup. For the size operation that OSSA has had, however, my own experience suggests that the undesirable features outweigh the desirable ones. Among the undesirable features is the tendency of separated groups to identify objectives and goals which are isolated from the realities of the total picture. The science group tends to become uncompromising in placing unreasonably difficult demands on the engineering group. The separated engineering groups tend to seek perfection in the achievement of purely engineering goals, rather than the ultimate scientific objectives. The functionally-oriented administrative organization will tend to become disassociated from program and pursue administrative perfection as a self-sustaining goal. The fragmented approach, thus, tends to stimulate tradeoffs within science, within engineering, and within administration, instead of across the board. These isolations tend to grow, and in the palnning [sic] for the future, the developers of engineering tools begin to manufacture and invent artitificial [sic] science and applications objectives, whose main function is simply to use the new tools. On the other hand, the science and applications groups would tend to plan programs that are either too fast or too slow for the developing capabilities.

We have, therefore, chosen the other approach of a well integrated mixing of the various activities. At the same time, we recognize the need for each activity to have an identity of its own with sufficient challenge to stimulate continuing growth, and clear recognition and acknowledgement on the part of top management of the individual accomplishments within science, within engineering, and within administration, as well as of the total collective accomplishment. As so often happens, a delicate balance between two possible extremes seems to be the requirement.

Scientist-Engineer Pairing

One of the means that we have used in OSSA to achieve the total integration mentioned above, has been to organize so that scientists and engineers must work together in sharing the total management responsibility and workload. Customarily, where a division director is an engineer, his deputy or second in command is a scientist, and vice-versa. This type of scientist-engineer pairing extends to considerable depth in the organization. At the top level, we have a chief scientist who oversees the total scientific activity of the office. Similarly, we have a chief engineer whose task is to oversee our total engineering posture, to insure proper engineering [9] discipline and practices throughout our program.

The Space Science Steering Committee, chaired by the chief scientist, consists of the key scientists and engineers within the office, thereby forcing our engineering talent to work with the scientists in steering the science program planning and execution. The same group, now chaired by the chief engineer, becomes our Space Applications Steering Committee, thereby forcing the scientists to work with the engineers in steering the planning and execution of applications program and projects.

Administration

The same principle applies in administration. Each technical office within OSSA has a few administrative people, to work on program review and resources management. These people work closely with and are subject to oversight by the Program Review and Resources Management Division.

Total Commitment

To the fullest extent possible, we try to arrange for each individual to be associated with a project from start to finish. The type and intensity of involvement may well, and usually does, vary widely during the course of a project. For example, a scientist may be a primary figure in the initial stages of a project, when experiments are being planned, experimenters organized into a team, and a payload for a mission is being defined. Later, when the payload is functioning in orbit, the scientist may be very actively involved in assuring the analysis, interpretation, and dissemination of the data from the project. The maximum activity of engineers on the project takes place during the middle period, when development, assembly, test, and launch of the spacecraft involved takes place. Nevertheless, the engineer must be involved with the early planning phases in order to get the project started on a sound engineering basis. Likewise, the scientist must continue to be involved during the primarily-engineering phases to insure that the integrity of the scientific objectives is maintained.

These comments apply to both Headquarters and the field, that is, to both program and project management activities. This philosophy of start-to-finish involvement, is a strong motivating factor throughout a project. When an individual starts something, he knows he is going to be with it to the end, so he gives extra attention to the full consequences of his recommendations and decisions. Moreover, he gains an added sense of accomplishment in that he is a partner in the final achievement as well as in the initial and intermediate activities.

Clear-cut Assignments

It is a basic principle of our management pattern that everyone should know what he is supposed to do. This requires, on the part of management, a full understanding not only of the total job, but of the individual peices [sic], and of how they all fit together. This avoids false starts, wheel-spinning, and wasted effort. It avoids unproductive duplication of effort. It enables management and the individuals themselves to assess progress, and to estimate what remains to be done.

[10] Understudies

To the fullest extent that we can, we attempt to give each individual an opportunity to develop toward a higher level position. Rather than filling in on an ad hoc basis from time to time, deputies share on a continuing basis with their principals the management workload of an office, and in this way become candidates for the "boss" position, should the boss himself have an opportunity to move up to some other position.

This pattern of providing for understudies in various forms and arrangements throughout the Office organization, provides for strength in depth, even when staffing may be minimal. It enables one to promote from within to fill a reasonable share of the vacancies that occur in the Office, rather than always looking outside the organization, a pattern which is extremely discouraging to hard-working, competent members of an organization. It enabled us to choose John Clark as our first Director of Science, moving him up from the position of Director of Geophysics and Astronomy, where Mr. Mitchell moved up to replace him. It enabled us to move Vince Johnson in as head of Launch Vehicles and Propulsion when Dr. Morrison left.

Proper Delegation

It is a basic principle in the OSSA management pattern that problems should be handled at the level in the organization where the talent, competence, experience, and strength exist to handle the problem. Thus, what we call program management, which is the over-all management of planning, budgeting, total OSSA use of resources (relations with Congress, with other organizations including the scientific community, and with other offices), and the handling of institutional problems for those Centers assigned to us, is undertaken in Headquarters. Project management, on the other hand, which includes the execution of missions selected for the NASA program, is delegated to selected centers.

With this delegation of responsibility goes also the delegation of the necessary authority to carry out that responsibility. In other words, when an individual, a group, or a center is assigned a responsibility, it is intended that the assignee do the job.

One of the principal jobs of top management is to see that the assignee is not taken off the hook by undercutting of his delegated authority through improper management.

It is a particularly delicate matter for our program managers, for example, to carry out their assignments in such a way that they do not encroach upon or undercut the authorities and responsibilities of the project managers in the Centers. Where program and project managers have developed a clear understanding of the differences and relations between their two jobs, a smooth-working operation can be and has been achieved. Where such an understanding is lacking, problems can develop, which require careful attention of top management in both Headquarters and the Center. Mr. Cortright and Dr. Townsend will talk more about this subject.

Institutional Support

It has long been the view that when an individual is at a certain level in an organization, he has a certain number of people working for him. When he moves up to a higher level, more [11] people work for him, and so on. My view is exactly the opposite. It is that the individual who leads a group should view himself as working for the group. If he is able to provide adequately for the needs of the group, in the way of ideas, resources, support from outisde [sic], good decisions, etc., then he is a good leader. If he is able to work for more people, then he is ready to be promoted.

In this sense, then, a very important task of Headquarters is to work for the groups and institutions assigned to Headquarters. This is my view of the task involved in institutional management. It is OSSA's job to provide for the health, well-being, and necessary resources for the Goddard Space Flight Center, Wallops Station, and the Jet Propulsion Laboratory.

In this area we must improve. It is my belief that NASA Headquarters needs to give considerable thought to how best to handle the total job of institutional management for our Centers. In the case of OSSA, the task is particularly difficult since we have only one large Civil Service Center, and therefore, no basis for developing trade-offs and exchanges between and among Centers. We in OSSA welcome the creation of the Office of Organization and Management, and sincerely hope that one of the early problems that this new Office will tackle is that of strengthening the Agency's approach to the problems of institutional management.

CONCLUSION

The speakers to follow will now discuss how we turn our management philosophy into practice in such a way as to develop acceptable solutions to the problems facing us. Mr. Cortright will begin with a review of organization and program and project management....

Document I-25

Document title: Donald Wise, Chief Scientist and Deputy Director, Apollo Lunar Exploration Office, NASA, to Homer Newell, Associate Administrator, NASA, August 24, 1969.

Source: NASA Historical Reference Collection, History Office, NASA Headquarters, Washington, D.C.

Given its focus on achieving a manned lunar landing before the end of the decade, NASA had a difficult time incorporating the priorities and working procedures of the scientific community into planning for Apollo missions. This letter of resignation exemplifies the frustrations of many space scientists regarding this situation.

[no page number]

August 24, 1969

Dr. Homer Newell
Associate Administrator
National Aeronautics and Space Administration
Washington, D.C. 20546

Dear Dr. Newell:

The following is a discussion of the reasons behind my resignation as Chief Scientist and Deputy Director, Apollo Lunar Exploration Office:

I came to the Agency because the scientific advisory boards to NASA on which I sat seemed to have little influence on the manned lunar program. After working inside the system to give science a more effective voice, I became convinced that the system was equally refractory to internal scientific advice.

The unfortunate split within NASA between manned space flight and space sciences comes to a focus in the Apollo Lunar Exploration Office, a management monstrosity taking direct commands and supposedly serving simultaneously the Associate Administrator for Manned Space Flight, the Associate Administrator for Space Science and Applications, the Apollo Program Director and their various deputies. On several occasions all three groups gave the same command.

This tiny headquarters office including only 10 scientists (3 PhDs and 4 short-term detailees) was expected to be intimately involved in all phases of lunar exploration including astronaut activities, surface science, orbital science, rover science, ALSEPs, lunar sample analysis and the handling of the 140 PIs, the processing of experiment proposals, the SR&T program, writing of Congressional material, long-range planning, and the briefing of at least a half dozen scientific advisory groups. Even with a reasonable command structure this group would have been much too small to follow all the science avenues of the massive Apollo program. Repeated pleas for additional scientific manpower or even adequate secretarial help went unheeded. The fact that the office could be as effective as it was is a tribute to its Director, Lee Scherer.

[2] While following these varied problems, I watched a number of basic management decisions being made, shifting priorities, funds and manpower away from maximization of exploration capabilities of the present Apollo system toward the development of large new manned space systems.

Simultaneously a cumbersome bureaucracy utilized excessive time of both internal and external scientists in getting lunar payloads defined and approved. The fast moving Apollo system, on the other hand, had a tendency to outrun the ability to implement these science decisions for the missions originally planned.

Further, the lack of SR&T funds for a number of past years has starved new lunar instrument and experiment development such that the scientific community could not react effectively to supply "on the shelf" experiments demanded by the Apollo pattern of asking for science on very short lead times.

In evaluating the total picture I felt our office with its headquarters responsibility to get lunar science moving was largely wasting its time running in tight circles within the bureaucracy and among the various competing elements of NASA. Further, it seemed that the system and priorities were not likely to change significantly in the near future. Accordingly I decided to return to college teaching this September from a leave of absence. Until such time as the Administrator, together with the Associate Administrators, determines that science is a major function of manned space flight and is to be supported with adequate manpower and funds, any other scientist in my vacated position would also be likely to expend his time futilely.

Sincerely,

[signature]
Donald U. Wise
Chief Scientist and Deputy
Director, Apollo Lunar Exploration Office
NASA Headquarters

Document I-26

Document title: James C. Fletcher, Administrator, NASA, to George Low, Deputy Administrator, NASA, "Conversation with Gerry Wasserburg and Willy Fowler," October 20, 1971.

Source: George M. Low Collection, Archives of the Rensselaer Polytechnic Institute, Troy, New York. Reprinted with permission.

Document I-27

Document title: Homer Newell, Associate Administrator, NASA, to James C. Fletcher, Administrator, NASA, "Relations with the Scientific Community and the Space Science Board," December 3, 1971.

Source: University of Utah Library. Reprinted with permission.

Between the time of the Apollo 11 mission in mid-1969 and the end of 1971, NASA struggled to get Presidential approval for a major post-Apollo human space flight development program. Given this focus, the space agency tended to give secondary priority to the concerns of the space science community. By 1971, relations between NASA and the science community had reached a low point. New Administrator James Fletcher, who had come to NASA in May 1971, was closely allied with the scientific community as the former president of the University of Utah. Fletcher in October 1971 met with two prominent university scientists, Gerry Wasserburg and Will Fowler, to discuss NASA-scientific community relations. Of the individuals mentioned in Fletcher's "eyes only" memorandum summarizing that meeting, John Naugle was Associate Administrator for Space Science and Applications and Rocco Petrone was Associate Administrator for Manned Space Flight. Later in 1971, as Fletcher addressed in particular NASA's relationship with the National Academy of Sciences (Philip Handler was NAS President; Charles Townes was Chairman of the Space Science Board), veteran space science manager Homer Newell gave Fletcher his views on the state of that relationship.

Document I-26

[no page number]

NATIONAL AERONAUTICS AND SPACE ADMINISTRATION
WASHINGTON, D.C. 20546

OFFICE OF THE ADMINISTRATOR

EYES ONLY

October 20, 1971

MEMORANDUM

TO: AD/Dr. Low

SUBJECT: Conversation with Gerry Wasserburg and Willy Fowler

As you know, the conversation with Wasserburg and Fowler was held off-campus (in the Cosmos Club) in a somewhat secret manner. They were reluctant to have it known that they were needling us on organizational structure within NASA. In general, their conclusion, after some years of dealing with NASA, is that NASA has not yet given science a fair priority in its programs and that even now a number of science programs are in jeopardy, such as:

1. Viking, particularly the life science aspects.

2. The lunar science for Apollo 16 and 17 as well as the follow-on lunar science.

3. The Grand Tour.

They did feel that HEAO has enough pressure from the scientific community that despite "inadequate scientific management" HEAO would probably turn out all right.

They felt that the major difficulties within NASA centered around poor science management at all levels, and particularly they cited the following difficulties:

1. In many programs there is no project scientist, or if one exists he is only part-time. This often leads to primary focus on the project scientist's own interests and not to "tough" management of the science aspects of the project. What is needed at this level is not only a competent person (not necessarily a scientist) who feels responsibility for the science aspects of [2] the program but who has plenty of clout and can plug appropriately into the system. They also felt that this project scientist needed an appropriate staff. When I asked how large this staff should be, they cynically remarked, "At least one person."

2. They felt that scientists were needed at the program office level also, and that John Naugle needs a Deputy for Science as does Rocco Petrone. I asked specifically whether Lee Scherer wasn't doing a good job and they thought he was doing a good job but perhaps he wasn't plugged in as he should have been early enough. They didn't mention specific difficulties with John Naugle's office except to indicate that John was a capable administrator but simply didn't have the time, or wouldn't spend the time, to deal with science problems at the Center or project level.

3. They felt also that a good scientist was needed at each of the Centers to deal with science programs. Since I didn't know many of the Chief Scientists, I asked specifically about Don Rea at JPL and they said he was a welcome addition but it was too early to say whether he would have an appropriate impact on JPL's programs.

As to solutions, they were of course somewhat more vague. They thought that the Administrator should simply direct that a project scientist with appropriate staff be assigned to each project and that more senior scientists should be assigned to each of the various other levels of NASA management and that if I did so, it would happen. When I suggested there might be a shortage of competent scientists in this regard, they admitted to the shortage of scientists interested in administration; however, they felt it could be done but didn't volunteer to help.

They also gave opinions on various scientific personnel within NASA, when pushed. I will be glad to discuss them with you, if you so desire.

J. C. F.

Document I-27

[no page number]

NATIONAL AERONAUTICS AND SPACE ADMINISTRATION
WASHINGTON, D.C. 20546

OFFICE OF THE ADMINISTRATOR

MEMORANDUM

TO: A/Dr. Fletcher

SUBJECT: Relations with the Scientific Community and the Space Science Board

Following are some points to keep in mind when discussing the matter of NASA-Space Science Board relationships with Charlie Townes or Phil Handler, or when preparing a letter on the subject. John Naugle and Henry Smith have reviewed the material below, and are in general agreement with it. Henry Smith did have some additional comments which I thought you should see; they are attached in Enclosure 1. Both Naugle and Smith, however, are very much concerned that as we wrestle with the problem of maintaining good relations with the outside science community, we not forget that we have good scientists within NASA who also need to be heard, and who need to be assured of the opportunity to derive professional satisfaction from their work. Naugle and Smith are preparing a memorandum to you on this subject of the total problem, which they will be sending through shortly.

First of all some background. Relations with the Space Science Board, and also with our own Boards and Committees, began to come apart about the time the Space Task Group Report was published. Strains developed because the Boards and Committees felt they were not being effective or listened to by NASA. The budgets in the Space Task Group Report were regarded as appallingly high. The emphasis given to large-scale programs—space shuttle, space stations, space bases, lunar bases, nuclear shuttles, Grand Tours, and manned missions to Mars—had a very negative effect. Our own Lunar and [2] Planetary Missions Board threatened to resign en masse. The Lunar and Planetary Missions Board took strong exception to the order of choice of some of the near-term planetary missions, but their more serious concern was that if they had been aware of the total context developed in the Space Task Group Report, the entire scope of their advice on planetary program would have been different. It was this kind of concern, which was also expressed by the Astronomy Missions Board, that led us to form the Space Program Advisory Council with Chairmen of our advisory Committees as members, so that in the Council these Chairmen can get the kind of perspective that was not available to our former Boards.

It was in this period of turmoil and reaction to the kind of program proposed in the Space Task Group Report, that disenchantment with Viking and active concern over large scale planetary projects like Grand Tour began to develop. When Harry Hess died, and Herb Friedman took over as temporary Chairman of the Space Science Board, these difficulties were exacerbated. Friedman has long been a strong proponent of the smaller types of space projects, and very much wants to see the sounding rocket work enlarged, particularly astronomy. Additionally, Dr. Friedman is personally intensely interested in high-energy and infrared astronomy, and wishes to see them pushed, if necessary, at the expense of ultraviolet and other areas of astronomy. This hostility to the large-scale projects, including particularly the large-scale manned space flight programs proposed for the future, on the part of the Acting Chairman of the Space Science Board, naturally made itself felt, and carried over into the handling of the Woods Hole Summer Study of 1970, which Dr. Friedman chaired.

Charlie Townes, very much at our urging and the Academy's, took over chairmanship of the Space Science Board at the time that we in NASA were planning a restructuring of our advisory committees in order to eliminate weaknesses experienced in the earlier arrangement. When he took over, Charlie undertook to revamp the Board, instituting a policy of rotation in membership. Thus many new members were brought on, and many of the older [3] members rotated off. As a consequence, the Space Science Board, like our space program advisory council and its Committees, is going through something of a learning period. Having more continuity than SPAC and its Committees, the Space Science Board has less of a problem at present in this regard than do our Council and Committees, but we should also recognize that the small size of the Board means that some areas of space science may not enjoy the same degree of representation or advocacy as others. Thus some divergence of scientific opinion must be expected from time to time.

The Space Science Board has for the past year and a half clearly been feeling its way toward what it would like to regard as the proper relationship with NASA. The recent meeting at the Jet Propulsion Laboratory was something of a milestone in this regard, when we went through in detail for the Board the various problems and considerations involved in our FY 1973 budget decisions, and outlined in as much detail as was available various options for the future in planetary exploration. A large number of the Board members indicated satisfaction with that meeting. As Roman Smoluchowski put it, prior to the JPL meeting the relationship between the Board and NASA had been largely that of adversaries, while as the result of the meeting the relationship seemed more like that of partners. This brings me to the first major point that I think ought to be established between us and the Space Science Board.

1. There is a need for more exposure on both sides to the give and take of problems and alternatives being considered on the other side. To continue the development of a feeling of partnership between NASA and the Board, we in NASA need to give more attention to ensuring that the Board has a good insight into the budget problems, political pressures, technical tradeoffs, manpower restraints, etc., that we are wrestling with. The Board, on the other hand, needs to be more open with us. There is a great tendency nowadays for the Board to call executive sessions, excluding even senior NASA personnel from them, to debate points of [4] view, pros and cons for different alternatives, opinions regarding NASA plans and approaches, etc., and then after a position has been determined to present that final position to NASA people. In the days of Berkner and Harry Hess, Dryden and I were welcomed into such discussion meetings, where we heard all the give and take, and were consequently more sensitive and alert to the feelings of the Board and its members on objectives, relative priorities, approaches, etc. In those days I detected no hesitation on the part of the Board members to express concern or dissatisfaction over things that NASA was doing, and the ensuing discussions often led to constructive approaches to resolving problems. The Board should be urged to invite Naugle, Myers, Barry, myself, and yourself when you have time, to hear these very important discussions.

2. A second important way to enhance the feeling that we are working together is to involve at least the chairman of the Space Science Board, and perhaps the Chairman

plus some selected members, in close discussions with Naugle and the Administrator during those last weeks of decisions when our budget proposal is taking final shape. Such discussions will enable us to detect whether misunderstandings of Space Science Board positions or intensity of feeling about different alternatives are leading to decisions that might otherwise not be made. Such discussions, though informal and off the record, would help reassure the Board that its views are indeed being taken into consideration in the final decisions. Then, even if a decision goes contrary to or differs in some respect from a Space Science Board recommendation, at least there will be an understanding of the reasons, and a better awareness of the fact that the decision was not an arbitrary rejection of the Board's position.

[5]

3. The Board has felt of late that NASA tends to provide backup support—studies, analysis, budget estimates, etc.—for those missions or projects that NASA wants to consider, but not adequate support for those that the Space Science Board would like to consider. We should spend more time discussing with the Board the kinds of studies and analysis that they need in order to carry out their advisory role. This will, of course, take manpower and dollars, but a modest investment in this area should pay big dividends.

4. Finally, we need to develop some way of joining hands in support of long-term projects, so that programs that we start with the support and encouragement of the Space Science Board don't later founder because the support is withdrawn or watered down midway through the effort. This should be helped by a fuller and more open discussion between NASA and the Board in early phases of planning. NASA needs to be sensitive to the concerns which may appear mild in the early phases, but which may grow in intensity as the project proceeds until they become a serious stumbling block. The Space Science Board, on the other hand, must avoid underplaying concerns that are in their view serious. The Board also should be urged, as was done by you at JPL, to discipline itself to regard a recommendation to undertake a project as a commitment by the Board to stand by and work with NASA not only during the early days of selling the project to the Administration and the Congress, but also during the long period of carrying it out. With rotating membership the Board will need to develop some sense of continuity of opinion, so the one year's membership understand and support the recommendations made by a different group of people some years earlier.

The form of our relationship with the Space Science Board and the scientific community will be of no real significance unless it also produces a program that the Board and Scientific community can believe in and support. In this regard, there are [6] a few points that the Space Science Board has made over and over for many years. Some of these are summarized in the next paragraph.

5. For a long time the Board has consistently urged that NASA develop a balanced program, balanced not only with respect to different disciplines, but also balanced within each discipline between large and small projects. An essential element of such a balanced program is, in the view of the Board, the flexibility and quick turn-around

time afforded by smaller projects and smaller spacecraft. Short lead times, and the ability to follow up quickly new scientific discoveries, are viewed as essential to a good scientific program. To achieve this flexibility and follow-up capability, the Board has repeatedly recommended more sounding rockets, Explorers and Pioneers, for example. The serious concern of the Board over large projects like Viking and the Grand Tour is two-fold. First at some dollar level such projects become far more expensive than they feel they can in good conscience justify on scientific grounds alone; and, secondly, through their incessant demands for funds and manpower, the large projects tend to squeeze out the essential smaller projects. Moreover, with the smaller projects in the total program mix, the larger projects make their best contribution and can be accepted as scientifically fruitful and worthwhile; without the proper number and kinds of the smaller projects, the larger projects are regarded as not producing the best science for the money invested.

6. Over the years, the Space Science Board and our own committees have expressed deep concern about the long term involved in space missions, which in some cases take between five and ten years for accomplishment. If, as space experiments become more sophisticated and more complex, they also take more and more time, then [7] this time becomes a substantial fraction of a man's career, more of a fraction than most scientists would like to gamble on program as risky as is space experimenting. This again argues for an adequate number of smaller projects and experiments that scientists can use to generate a steady flow of results. But equally important is the fact that if a long period intervenes between the selection of the experiment and the time of its accomplishment, this virtually guarantees that the experiment will be out of date or not the best that could be done at the time it is performed. As a consequence the SSB urges every effort on NASA's part to shorten lead times for experiments, and to make it possible to update or replace experiments as close to launch time as can be managed. One suggestion is that a group of scientists be formed for each major mission to keep in touch with the payload and advise on its status, recommending on desirable update and improvement during the course of the project. In response to these recommendations, John Naugle suggests that for each project we consciously allow for some contingency in both payload weight capacity and project dollars to accommodate new experiments late in the game, and that we ask Fowler's Physical Science Committee to review periodically the project status and make recommendations. (This is a very difficult problem. Naugle is personally committed to doing as much in this direction as possible. However, it must be pointed out that it just does take a long time to prepare these experiments for flight, and that the longest lead times are usually associated with scientific experiments, not with the spacecraft and housekeeping hardware.)

It is very natural for those who are in the middle of carrying out space projects, and of planning for the future, should have the broadest range of alternatives in mind and have strong feelings about what can and can't be accomplished, and what ought to be undertaken next. This sometimes results in NASA's moving out on a project before the scientific community has a [8] full appreciation of what is involved. When this happens,

NASA leadership often has to work on the problem of organizing its following right at the time when strong support from the outside may be crucial to sustaining Administration and Congressional support for the project. My last point refers to this problem.

7. We need to restore a NASA posture of being urged by the scientific community to do things, rather than urged not to do them. In many cases this may well be a matter of how we work new projects into the program. To be specific, a possible program for the next decade is that of building a large astronomical orbiting facility for use by the astronomical community. In his recent oral report to you, Jesse Greenstein indicated that his Academy study committee has given the large space telescope an important place on the committee's list of priorities. Our Astronomy Missions Board regarded this as an important project and recommended that we move in that direction. Recently, Herb Friedman spoke to me urging that we use the large space telescope as the principal focus for our manned space flight program beyond Apollo. Thus, support for the building of a large orbiting space telescope is beginning to form. At this point, however, NASA could easily outrun its support and generate some undesirable resistance. It might be well at this stage to move more slowly, consolidating interest in and support for such astronomy projects step by step, until we arrive at a stage where pressure for the large space telescope is so great that we can hardly fail to accede to it.

[signature]
Homer E. Newell
Associate Administrator

cc: AD/Dr. Low
S/Dr. Naugle

Attachment as stated

****

[Enclosure 1]

[no page number]

From Dr. Henry Smith/SS

Comments on Draft: December 3, 1971

P. [paragraph] 3. The bias towards astrophysics is also dominant. Skylab, which carries the $300M ATM, has not been subject to the same degree of opposition as the other large programs. This suggests that the Board would be mollified if among the large projects there were better balance between the different disciplines.

P. 8. One of the most sensitive areas of Board relationships is program cost growth. Historically, almost every program has grown 200% to 500% between the time of conception and approval, and the last launch. We secure the Board's support in the early phase of low cost estimates, and suffer their disillusionment when the costs grow beyond this initial frame of reference. On the other hand we can't realistically go back to the Board on every overrun, schedule slip or experiment addition. Two steps are necessary to mitigate this problem: (1) NASA must define program runout costs more realistically during the initiation phase, and must abide within this envelope; (2) or be prepared to seek the Board's concurrence to adjust the cost ceiling significantly from 1973 to 1975, and is probably the main reason the SSB continues in accord on the Viking plan.

P. 11-12. Adding new experiments late in a mission is a good idea when we have reserved some contingency for it and have announced our intention before hand. However, to really capitalize on new knowledge acquired after the initial payload selection, it would be necessary to have an open competition via an AFO. In prior years the discipline ad hoc subcommittees performed the continuing mission review and suggested updating and improvements during the course of a project. The scientific community accepted that because we always applied the conflict of interest rules to the subcommittees, and people accepted their opinions as fair and impartial. I don't think we want to ask the Board or SPAC to be directly involved in experiment selection.

P. 14 The scientific community has expressed a strong desire to be involved in the initial conception and planning stages of space science missions, in order to optimize scientific cost effectiveness. We have successfully done this in recent years in the planetary program, where science steering groups worked with the project teams on Viking, Mariner Venus-Mercury, etc. [2] However, even this procedure has pitfalls, as we are observing in the conflict between the Fowler Committee (Physical Sciences Committee) and the Grand Tour SSG [Summer Study Group]. The outsiders in the scientific community, always the larger majority with any single project, feel some distrust of the minority "in-group" who are naturally most competent and best informed about a project, and we cannot afford to disregard their opinions.

Document I-28

Document title: Frank Press, Director, Office of Science and Technology Policy, White House, to the President, November 27, 1979.

Source: Jimmy Carter Presidential Library, Atlanta, Georgia.

As he was leaving office at the end of 1976, President Gerald Ford approved the start of two major space science missions, which became the Hubble Space Telescope and the Galileo mission to Jupiter. In his first year in office, President Jimmy Carter approved U.S. participation in the International Solar Polar mission. But there were no "new starts" in space science in 1978. In the fall of 1979, the White House Office of Management and Budget recommended, given other financial pressures including the costs of the Space Shuttle program, that once again there be no new space science missions approved. Presidential Science Adviser Frank Press expressed to the President his concern about this recommendation. Carter in December 1979 did approve a new start for the Gamma Ray (later known as the Compton) Observatory. The other technology development addressed by Press, the Solar Electric Propulsion System, was linked to a planned NASA comet mission which would both carry a probe to fly by Comet Halley in 1986 and rendezvous with another comet later in the 1980s. President Carter did not approve a new start for this system.

THE WHITE HOUSE
WASHINGTON

November 27, 1979

MEMORANDUM FOR THE PRESIDENT

FROM: FRANK PRESS

SUBJECT: Presidential Appeal – NASA Budget

I am seriously concerned about the effect of OMB's recommendation that there be no new starts in FY 1981 in space science. Over the last decade the US has averaged nearly two new starts per year in space science and applications. If the NASA appeal is denied, you would be the first administration in NASA's history not to initiate a new space science start for two years in a row.

Last October you unveiled your civilian space policy at the Kennedy Space Center. You directed a balanced strategy in space science, applications and technology development. Three months later the FY 1980 budget reflected no new starts in space science or application. In fact, GALILEO—the Jupiter mission—is the only approved planetary mission now funded. Accordingly, I strongly support FY 1981 funding for the Gamma Ray Observatory (GRO). It has been NASA's highest priority new start for two years. The GRO may well be for black holes what the telescope was for the planets and the stars. GRO is a soundly based program of relatively low cost and promises a high scientific return. Its initiation would give needed balance to your announced space policy.

Secondly, and in that order, I recommend funding in FY 1981 for the Solar Electric Propulsion system (SEPS) and related technology development—even at a reduced level. This would preserve the option for the Halley Comet mission in 1985, which will capture world attention and provide valuable insight into the origin of the planets. Moreover, this development has future applicability for both civilian and military programs, which would be in fulfillment of your direction to maximize technology-sharing among the space sectors.

Our leadership in space will depend on the Shuttle. It is imperative for both civilian and military defense requirements that we maintain its development schedule. The large increase of funds for ensuring that the Shuttle is ready to launch future national security payloads competes with space science. However, it would be a pity if the expensive technology development for the Shuttle drives out the relatively inexpensive space exploration.

The Gamma Ray Observatory, Solar Electric Propulsion System and associated technology development are research efforts worthy of a great nation. Their initiation would demonstrate your decision to invest in the future and a degree of optimism that would be uplifting in these troubled times.

Document I-29

Document title: Space and Earth Science Advisory Committee, NASA Advisory Council, *The Crisis in Earth and Space Science: A Time for a New Commitment,* November 1996.

Source: Office of Space Science, NASA, Washington, D.C.

Beginning in 1982, NASA undertook a major scientific initiative with respect to the study of Earth from space, and Earth science was added to space science as part of the responsibility of NASA's Office of Space Science and Applications. This change was reflected in the name of the top-level external advisory group for the office, the Space and Earth Science Advisory Committee (SESAC). Chairing SESAC in the early 1980s was Louis Lanzerotti of Bell Telephone Laboratories. Beginning in 1984, SESAC undertook a comprehensive examination of the state of the space and earth science enterprises. It was quite troubled by what it found; its concerns were amplified by the January 1986 Challenger accident and the resulting delay of all NASA programs.

[cover page]

THE CRISIS
IN
SPACE AND EARTH SCIENCE

A TIME FOR A NEW COMMITMENT

A REPORT BY THE
SPACE AND EARTH SCIENCE
ADVISORY COMMITTEE

NASA ADVISORY COUNCIL

NOVEMBER 1986

[iii] **Overview**

Results from the Space and Earth Sciences, in the last quarter century, have stimulated a profound curiosity about our universe and an awareness of our own planet. The astounding successes of science missions in space, ranging from weather satellites, to astronomical observatories, to planetary reconnaissance and surface sampling, have created a new sense of the wonder and unity of our environment and have produced an almost dazzling array of compelling new scientific questions yet to be answered. Science in space is an unparalleled intellectual adventure, a technological endeavor, and the necessary precursor to the next great journey of mankind envisioned by the National Commission on Space (1986).

The nation has also had an important emotional, as well as intellectual, investment in its successes in space and in space research. American leadership in penetrating the unknowns of our planet and solar system, and in unravelling the mysteries of the universe, is as important to the general public as to those directly engaged in the scientific endeavor. But preservation of a leadership position at the frontiers of science is precarious and can be maintained only through diligence and commitment; American preeminence is now in question. This report assesses the current health of the Space and Earth Science Program and identifies the requirements for a renewed commitment to excellence. It concludes that the program is facing grave difficulties and that specific steps must be taken to ensure its vitality and long-term future.

The Crisis in Space and Earth Science

Even before the Challenger accident, and the resulting hiatus in the space program, it was becoming clear that the nature of the Space and Earth Science Program was changing and that major stresses were developing as a result of those changes. Within the scientific community there was a growing sense of unease and frustration over the program's diminishing pace. As the result of a number of trends, it appeared that a major transition was taking place in the nature of the Space and Earth Science Program, but it seemed that this transition was occurring more by accident than as a matter of conscious policy. Decisions were being made that had long-term consequences on ways the program would be conducted, but the consequences of those decisions were largely unexamined. More and more missions were being identified as candidates for "New Starts" at a time when prospects for New Starts were becoming uncertain. The competition among prospective missions had escalated to a counterproductive level; there was a growing sense that the future vitality of whole fields of research depended on single decisions. The emergence of the Space Station as a major NASA initiative was raising questions as to whether NASA's science program would be reoriented around this facility. Questions concerning priorities were being raised without there being any obvious way to systematically address those questions. More and more scientific groups seemed to be competing for fewer and fewer flight opportunities. At the same time, the pressure to start major new missions seemed to be leading to an erosion of those smaller-scale, less glamorous, less visible activities that, in [iv] many ways, formed the foundation of the program and ensured that the scientific return from major missions really justifies the investment.

Other grave difficulties were also appearing in the implementation of the program. Delays and cost overruns drained away resources that could have supported additional major missions or other important research; projects cancelled or repeatedly deferred after scientists had responded to Announcements of Opportunity, wasted the efforts of talented individuals; dependence on the Shuttle as the single launch vehicle introduced human safety as a crucial consideration into the program even for those missions where less risky alternatives should have been available; erratic funding patterns and continually shifting priorities created uncertainty for all components of the space research community. In view of the uncertain future and the lengthening time scales for execution of programs, talented individuals began to seek other opportunities.

All of these difficulties were dramatically amplified by the *Challenger* accident and the subsequent turmoil in the U.S. space program. Scientific spacecraft ready for launch were grounded. Delays of two years or more are inevitable; maintaining scientific teams and spacecraft readiness until missions are launched and results are available will be both difficult and costly. The mixture of launch capabilities available for future programs is not yet clear and there is a fear that the costs of the replacement Orbiter could threaten other elements of the NASA program.

As a result of many of these perceptions, two years ago the NASA Space and Earth Science Advisory Committee (SESAC) embarked on a wide-ranging examination of the programmatic issues facing the U.S. Space and Earth Science Program. The intent of this study was to determine the nature of changes underway, to understand the implications of those changes, and to make recommendations to enable NASA to proceed with a long-term, productive program in the Space and Earth Sciences. A major goal of the work was to develop a more rational process for making decisions, especially decisions concerning major new initiatives. The fundamental task of this effort was to determine how to optimize the use of the limited available resources in such a way as to construct the best possible scientific program.

All of these concerns became even more urgent in the wake of the *Challenger* tragedy.

Given the current critical circumstances and the clear threats to the vitality of the future program, careful examination of the premises upon which the NASA Space and Earth Science Program is based, planned, and executed is clearly in order. This report proceeds from such a fundamental examination to a series of recommendations intended to guide the conduct of the program in the years ahead.

[v] **Vitality in Science**

Before arriving at any conclusions, we must first address what is required to ensure the vitality of NASA's program. Scientific vitality comprises many elements. They include:

- *Stimulating questions.* The success of the Space and Earth Science Program can be traced in part to the abundance of stimulating questions about our environment and place in the universe. Although some disciplines of space research are more mature than others, stimulating questions abound in all the disciplines.
- *Observations and experiments.* There must be a steady flow of experiments and observations, discovery, and reconnaissance. Scientists first search for new phenomena or for

new ways of viewing known phenomena. Once a discovery is made, reconnaissance, systematic observation, and analysis begin with the goal of acquiring more complete understanding.

- *Theory and models.* Comprehensive theories and useful models spring almost naturally from a carefully planned base of observations. Observations validate theoretical predications or lead to creation of new theories, which, in turn, must be judged in terms of additional data.
- *Talented and dedicated people.* Essential to scientific progress is the involvement of talented, dedicated people driven to satisfy their curiosity about nature. They acquire a command of existing knowledge in order to make new contributions. Dedication alone, however, is not enough. Aspiring young scientists must have the support of a strong educational system in which they can learn by working with established researchers on substantive scientific questions.
- *A perceived future.* Any healthy science must have goals and opportunities that are perceived to be exciting and important both by specialists in the field and by the public at large. There must also be favorable prospects for the continuing support of those endeavors in order to drive the development of new levels of technological sophistication and scientific understanding.

Our ability to meet some of these requirements is now questionable, and, as a consequence, the vitality and the future of the Space and Earth Science enterprise are threatened.

[vi] **Signals of Stress and Change**

The systemic difficulties which have developed in the Space and Earth Science Program have, at least in part, resulted from the facts that:

- *The Space and Earth Sciences have widened their horizons.* The accomplishments of science in space have opened a broad frontier of new and fundamental scientific possibilities, have prepared the way for a variety of practical benefits, and now promise even greater rewards from the continued exploration of the Earth and the heavens. New disciplines are realizing the benefits of science in space. The successes of the Space and Earth Sciences could be but the dawn of a bright future. However, there are many more worthwhile opportunities for exploration than can be accommodated by the resources expected to be available.
- *Space technology required for new advances is more sophisticated and more costly.* The advances in space research have mandated the development of observing systems capable of greater temporal, spatial, and spectral resolution. Such systems are technologically complex, heavier, require more power, and produce data at rates that challenge current capabilities. But if science is to advance, the technological pace must be maintained, and the resulting increased costs have to be accommodated.
- *Interactions between an increasingly constrained NASA and a larger and more diverse scientific community are creating serious stresses.* Within the space research community, with its many components and interests, there is intense competition and tension. Strong proponents see lost opportunities; they fear their future may be one of delay and decay

rather than stimulating accomplishment. The complexity of the current endeavors raises concerns about the retention and stimulation of the individual creativity and initiative essential to scientific progress.
- *Assured access to space is no longer obvious.* The number of flight opportunities for Space and Earth Science payloads has gradually decreased. While this trend is due in part to the widening scientific horizons and more diverse research community, it also is the result of not matching the launch vehicles to the purposes of the scientific missions. Space and Earth Science cannot advance without assured access to space.

The character of the Space and Earth Science Program is changing. If the program is to be guided properly, conscious steps must be taken to manage the change. Awareness of the issues and trends must be the key first step in proceeding in a more systematic fashion.

[vii] **Recommendations**

In order to foster the vitality that is at the heart of a productive Space and Earth Science Program, SESAC presents the following recommendations.

1. *The Space and Earth Science Program must continue to incorporate a diverse range of activities, participants, and facilities.* (Chapters 4, 5, and 7)* The vitality of the program conducted by NASA's Office of Space Science and Applications (OSSA) rests in the availability of a range of activities and facilities. Low-cost suborbital missions are essential for addressing certain scientific questions on a short time scale, for technology development, and for graduate education. Moderate scale missions focus on specialized scientific issues. Major facility-class missions have become essential for answering fundamental scientific questions in each of the Space and Earth Science disciplines and must be provided in turn on an appropriate schedule. The OSSA Research and Analysis program is the foundation on which the vitality of the Space and Earth Sciences depends. It must be strengthened in a number of significant ways and protected from funding fluctuations.

Cooperation and collaboration among all components of the Space and Earth Science community—NASA Headquarters, the NASA Centers, the universities, industry, other Government agencies and Federal laboratories, and international partners—are the key to effectively conceiving, planning, constructing, and managing space missions. Each component of the space science infrastructure provides unique capabilities and perspectives, and this diversity must be maintained. In this report we reaffirm the significance of the several roles played by each component of the space research community and recommend that increasing cooperation be promoted. NASA should, with the assistance of the entire research community, explore the potential advantages of new organizational structures, including consortia and formal academic and industrial partnerships. Increasing capabilities outside the United States are potential sources of new opportunities. We praise the ongoing efforts between NASA and the European Space Agency (ESA) to establish a policy of reciprocity of flight opportunities. We note the valuable opportunities offered by the Japanese Institute of Space and Astronautical Science (ISAS) for U.S. participation in the Geotail and High Energy Solar Physics missions. Other possibilities for bilateral and multilateral cooperation with other space-faring nations also exist. We urge

NASA to pursue and take full advantage of collaborative and reciprocal opportunities which may arise.

[vii] 2. *The scientific requirements of a particular mission must be the dominant factor in selecting the launch vehicle, instruments, and spacecraft to be employed.* (Chapters 4 and 7) It is imperative to adopt the most appropriate launch vehicle for each program. NASA must reintroduce expendable launch vehicles into the fleet. Manned space flight must be used only when a manned capability is essential for meeting scientific requirements. But having choices available for launch is only one step in optimizing the program. Proper matching of instruments with spacecraft capabilities must be done on the basis of the scientific needs of the mission, not on the basis of exploiting an available facility. There must not be confusion between ends and means. This will become an increasingly significant point as we move into the era of the Space Station. Thoughtful preparations must be made for the utilization of the Space Station. Use of the Station should begin with simple experiments, which then evolve toward more complex ones as the Station's capabilities become better understood. Science payloads should not be selected merely on the basis of the availability of space on the Station. There are established mechanisms for selecting payloads on the basis of their scientific merit, and this philosophy must be maintained for Station or platform manifesting. The Space Station will be only one of a range of tools available to OSSA. OSSA should select what science is to be done before selecting the most appropriate mode of performing the experiments, whether that be as a Station or Shuttle payload, or an instrument on a unique free-flying spacecraft, a spaceprobe, or a servicable, retrievable platform.

3. *All aspects of the Space and Earth Science Program, and their total requirements for resources, must be thoroughly and realistically understood through rigorous planning.* (Chapter 7) NASA management and the research community must make efforts to optimize the current utilization of resources and talents. OSSA should reexamine its approach toward implementation of flight projects with the intent of reducing overall mission costs. This effort should include use of similar, but appropriately modified, spacecraft for several missions; reducing requirements for documentation while reappraising the level of reliability needed for each mission; and more realistically matching mission needs with spacecraft and instrument capabilities. Once a project has been started it must be completed on the most cost effective schedule. A flight project should not be started until the launch or carrier vehicle is assured and a clear understanding exists of the risks associated with any necessary new technology connected with the carrier. OSSA should also consider broader implementation of the current funding process applied to the Explorer program in which missions are developed and launched a few at a time within a fixed funding envelope.

Especially for larger missions, runout costs, including operations and data analysis costs, must be well understood before a project is officially started, and if a major delay or descoping appears necessary, then OSSA must address the issue of whether the program is still viable and retains its original priority. Large cost overruns cannot be tolerated. Because of limited resources, careful choices should [ix] be made about the number of projects which are in the definition and design stage (Phase B) at any given time. Just as the number of Phase B projects should be limited to those with a reasonable expectation of being started, so should Announcements of Opportunity only be released for those projects that have a reasonable prospect of entering the development phase with a few years following investigator selection.

4. *Carefully specified criteria must be used in setting priorities and deciding among proposed major space research projects or missions.* (Chapter 6) The Space and Earth Science Program consists of a large number of research and data analysis projects, of suborbital experiments, and of a family of space missions ranging from the small and short-lived, to the very large, permanent facilities. All elements of this program must be melded into a coherent whole. Moreover, the selections of the major facility missions that become the center piece activities of the individual science disciplines are especially significant because such choices involve substantial near-term funding requirements, determine the long-term direction of whole fields of research, and obligate funds well into the future. Careful specification of the criteria for evaluating the scientific merit, programmatic implications, and societal benefits of proposed new Space and Earth Science projects or missions is essential to make effective decisions. We propose such criteria, formulated as questions, and urge that the criteria be applied by all who are involved in making the difficult decisions that shape the future of the Space and Earth Science Program.

A Broader Issue

The proud advance of the Space and Earth Sciences in the first quarter-century of the modern space age have created many more exciting opportunities for science in space than can be accommodated by the present budget of the NASA program. Thus, the critical question to be faced is whether the Agency should be reponsive [sic] to scientific imperatives or curtail its efforts to fit within a budget determined on the basis of extra-scientific criteria. In either case, maintaining the focus and effectiveness of the program is essential to providing the greatest possible scientific return, thereby justifying public support. Regardless of the size of the program, resources must be effectively utilized to produce the highest quality scientific results.

If additional resources are not available, then there are only two options: either progress in all of the Space and Earth Sciences must be delayed or else some of the disciplines must be assigned a substantially higher priority to proceed at an optimum pace. In either case, some disciplines will perceive a future that is bleak at best and will lose vitality.

[x] *The relevant elements of the Executive Branch and Congress must participate in continuing discussions on the future of the NASA Space and Earth Science Program in order to foster stability, predictability, and realistic expectations.* Decisions and choices must be made. Once decisions are made, programs should proceed on a firm schedule. The continued health of the research program requires predictability in continued support from year to year. Graduate students cannot be encouraged to select a career in space research if they see fluctuations in the research base or if projects are started, postponed, restarted, delayed, refocussed, and possibly canceled. Obviously, senior scientists also cannot function in such an unstable environment. We must promote a more rational use of human resources. Above all, whatever the actual levels of funding for the various programs, a certain level

of stability must be imposed across the spectrum of research activities in order to provide a predictable program with realistic expectations.

Restricting access to new knowledge through parsimony is not in the nation's long-term interest. Science, by its very nature, promotes progress. Progress in science necessarily leads to further scientific endeavors, greater achievement, as well as greater costs. The direct and indirect rewards of effectively conducted research provide the increased productivity to finance the continued growth of science. The past three decades have clearly shown that the Space and Earth Sciences, carefully managed and carefully nurtured, can be among the nation's most rewarding investments.

We must, therefore, move ahead with our voyage into space, to observe and measure our Earth and its environment from great heights, to visit and explore the distant planets, to probe the depths of our Galaxy where stars are born and stars die, to search the outermost reaches of the universe to learn about our cosmic origins, to fathom the deeper laws of nature, to investigate the origins of life, and thus, to find our place in the greater design of the world around us. This is where America has made major intellectual contributions in this century and should also continue to do so. Let us press forward.

* The chapters referred to following each statement contain the arguments and discussions which have led to these recommendations.

Document I-30

Document title: Office of Space Science and Applications, NASA, Strategic Plan 1998, April 1998.

Source: Office of Space Science, NASA, Washington, D.C.

One of the problems pointed out by the 1986 SESAC report (see preceding document) was the unproductive annual competition among potential major space science missions for "new start" status. Upon becoming head of the Office of Space Science and Applications in 1988, Lennard Fisk crafted a different approach to planning future space science missions and assigning priorities among missions for "new start" status. He took a strategic approach to planning for space science, with a five-year time horizon for mission approval. This was the first of a series of strategic plans issued by the Office of Space Science and Applications while Fisk was its leader from 1988-1992.

[cover page]

OFFICE OF SPACE SCIENCE AND APPLICATIONS

STRATEGIC PLAN 1988

A STRATEGY FOR LEADERSHIP IN SPACE
THROUGH EXCELLENCE
IN SPACE SCIENCE AND APPLICATIONS

[2] **PREFACE**

The program of the Office of Space Science and Applications (OSSA) today may be characterized as a program in transition—transition from the exhilarating pace of the 1960s, through the era of fewer, but more sophisticated, missions of the 1970s, to the current trend toward large, complex, long-duration missions that by their very nature require a more deliberate pace. Planning for major missions has evolved from discipline-specific research to a multidisciplinary approach to answering major scientific questions. Our progress over the last three decades has brought us to the point where the number and breadth of science and applications disciplines depending on OSSA resources have grown substantially, and technological advancement has generated great new opportunities that carry with them increasingly complicated methodologies. Space science and applications planning is evolving toward a new approach to the future—consolidated strategies to carry out flight research programs.

The Presidential Directive on National Space Policy, which was approved on January 5, 1988, reaffirms the long-standing call for U.S. leadership in space. With the number of spacefaring nations increasing, and with those nations making significant inroads into all areas of space science and applications, U.S. leadership is being challenged. Leadership requires that a nation have a clear vision of a desired future, articulated by specific goals and plans, and that it visibly demonstrate its accomplishments by the achievement of those goals and plans.

In the present environment, strategic planning of the OSSA program has become increasingly important. Integrating priorities across research disciplines while maintaining a balanced program of major, moderate, and small missions and supporting activities is imperative. In addition, plans for using the significant new opportunities offered by the Space Station Program must be developed. To establish a context for decision-making, to provide a common focus for all the elements of OSSA's program, and to maintain the program's viability, vitality, and flexibility, OSSA has initiated a strategic planning process.

[3] OSSA has formulated a strategy that makes an appropriate contribution to achieving overall NASA goals, and that also directs the energies of OSSA and the institutions with which OSSA collaborates toward the realization of the specific goals and objectives of the disciplines within OSSA. The strategy also must be responsive to the guidance of the NASA Administrator, Congress, and OSSA's advisory groups, and to realistic projections of technology readiness, budget allocations, launch windows, availability of appropriate launch vehicles, and other resources.

The strategy is constructed around five actions: (1) establish a set of themes; (2) establish a set of decision rules; (3) establish a set of priorities for missions and programs within each theme; (4) demonstrate that the strategy can yield a viable program; and (5) check the strategy for consistency with resource constraints. The outcome of this process is a clear, coherent strategy that meets both NASA's and OSSA's goals, that assures realism in long-range planning and advanced technology development, and that provides sufficient resiliency to respond and adapt to both known and unexpected internal and external realities.

[signature]
L. A. Fisk
Associate Administrator for
Space Science and Applications

April 6, 1988

****

[12] **THE OSSA STRATEGY**

To chart a course for an enduring program to make its vision a reality, OSSA has formulated a strategy that is the culmination of extensive interaction and collaboration with the scientific and applications communities, careful consideration of resource guidelines, and iterative reviews of pertinent issues and challenges.

The OSSA strategic approach is constructed around five actions:

1. Establish a set of programmatic themes.
2. Establish a set of decision rules.
3. Establish a set of priorities for missions and programs within each theme.
4. Demonstrate that the strategy can yield a viable program.
5. Check the strategy for technology readiness and for consistency with resource constraints, such as budget, manpower, and launch vehicle availability.

Each of these actions is described in more detail below.

Taken together, these five actions define a programmatic process by which OSSA will plan its activities and allocate its resources. The programmatic themes provide direction and balance to our program, the decision rules guide us in choosing efforts among and within programmatic themes, and the list of priorities determines the order in which we will pursue the missions and programs within each theme. By exercising these actions, various plans for an integrated OSSA program result and these plans can be checked to determine whether they yield a viable program and are consistent with our resource constraints.

An important point to note is that exercising the above actions does not, nor is it intended to, result in a single plan. Rather, these actions define a realistic and flexible process that will provide the basis for making near-term decisions on the allocation of resources for the planning of future efforts. The least certain constraint on our planning is the budget level that will be available to [13] OSSA. The process defined here allows us to adjust to varying budget levels, both those levels that provide opportunities for an expanding science and applications program and those that constrain growth.

In developing this strategy, we have assumed that the overall NASA budget will continue to grow to accommodate overall Agency plans for the Shuttle and expendable launch vehicles, the Space Station, and the basic research and technology program, and that OSSA will receive a proportion of the overall budget that is consistent with its historical allocation. Further, we assume the implementation of current plans for a mixed fleet of launch vehicles, with the launch rates presently projected for the Space Shuttle and for expendable launch vehicles. (In general, expendable launch vehicles will be used for payloads that do not require crew intervention or other capabilities unique to the Space Shuttle.) The level of availability of the Agency work force is assumed to be consistent with Agency plans.

Overarching Agency initiatives, such as geostationary platforms as part of the Mission to Planet Earth, extended-duration Space Station crew certification, and/or Mars Rover/Sample Return as a precursor to human exploration, are not considered in the base strategy. If such initiatives are approved, the appropriate resources must be added over and above the baseline.

Finally, in developing the strategy, no explicit assumptions are made about dependence on international cooperation. Instead, we intend to define our strategy and then to go forward and seek opportunities for international cooperation to fit our plans.

Within these guidelines and assumptions, five basic themes drive the development of OSSA's strategy.

Programmatic Themes

The Ongoing Program

First and foremost, for missions in the ongoing program, the scheduling, resource allocations, and manifested slots on the Space Shuttle or an expendable launch vehicle must be protected and assured. The same high level of priority applies to ongoing research programs and mission, operations and data analysis activities.

Leadership Through Major and Moderate Missions

OSSA plans to move boldly forward to make fundamental and visible advances in key areas of space science, to ensure that our world leadership is preserved in the future. Because they provide the largest quantum leaps in the advancement of scientific knowledge and technological ability, our pursuit of leadership is most conspicuous through major and moderate missions.

Increased Opportunity with Small Missions

Small missions are vital to the program because they can be accomplished relatively inexpensively, allowing the consideration of more innovative ideas, and they can be conducted on a short time scale, offering quick turnaround and continuing opportunity. The small missions are particularly important for the training of the next generation of scientists and engineers, since the missions are of a size that universities can develop, and the development and flight of small missions can occur in the same period of time as that required to earn a graduate degree.

[14] The Transition to Space Station

Beginning with Spacelab and other in-space facilities, it is time to move aggressively, but sensibly, to develop the principal areas of space science and applications that will take advantage of unique Space Station opportunities, such as pressurized laboratories for microgravity science and life sciences research, the multidisciplinary use of attached payloads, and polar platforms for Earth science research.

The Research Base

The research and analysis program provides base support for a vigorous and productive research community and it presents a special opportunity for students to develop the skills that will enable them to conduct the programs of the future. Parts of the program need early enhancement, especially in the replacement of aging laboratory equipment, in the increase of theory and data analysis funding in certain disciplines, and in the ground-based and suborbital programs.

Decision Rules

The first step in the process of determining mission priorities and sequence is the establishment of a realistic budget level. Then, the five themes described earlier provide a template on which the OSSA program is built for 1989 and succeeding years. Ideally, at least one new initiative for each theme, excepting the ongoing program, would be included each year, and we would systematically pursue each item under each theme, in sequence by priority. However, in the event that the budget or other aspects of the external environment do not accommodate simultaneous enhancements in all four areas, certain rules have been formulated to determine the mix of program elements.

Complete the Ongoing Program.

The completion of the ongoing program always has the highest priority; no resources allocated to those programs already under way will be sacrificed or postponed in order to pursue new starts.

Initiate a Major or Moderate Mission Each Year.

Major missions preserve and enhance U.S. leadership in key areas of space science and applications and we will pursue major missions whenever available resources allow us to do so. If an assessment of foreseeable expenditures for candidate missions, over both the near term and the lifetime of the program, indicates that our resources do not permit a major mission, we will pursue a moderate mission.

Initiate Small Missions in Addition to Major or Moderate Missions.

In all cases, we endeavor to start a small mission or a small mission program every year, in conjunction with either a major or a moderate mission.

Move Aggressively, but Sensibly, to Build Science Instruments for the Space Station.

Space Station initiatives are determined by scientific discipline pace and balance, relevance to Space Station, and technological maturity. We will move forward systematically to provide a complete set of fully developed facilities and instrumentation for the Space Station.

Research Base Augmentations Will Be Sought Whenever They Are Warranted.
They are determined by the impact of the rest of the program on discipline stability, progress, and future needs.

[15]
The Plan for 1989

The five programmatic themes and the rules for decision-making were followed in the construction of our plan for 1989, which is detailed below.

ONGOING PROGRAM

First, the 1989 plan includes sufficient resources to keep each of the ongoing flight programs on schedule for launch in their manifested slots on the Space Shuttle or an expendable launch vehicle. The long hiatus in space science and applications launches is drawing to a close. In what promises to be an exciting year, 1989 is expected to see the launch of the Cosmic Background Explorer, the Magellan mission to Venus, the Hubble Space Telescope, the Astro Spacelab mission and the Galileo mission to Jupiter. In August of that same year, Voyager 2 will encounter Neptune, a major milestone in outer solar system exploration, and one that will surely return a wealth of scientific information.

Development will continue on an impressive array of major, moderate, and small mission to be launched from 1990 through 1993, including:

Gamma Ray Observatory
Roentgen Satellite
Combined Release and Radiation Effects Satellite
Ulysses
Laser Geodynamics Satellite
Extreme Ultraviolet Explorer
Ocean Topography Experiment (TOPEX)
Upper Atmosphere Research Satellite
Advanced Communications Technology Satellite (subject to action on the FY 1989 budget)
Wind
Geotail
Mars Observer
Polar
Mobile Satellite
Spacelabs, including a series of Space Life Sciences missions; International Microgravity Laboratory and U.S. Microgravity Laboratory missions; several Atmospheric Laboratories for Applications and Science; two Astronomy Laboratory missions; and two flights of the Space Radar Laboratory.

In addition to the flight projects, resources that support ongoing program elements in research and analysis, suborbital observations, theory and modeling, laboratory and supporting observations, and mission operations and data analysis for ongoing operating missions will continue uninterrupted.

LEADERSHIP: MAJOR AND MODERATE MISSIONS

Our plan makes a bold statement that the United States will pursue world leadership in space science in 1989 through an initiative in astrophysics. Our nation is poised for an accomplishment unique in the history of humankind—to observe the physical universe with unprecedented completeness and resolution. We have the demonstrated capability to construct high-technology orbiting telescopes that can observe the universe in all forms of electromagnetic radiation, and we have the unique capability with the Space Shuttle, and eventually the Space Station, to maintain these telescopes in orbit.

The key to realizing this ambition is the Advanced X-Ray Astrophysics Facility—a telescope facility designed to observe the universe in the X-ray region of the electromagnetic spectrum. This [16] facility will be 100 times more sensitive and have 1,000 times more capability for spectroscopy than any previous or planned X-ray mission.

The Advanced X-Ray Astrophysics Facility is to fly in concert with the Hubble Space Telescope, which will observe the universe in visible and ultraviolet radiation; with the Gamma Ray Observatory, which will observe in gamma rays; and with the Space Infrared Telescope Facility, which will observe in the infrared region. These Great Observatories, operating together, will provide a comprehensive physical picture of the universe's most enigmatic objects, and will observe the full range of phenomena in the universe, from the most tranquil to the most violent.

The Advanced X-Ray Astrophysics Facility will also provide a scientific opportunity that is unlikely to be repeated for many generations. The closest supernova to occur near Earth since the invention of the telescope 400 years ago was seen last year, and it can be studied by the facility, provided that launch occurs by 1995, before the X rays fade. Supernovae are responsible for the origin of all the heavier elements in the universe, including those essential for life. In this era in which United States leadership in space is being challenged, our plans assert that in the premiere scientific discipline of astrophysics, we will be second to none.

SMALL MISSIONS

To maintain program continuity and vigor through frequent flight opportunity for small missions, our plan proposed an augmentation to the Explorer program that builds on the augmentation that Congress provided last year. A clear and present need exists to stimulate the research community, particularly at universities, with exciting new opportunities, which will attract new scientists and engineers to space science. Historically, the Explorer program has been one of the means by which we have provided such opportunities, through frequent launches of focused science missions.

Accordingly, we are planning to augment the Explorer program to allow for more small missions, which can be launched on Scout-class expendable launch vehicles. These missions are sufficiently small that they can be built and launched within three years, yet they are sufficiently capable to accomplish first-class scientific objectives in astronomy, space physics, and upper atmospheric physics. We anticipate shortly releasing an Announcement of Opportunity to select these missions, and we expect that this opportunity will provide yet another indication that the space program is moving forward again.

SPACE STATION UTILIZATION

The fourth theme of our 1989 plan concerns the Space Station. It is time to begin to aggressively develop the principal areas of space science and applications that will take advantage of the unique opportunities that the Space Station will provide us. Four such areas—microgravity science, life sciences, multidisciplinary attached payloads, and Earth science from the Polar Platform—are being developed, and we have a separate strategy for each area.

The Space Station will provide us with a laboratory in which, with continual human interaction, we can conduct a broad range of microgravity experiments in materials science, fluid physics, [17] and biotechnology. These experiments will advance our knowledge of basic physics, chemistry, and biology, and will have direct applications to improving our understanding of processes that occur on the Earth and in space.

To use the Space Station as a laboratory, we will develop six facilities: (1) a Space Station Furnace Facility, (2) a Modular Combustion Facility, (3) a Fluid Physics/Dynamics Facility, (4) a Modular Containerless Processing Facility, (5) an Advanced Protein Crystal Growth Facility, and (6) a Biotechnology Facility. We will fly elements of these facilities in advance of the Space Station, both to test and perfect the design of the facilities, and to provide new research results in the important discipline of microgravity science and applications.

The plan for microgravity science provides for the full development of all six facilities required for the Space Station, and allows for their test flight on a Spacelab mission and/or a commercially developed space facility. The program leads to the full instrumentation of the Space Station for microgravity science by the time of man-tended capability.

Life sciences research is also an important activity that we will conduct on the Space Station. We have ongoing studies on precisely how we will accommodate the life sciences research on the Space Station, and we are developing one of the facilities that we are certain we will require on the Space Station—the 1.8-meter centrifuge, which is essential to any biological research in space. As with the microgravity facilities, the centrifuge will be flown and tested on a Spacelab and/or commercially developed space facility, and will then be transitioned to the Space Station.

Attached payload opportunities, which can be used by a broad range of science and applications disciplines, are also provided by the Space Station. Our strategy in this area is to begin with attached payloads that are not overly demanding on the environment and pointing capabilities of the Space Station; then, as we learn to use the Station and its full capabilities, we will evolve into using more sophisticated attached payloads. We anticipate shortly releasing an Announcement of Opportunity soliciting proposals for attached payloads to be carried on the Space Station during its initial one to three years of operation, and proposals for the definition of more ambitious investigations for possible attached payloads to be flown at a later time.

The Polar Platform of the Space Station provides us with the opportunity to make detailed observations of the Earth, of how it is evolving on a global scale, and of how we humans are influencing that evolution. We have recently released an Announcement of Opportunity, jointly with the Europeans and the Japanese, to select investigations for the Earth Observing System, which is to fly on the Polar Platform, and to select potential manned base attached payloads in the Earth sciences discipline. Our plan also includes the resources to conduct advanced technology studies to define instruments and information systems for the Earth Observing System.

The four parts of our Space Station initiative—the development of microgravity facilities, the development of the centrifuge and the planning for other life sciences facilities, the development of attached payloads, and the selection and study of Earth observing instrumentation for the Polar Platform—form a comprehensive plan to begin to make full use of the unique opportunities that the Space Station will provide.

[18] **RESEARCH BASE**

The fifth and final theme of our plan for 1989 concerns the Research and Analysis program, which is the vital underpinning to our program. We are proposing to augment this program to continue our rocket and balloon campaign to understand the recent supernova, in advance of when it can be observed and studied by the Great Observatories; to complete the purchase of a new high-flying Earth remote sensing aircraft; and to provide additional resources to take maximum advantage of the upcoming encounter of the Voyager spacecraft with Neptune. The plan provides for continuing advanced technology development on the Mariner Mark II missions, Comet Rendezvous Asteroid Flyby and Cassini. We are also planning to begin development of a network of signal processing equipment to be attached to radio astronomy facilities to begin, in 1992, to search for other intelligent life in our galaxy. Detection of life elsewhere in the universe may be the most profound event to occur in human history.

With a clear eye toward the next five years, the plan for 1989 allows us to make significant progress toward achieving our ultimate goals. The U.S. space science and applications program has historically produced an outstanding scientific return on America's investment, and we expect this to continue and grow through the implementation of our five-year strategy, described next.

Five-Year Strategy

Beginning with the overarching goals of NASA as articulated by National Space Policy, and working through OSSA's goals and objectives, the themes and decision rules cited earlier form the basis for our strategy for the years 1990 through 1994.

ONGOING PROGRAM

Through each succeeding year, the flight projects and research programs started the previous year combine with those that are still under way to form the ongoing program. In all cases, the highest priority of OSSA's strategy is to carry out the ongoing program.

LEADERSHIP: MAJOR AND MODERATE MISSIONS

All the major flight projects in the 1989 ongoing program will be launched by 1993; a new major flight project requires four to six years to develop. Thus, to pursue leadership in key areas, the necessary next step is to select the successors to the ongoing program. Several criteria drive decisions about the selection and sequence of major and moderate missions. First, we want to pursue missions of the highest scientific priority, as identified by the National Academy of Sciences and the NASA Advisory Council. Second, the several candidate missions that fall within this category are assessed for the degree of technological readiness to pursue them; this determines the degree of understanding of cost and schedule risk for these candidates. Third, the order in which major and moderate missions are pursued is governed by the need to pace the implementation of discipline-specific plans at a rate of approximately one major or moderate new start every five years in each discipline; this pace keeps all the discipline programs moving forward and maintaining vigor. Finally, the missions are viewed in the context of the NASA Space and Earth Science Advisory Committee's recommendations for mission selection that are elucidated in the report *The Crisis in Space* [19] *and Earth Science.* The report describes guidelines for the following criteria: (1) scientific merit, (2) programmatic considerations, and (3) societal and other implications.

Another guideline is that there should be one major or moderate new start per year. While we recognize the fact that circumstances may present occasions where more than one new start is possible, and others where no new start is possible, an average pace of one per year is necessary to meet the goals of leadership in key areas and to assure vigor and continuity. On the other hand, given a realistic estimate of resource constraints, more than one new start per year cannot ordinarily be expected, because available resources for small missions and for research and analysis must be preserved. Accordingly, the sequence of major and moderate missions in our five-year strategy has been determined as detailed below.

According to the decision rules, whenever resources permit, we will pursue major missions in order to preserve and enhance U.S. leadership. These major missions, in order of priority, are described below.

JOINT INITIATION OF COMET RENDEZVOUS ASTEROID FLYBY AND CASSINI MISSIONS

The Comet Rendezvous Asteroid Flyby (CRAF) mission will include a close flyby of a main belt asteroid followed by an extended multiyear rendezvous with a short period comet, permitting detailed study of the comet's nucleus, dust, and atmosphere at close range under both quiescent and active conditions. The Cassini mission, a potential cooperative project with the European Space Agency, will conduct a comprehensive scientific investigation of the planet Saturn, its rings and moons, the surface and atmosphere of its principal moon, Titan, and the nature of fields and particles in Saturn's magnetosphere.

These two missions have long been established as endeavors of high scientific priority, because they combine to address the fundamental OSSA goal of determining the origin and evolution of the solar system and of life. The large planets preserve unprocessed elemental and isotopic abundances; scientists believe that in Titan's atmosphere, chemical and physical reactions similar to those that led to the origin of life on Earth may now be taking place. The primitive bodies, comets and asteroids, preserve relatively unprocessed molecular and organic material from the interstellar medium and the solar nebula. Studying the outer solar system and the primitive bodies provides information about the early history of the solar system, and about the origin, evolution, and distribution of prebiotic organic materials.

Because of this shared scientific goal and the complementary nature of the two missions' objectives, and also because both missions use the same Mariner Mark II spacecraft design, CRAF and Cassini are combined for a joint program. We place this initiative as the highest priority for major missions because development must be started now to assure a robust solar system exploration program at the end of this century. This approach assures the continued strength of the program, and continues our tradition of leadership in exploring the outer solar system.

[20] THE EARTH OBSERVING SYSTEM

The Earth Observing System will place a suite of instruments in low-Earth orbit to make comprehensive observations of Earth's atmosphere, oceans, land surfaces, and biota. An integral part of the program is the collection, processing, analysis, interpretation, and archiving of the resulting data. The Earth Observing System is the centerpiece of NASA's implementation of the Earth System Sciences Committee strategy for integrated study of the Earth and of global change. Long-term, consistent measurements are required to understand global changes, and so, for at least 15 years, the mission will study the global-scale processes that shape and influence Earth as a system.

This study of global change on Earth is fundamentally important to humanity's future on this planet. We must apply the capabilities that we have developed in space to understand our own world, and to safeguard that world for the coming generations.

The Earth Observing System is recommended for an early start in order to address the accelerating need for information about the rapid evolution of Earth's environment, and to prepare to make timely use of platforms provided by the Space Station. Definition studies are expected to be completed in 1990 and a development start should follow as soon as possible.

THE SPACE INFRARED TELESCOPE FACILITY

The fourth Great Observatory, the Space Infrared Telescope Facility is a long-lived, meter-class, cryogenically cooled, infrared observatory to study the very cold regions of space. It will be launched by the Space Shuttle and serviced by the Shuttle and the Space Station. Regions and objects the facility will study are: location where the cosmic gas and dust condense into stars; cool objects in the solar system—planetary systems, asteroids, and comets; and infrared-emitting extragalactic objects. It will be 1,000 times more sensitive than the Infrared Astronomical Satellite. One of its major applications will be to obtain detailed infrared spectrometry of the faint infrared sources that the Infrared Astronomical Satellite discovered but could not observe in detail. The Astronomy Survey Committee of the National Academy of Sciences has treated the Space Infrared Telescope Facility as a high-priority mission.

The four Great Observatories—the Hubble Space Telescope, the Gamma Ray Observatory, the Advanced X-Ray Astrophysics Facility, and the Space Infrared Telescope

Facility—will provide world-class facilities for observing in all the major wavelength bands. Together, the Great Observatories will ensure U.S. leadership in astronomy and astrophysics for decades to come.

THE SOLAR PROBE

The Solar Probe will be humanity's first direct exploratory venture to the vicinity of the Sun. It will study the unexplored region between 4 and 60 radii from the Sun, where the solar wind begins to flow at supersonic speeds. The Solar Probe will measure the electromagnetic fields and will study the particle populations in the region close to the Sun. It will make fundamental measurements relating to stellar internal structure, gravitation, and relativity, and it will observe the structure of the solar atmosphere from the photosphere to the corona with exceptionally high spatial resolution.

[21]Because the Solar Probe offers a unique opportunity for leadership in exploration of the heliosphere, and because it has been cited by the scientific research community as a high-priority objective, it has been established as the fourth major mission in our five-year plan.

In the event that resources do not permit the implementation of a major mission, the moderate missions described below will be pursued in order of priority.

HIGH-RESOLUTION SOLAR OBSERVATORY

The High-Resolution Solar Observatory is a scientific platform for performing investigations of the Sun's fine-scale magnetic structures. Its scientific objective is to study in visible light, and at the limits of spatial and temporal resolutions at which they actually occur, the fundamental magnetohydrodynamic processes of the Sun's surface atmosphere. This program has repeatedly been endorsed as the highest priority of the U.S. space solar physics discipline and, as such, has received the highest recommendations by the relevant committees of the National Academy of Sciences.

THE LUNAR OBSERVER

The second mission in the Planetary Observer program, the Lunar Observer will be constructed from Mars Observer spares to conduct a one-year polar mapping mission to measure the Moon's global surface mineral and elemental composition, to assess global resources (including frozen volatiles at the poles), to measure surface topography, and to measure magnetic and gravitational fields. In addition to the valuable scientific information that this mission will provide, the data from the Lunar Observer will contribute to the Agency goal of preparing the way for a possible human outpost on the Moon. In order to efficiently and cost-effectively make the transition to the Lunar Observer using spares from the Mars Observer, the Lunar Observer must begin in 1992. Therefore, at that time, the Lunar Observer will become the highest priority moderate mission, even if the High-Resolution Solar Observatory has not been started.

GRAVITY PROBE-B

Gravity Probe-B is designed to be a cornerstone test of general relativity. Einstein's universally accepted theory of special relativity ties together the structure of time and space. His theory of general relativity, which is far less thoroughly tested, ties together space, time, and gravity. This theory is on a much less secure experimental footing than the special theory, and alternative hypotheses exist. Gravity Probe-B will measure both the distortion of the "fabric of space time," imposed by the Earth's presence, and the subtle dragging of this fabric, predicted to result from the Earth's rotation. The influence of these effects will be seen in subtle precessional changes affecting the behavior of a set of four ultra-precision gyroscopes operating in a drag-free, superconducting environment. The required technology for this demanding undertaking has been under development since 1965. The key elements will be tested using a functioning prototype to be flown on a Space Shuttle flight prior to the science mission.

[22] **SMALL MISSIONS**

The missions in this category are essential to sustaining the vigor of our program. They can be launched more frequently than major or moderate missions, perhaps as often as every two years per discipline. The small missions can provide opportunities comparable to classical Explorers.

Currently identified small missions of the five-year strategy include:

EARTH PROBES

To complement the observations carried out by the Earth Observing System, we have defined a Series of Explorer-class missions in Earth science, called Earth Probes. We plan a continuing series of these missions to be launched at a regular interval. For example, the Tropical Rainfall [Measuring] Mission, the Magnetic Field Explorer, and the Geopotential Research Mission are concepts for small missions that may be selected as Earth Probes.

LIFESAT

Lifesat is a small, recoverable, and reusable orbiting spacecraft that can be used as an inexpensive platform for conducting life sciences (and possible other) experiments. The spacecraft can be launched on a variety of expendable launch vehicles and can provide up to 40 days of microgravity environment. This program provides a particularly attractive opportunity for multinational cooperation.

SPACE STATION UTILIZATION

For this segment of our five-year strategy, we wish to initiate the space biology counterpart to the 1989 microgravity initiative. The goal of space biology research is to use the unique characteristics of the space environment, especially microgravity, to increase our understanding of life and its processes, and to understand how gravity affects and has shaped life on Earth. The objective of the research, which encompasses both plants and animals, is to understand the mechanisms by which organisms perceive gravity and transmit the information to a responsive site, to determine the role of gravity in reproduction, development, maturation, and function, and to understand the mechanisms by which environments in conjunction with microgravity affect living systems.

The development of second-generation attached payloads for a variety of disciplines will also need to begin during this five-year period.

A key factor in OSSA's preparation for the Space Station will be the continued use of Spacelab, the Space Shuttle mid-deck lockers, and other appropriate carriers to develop, test, and verify new and improved instrumentation for subsequent use on the Space Station.

RESEARCH BASE

The highest priority in this area is to augment the research and analysis base that is essential to OSSA's program. In particular, laboratory equipment and facilities need to be upgraded, and enhancements in funding need to be provided for new instrument development, more capable information systems and computational facilities, data analysis, and theoretical studies. Further, the suborbital program needs enhancement in balloons and rockets and in areas such as those described below.

[23] STRATOSPHERIC OBSERVATORY FOR INFRARED ASTRONOMY (SOFIA)

SOFIA is a 3-meter-class telescope planned to be flown in a modified Boeing 747 airplane to observe the Infrared Astronomical Satellite sources with good angular resolution at infrared wavelengths inaccessible from the ground. A facility with tremendous potential for science, SOFIA can offer an improvement in resolution over the 0.9-meter Kuiper Airborne Observatory, it can be readily available to the scientific community (with a short turnaround time), and it can be flown on a reliable, reusable vehicle. Since the Space Infrared Telescope Facility will not fly until the late 1990s, SOFIA will allow us to follow up on the exciting discoveries of the Infrared Astronomical Satellite in the interim, and SOFIA will complement the Space Infrared Telescope Facility when it becomes operational

NEW AIRCRAFT FOR EARTH REMOTE SENSING

Observations of Earth from instrumented aircraft complement those taken from space and on the ground, and provide critical flight demonstration tests of advanced remote sensing technologies. Currently, four aircraft—one DC-8, two ER-2s, and a C-130—make up the means of conducting this program. Since aircraft observations provide a method for uncomplicated launch and fast turnaround, we wish to update the fleet with more capable and more sophisticated craft.

Summary

The chart below graphically represents the strategy that will guide OSSA's plans from 1989 through 1994. The strategic approach described earlier, including consistent programmatic themes and decision rules, will continue to provide a methodology for OSSA strategic planning in the future.

	Ongoing Program	Major & Moderate Missions	Small Missions	Space Station Utilization	Research Base Enhancements
1989	Research and Analysis Mission Operations And Data Analysis Flight Projects Spacelabs and Other Carriers	AXAF	Scout-class Explorers	Microgravity Facilities 1.8m Centrifuge Attached Payloads Eos Payload Definition	SETI Microwave Observing Project CRAF/Cassini ATD SN 1987a Suborbital Observations ER-2 Purchase
1990 T H R O U G H **1994**		CRAF/Cassini Eos SIRTF Solar Probe HRSO* LO* GP-B	Earth Probes Lifesat	Space Biology Facilities Second-generation Attached Payloads	Laboratory Facilities, Mission Operations And Data Analysis Theory Suborbital

* Moderate Mission

Document I-31

Document title: Solar System Exploration Division, Office of Space Science, NASA Headquarters, *Discovery Program Handbook*, November 1992.

Source: Office of Space Science, NASA, Washington, D.C.

Document I-32

Document title: Tony Spear, *NASA FBC (Faster, Better, Cheaper) Task Final Report*, March 2000.

Source: Jet Propulsion Laboratory, Pasadena, California.

A new Administrator, Daniel Goldin, took over NASA in April 1992. He had a mandate from the White House to "reinvent" the space agency, which critics believed had lost much of its technological and managerial leadership. One of Goldin's early reforms was to switch from a strategy for science missions that emphasized complex, expensive, long-in-development missions to one that emphasized more frequent, less expensive missions. This change was first instituted in NASA's solar system exploration program through a program named Discovery. The first document is a handbook that set out for potential program participants NASA's new approach to solar system missions, which went by the mantra "faster, better, cheaper."

One of the successes of the faster, better, cheaper approach was the 1997 Mars Pathfinder mission, which used an innovative technique to land on the Martian surface and deployed a small rover to explore the surface in the near vicinity of the landed spacecraft. In July 1999, Goldin asked Tony Spear, who had been project manager for Mars Pathfinder and subsequently had retired from the Jet Propulsion Laboratory, to undertake a personal evaluation of NASA's implementation of the faster, better, cheaper approach. This review took on added urgency with the failure of the Mars Climate Orbiter, Mars Polar Lander and Deep Space 2 missions in late 1999.

Document I-31

[no pagination]

DISCOVERY PROGRAM HANDBOOK

Solar System Exploration Division
NASA Headquarters

November 1992

INTRODUCTION

This handbook describes the Discovery Program, which is a series of low-cost planetary exploration missions with very specific scientific, technical, and programmatic guidelines. The goals of the Discovery Program are to increase flight rates and launch schedule certainty, complement large missions to keep a steady rate of incoming planetary data, broaden university and industry participation in solar system exploration missions, and increase public awareness of solar system exploration missions.

The purpose of this handbook is to respond to the large number of questions that have been posed regarding the Discovery Program. Information is presented in a question-and-answer format so that the reader can immediately locate the material in which he or she is most interested. Since the Discovery Program is still evolving, this handbook is to be considered a working document that will periodically be revised.

Why was the Discovery Program developed?

The Discovery Program is an outgrowth of an effort begun about 2 years ago to develop a program of small planetary missions that would complement larger missions and keep the scientific community involved with a steady stream of new planetary data. While this planning was under way, the Senate Appropriations Committee in April 1992 directed NASA to prepare "a plan to stimulate and develop small planetary or other space science projects, emphasizing those which could be accomplished by the academic or research communities." In addition, more constrained budgets have led to a call for "faster, better, cheaper" missions.

The Discovery Program is responsive to this environment. The time frame of small Discovery missions is consistent with academic degree programs, which makes these missions an excellent training ground for graduate students and post-doctoral researchers. Because small missions can be conducted relatively quickly and inexpensively, they provide frequent opportunity for access to space. In addition, small missions help sustain a vital scientific community by increasing the available opportunities for direct investigator involvement from just a few projects in a career to many.

What are the Discovery Program guidelines?

Discovery Program guidelines address each of five typical project attributes: scope, cost, risk, schedule, and management. Figure 1 [all figures omitted] compares the more traditional approach with the Discovery approach to these project attributes. More detail is provided below.

Scientific Content Guidelines

1. Missions must have focused, well-defined scientific objectives.

2. Targets of exploration are not limited; however, inner solar system objects are the most likely to satisfy programmatic guidelines.

3. The payload comprises a small number of instruments, usually three or fewer.

4. Full missions or collaborations (e.g., supplying an instrument) are within program scope.

Technical Guidelines

1. Instruments proposed for flight on Discovery missions must be sufficiently mature to minimize development risk. Instruments that are "off-the-shelf" or in advanced development are preferred.

2. Proposed spacecraft must be simple, cost-effective, proven designs.

3. Missions must use launch vehicles of capability no greater than the Delta II, although piggyback options on larger vehicles will also be considered. The use of launchers smaller than the Delta II is encouraged.

Programmatic Guidelines

1. Development time must be 36 months or less. Project costs must be rigidly controlled. Failure to meet actual or projected cost and schedule limits will result in the cancellation of the mission.

2. Total development costs for a particular mission can be no more than $150M. Lower-cost missions are preferred.

3. Discovery is a level-of-effort program to be funded at $85M/year in FY 1992 dollars. This amount includes $75M/year for current mission definition and development and $10M/year for advanced mission studies and instrument development.

4. Mission teams consisting of representatives from universities, industrial firms with flight hardware experience, and a NASA or other Government Center are encouraged. A single individual (the Principal Investigator) must be in charge of the team.

5. International collaborations, e.g., NASA provision of an instrument for a foreign mission, are within the scope of the Discovery Program. International members may be part of the mission team; however, no funds will be transferred to non-U.S. agencies.

What are the phases of Discovery missions?

Figure 2 summarizes the life cycle of a Discovery mission, which consists of four distinct phases:

1. **Advanced Studies** conducted through the Solar System Exploration Division's Advanced Studies Branch, which form a continuing base for potential Discovery missions.

2. **Project Definition**, which includes preliminary mission design, assessment of concept readiness, cost and schedule evaluation, and preparation for non-advocate review.

3. **Project Development**, which includes final design, fabrication, assembly, and test through launch plus 30 days.

4. **Mission Operations**, which includes all flight activities as well as supporting ground operations from launch plus 30 days to the end of the mission.

When is the Discovery Program expected to begin?

The Discovery Program has been proposed as a line item beginning with NASA's 1994 budget request.

What is included in the $150M cost cap?

Project development costs; that is, final design, fabrication, assembly, and test through launch plus 30 days. This amount also covers the development cost of ground systems.

What is not included in the $150M?

Advanced studies, project definition, launch vehicles, and mission operations/data analysis.

What are the first Discovery missions?

Two concepts now under study will become the first two missions of the Discovery Program. The first is a small lander called MESUR (Mars Environmental Survey) Pathfinder, which will demonstrate the technologies required for the MESUR multiple-lander program. The second Discovery mission will be a Near-Earth Asteroid Rendezvous (NEAR). Each of these two missions is briefly described below.

MESUR Pathfinder. The next scientific mission to Mars is a global network of about 16 small surface landers (MESUR Network). This mission will have a very different, and quite possibly more stressful, landing procedure than the one used by Viking. Therefore, it is important to demonstrate the cruise, entry, descent, and landing functions before the first flight of MESUR Network landers. This demonstration mission, designated Pathfinder, will send a single aerocraft to Mars in 1996. The objective of Pathfinder is to demonstrate the flight and landing systems required for the MESUR Network mission, while acquiring limited, but important, scientific and exploration data on the Martian surface.

The Jet Propulsion Laboratory is managing this project under the Discovery guidelines. Among the actions being taken to ensure that program guidelines are met are the following:

1. **Streamline management.**

    Establish agreements with customer early and keep to them.

    Prioritize objectives, make sure everyone understands the consequences, and stick to them (e.g., Pathfinder is first and foremost an engineering mission; science and technology elements will be included only to the degree allowed by technical and cost guidelines).

Use a fast-track Announcement of Opportunity process—6 months from release of the AO to instrument selection.

Use a small, collocated project team.

2. **Optimize design within schedule, cost, and performance constraints.**

   Clearly understand mission objectives.

   Design to cost.

   Rely on available hardware and software.

   Judiciously use testing and verification procedures.

3. **Plan for operations early.**

   Design flight system and mission operations and ground data systems concurrently to reduce system costs.

NEAR. The second Discovery project under study is a Near-Earth Asteroid Rendezvous (NEAR) mission. The NEAR project development will begin in Fiscal Year 1996. Development time will be 27 months with a launch in January 1998. The spacecraft will rendezvous with the asteroid 4660 Nereus and remain in orbit for at least 8 months. A backup launch opportunity to the asteroid 3361 Orpheus occurring 2 months after the nominal launch date provides the mission with additional schedule resilience.

Carrying between three and five instruments, NEAR will assess the asteroid's mass, size, density, and spin rate, map its surface topography and composition, determine its internal properties, and study its interaction with the interplanetary environment.

The NEAR mission is being managed for NASA by the Applied Physics Laboratory (APL) of the Johns Hopkins University. APL's management plan for NEAR assigns the prime responsibility for the spacecraft design, construction, integration, and testing to APL. The instruments will be procured through a NASA Announcement of Opportunity. The accelerated AO process to be used for MESUR Pathfinder will be applied to the implementation of NEAR, as will the other management techniques discussed in connection with MESUR Pathfinder.

Mission support in the areas of navigation and use of the Deep Space Network will be provided to APL through the Jet Propulsion Laboratory. However, the mission operations, including spacecraft command, will be performed at the Applied Physics Laboratory. This mingling of a university facility (APL) and a NASA Center (JPL) is one example of the way these capabilities can be merged in a Discovery mission.

How will Discovery missions after Pathfinder and NEAR be selected?

NASA plans to select future Discovery missions through Announcements of Opportunity. Details are provided later in this handbook.

DISCOVERY PROGRAM MANAGEMENT

Two tenets of the Discovery Program are streamlined management processes and minimum review and documentation requirements. The Principal Investigator is the single point of responsibility; he or she essentially "delivers" the mission. The management plan that the P.I. and his or her team develop will be an important criterion in selecting Discovery missions for definition and development. Figure 3 summarizes the Discovery Program management approach.

General guidelines exist for Discovery mission management, although some tailoring for specific missions is likely. In general, when concepts are at the advanced mission study or project definition phases, they all will be managed by NASA Headquarters. Missions that proceed to development most often will have a NASA Center involved as an interface for technical expertise and/or contract management. For example, if the P.I. is at a NASA Center, contract management will be performed by the Center Project office. A contract for a mission with a P.I. not at a NASA Center may be managed either by Headquarters or a Center.

What will be the role of NASA Headquarters in managing Discovery missions?

NASA Headquarters will:

1. Manage the overall program; i.e., Headquarters will select and fund specific missions and provide fiscal oversight, reviews, and other program control elements.

2. Manage the Announcement of Opportunity process and selection.

In a typical P.I/Industry/NASA Center mission team, what are the respective roles and responsibilities?

The P.I. is solely responsible for managing the planning, development, and execution of the mission. He or she selects team members, allocates roles and responsibilities, and ensures that activities assigned to team members are proceeding on-time and within budget. The P.I. serves as the interface with NASA Headquarters and is responsible for both meeting and reporting on cost and schedule objectives. He or she is also responsible for developing the scientific objectives and instrument payload for the mission.

Industry would typically be responsible for developing, building, or procuring spacecraft and instruments. During the study phase, the industry team member would be responsible for detailed hardware planning and costing and for ensuring that plans can be implemented within budget. During the development phase, the industry team member would oversee the hardware fabrication, integration, and testing and the integration of the spacecraft to the launch vehicle.

A NASA Center, either chosen by proposers or assigned by Headquarters, will provide contract management and oversight, launch support, and engineering/test support. Other responsibilities may be assigned to the NASA Center at the discretion of the P.I.

What is the rational behind the formation of mission teams?

One of the goals of the Discovery Program is to broaden the participation of industry and universities in the solar system exploration program. Additionally, one of the management tenets is that the P.I. essentially "delivers" the entire mission. Therefore, the P.I. must determine the mission implementers as well as the scientific requirements. The formation of mission teams pulls together all the requisite skills to perform a mission from concept development to end of mission.

Will there be Guest Investigator programs?

NASA Headquarters reserves the right to add Guest Investigators to a Discovery mission team toward the end of the development phase or during flight.

SELECTION PROCESS FOR DISCOVERY MISSIONS AFTER NEAR

The selection process for Discovery missions is in some areas a departure from the traditional method of selecting larger solar system exploration missions. The special character of Discovery mandates new methods of concept selection and streamlined management oversight. Of course, the input of the science community will still be solicited through the advice of NASA's advisory groups, particularly the Solar System Exploration Subcommittee.

Post-NEAR Discovery missions will be selected through Announcements of Opportunity (AO). The first Discovery Program AO is expected to be released in 1994 or 1995. Prior to that time, studies of future Discovery missions will be conducted through the Solar System Exploration Division's Advanced Studies Branch. Mission concepts will be selected for study with the advice of the Solar System Exploration Subcommittee. Unsolicited proposals for future Discovery mission studies may be submitted at any time. Results of workshops and other community interactions may also be used in the process of selecting concepts for study. Concepts will be selected by NASA Headquarters based on suitability factors such as the way in which the mission addresses scientific objectives and programmatic goals, technical readiness, feasibility within Discovery guidelines, and soundness of approach.

By about 1996, it is expected that advanced Discovery mission studies will begin to be funded through the Discovery program line item. Once the Discovery line item has ramped up to $85M/year (FY 1992 dollars), $10M will be devoted to advanced mission studies and advanced instrument development.

The first Discovery Program AO will lead to the selection of at least two to three proposed missions for a 1 to 2 year project definition phase (see Figure 2). The intent of this phase is to resolve any open issues or instrument uncertainties, and develop an approved project plan. At the conclusion of this phase, a non-advocate review (NAR) for each proposed mission will lead to a down-selection of one or more missions that will be continued into the project development phase for flight. It is expected that the NARs will lead to the elimination of some mission proposals selected through the AO.

Subsequent AOs will be released at approximately 2-year intervals. A number of variables will determine the exact schedule, including NASA budgets, status and cost of other missions in the Discovery queue, and the success of the program in general. A similar process of selection for definition phase (through the AO) and subsequent down-selection for development and flight will ensue for each AO.

When will the first opportunity for new missions occur?

For a very small mission or payload, the first opportunity could be as early as FY 1997. More likely, the next project development start after NEAR will be in FY 1998.

What is the expected frequency of Discovery missions?

Frequency will vary according to the availability of resources. The Discovery Program is proposed as a constant level-of-effort budget line item at an annual funding level of $85M. Therefore, the rate at which project concepts can be developed and flown is determined by the cost of each project (within the development ceiling of $150M). Figure 4 illustrates how the rate at which new projects can be started relates to the average cost per project. Note the various program support cost assumptions associated with this illustration. If the average project costs $65M to develop, a new project can be started every year. If all projects require the maximum development cost of $150M, then the interval between project development starts will average 28 months.

Will there be a continuing opportunity for unsolicited proposals?

Yes, for advanced mission studies. Beyond that level of maturity, all missions will be selected through the AO process.

What type of information will be required in the mission plan presented in response to the AO?

1. A Work Breakdown Structure (WBS) that captures all costs, including mission operations and data analysis throughout the mission lifetime. A consistent WBS will be required for all proposals to ensure a common basis for comparison.

2. Information that indicates how resilient the mission is; for example, available margin within objectives, scalability of instruments, and trade-offs between mission focus and robustness.

3. An assessment of mission risk that includes strengths and weaknesses and an evaluation of cost and schedule risk.

By what criteria will the AO proposals be evaluated?

AO proposals will be evaluated through standard NASA procurement procedures. A preliminary list of evaluation criteria (with no priority order implied) includes:

1. Compatibility with the scope of the Discovery Program, especially in terms of cost and schedule.

2. Overall scientific and technical merit.

3. Way in which the mission fits within established scientific strategies.

4. Capabilities and experience of mission team.

5. Technical readiness.

6. Strength of management plan.

7. Resilience and flexibility to address unforeseen issues within the scope and guidelines of the program.

8. Level of risk—technical, cost, and schedule.

How much flexibility in cost and schedule requirements exists for Discovery missions?
There is not flexibility in cost or schedule above the requirement caps cited earlier.
Missions that do not or cannot meet program requirements (e.g., those that overrun cost or schedule) are subject to termination at any time during the project development phase.

Document I-32

[no page number]

<u>NASA FBC TASK FINAL REPORT</u>

Preface

In the early 1990s, NASA adopted the Faster, Better, Cheaper (FBC) approach to project management of its Space and Earth Science Missions. The goal was to shorten development times, reduce cost, and increase the scientific return by flying more missions in less time. One of the first Faster, Better, Cheaper missions was Mars Pathfinder, which touched down on Mars on July 4, 1997, and became a resounding success. The Project Manager for Mars Pathfinder was Tony Spear, who recently retired after a long and successful career at the Jet Propulsion Laboratory. In mid-1999, Administrator Daniel S. Goldin met with Tony and asked him to undertake a personal study of the Agency s implementation of FBC. He was formally tasked in July 1999 to assess best practices through a series of interviews and workshops. These coupled with his personal experiences and expertise have led to the following observations.

Introduction

This report summarizes the results of the NASA FBC TASK conducted from July 1999 through February 2000. NASA Headquarters, ten NASA Centers, industry, and academia participated in interviews and lively discussions during visits and three workshops, one held in Maryland and two in California. Industry and academia participated in one workshop held in California.

Task Objectives:

1. Define FBC and develop "rules of engagement"
2. Identify major challenges for FBC
3. Identify what is most important for the future of FBC
4. Propagate FBC throughout NASA identifying more effective Center-to-Center teaming arrangements
5. Provide recommendations in response to 1 through 4 above
6. Identify good examples of FBC
7. Provide metrics for measuring FBC

Background

In 1992, NASA Administrator Dan Goldin challenged all of NASA, including its industry and academia partners, to do Projects in a Faster, Better, Cheaper (FBC) mode crystallizing what already was taking root in NASA's SMEX and Discovery Program thrusts, with BMDO s Clementine Mission, etc.

Moving into the last decade of the 20th Century, it was already clear that for NASA to remain viable and credible, it must become more business-like, treat cost and schedule as important as Mission performance, and deliver on time for the advertised cost.

Dan Goldin quickly followed with these important directives to facilitate FBC:

- Focus on smaller Missions; stop putting all of our "eggs in one basket"
- Incorporate advanced technology into Missions
- Reduce NASA HQ Management moving more Program responsibility to the Centers [2]
- Construct exciting VISIONS and ROADMAPS in aggressive planning for future Missions
- And, Dan Goldin said: "It's OK to fail!" (This was not well understood, but is clarified in this report.)

All of this triggered a major transition from an "old" to a "new" NASA, which is continuing to this day as we move into the 21st Century.

So how well are we doing?

Well, as with any major upheaval, NASA is being turned over in every sector like fields in spring to bring new growth. As with any major human endeavor, there have been successes and failures in the stress and strain of venturing onto new ground, trying new things, and taking risk to gain significant return.

However, of all the hundreds of people interviewed, outside and inside NASA, no one said we should turn back. All realized that NASA must continue to improve its performance if NASA is to stay a world leader.

All Government Agencies, industry, and academia are at it, too, realizing they need to improve if this Nation is to compete in the Information Age, in the new World Economy.

We must stay the course, benefit and grow from our lessons learned, and move on to the next level for FBC.

Just What Is FBC? Its Definition

Of all our sessions, the ones defining FBC were the most animated. Everyone had his or her pet definition, and it was difficult to get consensus.

Examples of favorite debates: Should the Better go before Faster and Cheaper? Or is [it] the other way around? Another argument was that you could pick two, but you can't have all three and on and on.

However, the FBC Team concluded it is not true that only two of three in FBC can be obtained. A good example is Lunar Prospector. It was certainly "Faster" and "Cheaper" and its "Better" was the ingenious simplicity of its spacecraft system design to make such important measurements. "Better" for future FBC Missions will be achieved from advanced technology and methods. We haven't scratched the surface yet.

However, most agreed that FBC boils down to two basic things:

1. FBC is simply attempting to improve performance by being more efficient and innovative, and it applies to everything and everyone.

and

2. There's an intangible element, there is a team spirit associated with doing FBC, and people are the most important ingredient.

You can imagine the argument between the strict technicians and the humanists in our debates as to whether item 2 above was an essential part of the definition, but over and over, every successful FBC Team told us that people are most important and close team work was essential.

The FBC Teams became a family with extensive communications among the members. They fussed and fumed like a family, but reached out to help each member, as well, while still doing their own job. They [3] became totally excited about their work and remarked they were having fun! And this extended outside of the home base team to the industry, university and other NASA Center partners. This "team spirit" became infectious. Some FBC Teams, however, reported that the fun had gone away after having their resources cut too deeply. Physical co-location is best, but virtual co-location via the Internet is working well.

FBC is not trying to fit a challenging Mission scope within arbitrary schedule and cost caps. For the first generation of FBC Projects, mission scope fit fairly well within the caps that is, for Clementine, Near-Asteroid Rendezvous, Mars Pathfinder, Mars Global Surveyor, Lunar Prospector and Stardust, for example.

However, in our zeal to do FBC, and in teaming to do Programs at the NASA Centers, the challenge bar was raised too high for some of the second-generation Missions. The cost cap challenges were made too great, along with a mix of unstable funding and escalating requirements.

We need to slow down some, not rush too quickly into important Programs and Projects, plan and implement them more carefully, and move away from fixations on cost and near term gain.

To assist the learning process of doing Programs at the NASA Centers, better reality checks of the feasibility of implementing the Mission set under the Program funding profile are needed.

An FBC Project in a pre-project phase must be permitted to develop correct cost and schedule caps for the Mission scope by working estimates from the "bottom up" with all members of the Project Team participating, who then own their Project Plan.

Or, if it's necessary to set cost and schedule caps at the outset, then:

An FBC Project Team must be given the flexibility to define the Mission Scope that fits properly within the given caps; this was the case with Mars Pathfinder when NASA HQ gave the project flexibility to adjust Mission scope to fit.

In one of our workshops, the definition of FBC was "nailed" by two separate presentations from an unlikely pair:

- In a presentation made by Col. Pete Rustan, Clementine Project Manager, he strongly made the point that a challenging Mission scope can't be stuffed into arbitrary caps, that careful FBC pre-project planning and costing are as important as ever before.

- After an exciting presentation on how to look for life on other planets, and giving his support to FBC, Ken Neilson, a Caltech Life Scientist, pointed out that an FBC Mission could take as long as 25 years citing this example: a carefully planned Planetary Life Detection Program, led by the proper group of scientists, could take as long as 25 years to complete its sequence of Missions, all in an FBC mode.

So FBC is not resting on your laurels, not just accepting past ways without good reason. It's questioning the reason for every practice, it's continuously looking for improvement, it's stepping out with new methods, new technology, and taking risk—prudent risk. FBC is not taking undue risk by taking shortcuts under pressure around important tests and qualification.

At the same time, it means discipline, doing careful upfront planning, design and implementation, keeping past lessons learned out in front at all moments, and being checked, balanced and mentored by those scarred by [4] experience; it's maintaining a delicate balance between old and new. It means being humble, respecting how hard this business is, not automatically discounting others and their methods, and never thinking you know it all.

As evident in the FBC Rules of Engagement in Attachment A [omitted], there isn't anything magic about doing FBC. It's back to basics, lots of hard work, follow-through on the details, working openly and candidly, and total dedication by the Team, as with winning the Super Bowl or starting up an Internet company.

And notice that the FBC Rules of Engagement include the need for important checks and balances such as are provided by Independent and Peer Reviews.

And it's important to get this straight:

The Project Manager is "Captain of the Ship." The buck stops with him or her. The Project Manager and Team are responsible for ensuring that all elements of a Project are being implemented with acceptable Risk for those Project elements under their control and also for those outside their immediate Project control possibly the launch vehicle, a major instrument, etc.

While Project risk at the outset may be high, it must be sufficiently assessed and mitigated throughout development and operations. Not having enough funding or schedule resources are never excuses for failure, and it takes a Project Manager with good judgment and courage to declare under pressure that the Project is not doable for the available resources. This ability to judge, to walk the fine line between challenge and risk, is even more important in today's environment for FBC Projects.

FBC equates to all of NASA, to all sizes and categories of Missions, robotic and human, large and small, and to the institutional support of Missions from re-engineering to human relations, from individuals to teams. NASA will continue to have a mix of large and small Missions, each employing FBC methods. An important Rule of Engagement for the institutional support to FBC Projects is ensuring a people-oriented environment that facilitates candid and thorough communications by the FBC Project Team to create an environment that encourages the Project Manager to speak up, to ask for help when needed.

An open, candid environment is important at NASA Headquarters, at the NASA Centers, and with the industrial and academic partners, where forums for healthy debate need to be established. As a rule, in this day and age, most NASA studies should be done openly by the Study Team placing its objectives and status on the Internet to give a wider audience the opportunity to participate electronically, to chip in with their ideas, recommendations and critique.

Have all Projects wear "Three Badges of Courage":

1. Certification of the Project Manager and Team as to experience and expertise
2. Programmatic and Mission Risk Signatures: the Project Risk "fingerprints"
3. Rules of Engagement Performance Metric: a periodic tally as to how well the Project is implementing and operating according to the FBC Rules of Engagement

Examples of items 2 and 3 are attached to this paper as Attachments B and C [omitted]. These "Project Badges of Courage" are powerful graphic measures of Project Performance, but must be implemented by the institution constructively to help, not hinder, the projects. This process, as with all NASA Policies and Procedures, needs [5] to be frequently checked as to effectiveness and evolved to make it even more relevant as being accomplished with NASA 7120.5A, as an example.

MAJOR CHALLENGES FOR FBC

1. The current Mission failure rate is too high and must be reduced. This is not due to the introduction, say, of exotic technology or due to a difficult, unknown space envi-

ronment. Most failures over the last decade can be attributed to poor communication and mistakes in engineering and management. This was not what was meant when Dan Goldin said, "It's OK to fail."

Failing due to mistakes is not tolerable. Dan Goldin's "Its OK to fail" statement was made to encourage Project Managers to step up bravely to difficult, risky, but potentially highly rewarding Missions. Failures here can be honorable, even if still traumatic to the Project Team.

2. In the current NASA transition, we have moved from few to many Missions requiring many more Project Managers, Project Teams and institutional support including Review Teams. Also, management attention has become diluted across these many Missions. As before with few missions, Project Managers worked through the ranks for many years to gain significant experience before they became Project Managers. Now with many missions this is not always possible, making training, mentoring, and peer review even more important.

3. At the same time there is a talent drain due to retirement, downsizing, and loss of people to Industry.

4. We must remake the NASA institution to match business in the 21st Century Information Age. Importantly, each NASA Center must focus on a few core competencies for which they are world-class, and rely on other NASA Centers, other government agencies, industry, and academia for other capabilities to paint the full picture. In addition, a long-term stable funding environment must be established to foster these world class centers. This is critical.

   In a management interview, PricewaterhouseCoopers, a worldwide management consultant firm, forecasts that sets of global companies networked together, each company with its particular specialty, will compete with other sets of networked companies. This too will be the future structure for NASA HQ and the Centers.

   NASA Mission Centers must retain the expertise to do in-house Projects. This "corporate history" represents a sustaining expertise that is the foundation for space exploration; it is an important national asset.

5. NASA must guard against any effort to shift from basic research to development solely in support of the near term Missions. While the near term development is very important, a better balance must be struck to ensure that some "seed corn" is left over for research and advanced development to trigger revolutionary approaches to space exploration. A better NASA Integrated Technology Plan is needed.

   NASA and other government development agencies must be on the lookout for potentially high payoff technology breakthroughs whose accelerated development by the nation in "mini-Manhattan-like projects," but run openly with peer review, could give this nation a decided competitive advantage. Some candidates: carbon nanotubes and quantum dots. Yearly "out of the box" technology workshops should be conducted nationwide giving awards for the best ideas.

[6]

6. Existing or soon-to-mature Information Technology (IT) can be used to develop important advanced, computerized, design visualization aids for the front end of project development. The designers walk around their "virtual spacecraft" as they design it.

   This same technology can then be used, also, at the other end of the project, to develop Visualization Domes which, through tele-presence, immerse the Flight Operations Team, the press, and the public on the distant planet, say, on the surface of Mars at the lander and rover site—the ultimate armchair adventure!

   Sadly, there exist no plans to produce these productive and exciting products any time soon. These are needed now and require the requested funding to bring them into being as quickly as possible. There are good but fragmented pockets of exciting IT developments in NASA, but presently the IT work is localized. It is not integrated into an effective, NASA-wide Development Plan with a schedule committing to a set of incremental demonstrations leading to delivery of a major Mission IT architecture. Current IT related activities like Intelligent Systems, Intelligent Synthesis Environment, Consolidated Super Computing Management Office, etc., need to be folded into this integrated plan.

7. All work in NASA can be treated like Projects with a task objective, a plan, a cost estimate, risk management and importantly, a schedule of delivery commitments. This goes for research and advanced development activities as well. There is a reluctance in this community to projectize, but more structure in these can make them much more productive in an FBC mode. That is not to say breakthroughs or inventions can be scheduled. However, a research schedule can be produced which targets periodic peer review assessing relevance and possible need for new direction.

   Early involvement of the project team with the technology team in technology development, with lots of good communication, is necessary to facilitate acceptance and ownership by projects to fly new technology. A good example of this process is being accomplished at JPL in their TEAM X pre-project planning function. It's now called TEAM XT.

8. WE MUST DRIVE DOWN THE COST OF LAUNCH! THIS MUST BE A MAJOR NATIONAL PRIORITY
   see below

WHAT IS MOST IMPORTANT FOR THE FUTURE OF FBC

Future FBC = PTM People Technology Methods

1. **<u>Acquiring, motivating and keeping good people.</u>** Generating interest in NASA must start early in the schools. While there is good work here, it needs higher priority. There is nothing better than involving students in real live Missions, with some managed by students, with strong, encouraging assistance and mentoring by NASA expertise to give them a good chance to succeed. Let them navigate rovers on the Moon and Mars.

2. **<u>Infusion of Advanced Technology.</u>** Soon Projects, who now develop their own uplinks and downlinks [sic], will be provided proven, advanced, low cost multi-mission data systems with "bug free" software; this will be like not needing to build your own phone every time you call home.

[7] Advanced micro-electronics will bring the cost of small but powerful spacecraft, matched to an automated, Internet driven ground data system, down to a few million dollars so that universities, the world's developing countries and companies can explore space on their own.

Electrons and photons cannot tell if they are participating in a reconnaissance mission or making noble scientific measurements at Mars. All spacecraft share most of the same equipment functionality. A common multi-agency, Internet Store for high quality, modular, advanced components, supplied by multiple vending sources, must be a top national priority. This store would be fed by multi-agency advanced developments.

And as already understood by many, an accompanying reduction in launch cost is essential and must be a top national priority to drive FBC to a higher level, as well as to keep this nation the space leader. This is the single most important factor standing in the way of a "big bang inflationary-like" expansion of the nation and the world into space.

This is what NASA in the FBC mode must be about—paving the way for others to do space explorations cheaply, reliably, and safely, by effectively accomplishing high risk, but high payoff, enabling, advanced developments.

3. **<u>Infusion of Advanced Methods.</u>** This deals with expanding the multi-mission infrastructure in support of FBC Project Teams.

Core FBC Teams, becoming smaller in size, will be supported by:

- Multi-mission pools of technical and management expertise for consulting and peer review
- Best computer aided tools, processes, templates, model-based design and management standards, training
- Lessons learned data bases
- Risk evaluation tools

as well as

- The advanced, multi-mission technology mentioned above

PROPAGATE FBC THROUGHOUT NASA IDENTIFYING MORE EFFECTIVE CENTER-TO-CENTER TEAMING ARRANGEMENTS

It's propagated! The message is there loud and clear at NASA Headquarters, every NASA Center and throughout industry and academia. The challenge lies with institutionalizing it.

It's one thing to do an FBC Project experiment, it's another thing to instill this cultural change throughout the complete organization.

But every organizational element is facing this challenge. There are many good people within NASA and its support partners. There is still a lot of the right stuff.

Dan Goldin is right on with his FBC thrust. He has set the stage, created the environment. Now all we need to do is follow through on implementation of the exciting Roadmaps and Visions that have been generated.

[8] The key word now is "implementation." This requires careful planning and lots of work in the trenches. No Mission before its time. This requires unprecedented teaming and open, candid communications. No one person has the answer. It takes a lot of debate and evolution of ideas to get there. It takes courage to admit a wrong path and the need to move in another direction.

And if we do, NASA will be even more important in the 21st Century, looking for life out there, building the bridge for humans to cross over to space.

But there are many challenges, and a lot of hard work to do, and a lot of teaming to do.

Here are three examples of cultural [change] that need more work to take FBC to the next level:

1. *How to motivate individuals to team?* Getting them to align their direction, their ego vectors, in the same direction of the Project, if only for a short while, for the duration of the Project.

2. *How to motivate hard-crusted Project Managers to accept a larger role outside their immediate project responsibility?* Getting them to cooperate with the institution re-engineering support and technology development activities. Getting them to infuse new technology into their Project. Getting them not only to accept, but own larger institutional and Program requirements.

3. *How to motivate NASA Headquarters and Centers to team better?* Dan Goldin's great desire is for the Enterprises and Centers to take the initiative on self-management as, for example, if a team of Centers come in with, say, an Integrated NASA-IT Implementation Plan showing a schedule of deliveries, who's in charge, and who does what. If this would happen, then one of Dan's major objectives for the "new" NASA would be fulfilled.

ON CENTER TEAMING

Currently, Center teaming exists in established roles where historically there's a clear advantage for each Center. But there are new Center Teaming initiatives being implemented effectively. Where things are working, don't fix them.

But in general, Centers are stand-alone and protective because of the downsizing they have had to fare, periodic threat of closure and the need to compete for scarce resources.

The solution to better Headquarters and Center teaming will come with Information Technology—the wiring of NASA into one electronically networked NASA Center. With workers from the Centers and Headquarters, networked together and cooperating to accomplish a Project, each contributing with its specialty, the lines of distinction for Headquarters and the Centers will become more and more blurred. They will begin to act more as a whole not as single entities. The "stovepipes," "fiefdoms," and "castles" will come tumbling down.

As with inside NASA, better outside partnerships with industry and academia need to be worked, including involving them in workshops and listening to their feedback.

Yearly performance evaluations must include how well all Enterprises and Centers are doing in affecting these teaming transitions.

[9] *Here's what has to happen to bring this into reality. We need to:*

1. Solve the long-standing NASA Center core competency problem, establishing what each Center does and motivating all Centers to utilize other Centers' competencies. Establish stable funding.
2. Motivate each Center to neck down to, focus on, and become world class in its smaller set of core competencies.
3. Promote mobilization of key personnel around NASA, both technical and management rotate them on assignments between the Center and Headquarters.
4. Develop a NASA Integrated Technology Development Plan which balances research and advanced development with Mission focused technology development.
5. Balance competition of Technology Development with stable Center funding for their world-class core competencies.
6. Develop a NASA Information Technology Plan encompassing the likes of Intelligent Systems, Intelligent Synthesis Environment, CoSMO and all related IT activities.
7. Interact more closely with Industry and Academia to establish more effective partnering arrangements. Listen to their feedback.
8. Balance the Leadership of Programs between NASA and Headquarters.

There is a need for more Leaders at NASA Headquarters, more of a balance of Program Management responsibility between NASA Headquarters and the Centers especially in resolving the Center core competency problem and developing more effective NASA Technology Plans.

Recommendations

1. **Place higher priority on people acquisition, motivation, training.**

- Develop incentives for attracting good people and well-respected Leaders to come to work for NASA

- Expand the role and clout of NASA's Academy of Program and Project Leadership – see below

- Certify Project Managers and Teams as to experience and expertise – *Badge of Courage #1*

- Continue symposiums on lessons learned, re-engineering, information technology, cultural change, teaming, etc., bringing in experts from within/outside NASA

- Acquire outside help on cultural change, core competency, and organizational issues

2. **Assign responsibility to NASA Chief Engineer for:**

- Consolidating the findings of this report with the Mars Program and Mars Climate Orbiter Investigation Reports, deriving composite FBC Project Lessons Learned, FBC Rules of Engagement and Project Implementation check lists

[10]

3. **Assign responsibility to NASA Academy of Program and Project Leadership for:**

- Generating training material for FBC Training workshops for FBC Project Team leaders and teams which is first subjected to a "dry run" in front of experienced FBC Project managers from each Center, Industry and Academia

- Conducting these FBC Training Workshops throughout NASA, Industry, Academia

4. **Take aggressive steps to effect better teaming among NASA Centers, industry, and academia.**

- Start with strengthening NASA HQ Management, providing the "champions" as designated below

- Implement more effective NASA HQ relationships with the Centers

- Form a NASA Center Teaming Office at HQ to bring NASA into the 21st Century –Assign a NASA Center Champion

- Resolve Center Core Competency and Center of Excellence role issues and operations

- Place higher priority on funding and supporting University research and advanced development and their space flight Missions

- Assign the HQ Safety and Mission Assurance Office the responsibility for an Industry/Academia

- Workshop to effect better NASA teaming arrangement – including contracting and incentives

- Assign JPL the responsibility of conducting a NASA-Wide Methods Working Group to share and to further evolve re-engineering products. Use the NASA FBC Task Center Representatives already established

5. **Place higher priority on Advanced Technology Development — Assign a HQ Technology Champion**

- Form a Technology Office led by a results-oriented Chief Technology Officer – must have as much stature/clout as Enterprises

- Balance research and advanced technology development with focused technology development

- Balance competition of technology development with placing stable technology development at NASA Centers of Excellence

6. **Move out more aggressively on Information Technology development – the most important NASA HQ and Center-to-Center teaming arrangement – Assign a HQ Info Champion**

- Form an Information Technology Program encompassing Intelligent Synthesis Environment, Information Technology, Intelligent Systems, Consolidated Super Computing Management Office into one integrated plan.

7. **Strike better balance between FBC Challenge and Risk**

[11]

- Initiate Program reality checks

- Implement FBC Rules of Engagement and the associated performance metric

- Ensure Project teams own their Project Plans built from the "ground up"

- Develop "Badges of Courage" for each Project

- Expand Safety and Mission Assurance responsibilities at NASA HQ and at the Centers for verifying:
    - Team Certification
    - Risk Signatures
    - FBC Performance Metrics
    - Project Readiness for Start, Launch, Flight Operations
    - Compliance to FBC Lessons Learned
- Give immediate relief to understaffed Mars Operations, Launch and Payload Services
- Consolidate all Independent Review objectives into one Independent Review per year for all Programs and Projects
- Continually evaluate the effectiveness of NASA policies, rules, procedures, etc. – like being accomplished for NASA 7120.5A
- Bring industry academia and outside consultants in to review NASA's approaches

<u>Some Good Examples of FBC</u>

A short set, not the complete set, of good examples of FBC is given in Attachment D [omitted]. This set illustrates the diversity of FBC activities throughout NASA. Hats off to all the FBC individuals and teams, who are taking FBC to the next level. The complete list starts with the young people and extends to Center Directors to Associate Administrators to the NASA Administrator.

[12] <u>Metrics for Measuring FBC</u>

How to measure the value of FBC is a much-debated subject, too, and there have been a number of attempts at constructing this metric.

Here's another list of measures:
- Mission Success Rate greater than 8/10
- The degree to which both launch and spacecraft costs are reduced
- Number of Peer-Reviewed Scientific papers published, resulting from NASA Missions
- The degree to which an effective, NASA-wide Technology Development is achieved, including IT
- The extent to which NASA HQ and the NASA Centers are teamed together as one NASA Center and teamed effectively with Industry and Academia
- The degree to which the public is excited with and involved in Space Missions

Document I-33

Document title: Committee on the Future of Space Science, Space Studies Board, National Research Council, *Managing the Space Sciences*, 1995.

Source: Space Studies Board, National Research Council, National Academy of Sciences, Washington, D.C.

As the International Space Station proceeded in the early 1990s, some space scientists believed that the only way to protect the space science budget from being reduced to pay for the station's costs was to make the space science program somewhat autonomous from the rest of NASA. Senator Barbara Mikulski (D–MD), who chaired the appropriations subcommittee that oversaw NASA's budget and who had both the National Institutes of Health and the Goddard Space Flight Center in the state she represented, suggested that NASA ask the National Research Council to examine the desirability of creating a "National Institute for Space Science." This report contained the views of the Space Studies Board on this question and other suggestions for improving the management of the U.S. space science effort.

[cover page]

MANAGING THE SPACE SCIENCES

Committee on the Future of Space Science
Space Studies Board
Commission on Physical Sciences, Mathematics, and Applications
National Research Council

NATIONAL ACADEMY PRESS
Washington, D.C. 1995

[1]

Executive Summary

In April 1994 the National Research Council received a request from NASA Administrator Daniel S. Goldin that the NRC's Space Studies Board provide guidance on several questions relating to the management of NASA's programs in the space sciences.[1] The issues raised in the Administrator's request closely reflect questions posed in the agency's fiscal year 1994 Senate appropriations report. These questions included the following:

- Should all the NASA space science programs be gathered into a "National Institute for Space Science?"

---

1. In this report, "space sciences" refers to all of NASA's science programs conducted in or from space, including space astronomy, space physics, planetary exploration, microgravity research, space life sciences, and Earth science.

• What other organizational changes might be made to improve the coordination and oversight of NASA space science programs?

• What processes should be used for establishing interdisciplinary science priorities based on scientific merit and other criteria, while ensuring opportunities for newer fields and disciplines to emerge?

• What steps could be taken to improve utilization of advanced technologies in future science missions?

Since the creation of NASA in 1958, space science has been a key element of its mission. Indeed, the Augustine Committee report,[2] submitted at the end of 1990, asserted that science was NASA's most important mission. The committee responsible for the present report has proceeded on the same premise. A balanced and healthy program of space science is crucial to the future of NASA, regardless of the overall level of support available to the agency.

The most important recommendations of this report are listed below. They are further elaborated following the list.

[2]

• NASA should not establish a "National Institute for Space Science" that would pull together the three present science program offices.

• NASA should augment the responsibilities and authorities of the NASA Chief Scientist.

• NASA should establish a set of fair, open, and understandable processes to be used in the prioritization of space science research. These processes will ensure that major project proposals considered at progressively higher levels within the agency have the heritage of scientific merit that comes from a successful confrontation with competing proposals at lower levels.

• NASA should create a comprehensive strategy and plan for the technologies that support the space sciences, with the responsibility for near-term technology development residing in the science programs to be served and the responsibility for longer-term technology strategy and development residing in the Office of Space Access and Technology.

• NASA should change the funding of its field centers to full-cost accounting ("industrial funding"). Cost accounting should be based on full program costs, including civil service salaries. The committee endorses NASA's intentions to move in this direction.

• NASA should exercise caution in downsizing its Headquarters staff and transferring functions to the centers; this process could be carried too far and have unintended consequences. The committee identified a number of areas where it believes control should be retained at Headquarters.

• NASA science budgets should include a limited amount of dedicated funding for innovative ideas in high-risk, high-return areas lying outside the current framework of inquiry or design.

• NASA should take a cautious approach to the recently proposed establishment of focused science institutes. There should be a well-defined process for their selection and creation, and a clear plan for the phased transfer of base funds to programmatic funding.

---

2. Advisory Committee on the Future of the U.S. Space Program, *Report of the Advisory Committee on the Future of the U.S. Space Program,* December 1990.

The following expands key recommendations of the report:

Institute for Space Science—In response to direction in the FY 1994 Senate appropriations report, the committee considered a space sciences umbrella organization within NASA to coordinate and oversee all space science activities, functioning like the National Institutes of Health (NIH) within the Department of Health and Human Services. The committee reviewed the advantages and disadvantages of such a model and concluded that the [NIH] model, while effective in the arena of health research, is not appropriate for the space sciences. NASA space science benefits from close coordination with other elements of NASA, such as hardware development, launch services, and tracking and data operations, which have no counterparts in the NIH model. The committee believes the required coordination would be hampered by the creation of a quasi-autonomous space science institute. The committee therefore does not recommend establishment of such an umbrella institute.

The Role of the Chief Scientist—The role of the Chief Scientist was found to be a critical one from many perspectives, leading the committee to recommend expanding the authorities and responsibilities of this position. Despite the central role of the science associate administrators in the management of their respective science areas, the committee finds a need for greater integration and coordination of these programs. To achieve this, the position of Chief Scientist should be strengthened, particularly by the addition of concurrence authority in key matters affecting space science. The Chief Scientist should be a person of eminent standing in the scientific community with a significant record of accomplishment. A proposed "functional statement" for the Chief Scientist is given in Chapter 4. A major component of this official's integration responsibility is coordination and oversight of the recommended science prioritization process. Another component is coordination of the technology development programs that support space science.

The Prioritization Process—The committee believes that peer review is the most effective form of merit review for the selection of scientific research. A clear set of criteria, known and understood by all parties, is crucial to the prioritization of scientific goals. The relative ranking of science and mission [3] plans will be most strongly affected by scientific factors at the entry level, where proposals from the same discipline or subdiscipline compete against one another. As the arena of competition broadens to the interdisciplinary and then to the agency-wide level, other programmatic and political influences become increasingly important. It is essential, however, that all proposals being considered at progressively higher levels retain the heritage of scientific merit that comes from successful confrontation with their peers at lower levels. The office of the Chief Scientist should oversee these prioritization processes, especially as they cross-disciplinary boundaries. NASA management should cancel those programs or projects that are failing or whose priority has dropped substantially in this prioritization process. The committee found that peer review and the above corollary principles apply generally to technology research as well.

Technology Planning—New technologies are important as agents of change, enhancing the quality of scientific output and the ability to accomplish more with less. Technology development is undertaken both by NASA's science program offices and by its Office of

Space Access and Technology (OSAT). The committee recommends that NASA establish an agency-wide strategy and plan for the technologies that support the space sciences. These technologies may be characterized as near-term or far-term technologies (the latter defined as requiring more than five years to be ready for flight demonstration). The space science offices should have primary responsibility for identifying and reviewing near-term technologies, giving them greatest control of the technologies that most immediately affect the success of their programs. Each science office should allocate a significant fraction of its resources to Advanced Technology Development activities and should be willing to pool resources to achieve shared objectives. Most importantly, the implementation of all categories of technology development should be undertaken by the best-qualified individuals or teams within NASA, other government laboratories, industry, or academia, as determined by peer review.

Promising far-term technologies should be identified, funded, and managed by OSAT. Projects in these areas should be reviewed jointly by the science offices and by OSAT. Like near-term technology development, far-term projects should be carried out by the best-qualified individual or teams, as determined by peer review. These projects should stimulate exploratory development of possibly unconventional technologies having the potential of producing breakthroughs in capability. Finally, a rigorous review process should be put in place to identify those projects that ought to be terminated in the present constrained budgetary environment.

"Industrial Funding"—The committee examined the advantages and disadvantages of an explicit full-cost accounting system in which all charges, including salaries and facilities, are charged against projects (so-called "industrial funding"). This approach permits ready assessment of comparative costs that might otherwise be hidden in an institutional funding environment. The committee endorses NASA's decision (stated in the "Zero Base Review" briefing to the Congress) to identify, budget, and manage by total program costs, including civil service labor costs. The committee recommends that NASA change the funding of its field centers to an industrial funding arrangement. The committee believes that decisions on program priorities and budgets would be more rational if based on full-cost accounting, and program accountability and discipline in personnel management would thereby be enhanced. A similar recommendation was made in the NASA Federal Laboratory Review report.[3]

The Downsizing of Headquarters—NASA is currently "re-engineering" its organization. This re-engineering entails a very large downsizing of its Headquarters staff and a concurrent transfer of functions to the centers. The result is expected to be the analog of a lean "corporate management" model. While the committee endorses the intent, it notes that an unintended consequence could be a center-dominated model as opposed to the desired enterprise-focused one. Several recommendations are offered to avert this outcome. Not all program management functions should be transferred to centers. [4] Those complex programs that cut across centers should be retained at Headquarters and integrated with enterprise management. Support of scientific disciplines, management of peer review, and oversight and integration across center boundaries should remain Headquarters functions.

3. Federal Laboratory Review Task Force, NASA Advisory Council, *NASA Federal Laboratory Review*, February 1995.

Likewise, creation of a strategy and plan for the technologies that support space science should be a Headquarters responsibility. The adoption of industrial funding will further emphasize the importance of a suitably strong Headquarters organization.

Research in New Fields—The committee recognizes the competitive obstacles faced by smaller, newer, or less well established fields of science. The committee recommends that NASA science budgets include dedicated funding for innovative, high-risk, high-return ideas falling outside current frameworks of inquiry or design. This research is highly important and deserves special management attention, including that of the Chief Scientist. This recommendation is not intended to allow circumventing of peer review for the major parts of any science program.

Science Institutes—Creation of contractor-operated institutes may be advantageous in specific instances. However, the committee recommends that, as NASA proceeds with arrangements for the first focused science institutes, it give due attention to the processes by which these institutes are selected and created and by which, over a few years, their guaranteed base funding will be transformed into competed programmatic funding. Further, there should be consideration of a review process that will ensure either (1) that they compete successfully to maintain or increase their size, or (2) if less successful, that they are phased down in an orderly fashion. The committee recommends that additional initiatives along these lines be deferred until the above processes have been defined and the success of the two proposed institute pilots can be evaluated.

The committee's recommendations are gathered together by main theme in Chapter 7.

The NASA space science programs, from the dawn of the space age to the present, have produced an unprecedented flow of discoveries. The fiscal, political, and technological environment of the agency is now in a state of rapid change. It is vital that NASA respond to its challenges and opportunities in the most constructive manner to ensure the success of its future space science endeavors. The committee believes that the recommendations made in this report, if accepted by NASA, will aid in this objective.

Document I-34

Document title: NASA/National Research Council, "The Search for Origins: Findings of a Space Science Workshop," October 28-30, 1996.

Source: Space Policy Institute, George Washington University, Washington, D.C.

In August 1996, NASA announced that a team of scientists had discovered in a meteorite of Martian origin what could be evidence of ancient fossilized organisms. This exciting possibility added emphasis to NASA's attempts to reformulate its science program around the theme "Origins." At the request of the White House Office of Science and Technology Policy, NASA and the National Research

Council in late October 1996 assembled a group of three dozen biologists, planetary scientists, astronomers, and cosmologists to discuss this theme as an organizing principle for the space science program. This document, which summarizes the results of their deliberations, was the basis of a presentation to Vice President Al Gore on December 3, 1996.

[cover page]

The Search for Origins:

Findings of a Space Science Workshop

October 28-30, 1996

[1] **Findings of the Space Science Workshop**

This document reports the findings of three dozen biologists, planetary scientists, astronomers, and cosmologists assembled by NASA and the National Research Council at the request of the White House Office of Science and Technology Policy. They met in Washington, D.C., on October 28-30, 1996.

Steering Group
Claude R. Canizares, MIT, *Workshop Co-Chair*
Anneila I. Sargent, CalTech, *Workshop Co-Chair*
David C. Black, Lunar and Planetary Institute
Roger D. Blandford, CalTech
Joseph A. Burns, Cornell
James P. Ferris, Rensselaer Polytechnic Institute
Christopher P. McKay, NASA Ames Research Center
Patrick Thaddeus, Harvard-Smithsonian Ctr. for Astrophysics

Spacecraft Technology
Glen Fountain, Applied Physics Lab
Peter Lynn, Naval Research Lab

Universe Subgroup
Roger D. Blandford, CalTech, *Co-Chair*
Patrick Thaddeus, Harvard-Smithsonian Center for Astrophysics, *Co-Chair*
John N. Bahcall, Institute for Advanced Study
Marc Davis, UCal Berkeley
Alan Dressler, Observatories of the Carnegie Institution
Sandra M. Faber, Lick Observ.
Wendy Freedman, Observatories of the Carnegie Institute [sic: Institution]
Steven H. Kahn, Columbia

Richard Kron, Yerkes Observatory
John Mather, NASA GSFC
Marcia J. Rieke, Steward Observ.
David N. Schramm, U. of Chicago

Planetary Subgroup
David C. Black, Lunar and Planetary Institute, *Co-Chair*
Joseph A. Burns, Cornell, *Co-Chair*
Bernard F. Burke, MIT
George D. Gatewood, Allegheny Observatory
Janet G. Luhmann, UCal Berkeley
Jonathan I. Lunine, U. of Arizona
Glenn Mason, U. of Maryland
Eugene N. Parker, U. of Chicago
Laurence A. Soderblom, U.S. Geological Survey
George W. Wetherill, Carnegie Institution of Washington
Maria Zuber, MIT

Life Subgroup
James P. Ferris, Rensselaer Polytechnic Institute, *Co-Chair*
Christopher P. McKay, NASA Ames Research Center, *Co-Chair*
Michael H. Carr, U.S. Geological Survey
Bruce M. Jakosky, U. of Colorado
Andrew H. Knoll, Harvard
David McKay, NASA JSC
Kenneth H. Nealson, University of Wisconsin
Mary Jane Osborn, University of Connecticut Health Center
Norman R. Pace, UCal Berkeley
Yvonne Pendleton, NASA Ames Research Center
Beverly K. Pierson, University of Puget Sound
J. William Schopf, UCLA
Steven W. Squyres, Cornell
O. Brian Toon, U. of Colorado

[2]

ORIGINS

The question of our ORIGINS is as old as human thought:

- How did the Universe come to be what it is today?

- What is the ORIGIN of life? What are the building blocks, the habitats and the conditions necessary for primitive life and how did these come about in the Universe?

- Is life unique to Earth? Can we find convincing evidence that simple life forms once existed or even now exist elsewhere in or beyond our solar system?

The study of ORIGINS follows the 15 billion year long chain of events from the birth of the Universe at the Big Bang, through the formation of the chemical elements, of galaxies, stars, and planets, through the mixing of chemicals and energy that cradled life on earth, to the earliest self-replicating organisms and the profusion of life.

[3] A CHALLENGE AND AN OPPORTUNITY FOR TODAY

Today, we are in a unique position to pursue the quest for our ORIGINS:

- For the first time in history, we have achieved the level of understanding and technical capability to press for answers to fundamental questions concerning our ORIGINS, our history, and our context in the Universe.

- Revolutionary advances are possible over the next 15 years if we respond to public enthusiasm and push forward with the ORIGINS program.

- The ORIGINS quest informs, excites, and inspires the public. Its outcome may well have as profound an effect on human thought as the Copernican and Darwinian revolutions.

[4] RECENT DISCOVERIES

Recent discoveries from an array of diverse disciplines attest to the fact that life is remarkably hardy and that each step in the chain of ORIGINS occurred surprisingly quickly.

Discoveries in just the past few years provide the first scientific basis for believing that life may be widespread in the Universe, in our solar system and beyond.

We have a new comprehension of the development of the Universe, its constituent galaxies and stars, the number and variety of planetary systems, and the processes that shape them.

These discoveries blaze the trail of ORIGINS from the earliest moments of the Universe to the present.

[5] Recent Discoveries: Life on Earth

- On Earth, life thrives wherever there are two key ingredients: liquid water and usable energy; this includes unlikely, harsh environments like hot deep-sea vents, cold Antarctic rocks, acidic hot springs and rocks many kilometers below Earth's surface.

- Microbial life originated very early and evolved rapidly. Life on Earth sprang into existence nearly 4 billion years ago, soon after the end of the most violent phase in the formation of the planet. Our understanding of the process is greatly enhanced by the newly completed "tree of life," the genealogical family tree that links all living organisms.

- Life, whether simple or complex, developed from ancestral microbes that lived at high temperatures in the absence of oxygen, conditions believed to exist on the early Earth but also elsewhere in the solar system. Laboratory studies suggest that early life was based on self-replicating molecules of RNA (ribonucleic acid).

[6] **Recent Discoveries: The Possibility of Life Elsewhere**

- We now know that the two key ingredients of terrestrial life, water and energy, are or have been present at multiple locations in the solar system.

- Studies of meteorites from Mars show evidence of liquid water, organic compounds and possibly fossilized microbes. Images of the icy surface of Jupiter's moon Europa, recently returned from the Galileo spacecraft, suggest the presence of an underlying ocean of liquid water. Meanwhile, we have learned that our young Sun was extremely variable, affecting conditions in the solar system throughout the period of life's evolution.

- For the first time we have detected companions that may be planets around other stars, giving substance to the hypothesis that the solar system is not unique.

[7] **Recent Discoveries: Back to the Beginning of the Universe**

- The COBE satellite detected, for the first time, evidence for the seeds of cosmic structure formed in the earliest moments of the Universe. These are likely the most primitive ORIGINS of every present-day galaxy, star, and planet.

- The Hubble Space Telescope has imaged the birthplaces of stars in our galaxy and has looked to great distances and, correspondingly, far back in time to see galaxies taking shape when the Universe was still in its infancy.

- The ORIGINS of the chemical elements necessary for life are now being traced back through the history of the Universe. We have yet to see their earliest origins but recent measurements from the ground and space show that some elements were already formed when the Universe was one-tenth its current age.

[8] **CURRENT QUESTIONS ABOUT OUR ORIGINS**

These astonishing discoveries are scientific markers that trace the chain of ORIGINS. They also challenge us to find the "missing links." We need to understand more about:

- The *processes of life*
- The *habitats suitable for life*
- The *building blocks of the Universe*

[9] **Current Questions: The Processes Leading to Life**

- What are the characteristics and limits of life on Earth in extreme environments? What is the full range of biological diversity?
- What chemical ingredients and processes lay the foundation for life?
- What are the evolutionary pathways that life followed, and what determined them?

[10] **Current Questions: Habitats for Life**

- What solar and planetary conditions led to life on Earth?
- Did simple life forms emerge elsewhere in the solar system, particularly on Mars, and does it exist at present?
- How do stellar and planetary systems form and evolve in our galaxy? Are there worlds around other stars that could harbor primitive life?

[11] **Current Questions: Building Blocks**

- How did the Universe transform so rapidly from a Big Bang of almost infinite density and temperature to a panoply of galaxies, stars and planets?
- How and when did the chemical elements form in the earliest Universe and throughout its history?
- What is the nature of the dark matter, which we believe constitutes at least 90% of the material of the Universe, and how did its gravitation seed all present structure in the Universe?

[12] **THE CHALLENGE AND THE OPPORTUNITY BEFORE US: THE NEXT STEPS**

We now have a unique opportunity to fill in the "missing links" along the chain of ORIGINS by exploring on Earth and outward in space, in the present and backward in time.

Answers to the questions of ORIGINS are within our grasp if we take the next steps. The current and planned science programs of NASA are essential as is technology to enable subsequent strides.

The ORIGINS challenge provides a unifying core for the space science program. But neighboring disciplines address important problems of their own, provide the context for ORIGINS, and may unexpectedly turn out to be key in finding the answers—as happened in the recent discoveries from Martian meteorites.

NASA's planning process with peer review defines the programs, sets the priorities, and coordinates with other agencies. This process should continue.

[13] **Next Steps: Study the Dimensions and Processes of Life**

- Continue the search for microbial diversity on Earth and extend the genetic "tree of life."
- Determine the limits of environmental conditions under which life exists on Earth.
- Determine the sources of organic molecules that began life on Earth and expand knowledge of the pre-biotic environment of the early Earth.
- Study the ORIGINS of the first genes, metabolism, and cells.

NASA's Exobiology Program will bring new understanding of Earth's early environment and the interplay of planetary and biological evolution. This includes the effects of extraterrestrial influences, such as meteorite impacts, field work in paleontology, and laboratory work in RNA sequencing seeking life's earliest ancestor. There is related work in other agencies, such as research on life in extreme terrestrial environments and on Earth's climate supported by the National Science Foundation.

[14] **Next Steps: Seek Evidence of Life Elsewhere in Our Solar System**

- Continue the search for and detailed analysis of Martian meteorites present on Earth and understand the exchange of materials between planets.
- Characterize Mars' chemical environment, climatic evolution and geologic history, and identify promising sites for sample collection.
- Return to Earth a suite of carefully chosen samples to elucidate Mars' past and to search for evidence of both fossil and extant life.
- Survey potential habitats for life on Jupiter's moon Europa and characterize pre-biological chemical processes on Saturn's moon Titan.
- Return samples from comets and asteroids, and examine the primitive remnants of planet formation at the edge of the solar system.

This year, NASA's Mars Global Surveyor and Pathfinder will begin the essential survey of Mars. Galileo is observing Europa close-up; Cassini-Huygens will probe Titan; and the Discovery missions—NEAR and Stardust—will visit a near-Earth asteroid and a comet. Future probes can then target the most promising sites for detailed observation and, eventually, return samples to laboratories on Earth.

[15] **Next Steps: Understand the Habitats for Life**

- Study the formation and evolution of planet-forming disks around stars.
- Discover and investigate planets around many nearby stars.
- Characterize the Sun's history and its influence on the planets.
- Develop the capabilities for the eventual detection of life-sustaining planets around nearby stars.

Over the next several years, the Stratospheric Observatory for Infrared Astronomy will give new power to study primitive stellar nebulae, and the unprecedented sensitivity of the Space Infrared Telescope Facility will capture images of faint protoplanetary disks. Our knowledge of the structure and dynamics of the Sun, our star, is being greatly expanded by Ulysses and the spacecraft of the International Solar-Terrestrial Program. On the ground, the Keck Interferometer and long-wavelength facilities sponsored by the National Science Foundation will find and characterize new planets around other stars and lay the basis for future space missions.

[16] **Next Steps: Search for Our Earliest ORIGINS**

- View in detail the birth of galaxies, stars and planets in the infrared and longer wavelengths.
- Study the production of the chemical elements soon after the Big Bang and throughout the history of the Universe.
- Trace the amount and distribution of the dark matter that shapes the Universe and the galaxies within it.
- Measure the seeds of structure in the earliest Universe.

In the next few years, the Hubble Space Telescope with upgraded instruments will make even more dramatic images of stars and galaxies being born, and the Advanced X-ray Astrophysics Facility will trace dark matter and the chemical elements in clusters of galaxies. Later, the Space Infrared Telescope Facility will observe young galaxies forming in the early Universe, and the Microwave Anisotropy Probe will isolate the still younger seeds of those galaxies. Future missions can build on and expand these capabilities.

[17] **WHAT WE CAN LEARN ABOUT ORIGINS**

The investment in a mixed portfolio ORIGINS program will yield a steady return of major discoveries and, inevitably, major surprises.

Over the next 15 years, scientists and the public can share the excitement of discoveries such as:

- When and how primitive life emerged and flourished on Earth;
- Whether the Martian meteorites found on Earth or the samples returned from Mars confirm that life existed on that planet;
- The presence on Jupiter's moon Europa of a liquid water ocean that could harbor life;
- The detection of dozens of planetary systems, including some which may be conducive to life as we know it;
- Sharp pictures of planet-forming disks, infant stars, and growing galaxies;
- Detailed histories of the early stages of the Universe, including maps of the dark matter "seeds" that grew to form galaxies.

[18]

Summary

The Hubble Space Telescope pictures of embryonic solar systems and the discovery of evidence for possible past life on Mars have created intense public interest in the ORIGINS of the Universe and its contents.

These breakthroughs are astonishing returns being reaped from years of investment in many scientific disciplines, a major legacy of the 20th Century.

Now is the time to leverage that investment into the 21st Century, to make this the generation that leaves behind a major legacy of answers to the age-old questions of ORIGINS.

Document I-35

***Document title: NASA, The Space Science Enterprise Strategic Plan: Origins, Evolution, and Destiny of the Cosmos and Life,* November 1997.**

Source: Office of Space Science, NASA.

This document articulates NASA's Origins initiative, aimed at increasing understanding of the origins, evolution, and destiny of the cosmos and life within it. It tied the scientific questions to be addressed by the initiative to both short-term (2000-2004) and long-term (2005-2020) space science missions, and provides an easily understood rationale for NASA's twenty-first century space science activities.

[cover page]

The Space Science Enterprise Strategic Plan

Origins, Evolution, and Destiny of the Cosmos and Life

National Aeronautics and Space Administration

November 1997

[1] **Part I: Origins, Evolution, and Destiny**

A. Introduction

We humans are players in the greatest drama of all, the story of cosmic Origins, Evolution, and Destiny. Now, for the first time, we truly have the opportunity to seek scientific answers to questions as old as humanity itself:

- How did the Universe begin?
- How did life on Earth arise?
- What fate awaits our planet and our species?

We have begun to assemble answers to these grand questions using remarkable new tools on Earth and in space. But, more importantly, our understanding is growing through the intellect and imagination of men and women who look up and wonder, who devise new means of gathering information that lead to the formulation and testing of theories to explain what it all means. This is a Golden Age of discovery as exciting and significant as the time when humans turned their first telescopes to the heavens.

In the past few years, we have seen faint folds in the fabric of the Universe, the most ancient ancestors of all the galaxies, stars, and planets that surround us. We have used telescopes on the ground and in space to discover disks of gas and dust surrounding young stars—nurseries of potential worlds—and to discern evidence for giant planets orbiting nearby stars. We have found living creatures in extreme environments previously not thought capable of sustaining life—the dark depths of Earth's oceans and the dry valleys of the Antarctic. We have studied meteorites from Mars, one of which shows evidence of the presence of ancient water and the chemical building blocks of life, and—possibly—tiny, fossilized microbes. Our spacecraft have returned images of what may be ice floes above a liquid water ocean on Jupiter's moon Europa, and made us wonder if life may begin on moons as well as planets. We have seen a comet collide with Jupiter and studied a super-nova from its initial explosion to an expanding gas cloud. We have learned that Earth's climate, biosphere, and the workings of our entire technological civilization are profoundly influenced by the behavior of our varying Sun, a star we can study close-up. We have detected giant black holes that may be as massive as a billion suns at the center of our galaxy and in other galaxies, turning centuries of theory into fact.

From the Big Bang to Biology

Some 15 billion years ago, matter itself came into being in the aftermath of the Big Bang, the event when space and time began. Mysterious forces sculpted the formless sea of particles, leading first to structure in the Universe and then giving birth to galaxies and stars. Some massive stars lived short lives of violent intensity and died in colossal supernova explosions. Their death throes scattered heavy elements produced in their interiors into interstellar space. Our home planet condensed from a cloud enriched with iron and silicon. Our lifeblood and the tools of our civilization are made of elements forged in supernovas long ago.

The early years of Earth were scenes of incredible violence as comets, asteroids, and eruptions tilled the cooling surface and built and blew away oceans and atmosphere. But within just a few hundred million years the first living organisms emerged: Life, it seems, is remarkably hardy and its origin on Earth seems to have occurred surprisingly quickly. In the nearly 4 billion years since, life on our planet has made its home in astonishingly extreme environments and diverse places, habitable so long as there is even a trace of water and useable energy.

And, so we humans, made of star-stuff, descendants of one common ancestor, cousins to all life on Earth, children of ages of evolution and adaptation—now equipped with tools of glass and metal and plastic and silicon to extend our sense beyond our ordinary grasp—are able to look out at the Universe around us and know our solar neighborhood, our intimate relationship to galaxies and stars, and our deep connection to the cosmos.

We have [2] seen bursts of gamma rays from distant reaches of space and time, momentarily more powerful than a million galaxies. Our understanding of the Universe has been altered forever.

We have learned much, but many questions remain to be answered. How could an ordered Universe emerge from a formless beginning? Is life in our solar system unique to Earth, or might there be evidence of past or present life on other moons and planets? Can we forecast space weather by better understanding the forces that drive our Sun? In so doing, can we better protect our astronauts and the orbiting satellites on which our global communications depend? Can we develop the scientific base of information necessary to save Earth from an incoming asteroid like the one we believe ended the epoch of the dinosaurs 65 million years ago? Will a "Big Crunch" follow the Big Bang, billions of years from now, or will our Universe expand endlessly?

In the decade ahead we have the opportunity to address many of these exciting and engaging issues, developing missions to gain new answers and enrich the story. There will be twists and turns along the way, unexpected discoveries that will show us the Universe is not quite the way we thought. And there will almost certainly be difficulties. Developing new tools to extend the frontiers of the known is always challenging. But a coherent, practical, and affordable strategy is feasible. Anchored by missions included in the Origins Initiative (see box on next page), NASA's Space Science Enterprise can provide more precise answers to fundamental questions about the formation and evolution of the Universe, how the Sun influences Earth, the history of planets and satellites in our solar system, and the occurrence of life either in our tiny region of space or in the larger neighborhood of our Galaxy.

[3]

The Origins Initiative

The Origins Initiative is a set of missions and enhancements to current programs (see below). It emerged from Space Science Enterprise strategic planning and President Clinton's call for a reconsideration of space science following the August 1996 announcement that a Martin meteorite contained possible evidence of ancient microscopic fossilized organisms. Enterprise planning, begun in early 1996, included hundreds of scientists, engineers, educators, and communicators of science. They developed science and technology "Roadmaps" for each of the four science "themes" (Structure and Evolution of the Universe, Astronomical Search for Origins, Solar System Exploration, and Sun-Earth Connection) around which the Office of Space Science (OSS) is organized.

The Roadmaps, together with National Academy of Sciences reports, provided the foundation for a workshop convened by the National Academy and a symposium chaired by Vice President Gore. These in turn led to a Presidential funding request for the Origins Initiative, aimed at following the 15-billion year chain of events from the birth of the Universe at the Big Bang; through the formation of the chemical elements, galaxies, stars and planets; through the mixing of chemicals and energy that cradled life of Earth; to the earliest self-replicating organisms and the profusion of life. Although the missions of the Origins Initiative are drawn primarily from the Roadmaps of the Astronomical Search for Origins and Solar System Exploration themes, the search for origins embraces elements of all four OSS themes.

The Origins Initiative includes:

- An enhancement to the Mars Surveyor Program enabling return of selected samples from Mars by a mission launched in 2005
- A series of missions to the outer planets including missions to Europa and Pluto and the Solar Probe, which first flies by Jupiter (first New Start in 2000)
- The Space Interferometry Mission (New Start in 2001)
- The Next Generation Space Telescope (New Start in 2003)
- The Terrestrial Planet Finder (New Start in 2007)
- Enhanced technology investments to enable the above missions

B. Fundamental Questions, Enterprise Goals, Science Goals, Science Objectives, Missions and Programs

Detailed Space Science planning begins with a set of Fundamental Questions (see box on "Fundamental Questions"). These questions—challenging and exciting to scientists and non-scientists alike and amenable to scientific progress—form the basis for our scientific program over the next several decades. To address these Fundamental Questions, the Space Science Enterprise—guided by the National Academy of Sciences, and in conjunction with the space science community—has laid out a series of broad Enterprise Goals (see box on "Enterprise Goals") and more specific Science Goals (see box on "Science Goals") to guide our activities over the next decade or two. Each Science Goal is addressed through more detailed Science Objectives (see box on "Science Objectives")—scientific investigations that can be accomplished within the next 5-6 years through one or more space missions and ground-based programs. (The relationships between Science Goals and Objectives are shown in Appendix A [omitted].) Success in achieving a Science Objective can be measured, permitting a clear assessment of progress. And achieving Science Objectives provides new knowledge from which we formulate new questions and science goals, completing the cycle of scientific advancement.
[4] Question—Goal—Objective—Mission and Program, successive layers revealing more and more detail; a logical way of understanding everything we do, of planning for the near-and longer-term, and of measuring our progress.

Fundamental Questions

1. How did the Universe begin and what is its ultimate fate?
2. How do galaxies, stars, and planetary systems form and evolve?
3. What physical processes take place in extreme environments such as black holes?
4. How and where did life begin?
5. How is the evolution of life linked to planetary evolution and to cosmic phenomena?
6. How and why does the Sun vary and how do the Earth and other planets respond?
7. How might humans inhabit other worlds?

Enterprise Goals*

1. Establish a virtual presence throughout the solar system, and probe deeper into the mysteries of the Universe and life on Earth and beyond.
2. Pursue space science programs that enable and are enabled by future human exploration beyond low-Earth orbit.
3. Develop and utilize revolutionary technologies for missions impossible in prior decades.
4. Contribute measurably to achieving the science, mathematics, and technology education goals of our Nation, and share widely the excitement and inspiration of our missions and discoveries.

* High-level Enterprise Objectives are shown as bullets on Figure 3 [all figures omitted], which is the Space Science Enterprise roadmap drawn from the NASA Strategic Plan.

Science Goals

1. Understand how structure in our Universe (e.g., clusters of galaxies) emerged from the Big Bang.
2. Test physical theories and reveal new phenomena throughout the Universe, especially through the investigation of extreme environments.
3. Understand how both dark and luminous matter determine the geometry and fate of the Universe.
4. Understand the dynamical and chemical evolution of galaxies and stars and the exchange of matter and energy among stars and the interstellar medium.
5. Understand how stars and planetary systems form together.
6. Understand the nature and history of our Solar System, and what makes Earth similar to and different from its planetary neighbors.
7. Understand mechanisms of long- and short-term solar variability, and the specific processes by which Earth and other planets respond.
8. Understand the origin and evolution of life on Earth.
9. Understand the external forces, including comet and asteroid impacts that affect life and the habitability of Earth.
10. Identify locales and resources for future human habitation within the solar system.
11. Understand how life may originate and persist beyond Earth.

[5]

This Plan proposes near-term (2000-2004) and long-term (2005-2020) missions and programs to address each of the Fundamental Questions, Science Goals and Objectives,

Science Objectives*

1. Observe the earliest structure in the Universe.
2. Observe the emergence of stars and galaxies in the very early Universe.
3. Observe the evolution of galaxies and the intergalactic medium.
4. Measure the amount and distribution of dark and luminous matter in the ancient and modern Universe.
5. Test the Theory of General Relativity.
6. Identify the origin of gamma-ray bursts and high-energy cosmic rays.
7. Study compact objects and investigate how disks and jets are formed around them.
8. Study the formation and evolution of the chemical elements and how stars evolve and interact with the interstellar medium.
9. Measure space plasma processes both remotely and in situ.
10. Observe and characterize the formation of stars, protoplanetary disks, and planetary systems, and detect Neptune-size planets around other stars.
11. Measure solar variability and learn to predict its effect on Earth more accurately.
12. Study the interactions of planets with the solar wind.
13. Characterize the history, current environment, and resources of Mars, especially the accessibility of water.
14. Determine the pre-biological history and biological potential of Mars and other bodies in the solar system.
15. Determine whether a liquid water ocean exists today on Europa, and seek evidence of organic or biological processes.
16. Investigate the composition, evolution, and resources of the Moon, small bodies, and Pluto-like objects across the solar system.
17. Complete the inventory and characterize a sample of near-Earth objects down to 1-km diameter.
18. Reconstruct the conditions on the early Earth that were required for the origin of life and determine the processes that govern its evolution.
19. Investigate the processes that underlie the diversity of solar system objects.

* Scientific investigations that can be accomplished within the next 5-6 years through one or more space missions and ground-based programs.

and to fill in many of the blanks in our understanding. It builds on the legacy of the pioneering decades of space exploration and the continuing achievements of America's current missions. It moves forward using advanced spacecraft that are smaller, less expensive, and developed more quickly than ever before. Enabled by revolutionary technologies, a dozen diverse missions may now set forth where before there was one. This way we establish a virtual presence throughout the solar system, with single spacecraft and fleets of visitors sailing to the most interesting, mysterious and instructive locations such as Mars and Europa. Orbiting observatories will probe the Universe beyond our local neighborhood—looking ever farther back in time at higher resolution to see new details, and peering through new windows of perception opened by technological innovation—scanning the entire electromagnetic spectrum from gamma-rays to radio wavelengths. Dedicated to finding new ways to deliver world-class science under constrained budgets, we will launch more—and more capable—spacecraft and support more powerful telescopes than during any previous decade in the Space Age.

Figure 1 illustrates some relationships among our Science Goals, and Table 1 [omitted] summarizes how the major missions and some continuing lines of smaller missions [8] expected to begin or be enhanced between 2000 and 2004 contribute to achieving those Goals. (A full acronym dictionary and descriptions of individual missions and programs are contained in the Appendices [all omitted].)

In addition to the major missions, continuing lines of smaller, more focused missions will also contribute to all Science Goals. Discovery-class missions, the Surveyor series of Mars missions, small and medium Explorers, the planned series of Solar-Terrestrial Probes, and an international Payloads line provide ongoing opportunities for frequent flight and the flexibility to pursue new scientific opportunities. The New Millennium program, a series of missions to test revolutionary technologies in flight, will also make important scientific contributions.

Other vital elements of the Enterprise include focused extensions of current missions, international collaborations, and non-spaceflight research programs. For example, the Voyager Interstellar Mission is using two spacecraft that were highly successful outer solar system explorers to probe the boundary between the solar system and interstellar space. The Galileo Europa mission will focus on what may be the only water world in the solar system besides Earth. The International Solar Maximum Program will focus on using current missions to chart the next solar maximum, greatly expanding our understanding of solar variability. All of these missions are highly economical and ensure that we get the most out of our valuable space assets. Other international collaborations in areas such as x-ray and infrared astronomy, solar physics, and solar system exploration will continue to enhance the overall scientific return on our investments. Non-spaceflight research programs such as those conducted from ground-based observatories (e.g., the Keck telescopes), aircraft (e.g., the Stratospheric Observatory for Infrared Astronomy—SOFIA), balloons, and sounding rockets—and field, laboratory, and theoretical programs (e.g., Astrobiology)—also make important contributions.

The same technological innovations that make possible smaller, cheaper, yet more efficient spacecraft, also enable the public—students and life-long learners alike—to participate in these missions as never before, interacting with front-line researchers via the Internet and sharing the excitement of learning new things. As we approach the Millennium, space science will provide awe-inspiring sights, drive progress in technologies which will support our entire society in the century ahead, and deliver priceless knowledge about who we are, how we got here, and what our fate may be—our Origins, Evolution and Destiny.

C. Technology: Enabling and Enhancing

A central element of the overall Space Science strategy is developing new capabilities and innovative techniques that will enable us to meet the challenges set forth in this Plan. The missions proposed here must be accomplished within fixed budgets that are dramatically lower than those of past generations of missions. In many cases they require fundamentally new observational and measurement techniques.

To meet these needs, the Space Science Enterprise is committed to an aggressive and carefully planned program of technology research, development and utilization in which mission concepts and supporting technologies are developed in synergism.

[9] This program will rest on four pillars:

1. An aggressive, long-range core technology program to enable the next generation of high performance and cost-effective Space Science missions.
2. A solid, mainstream set of focused technology programs to enable near-term emerging missions.
3. A flight validation program, complemented by advanced development to bring laboratory pre-prototypes to flight readiness.
4. A far-reaching mission studies and advanced concepts program to explore the full range of near- and long-term mission options and how to achieve them.

In parallel, the Space Science Enterprise will form partnerships with industry and universities. This will ensure that the revolutionary developments that result from this program will be infused into the American economy via a pathway of new products, new commercial applications, and enhanced competitiveness for the benefit of the entire nation.

The Space Science Enterprise is committed to the principles of open competition and merit review as a key to excellence. Early fundamental technology research will be selected through open, peer-reviewed competition. The later stages of technology development will be periodically reviewed for merit by independent panels of experts.

D. Education and Public Outreach

Technology Goals

1. Lower mission life-cycle costs and provide critical new capabilities through aggressive technology development.
2. Develop innovative technologies to address far-term scientific goals, spawn new measurement concepts and mission opportunities, and create new ways of doing space science.
3. Develop and nurture an effective science-technology partnership to help optimize mission concepts and infuse new technologies into science missions, with the goal of dramatically lowering mission cost and risk.
4. Stimulate cooperation among industry, academia, and government to ensure that the nation can reap the maximum scientific and economic benefit from its Space Science mission and technology programs.
5. Identify and fund the development of important "cross-cutting" technologies that support Space Science and other NASA Enterprises.

The NASA Strategic Plan mandates that we "involve the education community in our endeavors to inspire America's Students, create learning opportunities, enlighten inquisitive minds," and "communicate widely the content, relevancy, and excitement of NASA's missions and discoveries to inspire and to increase understanding and the broad application of science and technology."

The Space Science Enterprise has an extraordinary potential for contributing to NASA's total educational program through these and related goals, such as helping to ensure that a continuing supply of scientists, engineers, and technologists will be available to meet the needs of the twenty-first century.

[10]The unique contribution that Space Science can make to education and the public understanding of science rests on the discoveries and new knowledge coming from our missions and research programs. These have engaged people's imaginations, informed teachers, and excited students and the public about science and exploration.

To realize this potential more fully, we have developed a comprehensive, organized approach to making education at all levels and the enhanced public understanding of science integral parts of Space Science missions and research programs. We will work closely with the space science and education communities to develop a variety of long-term partnerships between educators and space scientists and to ensure that the information, ideas, and materials emerging from the Space Science program are developed in a variety of formats useful to educators and understandable by the public.

Education and Public Outreach Goals

1. Use our missions and research programs and the talents of the space science community to contribute measurably to efforts to reform science, mathematics, and technology education, particularly at the pre-college level, and to the general elevation of scientific and technical understanding throughout the country.
2. Cultivate and facilitate the development of strong and lasting partnerships between the space science community and the communities responsible for science, mathematics, and technology education.
3. Contribute to the creation of the talented scientific and technical workforce needed for the 21st century.
4. Promote the involvement of underserved/underutilized groups in Space Science education and outreach programs and their participation in Space Science research and development activities; and
5. Share the excitement of discoveries and knowledge generated by Space Science missions and research programs by communicating clearly with the public.

Education and Public Outreach Objectives

1. Have a substantial education and outreach program associated with every Space Science flight mission and research program.
2. Increase the fraction of the space science community directly involved in education at the pre-college level and in contributing to the broad public understanding of science.
3. Develop a presence in every state in the U.S. to serve as a focal point for encouraging and assisting scientists and educators to develop partnerships and, in so doing, contribute in a meaningful way to Space Science education and outreach.
4. Organize a comprehensive, national approach for providing information on and access to the results from Space Science education and outreach programs.
5. Continue, and refine or enhance where appropriate, programs dedicated to the development and support of future scientists and engineers.
6. Provide new opportunities for minority universities in particular and for underserved/underutilized groups in general to compete for and participate in Space Science missions and research programs.
7. Develop the tools to evaluate the quality, effectiveness, and impact of Space Science education and outreach programs.

Chapter Two

NASA and Planetary Exploration

by Amy Paige Snyder

Prelude to NASA's Planetary Exploration Program

Four and a half billion years ago, a rotating cloud of gaseous and dusty material on the fringes of the Milky Way galaxy flattened into a disk, forming a star from the innermost matter. Collisions among dust particles orbiting the newly-formed star, which humans call the Sun, formed kilometer-sized bodies called planetesimals which in turn aggregated to form the present-day planets.[1] On the third planet from the Sun, several billions of years of evolution gave rise to a species of living beings equipped with the intellectual capacity to speculate about the nature of the heavens above them.

Long before the era of interplanetary travel using robotic spacecraft, Greeks observing the night skies with their eyes alone noticed that five objects above failed to move with the other pinpoints of light, and thus named them *planets*, for "wanderers."[2] For the next six thousand years, humans living in regions of the Mediterranean and Europe strove to make sense of the physical characteristics of the enigmatic planets.[3] Building on the work of the Babylonians, Chaldeans, and Hellenistic Greeks who had developed mathematical methods to predict planetary motion, Claudius Ptolemy of Alexandria put forth a theory in the second century A.D. that the planets moved in small circles, or epicycles, around a larger circle centered on Earth.[4] Only partially explaining the planets' motions, this theory dominated until Nicolaus Copernicus of present-day Poland became dissatisfied with the inadequacies of epicycle theory in the mid-sixteenth century; a more logical explanation of the observed motions, he found, was to consider the Sun the pivot of planetary orbits.[5]

---

1. For a detailed description of the evolution of the solar system and individual planets, moons, and other bodies, see David Morrison and Tobias Owen, *The Planetary System* (New York, NY: Addison-Wesley, 1987). J. Kelly Beatty and Andrew Chaikin, eds., *The New Solar System* (Cambridge, MA: Sky Publishing Corporation, 1990) is a comprehensive guide to solar system bodies, their properties, and their postulated evolutionary histories.

2. Without the aid of telescopes, Mercury, Venus, Mars, Jupiter, and Saturn were the only planets visible from Earth, which then was not known to be a planet as well. Interestingly, all of the planets are named for Roman gods with the exception of Uranus, who was a Greek god.

3. For an excellent history of planetary studies leading up to and including the inception of NASA, see Ronald A. Schorn, *Planetary Astronomy: From Ancient Times to the Third Millennium* (College Station, TX: Texas A&M University Press, 1998). William Sheehan, *Worlds in the Sky: Planetary Discovery from Earliest Times through Voyager and Magellan* (Tucson, AZ: University of Arizona Press, 1992) describes the history of human study and knowledge of individual planets and other solar system targets.

4. Although Ptolemy is credited with the development of epicycle theory, Hipparchus was also responsible for its rise.

5. Copernicus' treaty, *De Revolutionibus Obrium Caelestium* (1543) was banned by the Catholic Church for nearly two hundred years beginning in 1616 due to its "heretical" idea of removing Earth from the universe's center. See Thomas S. Kuhn, *The Copernican Revolution: Planetary Astronomy in the Development of Western Thought* (Cambridge, MA: Harvard University Press, 1957) for more on Copernicus' contributions. It should be noted that the Grecian Aristarchus of Samos had proposed that Earth and the other planets revolved around the Sun even before Ptolemy put forth the theory of epicycles.

During the next 150 years, Johannes Kepler of Denmark deduced that planets moved around the Sun in elliptical orbits, and Isaac Newton of England identified the force that yielded these orbits and interactions between planetary bodies as gravity.[6]

By the early seventeenth century, people no longer had to rely only on their eyesight to study the heavens—the refractor telescope, comprised of glass lenses—made its debut in 1609 and thus marked the start of a new era of planetary discovery. Though not the very first to scan the night sky with a telescope, Italian mathematician Galileo Galilei worked hardest to perfect his refractors (his best telescope achieved a magnifying power of thirty) and reported details of the Moon's surface features, observed the phases of Venus, and discovered the four largest satellites of Jupiter. As subsequent generations of astronomers worked to improve the power of refractors, they had to build increasingly lengthy telescopes, separating the eyepiece from the objective lens, to combat the distortion in colors that occurred in telescopes with wider objective lenses. Newton's invention of the reflector telescope, which used a curved mirror in lieu of glass lenses, was not limited by this problem. Observatories throughout the Western world installed larger and more powerful telescopes of both types as they improved in capability over the next several centuries. The developments in telescopes led to the discovery of three more planets in the solar system—Uranus, Neptune, and Pluto—as well as numerous moons, asteroids, and comets previously unseen by the unaided eye.[7]

During the nineteenth century, the United States emerged as a player in the field of planetary astronomy. In 1840, New York University Professor John William Draper photographed the Moon for the first time, while William Cranch Bond used the 15-inch refractor at the Harvard College Observatory to discover satellites and rings around Saturn in 1848. But while American facilities like the Harvard College Observatory focused on visual studies of the planets during the nineteenth century, many of them turned their attention to stellar research by the turn of the century. With the advent of more powerful telescopes came the desire among many astronomers to look beyond the solar system and farther into the reaches of space. In addition to better telescopes came advances in photography and spectroscopy—techniques that were helpful to planetary astronomy to some extent but proved more appropriate to the study of more distant objects in the universe.[8] Developments in the theories of quantum mechanics, relativity, and cosmology further sparked interest in astrophysics rather than planetary studies. By the turn of the century, visual planetary astronomy was becoming a subject in actual disrepute, as when wealthy astronomer Percival Lowell made highly-publicized yet inaccurate claims that apparent lines streaking the surface of Mars were irrigation canals constructed by intelli-

6. Newton's famous work on gravity and other physical principles is *Philosophiae Naturalis Principia Mathematica* (1687).

7. For more on the development of the telescope, see Henry C. King, *The History of the Telescope* (London, England: Charles Griffin, 1955).

8. When used to record details on planets' surfaces, time-exposure photographic plates tended to register only blurs due to the atmosphere's movement, indicating that the human eye was still the better instrument for recording physical features. Spectroscopy was also more applicable to stellar astronomy, as the chemical elements producing spectral lines were easier to identify for stars and nebulae than for planets.

gent inhabitants.[9] By the early twentieth century, with many observatories following tacit rules that planetary studies could occupy no more than 10 percent of telescope time, astronomers had practically abandoned the objects that were once the focus of celestial studies.

The drought in U.S. solar system studies came to an end with the nation's experience in World War II.[10] Astronomers and other scientists with backgrounds in physics played a major role in the development of radar, instrumentation to explore infrared wavelengths, and means to better forecast weather. In addition to yielding new techniques useful to planetary astronomy, these efforts served the nation well in fighting the war; in return, the government increased its patronage of scientific studies across many disciplines, including all areas of astronomy. The war also gave rise to rocket and missile technology, advanced mainly by the Germans but then exploited by the victorious Allies. Bringing home leftover V-2 rockets and leading German rocket engineers, American military forces quickly went to work to study the technology of the vehicles that would soon forever change the way people understood the solar system.

The military's interest in the utility of planetary studies continued even after the war. The Army, Navy, and Air Force conducted and funded a number of projects and even built new observatories to perform planetary research to further their understanding of meteorology and radar. As early as 1946 the Army studied the Moon's thermal radiation using radar equipment, whose improvement led to more precise determination of distances to the planets and the nature of their surface features.[11] The Naval Research Laboratory began planetary radio astronomy work in 1947 to gather data on the Sun's radio emission as well as infrared radiometric properties of the planets, while the Office of Naval Research supported planetary work at several universities.[12] In an effort to better understand weather patterns and the atmosphere of Earth, the Air Force funded a project at Lowell Observatory to examine global atmospheric circulation on other planets and also erected a solar observatory in New Mexico to investigate the Sun's impact on Earth's atmosphere and ionosphere.[13] In addition to the scientific studies, the Army and Air Force both engaged in projects to develop rockets and satellites capable of traveling to the Moon and planets for both

9. As had others, Lowell mistakenly interpreted the *canali* ("channels") on Mars described by Italian astronomer Giovanni Schiaparelli as engineered waterways. Beginning in 1895 he published a series of books and articles based on this belief, including *Mars* (1895), *Mars and Its Canals* (1906), and *Mars as an Abode for Life* (1908). Although Lowell's writings incurred the scorn of many astronomers, he left a great legacy to planetary science in the Lowell Observatory, which he founded in Flagstaff, Arizona, for the primary purpose of planetary studies. The observatory became more reputable after Lowell's death in 1916 and with the discovery of Pluto there by Clyde Tombaugh in 1930. See William Graves Hoyt, *Lowell and Mars* (Tucson, AZ: University of Arizona Press, 1976) for more details.

10. See Ronald E. Doel, *Solar System Astronomy in America: Communities, Patronage, and Interdisciplinary Science, 1920–1960* (Cambridge, England: Cambridge University Press, 1996) for an excellent account of planetary studies performed between the end of World War I and the inception of NASA.

11. John H. DeWitt and E. King Stodola, "Detection of Radio Signals Reflected from the Moon," *Proceedings of the Institute of Radio Engineers* 37 (1949): 229–42. For a thorough examination of planetary radar astronomy, see Andrew J. Butrica, To See the Unseen: A History of Planetary Radar Astronomy (Washington, DC: NASA Special Publication (SP)-4218, 1996).

12. Joseph N. Tatarewicz, *Space Technology and Planetary Astronomy* (Bloomington, IN: Indiana University Press, 1990), p. 16; Doel, *Solar System Astronomy in America,* pp. 192–93, 236–40.

13. The major results of the Lowell Observatory project can be found in Earl C. Silpher et al., "The Study of Planetary Atmospheres: Final Report," US Air Force Contract AF 19(122)–162, Lowell Observatory, September 30, 1952.

military and peaceful purposes.[14] Responding to the project needs of the military, a number of commercial aviation firms also moved into the business of space vehicle and spacecraft development, which made the prospects of interplanetary travel even more realistic.

Indeed, advances in space technology and newly perceived advantages to knowing about the solar system had rekindled U.S. interest in the field of planetary science by the mid-1950s. Some astronomers distinguished for their work in stellar astronomy turned their attention to targets closer to home. Perhaps the most renowned, Gerard Kuiper of the University of Chicago, who researched double stars and stellar evolution before the war, used infrared spectrometry to confirm the presence of carbon dioxide in Mars' atmosphere and water at the polar caps in 1948.[15] Between 1953 and 1963, Kuiper compiled a photographic atlas of the Moon as well as a comprehensive, four-volume summary of human knowledge of the solar system.[16] During the decade, American as well as international astronomers also created organizations to plan and discuss research in planetary astronomy. The Mars Committee, for example, consisted of scientists that met annually to share the results of their observations of the Red Planet.[17] Even popular literature reflected the new preoccupation with the planets, with writers—including scientists and engineers—conveying to the public in simple words modern understanding of the solar system and their vision of human exploration of neighboring worlds.[18] Little did solar system enthusiasts know that before the next decade, national efforts in planetary astronomy would come together under a single organization and begin a new paradigm of operation as a reaction to a stunning space feat performed halfway around the world.

---

14. In 1952, top German rocket engineer Wernher von Braun expounded his vision of interplanetary vehicles that would transport humans to the surface of the Red Planet in Wernher von Braun, *The Mars Project*, English translation prepared by Henry J. White (Urbana, IL: University of Illinois Press, 1953). For Air Force space technology activities, see Nick A. Komans, *Science and the Air Force: A History of the Air Force Office of Scientific Research* (Arlington, VA: Office of Aerospace Research, 1966) and David N. Spires, *Beyond Horizons: A Half Century of Air Force Space Leadership* (Washington, DC: U.S. Government Printing Office, 1997).

15. Gerard P. Kuiper, "Planetary Atmospheres and Their Origin," in Gerard P. Kuiper, ed., *The Atmospheres of Earth and Planets*, 2nd ed. (Chicago, IL: University of Chicago Press, 1952), pp. 358–60.

16. Gerard P. Kuiper, ed., *Photographic Lunar Atlas: Based on Photographs Taken at the Mount Wilson, Lick, Pic du Midi, McDonald, and Yerkes Observatories* (Chicago, IL: University of Chicago Press, 1960); Gerard P. Kuiper, ed., *The Sun* (Chicago, IL: University of Chicago Press, 1953); Gerard P. Kuiper, ed., *The Earth as a Planet* (Chicago, IL: University of Chicago Press, 1954; second impression, 1958); Gerard Kuiper and Barbara M. Middlehurst, eds., *Planets and Satellites* (Chicago, IL: University of Chicago Press, 1961); Barbara M. Middlehurst and Gerard P. Kuiper, eds., *The Moon, Meteorites, and Comets* (Chicago, IL: University of Chicago Press, 1963).

17. For more details on the Mars Committee's activities, see, for example, E. C. Silpher and A. G. Wilson, "Report on the Conference of the Mars Committee" (held at Lowell Observatory, October 22–23, 1953) and "Minutes of a Meeting of the Mars Committee Held at the Headquarters of the National Geographic Society."

18. Examples of popular books on planetary exploration written in the 1950s include Willy Ley, *The Conquest of Space* (New York, NY: Viking Press, 1949); Joseph Kaplan, et al., *Across the Space Frontier* (New York, NY: Viking Press, 1952); Cornelius Ryan, ed., *Conquest of the Moon* (New York, NY: Viking Press, 1953); and Willy Ley and Wernher von Braun, *The Exploration of Mars* (New York, NY: Viking Press, 1956). The history of popular interest in Mars can be found in Martin Caidin and Jay Barbree, with Susan Wright, *Destination Mars: In Art, Myth, and Science* (New York, NY: Penguin Putnam, 1997). For more details on public interest in space exploration see Howard McCurdy, *Space and the American Imagination* (Washington, DC: Smithsonian Institution Press, 1997). McCurdy argues that human rather than robotic space exploration has been the predominant focus of imagination throughout history, and for this reason human space travel became the emphasis in the U.S. civil space program.

A Federal Home for Planetary Science

Both the United States and the Soviet Union had pledged to develop and launch scientific Earth satellites during 1957 and 1958 for the International Geophysical Year.[19] In addition to improving understanding of Earth's atmosphere and its relationship to the Sun during this worldwide research effort, the nations hoped to demonstrate the feasibility of launching and orbiting around Earth spacecraft that could serve scientific as well as other purposes. Prior awareness of the Soviet Union's project, however, did not placate the American public when news spread in October 1957 that the Communist nation had succeeded in lofting into orbit a beeping, basketball-sized satellite known as Sputnik. In reaction to the Soviet achievement, government and military officials quickly made plans to mobilize a major national space effort.

Solar system exploration played a significant role in the nation's earliest attempts to outdo the Soviet Union in space.[20] The first such scheme was put forth just three weeks after Sputnik's launch by William Pickering, director of the Jet Propulsion Laboratory (JPL). Calling his proposal Project Red Socks, Pickering envisioned sending robotic probes to the Moon.[21] Though the probes would be equipped with scientific payloads, Project Red Socks' main purpose was to demonstrate the United States' capability to reach Earth's satellite and travel beyond. Early in 1958, the Advanced Research Projects Agency (ARPA), which then had responsibility for the nation's space projects, considered the proposal. By March, Secretary of Defense Neil McElroy announced that the United States would attempt to send robotic envoys to explore up-close another body in the solar system.[22] [II-1, II-2]

Under the direction of ARPA, the Air Force, the Army, and JPL immediately went to work to develop the hardware for the project, renamed Pioneer. Some have argued that the differences between the two military branches' approaches to Pioneer doomed the project in its planning stages.[23] Whether actually due to such differences or simply to the fact that space launch was a very new activity, Project Pioneer encountered one failure after the next. An explosion of its Thor-Able launcher shortly after liftoff on

19. James C. Hagerty, The White House, "IGY Statement," July 29, 1955. This document appears as I-17 in John M. Logsdon, gen. ed., with Linda J. Lear, Jannelle Warren-Findley, Ray A. Williamson, and Dwayne A. Day, *Exploring the Unknown: Selected Documents in the History of the U.S. Space Program, Volume I: Organizing for Exploration* (Washington, DC: NASA SP-4407, 1995), pp. 200–01.

20. Two books authored by leaders of NASA's early space science program are particularly valuable in providing a comprehensive, inside look at the development of NASA's space science program. These books are Homer E. Newell, *Beyond the Atmosphere: Early Years of Space Science* (Washington, DC: NASA SP-4211, 1980) and John E. Naugle, *First Among Equals: The Selection of NASA Space Science Experiments* (Washington, DC: NASA SP-4215, 1991).

21. Jet Propulsion Laboratory, *Project Red Socks* (Pasadena, CA: Jet Propulsion Laboratory, California Institute of Technology, October 21, 1957), pp. 2–3; William Pickering to Lee DuBridge, with attachments, October 25, 1957. Unless otherwise noted, all unpublished documents cited in this essay may be found in the NASA Historical Reference Collection, NASA History Division, NASA Headquarters, Washington, DC. Founded in 1936 as the Guggenheim Aeronautical Laboratory of the California Institute of Technology, JPL started as a rocketry research and development center operated by the California Institute of Technology under contract from the Army Ordnance. The center began tinkering with space probes after World War II.

22. Roy Johnson, ARPA Director, to Commanding General of Ballistic Missiles Div, ARDC, "Order to Proceed with Development of Three Lunar Probes," March 27, 1958; U.S. Army Ordnance Missile Command, "Development and Funding Plan for Project One, ARPA Order 1-58, as Amended," May 15, 1958.

23. Edward Clinton Ezell and Linda Neuman Ezell, *On Mars: Exploration of the Red Planet, 1958–1978* (Washington, DC: NASA SP-4212), p. 25.

August 17, 1958, prevented the first lunar probe from even passing through Earth's atmosphere. Two months later Pioneer 1 was successfully launched and returned data on near-Earth space, but failed to reach the Moon because its second stage shut down prematurely. Pioneer 2 failed when its booster's third stage failed to ignite. While Pioneer 3 traveled away from Earth for 38 hours and discovered a second Van Allen belt of trapped energetic particles around Earth, it failed to arrive at its lunar target when the Jupiter launcher's first stage cut off prematurely. By the time Pioneer 4 was launched in March 1959, passing too far from the Moon to use its scanning instruments, the Soviets had already successfully flown Luna 1 by the Moon and would soon crash-land a second Luna on the Moon's surface. Three more Pioneers failed by 1960, and the project came to an unsuccessful end.[24]

As plans for Project Pioneer were getting underway, President Eisenhower proposed to Congress in April 1958 the creation of a civilian agency to begin handling the nation's activities in space. A peaceful approach to space operations, the President reasoned, was preferable in the eye of the national and global publics to allowing the military to continue responding to the Soviet space challenge.[25] Receiving congressional support for this proposal, Eisenhower approved the law establishing the National Aeronautics and Space Administration (NASA), which began operations on October 1, 1958.[26] From that point onward, the new agency was responsible for national programs of human spaceflight, passive communications, meteorology, aeronautics research, and space science.[27]

Not specifying particular space science disciplines or projects that NASA should pursue, the space agency's enacting legislation only noted an obligation regarding "the expansion of human knowledge of phenomena in the atmosphere and space."[28] This language gave NASA the responsibility to decide how it would design its space science program. Soliciting the advice of scientists renowned in a variety of fields, the agency began within its first few months to assemble a space science program that would lead to greater understanding of the Earth and the cosmos by conducting investigations with spacecraft as well as ground-based facilities.[29] With scientists expressing great interest in making solar system exploration part of the national space science effort, NASA managers began planning at once for a repertoire of missions that would travel into deep space.

While space science enthusiasts had little difficulty reaching the decision to make solar system studies a scientific priority, arriving at a consensus on where to go first proved formidable. It became clear early on that NASA officials, scientists, and even spacecraft engineers made a distinction between lunar and planetary exploration. With the Soviets aiming for the Moon, NASA's top administrators could not resist making a successful robotic visit to the Moon and its environs its first priority in the area of solar system explo-

24 In 1965, NASA revived Project Pioneer. The new series of Pioneer spacecraft complemented interplanetary data returned from the Mariner probes.

25. Alison Griffith, *The National Aeronautics and Space Act: A Study of the Development of Public Policy* (Washington, DC: Public Affairs Press, 1962), pp. 100–01; Newell, Beyond the Atmosphere, pp. 88–89.

26. NASA grew out of the National Advisory Committee for Aeronautics, which had been established in 1915. "National Aeronautics and Space Act of 1958," Public Law 85–568, 72 Stat., 426. Signed by the president on July 29, 1958.

27. Newell, *Beyond the Atmosphere,* pp. 95–101. The military retained authority over active communications and reconnaissance. Responsibility for several other relevant areas, such as launch vehicle development, was left to NASA and the Department of Defense to arrange.

28. "National Aeronautics and Space Act of 1958," Sec. 102(c)(1).

29. Newell, *Beyond the Atmosphere,* pp. 100–15.

ration. As the next section of this essay reveals, NASA pursued scientific exploration of the Moon with great vigor from the start, putting this goal ahead of sending spacecraft to the planets. An important first step in succeeding in this effort, however, was for the space agency to make clear its interests and delineate its authority in making such decisions to JPL, which had been transferred from the Army to NASA by executive order in December 1958.[30] Destined to become NASA's premier facility for managing solar system exploration missions, JPL preferred to bypass the Moon and take on the challenge of sending probes to worlds beyond the Earth-Moon system.

Believing that beating the Soviets to Venus or Mars would be a loftier triumph than reaching the Moon, JPL managers and engineers began thinking about solar system exploration missions NASA could perform—even before the Center was officially transferred to the agency. Interpreting a memo from NASA's Office of Space Flight Development Director Abe Silverstein asking JPL to consider future space projects as a request to devise a long-range program for the agency, JPL developed a preliminary five-year plan of solar system exploration in November 1958.[31] By April 1959, JPL scientists produced a final report that addressed detailed aspects of sending spacecraft to the planets. JPL advised taking every possible opportunity to send probes to Mars and Venus, while filling in the "down time" with missions to the Moon, launching them on Atlas-Vega and Saturn 1 boosters. The researchers also suggested that NASA undertake a complementary program of ground-based planetary studies.[32] [II-5]

JPL's report clearly expressed the Center's desire to focus on planetary missions, with lunar exploration as a secondary goal. But NASA had opted by mid-1959 to concentrate on lunar exploration as its venue of competition with the Soviets and to reject JPL's plans to develop probes bound for Mars and Venus, piquing Pickering's concern about JPL's involvement with the space agency. That December, officials from NASA Headquarters and JPL exchanged correspondences concerning JPL's role in planning and management of the solar system exploration program.[33] [II-6] On December 28, a delegation from NASA Headquarters visited JPL to discuss plans for solar system exploration and to better define the responsibilities of the two entities in mission planning and execution.

By the end of the meetings, the attendant officials resolved that NASA Headquarters would remain responsible for overall program planning, while JPL would lead the engineering and execution of lunar and planetary missions—a position that it has maintained for the most part through the present.[34] NASA officials assured JPL that while lunar exploration remained the agency's main area of solar system interest, planetary work would get underway soon, with launches to Mars and Venus when-

---

30. For more on JPL's involvement with NASA, see Clayton R. Koppes, *JPL and the American Space Program: A History of the Jet Propulsion Laboratory* (New Haven, CT: Yale University Press, 1982); Newell, Beyond the Atmosphere, pp. 258–73.

31. John F. Froehlich, "Minutes of Meeting on N.A.S.A. Space Program of October 27, 1958," October 28, 1958; R. Newburn and M. Neugebauer, "Preliminary Consideration of a Limited Class of Problems Suitable for Study by Interplanetary Probes and/or Satellites," November 13, 1958.

32. Albert R. Hibbs, ed., "Exploration of the Moon, Planets, and Interplanetary Space" (Pasadena, CA: JPL Technical Report 30–1, April 30, 1959).

33. Richard Horner to William Pickering, December 16, 1959; William Pickering to Abe Silverstein, December 17, 1959; Abe Silverstein to William Pickering, December 21, 1959.

34. NASA General Management Instruction 2–2–11, "NASA-JPL Relationships," August 2, 1960.

ever they were in "optimum position for a planetary mission." A NASA ten-year plan created just days before the meeting had already affirmed the agency's commitment to studying the planets.[35] Finally, NASA pledged to create a single working committee for lunar and planetary exploration in the NASA management structure.[36] [II-7] Soon thereafter, Homer Newell, assistant director for space sciences and one of the Headquarters delegates on the trip, created the Lunar and Planetary Programs Office, to be headed by NASA officials but staffed by outside scientists, to recommend to NASA what projects the agency should undertake. As Newell noted years after he left NASA, although the NASA-JPL entanglement required the two entities to wrestle with "knotty issues in human relations," the JPL staff was instrumental in "laying the groundwork for the phenomenal successes that were later achieved in investigating [both] the [M]oon and planets."[37]

While getting scientific instruments into space became the focus of NASA's solar system exploration program, supporters of planetary and lunar studies pushed the agency to fund ground-based planetary astronomy as well. With scientists still having much to learn about the solar system, planetary enthusiasts argued that a strong ground-based program would serve as an economical way to gain knowledge of the planets needed to prepare spacecraft bound for neighboring worlds. Throughout the early years of NASA's existence, military, commercial, and non-profit groups with interest in the budding space program completed studies on the feasibility and importance of a federally funded, ground-based planetary program.[38] In June 1960, the topic of ground-based observatories became the focus of a Space Science Board conference on planetary atmospheres, where some of the attendees passed a resolution that the Board recognize the importance of federal support for ground-based planetary research.[39] That same month, Kuiper, by then a consultant to NASA's Lunar and Planetary Program Office, stressed a ground-based program's merit, expressing to Newell that a ground program was "not merely a matter of economy," but also "a logical necessity" for obtaining an "integrated" understanding of the data.[40] [II-8] The following year, National Academy of Sciences President Lloyd Berkner sent to NASA Administrator James Webb the Board's recommendation that NASA fund a strong program of both space and ground space science research.[41]

The urging of these groups that NASA support a ground-based planetary astronomy program came to fruition almost as soon as they voiced their desires, as NASA immediately

---

35. NASA, Office of Program Planning and Evaluation, "The Ten Year Plan of the National Aeronautics and Space Administration," December 16, 1959.

36. Homer Newell, memo to file, "Trip Report for the Visit to the Jet Propulsion Laboratory on December, 28, 1959 by Homer E. Newell, Jr., Newell Sanders, J. A. Crocker, Morton J. Stoller," December 30, 1959, pp. 1–3.

37. Newell, *Beyond the Atmosphere*, p. 259.

38. For a more detailed description of the evolution of federally supported, ground-based planetary studies, see Joseph N. Tatarewicz, *Space Technology and Planetary Astronomy* (Bloomington, IN: Indiana University Press, 1990).

39. Space Science Board, "Minutes of the Eighth Meeting," June 25, 1960. NASA Deputy Administrator Hugh Dryden, who was present at the meeting, suggested to the Board that the resolution be passed along to President Kennedy's science advisor, George Kistiakowsky, to be considered as part of a new national science policy.

40. Gerard Kuiper to Homer Newell, "Need for a Ground-Based Lunar and Planetary Observatory," June 18, 1960.

41. Lloyd V. Berkner to James E. Webb, March 31, 1961; Space Science Board, "Support of Basic Research for Space Science," March 27, 1961.

began subsidizing new and current observatories and laboratories to study the solar system. One of NASA's earliest major contributions to ground-based solar system research was the funding of the University of Arizona's Lunar and Planetary Laboratory. In 1960, Kuiper relocated the lunar and planetary operations of the Yerkes Observatory to this new facility, which aimed to serve as a "research and teaching unit concerned with the study of the Moon and the planets."[42] Staff of the Laboratory have assisted in collecting and interpreting data from NASA's solar system exploration missions since the Laboratory's inception. Throughout the 1960s, NASA also funded upgrades of several ground-based telescopes to make them more suitable for planetary astronomy purposes. The space agency built an observatory on Mauna Kea in Hawaii that has specialized in planetary investigations. In addition, NASA began development in 1958 of the Deep Space Network—the first worldwide, civilian satellite communications network. Consisting of three radio antenna stations in California, Spain, and Australia, the Deep Space Network has the ability to continuously track robotic spacecraft and remains NASA's means for communicating with probes sent into the solar system.[43]

Within a few years of its inception, NASA had become the primary supporter and coordinator of solar system exploration activities in the United States. The creation of a national space agency equipped with millions of dollars of federal money for planetary and lunar projects and spurred by international competition provided the invigoration solar system astronomy needed to move forward after its decline in the early part of the twentieth century. Moreover, technological progress had equipped astronomers with the means not only to study but also to *explore* the solar system *in situ* with spacecraft, the "sine qua non" of space science.[44] Between NASA Headquarters, JPL, and the other NASA Field Centers, the federal government had created an institution that, beginning in the 1960s, transformed scientists' knowledge about the Moon and planetary system.

NASA Shoots for the Moon

While ARPA was striving to successfully deliver Pioneer spacecraft to the Moon's vicinity in the fall of 1958, Naval Research Laboratory theoretical physicist Robert Jastrow arrived at NASA Headquarters to head the agency's program of basic research in astronomy and planetary science. Within a short time he came across *The Planets: Their Origin and Development*, a 1952 book in which Nobel laureate Harold Urey put forth his theories of lunar evolution.[45] Fascinated by Urey's arguments that the Moon was geologically dead

42. Ewen A. Whitaker, *The University of Arizona's Lunar and Planetary Laboratory: Its Founding and Early Years* (Tucson, AZ: University of Arizona [sometime after August 1985]), p. 29. This document provides an excellent history of the Lunar and Planetary Laboratory.

43. The first components of the Deep Space Network were two antennas built by JPL in the Mojave Desert. These antennas were originally intended to track and receive telemetry from the military's Pioneer probes and to test the feasibility of long-range satellite communications. JPL later used the antennas for a ground-based Venus radar experiment. Butrica, *To See the Unseen*, pp. 36–38. For more on the Deep Space Network's history, see William R. Corliss, *A History of the Deep Space Network* (Washington, DC: NASA CR-151915, 1976); Nicholas A. Renzetti, ed., *A History of the Deep Space Network from Inception to January 1, 1969*, Vol. 1, JPL Technical Report 32–1533, September 1, 1971; and Craig B. Waff, "The Road to the Deep Space Network," *IEEE Spectrum* (April 1993): 53.

44. Newell, *Beyond the Atmosphere*, p. 133.

45. Harold C. Urey, *The Planets: Their Origin and Development* (New Haven, CT: Yale University Press, 1952). Urey, together with George M. Murphy and Ferdinand G. Brickwedde, won the Nobel Prize for Chemistry in 1934 for their discovery of the existence of heavy water, the molecules of which consist of an atom of oxygen and two atoms of heavy hydrogen or deuterium. For an overview of Urey's achievements, see Stephen G. Brush, "Nickel for Your Thoughts: Urey and the Origin of the Moon," *Science* 217 (1982): 891–98.

and that its interior recorded conditions of the early solar system, Jastrow contacted Urey to discuss the prospects for scientific exploration of the Moon. In January 1959, just after the Soviet Luna 1 had passed within 5000 kilometers of the Moon and the American Pioneer program had endured its third failure, Urey visited NASA Headquarters to share his views on lunar exploration's scientific value.[46] After talking together, Jastrow and Urey approached Newell about initiating a special effort to land on the Moon to catch up with the Soviets. Receptive to the idea, Newell asked Jastrow and Urey to draft a memo proposing that NASA institute a plan to crash-land spacecraft laden with scientific instruments on the Moon's surface over the next two years, with the goal of a soft lunar landing by 1961.[47] Serving as the first formal scientific rationale for lunar exploration, this memo proclaimed that NASA should undertake a program of lunar exploration in accordance with goals put forth by planetary scientists. Urey and Jastrow asserted in the memo, "It is our opinion that a study of the Moon is more important than a study of Venus or Mars, from the standpoint of the origin of the solar system."[48] [II-3]

During the course of 1959, NASA officials stirred by the Soviet Union's success in reaching the Moon took heed of the proposition and elevated lunar exploration to a very high priority of the national space program—putting it in a more prominent position than planetary exploration.[49] [II-4] Two meetings of the newly formed ad hoc Working Group on Lunar Exploration in February 1959 laid out the main lines of a proposed automated lunar program. In late May, Silverstein and Newell reprogrammed two Atlas-Vega flights as lunar orbiters; two months later Silverstein instructed JPL to cancel plans for some Venus and Mars missions and to redesign the Vega upper stage for a series of lunar orbiting missions.[50] By July, NASA Administrator T. Keith Glennan formally recommended to a group of top presidential advisors and security officials that the nation concentrate its solar system exploration program on the Moon because it best supported national security goals and was a more proximate, accessible target than the planets.[51] With the approval of Glennan's proposal, NASA Headquarters ordered JPL to cancel its January 1961 Venus mission, leaving the center to work only on lunar missions.[52]

NASA's lunar interest did not stop at orbiting science spacecraft around the Moon but extended to crashing them into its surface. At the very end of 1959, NASA Headquarters personnel asked JPL to begin planning for a hard lunar landing project.[53] Taking the name Ranger, the project would consist of two initial engineering flight tests that would perform experiments on fields and charged particles in Earth's upper atmos-

---

46. Robert Jastrow, *Journey to the Stars: Space Exploration—Tomorrow and Beyond* (New York, NY: Bantam, 1989), pp. 9–14.

47. Homer Newell, "Meeting of Harold Urey, Robert Jastrow, John O'Keefe, and Homer Newell," January 16, 1959. This meeting is recorded in the author's notebook.

48. Jastrow, *Journey to the Stars*, p.13; R. Cargill Hall, *Lunar Impact: A History of Project Ranger* (Washington, DC: NASA SP-4210, 1977), p. 15.

49. U.S. Congress, House, Committee on Science and Astronautics, *The First Soviet Moon Rocket*, Report of the Committee, 86th Cong., 1st sess., on H.R. 1086 (Washington, DC: Government Printing Office, 1959), p. 6; Newell to Silverstein, March 23, 1959.

50. William Pickering to Abe Silverstein, July 10, 1959, as cited in Hall, *Lunar Impact*, p. 20.

51. T. Keith Glennan, memorandum for the file, July 24, 1959, as cited in Hall, *Lunar Impact*, p. 20.

52. William Pickering to Abe Silverstein, August 4, 1959, as cited in Hall, *Lunar Impact*, p. 20.

53. Homer Newell, memo to file, "Trip Report for the Visit to the Jet Propulsion Laboratory on December, 28, 1959, by Homer E. Newell, Jr., Newell Sanders, J.A. Crocker, Morton J. Stoller," December 30, 1959.

phere and near-Earth space, and three subsequent spacecraft which would gather scientific data on the Moon before impacting it. NASA officials hoped Ranger would demonstrate the technology necessary for spacecraft bound for deep space as well as the abilities to deliver scientific payloads to a celestial target, position experiments, perform a proposed scientific program, and transmit the results to Earth.[54] Instruments planned for the crash landers included a television camera to return close-up photographs of the surface, a seismometer, a gamma-ray spectrometer to determine the surface's chemical composition, and radar for reflectivity measurements. Silverstein hoped JPL would complete the project in thirty-six months.[55]

The five originally scheduled Ranger missions did, in fact, make it off the launch pad within three years. All five, however, failed, preventing the return of virtually all of the planned science data. Booster failures and inaccurate launch trajectories contributed to the first three Ranger failures. Ranger 4 crashed without control on the far side of the Moon, while Ranger 5 experienced a power failure that ended the mission. NASA and JPL investigations of the series of Ranger failures revealed that failures specific to the spacecraft themselves resulted from the fact that the missions had become increasingly risky when engineers removed many of the spacecraft's redundant systems in an effort to meet the Atlas-Agena launch vehicle's weight limitations.[56] After a complete design review, changes in the project's management and development practices, and the addition of several redundant features, NASA attempted to send four more Rangers to the Moon. Ranger 6 launched in January 1964 and successfully reached the Moon, but failed to transmit any photographs from its six television cameras, leading NASA, JPL, and Congress to conduct further investigations into the management and engineering processes of JPL and the space agency.[57] The congressional report concluded that NASA Headquarters failed to provide enough oversight, while JPL did not adhere to NASA's directions.

The United States finally claimed a completely successful shot at the Moon on July 31, 1964, when Ranger 7 became the first American spacecraft to return meaningful data before striking the lunar surface. Returning more than 4000 high-quality photographs of the Moon's surface, the spacecraft's success after a long string of failures lifted the morale of space supporters in NASA, JPL, Congress, and the public at large.[58] [II-11] Two subsequent Ranger spacecraft proved equally successful, with the final mission, Ranger 9, carrying the last ten minutes of the spacecraft's journey to the surface on live television—a public-stirring feat the Soviets had not yet accomplished. With the end of the Ranger program, NASA had achieved the best view to date of the

---

54. The history of the Ranger program is exposed in great detail in Hall, *Lunar Impact*.

55. JPL's early Ranger planning efforts can be seen in JPL, "Ranger Project Development Plan, Revision," June 5, 1961.

56. In truth, it turned out, engineers had underestimated the carrying capacities of the launchers, and thus needlessly removed many vital redundant systems. Hall, *Lunar Impact*, pp. 65–67; JPL, "Ranger RA-5 Failure Investigation, Report of JPL Failure Investigation Board," November 13, 1962; NASA, "Final Report of the Ranger Board of Inquiry," November 30, 1962.

57. JPL, "RA-6 Investigation Committee Final Report" Engineering Planning Document No. 205, February 14, 1964; NASA, "Final Report of the Ranger 6 Review Board," March 17, 1964; U.S. Congress, House, Committee on Science and Astronautics, *Investigation of Project Ranger: Hearings before the Subcommittee on NASA Oversight*, 88th Cong., 2nd sess., no. 3 (Washington, DC: Government Printing Office, 1964).

58. For an example of the media's response to Ranger 7's success, see "Impact!" editorial in *The New York Times*, August 2, 1964, p. E1.

Moon and its craters, returning photographs revealing features as small as a meter in size, and had also developed technologies and spacecraft designs to use on future solar system missions.

By the time the first Ranger mission launched, however, NASA's lunar exploration program had begun to change in fundamental ways. When in May 1961 President John F. Kennedy made his landmark announcement of the U.S. intent to send humans to the Moon's surface and return them safely to Earth, the nation readily embraced this chance to make major strides over the Soviets in space.[59] Already underway, the Ranger program piqued the interest of supporters of the manned lunar landing project, dubbed Apollo.[60] Although Ranger originally had been conceived as a program of scientific exploration consisting of five probes, many NASA officials believed the missions could contribute to the understanding of the surface as well as the landing systems that Apollo required. [II-10] Congress willingly appropriated the necessary funds for NASA to fly Rangers 6 through 9 to return high-resolution photographs of the lunar surface.[61] At the request of NASA to find a way to improve the missions' reliability and ensure the success of Ranger's Apollo objectives, JPL removed all scientific experiments from the additional Rangers, leaving only the television cameras.[62]

Despite returning excellent photographs, the Ranger program did not allow scientists to draw many conclusions about the nature or evolution of the Moon because they lacked other vital data. To the chagrin of planetary scientists, NASA had prioritized lunar studies over other solar system targets and then essentially stripped science for its own sake from the lunar exploration program.[63] JPL Lunar Program Director Clifford Cummings made the point while briefing Vice President Lyndon Johnson on October 4, 1961: "Originally our lunar program had been oriented toward scientific and technological objectives. Now...the emphasis has been changed so that support of the manned operations is the primary objective, and space technology and lunar science are secondary."[64]

By November 1961 NASA Administrator Webb had reorganized the agency to create separate offices for space science and manned space flight. Within two years the new directors of the respective offices, Newell and D. Brainerd Holmes, formed a working group of representatives from both offices to recommend a program of space

---

59. John F. Kennedy, "Urgent National Needs," Speech to a Joint Session of Congress, May 25, 1961. See document III-12 in John M. Logsdon, gen. ed., *Exploring the Unknown*, 1:453–54.

60. Oran Nicks to Edgar Cortright, "Lunar Program Support to Manned Lunar Landing," December 6, 1961.

61. NASA Associate Administrator Hugh Dryden testified before the Senate Committee on Science and Astronautics that Apollo's success depended on an improved understanding of the Moon's surface, and requested that Congress extend the Ranger program to meet Apollo's needs. U.S. Congress, Senate, Committee on Science and Astronautics, *NASA Authorization for Fiscal Year 1962: Hearings before the Committee*, 87th Cong., 1st sess., on H.R. 6847 (Washington, DC: Government Printing Office, 1961), p. 56.

62. Oran Nicks to William Pickering, June 9, 1961.

63. Homer Newell realized the disappointment of planetary scientists regarding the change in focus of NASA's lunar exploration program. He expressed to scientists opposed to the burgeoning Apollo project that he expected NASA to reestablish a program that would better serve planetary science's interests in the future. Urey to Newell, October 24, 1962; Homer Newell to Harold Urey, November 15, 1962; Hall, *Lunar Impact*, p. 181.

64. Clifford I. Cummings, "The Lunar Program," *Minutes of Briefing on the Occasion of the Visit of Lyndon B. Johnson, Vice President of the United States of America to the Jet Propulsion Laboratory*, October 4, 1961 (Pasadena, CA: Jet Propulsion Laboratory, October 19, 1961), p. 1.

science data acquisition that would assist planning for Apollo.[65] The Office of Space Science carried out two additional robotic lunar exploration programs for the manned program's benefit.[66] The first of these programs, Surveyor, started as an effort both to softly land spacecraft on the lunar surface and to develop lunar orbiters that would make scientific measurements over several years.[67] Apollo's dominance soon curtailed Surveyor's long-term scientific objectives and modified the program to serve the former's needs. NASA canceled the lunar orbiter portion of the project after grappling with schedule delays and cost escalation in both the Ranger and Surveyor programs as well as problems in the development of Surveyor's launch vehicle, the Atlas-Centaur. JPL managed to launch seven Surveyor spacecraft between 1966 and 1967 with five successful soft landings on the Moon.[68] These probes landed on many types of lunar terrain and returned numerous photographs and data on the composition of the surface.[69]

NASA revived the orbiting spacecraft concept in its second robotic lunar program, Lunar Orbiter. Although designated as an Apollo support project from the start, Lunar Orbiter had the potential to return a significant amount of scientific data. As a result, the project appealed to NASA Headquarters officials favoring manned space flight as well as space science and met the approval of both groups.[70] With JPL already overwhelmed by Ranger and Surveyor, in March 1963 NASA assigned the Langley Research Center in Hampton, Virginia, the task of managing Lunar Orbiter. The objectives of the program were to launch five spacecraft that would return one-meter resolution photographs and other data about the Moon's surface from orbit to facilitate planning Apollo landing sites.[71] From its first launch on August 10, 1966, all five Lunar Orbiter missions successfully fulfilled their objectives. The Lunar Orbiter project provided Apollo with the best lunar surface maps to date and gave flight operators experience tracking spacecraft in orbit around the Moon. After the first three missions satisfied almost all of the Apollo requirements, photographing twenty potential landing sites, scientists were able to use the last two missions to image targets of their choice on the near and far sides of the Moon.

Although NASA's primary intention for Apollo was to demonstrate that the United States could trump the Soviets in engineering a manned lunar landing, many groups supportive of space science pushed the agency to have the Apollo astronauts conduct a program of scientific exploration during their lunar stays. As early as 1962,

65. Homer E. Newell and D. Brainerd Holmes, "Establishment of a Joint OSS/OMSF Working Group," October 22, 1962; "Memorandum of Agreement between Office of Manned Space Flight [and] Office of Space Sciences, Scientific Interfaces," no date, signed by E.M. Cortright, July 25, 1963, and J.F. Shea, July 26, 1963.

66. John M. Eggleston to Director, Manned Spacecraft Center, "Utilization of Orbiter and Surveyor in Support of Apollo and Apollo Applications Program Objectives," January 18, 1967.

67. Erasmus Kloman, *Unmanned Space Project Management: Surveyor and Lunar Orbiter* (Washington, DC: NASA SP-4901, 1972), no pagination.

68. Development of the Atlas-Centaur was eventually transferred from the Department of Defense to NASA.

69. NASA, Office of Space Science and Applications, Lunar and Planetary Division, Surveyor Program, *Surveyor Program Results* (Washington, DC: NASA SP-184, 1969).

70. Hall, *Lunar Impact*, p. 209.

71. Bruce Byers, Destination Moon: A History of the Lunar Orbiter Program (Washington, DC: NASA TM-3487, 1977), pp. 96–7.

the Space Science Board polled members of the scientific community for their opinions on possible landing sites for the Apollo missions and experiments the astronauts could conduct.[72] [II-12, II-13] In addition to suggesting what types of space science data robotic spacecraft needed to acquire for Apollo, Newell and Holmes's Joint Working Group also developed Apollo science objectives. Newell solicited the assistance of geologists from the United States Geological Survey to support studies relevant to their expertise.[73] NASA even conducted several conferences to gather scientists interested in the Moon to help prioritize scientific plans and select landing sites.[74] In 1968 NASA established the Lunar Science Institute, a lunar sample and data research facility to be used by university researchers and managed by a university-based consortium; it was located near the Manned Spacecraft Center and Lunar Receiving Laboratory in Houston.[75]

Despite the interest NASA showed, its actual actions regarding science on Apollo were only lukewarm in intensity. On the very first lunar landing mission, NASA officials ended up flying a smaller scientific package than it had intended due to weight requirements of the lunar module and because the larger payload proved cumbersome for suited astronauts to manage.[76] Scientists who had played integral roles in the development of the Apollo science program, including Gene Shoemaker, the geologist who headed the Joint Working Group, denounced NASA for neglecting science in the manned space program and failing to assign any astronauts with scientific backgrounds to Apollo crews as of 1969.[77] Cutbacks in NASA's FY 1971 budget leading to the cancellation of two Apollo missions (in addition to one already canceled earlier) further outraged scientists who counted on the potential scientific returns of those missions. Associate administrator of the Office of Manned Space Flight George Mueller recognized Apollo's weak commitment to science during the first few flights; the last three missions, Apollo 15, 16, and 17, thus carried significantly more scientific experiments aboard the command and service modules as well as lunar surface experiments than their predecessors.[78] Although scientists could not conclusively determine the Moon's origin and evolutionary history

72. Space Science Board memorandum, "Scientific Program for the Apollo Mission," April 20, 1962; "Summary of Responses to SSB Inquiry on Scientific Program for the Apollo Mission," in *Scientists' Testimony on Space Goals, Hearings before the Committee on Aeronautical and Space Sciences*, U.S. Senate, 88th Cong., 1st sess. (Washington, DC: Government Printing Office, June 10–11, 1963).

73. Robert Gilruth to Thomas Nolan, USGS Director, March 29, 1963.

74. William David Compton, *Where No Man Has Gone Before: A History of Apollo Lunar Exploration Missions* (Washington, DC: NASA SP-4214, 1989) provides an excellent overview of the role of science and scientists in the Apollo program. *NASA 1965 Summer Conference on Lunar Exploration and Science* (Washington, DC: NASA SP-88, 1965); George E. Mueller to multiple addressees, "Establishment of Apollo Site Selection Board," NASA Management Instruction 1152.20, August 6, 1965; *1967 Summer Study of Lunar Science and Exploration* (Washington, DC: NASA SP-157, 1967), pp. 3–6.

75. Newell, *Beyond the Atmosphere*, pp. 240–42; James Webb to Dr. Frederick Seitz, President, National Academy of Sciences, December 20, 1967. This document appears as III-19 in John M. Logsdon, gen. ed., with Dwayne A. Day and Roger D. Launius, *Exploring the Unknown: Selected Documents in the History of the U.S. Space Program, Volume II: External Relationships* (Washington, DC: NASA SP-4407, 1996), pp. 538–40.

76. See, for example, Edward M. Davin to Manager, Apollo Surface Experiments Program, "Contingency Science Payloads," July 3, 1968, and George M. Low to George W. S. Abbey, "Lunar Mission Planning," September 3, 1968.

77. Homer Newell, author's notebook, December 27, 1969; Newell, *Beyond the Atmosphere*, p. 292. The first scientist-astronaut, geologist Harrison Schmitt, was assigned in 1971 to the crew of Apollo 17, the last Apollo mission.

78. George Mueller to Robert Gilruth, September 3, 1969; Homer Newell, "Conference Report, February 5, 1970, Lunar Science Institute, Houston, Texas, Subject: Critique of Apollo Lunar Missions and the Maximization of Scientific Returns for the Remaining Apollo Flights," February 6, 1970.

from the 380 kilograms of lunar samples and other data returned to Earth, they could confidently posit that the Moon's surface was chemically different than Earth and was in fact as geologically dead as Urey had surmised.[79] [II-24]

Without question, Apollo dominated NASA's solar system exploration effort during the 1960s. The national goal to send humans to the Moon's surface drove the space agency not only to choose lunar over planetary exploration as the primary solar system emphasis during the decade, but also to design its program of lunar scientific exploration to support the human space program. No one at NASA had been directly opposed to science for its own sake; instead, this pressing national objective and relatively limited resources led NASA to exploit the solar system exploration program for reasons other than the pursuit of pure scientific knowledge. In effect, the scientific results of NASA's lunar science program emerged as a byproduct of the Apollo project. In contrast, NASA's efforts to study solar system bodies beyond the Moon, which began very modestly in the early 1960s due to Apollo's prominence, while also designed with an eye towards eventual human interplanetary travel, ended up serving scientific understanding more directly.

To Worlds Beyond Our Moon

Despite initially standing second to lunar exploration, voyaging to worlds beyond the Earth-Moon system was not absent from NASA's early solar system agenda. In December 1959, NASA officials had promised JPL's Pickering that the agency would support the development of probes to visit the planets. Within the next six months, NASA received data from the only successful Pioneer probe, which measured radiation levels and magnetic fields between Earth and Venus, and began planning for missions to Earth's nearest neighbors, Mars and Venus.[80]

Only familiar with solar system bodies via data accumulated from ground-based resources, at the dawn of the space age scientists knew relatively little about the planets. Thus, while from the very start NASA considered spacecraft as elaborate as landers capable of gathering samples and returning them to Earth, the agency recognized that its first missions would have to be devoted to reconnaissance of its targets. The most appropriate spacecraft for its earliest Mars and Venus missions, NASA reasoned, were modest probes that would gather data as they flew by the planets. Orbiting and landing spacecraft, though attractive, seemed far too complex—and risky—while NASA was in its planetary exploration infancy.

JPL designed one spacecraft, called Mariner, with two variations to complete the flybys: Mariner A would perform simple flybys of the planets while Mariner B would release a landing capsule above the planet during its flyby. Initial plans for both models called for launch on the Atlas-Centaur, but problems in the Centaur stage's development forced NASA to reconsider that intention for fear that delays would prevent the

79. H. H. Schmitt, Gary Lofgren, G. A. Swann, and Gene Simmons, "The Apollo 11 Samples: Introduction," *Proceedings of the Apollo 11 Lunar Science Conference, Houston, Texas, January 5–8, 1970, Volume I, Mineralogy and Petrology*, ed., A. A. Levinson (New York, NY: Pergamon Press, 1970), pp. 1–54; *Apollo 15 Preliminary Science Report* (Washington, DC: NASA SP-289, 1972).

80. NASA, Office of Program Planning and Evaluation, "The Ten Year Plan of the National Aeronautics and Space Administration," December 16, 1959. After the failure of Pioneer 4 in March 1959, the project's management was transferred from the Department of Defense to NASA.

United States from beating the Soviets to a planetary shot.[81] In August 1961, NASA's Office of Space Flight Development realized it would have to modify its launch plans to achieve a Venus flyby in 1962, and thus canceled Mariner A in favor of a new probe, Mariner R, that would be compatible with the less powerful, but ready for use, Atlas-Agena vehicle. Within a year, JPL planned for Mariner R's scientific capabilities and developed and built the spacecraft, a hybrid of the Mariner A and Ranger designs. Although the first of two planned Venus shots was lost due to failure of the Atlas, Mariner 2 was successfully launched on August 27, 1962. Three and a half months later the probe passed within 16,000 kilometers of Venus, becoming the first spacecraft to flyby another planet and return scientifically valuable data on it and interplanetary space.[82] For 130 days Mariner 2 beamed information to Earth on Venus' climate and clouds and properties of the solar wind.[83] [II-15]

Centaur difficulties eventually led to the cancellation of a 1964 Venus mission and modification of a Mars mission to be launched in the same year. Further delays and a lower than originally predicted lift capacity for the stage forced NASA to scale down and then ultimately abandon Mariner B in favor of a less ambitious design, Mariner C, that lacked a lander and could ride aboard the Atlas-Agena. NASA's first Mariner mission to Mars failed when the upper-stage fairing failed to separate after launch and the solar panels could not deploy. The next spacecraft, Mariner 4, was launched successfully and approached Mars in July 1965. Returning twenty-one television images of the Martian surface as it passed the planet, Mariner 4 showed scientists that Mars' terrain was barren and cratered, like the Moon's, with no apparent canals, water, or signs of life.[84] [II-9]

Not long after JPL began work on the Mariner reconnaissance missions, many scientists started pushing NASA to pursue more ambitious solar system exploration missions. While the early Mariner flyby spacecraft would—and certainly did—provide impressive first close-up views of Earth's nearest planetary neighbors, they were limited in capability and tended to raise more scientific questions than they answered. Solving the mysteries of the planets' origins and evolutionary histories, surface and atmospheric compositions, interior structures, and other properties required probes equipped with larger, more capable instrument suites that could operate around the planets or on their surfaces for extended periods of time. Such desires led scientists and engineers to favor missions relying on increasingly more massive spacecraft, more powerful launch vehicles, and of course, larger budgets. The costs and technical complexity associated with ambitious missions often ran these projects or the entire solar system exploration program into trouble.

As early as 1961, JPL had studied possibilities for Mars and Venus exploration to follow the never-materialized Mariner B. JPL's preferred concept, called Voyager, was an ambitious program that would consist of orbiting as well as landing spacecraft that would carry more scientific instruments, collect and return more data, and operate for much

81. Ezell and Ezell, *On Mars*, p. 35.

82. Although the USSR's Venera 1, launched February 12, 1961, became the first spacecraft to flyby Venus in May 1961, contact with the spacecraft was lost a week after launch. Mariner 2 thus became the first spacecraft to fly by and return data on the planet.

83. NASA, *Mariner-Venus 1962: Final Project Report* (Washington, DC: NASA SP-59, 1965), pp. 12–15, 87–120. This early planetary mission did not carry a camera.

84. NASA, *Mariner-Mars 1964: Final Project Report* (Washington, DC: NASA SP-139, 1967).

longer than the Mariner probes.[85] Among the instruments scientists hoped to send to Mars were elaborate experiments to detect the presence of life on the planet that had long been suspected to harbor living creatures.[86] [II-16, II-17] Initially intending to send an orbiter and lander pair to both Venus and Mars, NASA ended up approving in 1964 four Mars-only Voyager flights—two in 1971 and two in 1973—at a cost of approximately $1.25 billion. [II-18] Deciding that Mars was its primary target and Voyager was the spacecraft with which it wanted to achieve its scientific goals, the agency eliminated Venus from Voyager plans and canceled two Mariner missions to Mars for 1966 and 1969 to assure the availability of funds for Voyager.[87]

Voyager's ambitiousness, coupled with political and economic circumstances, brought about difficulties in mission planning and ultimately led to the program's demise. Early on, some scientists and engineers questioned the wisdom of undertaking such a costly, sophisticated project; money aside, they wondered whether NASA had enough data on Mars from its first Mariner mission to the planet to design a suitable lander and to select an appropriate landing site by 1971. Moreover, the Voyager Lander was large and engineers struggled to develop a means of sterilizing the spacecraft for landing on Mars without destroying the functionality of its systems.[88] [II-14] Delays in the planned launch date occurred when NASA discontinued Saturn 1B—the vehicle initially intended to launch the Voyager spacecraft—and announced that the missions would fly on Saturn V, which would not be ready until at least 1967.[89] These delays pushed back the missions' launch dates; NASA's cost projection of the entire Voyager program grew to $2.2 billion through 1977.[90] At the same time, national priorities such as the conflict in Vietnam and President Johnson's Great Society programs were competing for funds with Apollo, and as a result, NASA began in 1965 to transfer funds from space science projects, including Voyager, to support its highest-priority manned lunar project.

By the end of 1965, NASA officials decided to cancel the 1971 Voyager mission after receiving only $10 million in the FY 1967 budget to begin flight hardware development. When the 1973 mission received no appropriations from Congress in FY 1968 and the White House made no attempt to restore NASA's request for Voyager, NASA did not attempt to reinstate the project. In lieu of this project, NASA flew Mariner spacecraft less

---

85. Bruce Murray, California Institute of Technology Associate Professor of Planetary Sciences, to Lee DuBridge, President, California Institute of Technology, February 23, 1965; California Institute of Technology, "Suggestions for Martian Exploration Following Mariner 4," February 23, 1965. Voyager Mars should not be confused with the late 1970s' Voyager mission to the outer planets.

86. Since its establishment in 1959, NASA's ad hoc Bioscience Advisory Committee had urged NASA to search for life on Mars. The Space Science Board had also encouraged biological studies on Mars. See Space Science Board, *Biology and the Exploration of Mars: Summary and Conclusion of a Study Supported by the NASA* (Washington, DC: National Academy of Sciences, 1965); Space Science Board, *Biology and the Exploration of Mars: Report of a Study Held Under the Auspices of the SSB, National Academy of Sciences-National Research Council, 1964–1965* (Washington, DC: National Academy of Sciences, 1966).

87. Robert Seamans to Donald Hornig, December 14, 1964, and James Spriggs to Homer Newell, "Voyager 1969," December 7, 1964.

88. Since NASA's inception, scientists were concerned that unsterilized spacecraft sent to other planets could carry terrestrial microbes. Some claimed that such "biological contamination" of other planets would be unethical. Many feared that sending unsterilized spacecraft to the planets would compromise scientists' ability to identify a microbe on another planet as indigenous to that planet. Both concerns led NASA to take measures in preparing spacecraft to ensure "planetary protection." Incidentally, the Soviet Union did not completely sterilize its early Mars spacecraft. Ezell and Ezell, *On Mars*, pp. 55–6, 104.

89. Ezell and Ezell, *On Mars*, p. 105.

90. *Ibid.*, p. 113.

capable than Voyager but more sophisticated than the earlier Mariners to Venus in 1967 and to Mars twice in 1969.

Planetary Exploration After the Height of Apollo

The combination of Voyager's cancellation, disagreement among scientists on planetary science objectives, and the start in 1967 of a downward trend in space science funding—and for all space program budgets, as the Johnson Administration reduced NASA's budget after Apollo's development was nearly complete—led NASA Administrator Webb that fall to temporarily halt work on new planetary missions to force the agency to reassess its plans to explore the solar system. Managers in the Office of Space Science and Applications developed several options on the course NASA's planetary exploration program could take.[91] They decided that while NASA had no commitments to fly any missions after the 1969 Mars Mariners, the agency should continue space science technology development and have ready a "wish list" of mission concepts to pursue should more money become available.[92] [II-21]

Scientists had mixed ideas regarding the strategy NASA should take for planetary exploration. The Space Science Board, for example, advised NASA to begin a program of "Planetary Explorers"—small, low-cost planetary missions, akin to the existing small Explorer missions for astronomy and space physics, to ensure frequent launches of solar system missions.[93] In contrast, the Lunar and Planetary Missions Board, a group of advisors from academia, research institutes, and aerospace corporations established by NASA in 1967 to critique the scientific merit of the agency's solar system exploration missions, suggested a more ambitious planetary program, which NASA rejected due to its high estimated cost.[94] When Webb appeared before the Senate Committee on Aeronautical and Space Sciences in November 1967, he proposed a revised planetary program that included five Mariner missions between 1971 and 1976 and a less ambitious Voyager-type, orbiter-probe mission to Mars in 1973, possibly to be followed in 1975 with a soft-landing mission.[95] [II-22] The Space Science Board believed the scheme to be overly ambitious while the Lunar and Planetary Missions Board thought NASA was not asking for enough. Webb, however, believed his own plan represented a balance between the desires of both groups, including enough activity to keep the planetary program agenda full while not requiring unrealistic amounts of money. The soundness of Webb's decision became evident when the

---

91. The Office of Space Sciences was named the Office of Space Sciences and Applications (see Chapter 1, page 12) during Webb's reorganization of NASA Headquarters in 1967.

92. Ezell and Ezell, *On Mars*, p. 134; James Martin to Charles Donlan, "OSSA Proposed Planetary Programs," October 5, 1967; "Planetary Program Extension, FY 1968–1969: Program Issues and Options," October 9, 1967.

93. NASA's Goddard Space Flight Center developed the Planetary Explorer concept when it conducted its own study in 1968 to investigate the capabilities of small planetary orbiters using Explorer spacecraft. The 1978 Pioneer Venus mission became NASA's first and only such Planetary Explorer. For the Space Science Board's recommendation on Planetary Explorers, see Space Science Board, *Planetary Exploration: 1968–1975* (Washington, DC: National Academy of Sciences, 1968).

94. The Lunar and Planetary Missions Board's efforts eventually led to the successful completion of planetary missions throughout the 1970s. See Ezell and Ezell, *On Mars*, pp. 144–48 and Barry Rutizer, "The Lunar and Planetary Missions Board," HHN-138, August 30, 1976, at the NASA History Office.

95. U.S. Congress, Senate, Committee on Aeronautical and Space Sciences, *NASA's Proposed Operating Plan for Fiscal Year 1968, Hearing*, 90th Congress, 1st session (Washington, DC: Government Printing Office, November 8, 1967), p. 16.

plan met the approval of Congress and, most importantly, when President Johnson noted in his January 1968 budget address to Congress: "We will not abandon the field of planetary exploration."[96]

The Johnson Administration remained true to its pledge, and NASA's proposed missions received the funding and new starts they needed. [II-23] After the success of the two Mariner probes to Mars in 1969, NASA attempted in 1971 to send two more Mariners to the Red Planet, not to fly past but to achieve orbit around the planet in order to return data at close range and over several weeks. Based on the early Mariner spacecraft design and ground equipment but larger in mass and more complex, JPL's new Mariner probes would orbit the planet for at least 90 days apiece. Planetary scientists hoped that long-term study of Mars would reveal information about the planet's weather patterns, polar cap phases, its potential of ever having sustained life, and possible landing sites for the future NASA Mars Lander. After Mariner 8 ended in a launch failure, Mariner 9 successfully left Earth on May 30, 1971, and became the first probe ever to enter orbit around another planet. Scientists feared the spacecraft would return little data when it arrived in the middle of a dust storm that swept across the entire planet, but within a couple of months the dust settled to reveal the planet's colossal canyons and mountains. Contrary to Mariner 4's bleak portrayal of Mars, Mariner 9 returned images of ancient lava flows and waterways, suggesting that Mars had had a very active geological past.[97]

Two years later, NASA launched its final Mariner mission. A flyby of Venus and Mercury, Mariner 10 became the first spacecraft to visit more than one planet. It remains the only probe to have visited the closest planet to our Sun. The mission returned photographs of almost half of Mercury's surface and revealed that the planet once had an intrinsic magnetic field.[98]

Perhaps most significantly for NASA's planetary exploration program in the era of Apollo flights, the Johnson Administration and Congress also allocated ample funding for the agency to land spacecraft on the surface of Mars to examine the planet's surface environment and search for the possibility of life. As originally proposed by Webb, during the mid-1970s the agency would deliver both orbiting and landing spacecraft to Mars—all of which would be based on a less ambitious Voyager design. During the fall of 1968, NASA officials met with representatives from JPL and Langley, the two Centers that would manage the mission, and potential contractors to explore alternatives for orbiters, landers, entry modes, and launch vehicles for the missions.[99] Although NASA managers and the Langley team concluded that flying two orbiter and soft lander pairs would be the most

96. "Planetary Exploration Program: Collection of Comments, Policy Statements, etc. (excerpts from NASA Press Conference on FY 69 Budget, January, 29, 1968)," no date; Ezell and Ezell, *On Mars*, p. 135.

97. For Mariner 9's scientific achievements, see Jet Propulsion Laboratory, "Mariner Mars 1971 Final Project Report, Volume V: Science Experiment Reports" (Pasadena, CA: JPL Technical Report 32–1550, August 20, 1973) and William K. Hartmann and Odell Raper, *The New Mars: The Discoveries of Mariner 9* (Washington, DC: NASA SP-337, 1974).

98. For more on the achievements of Mariner 10, see Schorn, *Planetary Astronomy*, pp. 257–58, 260–61.

99. Though JPL had vied for complete control of the mission, NASA preferred Langley's proposed management scheme and awarded the latter Center primacy over the mission's development. "NASA-LRC/JPL Management Agreement for Advanced Planetary Mission Technology Mars Lander-Mission Study," August 1968; Watson, "Viking Project Phase B Report," M63–110–0 [Circa Nov 1968]; William Pickering to Charles Donlan, April 17, 1968; Eugene Draley, "Langley Research Center Management Proposal for 1973 Mars Mission," April 18, 1968; Ezell and Ezell, *On Mars*, pp. 148–49.

expensive and technically complex choice, space sciences head John Naugle presented this scientifically optimal option under the name Viking to NASA Acting Administrator Thomas Paine in November 1968.[100] The following month, Naugle and Paine—a planetary exploration advocate who was concerned about catching up with the 1967 Soviet landing of Venera 4 on Venus—selected a Viking mission scheme to send two orbiter-lander pairs to Mars for an estimated $415 million. Each mission would include a soft lander with a surface lifetime goal of 90 days; the lander would be released from a Mariner 1971-class orbiter that would provide wide-area surveillance and a communications link for the lander. A Titan III-Centaur combination would boost each of the two orbiter-lander pairs to Mars in 1973.

Work began on the project immediately, with JPL designing and building the orbiter, Langley supervising lander development and system integration, Martin Marietta constructing the lander, and Lewis Research Center in Cleveland, Ohio, overseeing launch vehicle development.[101] Within the early months of development, it became clear that the Viking project would surpass its originally estimated cost figure. While the orbiter borrowed heavily from Mariner technology, JPL engineers nonetheless had to make significant changes to the design to enlarge the orbiter and its systems so they could power the lander before its release. The lander's sophisticated computer and biology and gas chromatograph-mass spectrometer instruments further contributed to the quickly rising costs.[102] As Viking's price tag escalated, NASA's budget continued to shrink. After reviewing projections for NASA's FY 1971 budget, NASA opted in 1970 to postpone the Viking missions' launches until 1975, which increased costs as well.[103] By the time the spacecraft were launched in 1975, NASA had spent over $1 billion on what had been intended to be a more modest alternative to the overgrown Voyager concept.

The Viking 1 Orbiter-Lander pair was launched from Cape Canaveral aboard a Titan III-Centaur launch vehicle on August 20, 1975, followed less than three weeks later by the identical Viking 2. After arriving in orbit around Mars, the Viking 1 Orbiter began its first task: photographing the surface regions that the Landing Site Working Group had selected for the Viking Landers to visit based on Mariner 9 data.[104] Revealing surface features in unprecedented detail, the orbiter showed the early landing site choices for both landers to be hazards, covered with craters, depressions, grooves, and ridges. This discovery dismayed the Viking team, for they had hoped to make the United States' first landing on Mars on July 4, 1976, the bicentennial of the Declaration of Independence. Despite the intense desire to meet this target date, they decided to delay the landings while the landing site staff analyzed the orbiter data to make new

100. NASA had initially referred to the mission concept as Titan Mars 1973 because the spacecraft would launch on a Titan III-class booster. A. Thomas Young, "Titan Mars '73 Mission Mode Meetings Summary," November 14, 1968.

101. Viking was the first NASA planetary project in which multiple NASA Centers and contractors participated in the design, development, and operations phases. Ezell and Ezell, On Mars, p. 153; James Martin, "Procurement Planning, Mars '73 Mission," February 13, 1968; James Martin, "Mars '73 Statement of Work," June 13, 1968, with enclosures; Langley Research Center, Viking Project Office, "Viking Project Mission Definition No. 2," M73–112–0, August 1969.

102. Viking Project Office, "Viking Lander Science Instrument Teams Report," M73–112–0, August 1969.

103. John Naugle to Edgar Cortright, "The Cost of the Viking Project," August 26, 1969; John Naugle, memo for record, "Decision to Reschedule Viking to 1975," January 4, 1970.

104. Gerald Soffen to James Martin, "Landing Site Recommendation," April 3, 1973.

selections. The Viking team realized the prudence of their new choices—and of postponing the mission—when the two landers touched down successfully on the planet's northern hemisphere: Viking 1 at Chryse Planitia on July 20, 1976, and Viking 2 at Utopia Planitia on September 3, 1976.

Most planetary scientists agreed that Viking's returns made up for its high price tag. For six years, the thirty-four instruments of the orbiters and landers worked together to paint the most detailed picture of Mars that planetary scientists had to date.[105] The probes showed the Martian surface to be a cold, dry desert whose plains were strewn with rocks and sand dunes. The landers' color cameras—the first ever sent on a robotic spacecraft—showed Mars' iron-rich terrain to be a rusty red and the sky reddish-yellow from its high dust content. Vast canyons, tall mountains, and networks of tributaries were telltale signs that tectonic and volcanic activity and water and wind erosion had altered the surface over the planet's history.[106] The atmosphere, much less dense now than in Mars' past, contained trace amounts of water; scientists speculated that the planet must store more water below the surface or at the poles.[107] [II-26]

Scientists and the public did experience one major disappointment regarding the Viking mission: its failure to detect any unambiguous signs of life on the Red Planet. For a century, science fiction authors had created in the worldwide public's mind an image of Mars as a life-friendly planet. Scientists who had studied Mars also suspected that the planet once, and possibly still, harbored at least microbial life forms. But Viking turned up no signs of life: the landers' cameras did not photograph any living creatures and their highly sensitive life detection experiments found no evidence of microbial life in the Martian dirt.[108] Though some scientists argued that this negative result was just as informative as a positive one, the failure of these experiments—arguably the mission's most intriguing—to discover life dashed the hopes of both scientists and the public. With Viking's depiction of Mars as a cold and lifeless planet, NASA reduced its support for further robotic exploration of the Red Planet and redirected its focus to other areas of the solar system.[109]

Although NASA's funding was shrinking, the agency succeeded in gaining White House and congressional approval for yet an additional pair of major missions in the late 1960s. While Mars had been the primary target of planetary scientists and the agency because of its enigmatic history and accessibility using current launch capability, the outer solar system still beckoned. Both the Lunar and Planetary Missions Board and the Space Science Board maintained that Jupiter and the planets beyond were intriguing targets about which humanity knew very little, and that NASA ought to consider sending low-cost

105. The Viking 2 Orbiter ceased functioning in 1978, while both Viking 2 Lander and Viking 1 Orbiter continued operating until 1980. The Viking 1 Lander stopped working in 1982.

106. Viking Lander Imaging Team, *The Martian Landscape* (Washington, DC: NASA SP-425, 1978).

107. For a comprehensive overview of Viking's achievements during its first year of operation, see Gerald A. Soffen et al., "Scientific Results of the Viking Project," *Journal of Geophysical Research* 82 (September 30, 1977): 3959–70. This article introduces a full issue of *Journal of Geophysical Research* devoted to Viking's returns. See also Gerald A. Soffen and Conway W. Snyder, "The First Viking Mission to Mars," *Science* 193 (August 27, 1976): 759–65.

108. For results of Viking's search for life on Mars, see, for example, Norman H. Horowitz, "The Search for Life on Mars," *Scientific American* 237 (November 1977): 57–8 and "Life on Mars?..." *New York Times*, September 20, 1976.

109. Opposition to a robotic Mars sample return mission, which had been discussed at NASA throughout Viking's development, can be seen, for example, in Daniel Herman, Advanced Programs and Technology Manager, to Lunar and Planetary Programs Director, June 14, 1977.

spacecraft to explore them. The Space Science Board specifically recommended that such an objective could be achieved by sending two Pioneer-class probes to Jupiter at opportunities in 1972 and 1973.[110] In February 1969, NASA Headquarters embraced the Space Science Board's advice and approved a pair of missions to provide the first close-up look at the largest planet in the solar system. Congress and the White House approved the plan later that year.

As originally planned, the new Pioneer project would explore the interplanetary medium beyond Mars' orbit, investigate the asteroid belt, and explore the planet Jupiter and its environment.[111] Managed by Ames Research Center at Moffett Field, California, the Pioneer spacecraft were identical small, lightweight (258 kilograms) probes based on the modules used for interplanetary Pioneers 6, 7, 8, and 9.[112] Because they would have to endure long distances and traverse the asteroid belt, whose hazards were not fully understood, the spacecraft were very simple and boasted several redundant key subsystems. Two spacecraft were built due to the very fact that engineers feared that one of the probes would fail before reaching Jupiter.

In March 1972 and April 1973, Pioneers 10 and 11 were successfully launched on Atlas-Centaur vehicles to begin what would become the most distant voyages human-made probes had made to date. Pioneer 10 was a pioneer in the true sense of the word, for its experiences would tell NASA how successfully spacecraft could pass through the asteroid belt, endure Jupiter's intense radiation, operate using not solar power but onboard nuclear power sources, and communicate across extreme distances from Earth. This spacecraft and its twin proved their abilities to achieve all of the above feats in addition to collecting and returning phenomenal science during their travels to and flybys of Jupiter. Using a combined total of 23 instruments, the two spacecraft mapped the magnetic field and distribution of dust particles in interplanetary space while exploring how the interplanetary magnetic field interacted with the solar wind and cosmic rays. Travelling through the asteroid belt permitted the spacecraft to investigate properties of the objects scientists and engineers had feared could destroy their efforts to reach the outer solar system. In Jupiter's environs, the probes gathered data on the magnetic and gravitational fields, temperatures, and atmospheric properties of the planet and its four inner moons. After Pioneer 10 made the first successful Jupiter flyby, NASA made the decision as Pioneer 11 was en route that the latter spacecraft would continue on to explore Saturn after providing additional information on Jupiter. At Saturn the probe made measurements of the planet's physical and chemical properties while also discovering a new ring and new moon around the planet.

After the completion of their mission objectives at Jupiter and Saturn, the probes began their journeys in opposite directions to find the heliopause—the "envelope" around the solar system beyond which the Sun does not influence interplanetary space—

---

110. Space Science Board, *Planetary Exploration*, 1968–1975 (Washington, DC: National Academy of Sciences, 1968).

111. For a comprehensive overview of the history and achievements of Pioneers 10 and 11, see Richard O. Fimmel, James Van Allen, and Eric Burgess, *Pioneer: First to Jupiter, Saturn, and Beyond* (Washington, DC: NASA SP-446, 1980).

112. Ames Research Center had developed spin-stabilized probes for NASA's revived Pioneer program. NASA launched Pioneers 6, 7, 8, and 9 between 1965 and 1968 to investigate properties of the interplanetary medium and the effects of the Sun on the inner planets.

and eventually to leave the solar system. Though they no longer transmit data to Earth since NASA terminated the missions a few years ago, both carry gold-anodized aluminum plates showing their origins in the solar system as emissaries of humanity. Journalist Eric Burgess and Cornell planetary scientist Carl Sagan encouraged NASA to add these plaques to the probes to convey to an intelligent civilization, which might find and decipher the plaques millions of years from now, the desire of another species to leave its own planet and explore the universe.[113]

Planetary Exploration in the 1970s

NASA had managed in the late 1960s to gain authorization and funding from Congress to develop nine spacecraft to explore the solar system. But by the middle of the new decade, the planetary program was experiencing tremendous difficulty securing new starts for missions. NASA's total budget had been declining since Apollo's development funding peaked in 1966; by 1969 the space science budget reached a low that it had not seen since 1961. With the Apollo program to end in the early 1970s, NASA sought to start on a new human space flight program: a reusable human launch vehicle which came to be known as the Space Shuttle.[114] As Apollo had before it, the new human space flight project consumed a large proportion of the dwindling NASA budget. At the same time, Viking's complexity and price tag were escalating and Mariner 10 and Pioneers 10 and 11 were in development. In effect, the space agency could afford to initiate few planetary exploration missions in the 1970s. Thus, before the three projects NASA and Congress chose to begin supporting in the 1970s, which sought to help refine planetary scientists' understanding of the solar system, they endured debates and several modifications to meet the resource constraints of the times.

Throughout NASA's first several years, scientists and engineers only explored the solar system as far as the orbit of Mars. Assuming that they could only reach the outer planets by means of reaction propulsion, they could not devise propulsion systems powerful enough to achieve such distances. In 1961, Michael Minovitch, a graduate student from the University of California, Los Angeles, working at the Jet Propulsion Laboratory, discovered a method of propelling spacecraft through the solar system that would not rely exclusively on fuel but would leverage the gravitational pull of planets as they approached these bodies.[115] Minovitch postulated that gravity-propelled interplanetary space travel would limit the fuel required on spacecraft, thus making them easier to launch, while often shortening the time otherwise required for them to reach their destinations. [II-20] California Institute of Technology graduate student Gary Flandro attempted later that decade to apply this principle to develop

113. Fimmel, Van Allen, and Burgess, *Pioneer: First to Jupiter, Saturn, and Beyond*, p. 248; Carl Sagan, Linda Salzman Sagan, and Frank Drake, "A Message from Earth," *Science* 175 (February 25, 1972): 881–84; NASA News Release 72–32, February 25, 1972.

114. More details on the development of the Space Shuttle can be found in chapter two of John M. Logsdon, gen. ed., with Ray A. Williamson, Roger D. Launius, Russell J. Acker, Stephen J. Garber, and Jonathan L. Friedman, *Exploring the Unknown: Selected Documents in the History of the U.S. Space Program, Volume IV: Accessing Space* (Washington, DC: NASA SP-4407, 1999), pp. 161–404.

115. M. A. Minovitch, "A Method for Determining Interplanetary Free-Fall Reconnaissance Trajectories," JPL Technical Memo 312–130, August 23, 1961, pp. 38–44.

trajectories to reach planets beyond Mars.[116] [II-19] Starting in the late 1970s the outer planets would be aligned such that a probe launched to Jupiter could leverage that planet's gravity to boost it to Saturn, where it would receive another "gravity assist" to launch it to Uranus, which would slingshot it to Neptune.[117] A spacecraft built by that time could take a "grand tour" of all of the outer planets except Pluto.

In 1969, the same year that Congress approved Pioneers 10 and 11, NASA heeded these efforts and began designing a mission concept, called the Grand Tour, around this rare opportunity. Much more ambitious than the Pioneer mission, the Grand Tour called for dual spacecraft launches to Jupiter, Saturn, and Pluto in 1976 and 1977 and dual launches to Jupiter, Uranus, and Neptune in 1979, with an estimated total cost of $750 million. Even before receiving a new start, NASA selected about a dozen teams of scientists to develop the mission's scientific objectives, while JPL and industrial contractors proceeded to draw up designs for the advanced spacecraft that would carry the instruments.[118] Budget constraints, however, meant that the space agency could only allocate $10 million of the $30 million the Grand Tour's developers requested to complete the design phase of the mission in FY 1972.[119]

NASA significantly descoped the mission and received the approval of Congress and President Nixon for a new start in FY 1973. The agency revised its plans to take advantage of the unique alignment of the outer planets by dropping Uranus, Neptune, and Pluto from its targets and redesigning the mission to use the proven Mariner-class spacecraft to improve reliability and to lower the costs of development. The new $250 million concept began with the name Mariner Jupiter-Saturn, but in 1977 NASA renamed the project Voyager.[120] In 1972 NASA selected nine instruments from more than 30 proposed to satisfy Voyager's mission objectives to study Jupiter and Saturn, those planets' satellites, the interplanetary medium, and possibly Uranus. Two identical spacecraft weighing 815 kilograms and equipped with numerous redundant systems were built for the mission. Following the example of Pioneers 10 and 11, each also carried a special memento: a gold-plated copper phonograph record containing images, sounds, and spoken greetings representative of the diversity of life and cultures on Earth.[121]

The Voyager spacecraft were launched on August 20 and September 5, 1977, from Cape Canaveral on Titan III-E/Centaur vehicles.[122] Arriving at Jupiter in 1979, the two spacecraft sent back the best resolution images to date of the planet's enormous, turbu-

---

116. G. A. Flandro, "Fast Reconnaissance Missions to the Outer Solar System Utilizing Energy Derived from the Gravitational Field of Jupiter," *Acta Astronautica* 12 (July/August 1966): 329–37.

117. R. D. Bourke and G. A. Flandro to T. A. Barber and F. N. Haurlan, October 10, 1966. The outer planets achieve this alignment once every 176 years.

118. NASA, "Invitation for Participation in the Mission Development for Grand Tour Missions to the Outer Solar System," October 1970; David Morrison and Jane Samz, *Voyage to Jupiter* (Washington, DC: NASA SP-439, 1980), p. 24.

119. George M. Low, Personal Notes #40, January 22, 1971.

120. Grand Tour enthusiasts also had proposed a Mariner Jupiter-Uranus mission to be launched in 1979 for $400 million, but the costs of Voyager and the possibility of sending the spacecraft to Uranus after the successful completion of the primary mission precluded NASA from starting on such a mission.

121. The complete story of the Voyager records' conception and compilation can be found in Carl Sagan, Frank Drake, Ann Druyan, Timothy Ferris, Jon Lomberg, and Linda Salzman Sagan, *Murmurs of Earth: The Voyager Interstellar Record* (New York, NY: Ballantine Books, 1978).

122. Voyager 2 was the first to launch, but Voyager 1 was set on a shorter, faster trajectory that enabled it to reach Jupiter first.

lent atmospheric storm—seen from Earth as the Great Red Spot—and the vastly diverse terrains of the four inner moons. They also detected a faint ring of particles encircling the planet as well as a plasma torus produced by the moon Io. Voyager 1 flew by Saturn and its largest moon, Titan, in 1980 before proceeding on a trajectory that took it out of the solar system, while Voyager 2 reached Saturn the following year to yield new information on the planet's atmospheric dynamics, ring structure, and satellites.[123] With the original mission objectives completed and the spacecraft still healthy, Voyager's managers requested approval from NASA Headquarters to send Voyager 2 onward to Uranus.[124] [II-27, II-28, II-29] After a successful flyby of that planet in 1986, the spacecraft traveled on for a 1989 rendezvous with Neptune, making the only approach to these two planets of any spacecraft to the present.

NASA's second planetary new start of the 1970s took probes back to Venus. In the late 1960s, American planetary scientists wishing to catch up with the Soviets' success in releasing a probe into Venus's atmosphere began planning for their own Venus Orbiter and probe mission.[125] Such a mission would allow them to study the planet's surface using radar from on orbit and probing the atmosphere with *in situ* measurements. Scientists and engineers at NASA's Goddard Space Flight Center studied the feasibility of using the Planetary Explorer concept to develop a low-cost Venus Orbiter, and also examined a number of probe options.[126] Members of both the Space Science Board and the Lunar and Planetary Missions Board agreed that Venus was still an important scientific target and endorsed NASA's use of low-cost orbiters and probes to conduct in-depth investigation of the planet.[127] Although the scientists originally envisioned sending several orbiters and probes to the planet throughout the decade beginning in 1973, a $200 million mission cost cap forced the team to settle on a single orbiter and multiprobe—comprised of a large probe and three smaller ones—in the late 1970s. This mission took the name Pioneer Venus, and became the only mission NASA ever designed and executed around the Planetary Explorer concept. In 1974 Congress authorized this downscaled version of a Venus mission for a new start in the following fiscal year, and NASA awarded the Hughes Aircraft Company a contract to build the orbiter and probe.

The orbiter and multiprobe were launched on separate Atlas-Centaur vehicles in 1978. On December 4 of that year the orbiter entered orbit around Venus. The constituent probes of the multiprobe separated to make individual, hour-long descents through the planet's atmosphere five days later. Arriving in different zones around the planet, all of the probes successfully returned *in situ* data on the atmosphere's composition, structure, and temperature before impacting the surface.[128] The orbiter also

---

123. For the results of Voyager's flyby of Saturn, see David Morrison, *Voyages to Saturn* (Washington, DC: NASA SP-451, 1982).

124. Raymond L. Heacock to Frank Carr, October 21, 1980; Raymond L. Heacock to Frank Carr, November 24, 1980; Frank Carr to Raymond L. Heacock, December 12, 1980.

125. The story of Pioneer Venus's development and achievements can be found in Richard O. Fimmel, Lawrence Colin, and Eric Burgess, *Pioneer Venus* (Washington, DC: NASA SP-461, 1983).

126. R. M. Goody, D. M. Hunten, V. Suomi, and N. W. Spencer, *A Venus Multiple-Entry-Probe Direct-Impact Mission* (Greenbelt, MD: NASA Goddard Space Flight Center, 1969).

127. Space Science Board, *Venus: A Strategy for Exploration* (Washington, DC: National Academy of Sciences, 1970).

128. Although the probes were not designed to survive impact, one of the probes survived and continued to transmit data for over an hour.

gathered data on the atmosphere, but more importantly became the first spacecraft to "see" through the thick atmosphere using radar and to map the entire Venusian surface.

By the mid-1970s, NASA officials and planetary scientists were expressing their worries about the reduced frequency in new starts for solar system exploration programs. Referring to the decline in funding for lunar and planetary exploration since 1974, NASA Associate Administrator for Space Science Noel Hinners told a Senate committee during a NASA FY 1977 budget hearing that at the current rate of budget decline, the solar system exploration program was on a "going-out-of-business" trend.[129] Others voiced their concerns to top national science officials that the lack of new starts in favor of other NASA priorities would destroy the program of solar system exploration that NASA had worked so hard to build up over nearly two decades.[130] [II-25] Only after scientists and NASA representatives offered extensive testimony to Congress, and Congress took several votes, did the FY 1978 budget include a new start for a planetary program—the last the space agency would see until 1984.

During the course of Voyager's development, NASA had begun looking ahead to the possibility of sending to Jupiter a long-lived orbiter with a probe that could be released into the planet's atmosphere. Originally called the Jupiter Orbiter Probe, the project's name was changed to Galileo shortly after receiving its FY 1978 new start in honor of the discoverer of the planet's four largest moons. Slated to cost no more than $750 million (FY 00 dollars), the spacecraft was scheduled to launch in 1982 for a 1984 arrival at Jupiter. During the remainder of the decade, those involved in the program at NASA began orbiter development based on a Voyager-type design and a probe based on Pioneer Venus, selected the scientific experiments for the mission, and reached an agreement with Germany for that nation to develop the engine for the probe.

As the next section shows, Galileo nearly faced cancellation several times in the late 1970s and early 1980s due to competition with the development of the Space Shuttle and other space science projects for scarce budget dollars. Once the decision was made to preserve the project, problems with the development of the Inertial Upper Stage—the mechanism scheduled to deploy Galileo from the Shuttle, which was NASA's new vehicle of choice for launching all types of probes—pushed the Galileo craft to a 1986 launch date. The unexpected *Challenger* disaster that occurred early that year grounded Galileo for another three years, and only after the Shuttle program resumed was the spacecraft finally launched in 1989. By the time of its launch, the repeated delays had boosted the mission's price to well over $1.4 billion (FY 00 dollars). A long wait on the ground also unfortunately led to the wearing away of lubricant on the orbiter's high-gain antenna, which was supposed to permit the return of science at high data rates. As a result, the antenna was unable to open completely once in space, and so mission operators had to rely on the spacecraft's smaller antenna, which had slower data return rates. [II-37] The Galileo Orbiter and probe still managed to return a wealth of data, beginning in 1995, on the nature of Jupiter's atmosphere and magnetosphere as well as its four inner moons. Imaging with a solid-

---

129. U.S. Congress, Senate, Committee on Aeronautical and Space Sciences, *NASA Authorization for FY 1977, Hearings, Part 2* (Washington, DC: Government Printing Office, February 1976), p. 1138.

130. G. J. Wasserburg to H. Guyford Stever, June 8, 1976.

state detector represented a marked improvement in sensitivity and resolution over Voyager's vidicon television camera system, and enabled Galileo scientists to make stunning revelations about the features on the planet's moons.[131]

Keeping the Planetary Program Alive in the 1980s

Viking, Voyager, Pioneers 10 and 11, and Pioneer Venus were still operating at the end of the 1970s, but solar system enthusiasts were only partly consoled by their scientific returns. Looming large in their minds was concern for the planetary program's welfare beyond those projects.[132] NASA's budgets were tight, its program objectives were numerous, and the agency had only one planetary mission, Galileo, in the works. In 1979 the financial crunch induced NASA to abandon its plan to send a spacecraft to rendezvous and "fly in formation" with Halley's Comet, due to pass near the Sun during 1985 and 1986.[133] While the Soviet Union, European Space Agency, and Japan would all greet the comet that visited the inner solar system once every 76 years, lack of support for planetary exploration at the end of the 1970s meant the United States—then the world's major space player—would miss out on the opportunity.[134]

These tough times led to the appearance of two new organizations within the planetary science community. Having returned to NASA as Chief Scientist in the late 1970s, John Naugle recognized that NASA had abandoned its habit of developing long-term strategies for solar system exploration, and that the lack of an integrated strategy made missions vulnerable to descoping or outright cancellation when other projects took priority.[135] In response, Naugle formed the Solar System Exploration Committee (SSEC), an *ad hoc* committee of the NASA Advisory Council, to "review the goals of solar system exploration; identify the essential attributes of a viable program in planetary sciences; and define new ways to reduce costs.[136] In addition, JPL's Bruce Murray, Louis Friedman (formerly of JPL), and Carl Sagan founded The Planetary Society in Pasadena, California, to gain grassroots support for the endangered planetary program. Eventually attaining a membership of more than 100,000, The Planetary Society has become the most visible pro-space group in the world.[137] [II-30]

---

131. Galileo has returned high-resolution images showing the diversity and enigma of Jupiter's four largest moons: cratered Callisto, mottled Ganymede, volcano-pocked Io, and icy Europa.

132. For an excellent review of the difficulties NASA's planetary exploration program faced in the early 1980s, see John M. Logsdon, *The Survival Crisis of the U.S. Solar System Exploration Program* (unpublished), June 1989. This document was prepared for the NASA History Office.

133. For discussions of the United States' failure to conduct a Halley's Comet mission, see John M. Logsdon, "Missing the Comet: Why No U.S. Mission to Halley," *ISIS* 80 (June 1989): 254–80; Bruce Murray, *Journey into Space: The First Thirty Years of Space Exploration* (New York, NY: W. W. Norton and Company, 1989), pp. 253–75; and Schorn, *Planetary Astronomy*, pp. 289–90.

134. The United States still ended up being the first nation to visit a comet with a robotic probe. In late 1985, NASA redirected the International Sun-Earth Explorer 3, launched in 1978, to fly through the tail of the lesser-known comet Giacobini-Zinner.

135. The Space Science Board's Committee on Planetary and Lunar Exploration issued several reports in the 1970s on planetary exploration strategies.

136. NASA, "Purpose of Solar System Exploration Committee," November 10, 1980; "Summary Minutes of the SSEC," June 1–2, 1981. Solar System Exploration Committee of the NASA Advisory Council, *Planetary Exploration through Year 2000* (Washington, DC: NASA, 1983), p. 5.

137. Keay Davidson, *Carl Sagan: A Life* (New York: John Wiley and Sons, 1999), p. 348.

In the fall of 1980, the lame-duck Carter Administration included an additional NASA planetary mission in its proposed budget: a Venus Orbiter that would follow up Pioneer Venus by returning radar images of the planet's surface at even better resolution. Called the Venus Orbiting Imaging Radar (VOIR), the mission was Carter's more modest preference to the Halley's Comet rendezvous probe. But Ronald Reagan's triumph in the 1980 presidential election signaled an era of continued difficulty for the solar system exploration program. The Reagan Administration gave NASA $6.1 billion in FY 1982—$604 million less than President Carter had proposed. Office of Management and Budget (OMB) Director David Stockman opted to achieve this budget cut by rescinding the FY 1982 new start on the VOIR mission and instructing NASA to cancel or substantially descope one of its other major space science missions: Galileo, the Hubble Space Telescope, or the U.S.-European International Solar Polar Mission. Much to the chagrin of the Europeans, NASA's Acting Administrator chose to cut the last project, sparing the only planetary mission in development, Galileo.

When the Reagan Administration proposed an even smaller budget for NASA the following fiscal year, NASA's new administrator, James Beggs, announced that the agency would be willing to eliminate its solar system exploration program altogether as long as the Space Shuttle and other space science projects retained adequate funding.[138] [II-31] But after learning in November 1981 that the Administration had cut the agency's budget request by $1.3 billion and reduced solar system exploration funding to $118 million—leaving funds to continue operational missions but none for Galileo's development—Beggs appealed the allocations to a Budget Review Board.[139] [II-32] The White House, however, remained committed to the proposed budget.[140] [II-33] Only in response to a strong push by supporters of planetary exploration and JPL did the White House restore funding for Galileo.[141] Although no funds were restored for VOIR that year, further negotiations between NASA and OMB ultimately brought the FY 1983 solar system exploration budget to $154.6 million, with an additional $92.6 million for Galileo's continued development. NASA would at least be able to sustain a modest planetary program. [II-34]

In 1983 the SSEC recommended a new solar system exploration strategy for NASA. Keeping in mind the need to achieve planetary science goals at reduced costs compared

---

138. James Beggs to David Stockman, September 29, 1981. Human space flight projects had always been NASA's top priority; the agency was also willing to put space physics and astronomy projects ahead of solar system exploration. Planetary scientists were still a minority group in the space science community; moreover, they were divided on future mission priorities. Space physicists and astronomers, in contrast, agreed that the Great Observatories, such as the Hubble Space Telescope, were their priorities. Logsdon speculates in *The Survival Crisis of the U.S. Solar System Exploration Program*, p. 17, that Beggs said he was willing to cut the planetary exploration program because he figured that the White House would in fact not accept this option. Thus, this as well as other NASA activities would end up receiving funding.

139. NASA, "FY 1983 Budget Appeal," December 5, 1981.

140. Office of Management and Budget, "Summary of OMB/NASA Positions: Space Science and Related Programs (Including Planetary Exploration)," no date; White House, "Selected White House Views (on NASA's planetary exploration program)," December 8, 1981.

141. Many planetary program supporters played upon the political importance of keeping JPL alive as a means to sustain planetary program funding. California Institute of Technology President Marvin Goldberg met with senators interested in the space program in December 1981 and convinced Senate Majority Leader Howard Baker to express his support for planetary exploration in a letter to President Reagan. Baker's letter was influential in the decision to preserve Galileo and the planetary program. Logsdon, *The Survival Crisis of the U.S. Solar System Exploration Program*, p. 35–38.

with the past, the SSEC concluded that NASA should develop a solar system exploration program based on spacecraft in a variety of sizes, but that low- to moderate-cost probes should form the program's core.[142] [II-36] The SSEC believed NASA could sustain a basic planetary program using this strategy for $480 million (FY 00 dollars) per year. As envisioned by the SSEC, the core program would consist of a series of "Planetary Observers"—small spacecraft based on the designs of existing Earth-orbiting probes. These spacecraft would require little in terms of development time and cost, while ensuring that planetary scientists would receive a steady stream of data even if the space agency continued favoring the Space Shuttle or other programs. The first two Planetary Observer missions the SSEC recommended to NASA were a smaller version of the VOIR mission and a Mars orbiter that would focus on the planet's weather patterns. The SSEC also suggested using a modular spacecraft design called the Mariner Mark II for larger missions to the outer solar system, such as its proposed Comet Rendezvous Asteroid Flyby and Titan probe missions.

That year, NASA tried again to push a Venus radar mission through OMB and Congress. Having descoped VOIR and reduced its total cost estimate to under $300 million, the agency was able to gain a new start in the FY 1984 budget for the new mission, now called the Venus Radar Mapper but renamed Magellan in 1986.[143] [II-35] Maintaining the same scientific objectives as VOIR, Magellan would carry, along with two other instruments, a synthetic aperture radar instrument that would return data to make sense of the geological history of the planet's surface and interior. With its original 1988 launch date postponed due to the *Challenger* disaster, the Magellan probe launched from the payload bay of STS-30 on May 4, 1989. Upon arriving at Venus, Magellan embarked on a five-year mission that yielded outstanding scientific results. The spacecraft's returned data enabled scientists to create high-resolution gravity and surface maps of over 95 percent of the planet. Magellan revealed Venus' surface to be covered with volcanoes, faults, impact craters, and lava flows.[144]

The first solar system missions since Pioneer Venus' 1978 departure, the 1989 launches of Magellan and Galileo were the only two missions NASA sent to the planets in the 1980s. During the decade, however, NASA did begin developing three additional solar system exploration missions: one based on the SSEC's Planetary Observer concept, and the two others on the proposed Mariner Mark II spacecraft. The experiences of developing these missions once again indicated to planetary scientists that there still existed a disparity between their interests and the projects that the White House and Congress were willing to fund. In addition, they illustrated the technical and programmatic risks of pursuing very large and ambitious planetary science missions.

---

142. The SSEC made decisions about scientific priorities based on recommendations by the Space Science Board's Committee on Planetary and Lunar Exploration. The SSEC plan is detailed in Solar System Exploration Committee of the NASA Advisory Council, *Planetary Exploration through Year 2000* (Washington, DC: NASA, 1983).

143. NASA Venus Radar Mapper Project Initiation Agreement, October 20, 1982; NASA Program Approval Document for Magellan, September 2, 1988.

144. For more on the scientific achievements of Magellan, see Carolynn Young, ed., *The Magellan Venus Explorer's Guide* (Pasadena, CA: Jet Propulsion Laboratory, 1990) and Peter Cattermole and Patrick Moore, *Atlas of Venus* (New York, NY: Cambridge University Press, 1997).

NASA was able to get a new start for a Mars probe based on the Planetary Observer concept in the same fiscal year in which Congress approved the Venus Radar Mapper.[145] Slated to cost $250 million, the Mars Geoscience/Climatology Orbiter (MGCO) was intended to extend and complement data obtained from the Mariner and Viking Mars spacecraft on the Red Planet's surface composition, atmospheric structure and circulation, magnetic field, and volatile content. To achieve this mission at relatively low cost, the mission planners intended to use proven designs, off-the-shelf components, and simple instruments.

In the end, however, MGCO hardly adhered to the standards the SSEC envisioned for the mission and soon evolved into a Viking-class project.[146] Knowing the low priority NASA had placed on solar system exploration, planetary scientists feared that this Mars mission would be the last to the planet in a great while. They also reasoned that launching on the Space Shuttle would provide "substantial weight and performance margins," and thus spacecraft size and mass were not the concerns they would be if the probe was riding on its own rocket.[147] As a result, those responsible for mission planning selected the most expensive instrument package proposed for the mission. This choice raised the mission's price directly, while also adding to the cost and development schedule because the probe's engineers had to design a more elaborate spacecraft bus than originally intended to accommodate the payload and to reduce the risk of the mission's technical failure. At the time of its launch in 1992, the mission—known by then as Mars Observer—had grown not only in scientific capability but also in cost, to nearly $1 billion—a figure far from that approved years earlier by Congress.[148] Despite the extraordinary measures taken to boost scientific returns and to reduce risk of failure, Mars Observer's potential was never realized; after performing a maneuver to put the spacecraft into orbit around Mars, engineers failed to regain contact with it, making the mission NASA's largest robotic spacecraft failure in history.[149] [II-40]

The SSEC had also recommended in 1983 that NASA undertake the development of a modular spacecraft for outer solar system flight called the Mariner Mark II, whose chassis would contain common control, propulsion, and communications systems to reduce the design costs of missions using it. Beginning in the mid-1980s, NASA studied two Mariner Mark II-based missions—one that would travel to Saturn and release a probe toward its moon Titan, and another that would rendezvous with a comet. In 1989 Congress approved funding in the FY 1990 budget for two solar system explo-

---

145. The Planetary Observer concept was never implemented as a line item in NASA's budget and thus ended with the MGCO/Mars Observer mission.

146. For the history of Mars Observer's development, see Charles Polk, *Mars Observer Project History* (Pasadena, CA: JPL D-8095, December 1990).

147. Solar System Exploration Committee, *Planetary Exploration through Year 2000*, p. 21.

148. Part of the cost increase was due to the need to reconfigure the spacecraft for launch on a different vehicle: after the *Challenger* disaster, NASA opted to fly the Mars Observer on a Titan III booster.

149. The most plausible source of failure, according to the Mars Observer failure report, was that one of the spacecraft's fuel lines ruptured when operators attempted to pressurize the propellant tanks and that this action sent the spacecraft spinning out of control and thus out of communication. Mars Observer Mission Failure Investigation Board, *Mars Observer Mission Failure Investigation Report*, December 31, 1993; NASA, "NASA Response to Mars Observer Loss of Signal Failure Review Board Report," no date. After the report was issued, the investigation board admitted that a hasty management decision might have been the root of the problem. See, for example, Kathy Sawyer, "NASA Admits Oversight on Report," *The Washington Post*, January 11, 1994, p. A 3.

ration missions based on the Mariner Mark II: Cassini and the Comet Rendezvous-Asteroid Flyby (CRAF). The Cassini spacecraft would carry an instrument suite to perform an in-depth survey of the planet Saturn, its rings, and its moons in similar fashion to Galileo at Jupiter. The spacecraft would also release a probe, furnished by the European Space Agency, into the atmosphere of Titan to provide scientists with a first in situ glimpse of Saturn's largest natural satellite. A proposed eight-year mission, CRAF would execute a close flyby of at least one asteroid and then proceed to rendezvous with and fly alongside a comet for a three-year period. Collecting material from the comet's nucleus and dust from its tail, the probe would analyze samples in situ with the objective of characterizing the comet's composition. NASA estimated that developing the missions in tandem, using the Mariner Mark II bus for both probes, would save $500 million over the cost of doing the two separately and would ensure that the agency could meet the $1.5 billion price tag OMB and Congress had set for the two missions' development.[150]

While Cassini and Huygens, the Titan probe, left Earth in 1997 for a 2004 arrival at Saturn, CRAF did not survive beyond the planning stages.[151] [II-38] With tight overall budgets and increasing costs of space station development, NASA assessed its priorities in 1991 and chose first to cut costs by deleting two instruments from the CRAF spacecraft and then to readjust its mission profile and push back its scheduled 1996 launch date.[152] CRAF's doom arrived in FY 1993, when NASA's budget fell ten percent short of its request and the agency completely cancelled the mission in an attempt to save Cassini. The latter mission seemed to NASA to have greater public appeal, due to Saturn's photogenic rings, and more political importance, due to international involvement with the probe.[153] [II-41, II-42] Although NASA had intended to employ an efficient means of developing the two spacecraft, even this measure did not save one of them from cancellation. In spite of the attempt to descope the mission to achieve some of its objectives, this effort only drove CRAF's total cost higher, ultimately sealing its fate and limiting future funding for planetary missions.

Embracing a Leaner Approach to Solar System Exploration in the 1990s

The *Challenger* disaster of 1986 gave NASA the impetus not only to reexamine Space Shuttle policy but also to review its space science program. Reflecting on the elevated costs, delayed development schedules, and increased technical risks associated with recent planetary missions, a committee of NASA advisors concluded that expanding missions' scopes without heeding resource limitations tended in the long

150. Space Studies Board, "Scientific Assessment of the CRAF and Cassini Missions," March 30, 1992. The Committee on Planetary and Lunar Exploration wrote this document as a letter report to NASA's Associate Administrator for Space Science, Lennard Fisk.

151. Cassini's greatest opposition was perhaps that of people who protested NASA's launching of the spacecraft for fear that its plutonium power source could rain all over Earth should the spacecraft be destroyed during launch or in passing Earth on its flight path to Saturn.

152. Space Studies Board, "Scientific Assessment of the CRAF and Cassini Missions."

153. The Senate Appropriations Subcommittee decided to eliminate CRAF in the FY 1993 budget. The President's budget for that fiscal year did not include CRAF, and Congress never opted to restore funds for the mission.

run to devastate the solar system exploration program both financially and scientifically.[154] Clearly, the experiences of developing planetary missions in the 1970s and 1980s showed NASA that bigger was not necessarily better for achieving scientific returns, especially when the agency's funds were tight and its priorities were manifold. While in 1989 the Bush Administration endorsed human missions to the Moon and Mars, many of those involved with NASA's robotic solar system exploration program believed that the agency ought to turn to more modest spacecraft.[155] In light of the blows the planetary program had endured over the past years, in 1989 NASA's Space Science Advisory Committee rekindled the idea of a low-cost missions program to maintain the vitality of planetary science, and the space agency finally embraced the concept.[156]

That year, NASA's space science planning committees began serious discussions about a program for low-cost planetary missions.[157] Coming to realize the gravity of the problems facing solar system exploration and recognizing that NASA would endure a gap in the flow of planetary data between the 1997 end of Galileo's mission and Cassini's Saturn arrival in 2004, these groups felt they had little option but to make small planetary missions with short development times a priority.[158] Daniel Goldin's appointment as NASA Administrator in April 1992 lent further support to the concept. Goldin had been an advocate of small and inexpensive, yet potent, space science missions since his days as an engineer at TRW. Once at NASA, he began to preach the gospel of "faster, better, cheaper" missions for all space science disciplines, including planetary science.[159] At the request of the Senate Appropriations Subcommittee to "prepare a plan to stimulate and develop small planetary...projects, emphasizing those which could be accomplished by academic or research communities," NASA delivered a report claiming that solar system exploration missions with low price tags and short development times would become the centerpiece of the agency's new programs in the 1990s.[160] [II-39]

154. NASA Space and Earth Advisory Committee, *The Crisis in Space and Earth Science* (Washington, DC: NASA, 1986).

155. In 1987 the NASA Advisory Council recommended that NASA follow robotic exploration of Mars with human visits as a means for the United States to reassert its leadership in space. Shortly afterward, NASA established the Office of Exploration to begin planning the scientific objectives and technological requisites of human missions to the planets. President Bush announced the Space Exploration Initiative in 1989 to send humans back to the Moon and on to Mars. Anticipated to revive NASA as a goal akin to Apollo around which the nation could rally, the initiative failed because it could not conjure the support Bush desired.

156. A year earlier, the Office of Space Sciences and Applications announced its commitment to augment NASA's Explorer program for non-planetary space science missions with small missions. See NASA, *Office of Space Sciences and Applications Strategic Plan* (Washington, DC: NASA, 1988).

157. For an overview of NASA's development of small missions for planetary science, see Stephanie Roy, "The Origin of the Smaller, Faster, Cheaper Approach in NASA's Solar System Exploration Program," *Space Policy* 14 (August 1998): 153–171. For the utility of such missions, see Space Studies Board, *The Role of Small Missions in Planetary and Lunar Exploration* (Washington, DC: National Academy Press, 1995).

158. Galileo's primary mission was scheduled to end in December 1997, but NASA has kept the spacecraft in operation due to its sustained health and remarkable scientific achievements.

159. "Faster, better, cheaper" (or any permutation of the three words) became an important mantra of NASA beginning in the 1990s. In general, it has been used in reference to projects intended to achieve outstanding results using fewer resources and less development time than past projects with comparable objectives.

160. NASA, "Small Planetary Mission Plan: Report to Congress," April 1992.

The program NASA proposed, called Discovery, received a programmatic new start in FY 1994 as a line item in the NASA budget.[161] Similar to the "Small Explorer" concept that had been in place for a few years for small, Earth-orbiting astrophysics and space physics missions, Discovery became the first small planetary missions program to receive its own budget line. NASA Headquarters imposed strict guidelines on Discovery. Under the program, individual scientists could propose entire missions to explore targets in the solar system. Every one to two years, NASA would review the proposals and select one or two to fund, based on their scientific value, cost, technical feasibility, and other factors.[162] The selected missions could cost no more than $170 million (FY 00 dollars), take no more than three years to develop, and launch on a booster no more powerful than a Delta II.[163] As envisioned by NASA, Discovery would not preclude missions to the outer solar system, but would be most appropriate for allowing investigators to conduct missions with focused scientific objectives to small bodies and within the inner solar system. Table 1 lists all of the Discovery missions that NASA has selected to date. Thus far, Discovery missions have demonstrated a new technique to land on Mars and to test the possibility of controlling from Earth a roving vehicle on the surface, and also have returned new data on the Moon's gravitational field and repudiated speculation that water ice exists on its surface.[164] Stressing public education and outreach, mission planners have made efforts to put these missions in the public eye in order to help renew public enthusiasm for solar system exploration.[165]

---

161. During the same year, the Ballistic Missile Defense Organization, with NASA's assistance, launched and operated a small, low-cost spacecraft called Clementine in the vicinity of the Moon to test sensors using advanced technology. Although the mission failed in its second planned task to track a near-Earth asteroid, the mission did help build confidence in NASA that the low-cost mission concept was viable.

162. The Discovery program has followed a competitive selection process for all but its first two missions, the Near Earth Asteroid Rendezvous and the Mars Pathfinder, which were missions in development that were grandfathered into the program to get it started.

163. The cost figure includes the price of design, development, and construction of the spacecraft and the first 30 days of the mission's operation. It does not include the cost of launch. NASA, Office of Space Science, Solar System Exploration Division, *Discovery Program Handbook*, November 1992.

164. For the preliminary scientific results of Mars Pathfinder, see several articles in *Science* 278 (December 5, 1997): 1734–74. For more on Lunar Prospector's search for lunar ice, see NASA Press Release 99-119, "No Water Ice Detected from Lunar Prospector Impact," October 13, 1999.

165. People around the world raved as they saw the first color pictures of the rover taken by the lander after Mars Pathfinder's successful bounce-landing on the Martian surface on July 4, 1997. That JPL's Mars Pathfinder World Wide Web site received a then to-date record of 47 million hits on one of the days shortly following the spacecraft's landing shows that the world took great interest in the Mars mission.

TABLE 1: NASA-APPROVED DISCOVERY MISSIONS

	Selection Year	Launch Date	Mission Description	Status
NEAR (Near-Earth Asteroid Rendezvous)	1993	February 1996	The first spacecraft to orbit and study an asteroid	In progress
Mars Pathfinder	1993	December 1996	Demonstrated a low-cost method of landing a spacecraft and science instruments onto the surface of Mars and using a small rover to explore Martian terrain	Completed
Lunar Prospector	1994	January 1998	Offered insight on the Moon's origin and evolution; also sought to determine whether water ice exists at the Moon's poles	Completed
Stardust	1995	February 1999	Will be the first spacecraft to collect comet and interstellar dust particles and return them to Earth	In progress
CONTOUR (Comet Nucleus Tour)	1997	N/A	Will encounter and study at least three comets	In development
Genesis	1997	N/A	Will collect wind particles to improve understanding of the evolution of the solar system	In development
MESSENGER (Mercury: Surface, Space Environment, Geochemistry and Ranging)	1999	N/A	Will orbit and conduct scientific investigation of Mercury	In development
Deep Impact	1999	N/A	Will probe beneath the internal composition and structure of a comet by crashing a projectile into its surface	In development

NASA embarked on yet another mission series in the 1990s. While Mars Observer was under development, the space agency made plans to establish a long-range program of Mars exploration. Its early concept was the Mars Environmental Survey (MESUR), whose goal was to distribute globally sixteen small landers on the Martian terrain in order to make measurements of the planet's surface, interior, and atmosphere. NASA planners anticipated that emplacing the MESUR network would not only benefit Mars science but also provide experience useful to the agency for developing technology for future robotic and human missions to the Red Planet. The agency hoped to initiate the network in the mid-1990s with a demonstrator lander called MESUR Pathfinder.[166]

Changing programmatic objectives as well as economic constraints, however, led NASA to suspend the MESUR concept, as it seemed likely that the agency would only be able to secure funding in the near future for a few of the project's ground stations. Still determined to create a long-term U.S. presence on and around Mars, NASA proposed another program that would fit better with the "faster, better, cheaper" concept that was becoming more popular and necessary to its programs' sustenance. Called Mars Surveyor, the program would abandon the notion of an integrated network of ground stations; instead NASA would send two low-cost spacecraft—an orbiter and a lander—to Mars every 26 months over the course of ten years.[167] Each mission, cost-capped at $175 million (FY 00 dollars) and limited to three years of development time, would address science objectives centered on understanding Mars' climate, resources, and the search for water and life.[168] By 2005, the agency hoped to send a robotic envoy to the surface that would be capable of collecting samples of Martian terrain and returning them to Earth. With the capability to achieve the scientific objectives of the recently lost Mars Observer, the Mars Surveyor program, to be managed by JPL, won the favor of both the Clinton Administration and Congress. The program received a $77-million new start in the FY 1995 budget and was approved by Congress shortly thereafter.[169] In the government as well as the public, enthusiasm for Mars study burgeoned in August 1996 when a team of planetary science researchers funded by NASA reported that they had found the first organic molecules of Martian origin—possible evidence that life once existed on the Red Planet—in ALH84001, a meteorite found in Antarctica and thought to be descended from Mars.[170] [II-43, II-44, II-45] With a renewed focus on the search for life on Mars, at the end of that year NASA kicked off its new Mars program with the launches of Mars Pathfinder and the Mars Global Surveyor, which arrived at Mars in 1997.[171] [II-46]

---

166. NASA ultimately flew MESUR Pathfinder as a Discovery mission called Mars Pathfinder.

167. Earth and Mars are in a configuration that minimizes the length of travel between the planets once every 26 months.

168. Space Studies Board, *Review of NASA's Planned Mars Program*, (Washington, DC: National Academy Press, 1996), p. 13.

169. The approval of Mars Surveyor marked the fastest entry ever of a NASA program into the federal budget, occurring in less than six months.

170. David S. McKay, Everett K. Gibson, Jr., Kathie L. Thomas-Keprta, Hojatollah Vali, Christopher S. Romanek, Simon J. Clemett, Xavier D. F. Chillier, Claude R. Maechling, and Richard N. Zare, "Search for Past Life on Mars: Possible Relic Biogenic Activity in Martian Meteorite ALH84001," *Science* 273 (16 August 1996): 924–30.

171. Although Mars Pathfinder was funded through the Discovery program, its activities also supported the goals of the Mars Surveyor program.

While both of these spacecraft safely reached Mars and conducted successful missions, neither of NASA's next two probes in the series achieved any of their science objectives. Launched independently, the Mars Climate Orbiter and the Mars Polar Lander both disappeared as they made their final approaches to the planet. These mishaps have forced the space agency to cancel plans to send a lander to Mars in 2001 and to rethink its approach to managing the Mars program.[172]

Whither the Past and Future of Planetary Exploration?

Ronald Schorn notes in his book-length history of planetary astronomy that "[t]he American space program...galvanized the field of planetary astronomy, revitalized it, and reformed it."[173] Indeed, his words could not be closer to the truth. The establishment of a national space program managed by a civil agency transformed solar system study from what years before was an endeavor of amateurs, whose homemade telescopes only allowed them to see into the cosmos as far as the planets, into a full-fledged scientific discipline worthy of pursuing for its own sake. Providing an organizational structure and armed with abundant federal funds, NASA essentially institutionalized the study of the planets and thus was able to attract geologists, astronomers, and engineers alike to help build up the field. Having become a big science enterprise, planetary astronomy's operations moved into the realm of space—a feat about which scientists for millennia had only dreamed. From the time of NASA's inception, scientists no longer talked only about planetary astronomy or science but planetary *exploration*. Building on technological developments for more than forty years, the space agency has orchestrated a program of robotic explorers that have truly revolutionized human understanding of the solar system.

As this essay has shown, however, the road to scientific success had not always been a smooth one for NASA's solar system exploration program. The task of building spacecraft that can endure long journeys over millions of kilometers of the harsh space environment and successfully return scientific data to Earth indeed has been, and still remains, a daunting technical challenge. But perhaps the even greater challenge to the field has been the constant need to compete with other programs for political and public support to receive funding. Since the end of Apollo, NASA has had a difficult time securing the level of funding it requests each year. Of the programs the space agency manages, human space flight—first Apollo, then the Space Shuttle and the Space Station—has consistently constituted the highest priority. Among the space sciences, solar system missions have experienced greater threats of cancellation than space physics and astrophysics projects because the planetary science community in general has had more difficulty reaching consensus on what research to undertake. NASA planners and scientists have, over time, argued over the relative importance of studying the Moon versus the planets versus smaller bodies such as comets and asteroids. Often, NASA or national needs other than science have governed the activities of the solar system exploration program.

172. Mars Climate Orbiter Mishap Investigation Board, Phase I Report, November 10, 1999; Mars Climate Orbiter Mishap Investigation Board, *Report on Project Management in NASA*, March 13, 2000; Mars Program Independent Assessment Team, *Mars Program Independent Assessment Team Report*, March 14, 2000; JPL Special Review Board, *Report of the Loss of the Mars Polar Lander and Deep Space 2 Missions*, March 2000.

173. Schorn, *Planetary Astronomy*, p. 181.

With budgets diminishing, the Cold War over, and NASA's realization that it cannot afford to sustain ambitious planetary missions, the space agency has turned to small, low-cost spacecraft to perpetuate the program. In light of the recent failures of some Mars spacecraft as well as small spacecraft in other space science disciplines, some scientists and program analysts feel that NASA is jeopardizing missions by imposing overly stringent constraints, reducing oversight of development and operations, and accepting too much risk. Some also believe that the limited resources are forcing small missions to sacrifice scientific capability to ensure their technical integrity. Perhaps solar system missions cannot achieve their scientific potential under reasonable levels of risk on the shoestring budgets and tight development schedules prescribed by the "faster, better, cheaper" strategy.[174]

Now that the solar system exploration program is under scrutiny once again, the space agency must make important decisions to secure the future of its missions to the planets. With over forty years of scientific and technical experience making some of humanity's most extraordinary achievements in the twentieth century, NASA has both the inspiration and capability to ensure planetary exploration's prosperity as the new millennium begins.

****

UNITED STATES SOLAR SYSTEM MISSION SUCCESSES

Spacecraft	Launch Year	Object Studied
Mariner 2	1962	Venus
Ranger 7	1964	Moon
Mariner 4	1964	Mars
Ranger 8	1965	Moon
Ranger 9	1965	Moon
Surveyor 1	1966	Moon
Lunar Orbiter 1	1966	Moon
Lunar Orbiter 2	1966	Moon
Lunar Orbiter 3	1967	Moon
Surveyor 3	1967	Moon
Lunar Orbiter 4	1967	Moon
Mariner 5	1967	Venus
Lunar Orbiter 5	1967	Moon
Surveyor 5	1967	Moon
Surveyor 6	1967	Moon
Surveyor 7	1968	Moon
Apollo 7	1968	Moon
Apollo 8	1968	Moon
Mariner 6	1969	Mars

174. Tony Spear, "NASA FBC Task Final Report," March 2000. In this study requested by NASA Administrator Dan Goldin, Spear's task group expressed the sentiment that NASA's management techniques and resource constraints contributed to the recent solar system mission failures.

UNITED STATES SOLAR SYSTEM MISSION SUCCESSES (continued)

Spacecraft	Launch Year	Object Studied
Mariner 7	1969	Mars
Apollo 9	1969	Moon
Apollo 10	1969	Moon
Apollo 11	1969	Moon
Apollo 12	1969	Moon
Apollo 14	1971	Moon
Mariner 9	1971	Mars
Apollo 15	1971	Moon
Pioneer 10	1972	Jupiter
Apollo 16	1972	Moon
Apollo 17	1972	Moon
Pioneer 11	1973	Jupiter, Saturn
Mariner 10	1973	Venus, Mercury
Viking 1	1975	Mars
Viking 2	1975	Mars
Voyager 1	1977	Jupiter, Saturn
Voyager 2	1977	Jupiter, Saturn, Uranus, Neptune
Pioneer Venus 1	1978	Venus
Pioneer Venus 2	1978	Venus
Magellan	1989	Venus
Galileo	1989	Jupiter and its moons
Clementine	1994	Moon
NEAR	1996	Asteroid
Mars Global Surveyor	1996	Mars
Mars Pathfinder	1996	Mars
Cassini*	1997	Saturn, Titan
Lunar Prospector	1998	Moon

*still en route to destination

Document II-1

Document title: Roy W. Johnson, Director, Advanced Research Projects Agency, to Commanding General, Ballistic Missiles Division, Air Research and Development Command, March 27, 1958.

Source: NASA Historical Reference Collection, History Office, NASA Headquarters, Washington, D.C.

Document II-2

Document title: U.S. Army Ordnance Missile Command, "Development and Funding Plan for Project One, ARPA Order 1–58, as Amended," 15 May 1958.

Source: NASA Historical Reference Collection, History Office, NASA Headquarters, Washington, D.C.

Three weeks after Sputnik's launch, a Jet Propulsion Laboratory (JPL) group led by Director William H. Pickering developed "Project Red Socks," a mission proposal consisting of nine flights to the Moon, which they believed would be the United States' best means of rivaling the Soviet Union in space. Although initially rejected by the Department of Defense, the proposal appealed to the new Advanced Research Projects Agency (ARPA), which was made responsible for U.S. space projects in early 1958. ARPA approved a lunar program in March 1958, giving three flight opportunities to the Air Force and two to the Army. ARPA Director Roy Johnson ordered the Air Force to proceed with their probes in Document II-1, while Document II-2 is the Army's plan to develop lunar probes to test communications functions and take a photograph of the Moon. None of these probes succeeded in reaching the Moon.

Document II-1

[each page stamped "SECRET" and "CONFIDENTIAL"]

[no page number]

ADVANCED RESEARCH PROJECTS AGENCY
WASHINGTON 25, D.C.

ARPA ORDER #2-58
[stamped "MAR 27 1958"]

TO: Commanding General
Ballistic Missiles Division, ARDC
Los Angeles, California

1. Pursuant to the provisions of DOD Directive 5105.15 dated 7 February 1958, the Secretary of Defense has approved and you are requested to proceed at once on behalf of the Advanced Research Projects Agency with the projects specified below. Additional details and directives will be issued by ARPA from time to time and will become a part of this order when so specified.

2. Project One – Provide ARPA by contract or otherwise with technical, consulting, administrative, supervisory, inspection, reporting or management services in support of ARPA programs as may be required from time to time. These services may apply to general ARPA requirements over and beyond those required for the specific projects outlined herein or to other specific ARPA projects which may be assigned to BMD at a later date. Detailed requirements for services will be assigned from time to time as tasks under this project.

Project Two – Make necessary contractual arrangements and provide such support as required to proceed with the objective of three lunar probes to be launched as soon as possible consistent with the requirement that a minimal amount of useful data concerning the moon be obtained. The launch vehicle is to consist of the THOR as the first stage, the VANGUARD as the second stage, and a solid rocket as the third stage. The payload is expected to be at least 30 pounds, after allowing for a retrofiring fourth stage rocket.

This project should also include attempts to develop better third stage engines than those now available if a reasonable chance of timely application to this project exists. Up to $1,000,000 is to be allowed for this development.

The determination of the types of payloads is to be made later, but as soon as possible, by ARPA in consultation with the systems manager, and other interested parties, if any.

Maximum possible use is to be made of existing ground equipment in the data telemetry link. However, if necessary, special ground equipment may be developed and used.

[2] 3. You will submit, as soon as possible, for review and approval by the Advanced Research Projects Agency a detailed development and related financial plan covering each project specified. These data shall include a time-phased schedule of work and estimates for work to be performed (a) at BMD, (b) by contract, and (c) at other government facilities.

4. This order makes available $3,000,000 under appropriation and account symbol "97X0113.002 Salaries and Expenses, Advanced Research Projects, Department of Defense" for FY 1958 for obligation by the Ballistic Missiles Division on behalf of the Advanced Research Projects Agency only for purposes necessary to accomplish the work specified herein. These funds are immediately available for direct obligation and for use in reimbursing the Ballistic Missiles Division for costs incurred under this order. Upon approval of detailed development and financial plans, as required herein or in accordance with amendments to this order, these funds will be increased as appropriate.

5. The Director, Advanced Research Projects Agency, will provide policy and technical guidance, either directly or through designated resident representatives. The

Ballistic Missiles Division will be responsible for arranging for the detailed technical direction necessary to accomplish the specified objectives and to comply with ARPA policy and technical guidance. This general relationship may be specified in greater detail by amendment to this order if such action is necessary.

6. The Director, Advanced Research Projects Agency, and the Office of the Secretary of Defense will be kept informed by such management, technical and accounting reports as may be prescribed pursuant to this order.

7. The use of equipment and materials procured in connection with these projects is subject to direction of ARPA and all reports, manuals, charts, data and information as may be collected or prepared in connection with the projects shall be made available to ARPA prior to release to other agencies or individuals under procedures to be approved.

8. BMD shall be responsible for preserving the security of these projects in accordance with the security classifications assigned and the security regulations and procedures of the Department of the Air Force.

[3] 9. Notwithstanding any other provisions of this order, BMD shall not be bound to take any action in connection with the performance of this work that would cause the amount for which the Government shall be obligated hereunder to exceed the funds made available, and the obligation of the BMD to proceed with the performance of this work shall be limited accordingly. BMD shall be responsible for assuring that all commitments, obligations and expenditures of the funds made available are made in accordance with the statutes and regulations governing such matters provided that whenever such regulations require approval of higher authority such approvals will be obtained from or through the Director, ARPA, or his designated representative.

[signed "Roy W. Johnson"]

Document II-2

[each page originally stamped "SECRET" and "CONFIDENTIAL"]

[each page stamped "UNCLASSIFIED"]

[cover page]

DEVELOPMENT AND FUNDING PLAN
FOR
PROJECT ONE, ARPA ORDER NR 1-58, AS AMENDED (U)

15 May 1958

SECURITY NOTE

This document contains information affecting the National Defense of the United States within the meaning of the espionage laws, Title 18, U. S. C., Section 793 and 794. The transmission or the revelation of its contents in any manner to an unauthorized person is prohibited by law.

[1] I. STATEMENT OF MISSIONS

The missions as set forth in Amendment 1 of ARPA Order 1-58, dated 2 May 1958, are as follows: (a) a JUNO II to be launched in or about November 1958 as an escape guidance experiment with a payload of about 15 pounds and consisting of a trackable beacon and an optical apparatus capable of determining its location in space relative to the moon; (b) a JUNO II to be launched in or about January 1959 as an escape guidance experiment with a payload of about 15 pounds and consisting of a trackable beacon and an optical apparatus capable of taking, developing, and scanning a picture of the moon; (c) and a JUNO II to be launched in or about February 1959 as a satellite with a payload of about 60 pounds and consisting of the heavy cosmic ray experiment.

NOTE: The schedule contained in Section V of this plan indicates firing in December 1958, February and March 1959. This schedule is necessitated by the injection of the ARGUS program.

II. DESCRIPTION OF PAYLOADS

A. LUNAR PAYLOADS

1. The first lunar payload will consist of two optical scanning devices which will measure the range to the moon and the location of the payload in the trajectory relative to the earth, moon and sun. A transmitter within the payload will transmit trajectory information as the lunar payload nears the moon. Transmission at 960 MC, 100 MW, will be timed for reception at the Camp Irwin, California, tracking site. The primary purpose of this first payload is to test out the most critical communication functions and camera triggering operations to be used in the second lunar probe. This experiment has a high probability of success because it is simple and straightforward.

2. The second lunar payload will have the mission of taking a single shot photograph of the moon, using Land Camera techniques, with a resolution of 20 miles. An optical system will be used to take, develop and scan the photograph of the moon. The feasibility of vacuum development of a photograph has been demonstrated. The scanned photograph will be transmitted to the ground using a narrow band system with a relatively long time transmission. The picture will be taken in a fraction of a millisecond and transmitted over a period of about one half hour. …

[4] III. DESCRIPTION OF CARRIER VEHICLE

A. GENERAL

The JUNO II carrier vehicle is basically a JUPITER missile with extended tankage, slightly modified guidance, and designed for the mounting of the JUNO I upper stages to form a four-stage vehicle. The JUNO II vehicle for the lunar probe missions (Figure 3) and the satellite mission (Figure 4) will be identical, except for differences in the payloads, already described in Section II, and minor differences in the guidance system. The vehicle consists of the thrust unit, the aft unit, the cluster and shroud. The aft unit and the cluster and [5] shroud form the body of the missile. The overall length of the carrier vehicle is approximately 72 feet, its maximum diameter is 105 inches, and it has a dry weight of approximately 10,800 pounds.

B. THRUST UNIT

The JUPITER thrust unit has been elongated to allow for an additional 36 inches of propellant tankage. A 13.8 inch fuselage segment inserted in the fuel container and a 22.2 inch segment in the LOX container increase the capacity to 181 seconds burning time (with a 3 second reserve). As a result of this increase in container site the high pressure sphere capacity for fuel container pressurization has been increased from 3.5 cubic feet to 4 cubic feet. The liquid rocket engine utilizes liquid oxygen as its oxidizer and RP-1 as its fuel and furnishes a nominal thrust of 150,000 pounds. During the powered portion of the missile flight pitch and yaw control is provided for by the gimbaling of the rocket engine and roll control is provided for by the swiveling of the turbine exhause [sic] nozzle.

C. GUIDANCE

1. Lunar Probe. The guidance system to be used for this mission will be essentially the same as that of the JUPITER missile with the following exceptions:

a. Velocity guidance in only two directions (slant range and slant altitude) will be used.
b. Interim guidance computers will be utilized.
c. Coordinate resolvers will be placed in the slant range and slant altitude circuits to provide two additional angles of elevation for the slant range axis.
d. Vernier motor will not be used.
e. Only two presettings (slant range velocity, and a constant term to account for forces of separation, etc.) will be used instead of four.
f. Boom angle-of-attack meters will be used.
g. Proportional jet nozzles will be used for spatial attitude control of body after thrust unit-body separation instead of ON-OFF nozzles. ...

[6] D. UPPER STAGES

The JUNO II cluster is almost identical to the JUNO I cluster used in the EXPLORER launching. The major differences are that the JUNO II uses a shroud to protect the upper stages from aerodynamic heating during the powered portion of flight and the JUNO II uses high performance propellant in the third stage as well as the fourth stage (Figure 5).

The second, third, and fourth stages will a consist of eleven, three, and one solid propellant rocket motors respectively. The second stage will provide a nominal thrust of 14,533 pounds at a nominal burning time of 8 seconds, the third stage will provide a nominal thrust of 4,762 pounds at a nominal burning time of 8 seconds, and the fourth stage will provide a nominal thrust of 1,506 pounds at a nominal burning time of 8 seconds. As in the JUNO I, the third stage is positioned inside of the second stage and the fourth stage is mounted on top of the third stage. The spin launcher is mounted on top of the JUPITER aft unit taking the place of a tactical nose crane. The payload will be mounted on the fourth stage and will be separated from the fourth stage upon its burnout.

E. OPERATION

1. Lunar Probe. The missile will be fired from Cape Canaveral at an aiming azimuth and pitch program dependent upon the time of day and day of the year. Prior to launch the spin launcher will be brought up to the required spin rate by electric motors. Shortly after launch the missile is programmed into the required trajectory and will continue in powered flight for about 181 seconds at which time engine cutoff occurs. During powered flight the missile will use angle-of-attack control since it is aerodynamically unstable. Several seconds later the shroud is separated by standard explosive screws and springs. The shroud will be forced forward of the vehicle by the separation forces and will then be forced out of the flight path by a small rocket motor. Five seconds after engine cutoff, separation of the thrust unit from the body occurs by the action of explosive screws and springs. A proportional spatial attitude control system will then position the body into the required [7] attitude and approximately 50-55 seconds after thrust unit-body separation the second stage will be fired by a missile-borne timer. The altitude at which second stage ignition occurs depends upon the final selection of the trajectory. The third and fourth stages are fired by timers at 8 second intervals from the firing of the second stage. The payload will be separated from the fourth stage and will follow a hyperbolic trajectory to the vicinity of the moon. The minor modifications made to the guidance and control system will provide the ability to launch the missile at any one of two and possibly four times a day and on three consecutive days. ...

IV. SIGNIFICANT ENGINEERING TASKS

A. LUNAR PROBES

1. The optimum trajectory and flight path for the LUNAR approach must be determined.
2. The analysis, design, and modification of the JUPITER guidance system for a low flight trajectory must be accomplished.
3. The ability to change the aiming azimuth, pitch program, and cutoff angle within a short time period at the launching site must be achieved. All of these functions change with time if the pre-specified lift-off time is not met.
4. A separable upper stage shroud to provide aerodynamic heating protection and support a boom type angle-of-attack meter must be developed.
5. A fourth stage satellite separation device than does not cause displacement of the satellite must be developed.

[8] 6. The tracking stations at Cape Canaveral Missile Test Center and Camp Irwin and sufficient hardware for two overseas sites must be brought into being.
7. The design of the photographic developing technique must be accomplished by 1 July 1958.
8. The ability to launch the missiles within a very close tolerance (several minutes) of the specified time must be achieved. ...

Document II-3

Document title: Harold Urey and Robert Jastrow to Homer Newell, Assistant Director for Space Sciences, NASA, January 16, 1959.

Source: Robert Jastrow, Director, Mount Wilson Observatory, Pasadena, California.

Document II-4

Document title: Homer Newell, Assistant Director for Space Sciences, NASA, to Abe Silverstein, Director, Office of Space Flight Development, NASA, "Proposed NASA Project—Lunar Explorations," March 23, 1959.

Source: Robert Jastrow, Director, Mount Wilson Observatory, Pasadena, California.

In January 1959, Robert Jastrow, head of NASA's theoretical division of space sciences, invited University of California professor Harold Urey to NASA Headquarters in Washington to give lectures on his scientific understanding of the Moon and planets. After the lectures, Jastrow and Urey approached Homer Newell with a suggestion that NASA should undertake a science-focused, crash project to land on the Moon by 1961 to remain competitive in space with the Soviet Union. Document II-3 is a draft of the proposal composed by Urey and Jastrow; no final, official version exists. Newell approved the document and formed an ad hoc Working Group on Lunar Exploration made up of NASA and academic scientists to evaluate and recommend experiments to be sent to the lunar surface and environs. On March 23, 1959, Newell proposed to Silverstein that lunar exploration should play a major role in NASA's program. These documents led to the development of the Ranger project.

Document II-3

[1] INTRODUCTION

We believe that effort should be concentrated on a single project of great scientific value and impact. Successful and rapid completion of such a project will enhance the reputation of the United States to a degree that cannot be achieved by the execution of a conventional scientific program on a normal schedule. With this purpose in mind, we propose that a crash program be set up for the execution of a lunar soft landing. We believe that with a substantial concentration of effort it will be possible to effect a lunar soft landing and surface exploration in two years. We propose that the lunar surface vehicle shall contain three experiments—a seismograph, a detector of radioactivity [the third experiment anticipated was a television experiment]. This project is presently scheduled for the approximate period 1963-64. We presume that it occupies a similar place in the USSR schedule of space research.

A soft landing with performance of the experiments listed below will capture the imagination of the scientific community and the general public to a greater degree than any project of comparable scientific value.

[2] The moon is probably the last object of the entire solar system which has been modified least by all the forces that have acted since the solar system formed. It was used by Prof. Urey in this book on the planets as the subject of the second chapter for the reason that the marks and scars on the surface were probably produced there 4.5 billion years ago. Since that time it has become generally recognized as a very primitive object, not derived from the earth as has been supposed in the past but as an object that is probably older as a body than the earth. In fact it may be that it is one of a group of primitive objects of which the earth and the other planets are formed. Its surface has, to be sure, been somewhat modified by the fall of meteorites on its surface; but even nearly a century ago it was recognized that many of these meteorites were probably part of the stage of the formation of the solar system itself. This was put forward by a very eminent American geologist by the mane of G. K. Gilbert so long ago as 1893. There are [3] questions that we should like to know about this subject in connection with the subject studies. Also in every completed data I have seen, it showed high concentrations of uranium, thorium and plutonium in its surface. If it was only partly melted, then these elements should be concentrated in the surface regions to a high degree. If it was not melted the concentrations should be the same as they are in the meteorites. In fact, it is not at all impossible that the stone meteorites which we are working on in our laboratory came from the surface of the moon. Very excellent work on these objects has been done by a group of men in the U.S. since the war. It has been found that they solidified and were last degassed some 4.5 billion years ago. This same subject has been followed to some extent by the USSR but the work in the U.S. has been definitely superior.

There is much controversy over these subject[s]. Because all our information must be secured by looking at the object from a distance of some 230,000 miles, it seems fairly certain [4] that much of the controversy is due to the inability of men to interpret clearly what they see from this great distance even with our best telescopes.

It is our opinion that a study of the moon is more important than a study of Venus or Mars, from the standpoint of the origin of the solar system. We cannot expect to get information from these two objects that will bear so directly on this question because the surfaces have been modified by erosion of some kind.

It is true that if we could detect life on these planets it would be a piece of information of vast magnitude but outside this, the information that can be secured from the surface of these planets is highly inferior to a study of the moon.

In designing experiments it is well to remember that the easier things should be done first. The moon is nearest, the problem of transmitting information is less in this case, [5] and as a result of our experiments we would get great experience for the more distant purposes.

Document II-4

[no page number]

Office Memorandum — UNITED STATES GOVERNMENT

DATE: 23 March 1959

TO: Abe Silverstein

FROM: Homer E. Newell, Jr.

SUBJECT: PROPOSED NASA PROJECT –
LUNAR EXPLORATIONS

The Office of Space Sciences has completed the initial planning phase of a Lunar Explorations project to be conducted as part of the National Space Program. The present planning status of the project is described in the attachment to this memorandum. It is based on staff review, involving the selective adoption and synthesis of concepts suggested in preliminary proposals received from interested institutions and the results of two meetings of a Space Science Discussion Group held on February 5, and February 14, 1959, respectively.

The preliminary budget estimate totals $46.8 million. A breakdown of these figures by fiscal years and major categories is given on page 8 of the attachment. It appears that presently planned vehicle system will have adequate capability to perform the missions of this project.

Recommendation

It is recommended that the Director of Space Flight Development

a. Approve in principle the Lunar Exploration Projects as part of the national space program.
b. Authorize the Assistant Director for Space Sciences to submit research project proposals which are part of the initial phases of the project for early funding out of FY 59 allocations, up to a limit of $2 million.
c. Authorize the Assistant Director for Space Sciences to proceed with detailed planning, specifications, and activation of the project.

cc: Wyatt
Stoller
Schilling
Clark
Fuhrman
Jastrow

Attachment: Lunar Explorations

DS ______________________ DSA ______________________
Homer E. Newell, Jr. G.F. Schilling

COPY

****

[no page number]

23 March 1959

PROPOSED NATIONAL AERONAUTICS AND SPACE ADMINISTRATION PROJECT

PROJECT TITLE: Lunar Explorations
PROJECT OBJECTIVE: The objective of the lunar explorations project is the investigation of the lunar environment, surface, and interior. The project will include lunar probes, lunar orbiters, rough landings, and soft landings, in accord with the estimated order of availability of the necessary vehicles and guidance systems. The final achievement of the soft landing with survival of instruments on the surface of the Moon will constitute the first establishment of a fixed data link between the Earth and a celestial body.

Relatively firm decisions have been made on the type of scientific equipment constituting the payloads for the lunar impacts and basic lunar orbiter. Firm decisions have also been made on several experiments for the advanced lunar orbiter, the rough landing, and the soft landing. Preliminary proposals have been received from scientific groups and industrial laboratories for these experiments. The remaining experiments for the rough and soft landing payloads must be determined after further evaluation.

The lunar probes and orbiters will carry scientific payloads designed to obtain information on the lunar environment, including magnetic field measurements, the properties of the interplanetary plasma, fluxes of energetic particles, the intensity of gamma radiation originating from radioactive decay of materials in the lunar surface, and measurements of the extent of the lunar atmosphere.

[3] The advanced lunar orbiters will also give preliminary information on the detailed structure of the lunar surface, for application to the rough landing and soft landing phases of the project. The orbit analysis of the lunar satellite will provide a measurement of the mass and figure of the Moon, leading indirectly to improved values for the radius of the Earth.

The lunar orbiter should be considered as a dual purpose payload. First, it will survey the surface of the Moon. Second, it will serve as an anchored space probe, placed at a convenient distance from the Earth for the return of data.

The rough landing refers to instrumented capsules designed for survival after an impact at a velocity of a few hundred feet per second. The scientific payload of the rough landing will include seismographs, magnetometers, and basic instrumentation for the measurement of the properties of the lunar surface, and the detection and analysis of the lunar atmosphere. The soft landing payload will include the above instruments, an X-ray fluorescence experiment for the analysis of lunar surface materials, and advanced instrumentation for the measurement of lunar surface properties.

Institutions that have proposed the construction of specific experiments include:

California Institute of Technology (lunar seismograph)
Columbia University (lunar seismograph)
Texas Instrument Company (surface radioactivity)
Jet Propulsion Laboratory, NASA (radiation package)
Beltsville Space Center, NASA (lunar atmospheric density; hydrogen content)
[4] Massachusetts Institute of Technology (plasma measurements)
Naval Research Laboratory (beta and gamma detection)
Los Alamos Laboratories (surface radioactivity)
University of California at La Jolla (X-ray fluorescence)
University of Chicago (energetic particles)

In addition, many institutions have evidenced a general interest in participating in various phases of the project.

Project Management will be provided by the Office of the Assistant Director for Space Sciences.

SPECIFIC EXPERIMENTAL OBJECTIVES: The project will be implemented by the following payloads:

1. Lunar Impact I (Thor-Delta, January 1960). This payload will include an Rb magnetometer, and a Lyman Alpha measurement of hydrogen density and scale height in the vicinity of the Moon.

   Payload breakdown:

Rb Magnetometer	40
Hydrogen density	5
Batteries and solar cells	15
Verniers	15
Doppler	5
Structure	20
	100 lbs.

2. Lunar Orbiter I (Atlas-Delta, February 1960). The lunar orbiter payload will contain a radiation package designed to study the thermal and directed fluxes of the interplanetary plasma, the energetic particle fluxes in the kilovolt and mev regions, and the intensity of beta and gamma rays produced by lunar surface activity. If weight permits, an Rb magnetometer will be included.

[5] Payload breakdown:

Plasma Probe	20

Soft corpuscular radiation	15
Beta and gamma rays	15
Batteries and solar cells	75
Retro	125
Structure	35
Magnetometer	40
	325 lbs.

3. Lunar Impact II (Thor-Delta, September 1960). This payload will include a high resolution TV tube designed for two pictures, one from an altitude of 400 miles covering 1600 square miles with a resolution of 400 feet, the second from 200 miles covering 400 square miles with a resolution of 400 feet. Required bandwidth is 5 kc. The second experiment will be a density measurement in the lunar atmosphere, using a cold cathode discharge gauge.

   Payload breakdown:

TV (JPL)	30
Ionization Gauge	10
Batteries	20
Verniers	15
Doppler	5
Structure	20
	100 lbs.

4. Advanced Lunar Orbiter II (Vega, April 1961). The payload of the advanced lunar orbiter will include high resolution TV, gamma detection by a scintillation counter with pulse height analysis, a magnetometer, and a refined package for detection of isotropic and directed corpuscular fluxes.

5. Rough Landing (Vega, September 1961). The seismograph will be the major objective of the rough landing payload. Additional instrumentation will include an ionization gauge, a penetometer and possibly other simple instruments for the measurement of physical properties of the lunar surface.

[6]

6. The soft landing payload will contain X-ray fluorescence apparatus for the analysis of lunar surface materials, a seismograph, a magnetometer and advanced instrumentation for the measurement of conductivity and other physical properties. Sample return is also planned.

VEHICLES:

*1. January 1960: Thor-Delta (Lunar Impact I)
2. February 1960: Atlas-Delta (Lunar Orbiter I)
*3. September 1960: Thor-Delta (Lunar Impact II)

*4. April 1961: Vega (Lunar Orbiter II)
5. September 1961: Vega (Rough Landing I)
6. February 1962: Vega (Sfot [sic] Landing I)

Summary of Vehicle Requirements Through 1961:

Atlas-Delta:	1
Thor-Delta:	2
Vega:	3

* On present schedule.

[chart on page 7 omitted]

[8] BUDGET ESTIMATE (In millions $)

1. Project Breakdown:

	Vehicle	Payload and R and D	Totals
Lunar Impact I	2.4	2.5	4.9
Lunar Impact II	2.4	2.5	4.9
Lunar Orbiter I	3.0	3.0	6.0
Lunar Orbiter II	5.0	4.0	9.0
Rough Landing I	5.0	6.0	11.0
Soft Landing I	5.0	6.0	11.0
	22.8	24.0	$46.8

2. FY Breakdown:

	FY 1959	FY 1960	FY 1961	FY 1962	Total
Vehicles	2.4*	10.4*	10.0		22.8
Payload and R and D	2.0*	5.1*	10.0	6.9	24.0
	4.4	15.5	20.0	6.9	$46.8

*Either already funded or on present budget estimates for lunar probes.

Lunar Explorations
Astronomy and Astrophysics Programs
Office of Space Sciences
March 23, 1959

Document II-5

Document title: Albert R. Hibbs, ed., "Exploration of the Moon, Planets, and Interplanetary Space," JPL Technical Report 30–1, April 30, 1959.

Source: NASA Historical Reference Collection, History Office, NASA Headquarters, Washington, D.C.

Even before JPL became a part of NASA in December 1958, engineers and scientists at JPL professed an interest in leading the way in national exploration of the Moon and planets. During the fall of 1958, JPL undertook a study to develop a five-year plan for lunar and planetary exploration. The final report of the study, released in April 1959, was the first comprehensive plan for solar system exploration. With input from many scientists around the nation with planetary research interests, the JPL report called for the United States to take advantage of every opportunity to launch to Venus and Mars over the next five years, while supplementing gaps in the schedule with lunar missions.

[cover page]

National Aeronautics and Space Administration
Contract No. NASw-6

REPORT NO. 30–1

EXPLORATION OF THE MOON, THE PLANETS, AND INTERPLANETARY SPACE

Edited by Albert R. Hibbs

[signature]
Albert R. Hibbs, *Chief*
Research Analysis Section

Copy No. HC 348

JET PROPULSION LABORATORY
California Institute of Technology
Pasadena, California
April 30, 1959

[iii]

PREFACE

Portions of the following report were originated under studies conducted for the Department of Army Ordnance Corps under Contract No. DA–04–495-Ord 18. Such studies are now conducted for the National Aeronautics and Space Administration under Contract No. NASw-6. ...

[1] **ABSTRACT**

The Jet Propulsion Laboratory has undertaken a survey of possible objectives in a program of exploration of the moon, the planets, and interplanetary space. This has been combined with a survey of the feasibility of engineering developments which would be required by such an exploration program. The results of this study are presented in this Report.

The Report describes the basis on which the study was conducted, presents a review of current knowledge about the moon, the planets, and interplanetary space, gives a brief summary of the results of the Laboratory study on the feasibility of a program for the exploration of space, describes a program of lunar and interplanetary flights, and outlines the necessary development activities to support the exploration program. The time scale covered extends from 1959 through 1964.

I. INTRODUCTION

In December of 1958, the Jet Propulsion Laboratory was requested by the National Aeronautics and Space Administration to prepare a study of the space exploration program. In particular, the Laboratory was asked to describe those portions of the program wherein it felt it might make the greatest contribution. It was suggested that the period to be covered by the study extend through 1964. Thus, this study was intended to be an outline for a Laboratory program over the next 5 years.

Work on this study was broken down into several different areas; for example, vehicle development, guidance and control, tracking and communications, and so forth. One particular area so defined was the study of the scientific missions which might be undertaken in this program, in particular, those specific scientific objectives which might become the primary objectives of the Laboratory program.

The Laboratory program will consist of both the design, development, and operation of some of the rocket vehicles to be used in the space program and the design, development, and operation of some of the payloads which will carry the scientific measuring devices. In the area of payload development, it is the intention of the Laboratory to concentrate on those payloads designed for lunar and [2] planetary investigations, as contrasted to artificial earth satellites. The area of study of the scientific missions in space, like the other areas of the study program, was governed by this statement of Laboratory intention—concentration on the moon, the planets, and the space between them.

This study of the scientific missions in space had the following objectives: To tie together the important scientific missions with feasible technical developments and produce a realistic 5-year program for the scientific exploration of space. The results of this study are being used in the construction of a Laboratory program for the development of the necessary rocket vehicles and payloads to carry out this program. This Report presents the results of this study of scientific missions. It includes also some portions of other areas in the overall study program which were used to assess the feasibility of the proposed scientific program.

The results presented herein are not intended to be hard and fast design decisions on vehicles, payloads, or scientific instruments. Furthermore, the schedules presented herein are consistent with scientific potentialities and astronomical dates but are not to be interpreted as program commitments. The results are intended to be as realistic as possible on the basis of present knowledge, but it must be kept in mind that further developments will undoubtedly change many details of the program in a very significant manner. Thus, these results represent a typical program which can be used as a basis for program planning.

[3]

II. BASIC PHILOSOPHY

Three criteria have been selected as having the most important effects on the program:

1. Technical feasibility
2. Public reaction
3. Scientific and technical merit

These criteria have been listed, to some extent, in order of relative importance. The question of technical feasibility has been the primary consideration throughout this study. Before one can decide that a particular experiment should be carried out because it is either worthwhile or desirable or both, one must first make sure that it is possible at all.

It is more difficult to determine the relative importance of the last two criteria. Occasionally, they both lead one to the same conclusion. For example, the search for life on another planet is of the greatest scientific importance and, at the same time, is encouraged by a strong public interest. On the other hand, and in this same area, the problem of decontamination of planetary probes may be approached in quite a different manner by scientific and nonscientific groups. Scientific groups recognize the need for decontamination as primary for the success of future explorations for life forms. However, the public may question whether or not it is worthwhile to postpone a Mars shot, for example, for 2 years so that problems of decontamination may be fully solved.

In approaching this particular problem, we have taken the rather optimistic point of view that (1) the problems of decontamination can be successfully worked out in time to meet the proposed schedule, and (2) the public can be educated as to the importance of this problem so that they will neither begrudge the amount of money spent on its solution nor object to the limitation of experiments resulting from its possible lack of solution.

This example is characteristic of the manner in which the basic philosophy of this Report has been applied. An attempt has been made to select the possible, worthwhile, and desirable scientific experiments in the program for the exploration of the moon, the planets, and interplanetary space. A representative flight program has been constructed in which these experiments will be undertaken over the next 5 years. On the basis of the investigations involved in setting up this program, an attempt has been made to select the particular areas of the program which appear to need the most urgent attention.

In applying the criteria of technical feasibility, we have assigned specific missions to the launching vehicles and developed representative estimates of the payload weights which these vehicles can carry to the various objectives. It must be recognized that these weights are far from definite and are intended to be representative of the type of vehicle available at a particular time during the program. Furthermore, we have broken down the payload weights into the various major payload components and have listed representative weights which might be ascribed to each of these components. Here, again, these weight estimates are far from definite, but they are reasonable and serve to point out the problems of weight limitation.

We have also attempted to estimate the degree of complexity which can be assumed for each of these components, including the scientific instrumentation. This estimate has been made with the clear realization that the need for extreme reliability limits the degree of novelty which can be introduced into the various payloads.

In investigating the scientific objectives of the program, we have also considered the purely technological problems which must be solved if the whole program is to be successful. The telemetered information sent back from the payload must contain the results of measurements made for purely engineering objectives. In order to make possible the development of increasingly complex payloads, we must develop a background of engineering design data on the behavior of materials and components in the completely new environment of empty space and the atmospheres of other planets.

The program developed from this basic philosophy and the resulting investigations has been projected 5 years into the future. The validity of estimates of technical feasibility, public reaction, and scientific and technical merit is naturally degraded as we proceed further and further into the future. It was felt that, on the basis of present knowledge, a reasonable prediction could be made for a period of no more than about 5 years.

For this reason, present study does not include consideration of the man-in-space program. It is not reasonable [4] to assume that the man-in-space program would have any direct bearing on the technical problems involved in the exploration of the moon, the planets, and interplanetary space over the next 5-year period. Although many of the scientific and technological experiments which will be carried out during this program will have a definite bearing on the design of the vehicles which carry men to the planets, it is not felt that this objective is in any way inconsistent with the already stated criterion of scientific and technological merit.

Some time after the close of this first 5-year period, the man-in-space program and the interplanetary space program will gradually merge.

Certainly, a manned landing on another planet is one of the most important objectives of a long-range program. Regardless of how clever we become with remote measuring devices, one hard-rock geologist landed on the moon, for example, would be worth many tons of automatic equipment. The public interest in full-color photographs taken by a remote camera on the surface of Mars will be little as compared to the wild reception which will greet the first crew of astronauts which returns alive from that planet.

It is the basic philosophy of this study to develop [a] possible, sensible, and desirable beginning for a program which will eventually take man to the planets. ...

[93]

VI. SUGGESTED PROGRAM

A. Flight Schedule

The flight schedule for the exploration of the planets depends upon both the availability of the equipment and the availability of the planetary target. The program discussed in this Report covers a time period from 1960 through 1964. During this time period, the principal vehicles to be used in the program are the Vega and the Saturn.

The targets considered to be feasible objectives for this time period are the moon, Mars, and Venus. The moon is available as a target every month. However, the planets are available only once for each synodic period. For Venus, this is approximately once every 19 months, and for Mars approximately once every 25 months.

In general, the suggested schedule calls for 3 flights a year during the 5-year period from 1960 through 1964, except for the year 1962, in which 4 flights are suggested, two each at Mars and Venus. At first glance, this seems like a meager firing schedule as compared with the usual missile-test schedules. However, it must be remembered that these flights do not include all of the flights which will be made with these vehicles. Several of the vehicles will be committed to satellite launchings during this same time period. In addition, it must be remembered that these flights are being carried out early in the development history of the vehicles, at a time when flight testing must proceed on a fairly cautious basis in order that maximum advantage may be taken of each flight test in further development of the missile. Furthermore, each of the flight tests will continue for several weeks or months, rather than for just a few minutes as for a typical military missile. The data resulting from this extended flight time must be given at least a preliminary analysis before sensible design decisions can be made for the next shot at that particular target.

Even as it stands, a schedule calling for 3 lunar or planetary shots every year for the next 5 years, using vehicles which have not yet had their first flight test, is extremely optimistic. It would be desirable to have backup firings available in the event the scheduled test met with disaster.

Unfortunately, the concept of a backup test takes on a new significance for planetary experiments. In order to meet the requirements for launching at a planetary target, the backup flight must be made within a few days of the original flight, or else delayed for a complete synodic period. Furthermore, since it will not be known for several months whether or not the payload satisfactorily completed its design objectives, it is not really possible to know within the few days allotted whether or not a backup is necessary. Thus, it would seem desirable to fire the backup shot regardless of whether or not the primary shot appeared to be a success within the first day or two.

B. Description of Typical Payloads

The payload weights available from the vehicles considered in this study are not yet accurately known. Very approximately, the Vega might carry 400 lb to Mars, 700 lb to Venus, and 1000 lb to the moon. Payloads for the *Advanced Vega* might be anywhere from 50% greater to twice as great. For the *Saturn*, preliminary estimates give values approximately 10 to 15 times larger than the *Vega*.

For many missions, large fractions of the total payload weight (all weight forward of the propulsion system of the last stage in the launching vehicle) must be devoted to retro-rockets or aerodynamic heating protection. Furthermore, the weight of the guidance system to be carried by the last stage must be subtracted from the gross payload before the true interplanetary payload is obtained. The weights of such systems are estimated to be about 100 to 200 lb, including power supplies necessary for their operation.

The weights of retro-rockets, mid-course correction rockets, aerodynamic heating protection, and structure can be estimated in terms of % payload weight. Such estimates are given in Table 5. As an example, assume the gross payload of the *Advanced Vega* for a Venus satellite mission is 1200 lb. First, subtract 150 lb for guidance of the launching rocket. Second, subtract 5% of the remaining 1050 lb for the mid-course correction rocket and at least 50% for the retro-rocket (assuming a circular orbit around Venus is required). This leaves 470 lb. The structure requires 15% of this, leaving 400 lb for other requirements. These other requirements include (1) attitude sensing and control system, (2) mid-course and terminal guidance systems, with target sensor, (3) telemetering instrumentation, [95] (4) transponder circuitry,

[94] **Table 5. Suggested Lunar and Planetary Flight Schedule**

Flights	1960			1961			1962			1963			1964		
Lunar Flights															
Near miss		V[b]													
Orbit						V[b]									
Rough landing					V[b]										
Soft landing											S[b]	S		S	
Orbit and return										S[b]			S		
Mars Flights															
Near miss			V[b]												
Orbit									S[b]						
Landing									V[b]						S
Venus Flights															
Near miss				V[b]											
Orbit								V[b]							
Landing								V[b]					S[b]		

[a]Legend: V, Vega; S, Saturn.
[b]Typical payload weight breakdown and list of potential experiments are given in Sec. VI-B.

Table 6. Weight Requirements of Payload Components
% of interplanetary payload[a]

Mission	Retro-Rocket, Including Structure[b] (I_{sp} = 265)			Aerodynamic Heating Protection	Structure % of net payload (subtract rockets and heating protection)
	Capture by Target, 2 Radii from Center	Circular Orbit, 2 Radii from Center	Landing		
Planetary miss	0	0	0	0	15
Lunar satellite	>15	>45	0	0	15
Venus satellite	>20	>50	0	0	15
Mars satellite	>50	>70	0	0	15
Lunar landing	0	0	>70	0	20–25
Planetary landing	0	0	0	35	20–25

[a]Mid-course correcting rocket accounts for 5% of total payload in all cases.
[b]Critically dependent on approach trajectory, i.e., time of flight (see Figs. 24, 25, 27, and 28).

(5) communication power supply, (6) antenna, (7) miscellaneous power supplies, and (8) scientific instrumentation.

Since none of the components necessary for this particular payload has received an adequate design study, it is impossible to give meaningful estimates of their weights. On the basis of preliminary estimates, it is probable that between 50 and 100 lb. might be available for scientific instrumentation for such a payload.

In this manner, estimates have been made of the total available weight for instrumentation in several of the payloads suggested in the proposed program. Tables 6 through 17 [only Table 6 is reprinted here] present listings of scientific instrumentation which might be included within such weight estimates for these payloads.

These Tables also show (1) the weight and volume requirements for each instrument, (2) the information rate or total amount of information which must be communicated to earth to give the results of the measurements, (3) the expected duration of the experiment, (4) the distance of the payload from earth at the time when the experiment might be performed, and (5) special remarks which apply to certain of the instruments. These listings are not the result of any official assignment of experiments to the various payloads. They represent only an estimate of reasonable experimental objectives consistent with present knowledge about the moon, the planets, and interplanetary space, and consistent also with present estimates of feasibility for both vehicle and payload design and instrument development. Thus, the information presented in these Tables should be considered as a starting point for more detailed program planning.

1. Lunar miss (Payload No. 1, August 1960). The first firing, in the summer of 1960, should be a "moon-miss" development test. This firing should test the scientific instrumentation, communications, guidance system, power supply, and attitude control system to be employed in the two subsequent firings. Table 7 summarizes the important characteristics of this probe.

2. Escape toward Mars (Payload No. 2, October 1960). This payload (see Table 8) will be fired with sufficient velocity to carry it to the orbit of Mars and will be launched at the right time and in the right direction to intercept Mars on its orbit. The guidance system available for the probe should have the capability of placing it within 1,000,000 miles of Mars.

The primary scientific experiment to be performed by this payload is infrared scanning of the region between 2 and 4 microns. This scanning will be accomplished by a spectrophotometer. In addition, a camera device (e.g., a vidicon) will photograph the same areas of Mars scanned.

If additional payload weight is available, a space package designed to measure environmental conditions between the earth and Mars will be included. Measurements of cosmic rays, magnetic field intensity, micrometeorite density, meteoric frequency, and solar radiation would be made by such a package.

The cosmic-ray instrumentation is designed to measure counting rate and total ionization by means of Geiger-Mueller counters and integrating ionization chambers. Several scaling circuits should be provided in case the probe should enter a high intensity radiation belt around Mars, if such exists. Some discrimination between protons and electrons in such a belt could be obtained if two similar ionization chambers were used, one with a stainless-steel wall and the other with a low-atomic number material such as beryllium over stainless steel. If sufficient weight is available, information on the interplanetary primary cosmic-ray charge spectrums could be obtained using two pulse ionization chambers filled with gas at different pressure. This arrangement makes it possible to apply a correction for the background of nuclear stars produced in the gas. The charge spectrum and directionality of the relativistic radiation could be obtained by means of a counter telescope containing a Cerenkov counter, but it is doubtful that enough payload weight and communications capability will be available. The cosmic-ray detectors should preferably be on an arm out from the main body of the probe in order to minimize background due to showers originating in the material of the probe.

A magnetic-field measurement is useful both for itself and to help in the interpretation of the cosmic-ray data. A magnetometer of the alkali-metal-vapor type utilizing magnetic-resonance effects and able to measure fields between 1 and 10^{-4} gauss should be available. Such instruments available by this time will probably not be able to measure the direction of the field, but only its magnitude.

A measurement of micrometeorite density mill provide valuable environmental data. Instrumentation to measure this would include microphones and erosion gauges. Instrumentation should be included for determining the...[98] danger due to meteors. Such instrumentation should be able to determine the probability of encountering a meteor large enough to penetrate a wall thickness comparable to that of a manned capsule. A gas-filled chamber of the proper wall thickness could be used for this purpose. A hit by a penetrating meteor could then be detected by monitoring the gas pressure in the chamber. A difficulty here is that in order to gather sufficient statistics such a chamber would probably have to be quite large. The cosmic-ray ionization chambers (which contain argon gas under pressure) might double as such detectors, but would be inefficient because of their small size.

3. Escape toward Venus (Payload No. 3, January 1961). This payload (see Table 9) will include all of the experiments described for Payload No. 2. The spectrophotometer, however, will be scanning a different region. The payload will also include a passive radar scanning in the far infrared region. This measurement will provide a rough measure of the surface temperature of Venus. Since the subsequent design of hard- and soft-landing probes for Venus is highly dependent on the planet's surface temperature, this measurement is given a high priority.

In addition, instrumentation will be carried for measuring the effects of the solar ionosphere on the transmission of radio signals. This experiment will continue indefinitely or until failure of the solar cells supplying the necessary power for transmission. This experiment will begin after Venus passage when the probe enters its final heliocentric orbit. The experiment will transmit a signal through the sun's ionosphere.

The characteristics of this probe are summarized in Table 9.

4. Lunar rough landing (Payload No. 4, June 1961). Sufficient gross payload weight should be available by this time to permit the use of a retro-rocket to slow the lunar rough-landing probe (see Table 10) to a velocity of about 100 to 200 ft/sec before impact and still leave ample payload available for instrumentation. It is not anticipated that a complete terminal-guidance system required for a soft landing (less than 50 ft/sec) will be available at this time. A cone penetrometer containing accelerometers can be used to measure surface hardness. This will require attitude control of the vehicle. The impact transmitter will transmit the impact data until it is destroyed by the impact. Such a surface-hardness measurement will provide valuable data for the lunar soft-landing probe. The cone penetrometer should be able to distinguish a surface of dust or sediment from one of hard rock.

If the landing is made on the dark side of the moon, the impact location could be marked by flash powder. The 200-in. telescope should be able to detect the flash of about 10 lb of powder if the telescope is aimed at the proper point. If the landing is made on the light side of the moon, close-up pictures of the lunar surface might be taken on the way in, giving an excellent view of surface features. A difficulty here is that large transmission bandwidth antenna would then be required to give sufficient data to transmit the pictures before the camera is destroyed on impact.

Experiments to be carried out after landing depend upon the development of instrumentation sufficiently ruggedized to survive the impact and the harsh temperature environment on the lunar surface. A ruggedized beacon and telemetry system with power supply, ruggedized temperature sensors, and a ruggedized seismograph can probably be developed. Also, it would be highly desirable to develop a radiation monitor and magnetometer to survive the impact. The lack of atmosphere and the probably small magnetic field on the moon should make it an excellent base from which to monitor cosmic radiation.

5. Lunar satellite (Payload No. 5, September 1961). This payload (see Table 11) will be placed in a well-controlled orbit around the moon using terminal guidance. Sufficient payload weight will be available for some rather elaborate instrumentation.

A gamma-ray spectrograph[9] will be able to compare the abundances of uranium, thorium, and potassium in the lunar crust. This measurement should make it possible to determine whether the surface of the moon is composed of granite, basalt or meteoric material. For a trajectory within 100 miles of the lunar surface, variation in surface composition could be determined with a resolution of approximately 100 miles.

High-resolution photographs of the surface of the moon will be taken at various wavelengths and polarizations. These photographs should provide information on the surface characteristics of the moon that will be valuable for choosing a site for a lunar soft landing… [102] A mapping of the lunar magnetic field and cosmic-ray measurements will be made to measure the radiation present in the vicinity of the moon. The cosmic-ray instrumentation chosen will depend largely on the information obtained from previous shots and upon whether or not the moon has been found to have a high-intensity radiation belt.

The mass spectrograph will attempt to determine the presence of a lunar atmosphere. The mass spectrometer will need a gas collection device in order to make a measurement, and even if such a collective device were developed, the lunar atmospheric density may be too low to be detectable in the background of gases produced by the outgassing of the probe itself. Such a mass spectrograph and associated collecting device should be tested in an earth satellite before this shot.

Observations of the orbit of the satellite will give a better determination of the mass of the moon.

6. Venus satellite (Payload No. 6, August 1962). The mission of this Venus satellite (see Table 12) and of the Venus entry shot which follows it are not only to acquire scientific data but to obtain environmental and engineering data that will be necessary for the design of the Venus soft-landing probe scheduled for 1964.

Ionospheric soundings of the Venusian atmosphere will be carried out with transmitters and receivers designed to operate at various frequencies between 500 kc and about 2000 mc. The main purpose of these soundings will be to find a radio "window" in the atmosphere and identify ionospheric layers. The receivers in this system will also detect sources of electromagnetic radiation (e.g., thunderstorms) generated on the planet.

Radar operating at the 3, 1.35, and 0.5 cm wavelengths will determine the relative abundances of oxygen and water vapor in the atmosphere. This equipment can also be used to give low-resolution radar mapping of the surface, with the resulting roughness measure a possible indication of the presence of life forms.[10]

Infrared devices will be employed to analyze the atmosphere and also to attempt infrared photography of the surface of the planet, particularly on the dark portion where the white vapor in the atmosphere may be absent. Photography will also be carried out in the visible and ultraviolet regions.

The cosmic-ray package will detect the extent of any trapped radiation in the vicinity of Venus and measure the primary radiation en route to Venus. A mass spectrograph will determine the constituents of the Venusian atmosphere at altitudes along the satellite orbit.

A magnetometer, probably of the magnetic-resonances type, will be included to give a mapping of the magnetic field of Venus.

7. Venus entry (Payload No. 7, August 1962). The Venus-entry package (see Table 13) will be equipped with sufficient protection against aerodynamic heating to provide a good chance of landing an instrument package on the surface. The entry vehicle will also include instruments to measure the characteristics of the Venusian atmosphere during descent.

A mass spectrograph will measure atmospheric composition. Thermocouples in the skin of the descending vehicle will give data making it possible to calculate the atmospheric temperature if the atmospheric composition and the Mach number are known. Light intensity near the surface should be measured by a photodiode to determine whether there is sufficient intensity to operate solar cells in the soft-landing probe.

At impact, a cone penetrometer (requiring attitude control) could determine the surface hardness. A ruggedized beacon with power supply designed to survive the impact and surface environment will be deposited on the surface. Tracking this beacon will give information on the rotation rate of the planet. If the beacon operates a sufficiently long time it can be used as a terminal-guidance device for the Venus soft-landing probe. This beacon will also contain sensors to give surface temperature.

Instrumentation to measure the moisture content of the atmosphere and to indicate whether or not the probe impacts on water should be included. This might consist of a hygroscopic resistor. An instrument for measuring the amount of dust in the atmosphere would also be valuable. A light source with mirrors to give a long light path, used in conjunction with a phototube to measure light attenuation, might be developed for this purpose.

Communications problems on this shot are particularly formidable and require considerable study. It is likely that a large vehicle antenna cannot be landed or even carried into the lower reaches of the atmosphere, which severely limits the available communications bandwidth...

[105]Some interplanetary experiments could be performed en route to Venus, if weight limitations permit.

8. Mars satellite (Payload No. 8, November 1962). The payload and objectives of this shot (see Table 14) are very similar to those in the Venus satellite. The problem of finding a radio "window" in the atmosphere of Mars is probably much simpler than is the case with Venus. Ionosphere soundings can be made at lower frequencies. Also, more emphasis can be placed on obtaining photographs in the visible region because of the transparency of the atmosphere.

9. Mars entry (Payload No. 9, November 1962). The Mars-entry payload (see Table 15) will be similar to that of the Venus-entry payload. If sufficient communications bandwidth can be obtained, photographic equipment might be included to allow close-up photographs.

10. Lunar orbit and return (Payload No. 10, February 1963). The purpose of this test will be twofold: (1) It will demonstrate the capability of performing such a mission as flying around the moon and returning to make a safe landing on earth. (2) It will be a full systems test of the Venus-landing payload (payload 12).

The instrumentation will be similar to that scheduled for the Venus-landing package, except that it will be modified to the extent necessary for operation on the surface of the earth. If the Venus-entry shots or Venus satellite shots have determined that the surface of Venus is either largely water or land, this payload will be designed to operate accordingly, and will be brought into earth over either ocean or land as required. If this question is still unanswered, the selection of land or water operation will be based on the most reasonable estimate of Venus surface conditions from the available evidence. Experiments in the vicinity of the moon will be those which can be carried out with the equipment available as part of the system test.

11. Lunar soft landing (Payload No. 11, June 1963). This payload of the lunar soft-landing probe (see Table 16) is intended principally as a geological exploration of the moon's surface and would include a complete photographic survey of the landing area, the determination of the surface texture and composition (mineralogical and chemical) at various selected spots around the landing site, and a measurement of the more general environmental factors such as weathering (i.e., by radiation), seismic activity, and magnetic fields.

If possible, the surface exploration would utilize a mobile vehicle whose motion and experimental program would be controlled by commands from earth. First the camera would take a complete series of color photographs, ranging from long-range panoramas to microscopic examinations of the immediate surface. On the basis of these photographs, a suitable sampling site would be chosen and the vehicle ordered to proceed there. A vibrator probe would then determine the thickness and texture (i.e., dust or debris) of the surface layer and measure the thermal conductivity of the material by heating a thermistor at the probe tip and recording the subsequent cooling curve. Chemical analysis would be attained by X-ray fluorescence, ultraviolet emission spectra, and neutron activation, so as to yield data on both the light and heavy elements. A determination of the natural radioactive species will give some information as to age and geologic history. X-ray diffraction and infrared luminescence will provide a fairly complete mineralogical analysis. Surface density (by beta and gamma scattering), thermoluminescence, and volatile constituent analysis (by mass spectrometry of a pyrolized sample) will indicate the degree and type of weathering.

Some of these experiments can be performed merely by lowering the instrument onto the lunar surface, others will require sample collection and manipulation. The latter, although more accurate and unambiguous, are more subject to malfunction and unforeseeable hazards. Therefore a certain amount of experimental redundancy is desirable, especially in the chemical analysis.

The lunar environment presents some unique difficulties, chief of which are the high vacuum and temperature extremes. Not only will it be difficult to design moving mechanical systems which can function without lubrication in high vacuum, but it is highly probable that the dust will adhere to any surface with which it comes in contact, e.g., camera lenses. The wide range of temperature extremes makes it seem unlikely that all of the payload can perform satisfactorily during both the lunar day and night. Considerations of power dissipation and moving mechanical parts tend to favor the lunar night, whereas the presence of light for the all-important photographic survey and for solar cells (which could greatly reduce the weight of the power supply, and thereby make a large roving vehicle more feasible) makes the lunar day seem more favorable. ...

[110]The low gravitational field of the moon makes it appear feasible to use small rockets to send samples of the lunar surface back to earth (although the problem of locating and recovering the returned sample is formidable) for detailed chemical and petrographic analysis. It should be noted, however, that such data, on isolated samples, would only have great value in the context of an area survey, and therefore a sample-return complements, but does not supplant, the in situ experiments.

From a scientific standpoint, the ideal configuration would utilize a completely self-sufficient roving vehicle, containing all of the surface exploration equipment together with the command receiver, program control, logic circuits, power supply, and transmitter and antenna for telemetering the data back to earth. Using this configuration, the roving vehicle would have a hypothetically unlimited exploration range, which might greatly increase the value of the data. Sample collection could be simplified or even eliminated for most experiments. On the other hand, such a vehicle would be quite large and would require a powerful motive system, the largest source of power consumption in the whole payload. Moreover, since the operation of any mobile vehicle will be somewhat risky and subject to unforeseeable hazards, it will place the experimental package in jeopardy.

The stationary package is by comparison quite small and low in power consumption. A seismograph, magnetometer, gravitometer, and temperature probe could operate continuously on 25 watts. It therefore seems feasible to relay this data via the moving vehicle or, better, to telemeter it direct to earth by a bandwidth transmitter.

Another configuration might involve a much smaller roving vehicle containing only camera, sample collectors, and surface-texture probe. This would probably be connected by power and communication cables to the main payload package which would contain the sample receivers, manipulators and all of the experimental instruments. The main payload package might also contain a few sample collectors, so that in the event of loss or failure of the roving vehicle, some data could be obtained. Such a configuration would effect some saving in power supply and perhaps also in structure weight. It would involve less of a risk of locomotion failure but, on the other hand, it would necessitate elaborate sample manipulation. Moreover, the exploration range would be limited by the cable.

If sample collection and manipulation involves insuperable problems, it will be necessary to resort to the larger roving vehicle containing no sampling experiments. Conversely, if vehicle locomotion is unfeasible or very risky, a completely stationary payload may be necessary. (In the latter event, the sample-return rockets would become even more important.) The choice will therefore depend not only on the available payload weights and power supplies, but on the satisfactory development of a reliable mobile vehicle and sample collection system and also on the surface texture and terrain data obtained from the lunar rough-landing probe.

12. Venus soft landing (Payload No. 12, March 1964). The mobile surface-exploration vehicle, proposed for the lunar soft landing, constitutes only part of a satisfactory exploration program for Venus or Mars. A thorough investigation of the weather and atmospheric conditions is of even greater importance, and the question of life on other planets is of such universal concern that some sort of biological experiment, however rudimentary should be included in the first soft landing.

In addition to the usual weather data—pressure, temperature, humidity, wind direction and velocity—the stationary instrument package should provide data on the Venusian daylight spectrum, and its variation with direction and time. Atmospheric composition can be determined by a mass spectrometer and long light-path spectrophotometer. The latter will also give some data on the presence of dust, fog, etc. A microphone will detect atmospheric noises. In addition, a focused sounder in conjunction with a microphone on the moving vehicle will give data on atmospheric sound propagation. The stationary structure would also include a seismograph, magnetometer, gravitometer, and soil probe.

The stationary structure will have provision for launching radiosonde balloons so as to obtain high-altitude weather data. In addition, most of the stationary instruments in the stationary package can be designed to function during entry.

The surface-exploration vehicle will contain, in addition to the all-important color television camera, a small radar scanner so that some topographic data can be obtained even if there is atmospheric interference with photography. The remaining experiments will be similar to those of the lunar soft-landing vehicle. However, since the Venusian surface will be far more unpredictable than [111] the lunar surface, it seems expedient to avoid experiments involving sample collection and manipulation. The sole exception would be a suction pump and filter for the collection of airborne solid particles; this has very high priority, since if any form of life were present, it would probably include airborne microorganisms. This sampler would be used in conjunction with a special microscope (also usable for soil and rock examination) and some simple biological experiment, as yet unspecified.

As with the lunar soft landing, the mobile vehicle should be capable of being directed to optimum sampling spots by command from earth, these commands being decided upon on the basis of the photographic data.

The communication and telemetering system is as yet undetermined. Radio communication through the Venusian atmosphere may be poor enough to warrant telemetering data from the surface-exploration experiments to a simultaneously launched satellite containing a powerful transmitter for relay to earth.

The experimental program proposed here is admittedly ambitious, not only because it involves a complex network of instruments, logic circuits, and program controls, but because it presupposes that the earlier satellites and entry probes will send back a maximum amount of reliable information and that this information will be favorable, i.e., that an atmospheric radio "window" exists, that photographic data has been obtained and discloses suitable soft-landing sites, etc. Should the data be ambiguous or unfavorable, it would be best to simplify the payload by incorporating the mobile-vehicle experiments into the stationary structure, though this would still probably require a smaller, sample collecting, roving vehicle.

It should be borne in mind, however, that the optimum planetary payload (i.e., with mobile vehicle) is readily adaptable to a Mars soft landing and that an opportunity for a Mars soft-landing shot occurs a few months after the Venus soft landing proposed here. Our present knowledge indicates that a Mars soft landing would have few of the environmental problems encountered on Venus or the moon, and would have a considerably greater chance of success. Therefore, during the development of the lunar and planetary soft-landing payloads, the possibility of Mars landing should always be kept in mind and, if necessary, given preference.

There is some doubt as to the nature of the surface of Venus. In the event that the entry probe indicates a water surface, a special water-landing payload could be developed, quite rapidly and easily, which would have a high probability of giving ample useful data. If, at the time of the Venus soft landing, there is still some uncertainty as to the existence of water-covered areas, it may be possible to partially ensure the regular soft-landing payload (as described above) against this hazard. The radiosonde launcher would contain a sufficient gas supply to inflate pontoons adequate for floating the entire payload; however, the complications of waterproofing, orientation, etc., may make this unfeasible.

The Venus-entry payload (Payload No. 7) and this Venus-landing payload presents [sic] serious communication problems. In both cases, it cannot be considered feasible to bring down a large parabolic antenna through the atmosphere of Venus. Since the information to be transmitted to earth from the simpler Venus-entry payload does not require a large bandwidth, that earlier payload might be able to use a simple dipole as a transmitting antenna after it arrives at the surface of Venus. But this situation does not hold for the Venus-landing package.

For the Venus-landing package, large volumes of information will be collected, including picture-type information. Thus, a large communication bandwidth is required. This, in turn, implies the use of a comparatively large parabolic reflector for the transmitter antenna.

Two solutions suggest themselves for accomplishing this requirement: (1) a foldable antenna could be included with the stationary instrument package. After a safe landing on the surface of Venus, this antenna would be deployed and swiveled to aim at earth. If the planet rotates, the antenna would have to track the earth in a manner consistent with the rotation rate of Venus. Such an antenna would have to be strong enough to withstand possible surface winds and rigid enough to support itself against the Venusian gravity. Clearly a large amount of the payload weight would be devoted to such an antenna if this solution were adopted. (2) The total payload injected into the Venusian orbit from earth might consist of two separate pieces which could be separated from each other after the final course-correction guidance maneuver prior to arrival at the target planet. One package, the actual landing package, would be provided with an additional course-correction rocket to direct it into the atmosphere of Venus. The portion left behind would follow the trajectory previously established to a point 1000 miles or so above the surface of Venus, where it would, with the help of a retro-rocket, become a satellite...[114] of that planet. The landing package would be equipped with a parabolic reflecting antenna having a diameter small enough so that it could be enclosed within the protective housing of the landing package during entry through the atmosphere and simply erected after arrival on the planet's surface. This antenna would serve to communicate the results of the experimental findings on Venus to a receiver located in the orbiting portion of the payload. This Venus satellite would then act as a relay station equipped with necessary antennas for receiving the information from the surface and transmitting it back to earth. If this solution were employed, a large fraction of the gross payload weight would have to be devoted to the satellite portion of the dual payload.

The answer as to which of these two solutions is the most feasible, or whether or not still another solution exists, must await a more thorough design study.

The characteristics of the Venus soft-landing probe are listed in Table 17.

C. Development Schedule

1. Procurement of engineering design data. The engineering design of a space probe is a task which draws upon knowledge at the very frontiers of many branches of technology. It is a difficult job even if the entire environment with which the probe has to contend in order to accomplish its mission is well known. There are, in fact, many aspects of the space environment about which little or no information exists. A good example is our lack of knowledge of the type of surface and atmospheric environment in which a Venus soft-landing probe would be required to operate. For these reasons any rational, long-range space-exploration program must be a carefully planned, step-by-step procedure in which engineering design data for the later vehicles is obtained as fully as possible by earlier vehicles. If the over-all program is to be a coherent one, the earlier vehicles must therefore be instrumented with this requirement in mind. If the eventual goal is to put a man into deep space, it is obvious that the necessity for accurate information concerning the environment he must face is critical.

As an example of the type of knowledge of the space environment that is necessary for engineering design purposes, the problem of temperature control should be considered. The instrumentation in a space probe operates efficiently only over a restricted temperature range, which makes it necessary to hold the temperature within fixed limits. Considerable experience in this area has been gained from the satellite and lunar vehicles that have already been launched. The temperature-control methods hinge on control of surface emissivities. The long-term effects on these emissivities of such space conditions as vacuum, micrometeorite impacts, corpuscular radiation, gamma radiation, and ultraviolet radiation are not well known. Two things are obviously necessary in order to acquire this needed information efficiently. The first is to determine what the characteristics of the radiation, vacuum and micrometeorite environment are, and the second is to develop ground testing facilities for simulating this environment, whenever possible. The success of the temperature-control system must then be put to the final test by monitoring and telemetering back temperatures at various points in the vehicle during its actual flight.

In organizing the projected NASA program, considerable attention has been given to the preceding requirements. The early interplanetary shots devote much of their instrumentation to the measurement of such space environmental conditions as micrometeorite erosion and cosmic radiation. Meteor detectors might be included to determine the meteor hazard to manned probes. The planetary satellites and entry probes are equipped to obtain atmosphere and surface data for use in designing soft-landing probes.

If an accurate measurement of an important unknown quantity is planned for a certain shot, it is well to obtain at least a rough idea of the value of this quantity from an earlier shot. (This at least determines the order of magnitude of the quantity and makes it unnecessary to design the instrumentation to measure over a range of several orders of magnitudes. Measuring over a large range often involves the use of logarithmic amplifiers with their associated inaccuracies.) Whenever possible, this procedure has been applied in the present program.

2. Ground test requirements. The ground testing of payloads associated with this program must be more extensive and thorough than test programs employed in the development of a missile weapons system, since the number of vehicles and shots involved is severely limited.

Some of the difficult testing problems which must be solved as soon as possible are listed as follows:

[115]1. In the past most airborne equipments have been required to operate for relatively short periods—minutes, hours, a few days, or a few weeks. Many of the probes considered here require satisfactory operation for periods of several months. Present schedules do not permit realistic life tests; therefore it will be necessary to establish meaningful accelerated life tests on critical components and component parts.
2. The evaluation of attitude control and sensing systems in a gravity-free environment poses special problems as yet unsolved.
3. The effects of radiation on certain classes of component parts have not been determined. Radiation-simulation test facilities should be made available as soon as possible.
4. One of the major problems associated with the design of deep-space probes is the control of payload temperature. There is need for facilities to accurately determine the reflectance characteristics of proposed external surfaces and materials. Further, there is need for a vacuum facility which incorporates radiation sources and sinks which will to some degree verify expected thermal time constants and payload operating temperatures.

Considerable work has been done in establishing test equipments and procedures for the environmental simulation of: vibration, shock, high and low temperature, humidity, sand and dust, linear or static acceleration, and spin.

The capabilities of present test equipments will probably be adequate for testing the early probes. However, it is reasonably certain that some equipment capabilities (vibration, shock, linear acceleration) will have to be doubled or tripled as larger payloads are considered.

It is important that environmental specifications be established early in any development program. Every effort should be made to provide adequate instrumentation for the determination of vibration and shock environments based on static firings of proposed propulsion systems. Specifications based upon inadequate information result in inadequate or over-designed components.

3. Typical schedules. The development of payloads for lunar and planetary exploration is a problem of greater complexity than the development of the missile systems which have been carried out to date. The development schedules reflect this fact. In particular, the requirement for a long reliable lifetime implies the necessity for a long environmental testing program preceding the actual launching. It is not likely that this testing period can be made a great deal longer than the actual flight time for some of the planetary missions, and in some cases it may be difficult to make it even equally long. Thus, if any component fails during this life test and indicates the necessity for a redesign, the redesigned version cannot have a life test of really adequate duration. This fact implies a great emphasis on the need for reliable subcomponents and elements. It would not be advisable to introduce a new type of transistor, for example, into the payload design at a late stage in the development program. It would be much more desirable to rely only on transistors which have been in use for a long time before their incorporation into the planetary payload, transistors for which a large background of reliability experience is available.

This conclusion implies that innovations in elements and component design cannot be readily introduced into the program. Although this may place a limitation on the versatility and capability of the payloads, it is a necessity in order to assure any degree of confidence in the final success of the mission. Another fact which makes the payload development problem more complex than the development of a missile system is the rigidity of the flight schedule. If, because of development difficulties, the flight schedule has to be delayed for as much as one week, then actually the flight time will be delayed for a year and a half to two years. Still another fact which adds to developmental difficulties is that each of the payloads developed for the planetary exploration program is different from its predecessor, in some cases to a very large degree.

All of these developmental problems are reflected in the development schedule. An example of a development schedule is given in Fig. 64 [omitted]. This is the schedule for the...[117] development of the first three payloads listed in the suggested program.

Some elements of this schedule are worth special attention. First, notice that the experimental objectives for Payload No. 2, listed as an escape toward Mars, must be defined by the end of April 1959. Furthermore, for this same payload, it should be noted that component testing must begin November 1959 and continue through the month of May 1960. The prototype model of this payload will be available in mid-January of 1960. Thus, the life testing of the prototype model can be carried out only for four and one half months, whereas the actual flight time of this payload will be at least five months, or longer.

The development program for the Mars escape payload is further complicated by the fact that the Lunar-Miss payload (Payload No. 1) and the Escape-Toward-Venus payload (Payload No. 3) must also be designed and developed during the same time intervals as those for the Escape-Toward-Mars payload. This duplication of effort can be carried out successfully only if all three payloads are basically similar.

The development schedule for the Mars payload allows 17 months between the definition of the experimental objectives and the launch date. Although this may seem like quite an adequate time span, inspection of the various portions of the schedule will show that actually the time allowed is quite short.

An alternate schedule for the development of such a payload is given in Fig. 65 [omitted]. This schedule has been constructed with maximum emphasis on reliability of the final payload, and is thus very conservative. From the point of view of the development engineer, however, this conservative schedule is quite realistic, and desirable. Unfortunately, this schedule calls for nearly 4 years between the definition of experimental objectives (which would have had to occur at the end of January in 1957) and the launch of the final payload in October of 1960.

A realistic schedule which permits the incorporation of new ideas and information which might be gained by other flights in the planetary exploration program, but also permits adequate testing for the assurance of reliable operation, would lie somewhere between the comparatively tight schedule shown in Fig. 64 and the very conservative schedule shown in Fig. 65.

As the payloads become more complex, laboratory environmental testing will not be adequate to assure the successful operation of the payload system. Flight tests of the complete system will be required. Figure 66 [omitted] shows a proposed schedule of major payload system tests associated with the development of several of the payloads listed in the proposed program. These system tests include air drops to check out the behavior of payload systems destined for planetary landings during the final phases of their landing and surface operation. Tests in sounding rockets as well as tests in earth satellites are indicated for payloads which must behave as satellites of other planets.

For payloads designed to land on the surface of the moon, high-speed-impact tests are shown. It is the function of these tests to launch the payload toward the earth...[119] at speeds comparable to those which will be attained by the payload on its approach to the moon. Thereafter, the retro-rockets will be ignited to slow the payload down for a landing on the surface of the earth. The function of the payload after the landing will also be checked.

For payloads destined to enter the atmosphere of other planets, the preliminary airdrop tests will be followed by high-speed re-entry tests into the earth's atmosphere. The speeds obtained during these tests must be in excess of the escape speed of the particular planetary target. For Venus, the required re-entry test speed is between two or three times the speed attained by an ICBM on re-entry into the earth's atmosphere.

To meet the schedule for such system tests, a complete prototype payload must be available in time for the earliest system test. In some cases, this is several months before the actual launch date. For example, for the lunar soft landings, high-speed-impact tests should begin late in 1962, although the actual launching of this payload does not occur until mid-1963.

The lead times necessary to meet the development and testing schedules for lunar and planetary payloads may seem unnecessarily long to those outside of the missile development industry. However, such lead times must be allowed if the program is to meet with any degree of success.

It is in the area of scheduling that the effect of public reaction makes itself felt most keenly. The public demand for dramatic "firsts" overrides the public concern about test failures. It is in response to this public demand that development schedules cannot be made as long as might be desirable if reliability were the sole objective.

On the other hand, neither the public nor the scientists nor the engineers would be satisfied with a program consisting only of a monotonous series of failures. Somewhere between the two extremes, long developmental testing on one hand and a completely crash program on the other, lies the desirable scheduling philosophy. The schedule presented in Fig. 64 for the development of the first 3 payloads in the series is perhaps a realistic picture of such a middle course for these comparatively simple payloads.

[120]

VII. CONCLUSIONS

The development of a typical payload considered in this Report is in many ways analogous to the development of a complete guided-missile system. The payloads contain guidance and control devices, communication devices, telemetry equipment and measuring devices, and in many cases a major portion of the payload weight is devoted to a rocket propulsion system.

Actually, the development problems associated with these exploration payloads are much more difficult than the development programs associated with the typical guided missile. Each of these payloads will be different from its predecessor both in weight, in objectives, and in component design. Furthermore, each payload must be capable of operating without the benefit of last minute checkout or adjustments, after having traveled for days, or perhaps months, through the vacuum of space. Some of the payloads must successfully enter an atmosphere of only partially known characteristics traveling at a speed of more than twice that attained by an ICBM on re-entry into the earth's atmosphere. Their "warhead" will not be a single device designed to operate only once, but rather a whole array of devices, some of which may operate only once, some of which will operate continuously, and others periodically at intermittent intervals.

Many of the mechanical devices must automatically carry out a program of exploration and analysis without the help of human maintenance or human direction. In many cases, they must perform this test in a completely unknown environment, for it will be their job to discover the properties of this new environment. If these devices should fail to operate properly, it will be difficult, if not impossible, to detect the cause of their failure and so make the necessary design changes before the next attempt.

At the present time, we have no technology which would permit the design and construction of automatic devices to be carefully carried out to some portion of the earth's surface, and then left to perform their function of analyzing the characteristics of the earth. And yet, in a few short years, we must design such devices to operate on the moon and the near planets.

The public demands sudden and spectacular achievement in their space program. This demand cannot be ignored nor relegated automatically to second place in comparison with the demands of reliability. Both demands must be met.

The development of the payloads for the exploration of space is a task which will test the limits of our ingenuity; but most of all, it is a task which must begin immediately if it is to have any hope of success within the proposed time scale.

---

9. Proposed by James Arnold, University of California, La Jolla, California.
10. Suggested by Gold, Space Science Board, National Academy of Sciences, Washington, D.C.

Document II-6

Document title: Abe Silverstein, Director, Office of Space Flight, NASA, to William Pickering, Director, Jet Propulsion Laboratory, December 21, 1959.

Source: NASA Historical Archives, Jet Propulsion Laboratory, Pasadena, California.

Document II-7

Document title: Homer Newell, Assistant Director for Space Sciences, NASA, "Trip Report for the Visit to Jet Propulsion Laboratory on 28 December 1959 by Homer E. Newell, Jr., Newell Sanders, J. A. Crocker, Morton J. Stoller," December 30, 1959.

Source: NASA Historical Reference Collection, History Office, NASA Headquarters, Washington, D.C.

Despite JPL's desire to focus on the planets and manage NASA's solar system exploration program, NASA Headquarters instead opted to make lunar missions the agency's priority. On December 16, 1959, NASA Associate Administrator Richard Horner wrote to JPL Director William Pickering stating that NASA, and thus JPL, should concentrate on lunar rather than planetary exploration. In a December 21, 1959, letter Abe Silverstein, Director of the Office of Space Flight, informed Pickering of how JPL should proceed with solar system missions over the next three years. That letter became the basis for a meeting between Headquarters and JPL officials at JPL on December 28, during which the Headquarters representatives assured JPL that although a lunar program was NASA's first priority, planetary exploration would get underway shortly and JPL would be its leader.

Document II-6

["CONFIDENTIAL" stamped on each page and crossed out]

["UNCLASSIFIED" stamped on each page]

[no page number]

December 21, 1959

Dr. William H. Pickering
Director, Jet Propulsion Laboratory
4800 Oak Grove Drive
Pasadena, California

Dear Dr. Pickering:

Based on a study by the several groups in the Headquarters staff participating in the lunar and deep space program, the following tentative flight program and mission designations have been established as a starting point for determining a post-Vega program.

(1) The lunar and planetary program for the next 36 months is tentatively established as follows:

Round	Date	Vehicle	Mission
1	2nd qtr. '61	Atlas Agena B	Lunar reconnaissance
2	3rd qtr. '61	"	"
3	4th. qtr. '61	"	"
4	1st qtr. '62 (probable)	"	"
5	2nd qtr. '62 (possible)	"	"
6	3rd qtr. '62	Centaur	Venus probe
7	4th qtr. '62	"	Mars probe

Discussion of longer range programming will be deferred to a later date.

(2) The following comments on the mission are pertinent:

(a) The lunar reconnaissance mission has been selected with the major objective in mind being the collection of data for use in an integrated lunar [2] exploration program. Of the several specific experiments reviewed for assignment to the early flights, the transmission of high resolution pictures of surface detail appears to be the most desirable. It is therefore requested that your program be directed to the consideration of a payload containing a picture transmission system which will acquire and transmit a number of images of the lunar surface. The system should have an overall resolution of sufficient capability for it to be possible to detect lunar details whose characteristic dimension is as little as ten feet. It is recognized that this resolution limit will be obtainable only if the overall field of view of the imaging system is reduced to the extent necessary to satisfy technological limitations.

It is requested that a technical study be initiated to arrive at the necessary decisions as to the focal length and aperture of the optics, the requirements for spacecraft attitude control and stabilization, the specific telemetry and television techniques to be used and the possibility of use of retro-rockets.

It has been suggested that the data system transmit several pictures in the period immediately preceding impact by the use of a relatively high powered transmitter which is operable only during the terminal phase of the lunar approach. Trade-offs between retro-rocket weight and telemetry system weight may make it possible to extend the picture transmission time by taking advantage of the longer time to impact resulting from the application of appropriately programmed retro-rocket impulses. Such trade-offs should be investigated. It has also been suggested that the use of a retro impulse great enough to over correct for the vertical (with respect to the moon) component of impact velocity, might result in an appreciable increase in the number of pictures which may be transmitted. Also, during the terminal phase a portion or element of the spacecraft might be made to impact on [3] the lunar surface in the field of view of the optical system, and so provide an additional means of estimating the lunar surface properties. JPL should include in its technical evaluation the pros and cons of including any of these or similar variants of the basic impact mission.

On the assumption that the terminal impact mission, as generally described, can be acceptably engineered within the existing restraints on time and state of the art, it is also requested that your staff evaluate the feasibility of carrying on the spacecraft, for use during the transit to the moon, a basic group of instruments for the evaluation of the properties of cislunar space. These instruments can be read out over a narrow bandwidth, low power drain, sampling telemetry system which can be cut off when the terminal phase of the mission is initiated. The instruments considered should include a cosmic ray telescope and ionization chamber similar to those flown in Explorer VI, a plasma probe or low energy proton analyzer, a rubidium vapor magnetometer, and a micrometeorite detector in that order of priority. First priority should be assigned to the image transmission system. Any extra weight power supply capacity may then be used for the other experiments.

If additional instrumentation will not compromise the prime objective, transmission of impact acceleration is desired. Also, if a satisfactory technique can be devised in the time available, a penetrometer measurement would be very desirable.

(b) The rough landing mission that has previously been discussed with JPL, with the objective of depositing an instrumentation package which will survive the impact and then transmit significant data, should be reexamined by JPL. This reexamination should evaluate the probability of useful data return from a survivable package [4] incorporating with a lunar seismometer of the type now being developed for NASA, thermal measurement equipment, microphones or micrometeorite detectors or such other instrumentation which JPL believes will be likely to function after a rough landing. The objective of the review shall be to establish whether or not a survivable payload package should be actively developed as an alternate or backup for the image transmission payload.

(c) You will note in the schedule of paragraph 1 that rounds 4 and 5 are listed as probable and possible respectively. Our programming presently calls for full funding of the first three Agena vehicles (1960 Budget) and partial funding for the fourth. Completion of the funding for the fourth and fifth rounds will be influenced by the appropriations actually made available for FY 1961 and FY 1962, but we plan to include these additional required amounts in the FY 1962 budget. Variations to the instrumentation details of the payloads for the last two lunar reconnaissance missions are possible if the initial rounds satisfy their mission objectives.

(d) In view of the inclusion in the program of the two Centaur planetary probes (rounds 6 and 7), JPL's evaluation of the necessity for a preliminary Atlas-Agena B flight to acquire technical data on systems and/or components to be used on the Centaur spacecraft is requested. If such a flight is considered essential to the development of the Centaur spacecraft, your position as to which round of the Atlas-Agena B series should be directed to this purpose should be stated. In general, this office considers the preparation of a diversity of spacecraft undesirable, but it is realized that a detailed engineering review may justify the preparation of a unit to prove out systems applicable to the Centaur mission which cannot otherwise adequately be tested on the lunar reconnaissance missions.

(e) The probable payload for the planetary missions, assuming miss distances in excess of 100,000 miles, would [5] include detectors of magnetic fields, cosmic rays, soft protons, plasmas, micrometeorites, and Lyman alpha radiation. Television and spectrometry should also be evaluated for inclusion. In the event that terminal guidance could be provided to insure a near miss or orbit, the payload content would be reviewed to establish the preferred experiments.

(f) It is expected that follow-on Centaur programs will include initial efforts at lunar soft landings. In the design of the interplanetary Centaur spacecraft, it is hoped that some consideration will be given to the requirements of the follow-on lunar missions.

(3) The current funding for the spacecraft and associated tracking and data acquisition will be the amounts presently authorized in the budget for 1960 as transmitted to you on December 7 by Dr. Glennan.

(4) Dr. Newell, and Messrs. Sanders, Stoller, and Cortright of my staff will visit JPL on December 28 to discuss with you and your staff the preliminary program layout and possibilities for its accomplishment. It is hoped that steps can rapidly be taken to firm up a program that can be implemented according to the schedule.

Sincerely,

[signature]
Abe Silverstein
Director of Space Flight Development

Document II-7

INTERNAL USE ONLY [appears on each page]

[no page number]

30 December 1959

MEMORANDUM FOR THE FILE

SUBJECT: Trip Report for the Visit to Jet Propulsion Laboratory on 28 December 1959 by Homer E. Newell, Jr., Newell Sanders, J. A. Crocker, Morton J. Stoller

The principal purpose for the visit to JPL was to discuss with Laboratory personnel the letter of December 21, 1959, in which Dr. Silverstein set forth Headquarters guidance on lunar and planetary missions for the next three years. The plan was to answer any questions that JPL might have concerning the intent of the letter and its content, and to discuss with JPL their initial thoughts in connection with the proposed missions.

Dr. Pickering welcomed the NASA representatives and turned the meeting over to them. Newell Sanders suggested that we begin by using the letter itself as a basis for the agenda. He proceeded to read the letter paragraph by paragraph, and discussion developed as the letter was read. When the suggested schedule of lunar and planetary missions had been set down, the JPL representatives wished to know how this proposed program fit into the long range plans, particularly the follow-on uses of Centaur and Saturn. In partial reply Crocker indicated that the program should begin to develop:

a. A spacecraft for use with the Agena on lunar work,
b. a spacecraft for use with Centaur for planetary and lunar orbit work, with perhaps a modification for soft instrumented landings,
c. a spacecraft for use with Saturn on planetary work with some modification, perhaps for instrumented landing of lunar roving vehicles, and finally,
d. a spacecraft for use with the Saturn for unmanned circumlunar missions and return leading to perhaps some modification for manned circumlunar missions and return.

Crocker emphasized that we were prepared to discuss only the Agena spacecraft and the planetary Centaur spacecraft in this discussion, but that the remaining spacecraft had been introduced to provide the necessary background for the present discussions.

At this point the JPL people asked for some clarification on the philosophy that NASA was going to follow in these matters. Specifically, the JPL members wanted to know if each spacecraft mentioned by Crocker was to be developed more or less independently of each other, or if the spacecraft developed for NASA were to form a family, with the advanced ones growing out of the experience gained in designing the preceding spacecraft. The NASA representatives indicated that the latter approach was basically the one that NASA wished to follow. It was emphasized that this would mean that a close working relationship would be required between JPL and [2] the other NASA activities. JPL agreed that this is the proper approach to take.

Dr. Pickering pointed out that the JPL recommendation had been to emphasize the planetary work rather than the lunar, whereas the present letter indicates a NASA decision to emphasize the lunar work. It was stated that NASA had indeed decided to emphasize lunar work, but it was pointed out that the planetary work should get underway at once and that there would be a planetary try every time the near planets, Mars, and Venus, were in optimum position for a planetary mission. It was also pointed out by NASA that the planetary program may have to be based on the Centaur for some time to come, instead of Saturns.

The JPL representatives asked whether or not NASA had in its planning taken into consideration the question of competition with Russia, scientific objectives, technological objectives, and the matter of organization. In reply it was stated that all of these had been taken into account. It was pointed out that the overall objectives of the NASA program in space flight had two very important aspects:

(1) The extension of the domain over which man may move and be active; and
(2) The extension of human knowledge about the earth, its environment, and space and the objects of space.

Both of these objectives are regarded as very important, and the NASA program planning is designed to support both of them strongly. In the matter of Russian competition, it is clearly understood that whether it be stated openly or not, the United States is in competition with Russia, and the stakes are very high indeed. It is further understood that the loss of the space race would be of great seriousness to the United States, economically, culturally, and politically.

But, it is felt that our competition with the Russians must be based on a sound program of science and technological development, and not on the performance of what may be called stunt-type missions. If the latter approach were taken, we would be in danger in every case of being scooped or bettered by the Russians and made to look even worse than we are, and in the long run we would lose out by not properly developing our ability to compete. It is felt, therefore, that on the technological side we must turn to building up our technological strength and capability. On the scientific side, we should undertake missions that are scientifically worthwhile, and that integrate into a sound long-range program. In the question of the organization, NASA plans to take a unified NASA approach to the problem of spacecraft and boosters, and in the area of scientific research to develop a broad praticipation [sic] of the scientific community with the NASA research centers and laboratories. The JPL people appear to be in agreement with this approach to things.

Dr. Pickering then raised the question of overall management of the Agena vehicle. Likewise the Centaur. With regard to the first vehicle, Pickering asked what were NASA's plans for the use of Agenas. The [3] following table was given to him:

No Agenas in 1960.
The JPL Agenas in 1961.
2 Thor-Agena satellites in the interval from late 1961 to early 1963.
2 Atlas-Agena heavy satellite observatories about 1963.

Pickering commented that this would indicate to him that the management problem of establishing NASA-Air Force-Lockheed relations would rest on the JPL lunar program.

Pickering then raised the question of whether or not it was sensible to consider Centaur for planetary missions in 1962. It was pointed out that these would be Centaurs after Centaur No. 6, perhaps Centaurs 7 and 8.

After the NASA letter had been read through paragraph by paragraph and discussed in general terms, and JPL questions answered, Pickering then proceeded to make some specific comments. He stated that it was questionable that mid-course guidance would be available on the first two Agenas to make a valid lunar mission possible on the schedule indicated by NASA. It is thought by JPL that a mid-course guidance system will work out appropriately for such missions, but will take sufficient time that one should not plan on having it until about the third Agena in the NASA schedule. This would indicate either of two approaches; (1) to have two firings in advance of the lunar missions devoted to engineering tests of attitude control and communications, and to scientific observations of the interplanetary medium, then to follow with the lunar missions as indicated in the NASA letter; or (2) to delay the first firing so that it could be a lunar mission. This might advance the first lunar mission by a few months over the presently scheduled third flight on the NASA list. The JPL people appeared to favor approach No. 1, and this in fact was further developed in afternoon discussions.

Pickering indicated that the Agena payload would probably be appreciably less than that for Vega. He pointed out that in order to adjust the Vega trajectory so that Goldstone would see it at the time of impact a penalty of about 18 percent was incurred; JPL thinks that the penalty in the case of Agena may be somewhat larger.

Pickering than [sic] raised the question of work on ion propulsion. He indicated that JPL would, of course, be a customer, perhaps the first customer, for ion propulsion vehicles, in view of their concern with spacecraft. Sanders replied that work is going into the area of electrical propulsion, but that he felt such propulsion may not be available in practical form for the next decade.

Pickering then raised the question of the availability of Saturn. It was pointed out that we do not yet have the story on the Saturn program. However, it may be considered that Saturn vehicles for lunar work would not be available until after 1964.

[4] This terminated this phase of the discussion of the Silverstein letter.

* * * * * * * * * *

The JPL staff then reviewed the Agena situation as they now see it. The following facts emerged from their discussion:

(1) Headquarters will in the next week receive from Lockheed a proposal for some Atlas-Agena-B lunar flights.

(2) Of the 60 to 70 rounds of Agenas scheduled to have been launched by the end of 1961, only two are listed for AMR (February and April on pad 14). This leaves AMR open for NASA-JPL use, and the problem of adapting the stand for Agena-B work should be minor according to some. However it will be necessary to act

immediately to salvage the experience of the personnel involved and the equipment at AMR for use on NASA Agena-B flights.

(3) JPL is approaching the Agena-B program from the point of view that the minimum additional development should be done. They emphasized that one must be careful not to buy the Vega program and its development time lags all over again, in placing too ambitious requirements on the Agena vehicle.

(4) JPL plans, in approaching its spacecraft work, to study carefully the vehicle to be used, the trajectory to be followed, and the percentage of payload weight that must be required in the mid-course system. JPL feels that it is important for Headquarters to name vehicle people who are going to handle the selection of the Agena vehicle, so that JPL can be in contact with them. JPL emphasized the need for them to know what vehicle is to be used, but also that the spacecraft once launched will be entirely separate from the launch vehicle, and will be complete in itself.

(5) Although it is a little early to speculate on the Agena-B performance, JPL tends to think that space payload weights will be substantially smaller than for Vega.

(6) It was noted that the philosophy in the Agena-A program was to put as much as possible of a standard nature in a booster, either Atlas or Thor, so that the upper stage would be independent of whether the Thor or Atlas was used as booster. It is understood that this approach will carry over to the Agena-B. There are some plans to carry out tests in the DOD program involving Agena-B restarts. For those who wish to take it up, Lockheed has a third stage design on paper for use with the Agena-B.

[5] This closed this phase of the discussions.

* * * * * * * * * *

Following the above discussions, the group broke up into several smaller groups for separate discussions. A meeting was held between Pickering, Goddard and Hibbs of JPL, and Newell of NASA to discuss a number of policy questions.

First, the question of an overall guiding policy for the space science program activities involving particularly other scientists from other organizations was discussed. This had been brought up in a letter of December 14 to Dr. Silverstein from Dr. Pickering. The following statement of policy was agreed upon:

"1. After selection of basic missions for the flights, tentative selections of responsible scientists and scientific instrumentation for the spacecraft will be carried out by the Office of Space Sciences, NASA, in collaboration with the Jet Propulsion Laboratory. This tentative selection will be based on programmed capabilities, missions, schedules of the flight program, and the scientific desirability of various experiments. It is anticipated that this initial step will involve more experiments then [sic] will finally be carried on the flights, that is, some future weeding-out will be necessary prior to the development of flight equipment.

"2. Prototype models or design concepts of scientific instruments for the experiments tentatively selected will be constructed under the direction of the responsible scientists with funds provided either directly from NASA or by the Jet Propulsion Laboratory with concurrence of the Office of Space Sciences, NASA. The schedule for completion of such prototypes will be established by the Laboratory to be consistent with the spacecraft development program.

"3. Completed prototypes or design concepts will be delivered to the Laboratory and evaluated by Laboratory personnel in collaboration with the responsible scientists.

"4. On the basis of such prototype evaluation and any other applicable information which is available concerning the experiment and the instrumentation, JPL with the advice and concurrence of the Office of Space Sciences, NASA, will make the final selection of instruments and responsible scientists for each flight. After this selection, additional development of the selected instruments will be carried out under the technical direction of JPL, and with the collaboration of the experimenting scientists.

[6] "5. The fabrication, testing, calibration, checkout and launching of flight instruments, and the acquisition and reduction of data from measurements taken in flight will be carried out by or under the direction of the Jet Propulsion Laboratory based on functional specifications determined by the responsible scientists and with the assistance of the responsible scientists. If the Laboratory determines during the course of fabrication of the flight instruments that modification of the functional specifications are required in order that the instruments operate reliably in the overall system, such modifications will be made on the basis of agreement between the Laboratory and the responsible scientists. The concurrence of NASA Headquarters will be obtained for any modifications which imply major changes in the scientific objectives of the experiment."

With respect to this item it was agreed that the NASA Headquarters would answer the December 14 letter with the reworded version of the policy, confirming the present agreement.

Dr. Pickering then brought up Mr. Horner's letter of December 16 concerning the role and responsibilities of the Jet Propulsion Laboratory in the NASA program, and a letter of Dr. Pickering's to Dr. Silverstein of December 17 following up on part of Mr. Horner's letter. Both of these letters are attached. Mr. Horner's letter mentioned the responsibility of the Jet Propulsion Laboratory for detailed mission planning, and in this connection Pickering's letter proposes the creation of a committee for lunar and interplanetary space exploration under the chairmanship of Dr. A. R. Hibbs. This committee is recommended "in view of the recent decision of NASA Headquarters that the Jet Propulsion Laboratory undertake responsibility for the planning and execution of lunar and interplanetary space exploration program," to use JPL's words. It was further recommended that the current lunar committee chaired by Jastrow, be absorbed into the new committee.

In reply I stated that it was my feeling that JPL had gone further than Mr. Horner had intended, and that the overall program planning was a Headquarters activity, with the detailed engineering planning and execution being a Jet Propulsion Laboratory responsibility. With regard to the need for a committee such as that recommended, I pointed out that we had already taken steps to set up such a committee, and that an invitation to JPL to name a member to the committee would be forthcoming. This new committee would be chaired by myself, and the formal membership would be drawn internally from NASA and its Centers. The broad contact with the scientific community can then be had by two devices:

(1) By inviting interested scientists to attend appropriate meetings of the new committee, and to participate in the discussions; and

(2) by requesting from the Space Science Board their thinking on lunar and planetary programs.

[7] The former arrangement has the advantage that meetings of the formal committee itself may be held either with or without the invited attendance of outside members; in cases in which internal budget and programming crises must be discussed and aired, it is highly desirable to be able to meet without generating discouragement in the scientific community. With regard to the Space Science Board, an informal request has already been made to the Board to provide NASA with its input on what is desirable scientifically in a planetary program.

Dr. Pickering and Dr. Hibbs then reemphasized the desirability of pulling the Lunar Committee in under this new committee. In particular they were concerned about having a Goddard man chairman of a committee that is so vitally concerned with a matter that is assigned to JPL as a primary responsibility. The fact that Jastrow acts as a NASA Headquarters man in chairing this committee is completely lost on the scientific community, they felt, and from JPL's point of view the situation is undesirable. After a brief discussion the following points of agreement developed:

(1) A lunar or planetary committee should not have a Goddard man as chairman, since JPL has been assigned responsibility for these areas.

(2) The incentive in proposing the committee with JPL chairmanship was derived from the Horner letter; however,

(3) Headquarters chairmanship is certainly fully acceptable.

(4) JPL agrees that for the present having the committee internal to NASA is probably the correct approach.

(5) There must, however, be broad input from the scientists.

(6) The Lunar Committee should be melded into the new committee which would then be concerned with lunar, planetary and interplanetary work.

The discussion then turned to the mechanics of working with Headquarters on the lunar exploration program. I indicated that the man in overall charge of the project activities was Ed Cortright, and that the material should either go through him or information copies should be sent to him. With regard to the space science activities these would be handled through my office, and in particular the lunar work would be handled by Schilling and the planetary and interplanetary by Clark. Hibbs wished to get on with the matter of firming up instrumentation for the forthcoming missions. I suggested that he get in touch with Clark and Schilling immediately to discuss the contracts that we have in existence for instrumentation and to work out a plan for the forthcoming flights. He said he would do so.

[8] I then asked Pickering if he had given further thought to sending a man to Headquarters to work with Clark on the planetary and interplanetary programs. Pickering stated that their problem was, like everyone else's, that of manpower. Since a man has been sent to work with Schilling, he can not at the present time spare someone else to work with Clark. He thought that perhaps the man assigned to Headquarters might be assigned one time to Schilling, the next time to Clark, and so forth. This ended this phase of the discussions.

* * * * * * * * * *

A meeting was held in Dr. Hibbs' office to discuss the space sciences program. The following were in attendance:

Dr. Hibbs, JPL	Dr. Newell, NASA
Dr. Eimer, JPL	Mr. Stoller, NASA
Dr. Davies, JPL	Mr. Sanders, NASA
Dr. Richter, JPL	Mr. Crocker, NASA

The JPL people reviewed the thinking that has gone into space missions and to the scientific experiments to be conducted on those missions. Relating the discussion to the recent letter from Silverstein to JPL, the JPL approach appears to shape up as follows:

(1) Two Atlas-Agena test missions to check out attitude control and communications, and to make interplanetary medium measurements.

(2) A series of lunar missions including television with a 200 line field, and a three meter resolution at the time the closest picture is taken, including a split capsule, the survivable portion of which would carry a seismometer, a temperature measuring device and accelerometers. The television would be carried in the portion that is not intended to survive, and along with the television there would be equipment for radioactivity measurements.

(3) Venus and Mars planetary missions.

For the first two test flights the scientific measurements would include instruments to measure electrons from 0 to 100 electron volts, and protons in a large number of bands from 0 to above 75 million electron volts. The JPL electrostatic analyzer, Goddard or SUI scintillation counters, the Simpson proportional counter, the Neher ionization chamber, and some magnetometer (perhaps STL's) would be included. The magnetic field measurements are necessary to support the plasma particle measurements. In addition a Lyman-alpha hydrogen radiation scanner would be used to look for an envelope of neutral hydrogen about the earth. Finally appropriate micrometeorite detectors would be included. The planned spacecraft for such a mission is shown in the attached drawing. A summary JPL review of experiments and experimenters that might be used in the proposed mission is also attached.

[9] The mission for the test vehicles contains some highly desirable scientific experiments, and would be most worthwhile in the space science program. Likewise their proposed follow on missions appear to be well thought out from the scientific point of view. It is, however, quite clear that the whole approach is still a continuation of the development plan that was going on in connection with Vega, and is largely slanted toward ultimate planetary work. It would seem that NASA should go along with the proposed missions. Otherwise a considerable delay would be required and a considerable amount of work that has already been done would be put on the shelf.

In connection with the proposed missions, the schedule for the two initial tests, with the interplanetary scientific measurements, was brought up. Richter pointed out that it was necessary to have the flight type hardware in hand by March in order to meet the schedule. This is a full year in advance of the actual flight. The NASA people raised the question of why such a long lead time was required for availability of the final working flight equipment. The NASA people wondered why mechanical mock-ups of the actual flight equipment wouldn't be adequate, leaving perhaps as much as another half year to the scientists to continue the work on the electrical aspects. It was granted that this whole spacecraft was much more complicated than anything we had ever attempted before, and that therefore lead times should be stated on a conservative basis. The NASA people felt, however, that the lead time suggested may well be much too conservative, and that therefore the choice of scientific experiments, instruments, or experimenters may be suffering because of this. That the scientific work is actually suffering because of this type scheduling, was brought out by the JPL statement that it was not possible to bring Bruno Rossi, for example, into the planning and design work since equipment had to be frozen at what was essentially available at the present time. The NASA representatives pointed out that if even another three months were available it would be possible to profit by the thinking of Rossi and his co-workers in his particular area. The NASA representatives then raised the question of whether the same is true of the other instrumentations.

Since Stoller was planning to remain another day, it was agreed that he would discuss this problem further with JPL people on Tuesday.

[10] Stoller and Newell agreed to convey these comments and thoughts to Headquarters, and in particular to Clark and Schilling so that they would have as much background as possible by the time Hibbs and his people got in touch with them.

This concluded the discussions for the day.

[signature]
Homer E. Newell, Jr.
Assistant Director for
Space Sciences

Enclosures:
1. Ltr. 16 Dec 59 - Horner to Pickering
2. Ltr. 14 Dec 59 - Pickering to Silverstein
3. JPL Spacecraft
4. JPL Summary Review
[enclosures omitted]

CC: Dr. Glennan
Dr. Dryden
Mr. Horner
Dr. Silverstein
Mr. Stoller
Mr. Sanders
Mr. Crocker
Mr. Cortright
Dr. Schilling
Dr. Clark

Document II-8

Document title: Gerard P. Kuiper, "Need for a Ground-Based Lunar and Planetary Observatory," June 1960.

Source: NASA Historical Reference Collection, History Office, NASA Headquarters, Washington, D.C.

Just as NASA was embarking on a solar system exploration program focused on sending robotic spacecraft to lunar and planetary targets, some scientists believed that ground-based studies of the solar system were still critical, allowing for the acquisition of more data for researchers to integrate with that returned by spacecraft. Such studies would also ensure a flow of scientific data between space missions. University of Chicago astronomer Gerard Kuiper was one of the leading advocates for the continuation—and expansion—of ground-based lunar and planetary studies. In June 1960 he wrote this memorandum to NASA's Subcommittee on Planetary and Interplanetary Sciences, insisting that ground studies were still imperative, but that most U.S. observatories were devoted to stellar astronomy. Kuiper made a strong case for federal support of ground-based solar system studies, and succeeded in relocating his work and colleagues from Chicago to a new facility—the Lunar and Planetary Laboratory—that he established at the University of Arizona with support from NASA as well as from other R&D-oriented federal agencies.

[no page number]

NEED FOR A GROUND-BASED LUNAR AND PLANETARY OBSERVATORY
(Memo prepared by Gerard P. Kuiper, U. of Chicago for the NASA Subcommittee on Planetary and Interplanetary Sciences—June 1960)

1. THE PROBLEM

Are the present and projected astronomical facilities and capabilities in the U.S. adequate to support the NASA Space Program for the Period 1960-1970?

2. GENERAL CONSIDERATIONS

(a) Ground-based astronomy is inexpensive compared to rocket-based astronomy and should, therefore, collect all useful data that can be obtained from the ground, and develop a body of theoretical knowledge as a basis for further planning.

(b) Ground-based observations can be continued over long intervals of time and have a known degree of uniformity, stability and reproducibility. This aspect is important in the determinations of dynamical properties (orbital planes, direction of axes of rotation, oblateness of planets, moments of inertia) as well as physical properties (time variations of atmospheric conditions, seasonal effects on surface, variable radio emissions).

(c) There is a surprising difference between planetary and stellar astronomy; training for the latter does not fully prepare one for the former. Planetology is allied to and derives major support from various geophysical disciplines (meteorology, aeronomy, geophysics, geochemistry) as well as from classical celestial mechanics; but it also requires specialized astronomical techniques (efficient use of telescopes, trained visual observation, astronomical spectroscopy, astrometry, photometry, polarimetry, etc.).

(d) The major U. S. observatories have been designed and built for stellar work and are used only incidentally and for a small fraction of the time for planetary and lunar studies. Very few U.S. astronomers are engaged in planetology and then in most cases only part-time. Training of students in planetology to the Ph.D. level has for the past 20 years been at the rate of perhaps one Ph.D. per three years.

[2] (e) As an astronomer I find myself in full accord with the identification of the three main problems of Space Research, as listed in the NASA study of April 1960 (NASA Program Planning in Space Sciences, p. 2). I believe that planetary astronomers have long felt that these are indeed the sustaining forces behind their programs.

3. SCIENTIFIC PROBLEMS

The chief problems of planetology are, briefly:

(a) Discovery of planetary and interplanetary objects. This includes the continued search for intra-mercurial planets, of natural satellites (now 31 known); of asteroids, including especially those with orbits near the earth; of comets, and of solid interplanetary particles, including those causing the zodiacal light and the counterglow.

(b) Planetary and satellite atmospheres, discovery of presence and subsequent studies of composition, density and temperature regimes, circulation, condensation products and atmospheric dust, radiative transfer, photochemistry, contributions by volcanism, escape to space, interpretation in terms of planetary evolution, possibility of sustaining life.

(c) Measurements of mass, diameter, obliquity, oblateness, rotation, and moments of inertia of the planetary bodies.

(d) Construction of surface maps with all attainable resolution, referred to a coordinate system derived from the planetary rotation.

(e) Physical Studies of the surface texture and composition by means of the properties of the reflected light, including spectrophotometric analysis, polarization measurements at different wavelengths, thermal and radio emissions. Radar probes. Contact studies of the surface materials by landings (geophysical, geochemical, radiation studies). Organic materials.

(f) Satellite studies. Magnitudes, variability of light due to rotation, diameters, masses, surface markings, spectrophotometry, composition. Relation of satellite mass and composition to planetary body. Irregular satellites. Problems of origin.

(g) The study of the gas and radiation content of interplanetary space and interplanetary magnetic fields; action on comets, interplanetary particles, [3] planetary atmospheres. Radiation belts.

(h) Laboratory studies. Identification by laboratory experiments of atmospheric and surface absorptions (at a wide range of gas temperatures) and of polarization properties; laboratory model-studies of ridges on lunar maria, impact craters, etc., to determine formation processes.

(i) Theoretical studies. Studies of thermal balance and radiative transfer, escape of atmospheric constituents to space, evolution and origin of planetary atmospheres; bearing on general problem of planetary origin and evolution and the development of life. Models of planetary interiors. Origin of solar system.

4. EARTH-BASED VERSUS ROCKET-BASED RESEARCH

Of the problems listed in Section 3 several can best be dealt with from earth-based observatories, while others can be done only from rockets or instrument packages, and some may be attempted by either method. Examples are given below.

The dynamical and metric properties of the moon, the planets and the satellites (c, f above) on the whole are best dealt with from observatories, though some supplementary and much-desired information could be contributed from rocket trajectories (such as on the figure of the moon, the position of the pole and the oblateness of Venus). On the whole this subject is in a reasonably satisfactory state, though much is left to be desired (e.g., satellite masses).

Data on atmospheric compositions (b) are very fragmentary and our knowledge is unsatisfactory. This is due largely to the ultra-violet cut-off of our atmosphere. For the brighter planets this limitation may be overcome by observations from artificial earth satellites although ultimately spectroscopy from probes near the planets may prove more powerful. The spectroscopy from the earth has, however, by no means been exploited fully. Examples are listed in section 5.

Maps (d) of the visible surface of the moon up to scale 1:1,000,000 should be made from the earth because the coordinate system is based on the rotational properties of the moon which can be derived only by observation over several years (the rotational properties are summarized by Cassini's laws and the constants of the physical [4] libration). Approximate maps of Mars can and have been made also from the earth.

High-resolution photography of the moon and planets (d) will require the use of instrumented rockets near these bodies. This will undoubtedly be one of the most significant contributions by Space Research, but in order to be of optimum value the photography so obtained must be locatable on lunar or Martian maps previously made (since different types of terrain are seen from the earth).

The surface properties of the moon, Mars and Mercury (e) are accessible from the earth through photometric, polarimetric, and thermal measurements, carried out as functions of phase angle and of position on the planet. Vastly more can be done from the earth than has been done so far. However, the methods are indirect and the results lack the precision of prospective physical and chemical studies made directly on surface materials. Nevertheless, much is to be gained by more fully exploiting the rich optical data already accessible and matching them through appropriate laboratory studies (see Section 5).

Theoretical and Laboratory Studies (h, i) must be made on a greatly enlarged scale regardless whether the planetary data are obtained in observatories or from rockets.

Discoveries from rockets of objects 1 km or less in size around Mars and Venus (a) seem a distinct possibility. Mars could well have a ring or swarm of small objects moving about it, with Phobos and Deimos being merely the largest of that class. A tenuous ring of such bodies may exist also around Jupiter. Around Saturn two systems of rings are known, the outer, bright ring being composed apparently of H_2O snow or snow-covered particles; and the inner or crepe ring being almost certainly composed of dark silicate material.

Satellite studies (f) can be vastly improved on, particularly for the Mars, Saturn and Uranus systems.

Studies of Interplanetary Space (g) have already been extremely rewarding. Clearly, rocket-borne research is vastly more powerful here than earth-borne research, although ground-based observations of comets, comet tails, the zodical [sic] light, and aurorae and magnetic storms are very important sources of information.

[5] 5. NEED FOR A NEW LUNAR-PLANETARY OBSERVATORY

Section 4 shows that in a vigorous Space program, such as out-lined by NASA for the period 1960-70, certain problems are best dealt with from the ground (usually with a powerful telescope), while others require the use of rockets. The inclusion of ground-based observations in an over-all program is not merely a matter of economy; it seems a logical necessity, by the nature of the problems themselves.

The writer, during an after-dinner address at the Denver Meetings* of April 30, 1958, stated: "To send up space rockets for planetary observations without adequate observatory facilities and personnel on the ground would be a waste; we would get data, but no integrated science." A list of some of the more promising problems of planetology that can be solved from expanded ground-based observations follows.

(a) Venus. The apparent CO_2 abundance, as measured by the strength of the CO_2 band at $\lambda 8689$ A, varies drastically from day to day, from place to place on the planet, and also systematically with phase angle for the planet as a whole, being much larger at full phase than for the crescent. ** These variations, which have a range of a factor 10, must be connected with the variable cloud layer and the convection in the Venus atmosphere.

The areas of different strength of λ8689 may be the same as the belts seen on ultraviolet photographs; similar dimensions appear to be involved. Clearly, it is important that a full-time study be made of this phenomenon over a period of at least a year or two.

The reflection curve of Venus is strongly wavelength-dependent and an accurate measurement of this curve may lead to an identification of yellow coloring material in the Venus clouds and possibly to the identification of new atmospheric constituents (several unsupported claims have been made by Kozyrev and others in this direction). A ratio-spectrograph must be constructed which expresses the intensity of the Venus spectrum (cut by thousands of Fraunhofer lines) [6] directly in terms of the solar spectrum, with a precision of 1 percent or better. This requires some major instrumentation (a budget of perhaps $20,000). Very similar problems exist for the Jupiter satellites, Jupiter's Red Spot, and Saturn's satellite Titan, which have characteristic colors whose origins are at present unknown.

(b) Mars. When Ektachrome film is cooled to dry-ice temperature, it becomes nearly ten times more sensitive and gives a very good color balance. This material should be used for high-resolution color photography of Mars and Jupiter, on a continuing basis. Also, extensive studies of polarization of Martian surface detail must be made as a function of wavelength and time, to cover the seasonal variations of the ground features and the variability of the atmosphere with its clouds and haze content, using the full power of the polarization method by covering the entire accessible interval from 3,000 to 23,000 A. Further, a much more detailed study must be made of the motions of the Martian satellites, whose orbits give important information on the oblateness and pole of the planet; the same is true for the acceleration of the motion of Phobos (found at the Naval Observatory around 1940 but since questioned), because of its importance for the theory of bodily tidal friction.

The only gas so far identified on Mars is CO_2 but the polar caps were found to be H_2O snow (both 82-inch). Sinton with the 200-inch discovered surface absorptions near 3 1/2 μ, that are of very great interest.

(c) Jupiter. The causes of the remarkable cloud colors (white, cream, yellow, brown, brick-red, black, light blue) must be sought on the basis of extensive high-quality color photography and laboratory work. The observation with the 82-inch telescope, of snow deposits on the satellites II and III but not on I and IV, must be made more quantitative with better equipment. The nature of the bright and dark spots on all four satellites must be cleared up.

(d) Saturn. The very strong dependence of the angular rotation on latitude, amounting to at least 11 percent, must be studied in connection with the appearance and dissolution of spots, bands, colored zones, etc. The snow cover of the Ring, found spectroscopically with the 82-inch telescope, must be studied further to get the particle size. [7] (e) Titan. This is the only satellite for which an atmosphere has been found and a detailed comparison of its spectrum must be made with laboratory spectra of methane obtained at liqued-nitrogen [sic] temperatures. It is possible that in this manner differences will be found between the spectra leading to new discoveries.

(f) Uranus and Neptune. Studies of these planets with the 82-inch led to the discovery of a band near 8270 A which was later attributed to the pressure-induced dipole spectrum of H_2 by Herzberg. With the 82-inch also a rich system of absorption features was found between 7400-7600 A, which so far has defied all efforts to identify it.

(g) The Asteroids are the parent bodies of the meteorites, and we are thus in indirect contact with them. The space density of asteroids is 3 or 4 orders of magnitude greater at 3 astronomical units than near the earth; this region is the birth place of the meteorites (by asteroid collisions). The McDonald asteroid survey (Ap. J. Supplements No. 32) must be extended to fainter limits with the 48-inch Palomar Schmidt (we have scheduled such observations for the Fall of 1960). But photometry of asteroids, leading to the periods of rotation, the shapes of these bodies and the stability of their rotational pole, must be carried out on a vastly greater scale than we have been able to do with the 82-inch telescope. This subject is full of promise for clarifying several aspects of the origin of the solar system; while it has also a direct interest to NASA in giving data on the collisional hazards beyond the planet Mars, hazards that are much greater there than near the Earth.

(h) The Moon. Here ground-based research can make very extensive contributions. Detailed maps and three-dimensional models of representative lunar formations can be made that have 4 times the resolving power of current photography, while the best photography available by late 1959, collected in the Photographic Lunar Atlas, is for some fields already superseded by new photography with the 82-inch telescope.

Laboratory model studies must be made to elucidate the basic lunar processes of crater formation, the central peaks, the pressure ridges with their dykes, the tension rilles [sic] with their central extrusion dykes, the lunar volcanoes with their very low slopes ($\leq 5°$), the lava domes, lunar isostasy affecting the shapes of crater bottoms, etc.

[8] (i) Radio Telescopes have made incisive contributions to our knowledge of Venus, the Moon, Mars, Jupiter, and Saturn, and are expected to do more in the future, as microwave receivers and masers improve. Particularly important are the studies of Venus and Jupiter. Radars have made some contribution to measuring the smoothness of the lunar surface.

Examination of the above incomplete list will show that the numerous problems that can be tackled from the earth are:

(i) scientifically important;
(ii) entirely relevant to future rocket-borne research and excellent preparation thereto; and
(iii) far too extensive for the comparatively small effort that is now made with limited telescope facilities, very inadequate instrumentation, small funds, and lack of adequate assistance.

These conclusions show that a Lunar and Planetary Facility—an organization not now existing in the West—would aid NASA's objectives immeasurably. It would enhance NASA's impact on cosmic science, greatly increase NASA's scientific output, and provide national and international leadership in an area of great current interest; an area now neglected in the West because it falls between established disciplines (geophysics and astronomy), is not part of University corricula [sic], and requires facilities quite beyond the powers of Universities, even if supported by modest research grants.

6. SCOPE OF THE NEW FACILITY

In addition to the obvious requirement of a competent scientific staff of varied background, the Facility would need:

(a) a spectroscopic laboratory, including a long absorption tube for the study of atmospheric gases, a second tube coolable to liquid N2 temperatures;

(b) a small geophysical laboratory for dynamical studies of lavas, model studies of the lunar maria, impact craters, central peaks, lunar ridges and rills, a model shop, facilities for polarization studies of surface materials and fogs to simulate the observed curves for the Moon, Mars, Mercury, and Venus.

(c) cartographic facilities, largely directed toward lunar mapping and research

[9] (d) various telescope attachments (spectrographs, infrared recorders, polarization equipment for $3000 < \lambda < 24000$ A, colorimetric devices, color photography, etc.);

(e) adequate darkroom and copying facilities;

(f) measuring machines, special photogrammetric equipment, computing machines;

(g) library.

Two crucial questions remain:

(I) Does the Facility need one or more optical telescopes?

(II) Does it need its own Radio Astronomy Facility?

The latter question is readily answered.

Large Radio Telescopes are extremely expensive and require a vast organization for their operation (the 600-foot Sugar Grove steerable reflector costs around $65,000,000). Their use in planetology, while of singular importance, is limited in scope and the observing time required for the study of the planets is very modest. The large U.S. radio facilities are government-owned and their part-time availability may be regarded as assured. Furthermore, the planetary programs possible with these instruments already figure prominently in their planned observing schedules. It is concluded that the vital radio data will be forthcoming without NASA entering the field.

By contrast, the large optical telescopes are privately owned (except for the 36-inch at Kitt Peak National Observatory, and the 84-inch there under construction). These telescopes are fully booked with stellar programs by their staffs, though some small planetary and lunar programs can usually be worked in. Additional facilities seem very desirable.

There appears to be no sufficient reason to duplicate the 200-inch telescope (at an estimated cost of $10,000,000 and an estimated delivery time of 6-10 years), because only some of the more difficult planetary programs require this large aperture, and in the past the 200-inch has been made available when smaller telescopes were found inadequate (Kuiper's measures of the diameters of Pluto and the Saturn satellites; Strong and Sinton's measures of the infrared spectra of Mars and Venus).

Photography and visual observation of the moon and planets will [10] demand much observing time; together with auxiliary programs it could effectively use a large telescope full time on a continuous basis. This work requires the very best image steadiness, which on 99% of the good observing nights is obtained with apertures not over 60 inches (larger apertures collect more turbulence).

The practical question is therefore: can the U.S. telescopes of apertures 36-100 inches be counted on to provide the necessary observing time to match the NASA space program?

The answer to this question is definitely "No." The 36-inch of Kitt Peak is already signed up for 18 months in advance. The 60- and 100-inch Mt. Wilson telescopes are regularly booked full except in the past for some occasional nights near full moon on the 60-inch. Occasional short periods (of a few nights each) can probably be obtained on these instruments (this is based on advice I recently received in Pasadena) but it is impossible to expect more. The new 120-inch at the Lick Observatory, under construction for some 10 years, was eagerly awaited by the staff, and is now fully occupied by the California astronomers on stellar programs. The 82-inch telescope at McDonald Observatory has for years been under heavy competition among the Chicago and now also the Texas astronomers, with Indiana purchasing 100 hours of clear time per year (but in danger of being crowded out). This listing exhausts the suitable telescopes (the Eastern and Midwestern Observatories have smaller telescopes and poorer observing conditions).

A 60-inch reflector placed at a first-rate location (presumably a mountain site in the Western U.S. or possibly Hawaii) would be expected to be most useful to the NASA program because:

(i) it would be optimum for 80-90 percent of the ground-based programs (the remaining programs could probably be carried out at existing facilities, at least initially);
(ii) delivery time would be reasonable (about 2 years);
(iii) the cost would be reasonable (about $500,000, incl. dome if the design is held simple); a small dormitory would add $100,000; additional buildings could be added later;
(iv) a smaller telescope, 24-36 inches in diameter, should probably be added; it could be completed even more rapidly and would [11] later be a useful standby for many programs.

If the instrument could be placed on an existing observatory site it would save road building and provision for utilities, which might for isolated mountains be very expensive.

The laboratory would presumably be best off the mountain (as Pasadena is for Mt. Wilson and Palomar, San Jose or Berkeley for Lick Observatory, Tucson for Kitt Peak). The initial cost of the laboratory might be about $300,000. The initial equipment outlay [would be] $100,000 with perhaps $100,000 additions per year.

If NASA decides to explore the addition of a telescope to their facilities, it is recommended that advantage be taken of the experience on site surveys now accumulating in the Chile Observatory Project. (This project is carried out jointly under an Agreement which I negotiated in Santiago in March 1959. In connection with my scheduled departure from Chicago it is planned to transfer the Project to AURA in August 1960.) Dr. Jurgen Stock is in charge of the field work and he has shown unusual ability to get at the significant facts under sometimes trying field conditions. Dr. Stock has found some sites in Chile that are extraordinarily good, perhaps better than anything we have in the U.S. At any rate, he will be returning to the U.S. around August 1, 1960 and would probably be available to make some comparative site tests in the Western U.S. (Kitt Peak, Mt.

Wilson, Palomar, Lick) and possibly Hawaii. I do not know of a more competent person to make such tests. The cost of this site survey (extended for 2-3 months next fall) would be around $10,000 and it would be an excellent investment even if no construction would immediately follow.

---

* Sponsored by the Air Force Office of Scientific Research and Institute of Aeronautical Sciences, with most of the Proceedings published in Vistas in Astronautics, Vol. II, 1959.
** The Atmospheres of the Earth and Planets, pp. 370-71, 1952.

Document II-9

Document title: Homer Newell, Assistant Director for Space Sciences, NASA, to Robert Jastrow, Director, Theoretical Division, Goddard Space Flight Center, NASA, with enclosure, April 7, 1961.

Source: Robert Jastrow, Director, Mount Wilson Observatory, Pasadena, California.

In 1960, JPL engineers initiated development studies for NASA's first interplanetary spacecraft, slated to be used for missions to Venus and Mars. JPL's concept was the Mariner spacecraft, which would follow two models: the A model would fly past planets while the B model would be capable of releasing a capsule toward Mars or Venus. In this document Homer Newell informed Robert Jastrow of NASA's plans to deliver a Mariner B probe to Mars in 1964 and provided guidelines for proposals for mission experiments. Mariner B's ambitiousness and complications with launch vehicles precluded this model from ever flying. Instead, NASA's first successful Mars mission, Mariner 4, was launched as a flyby mission of a Mariner A spacecraft on November 28, 1964.

[no page number]

[on NASA letterhead]

IN REPLY REFER TO DL (CPS:asm)

April 7, 1961

Dr. Robert Jastrow
Theoretical Division
Goddard Space Flight Center
Greenbelt, Maryland

Dear Sir:

The National Aeronautics and Space Administration is preparing plans for a planetary probe to Mars in 1964. This probe is identified tentatively as Mariner B.

The basic purposes of the 1964 Mars mission are to study the planet Mars from a close hyperbolic orbit about the planet, and to study the physics of the interplanetary medium through which the probe would pass. A test flight of the spacecraft into interplanetary space may be made prior to the firing. This flight would probably carry the same experiments as the planetary firing. This letter is an invitation to submit preliminary proposals for experiments to be carried on the Mars mission, and to provide you with planning information to help in formulating such proposals.

It is suggested that proposals contain the following information:

1. A description of scientific objectives, including possible results peculiar to the proposed experiment;
2. A description of instrumentation, including an estimate of developmental requirements;
3. Funding and personnel requirements, the former for 1 July 1961 to 30 June 1962.

It is likely that the number of preliminary proposals will far exceed the capability of the spacecraft to accommodate them. Thus, proposals submitted for the Mariner B scientific payload will be reviewed carefully on the following bases:

[2] 1. Scientific content and consistency with the overall scientific objectives of the mission as stated in the appended document;
2. Compatibility with the spacecraft power supply, structure, and telemetry;
3. Probability of meeting spacecraft construction schedule.

Evaluation of proposals will be made by the NASA, JPL, and, in special circumstances, by recognized authorities in the appropriate scientific disciplines. Final selection of the payload will be made by the Space Sciences Steering Committee of NASA Headquarters. Upon acceptance of a proposal, appropriate funding arrangements will be made.

The NASA has assigned to JPL the responsibility for the preliminary design of the spacecraft for this flight. The design of the spacecraft for this mission, and the integration of the most meaningful set of scientific experiments, require the close collaboration of the scientists with JPL. Based on previous experience in the design of similar spacecraft and experiments, it is suggested that experimenters keep in close contact with Dr. A. R. Hibbs or G. Neugebauer, of the Division of Space Sciences, Jet Propulsion Laboratory, during the preparation of proposals. In the case of approved proposals, JPL will provide liaison with the experimenter through a project scientist to assist in the design and development of instrumentation, and to help in the integration of the equipment into the spacecraft. Previously, development and construction of flight instrumentation have often been provided by industrial organizations selected on a competitive technical basis with specific information developed by the experimenter assisted by JPL.

To meet the spacecraft development schedule, it is necessary that proposals be received by 15 May 1961. For specific experiments, primarily in the interplanetary area, utilizing developed and proven techniques which can be readily integrated into the spacecraft, the proposal due date is 15 June 1961, subject to negotiation with JPL. Subsequent to selection of experiments and experimenters, collaboration with JPL will help to define the experimental details and interface specifications so that instrument development for all experiments may begin 15 July 1961. Instrumentation "breadboards" are scheduled for completion by 1 November 1961.

[3] The establishment of experimental groups combining both theoretical and experimental talent is acceptable, and proposals may be submitted by such groups. The success of the Mariner B program depends on a broad participation throughout the scientific community. The participation of your colleagues, who may not have been contacted directly, is solicited. In the event that you know of an interested party, his name should be submitted to NASA Headquarters.

JPL has been in the process of developing some infrared and ultraviolet instrumentation which might be applicable to Mariner B. A document describing this instrumentation can be obtained from Dr. A. R. Hibbs, of the Jet Propulsion Laboratory. It is acceptable to utilize these instrument developments in the preparation of proposals.

Proposals (10 copies) should be directed to Dr. Homer E. Newell, Deputy Director, Space Flight Programs, NASA Headquarters, 1520 H Street N.W., Washington 25, D.C. Three copies of proposals also should be addressed to Dr. A. R. Hibbs, Chief, Division of Space Sciences, Jet Propulsion Laboratory, 4800 Oak Grove Drive, Pasadena 3, California.

Sincerely yours,

[signed: "Edgar M. Cortright for"]
Homer E. Newell
Deputy Director
Space Flight Programs

Enclosure

****

April 7, 1961

TENTATIVE MARINER B DESIGN CONCEPTS
FOR USE IN EXPERIMENT PROPOSAL PREPARATION

The following specifications for this mission are the result of a study conducted at JPL. These specifications are to be used for planning purposes and do not necessarily represent a final design of the spacecraft. Two possibilities have been considered. The first consists of a spacecraft with a planetary entry capsule to be separated near Mars. The design distance of closest approach of the spacecraft is 8,000 kilometers, at which time the velocity will be 5 kilometers per second in Martian coordinates. The capsule will land slightly north of the Mars equator, with descent to be accomplished by ablating materials until approximately Mach one velocity is reached, and subsequently by parachute. The parachute descent time is estimated at 10 to 15 minutes. Planetary experiments can be conducted from the capsule or bus. Data from the capsule will be relayed to the bus for retransmission to earth. During capsule descent, and for approximately 30 minutes after landing, the capsule-bus transmission rate is estimated at 2,000 bits/sec. Following a 2 1/2-hour eclipse, communications will be reestablished at 20-200 bits/sec for 10 hours. Data automation storage in the bus is estimated at $2x10^7$ bits with an acceptance rate into the storage unit of $2.4x10^3$ bits/sec from all experiments. In the Martian vicinity, communication to earth is estimated at 100-200 bits/see. This necessitates 10 days of subsequent transmission to empty the data automation system.

Weight allocation for experiments on the bus is estimated at 140-180 pounds, exclusive of power and data handling. Forty to fifty pounds is estimated for the capsule. Power estimate for interplanetary cruising is 15 watts for all experiments. During planetary encounter, the power estimate for the bus experiments is 25 watts. Capsule instrumentation power is estimated at 10 watts for 12 hours subsequent to parachute opening.

The above figures represent the most crucial design parameters. Abandonment of the capsule concept, if necessitated by design considerations, would modify the experimental arrangements. An increase of bus experiment instrumentation from 140-180 pounds to 240-280 pounds may occur if the capsule is not feasible. The trajectory would be approximately the same as in the capsule-bus configuration.

The above information has been included to give the experiment proposer some idea of the need on his part to seriously consider minimal weight, power, and communication so that these factors do not unduly influence the final selection of experiments.

Document II-10

Document title: Owen E. Maynard, Spacecraft Integration Branch, Manned Spacecraft Center, Langley Air Force Base, Virginia, "Comments on Mr. Frank Casey's visit to J.P.L. to discuss Ranger and follow-on programs which could provide information pertinent to Apollo missions," February 1, 1962.

Source: NASA Historical Reference Collection, History Office, NASA Headquarters, Washington, D.C.

Document II-11

Document title: Donald A. Beattie and Paul D. Lowman to Director, Advanced Manned Missions Program, NASA Headquarters, "Summary of Preliminary Ranger Photo Analyses," August 14, 1964.

Source: NASA Historical Reference Collection, History Office, NASA Headquarters, Washington, D.C.

In 1959, NASA began planning for its first lunar project, Project Ranger, which consisted of a series of probes targeted at crash landing on the Moon's surface. In addition to returning data about the Moon and its vicinity for purely scientific interest, NASA officials after 1961 anticipated that Ranger would provide environmental data valuable to the preparation of hardware for the manned Apollo missions scheduled for the end of the 1960s. Document II-10 is a memo written by a NASA engineer noting the ways in which Ranger probes could aid the Apollo missions. Document II-11 summarizes some of the early scientific results of Ranger 7, the first Ranger probe to successfully collect data on and near the Moon after six prior Ranger mission failures.

Document II-10

[no page number]

NASA – Manned Spacecraft Center
Langley AFB, Virginia
February 1, 1962

MEMORANDUM for Associate Director

Subject: Comments on Mr. Frank Casey's visit to J.P.L. to discuss Ranger and follow-on programs which could provide information pertinent to Apollo missions.

1. During a recent visit to J.P.L. at Pasadena, California, a group of NASA employees from Langley Research Center, Ames Research Center and Manned Spacecraft Center had an opportunity to discuss the Ranger program and its follow-on programs with the J.P.L. staff. The purpose of this meeting was to determine if the present series of Ranger payloads and the follow-on payloads could be of value to the Apollo mission.

2. Since both the time and experiments available for obtaining further engineering data for design of Apollo systems and components is [sic] limited when viewed in terms of the unknowns, the following question was posed within the NASA group as a basis criterion for the planning of payloads to obtain further information on environmental data for the Apollo program:

"What are the environmental parameters for which additional data must be obtained before the Apollo missions will be attempted?"

In consideration of three Apollo phases, this criterion leads to the following conclusions:

Apollo Phase A No further environmental data required.

Apollo Phase B	Possibly additional data on radiation and meteoroids in cislunar and lunar space.
Apollo Phase C	The above comments on radiation and meteoroids is [sic] appropriate. In addition, more definite data on both the large and small scale lunar surface features, the existance [sic] and nature of lunar surface dust, and the physical properties of the lunar surface which constitute its ability to support a vehicle.

3. It was recognized that the limits, accuracy and coverage of environmental data to better establish the physical nature of the lunar surface in terms of Apollo missions requirements are incomplete, and that further inputs on these are necessary from MSC to J.P.L. Cognizant personnel at MSC should be reminded of this need and attempts should be made to supply available information to plan instrumentation of Ranger follow-on payloads.

[2] 4. On the basis of current knowledge and thinking relative to the nature of the lunar surface environment, and the need for engineering data for the design of Apollo systems and sub-systems, it appears that the selection of Ranger follow-on payloads should be directed primarily on the ability of these payloads to yield data which would permit a better evaluation of:

a. The large scale features of the lunar surface such as the locations, magnitude, and slopes of mountains, craters, and protuberences [sic];

b. The existance [sic] and distribution of small scale features of the lunar surface such as roughness, slopes, faults, sharpness, and vesicularity which will aid in the evaluation of the extent to which the Apollo vehicle must be able to hover and translate prior to landing;

c. The existance [sic] of a dust layer on the lunar surface and the properties of this layer which will permit it to be entrained in the jet exhaust and form clouds which may foul systems components and obstruct optical and R.F. transmission from the vehicle to space and the earth;

d. The ability of the lunar surface to support the Apollo vehicle including the existance [sic] and bearing strength of dust layers in excess of six inches in depth and the bearing strength and hardness of sub-surface material.

Secondary consideration should be given to the measurement of meteoroid and radiation parameters.

5. In consideration of the difficulty associated with obtaining environmental information over a substantial portion of the lunar surface to the accuracy required by Apollo C missions, it would be extremely helpful in the selection of Ranger and follow-on experiments if MSC and J.P.L. could agree on the landing site. It is not possible to get Ranger

payloads over to the western limb of the moon where the sea of tranquility is located. This would allow the maximum Ranger payload weight to be used to advantage.

6. Since the design freeze date for Apollo occurs in 1964, it is imperative that lines of communication be established immediately if Apollo is to have an input from Ranger and follow-on programs in time to be used as design criterion.

7. J.P.L. is presently investigating the problems of conducting experiments to obtain direct design data for Apollo. They will investigate [3] launch vehicle capabilities to implement the investigations and report their findings to NASA Headquarters about February 8, 1962.

Owen E. Maynard
Spacecraft Integration Branch

OEM:amm
CWM
MAF

Document II-11

[no page number]

UNITED STATES GOVERNMENT
Memorandum

M-C MT 8122
Date: August 19, 1964
MT-1 (DAB:mls)

TO: MT/Director, Advanced Manned Missions Program

FROM: MT-1/Donald A. Beattie
SM/Paul D. Lowman

SUBJECT: Summary of preliminary Ranger photo analyses

1. All conclusions presented below are the results of very preliminary analysis, "instant science," and will undoubtedly be revised to some extent. The complete set of pictures was not studied at the time these opinions were voiced. However, it was stated that "in general no new or unexpected knowledge developed from the pictures." This is due in part to the fact that at one time or another probably all possible lunar models were hypothesized by the scientific community and

using these pictures proponents of the various models will be able to find evidence to at least partially support their theories.

2. The maria appear favorable for landing although they contain many hazards. Light colored zones on the maria classified as ray (ejecta) material are probably extremely rough containing a great number of secondary craters. Kuiper now thinks the cratered area seen on photo 373 was formed by Tycho ejecta; Shoemaker apparently still favors Copernicus as the source. Other sources such as Bullialdus should not be ruled out. No statements have been made on the highlands since close-ups of such regions were not obtained.

3. As Ranger approached the surface, the added detail increased the observed surface roughness. At a point where craters of 150 feet or less could be resolved, (photo 379 and lower), the surface began to appear smoother and the density of preserved craters less than 150 feet in diameter was less than expected. Shoemaker ascribes this to a surface supersaturated with small impacts, both primaries and secondaries, that tend to obliterate smaller features.

4. The nature of the surface material in which the craters were formed is still not agreed upon despite the Ranger pictures. The subdued appearances of the terrain and the gently rounded contours of the craters are probably the most controversial features shown by the pictures. There are two major schools of thought at the present time:

[2] (1) The craters represent the last formed features of an originally solid rock surface which has been thoroughly pulverized or buried by several meters of fine ejecta. This would account for the apparent absence of large ejecta blocks around the craters. The erosion of surface forms during this process would also account for the soft contours of the observed features.

(2) The craters were originally formed in unconsolidated material such as volcanic ash. This would explain the absence of large ejecta blocks since the original surface would then be rather soft. Whether this would also account for the subdued surface appearance is not clear and to our knowledge no experimental work has been done that would support or refute this model.

5. Shoemaker's group points out that except for the large blocky feature in the small crater in photo 381 very few blocks can be seen on the surface. Most of the small objects on this same photo that appear to be blocks are really noise. The crater with the "blocks" in the upper left corner of photo 381 is, however, another controversial feature. It is considered by some to indicate ejecta blocks within the crater formed by their impact. Some at the Astrogeology Branch think that this may be in part a slide block which has become detached from the crater rim. As supporting evidence they point to other craters, such as the crater to the right of the center cross in the same picture, which seem to have slumped along their rims. Undoubtedly the "volcanologists" will claim this indicates small scale volcanic action and is a small cinder cone or other similar feature.

It is of interest that to date there has been little activity from this group so we can expect that a well-thought-out argument supporting their views will probably be forthcoming.

6. From the views expressed thus far there appears to be general agreement that the areas photographed at low altitude by Ranger do not look like lava fields. There is no sign of flow structure, contacts, spatter cones, or flow front scarps. A possible lava field structure may be the mare ridges such as seen running north from the large crater in photo 372. Except for this possible exception, it can be concluded that if the maria are lava fields either their characteristic terrestrial topography does not develop in the lunar environment or has been completely masked by some sort of cover or destroyed by repeated impact.

7. The following tentative inferences may be drawn from the pictures concerning post-Apollo missions:

   (1) Greater mobility than presently provided for by Apollo will be necessary for even preliminary reconnaissance investigations. The surface appeared monotonously uniform with no indication of rock changes or structural complexity within walking range of virtually any point shown within the [3] mare. It is possible that if the photographed area is blanketed ejecta, several samples would give a good composite picture of a large area. However, ejecta samples will have decreasing scientific value after the first manned landing as there will be little chance of understanding their point of origin or the degree to which they are representative of any given area.

   (2) Surface traverses in wheeled vehicles appears practical. This conclusion is independent of the correctness of either school of thought expressed in paragraph 4 above. If it is volcanic ash, terrestrial experience indicates that it will be easily trafficable. If it is lava, it has been so thoroughly pulverized that the usual obstacles to surface travel, such as aa [sic] blocks, pressure ridges, flow fronts, and lava tunnels have been obliterated. The view advanced by Gold in 1955 that the maria might be extremely soft, electrostatically suspended dust, seems decisively contradicted by the apparent permanence of terrain features. Nevertheless, a thin layer of fine powdery material may still present some difficulty which should not be overlooked.

   (3) Excavation for road building, and base or instrument emplacement should be practicable for the same reasons mentioned in (2).

cc:
SM/W. Foster
SM/V. Fryklund
MT-1/W. Taylor
MTF/T. Evans

Document II-12

Document title: R. C. Peavey, Secretary, Space Science Board, to Space Science Board Members, April 20, 1962, with enclosure, "Scientific Program for the Apollo Mission," April 20, 1962.

Source: Hearings before the Committee on Aeronautical and Space Sciences, United States Senate, Eighty-eighth Congress, First Session, "Testimony of Scientists on Goals of the Nation's Space Program," June 10 and 11, 1963.

Document II-13

Document title: W. W. Kellogg, "Summary of Responses to Space Science Board Inquiry on Scientific Program for the Apollo Mission," circa May 1962.

Source: Hearings before the Committee on Aeronautical and Space Sciences, United States Senate, Eighty-eighth Congress, First Session, "Testimony of Scientists on Goals of the Nation's Space Program," June 10 and 11, 1963.

In 1962, NASA's Office of Manned Space Flight turned to the National Research Council's Space Science Board, an advisory group of space scientists from around the nation, for input in planning for Apollo's scientific program. The questions posed to the scientists sought their opinions on lunar landing sites, mission duration, mission objectives, mission timing, scientific experiments, mobility on the lunar surface, and the prospect of including scientists among the Apollo crews. Document II-13 conveyed the Space Science Board's vision that a scientist, indeed, should be part of the Apollo astronaut group starting with the very first crew. NASA did not fly a scientist on an Apollo mission until the final one, Apollo 17, in 1972.

Document II-12

[213] Attachment C

NATIONAL ACADEMY OF SCIENCES
2101 Constitution Avenue
Washington 25, D.C.

April 20, 1962

S P A C E S C I E N C E B O A R D

MEMORANDUM - SSB-264

TO: See Distribution List

FROM: R. C. Peavey, Secretary

SUBJECT: Scientific Program for the Apollo Mission

The Office of Manned Space Flight of the National Aeronautics and Space Administration has asked the Space Science Board for assistance in defining the scientific program for the Apollo mission (i.e., the 3-man flight to the Moon and return, established for the 1960 decade by President Kennedy as a primary goal in the nation's space research program).

Although the NASA has itself established an ad hoc Working Group on Apollo Scientific Experiments and Training, the NASA has particularly requested the assistance of the scientific community in the planning for this program. Consequently, the SSB Executive Committee regards this request for assistance as of critical importance.

Attached is a memorandum raising specific questions regarding scientific aspects of the Apollo mission. We should greatly appreciate receiving your initial recommendations on questions raised therein on or before May 7, 1962; replies received by this date will be considered by the SSB Executive Committee at its meeting on May 14. Your subsequent thoughtful consideration and further recommendations on all matters raised in this memorandum are also requested. Your recommendations now will also materially assist a more detailed consideration of these topics during the Space Science Summer Study of the Board in Iowa this summer.

Distribution

All SSB Members	T. Gold	E. M. Purcell
	H. H. Hess	N. F. Ramsey
H. Odishaw	P. Morrison	C. E. Sagan
A. H. Brown	N. Nelson	J. W. Senders
M. Calvin	A. Novick	H. C. Urey
G. de Vaucouleurs	F. Press	F. L. Whipple

****

[214]
SPACE SCIENCE BOARD
NATIONAL ACADEMY OF SCIENCES
2101 Constitution Avenue
Washington 25, D.C.

April 20, 1962

Scientific Program for the Apollo Mission

The Office of Manned Space Flight of the National Aeronautics and Space Administration has asked the Space Science Board to assist in defining the scientific program for the Apollo mission (i.e., the 3-man flight to the Moon and return, established for the 1960 decade by President Kennedy as a primary goal in the nation's space research program). In this connection we should like to have your thoughtful consideration and suggestions regarding the following questions:

(i) Are there preferred sites (or lunar areas) toward which the Apollo mission should be directed? If preferred landing sites can be specified, can priorities be established in terms of scientific preference? It should be recognized that selection of preferred lunar landing sites may in part also determine selection of landing areas and the scientific objectives for the unmanned lunar exploration program which precedes Apollo. Please give the arguments favoring preferred locations.

(ii) The Apollo mission is presently defined as a 7-day mission (from launch to return to Earth) with a maximum capability of a 14-day mission. What can be said about preferred duration of the stay on the Moon in terms of scientific objectives? What should be the primary scientific objectives of a brief (e.g., one hour) stay; a one-day visit; a prolonged stay (e. g., several days)? Please suggest primary research objectives separately depending upon the allowable period of stay. (Consideration should also be given here to the possibility of setting up scientific equipment which could continue to telemeter back useful scientific information for a prolonged period.)

(iii) In terms of basic research objectives, should the Apollo mission occur during the lunar day or the lunar night? Please give reasons for your suggestions.

(iv) What recommendations do you have regarding the scientific payload for the mission if it is possible to transport only 100 pounds of scientific equipment? 300 pounds? 500 pounds? 1000 pounds? 1500 pounds? 2000 pounds? The possibility of providing a lunar supply payload prior to the manned mission should not be overlooked.

(v) What requirements for mobility of man can be specified for the Apollo mission? Here mobility should be considered in two contexts: (a) What is the desirable range over the lunar surface which scientists should be expected to undertake? (b) To what extent is individual dexterity required to be built into the protective suit with which the scientist will be equipped?

[215] (vi) What are your recommendations regarding the scientist as a crew member in the Apollo mission? For example, what observations should he be asked to make en route to the Moon and on the return trip to the Earth? Should he be a fully trained astronaut also capable of navigating the spacecraft and, if so, how should this be expected to modify his duties as a scientist during flight?

Your replies to the foregoing questions should be given in the realization that scientific exploration of the Moon will be undertaken in progressive steps. Manned exploration will be preceded by Ranger and Surveyor scientific missions. The manned exploration program should be designed to undertake scientific investigations not capable of being accomplished by instruments alone. The first Apollo mission will be followed by others whose program of scientific investigations should be designed to capitalize on all preceding missions.

Document II-13

[232] Appendix I

Summary of Responses to Space Science Board Inquiry on Scientific Program for the Apollo Mission

Prepared by W. W. Kellogg

Some 14 replies to the SSB Questionnaire of April 20 are pertinent to the Summer Study discussion of Man as a Scientist in Space Exploration. In undertaking to summarize these responses, I have taken the liberty of reformulating the original questions so that they will apply more closely to the set of questions posed to the Summer Study, and then giving a composite answer drawn from these letters. This is a dangerous process, but it turns out that there was a good deal of unanimity in the attitudes expressed, making it possible to find a consistent story by piecing the various contributions together.

I have not covered the comments about where to land, whether the dark or light hemispheres would be better for scientific work, or the discussions of en route experiments. The summary relates entirely to the scientist who will some day set foot on the Moon with the terrible responsibility of reporting back to the world the characteristics of that strange world.

1. Should a scientist (at least one) be a member of the first Apollo crew?

   Emphatically, "Yes!" (See under No. 5 for justification.)

2. What should his scientific training be?

   Physics, chemistry, and field geology.

   More specifically, he should have studied and be familiar with the following subjects:

   Lunar observations, not only the traditional ones now done from the ground (telescopic, radiometric, radar, etc.), but also the latest results of unmanned landings.

   Laboratory studies of materials in a vacuum, under various conditions of irradiation, proton bombardment, etc.

Field geology, to gain intuitive insight into various types of formations (volcanic, impact), and to develop the ability to perceive important features of his surroundings and to describe them clearly and objectively. However, he should be cautioned from depending too much on terrestrial analogies when studying the Moon.

[233]

3. Should he also be trained as an astronaut?

Of course. He should be familiar with all aspects of the spacecraft and be able to take over in an emergency. However, his qualification as crew member would not depend so much on his ability as a space-pilot as on his scientific aptitude.

4. What should this scientist do on the Moon?

The number one job would be the acquisition of representative samples of lunar rock and dust for return to Earth. If time on the Moon is short, he will be able to do little else.

His next most important job will be to note carefully his surroundings, spot unusual features, make verbal and photographic records of his observations. He should be permitted to travel as freely as possible.

If time and space permit, he would install apparatus to measure, record, and telemeter back to the Earth such things as seismic activity (3-axis seismometer), gravity and tilt, magnetic field, temperature of sub-surface, radiation environment, etc. This part of the job could also be done by other members of the crew who were not scientists, if they were available; in fact, this would probably be preferable, leaving the scientist member more time for his special investigations.

It is not necessary to spell out in detail what he will do at this stage. One of the chief advantages of a man over a machine is that he does not have to be programmed far ahead of time, and his design is fixed. We will learn much from our first unmanned landings that will suggest what he will need to do.

5. Why should a scientist be included in the crew?

Because the chief justification for going to the Moon (beyond the simple realization of a "National Goal in Space") is to acquire knowledge about the Moon.

Because a trained scientist has the ability and the motivation to gather the most significant information available during his short stay in that strange environment.

Because, in particular, the gathering of rock and dust samples for return to Earth must be done with the greatest discrimination, and with a scientist to screen quickly samples for significant characteristics.

Because there may be unexpected emergencies, or unanticipated phenomena on the Moon that may endanger the expedition, and an alert and knowledgeable scientist in the crew might make the difference between success and failure.

6. How do we develop our astro-scientists?

They should be selected from scientists at the graduate student or early postdoctoral level. Perhaps four would be selected immediately, more later.

[234] (The maximum age that he could be at the time of the flight must be decided by the human factors experts.)

Their training scientifically should be in the areas listed in No. 2. They should demonstrate ability to do original research in at least some of these areas. Facility with scientific instruments and measurements will be essential. They should be able to relate quickly new ideas, and be good at order-of-magnitude calculations. Training should start now and last for at least four or five years.

They should go through astronaut training for part of each year to become familiar with problems of space flight. It is hoped that this would not involve too large a fraction of their time, since emphasis should be on their development as scientists.

Document II-14

Document title: H. H. Hess, Chairman, Space Science Board, to James E. Webb, Administrator, NASA, August 5, 1963, with enclosure, "Space Probe Sterilization," August 5, 1963.

Source: Historian's Source Files, Washington National Records Center, Suitland, Maryland.

Shortly after the launch of the first Earth-orbiting satellites, University of Wisconsin biologist Joshua Lederberg expressed concern to his colleagues interested in space biological research that organisms from Earth could ride aboard spacecraft bound for the Moon and planets, contaminate these targets, and thus destroy scientists' chances of discovering indigenous life forms. In 1959, the Space Science Board and the International Council of Scientific Unions studied the problem, and concluded that the United States and the Soviet Union should take measures to avoid contaminating the solar system with Earth organisms. As a result, NASA issued its own policy pledging to sterilize spacecraft before sending them toward celestial targets. Document II-14 reaffirms the Space Science Board's belief in the importance of space probe sterilization with specific regard to the Moon and Mars.

[no page number]

2101 CONSTITUTION AVENUE
WASHINGTON 25 D.C. 20418

CABLE ADDRESS: NARECO
WASHINGTON, D.C.

NATIONAL ACADEMY OF SCIENCES
NATIONAL RESEARCH COUNCIL
OF THE UNITED STATES OF AMERICA
SPACE SCIENCE BOARD

August 5, 1963

Mr. James E. Webb, Administrator
National Aeronautics and Space Administration
Washington, D.C. 20546

Dear Mr. Webb:

You will recall that during the course of the sessions in Iowa last summer a working group of both scientists and engineers considered the subject of space probe sterilization. This study resulted in an appreciation on the part of biologists for some of the engineering considerations and I think a measure of understanding by the engineers for the concerns of biologists.

Since the Summer Study, the Board has continued to consider the question and has concluded that a restatement of its views with regard to space probe sterilization is required. The enclosed paper reports the views of the Space Science Board on this topic and I believe will permit NASA to issue publicly a statement on space probe sterilization which, when implemented, can be supported by the biological community.

The extent of interest in this subject among scientists suggests that our recommendations be made known to them, as discussed with your staff.

Sincerely,

H. H. Hess
Chairman

enc.

cc: Dr. Seitz
Dr. Dryden
Dr. Newell
Dr. Odishaw

****

[1]
SPACE SCIENCE BOARD
National Academy of Sciences
2101 Constitution Avenue
Washington, D.C.

August 5, 1963

SPACE PROBE STERILIZATION

During 1959 a special ad hoc group of the Space Science Board considered the subject of contamination of probes likely to impact the Moon or the planets. These considerations resulted in the adoption of a two-part recommendation which was transmitted to the Government in September 1959:

> "(1) that an immediate study program be undertaken to determine sterilization requirements for space probes and to develop recommendations, compatible with present design and assembly processes, regarding necessary sterilization procedures;
> (2) that procedures be immediately established and implemented to insure a complete inventory of all components of all space probes."

These recommendations formed the basis for studies of space vehicle sterilization by the National Aeronautics and Space Administration and resulted in NASA's general policy of "sterilizing, to the extent technically feasible, all space probes intended to pass in the near vicinity of or impact on the Moon or planets."

During the course of the 1962 Space Science Summer Study a working group of scientists and engineers together reviewed the three years of experience and reported its findings in "A Review of Space Research" (NAS-NRC Publication 1079, Chapter 10). As a result of this report and with additional information available to it, the Space Science Board has adopted a restatement of policy with regard to (1) lunar probes and (2) Mars probes.

The Moon

The lunar surface with its high temperatures, intense ultraviolet radiation, paucity of moisture, and high vacuum is a most unfavorable environment for proliferation of terrestrial organisms. Although some forms could survive in protected places, they would be relatively immobile. Lunar subsurface conditions, in contrast, are relatively unknown. However, except at the site of impact, deep [2] subsurface contamination from a lunar landing appears highly unlikely. Even so, the lunar exploration programs to date, both U.S. and Russian, have undertaken to minimize contamination in order to avoid depositing terrestrial organisms on the Moon; both of the probes which have impacted on the Moon are believed to have carried only a relatively small number of microorganisms. Nevertheless the deposition of terrestrial contaminants (viable or not) over portions of the lunar surface seems nearly certain. While this introduction of organic substances of terrestrial origin into the lunar surface seems at present unavoidable, we believe it continues to be undesirable. Minimizing contamination by future lunar impactors remains an important consideration from the scientific viewpoint. The chief purposes here are to avoid possible distortion of chemical evidence (e.g., by microbial action) which may bear on conditions which preceded the evolution of life and to preserve the deep layers uncontaminated for subsurface life-detection experiments.

In view of these considerations the Space Science Board recommends that the following policy be considered for spacecraft programmed to land on the Moon:

(i) Minimize contamination to the extent technically feasible. By appropriate selection of components (favoring those which are inherently sterile internally) and the use of surface sterilants it should be possible to achieve a cleanliness level to approximate that which prevails in most hospital surgery rooms.

(ii) Inventory all organic chemical constituents. This will permit the interpretation of analytical results from future collections of lunar material.

(iii) Accord a low priority to life-detection experiments by remote devices on the lunar surface. A high priority should be attached to sampling the subsurface at points removed from the immediate vicinity of any landing site.

(iv) Undertake the development of a sterile drilling system to accompany an early Apollo mission to return an uncontaminated sample of the lunar subsoil. Samples aseptically collected from this subsoil will be of both biological and geochemical interest. Should life exist on the Moon, it might be expected at some depth below the surface where temperatures never exceed 100°C and below the zone of ultraviolet radiation. Every effort should be made to keep this level free of contaminants until it can be sampled by drilling.

[3] Mars

The planet Mars is by far the most probable extraterrestrial body in the solar system to be populated by forms of life. One of the most significant possible discoveries in space research, and perhaps even the most important, would be the finding of extraterrestrial life. Discovery of living organisms on Mars must depend on means of detection which could not be expected to distinguish between terrestrial contaminants and members of an indigenous Martian biota. Some terrestrial microorganisms are known to survive simulated Martian environmental conditions. Therefore the contamination of Mars through the impacting of nonsterile probes from the Earth could destroy an opportunity to carry out a meaningful search for life forms on Mars with remote detectors. This opportunity is unique and its loss would be a catastrophe: it is essential to preserve Mars until complete sterilization of the probes to land there has been achieved.

Moreover, should the initial life-detection experiments to be sent to Mars yield negative results, sterilization of Martian probes should not be abandoned automatically. There will remain scientific reasons for continuing to adhere rigidly to a policy of sterilization during the initial phases of sample collection from Martian surface and subsurface. If sterile, Mars will provide a unique opportunity to detect and analyze organic compounds of nonbiological origin in the Martian soil. Such studies of prebiological geochemistry, free from interference by living organisms, can supply important and otherwise not directly attainable information concerning the origins of life. Therefore contamination (introduction of viable terrestrial microorganisms) and pollution (introduction of significant amounts of terrestrial, albeit sterile, organic matter) are to be avoided until adequate soil sampling can be accomplished even if initial results from remote detectors suggest that Mars may have no biota.

In view of these considerations the Space Science Board recommends that the following policy be considered for spacecraft programmed to land on Mars:

(i) Accord the highest priority to the prevention of the biological contamination of Mars until sufficient information has been obtained about possible life forms there so that further scientific studies will not be jeopardized. Recognition of this priority on the part of launching nations is in accord with their main scientific objectives, in contrast to a competition to be first in which these objectives might be forever sacrificed.

(ii) Establish and provide adequate support for an augmented research program to develop agents, methods and [4] techniques for the sterilization of Martian probes. Such a research program should mobilize both biologists and engineers to insure [sic] successful development of practical sterilization procedures.

(iii) Inventory all organic chemical constituents. This is precautionary, but the lack of an inventory might make impossible the interpretation of analytical results from future collections of Martian material.

(iv) Cooperate fully with all other nations in the protection of Mars against premature biological contamination. The exchange of information and the possibility of a joint research project between scientists of the USSR and the U.S. should be explored.

(v) Strengthen the current research program for the development of the best possible life-detection experiments to insure the incorporation of a life-detection experiment in the first Mars lander. This is of extreme importance for otherwise we may succeed in the sterilization of Mars probes but fail to accomplish our true objective.

Document II-15

Document title: "Mariner-Venus 1962: Final Project Report," NASA SP-59, 1965.

Source: NASA Historical Reference Collection, History Office, NASA Headquarters, Washington, D.C.

NASA's objectives for Mariner R, a spacecraft combining features of the Ranger and the cancelled Mariner A, were to launch two spacecraft to the vicinity of Venus in 1962, maintain two-way com-

munication with the probes in flight, and obtain scientific data on the interplanetary medium and Venus. The first probe was destroyed after its launch vehicle deviated from the planned trajectory. On December 14, 1962, Mariner 2 became the first space probe to successfully fly past another planet, and return meaningful scientific data to Earth. This report details the mission's development and key scientific results.

[title page]

NASA SP-59

MARINER-VENUS 1962
FINAL PROJECT REPORT

Prepared under contract for NASA by
Jet Propulsion Laboratory
California Institute of Technology

Scientific and Technical Information Division 1965
NATIONAL AERONAUTICS AND SPACE ADMINISTRATION
Washington, D.C.

[v]

Foreword

The flight of Mariner II to the planet Venus represents the first time that man has sent his instruments close to another planet and received meaningful data from them back on Earth. The following pages describe the Mariner R project and its problems, solutions, difficulties, and successes. Perhaps a part of the drama and some of the tensions associated with this kind of mission will be felt and understood.

From the successful launching of Mariner until its arrival at the planet 109 days later, the spacecraft was under continuous observation as it transmitted data back to Earth. In the control room in Pasadena, men who designed and built the instruments hovered over its readings as anxiously as any parent. As the days grew into weeks and the distances into millions of miles, the tensions in the control room mounted steadily. Finally, the spacecraft flew past the planet and its instruments made the first close-up observation of Venus. Three weeks later, on January 3, 1963, with the spacecraft 9 million km (6 million miles) beyond Venus and 87 million km (54 million miles) from Earth, the signals stopped.

The journey of Mariner was a spectacular achievement of modern science and technology. It was made possible by the coordinated efforts of many people and many organizations. The National Aeronautics and Space Administration selected the Jet Propulsion Laboratory to manage the project. The spacecraft was designed and built by JPL with the assistance of numerous industrial subcontractors. The launch vehicle was the responsibility of the Marshall Space Flight Center of NASA and the launchings were conducted at the Atlantic Missile Range with the support of the United States Air Force. Tracking of the

Mariner was assigned to the Deep Space Network, operated by the Jet Propulsion Laboratory. The scientific experiments were selected from proposals by scientists from many universities.

Several thousand men and women had a direct part in the Mariner R project. It would be impossible to list all of those who made some special contribution in their specific task, but every member of the project performed his job accurately, on time, and in a superior manner. The failure of any one of the many thousands of components would have invalidated the mission. Design, manufacture, and testing all demanded the very highest standards to achieve the necessary reliability.

[vi] While Mariner is the first interplanetary spacecraft and has opened a new era of planetary exploration, it is only a beginning. Soon there will be spacecraft flying by other planets, then orbiting the planets, and finally landing instruments on their surfaces. Exploring the solar system, becoming acquainted with the planets, answering questions about extraterrestrial life—these are the challenges that lie ahead.

W. H. PICKERING,
Director, Jet Propulsion Laboratory,
California Institute of Technology.

[The first pages of Chapter 1 are omitted]

[8] SCIENTIFIC EXPERIMENTS IN MARINER II

The Mariner II spacecraft carried six scientific experiments representing the efforts of scientists at nine institutions: the Army Ordnance Missile Command, the California Institute of Technology, the Goddard Space Flight Center of NASA, Harvard College Observatory, the Jet Propulsion Laboratory, the Massachusetts Institute of Technology, the State University of Iowa, the State University of Nevada, and the University of California at Berkeley. Table 1-I lists the original experimenters and their affiliations.

The two primary planetary experiments were a microwave radiometer and an infrared radiometer. They were designed to operate during a period of about 45 minutes while the spacecraft passed the planet at distances ranging from approximately 12,874 to 64,372 km (8000 to 40,000 miles). These radiometers obtained information about the planet's temperature and the nature of its atmosphere.

The other four experiments made scientific measurements during the cruise through interplanetary space and in the near vicinity of Venus. They were: a magnetometer; charged-particle detectors, including an ionization chamber and several Geiger-Mueller counters; a cosmic dust detector; and a solar plasma detector. They also gathered data in the immediate vicinity of Venus.

One of the important considerations in choosing these experiments was the compromise between what scientists would like to measure during the mission and what was technologically possible. For example, of the 447 pounds that could be placed in a Venus trajectory with the available launch-vehicle thrust, only about 40 pounds could be allocated to scientific experiments. In addition, engineers and scientists designed Mariner II to convert electrical power from the sunlight, report its findings from as far as 57,934,800 km

(36,000,000 miles), and, although sensitive and unattended, remain in precise working order for 3 to 5 months in outer space. Another restricting factor was time. Venus is in a favorable position for the launching of a Mariner-type spacecraft only during a period of a few weeks every 19 months.

Several theories concerning the nature of Venus' atmosphere and surface [9] have already been advanced in this chapter. One of the missions of Mariner II was to make scientific measurements in the vicinity of the planet which might substantiate one of these theories, or call for the formulation of a new one.

Table 1-I.—Mariner experimenters

Experiment	*Description*	*Experimenters*
Microwave radiometer	Determine the temperature of the planet surface and details concerning its atmosphere.	Dr. A.H. Barrett, Massachusetts Institute of Technology; D.E. Jones, JPL; Dr. J. Copeland, Army Ordnance Missile Command and Ewen-Knight Corp.; Dr. A.E. Lilley, Harvard College Observatory.
Infrared radiometer	Determine the structure of the cloud layer and temperature distributions at cloud altitudes.	Dr. L.D. Kaplan, JPL and University of Nevada; Dr. G.Neugebauer, JPL; Dr. C. Sagan, University of California, Berkeley, and Harvard College Observatory.
Magnetometer	Measure planetary and interplanetary magnetic fields.	P.J. Coleman, NASA; Dr. L. Davis, Caltech; Dr. E.J. Smith, JPL; Dr. C.P. Sonett, NASA.
Ion chamber and matched Geiger-Mueller tubes	Measure high-energy cosmic radiation.	Dr. H.R. Anderson, JPL; Dr. H.V. Neher, Caltech.
Anton special-purpose tube	Measure lower radiation (especially near Venus).	Dr. J. Van Allen and L. Frank, State University of Iowa.
Cosmic dust detector	Measure the flux of cosmic dust particles in space.	W.M. Alexander, Goddard Space Flight Center, NASA.
Solar plasma spectrometer	Measure the spectrum of low-energy positively charged particles from the Sun.	M. Neugebauer and Dr. C.W. Snyder, JPL.

During the cruise and Venus encounter phase of the mission, Mariner II telemetered information to Earth. As the sensors of the six experiments received information, they fed it to a data conditioning system (DCS) which was located in one of the modules in the hexagonal base of the spacecraft. The DCS prepared information for transmission to Earth in the form of a digital code.

Since all of the data collected by Mariner II could not be transmitted at the same time, an electronic clock was built into the DCS. This clock controlled the equipment so that the receiver "listened" to one experiment at a time for about 1 second. After 20.16 seconds, the DCS switched off the scientific data and then the telemetry system sent spacecraft engineering data for 16.8 seconds. This cycle was continued during the cruise in interplanetary space. When the spacecraft was switched to the encounter mode, however, the spacecraft devoted its telemetry system to the full-time transmission of scientific information from its six experiments.

[11] CHAPTER 2
Project Organization and Management

EARLY MARINER PROJECTS

The Mariner A and B projects were intended to launch spacecraft in the 1250-pound class that were designed to make scientific investigations in inter-planetary space and in the vicinity of Venus and Mars, respectively, during the 1962-1964 launch opportunities. Both types of spacecraft were to be launched by a vehicle consisting of a modified Atlas D first stage, and a Centaur liquid-hydrogen/liquid-oxygen, high-energy second stage.

The Centaur vehicle, under development by General Dynamics /Astronautics at San Diego, Calif., had two gimbal-mounted engines, each capable of generating 15,000 pounds of thrust. Ten small hydrogen peroxide monopropellant engines were provided for attitude control, consolidation of main propellants, and final velocity correction.

The Mariner A configuration was scheduled to fly the NASA P-37 and P-38 missions to Venus in the summer of 1962 as a developmental spacecraft on Centaurs 7 and 8. However, slippages in the Centaur schedule began to compromise the Venus launches and the missions were forced into rescheduling.

By the second week of August 1961, it was generally recognized that the Centaur would not be available for the 1962 Venus launch period. Consequently, in mid-August, JPL discussions with NASA explored the possibility of using lightweight, attitude-stabilized spacecraft for the P-37 and P-38 missions, since it was considered most important that the United States launch probes to the planets in 1962 if at all possible.

On August 28, 1961, in a letter to NASA Headquarters, JPL proposed the feasibility of a 1962 Venus mission, based on an Atlas-Agena launch vehicle and the use of a hybrid spacecraft combining features of the Ranger and Mariner A designs. This proposed Mariner R spacecraft could carry 25 pounds of instruments (later increased to 40 pounds). Only one launch could be guaranteed, but two were possible within the July-September 1962 period if the Agena weight could be reduced. The project would not require significant changes in the Ranger schedule, but would necessitate the transfer of certain launch vehicles.

[12] In addition to the activation of a Mariner R project, JPL would proceed with the design and development of the Mariner B spacecraft, scheduled for launch by Atlas-Centaur with dual Mars-Venus capability in 1964 and beyond. Coincidental with the imple-

mentation of the Mariner R project and the shift of emphasis in Mariner B, the Mariner A project was to be canceled.

Accordingly, NASA authorized cancellation of Mariner A, activation of the Mariner R project, and establishment of the dual capability for the Centaur-based Mariner B in 1964.

MARINER R (1962) PROJECT

The purpose of the Mariner R (1962) project was to perform the National Aeronautics and Space Administration's P-37 and P-38 missions to Venus during the third-quarter launch opportunities in 1962.

The primary objective of the Mariner R (1962) project was to develop and launch two spacecraft to the near-vicinity of the planet Venus in 1962, to establish and maintain two-way communication with the spacecraft throughout the flight, to obtain interplanetary data in space and during the Venus encounter, and to perform scientific surveys of the planet's characteristics. The launch vehicle used in this project was to be the Atlas D-Agena B (fig. 2-1) [omitted], [13] permitting a spacecraft weight of approximately 460 pounds and including about 25 pounds of scientific instruments.

Two spacecraft were scheduled for launch in order to increase the probability that at least one would accomplish its objective. Launch schedules for the two probes were arranged to take maximum advantage of the limited 56-day launch period available from July 18 through September 12, 1962. By June 11, 1962, the firing dates had been established and both spacecraft were ready for launching. The minimum separation between the two launch dates was established as 21 days.

Since the time from the first consideration of the Mariner R mission to the initial launch date was less than a year (mid-August 1961 to mid July 1962), to achieve the objectives in the limited time available it was necessary to make decisions quickly, to "freeze" the design at the earliest feasible point, and to meet all schedule milestones on time.

An all-out effort was initiated to design, develop, and procure components, and to test and launch the two spacecraft in an 11-month period. The many associated activities, such as trajectory work, preparation for launch and flight operations, and design and fabrication of ground support equipment, were pursued on a "crash" basis, concurrently with a major launch-vehicle effort involving design and manufacturing changes.

To take advantage of experience already gained, use was made of existing Mariner A and Ranger hardware and procedures wherever possible. The resultant design, limited by the many time and weight restrictions, produced a spacecraft with little or no redundancy.

Project planning specified delivery of two spacecraft (Mariners R-1 and R-2) and one set of spares (later assembled and designated as Mariner R-3), two sets of system test complex equipment and one set of spares, and two sets of launch complex equipment in support of the spacecraft.

All major milestones were met on time, including arrival of equipment at Atlantic Missile Range and the subsequent launches on July 21 and August 27, 1962, respectively, for the P-37 and P-38 missions.

Because of a launch-vehicle deviation from the planned flight path, Mariner R-1 was destroyed by the range safety officer after approximately 290 seconds of flight. Measures taken to correct the difficulties experienced in this launch included a more rigorous checkout of the Atlas rate beacon and a revision of the [14] data-editing equation, designed as a precaution against acceptance of faulty data by the ground guidance equipment.

[15] The subsequent launching of Mariner R-2 on August 27, 1962 (fig. 2-2) [omitted] initiated a space flight in which the project objectives were met with a high degree of success. A vast quantity of valuable scientific and engineering data was telemetered to Earth from the spacecraft throughout its flight, up to the time of its final communication on January 3, 1963.

Because of the high quality and great amount of the data transmitted by Mariner II, it was decided to terminate activity on the Mariner R spacecraft schedule for the 1964 Venus attempt—a spacecraft which would have carried a nearly identical set of scientific experiments. The cancellation decision was made in January 1963, and all Mariner R effort after that time was directed toward termination of the project. Much of the work underway was directly or partially applicable to the Ranger or Mariner Mars 1964 programs, and appropriate transfer of effort in these areas was made. ...

[33] **SPACECRAFT DESCRIPTION**

As discussed earlier in this chapter, the Mariner R spacecraft (fig. 3-1) [omitted] utilized many of the design principles and techniques developed for the Ranger program. The basic structural unit of Mariner R was a hexagonal frame made of magnesium and aluminum to which was attached an aluminum superstructure, a liquid-propelled rocket engine for midcourse trajectory correction, six rectangular chassis mounted one on each face of the hexagonal structure, a high-gain directional antenna, the Sun sensors, and gas jets for control of the spacecraft's attitude. The spacecraft configuration is shown in figures 3-2 and 3-3 [omitted].

The tubular, truss-type superstructure extended upward from the base hexagon. It provided support for the solar panels while latched under the shroud during the launch phase, and for the radiometers, the magnetometer, and the omnidirectional antenna, which was mounted at the top of the structure. The superstructure was designed to be as light as possible, yet be capable of withstanding the predicted load stresses. The six magnesium chassis mounted to the base hexagon housed the following equipment: The electronics circuits for the six scientific experiments; the communications system electronics; the data encoder and the command electronics; the attitude control and CC&S circuits; a power control and battery charger assembly; and the battery assembly.

The Mariner R spacecraft was self-sufficient in power. It converted Sun radiation into electrical energy through the use of solar panels composed of photo-electric cells which charged a battery installed in one of the six chassis on the hexagonal base. The control, switching, and regulating circuits were housed in another of the chassis cases. The battery operated the various spacecraft subsystems during the period from launch until the solar panels were faced into the Sun. In addition, the battery supplied power during trajectory maneuvers when the panels were temporarily out of sight of the Sun, and shared the demand for power when the panels were overloaded. The battery furnished power directly for switching various equipment in flight and for certain other heavy loads of brief duration, such as the detonation of explosive devices for releasing the solar panels. The

Mariner R battery used sealed silver-zinc cells and had a capacity [35] of 1000 watts/hr. It weighed 33 pounds and was recharged in flight by the solar panels. The two solar panels, as originally designed, were each 60 in. long by 30 in. wide and each panel contained about 4900 cells, or approximately 9800 solar cells in a total area of 27 sq ft. Each solar cell produced only about 230 one thousandths of a volt. The entire array was designed to convert the Sun's energy to electrical power in the range between 148 and 222 watts. When a later design change required a 2.5-sq-ft extension of one panel in order to add about 910 more solar cells, it was necessary to add an extension (Dacron impregnated with silicone rubber) to the other panel in order to balance the solar pressure on the spacecraft. In order to protect the solar cells from the infrared and ultraviolet radiation of the Sun, which would produce heat but no electrical energy, each cell was shielded from these rays by a glass filter that was transparent to the light which the cells converted into power. The power subsystem electronics circuits were housed in another of the hexagonal chassis cases. This equipment was designed to receive and switch power either from the solar panels, the battery, or a combination of the two, to a booster-regulator.

Mariner R was stabilized in space by the attitude-control subsystem. The roll axis was pointed at the Sun, providing stability about the pitch and yaw axes. Roll stability was achieved by keeping the Earth sensor, mounted on the directional antenna, pointing at Earth, in order to maintain continuity of communications. Pointing the roll or longitudinal axis at the Sun allowed the maximum amount of solar energy to strike the solar panels and aided the thermal control of the spacecraft by maintaining the Sun at a constant known attitude relative to the spacecraft.

The beam width of the high-gain antenna was 16.3° at half-power and, consequently, the antenna had to be pointed at Earth. This requirement was used to roll-stabilize the spacecraft, thus providing a stabilized platform for the science experiments. The Sun and Earth acquisitions were achieved through a series of sensors, gyros, and internal logic circuits which caused actuation of cold-gas valves. Expulsion of gas in preferential directions provided desired rates about the various axes to bring the spacecraft into the desired stable attitude.

The central computer and sequencer (CC&S) subsystem supplied timing, sequencing, and computational services for other subsystems of the Mariner R spacecraft. All events of the spacecraft were implemented in three distinct sequences or "modes": (1) the launch sequence controlled events which occurred during the launch phase; (2) the propulsion sequence controlled the events necessary to perform the midcourse maneuver; (3) the encounter sequence included [37] all CC&S commands required in the vicinity of Venus. A highly accurate electronic clock (crystal-controlled oscillator) scheduled the operations of the spacecraft subsystems. The oscillator frequency of 307.2 kc was reduced to the 2400- and 400-cps output required for the power subsystem. The control clock also timed the issuance of commands by the CC&S in each of the three operating modes of the spacecraft.

Mariner R used a technique for modulating its radio carrier with telemetry data known as phase-shift keying. In this system, the coded signals from the telemetry measurements displaced another signal of the same frequency but of a different phase. These displacements in phase were received on Earth and [38] then translated back into the codes, which indicate the voltage, temperature, intensity, or other values measured by the spacecraft telemetry sensors or scientific instruments. A continually repeating code was used for synchronizing the ground receiver decoder with the spacecraft. The decoder then deciphered the data carried on the information channel.

This technique was called a two-channel, binary-coded, pseudo-noise communication system and was used to modulate a radio signal for transmission.

Radio command signals transmitted to Mariner R were decoded in a command subsystem, processed, and routed to the proper using devices. The subsystem was used to receive the commands, send back confirmation of receipt to the Earth, and distribute them to the spacecraft subsystems.

Mariner R used four antennas in its communication system. A conelike nondirectional (omni) antenna was mounted at the top of the spacecraft superstructure and was used from injection into the Venus flight trajectory until Earth acquisition and during the midcourse maneuver (the directional antenna could not be used until it had been oriented on the Earth). A dish-type, high-gain, directional antenna was used following Earth orientation, and after the trajectory correction maneuver was completed. The directional antenna was located beneath the hexagonal frame of the spacecraft while it was in the nosecone shroud. Following the unfolding of the solar panels, it was swung into operating position, although it was not used until after the spacecraft locked onto the Sun and the Earth. The directional antenna was equipped with flexible coaxial cables and a rotary joint. It was moved in two directions: one motion was supplied by rolling the spacecraft around its long axis. In addition, two command antennas, one on either side of one of the solar panels, received radio commands from the Earth and were used for measuring spacecraft velocity and annular position in the two-way Doppler mode.

The Mariner R propulsion subsystem for midcourse trajectory correction employed a rocket engine that weighed 37 pounds with propellant and a nitrogen pressure system, and developed 50 pounds of thrust. The system was suspended within the central portion of the basic hexagonal structure of the spacecraft, with the thrust axis parallel to the roll axis of the spacecraft. The rocket engine used a type of liquid propellant known as anhydrous hydrazine and it was so controlled that it could burn from as little as 0.2 of a second to a maximum of 57 seconds, and increase the velocity of the spacecraft from as little as 0.7 ft/sec to as much as 200 ft/sec. The hydrazine was stored in a rubber bladder inside a doorknob-shaped container. At the ignition command, nitrogen gas under a [39] pressure of 3000 lb/sq in. was forced into the propellant tank through explosively activated valves. The nitrogen then squeezed the rubber bladder, forcing the hydrazine into the combustion chamber. (Hydrazine, a monopropellant, requires ignition starting for proper combustion.) In the Mariner subsystem, nitrogen tetroxide starting or "kindling" fluid was injected into the propellant tank by a pressurized cartridge. Aluminum oxide pellets in the tank acted as catalysts to control the speed of combustion of the hydrazine. The burning of the hydrazine was stopped when the flow of nitrogen gas was halted, by explosively activated valves.

The spacecraft's temperature control system was made as thermally self–sufficient as possible. Paint patterns, aluminum sheet, thin gold plating, and polished aluminum surfaces reflected and absorbed the amounts of heat necessary to keep the spacecraft and its subsystems at the proper temperatures. Thermal shields were used to protect the basic hexagonal components. The upper shield, constructed of aluminized plastic on a fiberglass panel, protected the top of the basic structure and was designed for maximum immunity to ultraviolet radiation. The lower shield was installed below the hexagon: it was made of aluminum plastic faced with aluminum foil where it was exposed to the blast of the mid-course rocket-engine exhaust.

The six electronics cases on the hexagon structure were variously treated, depending upon the power dissipation of the components contained in each. Those of high power were coated with a good radiating surface of white paint; assemblies of low power were provided with polished aluminum shields to minimize the heat loss. The case housing the attitude control and CC&S electronics circuits was particularly sensitive because the critical units might fail above 130° F. A special assembly was mounted on the face of this case; it consisted of eight movable, polished aluminum louvers, each activated by a coiled, temperature-sensitive bimetallic element. When the temperature rose, the elements acted as springs and opened the louvers. A drop in temperature would close them.

Structures and bracket assemblies external to the basic hexagon were gold plated if made of magnesium, or polished if aluminum. Thus protected, these items became poor thermal radiators as well as poor solar absorbers, making them relatively immune to solar radiation. External cabling was wrapped in aluminized plastic to produce a similar effect. The solar panels were painted on the shaded side for maximum radiation control properties. Other items were designed so that the internal surfaces were as efficient radiators as possible, thus conserving the spacecraft heat balance. ...

[327] **VENUS MEASUREMENTS**

During the Venus encounter phase of the Mariner II flight, the response of the 19-mm microwave radiometer indicated roughly equal temperatures on the light and dark sides of the planet; three scans (dark side, terminator, light side) indicated temperatures of 460°, 570°, and 400° K, respectively. Limb-darkening, observed with both microwave and infrared radiometers, is consistent with the high temperatures originating deep in the atmosphere or at the surface of the planet. In both the 8μ and 10μ channels of the infrared instrument, the central radiation temperature was of the order of 240° K. The apparent equality suggested that there was little carbon dioxide absorption in the light path, implying that the measured temperatures were those of thick clouds. Again, light- and dark-side temperatures were qualitatively the same. A region 10° K cooler than the rest was indicated on the southern part of the terminator scan. Near Venus, there was no indication of a magnetic field or of appreciable change in the solar plasma flux or the charged particle flux.

These data are consistent with the premise that Venus has no magnetic field. The charged-particle data, interpreted as indicating that the Venus magnetosphere did not extend out to the spacecraft orbit, suggest that its dipole moment [328] is less than 18% of that of the Earth. The magnetic field and solar plasma data, interpreted as showing that the spacecraft did not penetrate the transition region outside the Venus magnetosphere, indicate that its dipole moment is less than 10% of that of the Earth. No cosmic dust particles were detected, suggesting that their density near Venus is less than 2×10^{4} of their density near Earth.

Microwave Radiometer

Earth-based measurements of the radio emission of Venus have indicated that the planet's temperature is approximately 600° K for wavelengths in excess of 3 cm. This temperature may be contrasted with infrared measurements of Venus which yield values somewhat less than half those obtained by radio. The radio data, which are critical to our understanding of the Venusian environment, rest on terrestrial observations which suffer from lack of spatial resolution and insufficient precision. Flyby planetary probes offer the possibility of precision and resolution with modest radiometers. Accordingly, the Mariner II spacecraft was instrumented with a two-channel microwave radiometer operating at wavelengths of 13.5 and 19.0 mm.

The pertinent equipment performance parameters are given in table 8-II. The effective antenna gain was calibrated by using a black disk of known temperature, whose angular size was designed to be approximately the size of Venus [329] at encounter. This calibration was performed on Table Mountain near Wrightwood, Calif., in March 1962.

Table 8-II.—Microwave radiometer characteristics [328]

Parameter	Channel	
	1	2
Center wavelength, mm	19	13.5
Center frequency, Gc/sec	15.8	22.2
Predetection bandwidth, Gc/sec	1.5	2.0
Sensitivity, rms, °K	15	15
Calibration signals, °K	1500	800
Time constant, sec	40	40
Beamwidth, deg	2.5	2.2
Side lobes, db	-23	-23
Reference frequency, cps	950	1050

During the 109-day flight, 23 noise calibrations were made; thus, the gain, base-level, and time-constant performance of the radiometers could be monitored en route.

The radiometers were energized, and the antenna scan motion was activated about 6 1/2 hours before encounter. The scan motion had an angular extent of 123.5° and a nominal scan rate of 0.1 deg/sec. The microwave radiometer first made contact with the planet Venus at 18:59 GMT (spacecraft time) on December 14, 1962. During the next 35 min, three scans across the planetary disk were obtained, as follows:

Scan	Approx. angular extent, deg	Alt. at mid-scan, km	Location
1	10	40,200	Dark side.
2	15	37,750	Near terminator.
3	10	35,850	Light side.

Telemetered digital data points constituted the basic data, which had to be corrected for a number of effects before they could be considered as yielding the microwave temperature distribution across the planet. Among these corrections were the more important effects of the post-detection time constant and a detailed consideration of the antenna pattern.

The noise tube calibrations obtained en route to Venus made it possible to determine the in-flight time constant and gain of the radiometers. The gain of both channels decreased during the cruise, and the zero levels had systematic variations. These effects were more serious in the 13.5-mm radiometer.

Preliminary estimates of the peak-brightness temperatures of the three scans were: Scan 1 (dark side), 460° K; scan 2 (near terminator), 570° K; scan 3 (light side), 400° K. The temperatures are based on calculations which account for the effects of the antenna beam and the postdetection time constant. The errors of the quoted temperatures are estimated to be 15%. The analysis of the preliminary results suggests that there is no significant difference in the microwave temperatures on the light and dark sides of the planet. The results suggest a limb-darkening, an effect which represents cooler temperatures near the edge of the planetary disk. The ionosphere model of the Venus atmosphere, which [330] permits Earth-like temperatures, appears to be ruled out by these observations. On the other hand, the observed limb-darkening is consistent with a model of the Venusian environment which has high temperatures originating deep in the atmosphere or at the surface of the planet.

Thus, Mariner II found an unquestionable limb-darkening and also found that there is little difference in temperature on the dark side compared with the sunlit side of the planet. On the basis of the radiometer scans, the surface of Venus, where the 19-mm radiation originates, appears to have a temperature of about 400° K.

Infrared Radiometer

The infrared radiometer which was flown on Mariner II in conjunction with the microwave radiometer was designed to measure, with high geographical resolution, the infrared radiation from Venus in two wavelength regions. One of these was centered on the 10.4 carbon dioxide band, while the other was selected to correspond to an infrared window centered at 8.4μ. The infrared radiometer was mounted upon and boresighted with the microwave radiometer described in the preceding section. Both instruments, therefore, executed the same scan pattern caused by the combined effects of the probe motion and a rotation of the radiometers in a plane normal to the probe-Sun line. From the three scans of the planet, five pairs of radiation temperatures were obtained on the dark side, five on the sunlit side, and eight along the terminator.

The radiometer was calibrated at the Jet Propulsion Laboratory by using two cylindrical blackbodies; one was maintained at liquid nitrogen temperature, while the other was varied over the expected planetary temperature range. In addition, a one-point check was obtained during encounter by causing the radiometer to view a plate, located on the spacecraft structure, whose temperature was independently measured.

The data are consistent with an equality of the 8μ and 10μ radiation temperatures. This apparent equality would indicate that there was little carbon dioxide absorption in the light path. The implications are that the measured temperatures were cloud temperatures, that the clouds were quite thick, and that essentially no radiation was transmitted from the surface.

A definite limb-darkening was observed in both spectral channels; the radiation temperatures showed a monotonic decrease of approximately 20° K between the central region and the limbs. Central radiation temperatures are [331] estimated to be on the order of 240° K. The data do not show any clear-cut evidence of asymmetry in the limb-darkening, except for an anomaly on the southern part of the terminator scan. In particular, the light- and dark-side temperatures were qualitatively the same. The anomaly was about 10° K cooler than expected on the basis of symmetrical limb-darkening. One obvious interpretation of this temperature anomaly is that the clouds were locally higher, or more opaque, or both. ...

[336] MASS OF VENUS AND OTHER SOLAR CONSTANTS

The orbit of the Mariner II spacecraft is unique in that it was dominated first by the Earth, then the Sun, and finally the planet Venus. On December 14, 1962, the spacecraft came within about 35,000 km (22,000 miles) of Venus, and its flight path was deflected by about 40° because of its close encounter. In addition, precise two-way Doppler data were obtained throughout the 129-day period from launch to January 3, 1963. Fortunately, prior to the launch of Mariner II an atomic reference had been installed at the Goldstone tracking station of the Deep Space Instrumentation Facility, and this piece of equipment allowed the transmitter frequency to be held to better than one part in 10^{11} over a period of an hour. Equivalently, the Doppler shift in the received signal was measured to an accuracy of about 0.03 cps. In terms of velocity units, the corresponding accuracy of the range rate between the probe and station was on the order of 0.5 cm/sec at a received frequency of 960 Mc and a cycle count time of 60 sec.

A combination of the aforementioned factors allows the determination of certain constants of the solar system to an accuracy that has been unobtainable in the past. In particular, at least an order of magnitude improvement in the mass of Venus appears possible because of the close approach to the planet. The same order of improvement should be obtained for the mass of the Moon because the periodic component in the data which results from the motion of the Earth about the center of mass or barycenter of the Earth-Moon system is appreciable. The astronomical unit can also be obtained from the data, and its accuracy should eventually be comparable to that of the recent NASA/JPL radar determinations from the tracking of Venus (149,591,412 ± 482 km).

[337] Correlations of the mass of Venus with the position of the probe are fairly high. However, inaccuracy in the calculation of the trajectory near Venus is a serious matter, and the value of the mass determined with a Venus-centered integration could easily fall outside the probable error as computed from the normal equations associated with this solution.

The full scientific value of the Mariner II tracking data will not be realized until the heliocentric and encounter data are combined in one least-squares reduction. This is impossible at the present time because: (1) the low-thrust forces are neglected, (2) the calculation of the trajectory is inaccurate in the vicinity of Venus, and (3) the effects of uncertainties in the ephemerides of the Earth and Venus are unknown. It is unacceptable to be satisfied with the results without a detailed investigation of these three sources of error. It is expected that the final reduction will be accomplished with the inclusion of a physically reasonable low-thrust model, a Venus-centered integration of the equations of motion during encounter, and an inclusion of orbital elements of the Earth and Venus as additional free parameters in the solution.

Pending final reduction of data, a preliminary calculation of the mass of Venus is 0.81485 that of Earth, with an error probability of 0.015%. Since the Earth's mass is known to be approximately 5.977×10^{24} kg, Venus' mass becomes approximately 4.870×10^{24} kg.

---

[1] Mariner R-1 and R-2 are spacecraft serial numbers, which are used in all prelaunch references and until injection into planetary transfer trajectory. Thereafter, the references become Mariner I and Mariner II.

Document II-16

Document title: California Institute of Technology, "Suggestions for Martian Exploration Following Mariner IV," February 23, 1965.

Source: NASA Historical Reference Collection, History Office, NASA Headquarters, Washington, D.C.

Document II-17

Document title: Homer E. Newell, Associate Administrator for Space Science and Applications, NASA, to Lee A. DuBridge, President, California Institute of Technology, August 16, 1965.

Source: NASA Historical Reference Collection, History Office, NASA Headquarters, Washington, D.C.

Document II-18

Document title: National Aeronautics and Space Administration, "Project Approval

Document (Voyager Project)," January 27, 1967.

Source: NASA Historical Reference Collection, History Office, NASA Headquarters, Washington, D.C.

Even before NASA launched its first probe to Mars, Mariner 4, in 1964, agency officials and scientists had already begun to think about more advanced Mars mission concepts. In the early 1960s, JPL had proposed Voyager, a program of orbiting and landing spacecraft that would exceed the scientific capabilities of the Mariner probes in performing investigations of Mars and Venus. NASA approved a Mars-only Voyager program for the 1970s, as detailed in Document II-18. Documents II-16 and II-17 reveal the beliefs of NASA and JPL that the search for past or present life on Mars should be Voyager's priority. These documents also show that although JPL urged NASA to complete a series of flights prior to Voyager to provide environmental data on Mars and to test advanced spacecraft, Headquarters officials instead cancelled plans for Mariner missions to Mars in 1966 and 1969 to ensure funding for the ambitious Voyager program. Despite NASA's intentions to conduct a grand Mars program, budget pressures in other NASA programs weighed on Voyager and ultimately led the agency to cancel the program in 1967.

Document II-16

[cover page]

SUGGESTIONS
FOR MARTIAN EXPLORATION
FOLLOWING MARINER IV

*
*

California Institute of Technology
Pasadena, California

February 23, 1965

[1] SUGGESTIONS FOR MARTIAN EXPLORATION FOLLOWING MARINER IV

Introduction

The step from Mariner IV to Voyager represents an increase of an order of magnitude in payload and cost per planetary opportunity and a probable gap of seven to nine years in the return of significant new data. In addition, the design of landers and, to a lesser extent of orbiters, is extremely sensitive to the nature of the Martian atmosphere in whose description large uncertainties exist at present. The sensitivity is such that any initial design must be a compromise. Only when more information and more precise data are available can efficient designs of both spacecraft and experiments be undertaken. The recognition of the vast complexity and cost of the exploration of Mars is evidenced by the present interest in preliminary funding for a Voyager class program to be initiated in the early 1970s.

Accordingly, a critical question arises concerning the nature of the transition program covering the next few flight opportunities, regardless of the scale of the program undertaken. Numerous considerations indicate that an intermediate Mars program will lead to a greatly increased scientific benefit for a given expenditure on a Voyager program; indeed, the cumulative cost to resolve the question of life on Mars may well be less if the program includes well-designed preliminary investigations.

One rational intermediate step would appear to be a limited flight program based on the Atlas-Centaur booster system, whose payload is about three times that of the Atlas Agena used with Mariner, combined with a thorough ground- and balloon-based telescope program in certain areas.

Further discussion of these points is given below, with emphasis placed on engineering, scientific and earth-based observational considerations in turn.

Engineering Considerations

When it became apparent that the Atlas-Centaur launch vehicle would not be [2] available to send a Mariner B spacecraft towards Mars in 1962, serious consideration was given to whether an Atlas-Agena mission should even be attempted in 1964 in view of the necessarily limited scientific capability available with the Mariner C spacecraft. It is clear now that the decision to go ahead with a minimal fly-by mission was indeed a very wise one whether or not Mariner IV actually returns significant scientific data on Mars. The engineering design, testing, and flight experience of Mariners II and IV provide a firm engineering basis for the development and deployment of a considerably more complex fly-by bus of the Voyager class. However, the biological exploration of Mars also requires the deployment of sophisticated landers and probably orbiters under far more difficult operational conditions than those involved in lunar exploration. Early development and deployment of a simple capsule that survives impact on Mars can be expected to result in greatly improved Voyager lander performance. The history of the whole space program demonstrates the necessity of avoiding excessively large steps in progressing from one stage of the program to the next.

In addition, present uncertainties in knowledge of the Martian surface and atmospheric parameters clearly inhibit efficient scientific utilization of Voyager class payloads. For example, in one design of a lander utilizing two-stage parachutes and a small retrorocket the useful payload landed is almost halved if allowance must be made for maximum wind velocities of 80 mph (for 30 mb surface pressure). As a contrasting example, another design utilizing a crushable material and no parachutes or retrorockets is relatively insensitive to winds up to 100 mph, but, like all such landers, the payload is quite dependent on surface pressure, being twice as great at 40 mb as at 20 mb. Thus, due to our present ignorance of Mars, a lander must be designed for environmental extremes far greater than probably really will be encountered there.

There is an inverse relationship between uncertainty in critical environmental parameters and available scientific payload. In the case of an unsterilized orbiter, such [3] as might be desirable for mapping seasonal variations, our knowledge of the temperature and composition of the Martian exosphere is so poor that current estimates of scale height differ by factors of 5 or more. Because the sterilization requirement may be translated into the requirement of a 50 year lifetime, the periapsis height must be several thousand miles above the surface to insure [sic] an adequate orbiter life. Such an orbit is an order of magnitude too far away from the planetary surface for efficient mapping.

Unfortunately, neither the upper atmospheric density nor the surface wind velocities, among other parameters, appear to be determinable from the earth in the next 5 to 10 years to an accuracy required for design purposes. Accordingly, efficient Voyager landers and orbiters must be preceded by sterilized capsules and orbiters. Because of the long lead times associated with the design of spacecraft and experiments, it would appear that a landing capsule designed to investigate critical data at the Martian surface should be flown at as early a date as is technically feasible, even though the early capsule experience gained at one planetary encounter cannot be incorporated effectively in the following opposition. However, the atmospheric drag observed on a simple sterilized orbiter in 1971, for instance, could probably determine the lowest safe altitude at which an unsterilized Voyager orbiter could be deployed in flights in 1973 and later.

Scientific Considerations

The problem of biological exploration breaks down into two questions: (1) the positive determination of whether or not life exists (or existed) on the planet, and (2) the investigation of any such life in detail, particularly with regard to its chemical nature, but also in respect to its physiology, morphology, genetics, etc. The second task may be so difficult, in our opinion, that the eventual return of samples to the earth may be required in order to obtain complete answers to all the questions we would have about Martian life. In any case, it is unlikely that we shall make much progress on it in the 1969-1971 period. The design objective of the intermediate program, as well as [4] of the initial stages of the Voyager program proper, should be to detect Martian life, if it exists, and to determine enough about the surface environment to permit the efficient design of the more complex systems that would be required to investigate such life. It is essential that the program of Voyager life detection experiments be of sufficient credibility that cumulative negative results will indicate the absence of life on the planet with a reasonably high degree of confidence.

The information obtained from early experiments primarily directed toward the efficient design of later spacecraft and biological experiments will be, of course, of scientific value in its own right since it will answer many of the questions relating to the planet and its environment.

It is for these reasons we feel it is most important to place high priority on the carrying out, at an early date, of technologically feasible and highly significant observations directly related to biological exploration. Additional information could also be obtained without detracting from the main effort. For example, substantial increases in our knowledge of the mass, radius, geometric figure and magnetic field, which are of paramount importance in understanding the internal constitution of Mars and its relation to the other planets, very probably could be acquired by 1969 from an intermediate program with little effect on the primary biological mission of such a program. Other potential experiments which may place constraints on the mission such as photography, infrared spectroscopy and radiometry, mass spectrometry of the atmosphere, etc., must be evaluated on their contribution to the overall biological exploration problem, but still with some weight based on their relative importance to non-biological questions.

An Atlas-Centaur system, for example, appears capable of providing the necessary data in time (1969) to be of significance to the later Voyager programs. In addition it would appear feasible to fly a life-detection experiment in a simple capsule in 1971, so that the possibility of common and easy-to-find life could be investigated in a timely [5] way, leaving to Voyager the real challenge of the definitive life detection mission. Finally, in carrying out these necessary preliminary steps for biological exploration, a successful intermediate program could reap a rich harvest of the geological, geophysical, and meteorological exploration of Mars, rich enough in all probability to have justified such a flight program in the first place.

It seems to us that the proposed pioneering flights could be carried out by systems of at least the payload capability of Atlas-Centaur both to avoid the necessity of "wasting" to some extent early Voyager payloads, and to acquire the necessary information as soon as possible. In particular, we suggest consideration of Atlas-Centaur fly-by plus capsule missions in 1969 and 1971 as a first priority and Atlas-Centaur fly-by plus a very simple sterilized orbiter mission as early as funds permit (perhaps 1971). Pioneering flights could, of course, be carried out by larger vehicles; we have suggested the Atlas-Centaur system as representing a lower limit in payload capability.

Duplication of flights is necessary to increase the probability of successful return of data.

Ground-Based Observations

Ground-based optical and radio observations and balloon-borne infrared observations, combined with certain laboratory and field studies, can provide some of the environmental data needed for Voyager engineering as well as make major scientific contributions during the 1965, 1967 and 1969 oppositions. For instance, the surface pressure of the Martian atmosphere should be known to within perhaps 25% after the 1965 opposition from ground-based spectroscopic observations in the photographic infrared combined with improved laboratory investigation of CO_2 absorption. If the Mariner IV occultation experiment is successful, an independent check may exist of similar precision. Accordingly, surface pressure probably will be known well enough, and soon enough, so as not to represent a major obstacle to efficient design of either intermediate or Voyager landers.

[6] Ground-based and, particularly, balloon observations can produce considerably improved infrared spectra from either the entire planet or from large (1000 km diameter) areas of it over the coming years. Indeed, the spectral reconnaissance of the planet in the infrared clearly should be carried out in this way, leaving to future fly-bys and orbiters the task of following up any interesting spectral features on a higher geographical resolution. However, there is very little terrestrial experience in the interpretation of infrared reflection and emission spectra from naturally-occurring soils and it seems doubtful that even good spectra of the planet would lead to significant and widely-accepted composi-

tional interpretations at present, so that supporting research involving the earth should be encouraged. Thus, we assign a low priority to most kinds of infrared spectroscopy and interferometry experiments considered for use on probes until such time as a good spectral reconnaissance has been completed from the earth and a considerable background of supporting studies has been carried out under field as well as laboratory conditions. However, certain kinds of probe-experiments utilizing known H_2O or CO_2 infrared spectral features in order to map geographic variations in those gases (presumably associated with differing surface conditions) may be of value. We anticipate no advances of sufficient magnitude in high resolution photography from the ground or balloons to reduce in any way the extreme importance of probe-acquired photography for both scientific and engineering needs.

It is technologically feasible for ground-based radar studies of Mars to reach by 1969 the level of sensitivity presently utilized so successfully with Venus. Under such conditions many planetary characteristics would be determined including the degree of surface roughness, ionospheric electron densities, and a radius value to perhaps ±4 km. However, the most important possibility of all, from the point of view of aid to Voyager capability, would be the capability of discerning the presence or absence of, and seasonal variations in, soil moisture in the equatorial regions of the planet. Such information would be of the utmost significance to both the design and deployment of life [7] detection experiments.

Furthermore, it is also technologically feasible to utilize passive radio techniques, including interferometry, to investigate the presence of radiation belts and to make measurements from which surface parameters such as dielectric constants, thermal conductivities and densities can be determined to a depth on the order of a meter. Radar polarization measurements in combination with passive radio observations may measure the Martian magnetic field and the orientation of the magnetic poles.

Experiment Design

For the proposed intermediate program the design of suitable experiments would have to be initiated soon. Since only the crudest information on the Martian environment is available for use in design, those particular experiments in any branch of the investigation which depend least on the precision of present knowledge appear to be most suitable and should be encouraged. This is especially true of biological experiments, which tend to be extremely dependent on environmental assumptions at present.

Conclusions

A. Voyager Program
   1. The primary objective of the Voyager Program should be the determination of the presence of life on Mars and the nature of that life; cumulative negative results from the Voyager life detection program must likewise be indicative, with a reasonably high degree of confidence, of the absence of life on Mars at present. The secondary objective should be the acquisition of general scientific knowledge of the planet, a goal of major scientific importance in itself.
   2. Certain environmental data which are critical to the efficient design and deployment of landers and orbiters as well as to the scientific experiments themselves must be obtained at the planet.

3. If this information is not acquired in time to guide the design of the Voyager spacecraft and associated experiments, those spacecraft will be significantly limited [8] in their scientific capability. In such a case, both the total time and accumulated costs required to reach a reasonably confident answer regarding life on Mars can be expected to be greater than would have been the case had the necessary data been available in time.
4. We recommend consideration of an intermediate flight program to obtain needed environmental data concerning Mars and to ensure early and efficient scientific exploration of that planet.

B. Intermediate Flights
1. With launches timed for the 1969 opposition, the Atlas-Centaur system appears capable of projecting payloads which are adequate to collect and return needed information about the Martian environment. Together with the additional geological, geophysical and meteorological data that could also be obtained, these results would probably be of such significance as to justify the costs of the intermediate program apart from subsequent events. Early emphasis should be placed on simple experiments, particularly of a biological nature, which make the fewest assumptions about the nature of the Martian environment.
2. An intermediate flight program would permit the step-by-step development and testing of more advanced spacecraft and scientific experiments on a sound base of known data.
3. The following approach appears to be one suitable evolutionary program:

1969
a. Fly-by plus Simple Sterilized Impact Capsule
Capsule aimed at collection of surface environmental data pertinent to design of Voyager lander and of life detection experiments. Fly-by bus to carry out photographic and other reconnaissance. Either two or three launches of identical systems to be made to increase probability of success.

[9] 1971
a. Fly-by plus Simple Sterilized Impact Capsule
Same system as in 1969 except for the inclusion in the capsule of a simple life detection experiment and other equipment to collect additional environmental data of biological significance.
b. Fly-by plus Sterilized Orbiting Test Body
Orbiter containing a minimal payload primarily for the determination of exospheric density in order to establish the minimum acceptable altitude for a 1973 or later Voyager orbiter.

C. Observations From the Ground and Balloons

Certain ground-based programs are of special significance to the effectiveness of Voyager and warrant particular attention and encouragement by NASA. In particular, serious encouragement should be given to the increase of sensitivity of suitable ground-based radar systems to be operated as planetary radars with Mars in 1969. Infrared spectral reconnaissance of Mars from balloons during the 1967 and 1969 oppositions should be stimulated by NASA as well as an intensive study of infrared emission and reflection spectra from natural terrestrial soils and from simulated Martian surfaces.

Document II-17

[no page number]

[NASA emblem] NATIONAL AERONAUTICS AND SPACE ADMINISTRATION
WASHINGTON. D.C. 20546

IN REPLY REFER TO: SL (DPH:mem)

[stamped "AUG 16 1965"]

Dr. Lee A. DuBridge
President
California Institute of Technology
Pasadena, California

Dear Dr. DuBridge:

We have reviewed in detail the CIT report entitled "Suggestions for Martian Exploration Following Mariner IV," dated February 23, 1965. It is the purpose of this letter to amplify Mr. Nicks' letter of March 11, 1965, and our discussions of May 4, 1965, and June 14, 1965, relative to this report. The broader aspects of planetary exploration are covered in a separate letter.

The initiation of the Voyager Program indicates NASA's intent to accelerate the pace of the planetary program. The Voyager Program is to provide the means for the scientific exploration of the planets. Initially, the emphasis is being placed on Mars where the prime interest is the question of extraterrestrial life. However, it is our intention to investigate the planet's surface, atmosphere, and body characteristics, as well as the planetary environment. Thus, the exploration of Mars by the Voyager Program will involve many different scientific disciplines and will be as thorough an exploration effort as possible.

With this general objective in mind, it is important that the Voyager Program be planned as an evolving long-term program which will include flights during several opportunities. Thus, later missions will be based upon information from earlier missions. As the program evolves, significant changes in the scientific experiments are anticipated. These changes will result from scientific results from earlier missions, advances in scientific instrumentation, and increases in spacecraft and capsule capability.

[2] The landed capsule will probably vary significantly during the Voyager Program. We anticipate that information concerning atmospheric and surface characteristics gathered

in early missions will permit the efficient design and operation of large capsules and their experiments during later missions. Nevertheless, we are exploring the possibility of utilizing common hardware (such as the aerodynamic heat shield and structure) for several opportunities in order to conduct a more cost-effective program. In the case of the spacecraft bus, a reasonably common design can be used for all missions. While some changes will be desirable, it is our intention that the basic design will not be significantly modified. Consequently, the bus will be designed for flexibility; i.e., the ability to transport and support one or more capsules of various sizes to Mars, to accommodate significant changes in scientific instruments, and the ability to fly at many opportunities. In addition, it is felt that the bus should have orbiting capability. Along this line, current design studies indicate that orbiting capability must be designed into a spacecraft bus at the outset. Attempting to modify a flyby bus into an orbiter appears to be extremely difficult and a new development may be required.

It appears obvious that landing scientific instruments on the Martian surface is an absolute requirement for the primary objective (biology) and also for other important scientific objectives. It also appears obvious that a thorough scientific exploration program requires remote observation from Martian orbit. The need for Martian orbiters has been expressed by several representatives of the scientific community; for example, Dr. Donald F. Hornig, Special Assistant to the President for Science and Technology, and the Space Science Board of the National Academy of Sciences. Thus, it is our position that a long-term scientific exploration of Mars will require both lander and orbiter missions.

As you know, the Voyager Program is being conducted in phases. The initial phase, project and mission definition, will permit thorough definition of the spacecraft and mission characteristics in terms of mission profiles, spacecraft design and technological requirements, resource requirements, etc. It is fully expected that such an effort will provide the information required for a national commitment to the hardware and operations phase of the program.

[3] Based upon the general comments noted above, the following sequence of Voyager missions is being used in the definition phase:

1968/1969 ....Earth-entry flight tests of capsule models
1969 ....Bus flight tests
1971 ....Operational missions: Orbiter + survivable capsule
1973 ....Operational missions: Orbiter + survivable capsule
After 1973..........Operational missions at each opportunity with emphasis on "large" survivable capsules

It must be recognized that this mission sequence is the "Current Plan" and is being used as the basis for current project definition efforts. As noted below, various mission alternatives are possible and are under evaluation. However, I wish to emphasize that the mission sequence noted above appears to be the proper one, at this time, based upon all factors.

Two basic questions arise relative to the mission sequence noted above:

1. Should the bus be designed for orbiting capability and should such missions be performed in 1971 and 1973?

2. Are non-survivable atmospheric probes required to permit the design of reasonably efficient survivable capsules, if such capsules are "relatively simple?" If atmospheric probes are required, should the 1969 test flights be modified to be operational missions including such probes?

The answer to the first question will be strongly influenced by the need for designing orbiting capability into the bus from the very beginning. As noted above, this currently appears necessary to prevent the separate development of an orbiter at a later date. Whether or not orbiting missions (rather than flyby missions) are conducted in 1971 [4] and 1973 can be answered at a later date if the decision is made to design the bus as an orbiter. It is our current view, as supported by a large segment of the scientific community, that the bus should be designed as an orbiter and orbiting missions should be conducted in 1971 and 1973.

The second question is more difficult to answer. It will depend upon the Martian atmospheric uncertainty that remains following analysis of recent ground-based observations and following analysis of the data from the Mariner IV encounter. The effect of this uncertainty on survivable capsule design will then indicate the need for non-survivable atmospheric probe missions <u>prior</u> to the design of survivable capsules. Other factors, such as resources, schedules, and impact on later missions, particularly the large survivable capsules planned for 1975, must also be considered.

Our views on the specific conclusions of the February 23 report are as follows:

A. <u>Voyager Program</u>

1. We agree that the initial objective of the Voyager Program should be the determination of the presence of life on Mars, and the nature of that life if it exists. On the other hand, the biological objective should not assume a position of such importance that exploration in other scientific disciplines is overly compromised.

2. We agree that certain environmental data (which in itself is of scientific importance) is important to the efficient design of future missions. Certain of these data can only be obtained by placing scientific payloads on the surface of the planet and in orbit about the planet.

   If reasonably efficient orbiters and landers can be designed based upon knowledge obtained from Earth and from Mariner IV, and if the technologies developed in order to accomplish these initial missions contribute to future missions, then such initial missions should and will be included in the Voyager Program.

[5] 3. As noted above, it is recognized that certain environmental data can be critical fo[r] future missions. On the other hand, the implication that the initial spacecraft will be significantly limited in their scientific capabilities is an overstatement. In laying out the Voyager program, careful consideration is being given to our current knowledge of the Martian atmosphere and surface char-

acteristics. It is our intention to design the most efficient spacecraft for the initial missions possible within the limitations of our knowledge of the planet. It is not our intention to attempt to design the most sophisticated landed mission possible within the weight constraints associated with the selected launch vehicle until we have direct measurements on surface winds, topography, soil characteristics, etc. We are taking this approach since we agree on the possible pitfalls of being overly ambitious on the initial missions.

B. Intermediate Flights

During the past year, very careful consideration was given to the possibility of spacecraft flights using the Atlas/Centaur during the 1966 and 1969 opportunities. After carefully considering the planetary unknowns required to design a reasonably efficient, initial Voyager configuration and the total resources needed to conduct both programs, it was concluded that Mariner flights in the Atlas/Centaur class would dilute the effort that could be applied to the Voyager Program without a large enough payoff. It is felt that the Voyager Program itself would provide much of the same information. Admittedly, the bypassing of one or, possibly, two opportunities resulted.

The design of the spacecraft bus is not significantly influenced by the Martian unknowns. Additional information on radiation fields, micrometeorites, etc., would permit the bus design to be more highly optimized. Obviously its use as an orbiter is strongly influenced by the upper atmosphere of Mars. On the other hand, this does not influence the design of the spacecraft bus itself. Thus, such a bus can be used in either a flyby mode or as an orbiter selecting an [6] orbit based upon our best knowledge of the atmosphere at the time. Thus, as more is learned about the atmosphere, it may be possible to select orbits for the bus which may be closer to the planet.

The biggest effect of the Martian unknowns, of course, is in the design of the landed capsule. We are analyzing the results of the recent ground-based program and the Mariner IV occultation experiment to determine if these types of investigations have narrowed the uncertainty sufficiently well that a reasonably efficient landed capsule can be designed for the first opportunity. We are most anxious to conduct landed capsule missions to Mars since we recognize the scientific value of this type of mission. In addition, such missions will provide important data on the surface environment and characteristics. Thus, early landed capsule missions are required to obtain those data that will be required to design large, sophisticated, and expensive landed capsules and their experiments. If we find that the design of the initial landing capsules is overly compromised by the unknowns in the Martian atmosphere, some type of non-survivable atmospheric probe may be required early in the program. It is our intention to modify the program to include this type of experiment at the earliest possible opportunity if such an approach is absolutely required.

C. Observations from the Ground and Balloons

We agree with the conclusion that an effective and efficient planetary exploration program should include a significant ground-based and balloon activity to augment and supply information to the space flight activity.

During the Martian opposition of 1965, an extensive effort was made to make maximum utilization of available ground-based facilities to study Mars. A commendable effort was made to obtain further information on the Martian atmosphere, in particular, the surface pressure and on a direct photographic and visual patrol of the planet. Knowledge of the surface features at the time of Mariner IV encounter will assist in the interpretation of the Mariner IV photographs.

[7] Balloon observations of the planet during this opposition do not exist primarily due to the high failure rate that has plagued balloon astronomy. Continued efforts in the field of balloon astronomy will make the balloon a more useful tool to be used during the coming oppositions. These efforts are being sponsored both by NASA and other agencies.

Additional radar studies are being encouraged at the limited number of radar facilities capable of studying the planets.

The inadequacy of ground-based optical telescopes for planetary observations was evident in preparing for this opposition. To help overcome this difficulty in later oppositions, NASA is sponsoring the construction of a few telescopes designed and located principally for planetary observations. These instruments, along with others that are not NASA sponsored, should be ready in time to make optimum use of the next oppositions.

In summary, the Voyager Program has been configured as an evolutionary long-range program to obtain the maximum amount of scientific data at each opportunity while obtaining those environmental data which are required for the design of future Voyager missions. We consider exobiology of Mars as the initial objective of the Voyager Program. However, we do not intend to overly compromise our ability to conduct experiments in other scientific disciplines in order to emphasize biology. The recommendations concerned with intermediate flights between Mariner IV and Voyager were seriously considered by the NASA. As noted above, the Voyager Program is being planned to obtain those data required in its early missions for the design of spacecraft for the later missions. We feel that this approach presents the best opportunity to the scientific community for the exploration of Mars consistent with reasonable engineering design of spacecraft and reasonable resources.

Sincerely yours,

[signature]
E. M. Cortright
[for] Homer E. Newell
Associate Administrator for
Space Science & Applications

Document II-18

[1]

NATIONAL AERONAUTICS AND SPACE ADMINISTRATION
PROJECT APPROVAL DOCUMENT

January 27, 1967

RESEARCH AND DEVELOPMENT

Code Number: 84-840-818

Program: Lunar and Planetary Exploration

Project: Voyager Project

Objectives:

a. The Primary objective of the Voyager missions to Mars, beginning in 1973, is to obtain information relevant to the existence and nature of extraterrestrial life, the atmospheric, surface, and body characteristics of the planet, and the planetary environment by performing automated experiments on the surface of and in orbit about the planet.

b. A secondary objective is to further our knowledge of the interplanetary medium between the planets Earth and Mars by obtaining scientific and engineering measurements while the spacecraft is in transit.

Technical Plan:

a. Design, construct, test, and launch planetary vehicles in the 20,000 pound class to be utilized for unmanned scientific exploration of Mars, and to be launched by the Saturn V launch vehicle. Two planetary vehicles will be launched by a single Saturn V during both the 1973 and 1975 Mars opportunities. The planetary vehicle includes the spacecraft, including an orbiting bus module and propulsion module, and the capsule, including surface experiments. Necessary steps will be taken to insure that the requirements to prevent contamination of the planet will be met. All launches will be from AFETR.

[2]

b. The scientific payloads will be selected at a time consistent with the schedules for the 1973 and

1975 missions. In addition, advanced development of experiment systems (including the Voyager Biological Laboratory) for future missions will be undertaken.

Reliability and Quality Assurance:

The Jet Propulsion Laboratory and participating NASA Centers will establish and manage Reliability and Quality Assurance Programs (in-house and/or at the prime and subcontractors as appropriate) to the degree necessary to satisfy the overall mission requirements. Applicable provisions from NASA quality publications in the NPC 200 series will be invoked contractually in procurements. The basic elements of the reliability program will be compatible with applicable provisions of NPC 250-1. Status of these programs will be reported either separately or as part of periodic progress reports.

Management:

a. Program Management: Office of Space Science and Applications

b. System Management as follows:

(1) Spacecraft System Management: Marshall Space Flight Center

(2) Capsule System Management: Jet Propulsion Laboratory and Langley Research Center (Interface to be determined during Phase B)

(3) Tracking and Data Acquisition System Management: Jet Propulsion Laboratory

(4) Launch Vehicle System Management: Marshall Space Flight Center

[3]

Management Information and Control Systems:

a. NASA-PERT and the NASA Financial Management Reporting System for cost-type contracts will be used. Reporting categories shall be selected in a manner which will permit integrated time-cost management control and reporting. Reporting categories shall be consistent with Voyager Work Breakdown Structure.

b. Financial reporting will be against the code numbers prescribed in the Agency-Wide Coding Structure.

c. Reporting will include the monthly OSSA Project Management Report (MICS).

Procurement:

a. The procurement effort for the planetary vehicle will be handled as two separate procurement actions through Phase B. One effort will cover the Spacecraft System; the other effort will cover the Capsule System.

b. Phase B procurement of both the Spacecraft and Capsule Systems will be by the Jet Propulsion Laboratory. The Marshall Space Flight Center will review the results of the Spacecraft Phase B in preparation for the Phase C RFP. The Capsule Phase B will be conducted by the Jet Propulsion Laboratory working closely with the Langley Research Center.

SUBMITTED BY:

[signed "Homer E. Newell"]
Associate Administrator for
Space Science and Applications

[4] The plan as described above is approved in principle and for execution through the following:

Phase B of the Spacecraft System and Phase B of the Capsule System

Continuing design and supporting development of the Spacecraft System, Capsule System, and potential scientific payloads (including the Voyager Biological Laboratory) prior to the issuance of RFPs for Phase C

Continuing mission design

APPROVED:

[signed "Robert C. Seamans"]
Deputy Administrator
January 27, 1967

Document II-19

Document title: G. A. Flandro, "Fast Reconnaissance Missions to the Outer Solar System Utilizing Energy Derived from the Gravitational Field of Jupiter," April 18, 1966.

Source: *Astronautica Acta*, Vol. 12, No. 4 (April 18, 1966): 329–37. Reprinted with permission.

Document II-20

Document title: Dr. Michael A. Minovitch to Roger Launius, Chief Historian, NASA History Office, November 7, 1997.

Source: Dr. Michael A. Minovitch, Phaser Telepropulsion Inc., Los Angeles, California.

Until the 1960s rocket scientists relied on chemical engines to boost launch vehicles and their payloads toward and through space. Although adequate for near-Earth space exploration, chemical rocket propulsion was incapable of taking any but very small spacecraft to the outer solar system on a reasonable time schedule. While working at JPL in the summer of 1961, Michael Minovitch, a University of California, Los Angeles, graduate student, developed a method of propelling spacecraft to distant planets by sending them past intermediate planets and leveraging the gravitational interactions. That year Minovitch published his work in a JPL publication. Document II-19 contains excerpts from a paper by California Institute of Technology graduate student Gary Flandro that built on Minovitch's work by calculating gravity-propelled trajectories to the outer planets. In Document II-20 Minovitch clarified to NASA Historian Roger Launius nearly 40 years later his key role in the invention of this method.

NASA planned to use gravity-assisted trajectories as the basis for the Grand Tour, a mission to send spacecraft past all of the outer planets when they were to be aligned in the late 1970s. NASA eventually descoped this mission and renamed it Voyager. Gravity-assisted trajectories have also been used for the Galileo mission to Jupiter and the Cassini mission to Saturn.

Document II-19

[329]

California Institute of Technology, Pasadena, Calif., U.S.A.

Fast Reconnaissance Missions to the Outer Solar System Utilizing Energy Derived from the Gravitational Field of Jupiter

By

G.A. Flandro

With 16 Figures

(Received April 18, 1966)

Abstract

Fast Reconnaissance Missions to the Outer Solar System Utilizing Energy Derived from the Gravitational Field of Jupiter. Contrary to popular belief, indirect ballistic trajectories involving close approach to one or more intermediate planets need not require longer flight duration than is characteristic of direct transfer orbits. In fact, significant reduction of both required flight time and launch energy results if efficient use is made of the energy which can be gained during a midcourse planetary encounter. From the point of view of a passing space vehicle, the intermediate planet appears as a field of force moving relative to the inertial heliocentric coordinate system. Thus, work is done on the spacecraft, and its heliocentric energy may be increased or decreased depending upon the geometric details of the encounter. This paper describes the application of energy derived in this fashion, utilizing gravity perturbations from Jupiter, for reduction of required launch energy and flight duration for exploratory missions to all of the outer planets of the solar system. The latter half of the next decade abounds in interesting multiple planet apportunities [sic] due to the similar heliocentric longitudes of the major planets during this time period. Trajectories to Saturn, Uranus, Neptune, and Pluto using the midcourse energy boost from Jupiter are best initiated in the years 1978, 1979, 1979, and 1977 respectively. Flight time reduction ranges from one half the required direct trajectory duration for Earth-Jupiter-Saturn missions to as much as 85% of the direct transfer time for Pluto flights via Jupiter. Many multiple-target trajectories are also possible. Of particular interest is the 1978 Earth-Jupiter-Saturn-Uranus-Neptune "grand tour" opportunity which would make possible close-up observation of all planets of the outer solar system (with the exception of Pluto) in a single flight. ...

[330]

1. Introduction

Of crucial importance in the study of the origin, evolution, and structure of the solar system is the acquisition of close-up scientific data from the major planets (Jupiter, Saturn, Uranus, and Neptune) and Pluto. However, as indicated in Table 1, direct trajectories to these bodies are characterized by high launch energy and very long flight duration. At least the latter of these two factors must be reduced if practical exploration of the outer solar system is to be accomplished. A very attractive source of energy which can be tapped to bring about this reduction is the gravitational perturbation of an intermediate planet. The gravitational perturbation technique for trajectory shaping has been under intensive study recently [1-5]. However, due to the nature of the missions investigated, a widely held misconception has arisen to the effect that indirect multiple-planet trajectories in general require greater flight time than direct transfers to the same target bodies with the same launch energy. It will be shown here that significant reduction in flight duration results if efficient use is made of the energy which can be gained during a midcourse planetary encounter.

The latter half of the next decade abounds in interesting multiple-planet missions utilizing massive gravitational perturbations of Jupiter. Minovitch [2] has studied deep-space, out-of-ecliptic, and close solar probe trajectories via Jupiter. The goal of the present study was the determination of optimum launch opportunities and corresponding trajectory characteristics for flights to the outer planets of the solar system using energy gained during close approach to Jupiter. Three-dimensional conic computer programs were employed, this procedure having been verified by comparison of conic and integrated trajectory results [1, 3]. The most interesting mission possibilities and the corresponding launch years are summarized in Table 2. More detailed descriptions of these missions are presented later. ...

Table 1. *Characteristics of Direct Minimum Energy Trajectories to the Outer Planets*

Mission	Minimum Launch Energy, C_3 (km^2/sec^2)	Flight Duration, T (Years)
Earth—Jupiter	86.5	2.5
Earth—Saturn	108.8	6.1
Earth—Uranus	126.1	16.0
Earth—Neptune	135.0	30.7
Earth—Pluto	135.3	45.7

Table 2. *Multiple-Planet Trajectories to the Outer Solar System*

Mission	Launch Years
Earth-Jupiter-Saturn-Escape	1976, 1977, 1978*
Earth-Jupiter-Uranus-Escape	1977, 1978, 1979*, 1980, 1981
Earth-Jupiter-Neptune-Escape	1977, 1978, 1979*, 1980, 1981
Earth-Jupiter-Pluto-Escape	1975, 1976, 1977*, 1978, 1979
Earth-Jupiter-Saturn-Uranus-Neptune	1976, 1977*, 1978

* Optimum launch year

[337]

4. Conclusion

The 1975-1980 time period is characterized by an abundance of interesting multiple planet trajectories which efficiently utilize energy derived from a close approach to the planet Jupiter. The trajectories discussed here are characterized by very short flight times in comparison to those for direct flights from Earth to the corresponding target planets. Although higher launch energies are suggested for some of the multiple-planet flights, the additional expense of this energy might be offset by the great savings afforded by the short flight times. This is due to the expense of providing adequate vehicle reliability for the extended flight duration characteristics of direct trajectories, and to the high costs involved in maintaining tracking, orbit determination, and other flight related activities for protracted periods.

The great communications distances involved in outer solar system flights give rise to some difficulties with regard to antenna size and positioning, transmitter power requirements and so on. Signal propagation times are of course very long (about 4 hours one way from Neptune), and real-time control of the spacecraft, especially during the critical encounter sequences, would have to be relegated to automatic onboard control devices. An interesting discussion of long-distance communications problems is given by Kirsten [6].

The very important problem of guidance was not considered in the present study, but it is expected that development of planetary approach guidance techniques [7] coupled with improved Earth-based radio guidance should make the missions discussed herein entirely feasible. The large boost vehicles which should be available in highly developed form by that time should enable the spacecraft to accommodate the large supply of fuel required for necessary midcourse trajectory corrections, as well as a significant instrument payload.

Acknowledgments

The author is grateful to Messrs. E. Cutting, F. Sturms, and R. Richard for several interesting discussions, and to A. Joseph, D. Snyder, and Mrs. H. Ling for their assistance in computer programming and trajectory calculations.

References

1. Minovitch, M.A., "Determination and characteristics of ballistic interplanetary trajectories under the influence of multiple planetary attractions," Technical Report No. 32–464, Jet Propulsion Laboratory, Pasadena, Calif. (October 31, 1963).
2. Minovitch, M.A., "Utilizing large planetary perturbations for the design of deep space, solar probe, and out-of-ecliptic trajectories," Technical Report No. 32–849, Jet Propulsion Laboratory, Pasadena, Calif. (December 15, 1965).
3. Cutting, E. and Sturms, F.M., Jr., "Trajectory analysis of a 1970 mission to Mercury via a close encounter with Venus," AIAA Paper No. 65–90, AIAA 2nd Aerospace Sciences Meeting, New York, N.Y. (January 25–27, 1965).
4. Ross, S., "Trajectory design for planetary mission analysis," AAS Preprint No. 65–130, AAS/AAAS Special Astronautics Symposium (December 29, 1965).
5. Niehoff, J.C., Private Communication. ITT Research Institute, Chicago, Illinois (1965).
6. Kirsten, C.C., "Data acquisition at planetary ranges," *Proceedings of the NASA-University Conference on the Science and Technology of Space Exploration,* Vol. 1 (NASA SP-11, 1962).

7. Gates, C.R. and Gordon, H.J., "Planetary approach guidance," *J. Spacecraft and Rockets* 2, 160–166 (March-April 1965).

G.A. Flandro
Karman Laboratory of Fluid Mechanics and Jet Propulsion
California Institute of Technology
Pasadena, Calif. 91109, U.S.A.

(On leave of absence from Jet Propulsion Laboratory,
California Institute of Technology)

---

[1] This paper presents the results of one phase of research carried out at the Jet Propulsion Laboratory, California Institute of Technology, under contract No. NAS 7-100, sponsored by the National Aeronautics and Space Administration.

Document II-20

[On Phaser Telepropulsion Inc. letterhead]

[1]
Roger Launius
NASA History Office
Mail Code ZH
Washington, DC 20546-0001

November 7, 1997

Dear Mr. Launius:

As the Director of NASA's Historical [History] Office, I'm sure that you would like to know the details of the invention that made it possible to explore the entire Solar System with instrumented spacecraft, but was kept from NASA by erroneous explanations for over three decades. I believe that I, as the inventor, have a responsibility to give you this history, so that it can be made known, preserved, and made available to the American people as part of their rightful heritage.

A close examination of almost every book on the history of space travel reveals that there is essentially only one method for achieving interplanetary space travel, namely reaction propulsion generated by expelling mass at high velocity. This was the method proposed by all of the early pioneers, e.g., Tsiolkovsky, Goddard, and Oberth. However, it was recognized early on in studying the technical feasibility of interplanetary space travel that the velocity requirements for exploring most of the Solar System were so high that they were well beyond the reach of chemical rocket propulsion. This was due to the fact that the exhaust velocities of chemical rocket engines were limited to rather low values that could not be increased because of fundamental thermodynamic reasons. Thus, all of these early pioneers (and all of the theoreticians and propulsion engineers that followed them into the 1960s) believed that the only way that most of the Solar System could be explored was by developing advanced propulsion systems such as nuclear or electric systems. This was viewed as a mathematical certainty resulting from the "rocket equation" upon which the entire theory of reaction propulsion, and hence the technical feasibility of space travel rested. Reaction propulsion was taken for granted as the only practical method for propelling a space vehicle. However, after years of effort, these advanced high-specific impulse propulsion systems were found to be beyond engineering feasibility. Thus, by the mid-1960s, it became evident to many theoreticians and propulsion engineers that most of the Solar System would remain out of reach and unexplorable for a very long time. But most of the Solar System was explored. And this was achieved by the invention of an entirely new method of interplanetary space travel that was so radical that it was originally dismissed at JPL as violating the law of conservation of energy.

During the summer of 1961, while working at JPL as a temporary graduate student in mathematics from UCLA, I invented this new method for exploring the Solar System.[1] It was based upon replacing direct-transfer trajectories to a target planet using reaction propulsion—that was taken for granted as self-evident at that time—with indirect trajectories passing one or more intermediate planets so that the spacecraft could be propelled by the resulting gravitational interactions. But after 36 years, I have yet to see one book on the history of space travel that accurately describes this fact. However, my invention was, in fact, recognized by "peer-reviewed" professional literature and by JPL many years ago.[2-4] I have named the method "gravity propelled interplanetary space travel" or simply "gravity propulsion." It is popularly known as "gravity-assist trajectories."

[2] I believe that the American people would like to know that there were actually two fundamentally different methods proposed for exploring the Solar System. One method, reaction propulsion, proposed and formulated by Tsiolkovsky, Goddard, and Oberth (an "engineering method" using a lot of hardware and propellant that Goddard invented) and described in all the history books on space travel which could not generate the high velocities required for exploring most of the Solar System—and my method invented in 1961 (a "mathematical method" that I invented by solving a mathematical problem called the "Three-Body Problem") that did enable the entire Solar System to be explored, but is never mentioned in any history book.

This is what my invention represented and accomplished, and why it is important in the history of space travel. It literally opened up the entire Solar System for exploration with instrumented spacecraft, and it did it with relatively small launch vehicles propelled by ordinary chemical rocket propulsion, and no subsequent reaction propulsion. Thus, it achieved what was believed to be a physical impossibility in 1961. Very few innovations in the history of science have made it possible to break through a fundamental energy barrier, believed to be technically impossible to penetrate, and obtain so much new scientific information for mankind.

The technical and historical details of the invention are described in a paper that I am enclosing herein.[5] It contains 170 published references and verifiable documents. I'm sure you will find it very interesting. Since this history is so fundamentally important, I am hoping that you will send it (along with all of the enclosures) to the National Archives, where it can be properly catalogued and made available to the general public. I'm sure that NASA, various scientific organizations, aerospace societies, and the authors of scientific and historical books, as well as ordinary citizens, would want to know the true facts behind the invention that broke the classical high-energy barriers of interplanetary space travel and opened up the entire Solar System to exploration with instrumented spacecraft. Since the invention was made by an American citizen, it represents part of our technological history as a nation that NASA, and our country should be very proud of. I would be willing to donate original documents or artifacts for examination and/or display by NASA, or by the Air & Space Museum of the Smithsonian Institution.

Since the Voyager 2 mission was only one of several high-energy deep-space missions that was made possible by the invention, you should also make the information in this paper[5] available to Ray James (or to anyone else connected with the writing of the history behind the Voyager 2 mission). You will find a great deal of historical and technical information about it in this paper,[5] such as how and when the gravity propelled encounter sequence, Earth-Jupiter-Saturn-Uranus-Neptune was really "discovered," and how it was numerically determined. These are important facts that have never been accurately published in the popular literature.

If there is any doubt about the authenticity of any of the UCLA documents enclosed herein they can be verified by contacting Professor Michael Melkanoff through UCLA's Department of Computer Science. Melkanoff became Chief of Computer Operations in December 1962 and gave me copies of the documents in 1974. I believe that Professor Melkanoff retired from UCLA in 1994. If anyone wishes independent verification of the FORTRAN listing of the computer code for my gravity propelled trajectory program, or a description of its operation, they could call Dr. Lowell [3] Wood at the Lawrence Livermore Laboratory. I solved all of the mathematical problems (it became my Ph.D. Dissertation in mathematics at the University of California, Berkeley), wrote all of the computer codes, and key-punched the entire 1960-1980 planetary ephemeris for all nine planets myself. Lowell helped me debug the computer code several times in early 1962.[6] (But it was improved many times.) If anyone wishes independent verification of any other document enclosed herein, or cited in the two IAF papers (IAA-90-630 and IAA-91-677) giving more details of the invention,[7,8] this can also be arranged...

Sincerely,

[signature]
Dr. Michael A. Minovitch

[4] Enclosures (16)

1. Minovitch, M. A., "A Method for Determining Interplanetary Free-Fall Reconnaissance Trajectories," JPL, TM 312–130, August 23, 1961, pp. 38–44.

2. Hollister, W. M. and Menning, M. D., "Periodic Swing-By Orbits between Earth and Venus," *Journal of Spacecraft and Rockets*, Vol. 7, No. 10, October 1970, pp. 1193–1199.

3. Pickering, W. H., "The Grand Tour," *American Scientist*, Vol. 58, March/April 1970, pp. 148–155.

4. "Slingshot Magic" Chapter 7 in *The Voyager Neptune Travel Guide*, (ed. Charles Kohlhase) JPL Publication 89–24, June 1, 1989, pp. 103–109.

5. Minovitch, M. A., "The Invention Of Gravity Propelled Interplanetary Space Travel: A Technical and Historical Presentation to the Jet Propulsion Laboratory," Oct. 1997.

6. Reichhardt, T., "Gravity's Overdrive," *Air and Space Smithsonian*, Vol. 8, No. 6, February/March 1994, pp. 73–78.

7. Dowling, R. L., et al., "The Origin of Gravity-Propelled Interplanetary Space Travel," 41st Congress of The International Astronautical Federation, October 6–12, 1990, Dresden, Germany, IAA Paper No. 90–630.

8. Dowling, R. L. et al., "Gravity Propulsion Research at UCLA and JPL 1962–1964, " 42nd Congress of The International Astronautical Federation, October 5–11, 1991, Montreal, Canada, IAA Paper No. 91–677.

[remaining enclosures omitted]

Document II-21

Document title: NASA, "Planetary Program Extension: FY 1968–1969 Program Issues and Options," October 9, 1967.

Source: Historian's Source Files, Washington National Records Center, Suitland, Maryland.

Document II-22

Document title: "Discussion of Planetary Exploration Plans."

Source: Hearing before the Committee on Aeronautical and Space Sciences, United States Senate, Ninetieth Congress, First Session, "NASA's Proposed Operating Plan for Fiscal Year 1968," November 8, 1967.

Document II-23

Document title: Office of Planetary Programs and Office of Space Science and Applications, "Planetary Program Review," July 11, 1969.

Source: NASA Historical Reference Collection, History Office, NASA Headquarters, Washington, D.C.

By late 1967, NASA's solar system program was in disarray. The pressure of completing the Apollo mission on schedule, particularly after the January 1967 Apollo 1 fire that killed three astronauts, gave Apollo highest priority within the NASA budget. After reaching its highest level the preceding year, the NASA budget had decreased, primarily due to the overall budget pressures resulting from the military buildup in Viet Nam and the demands of various Great Society programs. Then Congress refused to approve development funding for the ambitious Voyager missions. In this situation, NASA Administrator James Webb in October 1967 brought together his top officials to consider how to proceed. Document II-21 lists the options presented to Webb, and suggested that a new "planetary extension" program be adopted that preserved the possibility of large planetary missions being approved as soon as the budget climate improved. Webb accepted this option as an interim measure and presented it to Congress in November 1967 [Document II-22], but he also directed his scientific managers to develop a fresh approach to planetary exploration. Taking over a year to formulate, the new approach formed the basis of the planetary program that NASA proposed in 1969 as part of its post-Apollo planning activities [Document II-23].

Document II-21

[no page number]

PLANETARY PROGRAM EXTENSION
FY 1968-1969 PROGRAM ISSUES AND OPTIONS

ISSUE: Should NASA plan any flight missions for planetary exploration in the 1970s?

OPTIONS:

1. Provide no funds for planetary program extensions in FY 1968 and 1969. Because of present budgetary constraints, the funding levels for FY 68 and 69 are insufficient to maintain any flight missions in the early 1970s. This Option will require the phase out of JPL after Mariner 69, the loss of the scientific support presently being provided to the planetary program, termination of all contractor efforts and the reassignment of all in-house personnel to other agency programs. When the nation feels that sufficient support should be given to the exploration of the planets, then the entire effort will have to be restarted.

2. Provide the planetary program extension with a sufficient SR&T budget to maintain technology and pools of scientific, technical and managerial talent to support future planetary missions (see attached Plan 5). This Option will reduce the JPL manpower level to approximately 3000-3500 through the end of FY 70, continue LaRC effort in support of planetary entry and landing technology, maintain a minimum effort related to planetary exploration at ARC and provide a reason-able technological and scientific foundation to enable the implementation of future planetary missions.

3. Include a Mariner flight extension consisting of two missions to Venus in 1972 and two missions to Mars in 1973 and possibly use spare Mariner 69 hardware for earlier missions. This Option would provide the technology, experience and motivation to continue the exploration of the planet with either Voyager or Mariner class orbital and landing missions in the mid or late 1970s. This Option (see Plans 2 or 3) [omitted] would result in a reduction in the JPL manpower level to about 3500 to 4150 by the end of FY 69. A reasonable project activity would be maintained at LaRC and a minimum planetary related SR&T effort would be conducted at ARC.

[2]

4. The Voyager Phase B activities will be completed by about November 1, 1967, and all existing Voyager project activities will be terminated. In FY 1970, if it appeared that the Voyager program would be supported, all activities would be restarted. Such a start in FY 70 would permit Voyager missions in 1975 similar to those previously planned for 1973.

5. Initiate the Voyager program in FY 68 or 69. The effort would be directed at the 1975 missions, thereby stretching the development cycle by two years. The effort could, however, be directed at orbital missions of Mars in 1973 to be followed by landing missions in 1975. This Option would require a Voyager committment [sic] this year.

RECOMMENDED POSITION:

It is recommended that an extension to the Mariner flights (Option 3) be selected and that Voyager not be initiated until 1970 (Option 4). This would provide for continuation of the planetary exploration program (without a Voyager committment [sic]) at a reduced level and more effectively use the scientists, engineers and administrative personnel by focussing their activities at specific missions which incorporate the technologies required for future detailed exploration of the planets. If it is not possible to include post 69 Mariner flights in NASA's FY 69 program, a $35M SR&T effort should be established (Option 2) in FY 69 to prepare for future planetary exploration.

[3]

PLANETARY EXTENSION SR&T EFFORT

PLAN 5

During the past 7 1/2 years teams have been formed of NASA Centers, JPL, Industries, and Scientists; technologies have been developed, and 25 flight missions have been conducted to the moon and planets. The question now being asked is, "What sort of SR&T effort can we outline that will enable the nation to keep that capability for the next 7 1/2 years, until 1975, without any planned flight missions to the planets?"

It is almost certain that the team capabilities and technologies necessary for planetary missions will advance more effectively if the NASA, industry and scientific communities can be applied to other facets of the space program, until needed for the planetary effort. However, if the probability is high that the call to conduct planetary missions will come within about two years, it may be that sufficient "in line" SR&T tasks can be performed to maintain the momentum that now exists, in a truly effective manner.

Toward this end, we have prepared a list of sample tasks and estimated budget needs for a program of this sort as objectively as possible. Tasks have been defined to cover the basic needs for planetary developments, ranging from the most fundamental studies to the operational aspects of given missions. While the coverage of tasks is broad, their content and estimated funding requirements have in no way been inflated—they are aimed at realistic applications of resources both in dollars and in manpower.

Document II-22

[title page]

SENATE COMMITTEE ON AERONAUTICAL AND SPACE SCIENCES
NASA'S PROPOSED OPERATING PLAN FOR FISCAL YEAR 1968
HEARING, 90TH CONGRESS, FIRST SESSION, NOVEMBER 8, 1967

[15] DISCUSSION OF PLANETARY EXPLORATION PLANS

Senator SMITH. Yes, Mr. Chairman; I am sorry to be late. We are on the continuing resolution of the appropriations conference, Senator Holland asked me to inform you he will come over as soon as he can from there.

Mr. Webb, with regard to planetary exploration, you state that you do not believe it's the policy of the Nation or the intent of the Congress that we abandon the field of planetary exploration. I heartily endorse your assessment of the situation, at least so far as the Senate is concerned, but I am not so certain about the majority of our colleagues in the House. You have indicated allocating $143.4 million to the lunar and planetary category, which is about $12 million above the authorization level, in order to seek reestablishment of planetary exploration in the fiscal year 1969 budget.

Could you tell us what types of flight programs you now envision will make up this program?

Mr. WEBB. Yes, Senator Smith. In answer to a question by the Chairman, I pointed out that we had built up through the expenditure of about $700 million and about 16 flights to the moon and planets a very real competence represented by Mariner flights, the Lunar Orbiter fights, the Surveyor flights, the Ranger flights, and the development of Voyager through phase B, which has cost about $37 million; that we had put in 20,000 to 30,000 man-years over the last several years in developing a planetary capability.

We have built up a great capability here and if we now start on the assumption that the action this year is final and there will be no planetary program for a long time after 1969, then we will disband those teams and keep the best people working in supporting research [16] and technology and begin to study new systems that might be useful, say, 10 years from now.

We would not be able to maintain teams that could fly again soon.

Now, before I take that final action, I would like one more opportunity to persuade the President that he should provide some funds in 1969 to maintain these teams in being and to fly certain flights that you have asked about which I would like to outline briefly, if I may.

The first plan—the plan that I believe would be most useful to the Nation would be one in which we revived a project that was not authorized this year; namely, a 1971 flight to Mars with the Mariner type of spacecraft. I believe we should follow that with four other Mariner-type flights within the early 1970s, perhaps flying the five-shot series by 1976. Then I believe we should have two Voyager-type Orbiter missions to Mars in 1973, which would fly on a Titan III booster and follow this, then, with a Saturn V flight in 1975 that would carry two Orbiters and two landers on the one large rocket, so that we would have a double Orbiter and a double lander capability for going to Mars in 1975.

So in essence, I would propose that we make a Centaur-boosted Mariner flight to Mars in 1971, a Centaur-boosted flight to Venus in 1972, or as soon as the opposition occurs after 1971.

Let me go to a table here just one moment. I would like to be sure I am correct.

I want to correct my statement by saying I believe the 1972 Mariner flight to Venus should fly on a Titan III-C, which would be the first of the Titans that we would use in our program. Also, we would like to have a Mariner flying on a Titan III-C that would fly by Venus and then move on out to Mercury, giving us information on both of those planets, in 1973, and follow this with a Voyager Mars orbiter flying on Titan III-C's. That would be two launches in 1973 to Mars—these might carry atmospheric probes, very likely would carry probes, but they would not be capable of landing—and have a Saturn V launch in 1975.

Senator SMITH. Do not those Titan III-Cs take us into the military side of the program?

Mr. WEBB. No more than the use of the Atlas to fly John Glenn or the Titan II for Gemini. The Titan III can lift a military payload. It is part of the standard launch vehicle program. If the Saturn I-B is to be phased out, we would wish to plan a mission that could fly on a Titan III-C. It would be a smaller vehicle than we could fly on the Saturn I-B, but it would be a good scientific mission and would keep these teams at work and would, I believe, reduce the unit cost of these boosters to the military.

Senator SMITH. Thank you.

Did you complete the other part of your statement?

I did not mean to interrupt.

Mr. WEBB. I think that covers the flights. I am thinking of a 5-Mariner flight series in the decade of the 1970s and a program for the orbiters in 1973, landing on Mars in 1975, and then a decision in that period as to what we do after that.

Document II-23

[cover page]

PLANETARY PROGRAM REVIEW

11 JULY 1969

PRESENTED BY:

OFFICE OF PLANETARY PROGRAMS;
OFFICE OF SPACE SCIENCE AND APPLICATIONS

GRAPHICS, COMPOSITION AND PUBLICATION BY:

PROGRAM AND SPECIAL REPORTS DIVISION
EXECUTIVE SECRETARIAT

NASA HEADQUARTERS, WASHINGTON, D.C. 20546

[1] INTRODUCTION
By: Mr. Donald P. Hearth

Our solar system stretches across a distance of some 8 or 9 billion miles. It includes nine planets with their 32 moons; thousands of asteroids, most of which are located in a belt between the orbits of Mars and Jupiter; and hundreds of comets. Some of the comets may possibly come from outside our solar system.

The planets range in size from Mercury which is a little more than a third of the diameter of Earth, to Jupiter which is over 11 times the diameter of Earth. Thus, Earth, on the scale shown in Figure 1 [all figures omitted], is a relatively small planet.

The statistics of nine planets and 32 moons may be wrong. Pluto was only discovered in 1930. In addition, the tenth moon of Saturn was discovered in December 1967.

Man has investigated Venus and Mars with spacecraft. In the process, he has examined the 80 million mile portion of the solar system between these two planets. This is a small portion when compared to the 3 to 4 billion mile distance to the planets Neptune and Pluto.

With telescopes man has been able to examine the planets out beyond Mars, although to a limited extent. As we review some or the mysteries of our solar system we will see that most of what we do know has come from man's observations with telescopes.

[2] Our start in the exploration of our solar system with spacecraft, illustrated in Figure 2, was the Mariner II flight to Venus in 1962 followed by two more Mariners—one to Mars in 1965 and one to Venus in 1967. The Mariner VI and VII spacecraft, launched earlier this year, are rapidly approaching Mars and will be there within 3 weeks. Mariner VI, for example is about 7 1/2 million miles from the planet and closing in rapidly.

During the past 4 years, NASA has placed a series of Pioneer spacecraft into orbit about the Sun to examine the interplanetary medium. The last of the current series of these spacecraft will be launched in August 1969.

Let me define now what we mean by terms like "on-going program," "this year's program," and the "approved program." By these terms we mean the program that President Nixon submitted to the Congress in the Fiscal 1970 budget. In the on-going program, we are broadening our horizons, as shown in Figure 3. We are going out beyond the 80 million mile region discussed earlier, and we're doing more with the nearby planets.

In 1971, for example, we'll be orbiting Mars; thus, extend[ing] our observing time of that planet from a matter of minutes, as associated with the flyby, to months and perhaps even longer.

In 1973 we will, for the first time, land on Mars and continue observation from orbit. This will give us our first opportunity to make direct observations on the surface of another planet, and perhaps give us the first clues on the existence of or lack of extraterrestrial life.
[3] We're moving outward in the solar system beyond Mars with the Pioneer F & G missions; modifications to the current Pioneer spacecraft [will be] used to examine the interplanetary medium. Pioneer F & G will go out to the planet Jupiter, a journey that will take 2 years to complete.

These spacecraft will carry instruments to measure the interplanetary medium. At Jupiter, Pioneer F & G will measure the radiation belts, the magnetic field, and make some measurements of the atmosphere.

We will be searching for an indication of the Helium content on Jupiter. The Helium/Hydrogen ratio is a very important parameter in understanding the planet.

The spacecraft will also carry visual imaging equipment to photograph Jupiter at resolutions considerably better than we get from Earth.

In addition, we're moving in toward the Sun. Our international project with West Germany (Helios) will give us the capability, by 1974, of sending spacecraft inside the orbit of Mercury to within about 30 million miles from the Sun.

In 1973 we also hope to examine, with a Mariner spacecraft, the planet Mercury for the first time. This will be a two-for-one flight in which we will actually view the planet Venus on the way into Mercury.
[4] This, then, represents our spacecraft missions in the on-going program. You'll hear more about some of the possibilities for the future later.

Now, this technique, that is, the examination of the planets by spacecraft, is only one means of exploration. Ground-based observations are the other technique.

There are two new telescopes that are coming into operation this year which are being devoted almost entirely to observing the planets (Figure 4): the 88-inch telescope at Hawaii and the 107-inch telescope at the University of Texas. Actually, a number of findings on the planets have been made from telescopes, and these will come out as we move along. One of the more interesting ones, just this year, was from an 82-inch telescope at the University of Texas which found the first concrete evidence of the existence of water vapor in the Martian atmosphere.

This, then, is the task we have. I'd like to now review the team that NASA has put together to do this job (Figure 5). I think it's a good team.

Within Headquaters [sic] we have some 23 professionals in our office that do the planning and provide overall direction and guidance to the program.

We depend very heavily on the management and technical capabilities of the NASA field centers (Figure 6). The Jet Propulsion Laboratory has been actively engaged in the planetary program dating back to Mariner II, and including the current Mariners that are on their way to Mars. I would expect that JPL will continue to play a major role in the future.

The Ames Research Center has been responsible for the Pioneer Project since about 1962 or 1963 and will be responsible for the modifications to the Pioneer for the Jupiter missions. [5] The Langley Research Center, which did an excellent job on the Lunar Orbiter Project, is responsible for Viking, the orbiter/lander combination to Mars in 1973. They will also have a major role in the future.

Our international project, Helios, is the responsibility of Goddard, which is also responsible for the Delta launch vehicle with which we have been launching our Pioneers. Our larger launch vehicles, the Atlas Centaur and the Titan family, are the responsibility of the Lewis Research Center in Cleveland. All of our planetary launches are from the Cape.

This, then, represents the six NASA centers involved in our flight projects. We are also assisted in the Supporting Research and Technology Program by three other centers: the Electronics Research Center, the Marshall Space Flight Center, and the Manned Spacecraft Center.

This is the NASA team involved in the program. NASA depends, of course, very heavily upon two other groups in the country (Figure 7).

Most of our scientific investigators come from the universities. There are 53 colleges and universities involved in our program at various geographical locations across the United States.

Most of the hardware is provided by industry, and there are literally hundreds of organizations involved. There are some 21 industrial organizations involved to a major extent—again, spread geographically across the United States.

Today you are hearing from four members of this team; we are only spokesmen for this team.

Before Dr. Rea reviews the scientific aspects of the program, I'd like to say a few words on the values of exploring our solar system.

The[re is] opportunity for scientific discovery—answering important questions such as does life exist elsewhere in our solar system, or answering the question as to why did life develop only on the Earth. The application of this scientific knowledge to a better understanding of our own planet, the Earth, is also important. But there are two other values which we sometimes fail to mention. They're what I call exploration and technology.

Picture the sense of exploring this vast exapnse [sic], of going out to planets such as Neptune and Pluto, 3 billion and 4 billion miles in space, and returning photographs of these planets back to Earth. It is very challenging and very exciting, and is the sort of thing that this nation can be proud of. If one looks back through history he will find that countries, and indeed whole societies, have wanted to explore the unknown and achieve the difficult, providing society with an opportunity to look upward. In its small way, I think planetary exploration fills that desire. [6] By its very nature, these missions are very difficult and force the growth of technology. Again, history has clearly shown the need to provide stimulants to technology if we are to make progress.

Why explore the solar system? There are three reasons—science, exploration, and technology.

Dr. Rea will not [sic: now] discuss the goals and objectives of our program and what I call the mysteries of our solar system. ...

Document II-24

Document title: Johnson Space Center, NASA, "Apollo Program Summary Report," April 1975.

Source: NASA Historical Reference Collection, History Office, NASA Headquarters, Washington, D.C.

While the Apollo program was primarily an attempt by NASA to demonstrate the technological ability to send people to the Moon and return them safely to Earth, the program also had significant scientific merit. From the early 1960s, scientists advised NASA about the types of experiments that astronauts should conduct on the lunar surface (see, for example, Documents II-12 and II-13). Document II-24 is an excerpt from the science summary found in NASA's final report on Apollo. The report integrated the results from all of the Apollo missions. Apollo returned to scientists the first surface samples from another solar system body and also provided the most and highest quality data to that date on the Moon's structure, composition, and evolutionary history.

[cover page]

JSC-0942

APOLLO PROGRAM SUMMARY REPORT

[Apollo program emblem]

National Aeronautics and Space Administration
LYNDON B. JOHNSON SPACE CENTER
Houston, Texas

April 1975

[3-1]

3.0 SCIENCE SUMMARY

3.1 INTRODUCTION

The reality of, and enthusiasm for, lunar science greatly increased with the safe return of the Apollo 11 astronauts from man's landing on the moon. Although serious effort in planning, designing, developing, testing, and training for the scientific aspects of the Apollo program had been started much earlier by NASA, the greater emphasis had been correctly concentrated on the accomplishment of the safe lunar landing and return of the crews. Early accomplishment of the spacecraft operational objectives opened the way for more attention to be focused on the scientific potential of Apollo missions. The operational and scientific success of each successive mission stimulated a more vigorous interest in the solar system and established the study of the moon as a modern interdisciplinary science.

Although a considerable amount of scientific data was obtained during the early Apollo missions (Apollo 7 through 14), a significantly greater amount of data was obtained as the result of the Apollo 15, 16, and 17 missions. For each of the latter missions, a diverse set of experiments was installed in the service module and collected data during lunar orbit. These experiments increased the scientific scope of the missions, and the data obtained complemented the data from the experiments being operated on the lunar surface. In addition, more extensive first-hand exploration of the lunar surface was accomplished by the crews on these missions because longer stay times were allowed, and because the addition of the lunar roving vehicle increased the range of travel on the lunar surface as well as the load of instruments, equipment, and lunar sample material transported on crew traverses. Also, more science data were provided by the lunar surface complement of experiments operated by the crews during the extravehicular activities and by the continuing post-mission telemetry from the science stations established at each site.

The large amount of data and material collected as the result of the lunar missions will continue to provide study sources for many years. The crews took thousands of science-quality photographs on the lunar surface and from lunar orbit. Approximately 380 kilograms of lunar soil and rocks were brought back to earth in the returning spacecraft. Five long-term science stations were established on the lunar surface with 22 operating experiments continuing to transmit science data to the earth. The Apollo 12 crew retrieved selected components of a previously landed Surveyor spacecraft. Many materials were transported to the moon, exposed in the lunar environment, and returned for analysis and study.

Findings resulting from the Apollo lunar science program are discussed in the following sections. Science hardware performance is also discussed in conjunction with each experiment. Much of the information in these sections was extracted from the Apollo Preliminary Science Report series. In some cases, publication of results was scheduled by NASA before sufficient data were available to the principal investigators for comprehensive analyses. Thus, results published in the early reports were not as complete as in later reports. In these cases, an attempt has been made to include the latest information. References 3-1, 3-2 and 3-3 provide reviews of the present understanding of the moon's composition and history.

3.2 LUNAR SURFACE SCIENCE

During each Apollo lunar landing mission, the crewmen emplaced and activated a lunar geophysical observatory to be controlled and monitored from earth, collected samples of lunar soil and rock, photographically documented the geologic features of the landing area, and performed other exploration activities. The locations of the Apollo landing sites are shown in figure 3-1 [omitted] and the lunar surface science activities (formal experiments and science detailed objectives) are identified in table 3-I. The Apollo missions during which the activities were accomplished are also indicated in the table.

[3-3] TABLE 3-I. — APOLLO LUNAR SURFACE SCIENCE SUMMARY

Experiment/objective	Experiment number	Mission 11	12	14	15	16	17
[a]Lunar geology investigation	S-059	X	X	X	X	X	X
Soil mechanics experiment	S-200	X	X	X	X	X	X
Lunar sample analysis	–	X	X	X	X	X	X
[b]Passive seismic experiment	S-031	X	X	X	X	X	
[b]Active seismic experiment	S-033			X		X	
[b]Sesimic [sic] profiling experiment	S-203						X
[b]Lunar surface magnetometer experiment	S-034		X		X	X	
Portable magnetometer experiment	S-198			X		X	
[b]Heat flow experiment	S-037				X	X	X
[b]Lunar surface gravimeter experiment	S-207						X
Traverse gravimeter experiment	S-199						X
Surface electrical properties experiment	S-204						X
Lunar neutron probe experiment	S-299						X
[b]Laser ranging retro-reflector	S-078	X		X	X		
[b]Charged-particle lunar environment experiment	S-038			X			
[b]Solar wind spectrometer experiment	S-035		X		X		
Solar wind composition experiment	S-080	X	X	X	X	X	
[b]Suprathermal ion detector experiment	S-036		X	X	X		
[b]Cold cathode gage experiment	S-058		X	X	X		
Cosmic ray detector (sheets) experiment	S-152					X	X
[b]Lunar dust detector experiment	M-515	X	X	X	X		
[b]Lunar ejecta and meteorites experiment	S-202						X
[b]Lunar atmospheric composition experiment	S-205						X
Surveyor III analysis	–		X				
Long-term lunar surface exposure	–						X
Far ultraviolet camera/spectrograph	S-201					X	

[a]Field geology activities included documentary photography, collection of lunar material samples, and crew observations.
[b]Part of an Apollo lunar surface experiments package.

[3-4] As noted in table 3-I, some experiments are part of the geophysical observatories called Apollo lunar surface experiments packages. Using a long-life self-contained power source (radioisotope thermoelectric generator) and communications equipment, each Apollo lunar surface experiments package operates as a remote science station to collect and transmit to earth scientific and engineering data obtained over extended periods of time. The system was flown on Apollo 12 and all subsequent Apollo missions. The aborted lunar landing of Apollo 13 resulted in the loss of the package of experiments; however, the overall program objectives were met by rear-

ranging the experiment assignments of the subsequent flights. A variation of the Apollo lunar surface experiments package, known as the early Apollo scientific experiments package, was flown on the Apollo 11 mission. This package was selected to minimize deployment time and to simplify crew tasks during the first extravehicular activity on the lunar surface.

Rock and soil samples have been collected from most of the major physiographic or photogeologic units identified on the lunar surface prior to the Apollo missions. This collection has and will continue to provide a steady flow of data on the history of the moon. The staggering amount of published material presenting the results of experiments and the analyses of lunar samples cannot be covered in this document. However, the major findings are briefly summarized.

The moon may have accreted to its present mass 4.6 billion years ago. Early activity may have included large-scale magmatic differentiation to produce an anorthositic crust. Throughout early lunar history until about 3.9 billion years ago, the lunar surface was subjected to intense bombardment which produced most of the large ring basins and the deposits of the lunar highlands. Samples from the highlands indicate a very complex history of shock melting and fracturing of the anorthositic crust. Fragments interpreted as plutonic rocks from the crust have been found in some breccia samples collected at highland sites.

Millions of years after the period of intense bombardment, volcanism along the margins of the large ring basins, such as Mare Imbrium, began to fill the basins with lava flows. In a period from about 3.8 to 3.1 billion years ago, these basins were filled with iron- and titanium-rich basaltic lavas; these are now the flat, dark colored mare plains.

Meteoritic bombardment of the lunar surface has continued to the present, although less vigorously than in the past, forming craters and covering the surface with loose debris or regolith. Studies of soil samples from the regolith sections (cores) reveal an incredibly complex history of bombardment by meteorites and galactic and solar radiation through time.

The moon is now inactive, having cooled to a state of inactivity more than 3 billion years ago, the time of formation of the youngest lavas. In contrast with the earth, there is no water and there are no life forms. The surface is, however, constantly changing due to bombardment by cosmic debris. ...

Document II-25

Document title: G. J. Wasserburg, Professor of Geology and Geophysics, California Institute of Technology, to Dr. H. Guyford Stever, Director, National Science Foundation, June 8, 1976.

Source: NASA Historical Reference Collection, History Office, NASA Headquarters, Washington, D.C.

Although NASA was able to launch six planetary missions during the first half of the 1970s, only two new starts for follow-on missions had been authorized by 1976. The Apollo program came to an end in the early 1970s but NASA had begun on its next space flight project, the Space Shuttle, and planetary exploration funding once again suffered because of its perceived lower priority. In 1976, California Institute of Technology professor Gerald Wasserburg, then chairman of the Space Science Board's Committee on Planetary and Lunar Exploration, wrote to Dr. Guy Stever, who was at the time serving as the President's Science Advisor as well as Director of the National Science Foundation to express his concern that solar system exploration was in jeopardy. He pointed out that the Office of Space Science was the only NASA office to have experienced serious budget cuts and urged NASA to consider developing a stronger commitment to deep space exploration. Wasserburg sent an identical letter to NASA.

[no page number]

CALIFORNIA INSTITUTE OF TECHNOLOGY
PASADENA CALIFORNIA 91125

DIVISION OF GEOLOGICAL
AND PLANETARY SCIENCES

TELEPHONE (213) 795-6811

June 8, 1976

Dr. H. Guyford Stever
Director
National Science Foundation
Washington, D.C. 20550

Dear Dr. Stever:

I am taking the liberty of writing a personal note to you and two or three others to convey to you my most profound concern with the status and future of space exploration. For the past year and a half I have served as chairman of the Committee on Planetary and Lunar Exploration (COMPLEX) of the Space Science Board. The report of the committee, which was adopted by the Board, is enclosed for your perusal. We have outlined a general strategy for the exploration of the outer solar system for the next decade. This year COMPLEX is charged to do a similar job for the inner solar system, the asteroids, and the comets. This is a challenging and exciting endeavor on which all of us can work with a great deal of enthusiasm and interest. It is our intent to provide guidelines for a rational strategy of planetary exploration that would constitute a baseline endeavor for the U.S. space program. The wisdom of the strategy which we have evolved is subject both to scrutiny and revision as the times demand it. My real concern is whether or not any of the substantial goals will be achieved.

Unless a sharp reversal of path from the program decisions of the past few years is made, it appears that the whole endeavor of planetary exploration is subject to the highest jeopardy. It is my judgment that unless the situation is drastically improved, the national capability to maintain any significant planetary program will be most seriously debilitated within the next eight months and will be lost within the next two years.

Subsequent to the Apollo missions, the manned space flight program was continued by the commitment to the space shuttle. This is a most challenging and ambitious endeavor even within the restricted framework which was finally accepted by the President and the Congress. However, the NASA funding base has remained essentially level and caused a large real dollar decrease. This has resulted in a serious squeeze of the major shuttle effort and an extreme constriction on the budgetary base for other programs. In particular, a review of the Office of Space Science budget for the current fiscal year shows it to be the only office in NASA to have suffered serious cuts. There were no major new starts in the whole OSS area and no new starts in the planetary program. To date, there has been a three-year period over which no new starts have been authorized in the planetary program. Since fiscal year 1971, there have been only two new starts in the planetary area. These are as follows:

[2]

Mission	FY New Start	Launch	Encounter
Pioneer Venus (PV) Entry	1975	1978	Late 1978
Mariner Jupiter Saturn (MJS) Fly-by	1972	1977	Jupiter 1979 Saturn 1981 Uranus 1984

From a superficial point of view, it may appear that the United States is continuing to fulfill its role as a leader in space exploration. In a few weeks we will have achieved our first soft landing on Mars. Our entry into the Venusian atmosphere will take place in 1978, and in 1981 we will have achieved an encounter with Saturn with a properly instrumented spacecraft. These efforts, if successful, would appear to indicate that a vigorous program is under way. However, a study of the time between the conception of a deep space mission, the authorization to proceed, the time of launch, and the final time of encounter with the planet is about a decade. All the exciting achievements that are now being carried out are the fruits of seeds planted and nurtured a long time ago. Because of these long lead times which are required to carry out deep space exploration for both the inner and outer solar system, it is necessary that a continuing program be carried forward at a steady and sensible pace.

The fiscal and manpower resources on which the program of planetary exploration has been based are greatly reduced so that at present there is only one center (JPL) with resident responsibility for planetary exploration, thus reducing the manpower assigned to this area. A variety of other positive steps are now being considered by NASA in order to minimize the expenditures necessary to accomplish adequate deep space missions. This will involve the use of more standardized and efficient spacecraft so as to obtain exciting results without major development costs for each mission. This more efficient use of manpower and funds could, in principle, yield a healthy and vigorous program without the excesses of feast or famine. However, if one uses the budgetary figures and the lack of new starts as a real measure of national commitment, the only reasonable conclusion is that the United States is at the verge of relinquishing its role as the leader in deep space exploration. It is my assessment that if there are no decisions made in FY 1978 for new starts in 1977, this would begin the disbanding of basic teams which are necessary to maintain a capability for deep space exploration. Such erosion has already started in private industry. By mid-1978 this process would have gone to such an extent that the capability for deep space exploration will have been effectively lost. The only remaining capabilities would be those associated with operations and data acquisition of the MJS mission and the Deep Space Network.

If the planetary team were thus disbanded, it would be exceedingly difficult in both cost and time to reform our capability to carry out deep space missions. The possible circumstance is further apparent in the discontinuation of the Titan Centaur and the concommitant [sic] lack of hard requirements for an interim upper stage (IUS) that can guarantee an adequate planetary capability.

It may be argued that a more "introspective" type of space exploration, confined to earth orbit, should be the primary space goal in the next decade. Certainly the Large Space Telescope should provide one major and exciting effort that would expand our outward view within such a limited framework. However, it is the view of the [3] Space Science Board that "planetary exploration will continue to be an area of major scientific importance over the next decade and that a continuing vigorous activity in this field is fully justified."

The field of deep space exploration has proven to be a most exciting endeavor and has brought this nation great respect and admiration from the whole world. The past accomplishments will be recorded by all peoples as major human achievements. These endeavors require the highest levels of technologic and management skills and exhibit the vigor of the nation. These manifestations of our high technology are more public than those which are carried out within DOD and more easily permit the transference of high technology to the public domain.

The costs of an ongoing planetary program need not be burdensome but do require adequately formed national goals to permit the fulfillment of such long-term endeavors. Considering that the capability exists, that it can be maintained without excessive cost or risk, and that it invigorates and extends our major resource of high technology, I urge you to give this pressing matter the attention and support which is needed at the highest councils of government.

Respectfully yours,

[signature]
G.J. Wasserburg
Professor of Geology and Geophysics

GJW:elb

Report being sent under separate cover

Document II-26

Document title: Noel W. Hinners, Associate Administrator for Space Science, NASA, to Administrator, NASA, "Viking 75 Mission: Assessment of Primary Mission," January 6, 1977.

Source: Historian's Source Files, Washington National Records Center, Suitland, Maryland.

In 1975, NASA launched two Viking missions to Mars with the objective of advancing human understanding of the Red Planet. In particular, scientists hoped to gather data on the past, present, and potential future existence of life on the planet. The missions, designated Viking 1 and 2, each consisted of an orbiting probe and a landing spacecraft. All four spacecraft succeeded in reaching Mars in 1976, with Viking 1 becoming the first U.S. probe to land on another planet on July 20. The probes collected unprecedented scientific data for several years. This document offered an overview of Viking's scientific findings as of 1977. Scientists were unable to determine from the Viking data whether life ever existed on Mars.

[no page number]

MEMORANDUM

January 6, 1977

TO: A/Administrator
FROM: S/Associate Administrator for Space Science
SUBJECT: Viking 75 Mission, Assessment of Primary Mission

Two Viking spacecraft were launched with Titan/III Centaur launch vehicles on August 20 and September 9, 1975, from the Air Force Eastern Test Range, Pad 4. Both launches occurred within their nominal launch windows. After launch, the spacecraft identifiers were changed from A and B to 1 and 2.

The spacecraft arrived at Mars and were successfully inserted into orbit on June 19 and August 7, 1976. The Viking 1 Lander successfully landed at 22.5° N latitude by 48.0° W longitude on July 20, 1976. The Viking 2 Lander descended safely on September 3, 1976, at 47.9° N latitude by 225.9° W longitude.

All the Viking scientific instruments, except for the seismometer on Lander 1, which failed to uncage, operated satisfactorily and continue to send data back to Earth. With the exception of the seismometer failure, it has been possible to work around the few anomalies that occurred by using alternate operating modes designed into the system.

Details of mission plans and operations have been reported in the Prelaunch Mission Operation Report (August 1, 1975), Post Launch Mission Operation Reports # 1 and #2 (August 28, 1975, and September 16, 1975), Pre-Orbit Insertion and Landing Mission Operation Report (June 9, 1976) and the daily Viking Status Reports issued from pre-encounter through the primary mission.

Based upon the results of the Viking 75 Mission, the primary mission is adjudged as successful.

[signature]
Noel W. Hinners

[2] NASA OBJECTIVES FOR THE VIKING PROJECT

OBJECTIVES

The purpose of the Viking missions is to significantly advance the knowledge of the planet Mars by means of observations from Martian orbit and direct measurements in the atmosphere and on the surface during the 1975 opportunity. Particular emphasis will be placed on obtaining biological, chemical, and environmental data relevant to the existence of life on the planet at this time, at some time in the past, or the possibility of life existing at a future date.

[signature]
Robert S. Kraemer
Director, Planetary Programs
Date: [handwritten "7-31-75"]

[signature]
Noel W. Hinners
Associate Administrator for Space Science
Date: [handwritten "July 31, 1975"]

ASSESSMENT OF THE VIKING 75 MISSION

Based upon the results of the Viking 75 Mission with respect to the approved prelaunch mission objectives, the primary mission is adjudged a success.

[signature]
A. Thomas Young, Director
Lunar and Planetary Programs
Date: [handwritten "1/3/77"]

[signature]
Noel W. Hinners
Associate Administrator for Space Science
Date: [handwritten "1/5/77"]

[3] PRELIMINARY SCIENTIFIC RESULTS

The Viking Project objectives were to significantly advance the knowledge of the planet Mars by means of observations from Martian orbit and by direct measurements in the atmosphere and on the surface during the 1975 opportunity. Particular emphasis was placed on obtaining biological, chemical, and environmental data relevant to the existence of life on the planet at this time, at some time in the past, or the possibility of life existing at a future data [sic].

Preliminary scientific results for the primary mission have been reported in:

Viking 1 Early Results	NASA SP-408
Science Vol. 193	27 August 1976
Science Vol. 194	1 October 1976
Science Vol. 194	17 December 1976

These results are summarized in the following brief statements.

Orbiter imaging shows the Mars surface to be much more heterogeneous than anticipated. Some of the surface features are very ancient while others appear to be of recent origin. Crater frequency and size distribution is being used as a basis for estimating the sequence in the formation of these features. The major volcanic piles are comparatively young although no present activity has been observed.

Water is more abundant than was suggested by earlier data. The residual polar caps are composed of water ice. Unique lobate crater ejecta and large areas of surface slumping also suggest subsurface water or permafrost. Atmospheric water vapor shows distinct diurnal and seasonal cycling.

Nitrogen, argon, krypton, and xenon were detected in the atmosphere and the isotope ratios for carbon, oxygen, nitrogen, and argon were established. The isotope ratios are in some cases significantly different from those observed on Earth. These observations will continue to be the subject of study and discussion relative to the evolutionary history of Mars.

The landing sites are surprisingly similar although they appeared to be quite different in character at the limit of orbiter image resolution. While there are a few notable differences in the two sites, they are generally rock strewn landscapes with fine wind-blown material interspersed. At site 1 there are drifts of fines that show evidence of stratification, suggesting cyclic episodes of deposition and erosion. At site 2 a small depression crosses the near field of view. This may be a part of the large scale polygonal fracture pattern covering much of the northern hemisphere at the latitude of the landing site.

[4] Elemental analysis of the surface fines show a high concentration of silicon, iron, magnesium, calcium, aluminium and sulfur in that order of relative abundance. The high Ca/K ratio, together with the elemental abundances, indicate the material to be basaltic rather than granitic in origin. This indication is supported by the Radio Science observation that the surface dielectric constant at the landing sites is consistent with that for pumic or tuff.

Rocks in the fields of view show a wide diversity of size, color and texture. However, attempts to obtain a sample of small rocks or coarse gravel for elemental analysis were unsuccessful during the nominal mission. This is due to an apparent characteristic of the sample sites where pebble-like features are really clods of adhesive surface material.

The Biology experiments have not unambiguously demonstrated the presence or absence of living organisms in the Mars surface samples. What appear to be positive indications have been received from two of the three experiments. The third experiment shows a high reactivity of the surface material when exposed to moisture. The chemistry of the soil is not understood and the response appears to be significantly different than tests with Earth and lunar soils. Experiment parameters will be modified during the Viking Extended Mission in an attempt to better understand this unique chemistry and to determine with more confidence the presence or absence of biological activity. Earth based tests are also underway to help understand the biology experiment data.

No organic compounds were identified in the surface samples acquired. The organic analysis data did reveal the presence of a hydrated mineral, stable at 200°C, but which released water equivalent to about 1% of the sample at 350° and 500° C. A somewhat higher water concentration was measured in a sample from beneath a rock.

The Physical and Magnetic Properties investigations show the surface to have good bearing strength. The fine surface has a cohesiveness of approximately 10^3 dynes/cm^2 and contains 3 to 7% of magnetic particles.

Meteorological observations show the weather to be mild during the northern summer and highly predictable from day-to-day. Temperatures range from a low of 187° K just before sunup to 242° K in mid-afternoon. The mean vector wind has been from the south at 2.4m/sec with steady state variations from near zero at midnight to 8-9 m/sec in midday. Atmospheric pressure on Mars shows a predictable semidiurnal harmonic. There was a small but steady drop in pressure from the time of landing to late in the primary mission. This drop was estimated to be the result of south polar deposition of CO_2 from the atmosphere. A similar conclusion was drawn from the infrared thermal mapper data taken from orbit. Thus Mars is expected to have a semiannual pressure cycle as the deposition and evaporation of CO_2 shifts from pole to pole with the seasons.

Document II-27

Document title: Raymond L. Heacock, Voyager Project Manager, Jet Propulsion Laboratory, to Frank Carr, Acting Voyager Program Manager, Office of Space Science, NASA, October 21, 1980.

Source: NASA Historical Reference Collection, History Office, NASA Headquarters, Washington, D.C.

Document II-28

Document title: Raymond L. Heacock, Voyager Project Manager, Jet Propulsion Laboratory, to Frank Carr, Acting Voyager Program Manager, Office of Space Science, NASA, November 24, 1980.

Source: NASA Historical Reference Collection, History Office, NASA Headquarters, Washington, D.C.

Document II-29

Document title: Frank Carr, Acting Voyager Program Manager, Office of Space Science, NASA, to Raymond L. Heacock, Voyager Project Manager, Jet Propulsion Laboratory, December 12, 1980.

Source: NASA Historical Reference Collection, History Office, NASA Headquarters, Washington, D.C.

When budget constraints prohibited NASA from starting on the ambitious Grand Tour mission to visit all of the outer planets except Pluto in the early 1970s, agency officials designed a simpler and less expensive mission to visit the outer solar system in the late 1970s. Called Voyager, the mission consisted of two identical spacecraft that would augment the scientific data returned by Pioneers 10 and 11 from Jupiter and Saturn. Voyager 2 also had the possibility of continuing on to explore Uranus and Neptune. In the first two documents, JPL Voyager Project Manager Ray Heacock attempted to persuade NASA's Acting Voyager Program Manager Frank Carr of the scientific community's support for and capability of Voyager 2 to continue to Uranus. Carr granted this permission in Document II-29. After reaching Uranus in 1986, Voyager 2 proceeded on to fly by Neptune in 1989. Still returning data on the interplanetary medium, this spacecraft is now travelling out of the solar system. Voyager 1 began its departure from the solar system after exploring Saturn by traveling in a direction perpendicular to the ecliptic, or the plane in which the planets orbit the sun.

Document II-27

[no page number]

[on Jet Propulsion Laboratory letterhead]

October 21, 1980

Refer to: 260-RLH:lw

Mr. Frank Carr
Code SL-4
NASA Headquarters
Washington, DC 20546

Dear Frank:

The purpose of this letter is to document the Voyager proposed Uranus confirmation criteria in compliance with milestone #6 of chart CEK-3 in the attached presentation package. After you have reviewed the enclosed material and rationale, we would appreciate a letter from NASA that closes milestones #5 and #7.

Over the past several months, the scientific interest in Titan has diminished in intensity, while the interest in the Uranus system has increased. The decline in Titan interest has resulted primarily from lower estimates of atmospheric pressure and surface temperature. It also appears that Titan is largely covered by clouds, making it difficult to see the surface. In the meantime, the Voyager 2 health and propellant supply have preserved the opportunity to continue past Saturn to the remote Uranus, and possibly even Neptune, planetary domains. This is an opportunity that may not occur again until after the turn of the century. For this reason, and because of the paucity of data about the Uranus and Neptune systems, there is strong support from the scientific community to continue to Uranus. We have also assessed our ability to operate Voyager 2 at these greater distances and have not uncovered any major problems.

It is therefore our recommendation that the current plan be maintained whereby Voyager 2 be allowed to continue past Saturn to Uranus unless both of the following adverse (and unlikely) circumstances should occur: Voyager 1 fails to return adequate Titan and ring science, and the Voyager 2 health is not capable of supporting a minimal Uranus science mission. The occurrence of only one of these conditions would not alter our nominal plan to continue to Uranus. Although the Voyager 2 Uranus trajectory at Saturn provides less Titan and ring science return than might be obtained with other trajectories, the Saturn science return with the Uranus trajectory is more than acceptable.
[2] The primary Uranus reconnaissance objectives as outlined by the Space Science Board included the following five items: (1) the planetary magnetic field, (2) Uranus' gravity field, [J2] and satellite masses, (3) gross morphology of the planet and satellites, (4) atmospheric structure and composition, and (5) planetary heat source. Although magnetometer results alone would address objective 1 directly, the approximate magnitude, tilt, and rotation period of the magnetic dipole can also be derived from PLS, PWS, LECP, or CRS data, provided that the spacecraft enters the Uranian magnetosphere, which may extend 25 to 50 R_U from the planet. The PRA may be able to remotely determine a more accurate rotation period without the requirement of entering the magnetosphere. Thus, the first objective should be achievable even with only three working fields and particles instruments. Determination of the magnetic rotation period is also directly related to objective 2. The atmospheric objectives (item 4) can be at least partially addressed by a single-frequency, radio occultation that will provide a temperature-pressure profile for two points in the atmosphere. With an appropriate trajectory and a working radio system, objective 2 can also be accomplished. Thus, significant contributions to objectives 1, 2, and 4 can be obtained with the instrumentation identified as the minimal Uranus science mission. These conditions are summarized in chart CEK-8 from the enclosed presentation package.

We have recently completed our analysis of the potential risks associated with the Voyager 2 E-ring crossing at 2.87 R_S. This is a very difficult problem to analyze completely, since we do not have a reliable assessment of the particle size distribution. If the particles are larger, for example, than 1 gm, the probability of spacecraft impact is nil. If the particles are smaller, for example, than 10^{-5} gm, then impacts are a certainty but can cause no harm. If particle distributions are assumed in between these two sizes, then hit probabilities exceed 50% for critical spacecraft subsystems. However, the probability of spacecraft failure is less than the hit possibility due to subsystem redundancy and adaptive workaround options. If one extrapolates the Pioneer 11 ring-crossing experience to the Voyager 2 conditions, then the odds of a safe crossing should be at least 80%. Unless Voyager 1 sheds new light on this subject, we would place the Voyager 2 survival odds at 80% or better.

Returning to our principle criteria of Voyager 1 science return and Voyager 2 health, the following meeting dates are planned. On 11/17/80, the Flight Science Office will report on the adequacy of Titan and ring science returned by Voyager 1. If acceptable, we would know at this time that continuation to Uranus would be recommended at the 11/24/80 confirmation meeting. However, independent of the outcome of this science return assessment, the Flight Engineering Office would still present on 11/24/80 a checklist status summary of the Voyager 2 subsystems and their [3] prognosis for being able to support the minimal Uranus science mission. At the culmination of the 11/24/80 meeting, the final project recommendation would be made to the appropriate NASA managers.

If you have any questions on the material contained in this letter or in the attached presentation package, please give me a call. Otherwise, I look forward to your response on milestones #5 and #7.

Very truly yours,

[signature]
Raymond L. Heacock
Project Manager
Voyager

RLH:lw

Enclosure [omitted]

cc: E.K. Davis
C.E. Kohlhase
R.P. Laeser
E.L. McKinley
B.C. Murray
R.J. Parks
C.H. Stembridge
E.C. Stone

Document II-28

[no pagination]

[on JPL letterhead]

November 24, 1980

Mr. Frank Carr
National Aeronautics and Space Administration
Code SL-4

Washington, DC 20546

Dear Frank:

We held our Uranus confirmation recommendation meeting at JPL on 11/21/80. The presentation material from that meeting has been enclosed with this letter. The Voyager Project recommendation was clearly for Voyager 2 continuation on to Uranus following the Saturn encounter next August. The principal rationale for this recommendation involved two key factors: the successful Titan and ring science return from Voyager 1, and the good health prognosis for Voyager 2 in terms of achieving a minimal Uranus science mission or better. Our Uranus recommendation is also consistent with the criteria that were transmitted to you in my letter of 10/21/80. If possible, please provide your verbal concurrence with our Uranus recommendation by 12/1/80, with a written follow-up by 12/5/80.

During the course of our meeting, the subject of Neptune continuation following the Uranus encounter was briefly discussed. It was recommended and agreed that no future actions should occur that would jeopardize or preclude the Neptune option. For example, this implies that we carefully manage such expendables as hydrazine, TWTA operating lifetime, and DTR cycles. In this regard, the Mission Planning Office will ensure that resource allocation guidelines are consistent with preserving a reasonable chance for a successful Neptune encounter in August of 1989.

I look forward to hearing from you during the week of 12/1/80. Please call me if you require further information before reaching your final decision.

Very truly yours,

[signature]
Raymond L. Heacock
Project Manager
Voyager

Enclosure [omitted]

cc: E. K. Davis
C. E. Kohlhase
R. P. Laeser
E. L. McKinley
B. C. Murray
R. J. Parks
C. H. Stembridge
E. C. Stone

Document II-29

[no pagination]

[stamped: "DEC 12 1980"]

SL-4

Mr. Raymond L. Heacock
Voyager Project Manager
Jet Propulsion Laboratory
Mail Stop 264-443
4800 Oak Grove Drive
Pasadena, CA 91103

Dear Ray:

The Voyager Project's recommendation for Voyager 2's continuation on to Uranus is approved.

This approval is based on the agreed criteria having been met, namely:

- the successful achievement of adequate Titan and Saturn ring science by Voyager 1, or
- the JPL assessment that the Voyager 2 health is capable of supporting a minimal Uranus science mission.

You are asked to submit a revised Project Plan which addresses the Voyager 2 Uranus Mission, including both nominal and minimal aspects.

Sincerely,

[signature]
Frank A. Carr
Acting Voyager Program Manager
Solar System Exploration Division
Office of Space Science

cc: CalTech/Stone
S/Stofan
SL-4/ Guastaferro
Robins
Mitz
S-6/ Panagakos

****

Voyager Project's Recommendation for Voyager 2's Continuation to Uranus:

Approved:

____

C.H. Robins, Jr.
Chief, Solar System Mission Operations
Solar System Exploration Division
Office of Space Science

____

A. Guastaferro
Director, Solar System Exploration Division
Office of Space Science

[signature]
A. J. Stofan
Acting Associate Administrator for Space Science

____

J. D. Rosendhal
Acting Assistant Associate Administrator for Space Science

Document II-30

Document title: Carl Sagan, President, The Planetary Society, to members of The Planetary Society, August 1, 1981.

Source: NASA Historical Reference Collection, History Office, NASA Headquarters, Washington, D.C.

Document II-31

Document title: James M. Beggs, Administrator, NASA, to the Honorable David A. Stockman, Director, Office of Management and Budget, September 29, 1981.

Source: NASA Historical Reference Collection, History Office, NASA Headquarters, Washington, D.C.

Document II-32

Document title: NASA, "FY 1983 Budget Appeal," December 5, 1981.

Source: NASA Historical Reference Collection, History Office, NASA Headquarters, Washington, D.C.

Document II-33

Document title: White House, "Selected White House Views (on NASA's planetary exploration program)," December 8, 1981.

Source: NASA Historical Reference Collection, History Office, NASA Headquarters, Washington, D.C.

Document II-34

Document title: Budget Review Board, "Budget Review Board Decisions," December 11, 1981.

Source: NASA Historical Reference Collection, History Office, NASA Headquarters, Washington, D.C.

The planetary exploration program once again was struggling for its survival in the late 1970s and early 1980s. New JPL Director Bruce Murray in 1976 had decided that getting approval for a U.S. mission to Comet Halley during its 1986 swing through the inner solar system should be a high priority objective, even though the Space Science Board of the National Academy of Sciences had assigned scientific priority to a radar mission to map the surface of Venus. That Venus mission was approved by the outgoing Carter administration in late 1980, but this approval was quickly rescinded by the new administration of President Ronald Reagan, which took office in January 1981. This meant that the only approved future planetary mission was the Galileo mission to Jupiter, and that mission was also threatened with cancellation.

As one means of rallying public support for the planetary exploration program, Murray, well-known planetary scientist Carl Sagan, who had become an articulate spokesman for the program, and JPL staff member Louis Friedman formed The Planetary Society in 1980. Membership in the Society quickly grew to over 70,000 people. As part of a last-ditch effort in mid-1981 to convince the White House to approve a Halley mission, Sagan asked Society members to make their views known [Document II-30].

NASA's new administrator, James Beggs, faced with a requirement to further reduce NASA's budget, in September 1981 proposed a draconic measure: terminating the planetary program and thereby making JPL surplus to NASA's needs. [Document II-31] This may have been a tactical gambit on Beggs' part, on the assumption that the White House would be unwilling to take such a step. However, the Office of Management and Budget seemed prepared to accept the notion of shutting down the program [Documents II-32 and II-33]. Only last minute political intervention by influential Caltech supporters reversed this position; finally, the White House Budget Review Board accepted a suggestion by Science Adviser George Keyworth that enough money be added to the NASA budget to maintain the Galileo mission and keep JPL an active part of NASA [Document II-34].

Document II-30

[no pagination]

[on letterhead of The Planetary Society]

[handwritten: "8/1/81"]

Dear Member of The Planetary Society:

A mountain of ice and dust and organic molecules is streaking from beyond Pluto towards the Sun. It is probably a fragment left over from the formation of the solar system. It may hold clues to the origin of life. The most famous visitor to Earth in human history, Halley's Comet will once more pass our planet in 1986. It cries out for exploration. All the other nations of Earth able to do so are sending a spacecraft to encounter the comet. Only the United States seems reluctant. And only the United States is able to muster a Voyager-class mission, thereby greatly enhancing the effectiveness of the first truly multinational exploration of space.

As the editorial from the Los Angeles Times (overleaf) [omitted] and others in many American newspapers indicate, a Halley mission is enormously popular. It is scientifically meritorious. It costs less than 1% of the NASA budget (tiny for federal spending, but far too much for public subscription). It will reinvigorate a U.S. planetary program sadly neglected for several Administrations. It is an enterprise to make us proud of our nation and our species. And it can still be done.

But only the White House can make an American Halley mission happen at this late date. The Planetary Society is not a political organization. We have never before made such an appeal to our members. Still, as the largest space interest group in the world, we recognize a special responsibility: an American letter or telegram to President Ronald Reagan (The White House, Washington, D.C. 20500) from each of our 70,000 members and from their friends might make the difference. It would certainly be hard to ignore. We urge you to take action without delay.

Such a Halley mission will be celebrated by our descendants in the far future. Thank you for helping to make this historic event possible.

Cordially,

[signature]
Carl Sagan

CS/pf

Document II-31

[no pagination]

[stamped "SEP 29 1981"]

Honorable David A. Stockman
Director
Office of Management and Budget
Washington, DC 20503

Dear Mr. Stockman:

We have reviewed the guidelines you provided for the NASA budget outlays for FY 1983 at $6,041 million and FY 1994 at $5,687 million. In examining the NASA program to determine what must be done to reach these ceiling levels, I have come to the firm conclusion that they take us past the point at which we can simply take percentage reductions out of each of our programs to meet the proposed guidelines. Rather, we are now at the point at which it becomes necessary, in order to maintain viable programs in some areas, to close down other major programs that NASA has operated since its inception.

The proposed guidelines represent real reductions over the current spending level in FY 1981 of 10% in FY 1983 and 20% in 1984 when the effects of inflation are properly taken into account. The conclusion I have reached is inescapable when added to the fact that the NASA budget has already been reduced by over 20% in real terms, in the 10 years since the Shuttle decision was taken and it was agreed between NASA and OMB (Cap Weinberger) that NASA's budget would remain constant in real terms at the $3.4 billion level in 1971 dollars.

What I am compelled to do to meet these guidelines is to delete the planetary exploration program to reach an intermediate $6.5 billion level, and then to wipe out the space applications program and to make significant reductions in the Space Shuttle program to meet the $6.0 billion level.

It is important to understand the reasons behind the decisions I have outlined. The planetary exploration program is one of the most successful and viable NASA programs. However, it is our judgment that in terms of scientific priority it ranks below space astronomy and astrophysics. Planetary exploration is much more highly dependent on launch vehicles, and it is our opinion that the most important missions that can reasonably be done within the current launch vehicle capability have, more or less, been done. The next step in planetary exploration is to do such things as landing missions and sample return missions, and these require full development of the capability of the Shuttle and the ability to assemble elements in earth orbit before sending the assembled spacecraft on its way. In our judgment, it is ultimately better for future planetary exploration to concentrate on developing the Shuttle capabilities rather than to attempt to run a "sub-critical" planetary program given the current financial restrictions we face. Of course, elimination of the planetary exploration program will make the Jet Propulsion Laboratory in California surplus to our needs.

In the case of the space applications program our intention is to attempt to retain the scientific aspects of the program but to delete the future efforts in weather, remote sensing and communications oriented toward commercialization and demonstration. We would make this decision recognizing that applications have been politically popular and that these are the areas that are being emphasized in the European and Japanese Space efforts. Finally, to reach the $6.0 billion level, we would be forced to reduce the flight rate of the Shuttle below the 32 flight level through 1985 recommended in the FY 1983 budget.

We believe that, while painful and unpopular, the above approach will provide a program which marginally meets the objectives of the basic NASA mission. Making cuts in all programs, with the result that none are truly viable, will not advance the interests of the United States.

Therefore the budget guidelines you have proposed go beyond simple cut backs and require policy decisions from the appropriate officials in the White House. A few weeks ago, I met with Admiral Garrick of Mr. Meese's staff and informed him that any NASA budget for FY 1983 much below the level of $7.0 billion would require such policy decisions. I am now requesting a meeting with you and the President's Counselor, Mr. Meese, to discuss the policy questions raised by the budget guidelines you have proposed. It is most important, in my opinion, that all involved clearly understand the political implications and the international complications that result from these decisions.

Sincerely yours,

[signature]
James M. Beggs
Administrator

Concurrence: "Hans Mark" [hand-signed]/AD

A: JMBeggs:A24716:tm:53918:9/29/81

[handwritten:

bcc: A
AD
ADB
B
R]

Document II-32

[no pagination]

[stamped "DEC 5 1981"]

FY 1983 BUDGET APPEAL

Agency: National Aeronautics and Space Administration

Appeal Item: PLANETARY EXPLORATION

Budget Impact:

	1982	1983	1984
	($ in Millions)		
OMB Passback			
Budget Authority	224	118	100
Outlays	222	136	111
Agency Position			
Budget Authority	351	394	353
Outlays	313	368	375
Amount Appealed			
Budget Authority	127	276	253
Outlays	91	232	264

Program Description: The Planetary program objectives are to understand the origin and evolution of the solar system, to better understand the Earth through comparative studies with other planets, and to understand how the appearance of life in the solar system is related to the chemical history of the solar system. The Nation's planetary program is concentrated at the Jet Propulsion Laboratory which has an extraordinary record of successes over two decades in the exploration of planets, their satellites, and the interplanetary medium. Following the grand achievements of Viking and Voyager, the current program is centered around the Galileo mission to Jupiter, preparation for the next planetary exploration step—a Venus Orbiting Imaging Radar mission, and support to the ESA International Solar Polar Mission.

Justification for Agency Position: The scientific return of this program has been extraordinary, and the implications for the future are boundless. Americans have taken enormous pride in the nation's planetary exploration endeavors which have been a true reflection of the greatness and vigor of the United States. Citizens of foreign nations regard the prodigious successes of the program—which have included the first landings on Mars, and the first missions to the Jovian and Saturnian systems—with awe and admiration. The planetary exploration program has, in addition, made a very real contribution to the strength and security of the nation through important technological advances that have been driven by the exceptionally demanding nature of deep space missions. The impact of the OMB reduction on Planetary Exploration would be termination of the Galileo mission, the Venus Orbiting Imaging Radar (VOIR) mission, phase-down of the Deep Space Network, termination of the construction of the 34-Meter Antennas, and termination of the Space Flight Operations Facility Modifications, while ostensibly continuing support to the European Space Agency's (ESA) International Solar Polar Mission (ISPM) using TDRSS backup Inertial Upper Stage. This reduction would also result in the termination of approximately 950 positions at the Jet Propulsion Laboratory, as well as termination of approximately 700 positions in aerospace industries primarily in California and at NASA Centers in California, Texas and Maryland. Other specific impacts inherent in the reduction include:

- Only one mission to the planets is currently under development—Galileo. Approximately $350 million has been committed to Galileo over the last four years to bring the mission to a considerable state of maturity and hardware readiness. The mission has been designed with the close involvement of the science community to undertake fundamental scientific inquiries into planet-formation processes and indepth [sic] studies of the Jovian atmosphere and magnetosphere. The probe will provide the first in situ measurements of the composition of Jupiter, the principal repository of solar system material outside the Sun. The orbiter is equipped with a new generation of cameras and instruments to study the Jovian meteorology and the four large moons, including volcanic Io. The planned tour of the Jovian magnetosphere will provide the systematic coverage needed to understand the complexities of Jovian plasma physics processes that could never be simulated in any laboratory. The Space Science Board of the National Academy of Sciences has given this mission its strongest endorsement. This reduction would result in the loss of the Nation's opportunity to further man's knowledge and understanding of the active processes of the planet Jupiter and the ... [illegible] ... and the Earth's formation.

- NASA's solar system exploration program has been an international endeavor from its inception and, as such, enhances this Nation's relations with our allies. The Galileo mission is being undertaken jointly with the Federal Republic of Germany which is contributing over $50M of hardware and $25M of science instrumentation, with $50M invested to date.

- The Venus Orbiting Imaging Radar (VOIR) mission would be abandoned, thereby losing the opportunity to gain detailed knowledge of the planet Venus; this would terminate our effort to further our understanding of the evolution of Venus. Because of the Earth's similarity to Venus, this loss would be a serious blow to studies of the evolution of the Earth.

- The International Halley's Comet Watch and Co-Investigator support to the Giotto mission would be terminated. Thus, the United States would have to forgo any active participation in the Halley's Comet scientific program.

- The Deep Space Network reduction would significantly impact several missions currently in flight. Some specific impacts include:
  - Voyager/Uranus encounter could not be satisfactorily supported.
  - Several missions require simultaneous support which cannot be provided.
  - The Voyager/Neptune encounter opportunity would be lost.
  - Critical and unique skills at JPL would be lost.

- The 34-meter Antenna reduction would substantially reduce quality of data for the Voyager encounter with Uranus and other missions which cannot be supported by the Tracking and Data Relay Satellite System (TDRSS).

- The Space Flight Operations Facility is the operational "nerve center" for all planetary flight control activities at JPL. This reduction would force continued hazardous and inefficient conditions in the facility.

- These appealed funding levels provide for use of a solid rocket Inertial Upper Stage to perform the planetary program. The high energy cryogenic Centaur would be the most effective stage to perform the Galileo mission to Jupiter providing for a much earlier arrival (1987 vs 1990) and a greatly enhanced science mission (11 vs 6 satellite encounters). The justification for Centuar [sic] is covered as a special appeal item.

Implications for National Capability

- The precipitous reduction in activity at the Jet Propulsion Laboratory risks loss of a major national asset. It is our understanding that DOD is planning to increase their reliance on the Jet Propulsion Laboratory for assistance in development of advanced sensor systems for national security applications. An unstructured phase-down of JPL would result in the loss of the most talented members of JPL's staff to the detriment of the planned DOD activities. In addition, this reduction would adversely impact National security by undermining the technological base of current DOD project support, for example the Deep Space Mission capability would be significantly reduced and DOD Project support would have its technological base undermined. Some examples of this support are: Autonomous Spacecraft Project; U.S. Army Tactical Command/Control/Communication systems; U.S. Army Mobile Field Instrumentation System; U.S. Army project management methodology; and Air Force Rocket Propulsion lab support.

Document II-33

[no pagination]

DATE: 12/8/81

DEPARTMENT: NASA

ISSUE: Planetary Exploration

SELECTED WHITE HOUSE VIEWS

Office of Policy Development:

The key items here are the decision whether to cancel the Galileo mission to Jupiter and the Venus orbiting imaging radar. These are the two main items whose continuation we touted as the justification for the cancellation of other items in the FY-82 budget.

If the United States is to continue a presence in planetary exploration these are the lead items which should be continued.

It should also be noted that the cuts in the deep space network and 34 meter antenna may make it difficult to obtain the payoff from the Voyager encounter with the outer planets. The modest additional expenditure would seem justified to garner knowledge from these additional planetary encounters.

Office of Science & Technology Policy (through Office of Policy Development)

Before addressing specific NASA budget appeal items, it is important to realize what the agency funding trends have been:

		Shuttle	Other	Total	
FY-81			2,729	2,789	5,518
82	(Sept. Req.)	3,067	2,684	5,751	
83	agency appeal	3,619	3,446	7,065	
	OMB recommendation	3,360	2,894	6,254	

The point of this summary table is that the space program has received preferred treatment in the budgetary process, that the shuttle is consuming an increasing fraction of the total NASA budget, and that generous support is retained for space science and application. In hardly unique fashion, the appeal centers around a desire to maintain all ongoing activities while attempting to make the Shuttle operational.

More specifically, the Budget Review Board will address the two major appeal items, aeronautics research and planetary exploration.

Neither OMB nor OSTP has questioned the need or utility of the NASA aeronautics program. It is essential to the successful evolution of our military aircraft as well as providing basic research upon which our aircraft industry depends. The issue is the proper level of support, and distinguishing between those elements that are appropriate for government versus those for the private sector or direct DOD funding. In FY-82, NASA aeronautics funding was $265 M for R&D plus $232 M for support of manpower and facilities. The FY-83 proposal is $139 M and $179 M, respectively. I believe these reductions are excessive and will likely result in loss of essential elements of a critical technology. However, OMB and OSTP have discussed this extensively during the OMB Director's review and subsequently concurred that an increase of $89 M, shared between R&D and manpower, would restore those critical elements and yet maintain the proper government role. The primary issue of concern to NASA is that the manpower be retained. The restoration of $89 M retains the essential manpower, but does not exclude aeronautics from the overall 10% personnel reduction in civilian agencies. In summary, the NASA request is "business as usual" whereas the OMB compromise position recognizes the importance of aeronautics research but requires that clear industrial subsidy be reduced and DOD directly fund some activities where feasible.

NASA is appealing, with priority second to its aeronautics appeal, a large proposed cut in planetary research, but is ignoring a commensurate increase in other space science. The OMB position, with which I totally concur, is that the planetary missions currently in initial development for mid-80s launches (i.e., Galileo to Jupiter and the Venus Orbiting Imaging Radar (VOIR)) are rapidly becoming so expensive as to require major increases in the space program by 1985. We further believe that astronomy and astrophysics (e.g., Space Telescope, Gamma Ray Observatory) represent better, more effective use of the newly acquired potential of the Shuttle as well as representing better science. An important point is that we have invested more than ten years of effort upon planetary exploration, have acquired a vastly improved understanding of the evolution of the solar system, and we have much existing data to digest. The new NASA missions revisit the planets at much higher cost without commensurate additional scientific payoffs. The Shuttle offers us a new capability to expand our horizons through the aforementioned new astrophysical initiatives.

NASA is not in principle opposed to this philosophy. Their basic concern is over continued stability at the Jet Propulsion Laboratory in Pasadena. JPL is a laboratory of some 4,000 people, only 1,000 of whom are directly affected by this program redirection. In addition, DOD has already committed support for approximately 250 of these, with an implied increasing commitment in future years. Although some political pressure may be exercised, the real impact on JPL is not devastating. <u>I believe the cut in planetary exploration represents an example of good management. If "business as usual" were to continue in planetary exploration, an unjustifiable increase in the overall space program would result.</u>

Document II-34

December 11, 1981

<u>Budget Review Board Decisions</u>

<u>NATIONAL AERONAUTICS AND SPACE ADMINISTRATION</u>

1. <u>Issue</u>: Aeronautics Research and Technology

   <u>Decision</u>: Dr. Keyworth suggested an alternative of adding $86 million in budget authority to the OMB recommendation, $43 million for research and development and $43 million for manpower and facilities. The BRB asked NASA to consider the alternative and report back as soon as possible.

2. <u>Issue</u>: Planetary Exploration

   <u>Decision</u>: Dr. Keyworth suggested an alternative of adding $80 million to $90 million to the OMB recommendation. This alternative would permit the stability and excellence of the Jet Propulsion Laboratory to be continued. The BRB asked NASA to consider this alternative and report back promptly.

The Board expressed the preference that both of these issues be settled by OMB and NASA and that an appeal to the President on them be avoided.

Document II-35

Document title: NASA, "Venus Radar Mapper Project Initiation Agreement," October 20, 1982.

Source: NASA Historical Reference Collection, History Office, NASA Headquarters, Washington, D.C.

In 1980, outgoing President Carter approved a NASA mission called Venus Orbiting Imaging Radar to follow the 1978 Pioneer Venus project and return images of the planet's surface at improved levels of resolution. The new Reagan Administration, however, imposed severe constraints on the solar system exploration program beginning in 1981, precluding a new start on the project until FY 1984. By that time the original mission had been reduced in scope; the new concept became known as the Venus Radar Mapper. This document formally put forth the plans for the mission, whose name changed to Magellan in 1986. The Magellan spacecraft was launched from the Space Shuttle in 1989 and conducted a five-year mission after its arrival in orbit around Venus. Covering 95 percent of the planet's surface, Magellan enabled scientists to construct the most comprehensive map of Venus they had to date.

[cover page]

VENUS RADAR MAPPER

PROJECT INITIATION AGREEMENT

[signature, hand-dated 9-16-82]
Charles H. Terhune
Acting Director
Jet Propulsion Laboratory

[signature, hand-dated 10-20-82]
Burton I. Edelson
Associate Administrator
Office of Space Science and Applications

[signature, hand-dated 9-23-82]
Jesse W. Moore
Director
Earth and Planetary Exploration Division

[1] VENUS RADAR MAPPER (VRM)
PROJECT INITIATION AGREEMENT

This initiation agreement sets forth the major responsibilities, inter-faces, procurement plans, and schedule and resources to be followed in the implementation of the VRM mission. The Project is planned for a FY '84 start leading to a launch in April of 1988 and mission completion in mid-1989. The Jet Propulsion Laboratory, as the implementing center, will manage the Project and conduct the flight operations and will contract with industry for the development of the spacecraft and radar.

I. Objectives and Requirements

The Venus Radar Mapper mission addresses fundamental scientific questions about the geology of Venus, such as: What are the processes that form and modify the surface and interior? Is there evidence of plate tectonic activity? Did Venus have water and oceans? How does Venus rid itself of internal heat? The objectives of VRM are to acquire data pertinent to such questions and make a preliminary determination of the geologic history of the planet. Specifically the objectives address:

- surface morphology and processes
- density distribution and interior dynamics
- small-scale surface characteristics

Images of the major physiographic features (mountains, craters, basins) must be obtained. Global coverage is required at a resolution sufficient to identify the features and characterize the geological processes.

Global topographic data are required for geophysical models and to aid in image interpretation by providing local relief, elevations, and slopes.

Gravity data are required for geophysical analyses and to model the interior state and dynamics of the planet.

II. Technical Approach

Mission

Launched by the Shuttle/IUS-II on a direct transfer to Venus, the spacecraft is inserted into a 3.7-hour near polar elliptical orbit by an offloaded solid rocket motor. If the Shuttle/Centaur is used, the spacecraft would include a fully loaded solid rocket motor for orbit insertion allowing more flexibility in orbit design and greater spacecraft weight. For the next 243 days the spacecraft maps the planet, recording data near periapse and playing back data during the apoapsis phase.

[2] Spacecraft

The VRM Spacecraft will be implemented to the maximum extent practical with existing designs and residual hardware. A single spacecraft will be assembled, tested, and flown. The level of achievable performance with existing designs will be established and traded-off against desired performance before new designs are considered. Trade-offs will address system performance cost, risk, and impact on other project elements.

The baseline design of the VRM Spacecraft will make maximum use of the Voyager and Galileo designs and residual hardware. Designs and hardware from other programs will also be employed as available and applicable. A minimum of new designs will be used.

The major structural elements of the spacecraft will be the residual Voyager equipment bus and high-gain antenna, a new forward equipment module for additional housing space and a new spacecraft adapter.

The Command and Data System (CDS) from Galileo will be used with a minor modification. Two Galileo digital tape recorders will be used, also with minor modifications. The Galileo CDS support equipment and its computers will be used in the spacecraft integration complex.

The Attitude and Articulation Control Subsystem (AACS) will be structured around the Galileo AACS processor, memory modules, and power supply assembly. The inertial reference unit will be a Kearfott SKIRU III, and the star scanner will be the same model used on the Boeing inertial upper stage. The Sun sensors and associated electronics will be existing Adcole flight-proven designs, and the reaction wheels will be a NASA standard design.

The Electrical Power Subsystem includes two single-axis solar arrays articulated by Galileo-residual actuators. The power distribution unit, inverter, and pyrotechnic switching unit will be Galileo residual hardware. The power control unit and shunt regulator will be flight-qualified hardware designs from an Air Force satellite program. The two 20-AH batteries will be a NASA standard design.

The Telecommunications Subsystem will use the 3.7 m residual Voyager high-gain antenna, two Viking low-gain antennas, a single Mariner medium-gain antenna design and the ISPM residual traveling wave tube amplifier assemblies. NASA standard deep space transponders shall be used with a Voyager-designed control unit.

A STAR-48 SRM will be used for orbit insertion. The trajectory correction maneuvers will be performed by four 178-N thrust hydrazine, mono-propellant engines. Orientation during the SRM thrusting will be controlled by these same four engines augmented by four 445-N engines and four 22-N engines, all of which have extensive flight history. Twelve residual 0.9-N thrusters from Voyager will be used for momentum dumping throughout the mission. The propulsion subsystem shall be operated in [3] a blowdown mode from a single Shuttle auxiliary propulsion unit tank. Additional pressurant will be stored in a small, Galileo residual tank for a one-shot recharge prior to orbit insertion. Flight-proven ordnance valves and latch valves will be used for propellant isolation to meet the Shuttle safety requirements.

Radar

The radar consists of both a synthetic aperture radar (SAR) and an altimeter sharing common radio frequency components but having separate antennas. The SAR antenna is a residual 3.7 meter antenna from the Voyager program and is shared with the telecommunication system. The altimeter antenna uses eight horns (one at a time, depending on the spacecraft roll angle) which are mounted around the circumference of the SAR/telecommunication antenna.

The SAR operates in the S-band (approximately 2340 Mhz) to allow use of the 3.7 meter antenna. The radar is designed to accommodate an elliptical orbit and provide a minimum of 500 meter radar resolution.

To meet the data rate constraint, the radar has two key characteristics: the first uses variable transmission groups or "bursts"; and the second uses an analog to digital converter with a block floating point quantizer. The bursting of the SAR transmission reduces the number of azimuth looks. The block floating point quantizer allows transmission of less bits per sample while still providing the required dynamic range for the received signal. Parameter adjustments of the radar will be controlled by the spacecraft Command and Data System.

The radar hardware design is strongly based on previous Hughes Aircraft Company designs. Much of the design requires little or no modification. A breadboard and an engineering model are planned for use in the development and pre-launch testing.

Ground Systems

The uplink and real-time telemetry processing are performed in the Project-dedicated Mission Support Area (MSA) on PDP/VAX family mini-computers with smart terminals. During non-critical mission periods, spacecraft performance analysis is carried out at Denver by MMC personnel. Non-real-time processing of SAR data is performed in the Radar Data Processor (RDP); the data is received on DSN generated tapes.

Limited use is made of the JPL Mission Control and Computing Center (MCCC). The Univac 1100 computers are used for navigation, and Flight Projects Support Office (FPSO) personnel are used to monitor project-driven (MCCC) telemetry displays during some nonprime shifts.

Development of the VRM ground systems is based on high inheritance of hardware and software. Inheritance is based on Galileo software in the navigation, sequence design and spacecraft monitor areas. The realtime telemetry system design is based on a proven project-dedicated PDP [4] minicomputer system similar to that used on a previous project. The RDP is developed around the OAST-funded Advanced Digital SAR Processor (ADSP).

Risk Assessment

The VRM mission will be implemented such that the technical risk is essentially equivalent to recent past planetary missions. The application of redundancy, the quality of parts, materials and proceses [sic], the thoroughness of testing, and the initial system margins are comparable. The payload classification is Class A.

Document II-36

Document title: Solar System Exploration Committee, NASA Advisory Council, "Planetary Exploration through Year 2000," 1983.

Source: Solar System Exploration Committee, NASA Advisory Council.

As part of the survival crisis of the solar system exploration program, NASA's leaders recognized the need for a new strategy to underpin the program. As a means of developing that strategy, NASA Chief Scientist John Naugle in 1980 created, as an ad hoc committee of the NASA Advisory Council, a Solar System Exploration Committee. The existence of that committee became even more important as

NASA recovered from the near-cancellation of the solar system exploration program by the Reagan administration at the end of 1981. The committee's activities involved most of those in and out of NASA with an interest in planetary exploration. The committee's deliberations were influential in shaping NASA's approach to the area from 1982 on.

[cover page]

PLANETARY EXPLORATION THROUGH YEAR 2000

A CORE PROGRAM

PART ONE OF A REPORT BY THE SOLAR SYSTEM EXPLORATION SUBCOMMITTEE OF THE NASA ADVISORY COUNCIL

EXECUTIVE SUMMARY

[4] **1. Overview**

During the 20 years between *Mariner 2*'s first flyby of Venus to *Voyager 2*'s final encounter with Saturn in 1981, planetary exploration experienced a Golden Age. Robot spacecraft were launched every few months, reaching outward first to the Moon, Venus, and Mars, then ultimately to every planet known to ancient peoples, from Mercury to Saturn. Most of these spacecraft were launched by the United States. They bore names symbolic of their exploratory missions: *Ranger, Surveyor, Pioneer, Mariner, Viking,* and *Voyager.* The other principal nation to contribute to this era of discovery, the Soviet Union, focused its efforts more narrowly on the Moon, Mars, and Venus. The [5] U.S.S.R.'s planetary missions also achieved remarkable successes. Thus, within less than a generation, humans discovered more than two dozen new worlds, and placed our planet for the first time into its proper context.

Toward the end of this period, constrained budgets for the space sciences resulted in increasingly lengthy intervals between the initiation of new planetary missions. Against this background, the Solar System Exploration Committee (SSEC), an *ad hoc* committee of the NASA Advisory Council, was established in 1980 to take a fresh look at planetary exploration. The resulting study is a joint effort by NASA and members of the scientific community to review the goals of solar system exploration; identify the essential attributes of a viable program in planetary sciences; and define new ways to reduce costs. Based on this intensive study, the SSEC has formulated a Core program which is intended to establish a long-term, stable base for the planetary sciences, in which scientifically exciting missions can be carried out within a framework of cost-saving innovations in both engineering and management.

The SSEC indentified [sic] four goals for solar system exploration. ***The primary goal continues to be the determination of the origin, evolution and present state of the solar sytem*** [sic]. Although great progress has been made in the last two decades in addressing this goal, there remain a number of questions of high scientific priority. Two additional goals include ***understanding the Earth through comparative planetary studies*** and ***understanding the relationship between the chemical and physical evolution of the solar system and the appearance of life***, both of which require intensive study of other solar system bodies. Finally, ***the survey of resources available in near-Earth space is a new goal*** identified by the SSEC; it is essential that the relevant research be done before actual use of such resources can be contemplated.

The Core program which addresses these four goals has a number of attributes identified by the SSEC as essential. First, the program is based on the science strategies developed by the Space Science Board of the National Academy of Sciences, updated by four working committees of planetary scientists including members of the Committee on Planetary and Lunar Exploration of the Space Science Board. Second, the Core program provides a balanced approach to solar system exploration with near-term missions to the terrestrial planets, the small bodies (comets and asteroids), and the outer planets. Third, the Core program reestablishes a critical level of flight activity that is necessary for a healthy scientific program. And fourth, the Core program is designed for a realistic, sustainable budget so that stability can be restored to the planning and implementation of new missions and the associated research and data analysis.

The SSEC recognizes that the achievement of these attributes within a Core program requires several new or revitalized approaches to reducing the costs of individual missions. These approaches include specifying focused, high priority science objectives for missions in the Core program and selecting missions which do not require the development of new technology and which have relatively unconstrained launch opportunities. Increased hardware inheritance is another key approach to reduced cost; this includes use of spare [6] hardware; derivatives of industry-built Earth-orbital spacecraft; and modular spacecraft design. Finally, the Committee underscores the importance of automation and sharing of mission operations as two additional cost-saving approaches.

The recommended Core program incorporates these new approaches to implementation and demonstrates that a viable level of scientific activity addressing high priority science can be achieved within a tightly constrained budget. Based on its current assessment of the various factors involved, the ***SSEC recommends an initial sequence of four Core missions***:

1) Venus Radar Mapper
2) Mars Geoscience/Climatology Orbiter
3) Comet Rendezvous/Asteroid Flyby
4) Titan Probe/Radar Mapper

The ***SSEC further recommends a balanced set of subsequent Core missions*** in the three areas of the terrestrial planets, the small bodies, and the outer planets.

In order to implement the Core program within a realistic budget, the ***SSEC recommends the establishment of a Planetary Observer Program***, a level-of-effort program similar to the *Physics and Astronomy Explorer Program*, consisting of low-cost, modestly scaled inner solar system missions, the first of which would be the *Mars Geoscience/Climatology Orbiter*. The *Planetary Observer* spacecraft would be inexpensive derivatives of existing Earth orbital spacecraft.

The ***SSEC also recommends the development of the Mariner Mark II spacecraft***, a simple modular spacecraft which could be inexpensively reconfigured for different missions beyond the inner solar system, the first of which would be the *Comet Rendezvous/ Asteroid Flyby*.

The ***SSEC further recommends the development of a Common Mission Operations System*** to be shared by all Core missions subsequent to *Venus Radar Mapper*.

In conjunction with the Core missions, the ***SSEC recommends strengthening the planetary research and analysis programs***, both to analyze currently available data and to develop instrumentation for flights in the Core program.

As an additional avenue to reduce cost and to increase flight opportunity, the ***SSEC recommends that vigorous efforts be made to seek mutually beneficial international cooperation in solar system exploration***.

Because there are major scientific objectives in the Space Science Board strategies which are not addressed by the Core missions, the ***SSEC recommends augmenting the Core program with technologically challenging missions as soon as national priorities permit***. Part Two of the Committee's report, to be concluded next year, will present recommendations for this Augmented program.

As discussed in more detail in the following sections, the SSEC believes that its recommendations comprise a healthy and stable program that addresses primary scientific questions within the constraints of realistic, sustainable funding. The results from the recommended Core program will contribute significantly to the nation's continued leadership in solar system exploration.

[7] **2. The Core Program**

The SSEC recommends that the United States planetary exploration program be based on a Core program of missions that meet the following criteria: high scientific priority; moderate technological challenge; and modest cost. By addressing highly focused scientific issues these missions, though restrained in scope, are capable of making substantial progress during the next two decades toward the goals previously identified by the National Academy of Sciences' Space Science Board. The Core program will provide a stable base within NASA and the universities to maintain the capability for planetary exploration. The program will draw significantly on the capabilities of the aerospace industry and will provide the basis for subsequent augmentation with more challenging missions.

[8] **Scientific Basis**

The basis upon which the missions of the Core program were selected has been the science strategies developed by the National Academy of Sciences' Space Science Board. These strategies have been brought up to date by four working committees of planetary science specialists including members of the Space Science Board's Committee on Planetary and Lunar Exploration.

Affordability

Throughout the past twenty years of planetary exploration the program has pushed technology in many areas: launch capability; telecommunications; spacecraft reliability and automation; and atmospheric entry techniques. Meeting these technological challenges has been expensive, but it also has endowed us with a strong technological base that can be exploited to support relatively low cost missions of high scientific priority. These are the missions of the Core program. To achieve the high inheritance required to keep costs down, the missions will use: available spare hardware; derivatives of industry-built Earth-orbital spacecraft; a new, simple, deep-space spacecraft of reconfigurable design; and an updated, multi-mission operations system.

Program Stability

It is proposed that the low cost, inner solar system missions—named *Planetary Observers*—be funded and managed as a continuing program analogous to the successful *Physics and Astronomy Explorers*. The science return from these missions is very high and, because of the accessibility of the inner planets, the data return essentially will be continuous. Allowance is also made in this *Observer* program for funding the U.S. participation in European Space Agency missions now that ESA has opened its payload selection to U.S. investigators in a manner that reciprocates U.S. payload selection policy. Furthermore, the *Observer* program could include joint U.S.-European missions.

The need for a launch capability able to satisfy all the needs of the Core program will be satisfied by the ongoing joint NASA/USAF development of the *Centaur* stage. The *Shuttle/Centaur* will also provide the capability to undertake an augmented program.

The Core program contains resources to analyze already available data and to develop instrumentation for flight on the Core program missions. Resources are also earmarked for the in-depth analysis of data to be returned by the Core missions. These augmented research resources will provide greatly increased stability for the national planetary exploration capability that lies within the universities.

Roles of NASA Center, Aerospace Industry, and Universities

NASA's Jet Propulsion Laboratory is expected to continue to be the major center for planetary exploration with responsibility for project management, for the development of a new, simple, deep-space spacecraft, and for mission operations. NASA's Ames Research Center is expected to continue to support the development of planetary [9] probes—*Galileo*-based atmospheric entry probes and surface penetrator probes. NASA's Goddard Space Flight Center is expected to continue to provide essential support for planetary flight instrumentation. The aerospace industry will supply spacecraft for the exploration of the inner solar system derived from highly capable scientific and commercial spacecraft already built for use in Earth orbit. The universities will continue to be involved in all aspects of the planetary sciences research programs and to provide most of the flight experiments.

Opportunities for International Cooperation

Several missions of the Core program can be carried out in coordination or collaboration with international partners. For example, the recommended *Mars Aeronomy Orbiter* is similar to ESA's candidate *Kepler* mission; the *Comet Rendezvous* mission of the Core program would be ideally carried out in association with a second core mission, the *Comet Atomized Sample Return*; two of the outer planet Core missions—*Titan Probe/Radar Mapper* and *Saturn Orbiter*—could be combined into a collaborative mission.

In addition, international cooperation might also provide an affordable means of augmenting the Core program with additional missions to the mutual benefit of both partners, by permitting joint undertakings that would strain available resources if attempted unilaterally. Discussions with European scientists are underway to examine all these possibilities.

[10] **3. Goals for Planetary Exploration**

It is proposed that the primary goal of the planetary exploration program continue to be the scientific exploration of the solar system. A start also should be made toward a scientific survey of the Moon and Earth-approaching asteroids that would lead to an understanding of their resource potential.

The primary purpose of the planetary exploration program is to achieve a deep understanding of the solar system. The motivations for attempting this insight are at least two-fold.

The first is to understand the origins of the solar system, one of the longest standing goals of human thought. The planetary research program's ultimate objective is to discover how the basic physical laws operate to produce the world in which we live. Such understanding in turn allows us to attempt to predict and to control those natural phenomena. Planetary science uses theory, experiment, and observation to turn knowledge of natural laws into understanding of the world. A major goal of this inquiry is an understanding of the origin and cosmic prevalence of life.

[11] The second motivation is the recognition that the solar system is the entire extended environment of Earth's inhabitants. These is no conceptual barrier to extending the sphere of major human activity ultimately to fill this environmental niche.

Stated specifically, the goals of the planetary program are:

- To continue the scientific exploration of the solar system in order to comprehend its origin, evolution, and present state;
- To gain a better understanding of the Earth by comparative studies with other planets; and
- To understand how the appearance of life related to the chemical and physical history of the solar system.

The program should also have a new, secondary goal:

- The survey of resources available in near-Earth space in order to develop a scientific basis for future utilization of these resources.

Therefore, the Core program includes missions which, together with suitable ground-based and Earth-orbital techniques, will acquire information characterizing the chemical, mineralogical, and physical properties of the Moon and the Earth-approaching asteroids to a level sufficient to provide a first order assay of these bodies.

Document II-37

Document title: Wes Huntress, Director, Solar System Exploration Division, Office of Space Science and Applications, to Lennard Fisk, Associate Administrator, Office of Space Science and Applications, and Al Diaz, Deputy Associate Administrator, Office of Space Science and Applications, "Office of Space Science and Applications Incident Report," April 15, 1991.

Source: NASA Historical Reference Collection, History Office, NASA Headquarters, Washington, D.C.

After being in development for well over a decade, NASA's Galileo probe was launched toward Jupiter in October 1989. Eighteen months into the spacecraft's voyage, flight controllers executed the planned procedure to unfurl Galileo's umbrella-like high-gain antenna, which was designed to send scientific data to Earth at significantly higher rates than the low-gain antennas that the probe had used since launch. As this document indicates, ground operators were unable to fully deploy the antenna. Over the next several months, the Galileo flight team attempted to coax the antenna to open by cooling and shrinking the antenna central tower repeatedly. Despite these efforts, the antenna never fully opened and thus Galileo could only return data to Earth using its low-gain antennas. An investigation into this anomaly found that the most likely cause was that lubricant on the antenna had worn away while the spacecraft remained on the ground for so many years. Although the antenna problem impaired the rate and thus volume of data the Galileo team could obtain, the spacecraft's instruments operated successfully and returned valuable scientific data on Jupiter and its moons for more than four years.

[no pagination]

OFFICE OF SPACE SCIENCE AND APPLICATIONS
INCIDENT REPORT

TO: E/ Dr. L. A. Fisk
Mr. A. V. Diaz

FROM: EL/ Dr. W. T. Huntress Jr.

DATE: April 15, 1991

COPIES TO: E/ J. Alexander; K. Schmoll
ESP/ A. Phillips (Files)

PROGRAM: Galileo
DATE OF INCIDENT: April 11, 1991
LOCATION OF INCIDENT: IN FLIGHT

DESCRIPTION OF INCIDENT:

On April 11, 1991, a real time sequence was uplinked to the Galileo spacecraft for the purpose of deploying the high gain antenna (HGA). Real time commanding was required due to a previous incident on March 26, 1991, which had resulted in the spacecraft executing a safing sequence. This safing process automatically aborted the executing VE-14 sequence which provided for automatic deployment of the HGA on April 10, a day previous to the final selected date. In preparation for the April 11th deployment, the spacecraft was commanded to a state identical in all pertinent respects to that provided by the VE-14 sequence for the planned April 10th deployment. The real time sequence energized the redundant deployment drive motors for eight minutes though in line dual redundant micro switches were expected to remove power at full deployment within approximatimately [sic] three minutes. The sequence was started by the on board software on schedule at 3:51 EDT, Thursday, April 11, 1991. Telemetry measurements of the spacecraft power bus ramped as expected for nearly one minute then reached levels well above nominal for the entire eight minutes that the motors were enabled by the sequence thus indicating that the in line micro switches had failed to activate. Further, a second independent microswitch activated telemetry event indicative of full deployment was not received. Additionally, the preliminary spacecraft spin rate telemetry measurement indicated that the spin rate had decreased only a fraction of that expected; 0.3 milliradians of the 1.3 milliradians expected. At 3:59 EDT eight minutes after energizing the motor and in accordance with the programmed sequence, the motors were successfully commanded off thus completing the functional commands critical to the deployment sequence.

Subsequent analysis has concluded that the antenna did not fully deploy. The present best estimate based on review of the telemetry data for the past three days is that the antenna is partially deployed; perhaps asymmetrically, to a maximum of 34 degrees. This conclusion is based on obscuration of the sun gate which is mounted on the bus sun shade truss with its optical axis parallel to the spacecraft spin (Z) axis.

Presently, three potential causes for the partial deployment have been postulated. Two of them relate to mechanisms; one involves the jack screw/carriage assembly used during deployment, the second deals with the antenna retention assembly which holds the antenna in place during the launch and deploy sequence. The third cause is attributed to improper curing of the epoxy used in securing the antenna ribs and the cord stringers used for shaping the antenna when deployed. Presently, these three theories are being given substantially equal weight by the project. No conclusive evidence has yet been assimilated with respect to any of the theories. The project continues to examine all available data in the hope of determining the exact cause of the partial deployment and thereby develop measures to correct the problem.

IMPACT ON PROGRAM/PROJECT AND SCHEDULE: Failure to fully deploy the HGA would substantially reduce the downlink capability. Capability with the HGA is 134,000 bits per second: without, 10 bits per second. Therefore, the planned Jovian 10 orbit tour would be seriously degraded.

Document II-38

Document title: Louis J. Lanzerotti, Chairman, Space Studies Board, National Research Council, to Dr. Lennard A. Fisk, Associate Administrator for Space Science and Applications, NASA, March 30, 1992, with enclosure, "Scientific Assessment of the CRAF and Cassini Missions," March 30, 1992.

Source: Space Studies Board, National Research Council, Washington, D.C.

The Solar System Exploration Committee recommended in 1983 that NASA should undertake, among other missions, projects to explore Saturn and to visit at least one asteroid and comet. These were to be moderate-sized missions based on the common use of a modular spacecraft called the Mariner Mark II. Congress approved funding in the FY 1990 budget for two missions to achieve these objectives: Cassini and the Comet Rendezvous-Asteroid Flyby (CRAF). When costs for the two missions began to escalate above the established $1.5 billion cost cap, NASA began to delete instruments from CRAF. Despite the Space Studies Board's conviction that CRAF was still a scientifically valuable mission, as stated in this 1992 letter and report, the White House proposed CRAF's cancellation in 1993. NASA proceeded only with the Cassini mission, abandoning the Mariner Mark II spacecraft and designing a less expensive, unique spacecraft for the mission.

[no pagination]

NATIONAL RESEARCH COUNCIL
COMMISSION ON PHYSICAL SCIENCES, MATHEMATICS, AND APPLICATIONS
2101 Constitution Avenue Washington, D.C. 20418

SPACE STUDIES BOARD

Office Location:
Milton Harris Building
Room 584
2001 Wisconsin Avenue, N.W.
(202) 334-3477
Fax: (202) 334-3701

March 30, 1992

Dr. Lennard A. Fisk
Associate Administrator for Space Science and Applications
Code S
NASA Headquarters
Washington, DC 20546

Dear Dr. Fisk:

At its meeting on February 18 and 19, 1992, a subpanel of the Space Studies Board's Committee on Planetary and Lunar Exploration (COMPLEX) chaired by Professor Peter

H. Stone, of the Massachusetts Institute of Technology, carried out a detailed review of the CRAF (Comet Rendezvous-Asteroid Flyby) and Cassini (Titan Probe-Saturn Orbiter) missions. This review was part of COMPLEX's continuing advisory program to assess the responsiveness of NASA missions to science objectives given in COMPLEX's published strategies for exploring the solar system. The results of this review were presented to the Space Studies Board at its meeting on February 26-28 for consideration in the broad context of the status and outlook of the U.S. civil space research program.

COMPLEX's review was planned well before the release of the President's budget message on January 29. That message proposed cancellation of CRAF and called for a reassessment of the technical and schedule risks in the Cassini program. Nevertheless, COMPLEX proceeded with its review to assist those who must respond to the President's proposal. We recognize that scientists, as others, are subject to the effects of large budget deficits and that research must compete with other national needs. We feel obligated, however, to articulate the consequences of budgetary decisions. This letter summarizes the Board's overall assessment of the two missions and is accompanied by a summary providing COMPLEX's detailed scientific evaluation.

In brief, the Board recognizes that the current and near-term national budget environment severely constrains the conduct of the nation's space research program. Therefore, the Board recommends that NASA carefully reevaluate the Cassini spacecraft and instrument complement with the objective of ensuring the mission's prospects for adequate and stable funding leading to the scheduled 1997 launch, while retaining the maximum science content possible. This reevaluation should take into account cancellation of CRAF, if this mission is indeed canceled as proposed in the President's FY 93 budget message. Based on COMPLEX's evaluation, it is the strong recommendation of the Board that a scientifically responsive Cassini mission, reconfigured if necessary, proceed to development and launch on the present schedule.

The Board is dismayed by the proposed cancellation of the CRAF mission, which would be of great scientific merit even without the comet penetrator experiment. However, the Board recognizes that present and anticipated resources are not likely to be adequate to successfully undertake both missions and to meet the science objectives of both at this time.

The U.S. program of outer solar system exploration has brilliantly demonstrated American vision and technical mastery. The United States, alone, has undertaken and completed the initial reconnaissance of the major planets of the outer solar system, visiting in turn Jupiter, Saturn, Uranus, and Neptune and obtaining revolutionary data about these planets and their atmospheres, moons and rings, and plasma environments. The Board believes that a vigorous program of outer solar system exploration is an essential part of a national space exploration agenda. Because of the very long travel times to the outer solar system, seven or more years, it is important not to interrupt development of our next mission or delay its launch. The Saturn system, with its complex interacting system of magnetic fields, plasmas, rings, and moons, is an ideal laboratory for many of the physical processes believed to be important in the formation and present-day dynamics of our solar system and of planetary systems of other stars. It is for these reasons that the Board believes that high priority within the broad civil space agenda should be attached to the ongoing U.S. Saturn exploration program. We further believe that the Cassini mission should proceed without delay in order to benefit from the extremely favorable orientation of Saturn's rings at the spacecraft's projected arrival in 2004.

This review of the Cassini mission was COMPLEX's first since the initial selection of the instrument payload and has been completed prior to the final confirmation of these instruments. Thus, COMPLEX's conclusions are based on the current state of definition of the mission. Given the pending confirmation of the payload, and the programmatic changes that could result from the proposed cancellation of CRAF and technical reassessment of Cassini, COMPLEX plans to reexamine Cassini at a later time.

Sincerely,

[signature]
Louis Lanzerotti
Chairman

Enclosure: Scientific Assessment of the CRAF and Cassini Missions, March 30, 1992

****

[1]

SCIENTIFIC ASSESSMENT
OF THE
CRAF AND CASSINI MISSIONS

March 30, 1992

Summary

At its meeting on February 18 and 19, 1992, a subpanel of the Space Studies Board's Committee on Planetary and Lunar Exploration (COMPLEX) chaired by Professor Peter H. Stone, of the Massachusetts Institute of Technology, carried out a detailed review of the CRAF (Comet Rendezvous-Asteroid Flyby) and Cassini (Titan Probe-Saturn Orbiter) missions. This review was part of COMPLEX's continuing advisory program to assess the responsiveness of NASA missions to science objectives given in COMPLEX's published strategies for exploring the solar system.

It is COMPLEX's opinion that Cassini is highly responsive to the scientific priorities set out in its report, *A Strategy for Exploration of the Outer Planets: 1986-1996*.[1] The instrument payload that has been tentatively selected, the mission plan that has been outlined, and the spacecraft that is being developed together provide an excellent opportunity to advance our understanding of Saturn and its satellites, rings, and magnetosphere. The Saturn system is unique within the solar system because of the wide variety of interactions—electrodynamical, hydrodynamical, and gravitational—among the system's different components. Improving our understanding of these interactions is important for developing better theories of evolution of the early solar system and of planetary and satellite systems in general. In addition, study of Titan's atmosphere is of high priority because it has a composition and chemistry that may be similar to Earth's early atmosphere. The

Cassini mission as currently configured is extremely responsive to the objective of studying the Saturn system as a whole.

COMPLEX notes with concern that present budget constraints are jeopardizing all of the planetary program's large missions, including Cassini. The recent reconfiguration of the Earth Observing System into a series of small spacecraft might be thought to provide a guide for the achievement of science goals outside the context of large missions. Such an analogy is inappropriate for Cassini. The long travel times between Earth and the outer solar system require long-lived components, specialized power systems, and long-distance communications fundamentally different from those required for Earth-orbital missions. With current technology, any mission sent past the asteroid belt must be more than a Discovery-class mission. While intermediate-size missions (larger than Discovery class, but smaller than Cassini) could undoubtedly achieve some of COMPLEX's objectives for the Saturn system, they could not achieve many others. For example, studies of the interactions between the different components of the system, and concurrent coordinated observations of Titan's atmosphere by both the Huygens probe and by remote sensing instruments, require large suites of instruments that place heavy demands on the spacecraft's resources. Thus COMPLEX believes that the Cassini exploration of the Saturn system cannot be fully accomplished by reconfiguration into one or more small spacecraft.

COMPLEX views with dismay the proposal to cancel the currently approved CRAF mission. This mission is the outcome of many years of planning by numerous groups of distinguished scientists, NASA centers, and competitively selected scientific instrument teams. In proposing planning CRAF, the research community fully recognized the importance of [2] assessing priorities in choosing to pursue this major endeavor.

COMPLEX has long articulated the unique scientific opportunities provided by the in situ study of cometary nuclei, believed to be the best-preserved relics of the earliest history of our solar system. The report *Strategy for the Exploration of Primitive Solar System Bodies—Asteroids, Comets, and Meteoroids: 1980-1990*[2] assigned highest priority to reconnaissance and initial exploration of comets, with special emphasis on the rendezvous mode planned for CRAF. This mode is essential for studying the sequence of events that occurs as a comet approaches and recedes from the sun. The CRAF mission has been developed in full accordance with the science objectives and recommendations of COMPLEX. In addition, the mission incorporates an excellent set of asteroid flybys, another high-priority recommendation of COMPLEX.

COMPLEX recognizes that budget constraints have forced significant changes in CRAF since its last review in June 1990. These changes were the deletion of the penetrator experiment (PENL) and of the Scanning Electron Microscope and Particle Analyzer (SEMPA) experiment, a launch delay, a change in the mission's targets, and an increase in the required lifetime of the mission. However, in COMPLEX's opinion, these changes do not invalidate its earlier judgments. CRAF remains a scientifically sound mission, responsive to COMPLEX's most important near-term priorities for the exploration of primitive solar system bodies. Cancellation of CRAF will not lessen the importance of these scientific objectives, which should be pursued at the earliest possible opportunity.

Cassini

COMPLEX's 1986 report, *A Strategy for Exploration of the Outer Planets: 1986-1996*, states that the highest priority for outer planet exploration in the next decade is intensive study of Saturn—the planet, satellites, rings, and magnetosphere—as a system. Specifically, the recommended exploration and intensive study of the Saturn system include the following objectives:

- Titan's atmosphere: Measure the composition, structure, and circulation of Titan's atmosphere, and characterize the atmosphere-surface interaction;
- Titan's surface: Carry out a reconnaissance of the physical properties and geographic variability of Titan's surface;
- Saturn's atmosphere: Determine the elemental composition, dynamics, and cloud composition and structure, to a level well below the H_2O cloud base;
- Saturn's rings: Measure particle composition and spatial distribution, determine the evolution of dynamic structures, and search for shepherding satellites;
- Saturn's small satellites: Make comparative determinations of surface composition, density, geologic history, and geomorphological processes;
- Saturn's magnetosphere: Specify the structure, dynamics, and processes, and the interactions of the magnetosphere with Saturn's atmosphere, rings, icy satellites, Titan, and the solar wind.

These objectives can be met with an appropriately chosen mission profile and complement of scientific instruments, mounted on a spacecraft with sufficient power and communications capabilities. COMPLEX is favorably impressed with the progress made by the Cassini Project in the design of such a spacecraft and mission. The spacecraft accommodates [3] the Cassini science requirements, while being flexible to mission changes. It appears to be a robust and capable carrier for the Cassini investigations. The level of maturity in the design is high for the current phase of development. It is clear that a number of difficult problems have been solved while maintaining prudent engineering margins. (In addition to adequately meeting the Cassini requirements, the spacecraft will also serve the needs of CRAF.)

The instrument payload selected for Cassini is highly responsive to most of the important science objectives for the Saturn system. Following is a summary of the information that the currently configured mission will be able to obtain for each of the major components of the system.

Titan

Many of the scientific goals for Titan will be addressed by the Huygens probe, the component of the Cassini mission supplied by the European Space Agency (ESA). Its current suite of instruments, complemented by spectroscopic and radar observations from the Cassini orbiter, will do an excellent job of fulfilling these goals by providing a first characterization of Titan's atmosphere and surface. The probe's instruments include gas and haze-particle analyzers, capable imaging and spectral radiometers, as well as atmosphere profilers to determine temperature and pressure. Doppler tracking of the descent will provide the first direct measurement of Titan's atmospheric circulation. Near the surface the probe instruments will measure the composition of the atmosphere, the shock of landing—different for a solid as opposed to a liquid surface—and the density and refractive index of a liquid surface, if present.

Although not in orbit around Titan, the Cassini orbiter will repeatedly pass over Titan's surface and will directly measure the composition of the upper atmosphere. The orbiter's infrared spectrometer will determine temperature and composition globally and as functions of time, complementing the measurements made during the probe's descent. Orbiter imaging at visible and infrared wavelengths will determine haze structure and variability. Properties of the upper atmosphere will be measured during Titan flybys by the orbiter's ion and neutral mass spectrometer. The complementarity of obtaining orbiter data coincident with Huygens probe data is an important advantage of the Cassini mission as currently configured.

As the Cassini orbiter repeatedly passes over Titan, its radar will yield further information on the nature of the surface in high-resolution strip scans. This will allow imaging of a significant fraction of the surface at a resolution of 1 km or better. The radar will provide information on the composition of the surface and the depth of hydrocarbon oceans or lakes, if they are present. The radar will also operate in a radiometer mode and map surface dielectric constant variations over the entire surface.

In sum, Cassini will represent a major step in achieving COMPLEX's objectives for Titan.

Saturn's Atmosphere

The Cassini orbiter will determine properties of Saturn's atmosphere at all latitudes and will monitor dynamical changes. Infrared spectra will yield composition and temperature throughout the stratosphere and upper troposphere and, when combined with near-infrared and imaging measurements of reflected sunlight, will determine the thermal energy balance both locally and globally. Cloud structure and horizontal atmospheric motions within the upper troposphere will be obtained from temporal imaging sequences. Temperature, pressure, and ammonia abundance will be determined with excellent vertical resolution to a depth [4] corresponding to a pressure of approximately 1 bar by radio occultations. When combined with infrared spectra, radio occultations will also provide an improved determination of the helium abundance.

The Cassini instruments, together with an orbital tour that includes high-latitude coverage, will address all the Saturn atmospheric objectives outlined by COMPLEX except those for inert gas composition and isotopic abundances.

Rings

Throughout the Cassini mission, the orbiter will take images of Saturn's rings at a full range of viewing angles in both reflected sunlight and the thermal infrared. These images will form the database for tracking dynamical effects, wave motions, and spoke kinematics. The orientation of the rings as seen from Earth is particularly favorable for the proposed orbital tour. According to current mission plans, the rings will occult the orbiter 25 times, providing excellent measurements of the rings' transmission and scattering properties at three distinct radio wavelengths. The radio science experiment and complementary ultraviolet stellar occultation data will determine the particle size and mass distributions in all of Saturn's rings as a function of their distance from the planet.

Cassini's complement of selected instruments is sufficient to achieve all the objectives for Saturn ring science set down by COMPLEX.

Saturn's Small Satellites

Besides Titan, Saturn's satellite system includes several small icy bodies, each displaying a variety of surface landforms and evolutionary histories. They have been affected by internal activity, possibly including tectonism and ice volcanism, and external processes, such as impact. The morphology and stratigraphy of the satellites will be evident from the data provided by the imaging system. These data will advance the understanding of satellite surface processes and history, in addition to addressing the thermal history and state of satellite interiors. The spectroscopic instruments will determine the chemical and mineralogical composition of satellite surfaces. The distribution of various compositional units defined by both spectroscopy and imaging will permit the three-dimensional reconstruction of the configuration of the outer crusts of the icy satellites. Cassini's current instrument payload and mission configuration will provide an unprecedented view of the composition, state, and geological evolution of the small icy satellites of Saturn.

The proposed Cassini mission configuration is fully responsive to COMPLEX's goals for the exploration of Saturn's small satellites.

Magnetosphere

The particles-and-fields instruments will be able to measure particle fluxes with good coverage and good resolution of energy, spatial orientation, mass, and time. Measurements of plasma waves will enable the determination of the sources and sinks of magnetospheric plasma. The particles-and-fields instruments also have an excellent capability to characterize the interaction between Titan and Saturn's magnetosphere. Cassini's magnetometer will be able to determine the configuration of Saturn's nearly axially symmetric magnetic field. This, when combined with the directional capabilities of the radio receiver, will allow determination of the origin of the kilometric radio emission modulation and, hence, the characterization of the nonsymmetric components of Saturn's magnetic field. The temporal and spatial (both radial [5] and latitudinal) coverage of Saturn's magnetosphere during the Cassini mission should clearly establish the nature and origin of temporal variations in the magnetosphere.

The Cassini payload and mission design appear to be fully capable of achieving the major scientific objectives of studying Saturn's magnetosphere.

COMPLEX's overall conclusion is that the Cassini mission, as currently configured, is extremely responsive to the highest-order priority for exploring the outer planets, i.e., intensive study of Saturn as a system.

CRAF

The primary objective of CRAF since its inception has been a comet rendezvous. Comets represent some of the least-altered material left from the formation of the solar system. Thus the study of comets yields important constraints on conditions in the early solar nebula. However, ground-based observations of comets are limited by interference from Earth's atmosphere and by the generally poor viewing geometry for comets when they are near the Sun. Furthermore, the presence of a cometary coma makes viewing the nucleus difficult. The first close observations of a comet were obtained in 1985 when the International Cometary Explorer encountered Comet Giacobini-Zinner. Later, in 1986, spacecraft from Europe, Japan, and the Soviet Union completed fast flybys of Comet Halley. COMPLEX concluded in a letter report[3] that these encounters left COMPLEX's objectives for the exploration of comets largely unchanged. These objectives, given in COMPLEX's 1980 report, *Strategy for the Exploration of Primitive Solar System Bodies--Asteroids, Comets, and Meteoroids: 1980-1990*, are as follows (in order of priority):

1. To determine the composition and physical state of the nucleus (determination of the composition of both dust and gas is an important element of this objective);
2. To determine the processes that govern the composition and distribution of neutral and ionized species in the cometary atmosphere; and
3. To investigate the interaction between the solar wind and cometary atmosphere.

COMPLEX has reviewed CRAF four times, and each time concluded[4] that the mission as configured at the time of the review was responsive to the above objectives. The present review considers whether the changes in the mission since the last review, in July 1990, invalidate earlier conclusions. The significant changes were the descoping of the instrument payload, in the fall of 1990, and the change in the mission profile, in the fall of 1991, which delays the date of launch.

The descoping of the instrument payload in the fall of 1990 was forced by a new NASA assessment of the costs and risks involved with the development of the CRAF penetrator (PENL). This assessment led to a programmatic decision to remove PENL and the Scanning Electron Microscope and Particle Analyzer (SEMPA) from the instrument payload. In its July 1990 review, COMPLEX had identified SEMPA as not being as effective as the Comet Ice and Dust Experiment (CIDEX) and the Cometary Matter Analyzer (COMA) in addressing the most important science objective of CRAF, namely, determining the composition and physical state of the nucleus. Thus at the time, COMPLEX stated[5] that SEMPA had lower priority than PENL and that, in spite of the loss of SEMPA, CRAF would remain responsive to COMPLEX's previously stated science goals.

[6] The loss of the penetrator experiment is much more serious. As stated in the same letter, "deletion of the penetrator would severely compromise the ability of the CRAF mission to address the highest-priority goals identified by COMPLEX." PENL was the only experiment that would have sampled the comet nucleus in situ.

As important as the loss of PENL was, however, COMPLEX continues to hold the view, stated in its 1980 report Strategy for the *Exploration of Primitive Solar System Bodies—Asteroids, Comets, and Meteoroids: 1980-1990,* that comet "science objectives can be met during the next decade without undertaking to land on or penetrate a comet nucleus." This view is based on the fact that the descoped CRAF will still be able to contribute many things to comet science in response to COMPLEX's primary near-term objectives. Following is a list of what CRAF would still be able to do:

- Measure the shape and size of the nucleus;
- Measure the mass to better than one percent;
- Accurately compute the bulk density from the mass and volume;
- Measure the mass distribution of the nucleus;
- Map the active and dormant regions of the nucleus to determine why they are different;
- Study the morphology and evolution of craters;
- Characterize the surface energy budget;
- Determine the surface composition;
- Determine the dust composition;
- Determine the gas composition—both neutral and ion species;
- Gain information about the onset of activity in the comet and the formation of the coma;
- Characterize jet features and the relation between dust and gas in jets;
- Study the magnetic field in the coma; and
- Study the tail, including the interaction with the solar wind.

Many of these anticipated results directly address the primary objective of characterizing the nucleus. COMPLEX therefore believes that the descoped CRAF mission is still responsive to its highest-priority near-term goals for comet science. At the same time COMPLEX reiterates its comments in *Assessment of Solar System Exploration Programs: 1991*[6]: "The 1990 deselection of the CRAF penetrator requires continuing attention to alternative means of directly sampling a comet nucleus. The completion of this goal may still require acquiring and returning a sample of a cometary nucleus in some future mission."

The change in the CRAF mission profile was forced by the Congressional budget decisions for FY 92. The change caused the CRAF launch to be delayed from February 1996 to April 1997. This slippage delays the comet rendezvous from 2003 to 2006, but accommodated a cut in the proposed funding for the CRAF/Cassini program in FY 92. This change does not per se cause any loss in the mission's anticipated return for comet science.

The launch slippage does, however, enhance significantly the results anticipated for asteroid science. The primary goals of asteroid exploration, set forth in COMPLEX's 1980 primitive bodies strategy, are to determine the composition, bulk density, and surface morphology of asteroids. Elucidation of the diversity of asteroids is an essential aspect of these goals. The earlier mission profile would have included flybys of only one or two small asteroids, whereas the new profile includes flybys of two large asteroids, 88 Thisbe and 19 Fortuna, and at least one small asteroid, 1084 Tamariwa. The large asteroids are particularly important since they are likely to be primitive, undifferentiated bodies that would provide information about processes of planet formation in the early solar system.

[7] In addition to the three asteroids, the baseline mission now includes a gravity assist from Mars. This will afford an important opportunity to augment knowledge about Mars, using ultraviolet, visible, and near-infrared imaging capabilities not included on currently planned Mars missions. Along with the opportunity to study Mars itself comes the chance to study Phobos and Deimos, its two satellites. CRAF will be able to compare and contrast these irregular satellites with the three asteroid targets.

The slippage of the launch by one year and of the comet rendezvous by three years does not come without drawbacks. In addition to the added run-out costs, there is additional risk of component failure. However, the lifetime of the reconfigured mission still falls within the design criterion originally specified for CRAF. Although reduction of power from the radioisotope thermoelectric generators is a concern, the rendezvous nature of this mission, with the spacecraft in prolonged close proximity to the comet, relaxes some of the constraints on power usage and makes power usage less a concern for CRAF than for a more rapid flyby mission. Therefore, COMPLEX concludes that the complications arising from the one-year launch delay for CRAF are more than offset by the enhanced opportunities resulting from NASA's redesign of the mission profile.

REFERENCES

1. Space Science Board, *A Strategy for Exploration of the Outer Planets: 1986-1996*, National Academy Press, Washington, DC, 1986.
2. Space Science Board, *Strategy for the Exploration of Primitive Solar System Bodies Asteroids, Comets, and Meteoroids: 1980-1990,* National Academy of Sciences, Washington, DC, 1980.
3. Space Science Board, letter to Geoffrey A. Briggs, May 27, 1987.
4. Space Science Board/Space Studies Board, letter to Geoffrey A. Briggs, May 31, 1985; letter to Geoffrey A. Briggs, May 27, 1987; letter to Geoffrey A. Briggs, September 1, 1988; letter to Lennard A. Fisk, August 10, 1990.
5. Space Studies Board, letter to Lennard A. Fisk, August 10, 1990.
6. Space Studies Board, *Assessment of Solar System Exploration Programs: 1991,* National Academy Press, Washington, DC, 1991.

Document II-39

Document title: NASA, "Small Planetary Mission Plan: Report to Congress," April 1992.

Source: Solar System Exploration Division, Office of Space Science and Applications, NASA Headquarters, Washington, D.C.

Despite the push by the Solar System Exploration Committee earlier in the decade for smaller missions, during the 1980s NASA only launched two large solar system exploration missions: Galileo and Magellan. The Mars Observer mission scheduled for a 1992 launch had also grown into a billion-dollar spacecraft. Realizing that missions had become increasingly ambitious and expensive and thus occurred less frequently, space agency officials decided that they needed to focus on small, low-cost planetary missions in the 1990s. NASA's response to the problem was Discovery, *a program under which scientists could propose solar system missions that could be developed under cost and time caps. NASA would select missions from the proposals every one or two years. This document was a report written by NASA at the request of the Senate Committee on Appropriations in 1992 outlining the agency's rationale and plans for small planetary missions in the 1990s. In 1994,* Discovery *became the first small planetary missions program to receive its own budget line.*

---

[1] **NATIONAL AERONAUTICS AND SPACE ADMINISTRATION**
SMALL PLANETARY MISSION PLAN
REPORT TO CONGRESS
April 1992

INTRODUCTION

In the Senate Report accompanying H.R. 2519, the Fiscal Year 1992 VA, HUD, Independent Agencies Appropriations bill, the Committee on Appropriations directed NASA to prepare "a plan to stimulate and develop small planetary or other space science projects, emphasizing those which could be accomplished by the academic or research communities." This document outlines NASA's small planetary projects plan within the context of overall agency planning. In particular, this plan is consistent with Vision 21: The NASA Strategic Plan, and the Office of Space Science and Applications (OSSA) Strategic Plan.

Over the last 5 years, OSSA's strategic planning process has provided for the introduction and implementation of space science projects ranging from small missions that proceed from concept to flight in less than 3 years to large flagship missions that span a decade. Currently, most science disciplines are pursuing a balanced program of investigations with a flagship flight project as a cornerstone.

However, a change is occurring in the character of the space science program: the aggressive emergence of small missions. The current budgetary environment challenges NASA's ability to sustain a program of high science value and opportunity. As part of an overall approach to maintaining the vitality and progress of the science community, NASA is emphasizing the introduction of small projects in its near-term plans. The budget dictates that all missions be made as [sic] robust and responsive to likely budget fluctuations. Over the next several years, this emphasis will become apparent in the new initiatives NASA brings to the Congress for approval through the budgetary process.

Nowhere is this shift in attention to small missions more apparent than in NASA's Solar System Exploration Division (SSED). Two years ago, small planetary missions were just beginning to be discussed by the scientific community. Today they are the centerpiece of NASA's new programs for the 1990s. As illustrated by Figure 1, SSED has examined each element of the traditional approach to planetary exploration and developed an innovative, small project-oriented approach for more streamlined, cost-effective missions. Wherever possible, this method has been applied to develop new approaches to planned missions and to plan new missions. This document discusses several mission concepts that have been developed using the small projects approach.

In addition, new small robotic missions to the Moon, managed by NASA's Office of Exploration, are being proposed as part of the Space Exploration Initiative.

ATTRIBUTES OF SMALL PLANETARY MISSIONS

Small planetary projects address focused scientific objectives using a limited number of mature instruments, and are designed to require little or no new technology development. Small missions can be implemented by university and industry partnerships in coordination with a NASA Center to use the unique services the agency provides. The timeframe for small missions is consistent with academic degree programs, which makes them an excellent training ground for graduate students and post-doctoral candidates. Because small missions can be conducted relatively quickly and inexpensively, they provide greater opportunity for increased access to space. In addition, small missions contribute to sustaining a vital scientific community by increasing the available opportunities for direct investigator involvement from just a few projects in a career to many.

Figure 1. The Solar System Exploration Division Has Developed an Alternative Small Projects Approach to Each Facet of Its More Traditional Approach.

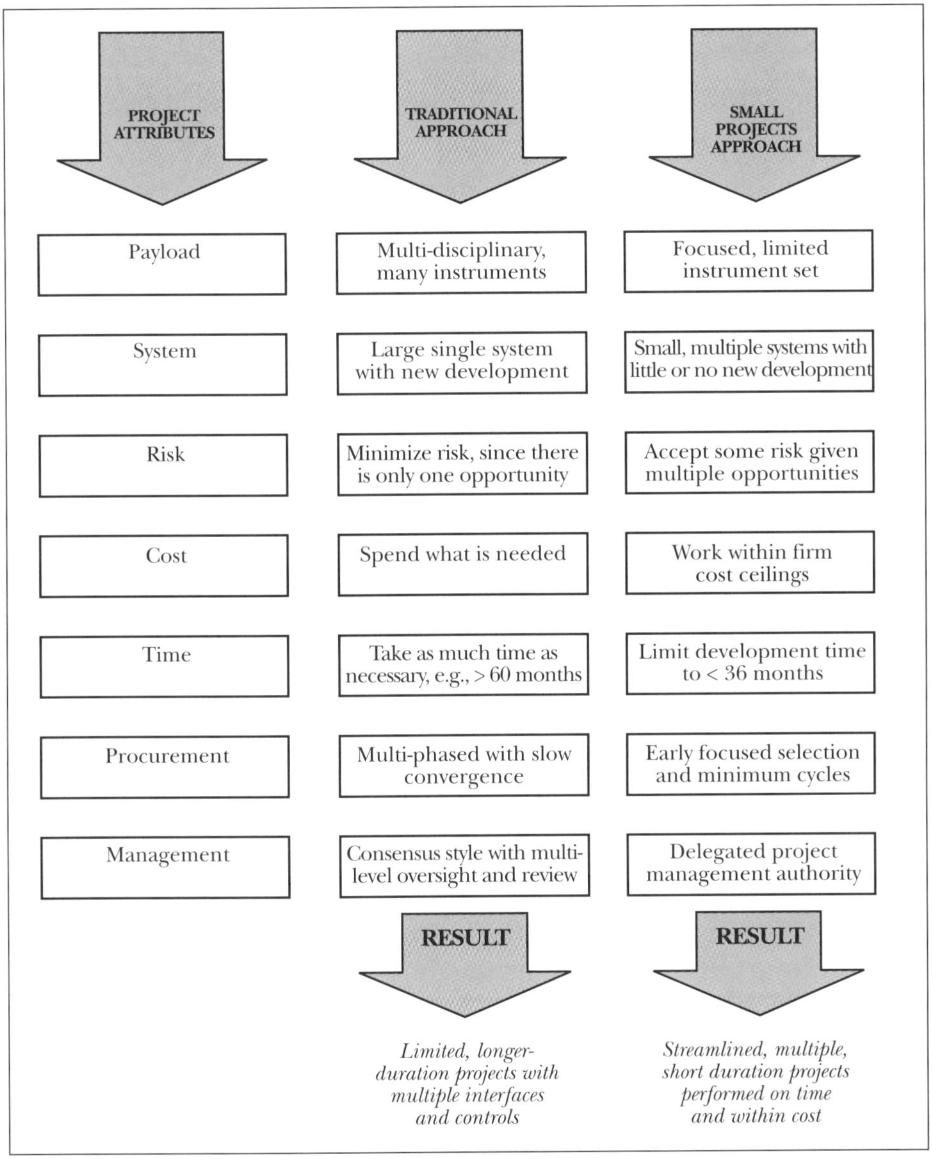

The planetary science community is entering another period of sustained planetary mission data with the success of the Magellan Venus mapping mission, the launch of the Mars Observer mission later this year, and the Galileo encounter with Jupiter in 1995. However, a significant gap exists between the end of Galileo's mission at Jupiter in 1998 and the beginning of Cassini's mission at Saturn, which is not scheduled to occur until well after the turn of the century. Figure 2 illustrates how the proposed small projects discussed below—Discovery, MESUR, and TOPS-0—are designed to sustain the flow of new data throughout this period.

Small missions have fixed cost ceilings and schedules, and are managed with considerably less oversight. NASA program office control is in the nature of assessing hardware and project readiness before approval and adherence to a strict implementation plan, with only ancillary [3] technical support. Once the project is under way, a strong resistance to change must be present to contain cost growth. The small scales and short lifetimes of these projects permit them to be terminated if they grow out of scope.

Figure 2. Planned Small Projects Will Maintain a Steady Flow of New Data to Complement Approved Larger Missions.

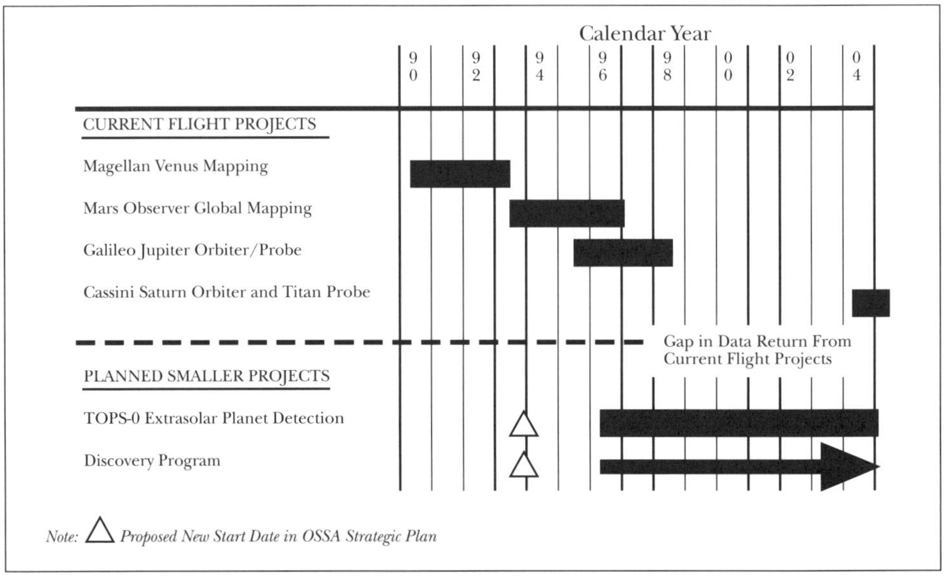

SMALL PLANETARY PROJECT OPPORTUNITIES

The design concepts under review for future small planetary missions draw heavily on NASA's extensive experience with small- and moderate-sized spacecraft, especially the Explorer and Earth Probes programs. Both of these on-going NASA programs emphasize focused science objectives and streamlined management structures, enabling important advances in scientific knowledge. Since its inception, NASA has launched 66 Explorer spacecraft, covering most major disciplines within space science. Some of the more recent Explorer missions include the Cosmic Background Explorer (COBE) launched in 1989, which [4] is revolutionizing our understanding of the early formation of the universe and the Infrared Astronomical Satellite (IRAS), launched in 1984, which provided the first all-sky survey of infrared radiation.

Explorer missions currently in development include the Extreme Ultraviolet Explorer (EUVE) scheduled for a 1992 launch, the X-ray Timing Explorer (XTE) which will follow in 1996, the Advanced Composition Explorer (ACE) planned for launch in 1997, and the Far Ultraviolet Spectroscopic Explorer (FUSE) which will fly early in the next century. NASA is also completing preparations for launch of the Solar, Anomalous, and Magnetospheric Particle Explorer (SAMPEX) later in 1992. SAMPEX was developed in under 3 years as the first Small-class Explorer (SMEX). The Earth Probes program was approved as a new start in the fiscal year 1991 budget. Missions under development include the Total Ozone Mapping Spectrometer (TOMS) scheduled for launch aboard a NASA Small Expendable Launch vehicle (SELV) in 1993; a separate TOMS instrument will be launched on a Japanese H-II launch vehicle in 1996; the NASA Scatterometer (NSCAT) scheduled for launch in 1996; and the Tropical Rainfall Measuring Mission (TRMM) planned for launch in the 1996-97 timeframe.

NASA has identified several exciting, scientifically important small mission candidates. NASA also has developed an innovative small-spacecraft approach to establishing a comprehensive network of small landers on Mars. To illustrate the rich variety and scale of small project opportunities, we briefly describe the following two initiatives:

(1) ***Discovery***, a program concept under study that would involve developing a series of scientifically exciting, cost-effective missions that can be carried out with small spacecraft, small launch vehicles, a constrained instrument payload, and highly focused science objectives for inner solar system objects. Discovery management will be modelled [sic] on NASA's existing Explorer and Earth Probe programs. Total mission cost is not expected to exceed $150 million, with an annual peak funding level of $85 million (all figures cited are in Fiscal Year (FY)1992 dollars).

(2) ***Toward Other Planetary Systems (TOPS-0)***, a program concept under study that would consist of a combination of ground-based observations, scientific research, and technical developments, including the Keck II observatory, to search for, identify, and examine planets around other stars. The cost of this program through the current decade would be less than $100 million.

In addition, the Office of Exploration will manage a lunar resources mapping mission and a lunar gravity-topography mission, both of which would feature small, inexpensive and quick concepts that would produce new lunar global scientific data sets while setting the stage for later human exploration. These missions are each characterized by a small instrument set (e.g. three), using off-the-shelf concepts or flight-proven instruments. The price range of these missions, including launch, is estimated at $100 million each. There will be significant academic involvement in these missions, particularly in the analysis of data sets, which will far surpass existing lunar data in coverage and resolution. ...

[7] SMALL PLANETARY PROJECTS PLAN

NASA has defined and begun to implement a Small Planetary Projects Plan. Work on this plan has been in progress for more than a year, initially stimulated by the creation and structure of the present OSSA strategic planning process, which recognizes the important role of small projects. The Report of the Advisory Committee on the Future of the U.S. Space Program ("Augustine Report") further underscores the importance of this effort. The Augustine Committee urged "that universities, other [8] organizations, and their investigator teams be used increasingly as 'prime' contractors for space research instruments and projects."

Figure 4. Roles and Responsibilities for Small Projects Aare Clearly Defined

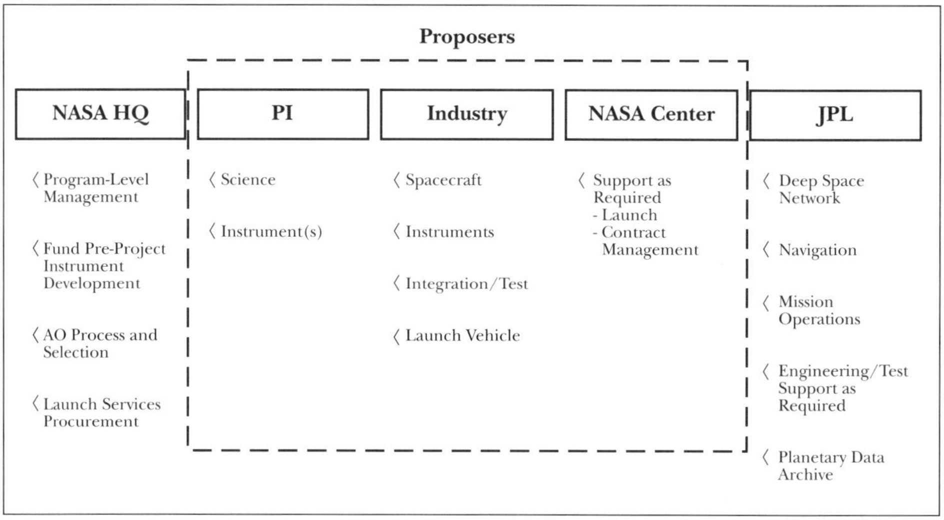

Encouraged by these actions and observations, our plan for small missions has taken shape with the following characteristics:

- Opportunities have focused, well-defined objectives, and use well-developed instruments and flight systems.
- Missions are implemented by qualified teams, preferably with substantial academic representation.
- Strictly observed ceilings will be set on cost, which must include reasonable contingencies.
- Risk will be tailored to each mission, generally at a level moderately higher than for larger projects.
- Project development time will be less than 3 years to ensure cost control and to maintain a steady flow of small projects within the overall program.
- Project oversight will be kept to a minimum, with implementation authority passed down to the Project Manager.

- The project procurement process will be streamlined to minimize the time and effort required to quickly establish each small mission project.

Using these characteristics and focusing on the Discovery Program, NASA has defined its approach for developing and operating small missions. The process begins with the definition of candidate concepts that meet small mission cost and schedule guidelines. This definition serves two purposes: identifying long-lead-time instrument development requirements, and stimulating the interest of industry and academia in developing these concepts into flight hardware. On the basis of scientific priority and concept readiness, a design is selected, and implementation is initiated. Procurement of selected Discovery concepts is expected to be performed using the existing Announcement of Opportunity process. Procurement will be at the flight project level with the selected consortia (academia and industry) of implementers given the responsibility to acquire or develop and integrate the science instruments and flight system based on their winning proposal. [9] Figure 4 illustrates the expected roles and responsibilities of team members in Discovery and other small planetary projects.

Several activities were conducted during the past year to further refine the plan. A Discovery Program working group was formed with strong representation from the academic community to identify science objectives, instrument payloads, and candidate missions. A Discovery Cost and Management Team of Advisors was assembled to develop management and control principles for effective administration of small planetary projects. This team was made up of past planetary project managers, program managers, and individuals with direct experience in the development of fixed cost spacecraft. Their recommendations are summarized in Table 1.

Table 1. A Cost and Management Team of Advisors Made These Recommendations for Small Planetary Projects Management.

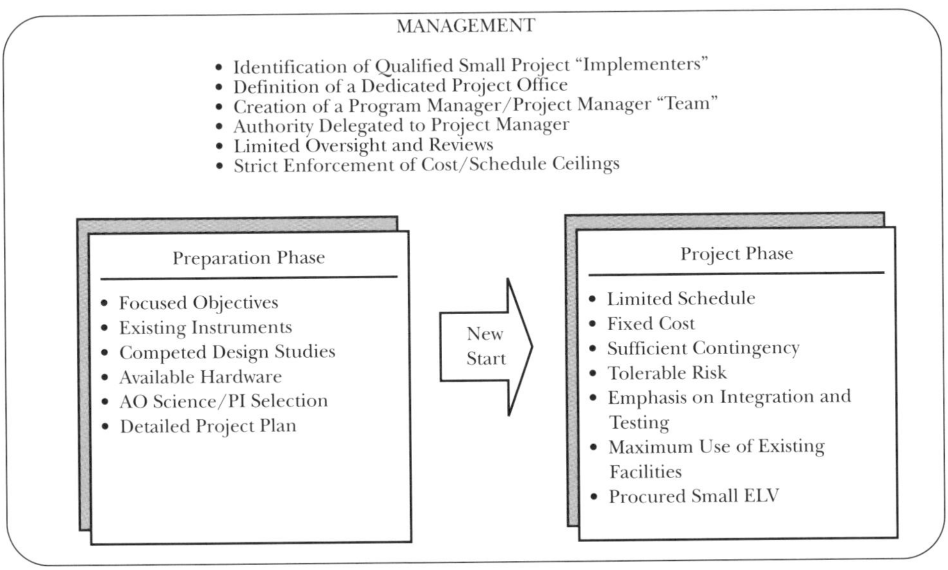

Near-term activities associated with the Small Planetary Projects Plan include the preparation of new start proposals and further planning. Both of these activities will draw on the $10 million requested by the President in Fiscal Year 1993 for small planetary project development. NASA plans to work with the science community to determine the priority of TOPS-0, MESUR Pathfinder, and NEAR.

NASA will continue to seek new candidate small projects. Emphasis will be on developing additional Discovery missions and on the study of scientifically focused, small missions to the outer planets. Outer planet concepts will have to have strictly limited scientific objectives to meet small mission cost criteria. These studies will involve academia and independent research institutions so that their interests and capabilities can continue to be appraised.

[10] **SUMMARY**

NASA shares the Committee's view of the importance of small planetary projects. We are well along in small project development and are paying particular attention to the involvement of the University and Independent Research & Development communities. We intend to implement small planetary projects as soon as possible, consistent with good preparation, sound management practices, and within available resources. Cost containment of these initiatives will make small projects a viable, enduring element of the overall space science program. We intend to rigorously enforce our cost objectives so that as many opportunities as possible will be realized by a community of qualified and enthusiastic scientists and engineers. These small projects can afford timely new opportunities to many investigators and institutions of our science community, fill in important gaps in our planetary exploration program, and revitalize educational interest in the space sciences.

Document II-40

Document title: Mars Observer Mission Failure Investigation Board, "Mars Observer Mission Failure Investigation Board Report," December 31, 1993.

Source: NASA Historical Reference Collection, History Office, NASA Headquarters, Washington, D.C.

NASA developed the Mars Geoscience/Climatology Orbiter in the mid-1980s to build upon the data on Mars' surface composition and atmospheric structure, circulation patterns, and composition received from past missions to the planet. Although started as a low-cost project, the mission's price escalated when scientists chose expensive instruments and engineers faced launch vehicle difficulties. When mission operators attempted to put the $1 billion spacecraft, which came to be known as Mars Observer, into orbit around Mars, they lost contact with the probe. This document reported an independent panel's finding that the loss of communication most likely resulted from a failure in the propulsion system when operators attempted to pressurize the propellant tanks. Shortly after the report's release, the panel admitted that the report should have emphasized that the failure may have been directly linked to a management decision to postpone fuel tank pressurization until reaching Mars instead of completing this procedure five days after launch, as originally planned.

[cover page]

VOLUME I

MARS OBSERVER
Mission Failure
Investigation
Board Report

A report to the Administrator, National Aeronautics and Space Administration on the investigation of the August 1993 mission failure of the Mars Observer spacecraft.

Submitted by the
Mars Observer Mission Failure
Investigation Board

31 DECEMBER 1993

[no page number]

List of Members, Advisors, Observers and Others

Members
Timothy Coffey, Director of Research, Naval Research Laboratory (Chairman)
Thomas C. Betterton, Rear Admiral, USN
Michael D. Griffin, Chief Engineer, National Aeronautics and Space Administration (NASA)
Joseph F. Janni, Chief Scientist, Air Force Phillips Laboratory (AFPL)
Kathryn D. Sullivan, Chief Scientist, National Oceanic and Atmospheric Administration (NOAA)
Peter G. Wilhelm, Director, Naval Center for Space Technology, Naval Research Laboratory (NRL)

Ex-officio Members
Leven Gray, Office of Safety and Mission Assurance, NASA (ex officio)
Gerald Hoskins, NRL (executive secretary)

Advisors, Technical Team Members, and Observers
William Panter, Program Manager - Mars Observer, NASA
Paula Cleggett-Haleim, Public Affairs Office, NASA
Christine Anderson, AFPL
William Baker, NRL
Daniel Barney, NASA
Robert Beal, NRL
Jay P. Boris, NRL
Michael Brown, NRL
Larry Burman, NRL

Robert Eisenhauer, NRL
Joe Foreman, NRL
Andrew Fox, NRL
Henry Hoffman, NASA Goddard Space Flight Center
Samuel Hollander, NRL
Nelson Hyman, NRL
Marvin Levenson, NRL
Christopher Lichtenberg, NRL
Porter Lyon, NRL
Joseph Mattaino, NRL
Wilfred Mazur, NOAA
William McIntyre, NRL
James Nichols, AFPL
William Purdy, NRL
James Rochier, Captain, USAF, Defense Meteorological Satellite Program (DMSP) Office
David Schriftman, NRL
D. Jeffrey Shortt, NRL
Robert Skalitzky, NRL
Gregg Snyder, NRL
Kurt Weiler, NRL

Report Editor
Kenneth Lackie, NRL

[B-2]

PART B
EXECUTIVE SUMMARY

The Mars Observer Mission Failure Investigation Board was established by Mr. Daniel S. Goldin, Administrator, National Aeronautics and Space Administration. The Board was charged to review, analyze, and evaluate the facts and circumstances regarding the loss of spacecraft communications and the failure of the Mars Observer mission; determine the cause of the failure; and report the results to the Administrator.

The Mars Observer program, originally named Mars Geoscience Climatology Orbiter program, was recommended and developed by the Solar System Exploration Committee of the NASA Advisory Council during the period 1981-1983. The spacecraft, orbit, and instruments were to be designed to maximize the scientific return within a modest cost framework. Given approval for a program start in fiscal year 1985, the Jet Propulsion Laboratory (JPL), acting as the implementing Field Center for NASA, was assigned responsibility for managing the program, including contracting with industry for the build and test of the spacecraft bus, acquiring the science instruments, and conducting the flight operations.

Mars Observer was launched from the Cape Canaveral Air Force Station on 25 September 1992. Both the Titan III and the Transfer Orbit Stage vehicle worked well, although the first stage of the launch vehicle suffered a fuel-depletion shutdown during launch. The cruise phase from Earth to Mars was relatively trouble-free, with only a few anomalies noted. The first of a series of maneuvers designed to insert the spacecraft into an orbit around Mars had been planned to take place on 24 August 1993. The sequence of events leading to the first maneuver began as scheduled on 21 August. The first action in this sequence involved pressurization of the propulsion system, initiated and controlled by a sequence of software commands previously stored in the spacecraft computers.

In accordance with the mission's published flight rules, the transmitter on the spacecraft had been turned off during the propellant-tank Pressurization Sequence on 21 August; as a result, there was no telemetry during this event. No data from the spacecraft have been received since that time. This lack of telemetry has seriously hampered an unambiguous determination of the cause of the mishap. The Failure Investigation Board therefore adopted an approach that first identified technically possible failure scenarios, eliminated those deemed implausible, and then categorized the remaining scenarios as either "possible" or "most probable." These scenarios were developed for each spacecraft system. They are included in Part F of this report.

[B-3] To carry out the investigation, the Board established technical teams corresponding to the major subsystems of the spacecraft. The specific technical teams established were:

- Electrical Power
- Attitude and Articulation Control
- Command and Data Handling
- Telecommunications
- Mechanical, including Propulsion
- Software

The teams included representation from NASA, NOAA, AFPL, NRL, and the DMSP Program Office. Each technical team member was required to have significant hands-on experience in areas related to the team's assigned system.

The investigation process involved briefings to the Board and the technical teams by JPL and Martin Marietta Astro Space (MMAS) to establish a baseline understanding of the Mars Observer spacecraft and the ground system supporting its mission. Team visits were made to MMAS and JPL for detailed subsystem reviews and for the development of failure scenarios. The Board was also briefed on a range of related topics, including the NOAA-13 spacecraft failure investigation; the LANDSAT-6 satellite failure; and lessons learned and observations from the LANDSAT-7 program and from the Global Geosciences (GGS) program. Additionally, the Board and the teams closely monitored the progress of independent JPL and MMAS investigation boards.

The Board and the technical teams began by identifying credible failure modes and design weaknesses in the spacecraft. Approximately 60 scenarios were developed and assessed in terms of:

- Consistency with observables;
- Probability of occurrence; and
- Correlation with Pressurization Sequence events.

Four stages of filtering were involved in this process. The first stage identified those failures that could lead to the immediate loss of telecommunications downlink. The second stage eliminated all random failures, since the circumstances of this mishap required a failure to have occurred during the specific 14-minute period (ten minutes during which the transmitters were off, plus four minutes of tube warm-up time) without telemetry. The third stage consisted of identifying the subset of single failures from the second stage that could lead to extended loss (hours to days) of downlink. The fourth stage was focused on the subset of the third-stage failures [B-4] that could be correlated with the Pressurization Sequence, which included several commands and activities that were being executed for the first time during the mission. Those failures that survived through the fourth stage were then examined with respect to supporting test data, analyses and failure history. Specific tests and analyses were identified and performed to validate or invalidate postulated scenarios. This process permitted the Board to classify the failures as to the most probable cause and potential causes.

As a result of these studies, analyses and tests, the Board was led to three principal conclusions:

- *First Principal Conclusion*

Despite extensive analysis of the circumstances surrounding the mission failure of the Mars Observer spacecraft, the Board was unable to find clear and conclusive evidence pointing to a particular scenario as the "smoking gun." Most of the failure scenarios were determined to be implausible or extremely unlikely. The Board was, however, unable to eliminate several failure scenarios. From these remaining scenarios, the Board concluded through a process of elimination that the most probable cause of the loss of downlink from the Mars Observer was a massive failure of the pressurization side of the propulsion system. The Board also concluded that the most probable cause of that failure was the unintended mixing of nitrogen tetroxide (NTO) and monomethyl hydrazine (MMH) in the titanium tubing on the pressurization side of the propulsion system. This mixing was believed by the Board to have been enabled by significant NTO migration through check valves during the eleven-month cruise phase from Earth to Mars. This conclusion is supported (but not proven) by NTO transport-rate data acquired by JPL, by NTO/MMH reaction simulations performed by NRL, and by NTO/MMH mixing tests performed by AFPL.

- *Second Principal Conclusion*

The Board concluded that the Mars Observer spacecraft design is generally sound. The investigation did, however, identify issues (some unrelated to this failure) that should be addressed and corrected prior to any flight of the same or derivative-design spacecraft.

[B-5]

- *Third Principal Conclusion*

The Board concluded that, although the result was a very capable spacecraft, the organization and procedural "system" that developed Mars Observer failed in several areas. In particular, the system failed to react properly to a program that had changed radically from the program that was originally envisioned. Too much reliance was placed on the heritage of spacecraft hardware, software, and procedures, especially since the Mars Observer mission was fundamentally different from the missions of the satellites from which the heritage was derived. The complementary strengths of JPL and Martin Marietta

Astro Space (formerly RCA Astro-Electronics and General Electric Astro-Space Division) were not used by NASA as effectively as they should have been.

Secondary Conclusions

In addition to its assessment of the most probable failure presented earlier, the Board found that the following failures must also be considered as potential causes of the loss of downlink:

- Electrical Power System failure resulting from a regulated power bus short circuit.
- Regulator failure resulting in NTO and/or MMH tank over-pressurization and rupture.
- Ejection of a NASA Standard Initiator at high velocity from a pyro valve, puncturing the MMH tank or causing severe damage to some other spacecraft system.

The Board was generally impressed with the spacecraft that was developed for the Mars Observer mission. However, considering the potential for reflight of an identical spacecraft, or the use of derivative designs or hardware in spacecraft currently in development or planned for future similar mission requirements, a number of specific concerns were noted:

- Propulsion System
  - Inappropriate isolation mechanisms between fuel and oxidizer for an interplanetary mission.
  - [B-6] Lack of post-assembly procedures for verifying cleanliness and proper functioning of the propellant pressurization system.
  - Current lack of understanding of the differences in pyro-initiator characteristics between European Space Agency initiators and NASA Standard Initiators.
  - Inadequate thermal instrumentation, control, and modeling for the mission profile.

- Electrical Power System
  - Potential power bus short circuit susceptibility, due to improper assembly, single component failure, or insulation failure.

- Command and Data Handling System
  - Critical redundancy control functions can be disabled by a single part failure or logic upset.
  - Redundant crystal oscillator (RXO) can lose one of its two outputs without remedy of fault protection.
  - The actual state of the backup oscillator in the RXO is not available in telemetry.

- Software/Fault Protection
  - A top-down audit of fault protection requirements, implementation, and validation is needed.

- Systems Engineering/Flight Rules
  - The flight system should be qualified and capable of providing insight into critical mission events. An example of this would be the availability of telemetry during critical events.
  - The flight system should be allowed to maintain attitude control during critical operations.
  - If any rebuild or modification of the spacecraft is anticipated, the documentation should be updated to reflect the as-built/as-flown configuration.

The Board noted that the Mars Observer that was built departed significantly from the guiding principles originally established for the program, yet the acquisition and management strategy remained unchanged. The role of JPL in this fixed-price procurement was, at best, cumbersome, and did not appear to make the most effective use of the unique resource represented [B-7] by JPL. In any event, the use of a firm, fixed-price contract was inappropriate to the effort as it finally evolved. The original philosophy of minor modifications to a commercial production-line spacecraft was retained throughout the program. The result was reliance on design and component heritage qualification that was inappropriate for the mission. Examples of this reliance were the failure to qualify the traveling wave tube amplifiers for pyro firing shock; the design of the propulsion system; and the use of a fault-management software package that was not fully understood. The Board also noted that the discipline and documentation culture associated with, and appropriate for, commercial production-line spacecraft is basically incompatible with the discipline and documentation required for a one-of-a-kind spacecraft designed for a complex mission. Mars Observer was not a production-line spacecraft.

While the Board can find no direct linkage between the mishap and these systemic weaknesses observed in the Mars Observer program as it evolved over the years, these weaknesses, nevertheless, remain a significant concern for future programs.

The Board would like to express its appreciation for the support provided to the investigation by the six technical teams, the other NRL and AFPL personnel who supported it, the NASA representatives, the JPL Project Team and Investigation Board, and the MMAS Technical Teams.

Document II-41

Document title: Office of Space Science, NASA, "Final Environmental Impact Statement for the Cassini Mission," June 1995.

Source: NASA Historical Reference Collection, History Office, NASA Headquarters, Washington, D.C.

Document II-42

Document title: Office of Space Science, NASA, "Supplemental Environmental Impact Statement for the Cassini Mission," June 1997.

Source: NASA Historical Reference Collection, History Office, NASA Headquarters, Washington, D.C.

In compliance with the National Environmental Policy Act, NASA must issue an Environmental Impact Statement for every spacecraft launch it conducts. Inner solar system probes, whose systems operate using solar power, generally have been approved readily. Spacecraft that travel greater distances, however, cannot depend on the sun but instead use radioactive materials for power. Such missions have been the focus of substantial public opposition before launch due to fear that a launch failure of a spacecraft containing radioactive materials could devastate the environment and life on Earth. These documents contain NASA's analysis of the potential effects on Earth of a catastrophe involving the plutonium-laden Cassini spacecraft during either launch or passage by Earth en route to its final destination, Saturn. The low probabilities of devastating consequences allowed the launch to proceed, despite intense public protests.

Document II-41

[cover page]

Final Environmental Impact Statement for the Cassini Mission

Solar System Exploration Division
Office of Space Science
National Aeronautics and Space Administration
Washington, DC 20546

June 1995

[v] EXECUTIVE SUMMARY

This Final Environmental Impact Statement (FEIS) has been prepared in accordance with the National Environmental Policy Act of 1969 (NEPA) (42 U.S.C. 4321 et seq.), as amended; the Council on Environmental Quality Regulations for Implementing the Procedural Provisions of NEPA (40 CFR Parts 1500-1508); and the National Aeronautics and Space Administration's (NASA) policy and regulations (14 CFR Subpart 1216.3) to support the decision-making process concerning the Proposed Action and alternatives for NASA's Cassini space exploration mission.

PURPOSE AND NEED FOR THE ACTION

The Cassini mission is an international cooperative effort being planned by NASA, the European Space Agency (ESA), and the Italian Space Agency (ASI) to explore the planet Saturn and its environment. The mission would involve a 4-year tour of Saturn, its atmosphere, moons, rings, and magnetosphere by the Cassini spacecraft, which consists of the

Orbiter and the detachable Huygens Probe. The Huygens Probe would be released from the Cassini Orbiter to descend by parachute through the atmosphere of Saturn's largest moon, Titan. During the descent, instruments on the Probe would directly sample the atmosphere and determine its composition. The Probe would also gather data on Titan's landscape.

The Cassini spacecraft would carry three radioisotope thermoelectric generators (RTGs) that use the heat from the decay of plutonium (Pu-238) dioxide fuel to generate electric power for the spacecraft and its instruments. The spacecraft would use radioisotope heater units (RHUs) (157 are planned), also containing plutonium dioxide, to generate heat for controlling the thermal environment onboard the spacecraft and several of its instruments. The U.S. Department of Energy (DOE) would supply the RTGs and RHUs to NASA.

NASA would provide the ground communications network and two scientific instruments for the Huygens Probe. ESA would provide the Huygens Probe, and ASI would provide major elements of the Cassini Orbiter's communications equipment and elements of several science instruments.

The Cassini mission is part of NASA's program for exploration of the solar system. The goal of the program is to understand the birth and evolution of the solar system. Initially, this program concentrated on flyby or reconnaissance-type missions to the outer solar system. With the launch of the Galileo spacecraft in 1989, the program began its transition to exploration-type missions to the outer planets using orbiters and atmospheric probes. The Cassini spacecraft would make remote and close-up measurements of Saturn, its atmosphere, moons, rings, and magnetosphere. This information could also provide significant insights into the formation of the solar system and the conditions that led to life on Earth.

[vi] ALTERNATIVES EVALUATED

The Proposed Action addressed by this FEIS consists of preparing for and implementing the Cassini mission to Saturn to conduct a 4-year scientific exploration of the planet, its atmosphere, moons, rings, and magnetosphere. NASA proposes to launch the spacecraft from Cape Canaveral Air Station (CCAS) (formerly Cape Canaveral Air Force Station [CCAFS]) in October 1997 using a Titan IV (Solid Rocket Motor Upgrade [SRMU]) and a Centaur upper stage (i.e., Titan IV (SRMU)/Centaur) to place the Cassini spacecraft into a 6.7-year Venus-Venus-Earth-Jupiter-Gravity-Assist (VVEJGA) trajectory to Saturn. The SRMU is the most recent upgrade of the solid rocket motor [SRM] used on the Titan IV. If the October 1997 launch opportunity were missed, a secondary launch opportunity exists in December 1997 using an 8.8-year Venus-Earth-Earth-Gravity-Assist (VEEGA) trajectory and a backup launch opportunity exists in March 1999 using a 9.8-year VEEGA trajectory. In the event that the Titan IV (SRMU)/Centaur were not available, a Titan IV (SRM)/Centaur would be used. The launch opportunities would remain the same.

The alternatives to the Proposed Action evaluated in detail are a 1999 mission alternative, a 2001 mission alternative, and the No-Action alternative (i.e., the cancellation of the mission). The 1999 mission alternative would entail dual Shuttle launches from the Kennedy Space Center (KSC), separated by 21 to 51 days, to deliver the Cassini spacecraft and the upper stages) into low Earth orbit. An on-orbit mating of the upper stage(s) and the spacecraft would be performed by astronauts followed by insertion of the spacecraft in March 1999 into its 9.8-year VEEGA interplanetary trajectory to Saturn. A backup launch opportunity, a 9.4-year VEEGA, occurs in August 2000. The 2001 mission alternative would use the Titan IV (SRMU)/Centaur to launch the Cassini spacecraft into a 10.3-year Venus-Venus-Venus-Gravity-Assist (VVVGA) trajectory to Saturn. The spacecraft would require 20 percent additional propellant, as well as completing development of and flight testing a high performance rhenium engine for spacecraft propulsion to accommodate the amount of maneuvering associated with the VVVGA trajectory. An 11.4-year VEEGA backup launch opportunity occurs in May 2002. The No-Action alternative would cancel the mission.

In developing the alternatives (i.e., the Proposed Action and the 1999 and 2001 missions), the available options for the following key components of the mission design were evaluated: launch vehicles, interplanetary trajectories, and power sources for spacecraft electrical needs.

Several criteria were used to evaluate the options: technological feasibility and availability of the option for implementing the mission at the earliest opportunity, impact of the option on the ability of the spacecraft to achieve the mission science objectives, and potential of the option for reducing or eliminating environmental impacts that could be associated with the mission. The evaluation provided the following results: (1) the Titan IV /(SRMU)/Centaur is the most capable U.S. launch vehicle available to implement the mission; (2) the Cassini mission to Saturn requires planetary gravity-assist trajectories; and (3) the spacecraft requires the use of RTGs to satisfy the mission electrical power needs.

[vii] The overall result of the options evaluated indicates that implementation of the Proposed Action, with its three launch opportunities (i.e., primary in October 1997, secondary in December 1997, or backup in March 1999), provides the greatest opportunity to achieve the mission science objectives. The 1999 mission alternative and the 2001 mission alternative also are technically feasible and provide opportunities to achieve most of the science objectives planned for the mission but with less science return (i.e., data).

ENVIRONMENTAL IMPACTS

The only expected environmental impacts of the Proposed Action, as well as the 1999 and 2001 mission alternatives, would be associated with the normal launch of the Cassini spacecraft on the Titan IV (SRMU or SRM)/Centaur or the Shuttle. These impacts have been addressed in previous NEPA documents prepared by the U.S. Air Force (USAF) for its Titan IV launch operations at the CCAS (USAF 1986, USAF 1988a, USAF 1988b) and for the Titan IV using the SRMU (USAF 1990) and prepared by NASA for the Shuttle launches (NASA 1978, NASA 1979, NASA 1988b, NASA 1989b, NASA 1990). The evaluation of these alternatives also used other NEPA-related documentation, including the EIS for the Kennedy Space Center (KSC) (NASA 1979) and the KSC Environmental Resources Document (NASA 1994).

For the Proposed Action, the environmental impacts of a normal launch of the Cassini spacecraft on a Titan IV (SRMU or SRM)/Centaur would result from exhaust emissions (i.e., the exhaust cloud) from the two solid rocket motors (principally aluminum oxide particulates [A12O3], hydrogen chloride [HCl], and carbon monoxide (CO]), which would have a short-term impact on air quality in the vicinity of the launch site; noise from the

SRMUs or SRMs, which would not adversely impact the nearest unprotected person for the general public); deposition of acidic SRMU or SRM exhaust products, largely on the launch complex itself, but which could reach nearby marsh and surface water areas where natural buffering would substantially reduce any impacts; and short-term impacts on stratospheric ozone along the launch vehicle's flight path from the SRMU or SRM exhaust products. No substantial long-term environmental impacts would be associated with a normal launch of the Cassini spacecraft for any of the launch opportunities.

The radiological concern associated with the mission is the potential release of some of the approximately 32 kg (71 lb) of plutonium dioxide (consisting of around 71 percent by weight Pu-238 at launch) in the RTGs and RHUs onboard the spacecraft. In the unlikely event that an accident were to occur during the launch of the spacecraft (i.e., from the time of ignition of the SRMUs or SRMs, through the insertion of the spacecraft into its interplanetary trajectory), the safety features incorporated into the RTGs and RHUs, in most cases, would limit or prevent any release of the plutonium dioxide fuel. However, in the unlikely event of a launch phase accident causing a release of plutonium dioxide fuel, no health effects (i.e., excess latent cancer fatalities [above the normally observed cancer fatalities]) would be expected to occur if members of the population were exposed to the released radioactive fuel.

For launch Phases 1 through 6 on the Titan IV (SRMU)/Centaur, four accident scenarios were identified as representative of the categories of failures that could release [viii] plutonium dioxide fuel to the environment. In addition, two postulated very low probability (i.e., much lower than the probabilities for Phases 1 through 6) accident scenarios that could occur during the interplanetary portions of the VVEJGA and VEEGA trajectories were identified as the short-term and long-term inadvertent reentry scenarios. The short-term scenario would involve the inadvertent reentry of the spacecraft into the Earth's atmosphere during a planned Earth swingby, and the long-term scenario would involve a spacecraft failure that leaves the spacecraft drifting in an Earth-crossing orbit and potentially reentering the Earth's atmosphere a decade to millennia later. Preliminary estimates for a Titan IV (SRM)/Centaur launch indicate that the radiological consequences and the risk would be similar to those for the Titan IV (SRMU)/Centaur.

Depending on the accident scenario, the CCAS/KSC regional area, limited portions of the African continent under the vehicle flight path, or indeterminate locations within the global area could be impacted by plutonium dioxide fuel releases. The CCAS/KSC regional area could be impacted if a Phase 1 accident were to result in a release. Areas outside the region (i.e., portions of the African continent; areas elsewhere around the world) could be impacted if an accident resulting in a release were to occur in Phase 5 or 6. Considering potential accidents that could result in a release across all launch phases, no excess cancer fatalities would be expected in the exposed population. No releases of plutonium from the RTGs to the environment are postulated if any of the representative accident scenarios occurred in Phases 2, 3, or 4.

During the interplanetary portions of the mission, postulated short- and long-term inadvertent reentry accident scenarios could result in releases of plutonium dioxide to the environment. However, NASA is designing the mission to avoid the potential for such accidents. The mission's design ensures that the expected probability of an inadvertent reentry would be less than one in a million. If such an accident were to occur, plutonium dioxide could be released in the upper atmosphere and/or scattered in indeterminate locations on the Earth's surface. Within the exposed population of 5 billion people, approximately 1 billion people (i.e., 20 percent or 1 /5 of the population) would be expected to die of cancer due to other causes. The estimated fatalities that could result from an inadvertent reentry with release would represent an additional 0.0005 percent above the normally observed 1 billion cancer fatalities.

The principal method used in this document for characterizing the radiological impacts of each alternative evaluated is health effects risk. Health effects are expressed as the number of excess latent cancer fatalities (above the normally observed cancer fatalities) caused by exposure to the plutonium dioxide fuel. As used in this FEIS, health effects mission risk is the probability of an accident with a plutonium dioxide fuel release (i.e., the probability of an initiating accident times the probability of that accident causing a release of plutonium dioxide, since not all accidents would result in a plutonium dioxide release) multiplied by the consequences of that accident (i.e., the health effects that could be caused by the exposure of individuals to the plutonium dioxide), summed over all postulated accidents. Estimates of health effects mission risk, as discussed in this FEIS, represent the expectation latent cancer fatalities. The expectation health effects mission risk over all mission phases (i.e., the total or overall health effects mission risk) does not include contributions to risk from the long-term reentry scenario.

[ix] For the Proposed Action, the health effects mission risk considering all launch phases for the primary launch opportunity would be 8.4 x 10^{-7}. The health effects mission risk from the short-term inadvertent reentry accident during the Earth swingby portion of the primary launch opportunity's VVEJGA trajectory would be 1.7 x 10^{-3} and for the secondary and backup opportunities' VEEGA trajectories would be 1.8 x 10^{-3}. The total health effects mission risk (considering all launch phases and the Earth-Gravity-Assist trajectories) from the primary launch opportunity would be 1.7 x 10^{-3} and from the backup launch opportunity would be 1.8 x 10^{-3}. The health effects mission risks from the Cassini mission would be small and less than the total health risks faced by the public from construction and/or operation of large industrial projects.

The environmental impacts of a normal launch of the 1999 mission would be associated with the normal operations of the Shuttle. These Shuttle operations would result in temporary impacts on air and water quality near the launch site. Because this alternative would require two Shuttle launches, impacts would occur two times separated by 21 to 51 days.

During the second Shuttle launch for this mission alternative, certain accidents that may occur could result in a release of a portion of the plutonium dioxide from the RTGs to the environment. The local CCAS/KSC regional area could be impacted if a Phase 1 accident resulted in a release. Limited portions of the African land mass could be impacted by a Phase 2 accident, and Phases 3 and 4 accidents could impact indeterminate locations within the global area. In addition, releases could occur from an accident occurring during a short-term inadvertent reentry.

Potential failures and radiological consequences associated with the Earth swingby portions of the VEEGA trajectory would be expected to be identical to those analyzed for the VEEGA swingbys for the 1999 backup launch opportunity of the Proposed Action.

Using estimation methods similar to that for the Proposed Action, the health effects mission risk over all the mission launch phases for the 1999 mission alternative is 2.1 x 10^{-6}. The corresponding risk from a short-term inadvertent reentry during the Earth swingby portion of the VEEGA trajectories would be 1.8 x 10^{-3}, and the total health effects mission risk would be 1.8 x 10^{-3}.

The environmental impacts of a normal launch of the 2001 mission alternative would be similar to those estimated for the Proposed Action. The spacecraft with a high performance rhenium propulsion engine would be launched on the Titan IV (SRMU)/Centaur. The launch accident scenarios that could result in a release of plutonium dioxide fuel and the associated consequences and risks would be identical to those evaluated for the Proposed Action. The overall health effects mission risk from the launch phases is 8.4 x 10^{-7}. The primary launch opportunity of this 2001 mission alternative would not use the Earth for a gravity-assist (the trajectory is a VVVGA); subsequently, there would be no consequences and health effects mission risks associated with a short-term inadvertent reentry. Because there is no non-EGA backup launch opportunity for the 2001 mission alternative, the backup opportunity would use a VEEGA. The health effects mission risk from the backup short-term inadvertent reentry is 1.8 x 10^{-3}. The overall [x] health effects mission risk from the primary opportunity is 8.4 x 10^{-7} and from the backup is 1.8 x 10^{-3}.

For all launch opportunities, should the spacecraft become uncommandable any time after injection into its interplanetary trajectory and before the final planetary gravity-assist, the spacecraft could eventually reenter the Earth's atmosphere a decade to centuries later (i.e., long-term inadvertent reentry scenario). The health effects mission risk of such an event is assumed to be similar (i.e., same order of magnitude) to that estimated for the short-term inadvertent reentry for the primary launch opportunity associated with the Proposed Action.

No environmental impacts would be associated with the No-Action alternative.

MISSION-SPECIFIC CONSIDERATIONS

The Proposed Action has the greatest potential to accomplish the mission and its scientific objectives. In addition, because the Proposed Action would ensure that adequate performance margins are available (e.g., spacecraft propellant available for maneuvers during the Saturn science tour), it would have the greatest likelihood to take advantage of both planned and unplanned opportunities for science return. The expected science return for the Proposed Action's December 1997 and March 1999 contingency launch opportunities would be less due to the later arrival time at Saturn. For similar reasons, the expected science return for the 1999 mission alternative using the two-Shuttle launch would be less than the return obtained from the Proposed Action.

Although the 2001 mission alternative would achieve most of the planned science objectives, it would not return as much science as the Proposed Action. The larger propellant tank and propellant load would reduce the overall mission performance, requiring the use of a specially developed rhenium spacecraft propulsion engine. Even with the use of this more efficient propulsion engine, the number of Titan flybys would be reduced from 35 to 21. Other trajectory adjustments would be necessary to conserve propellant. In addition to reducing the opportunity for obtaining the planned science return, the ability of the spacecraft to take advantage of unplanned discoveries would be limited. Because this alternative requires a longer flight time than the Proposed Action, and the launch would be delayed relative to the primary launch opportunity, the international partnerships formed to develop the Cassini spacecraft, Huygens Probe, and other space-related projects could be disrupted.

Because the No-Action alternative would cancel the mission, the science return would be lost, and the ability of the United States to enter into future international agreements for cooperative space activities could be impaired.

Document II-42

[cover page]

FINAL

SUPPLEMENTAL ENVIRONMENTAL IMPACT STATEMENT
FOR THE CASSINI MISSION

Office of Space Science
National Aeronautics and Space Administration
Washington, DC 20546

June 1997

[iii] EXECUTIVE SUMMARY

This Supplemental Environmental Impact Statement (SEIS) has been prepared in accordance with the National Environmental Policy Act of 1969 (NEPA), as amended (42 U.S.C. 4321 et. seq.); the Council on Environmental Quality Regulations for Implementing the Procedural Provisions of NEPA (40 CFR Parts 1500-1508); and the National Aeronautics and Space Administration's (NASA's) policy and procedures (14 CFR Subpart 1216.3) to support the decision-making process concerning the Proposed Action and alternatives for NASA's Cassini space exploration mission.

NASA completed development of the Cassini mission Environmental Impact Statement (hereafter denoted 1995 Cassini EIS) with distribution of the Final EIS to the public and other interested parties in July 1995. The Record of Decision (ROD) was rendered in October 1995. The 1995 Cassini EIS contained NASA's evaluation of the potential impacts of completing preparations for and implementing the Cassini mission, with particular emphasis on accidents that could potentially occur during launch and cruise phases of the mission, and which could impact human health and the environment. While the 1995 Cassini EIS analyses used the best information available at that time, the 1995 Cassini EIS noted that NASA and the U.S. Department of Energy (DOE) were continuing to analyze and evaluate additional accident scenarios specific to the Cassini spacecraft and its launch vehicle and trajectory. In both the 1995 Cassini EIS and the ROD, NASA made the commitment that, should significant differences arise between the results of the ongoing analyses and the 1995 Cassini EIS, NASA would evaluate the information and make a determination regarding the need for additional NEPA documentation, including supplementing the 1995 Cassini EIS. Updates of the safety analyses in support of the 1995 Cassini EIS were recently made available to NASA. NASA has evaluated those analyses accordingly, and has determined that the purposes of NEPA are furthered by preparation of this SEIS.

PURPOSE AND NEED FOR THE ACTION

The Cassini mission is an international cooperative effort of NASA, the European Space Agency (ESA), and the Italian Space Agency (ASI), to explore the planet Saturn and its environment. Saturn is the second-largest and second-most massive planet in the solar system, and has the largest, most visible, dynamic ring structure of all the planets. The mission is an important part of NASA's program for exploration of the solar system, the goal of which is to understand the system's birth and evolution. The Cassini mission involves a four-year scientific exploration of Saturn, its atmosphere, moons, rings and magnetosphere. The Cassini spacecraft consists of the Cassini Orbiter and the detachable Huygens Probe.

The Cassini mission represents an important step in the exploratory phase of planetary science, with the detailed data that would be obtained from the mission providing an important basis for continuing Earth-based studies of the planets. There are five major [iv] areas of investigation planned for the Cassini Mission. An overview of each area of investigation follows:

- The previous Pioneer and Voyager swingby missions to Saturn obtained only short-duration, remote-sensing measurements of the Saturnian atmosphere. These measurements have been sufficient to generally determine the basic composition, energy balance, temperature profile, and wind speeds in the planet's upper atmosphere. Cassini would further investigate cloud properties and atmospheric composition, wind patterns, and temperatures, as well as Saturn's internal structure, rotation, ionosphere, and origin and evolution. The missions would involve orbits near the equator and the poles of Saturn so that the entire planet could be studied.

- Titan is shrouded by dense clouds; therefore, little is known about its surface. Data collected by the instruments onboard the Cassini orbiter and the Huygens Probe would provide a better understanding of the abundance of elements and compounds in Titan's atmosphere, the distribution of trace gases and aerosols, winds and temperature, and surface state and composition. In particular, the spacecraft's radar would penetrate Titan's dense atmosphere and reveal the moon's surface characteristics. The Huygens Probe, carrying a robotic laboratory, would perform chemical analyses of Titan's atmosphere and clouds. As the Probe descends, the onboard instruments would measure the temperature, pressure, density, and energy balance through the atmosphere to the moon's surface. The surface properties would be measured remotely, and a camera would photograph the Titan panorama and relay the images to Earth via the Cassini Orbiter.

- Saturn's other satellites (i.e., moons) are ice-covered bodies. Cassini would investigate their physical characteristics, the composition and distribution of materials on their surfaces, their internal structure, and how they interact with Saturn's magnetosphere. Of particular interest is the half-dark and half-light moon, Iapetus. The light side of the moon is believed to be composed of ice and the dark side possibly of some organic material. The data obtained by Cassini would assist in determining the geological histories of the satellites and the evolution of their surface characteristics.

- The Voyager swingbys in 1980 and 1981 proved Saturn's ring system to be much more complex than previously realized, with intricate dynamic interactions in most parts of the system. The short-term Voyager studies showed a wide range of unexplained phenomena in the rings, including various wave patterns, small and large gaps, clumping of material and small, so-called "moonlets" embedded in the rings. Long-term, close-up observations of the rings by Cassini could help resolve whether the rings are material left over from Saturn's original formation, or whether they are remnants of one or more moons shattered by comet or meteor strikes. Applied to larger-scale disk-shaped systems, the detailed studies of Saturn's rings proposed for Cassini would provide important contributions to theories of the origin and evolution of the dust and gas from which the planets first formed.

  [v] The tilt of Saturn's ring plane changes as the planet orbits the Sun and the changing angle of sunlight illuminating the rings dramatically alters their visibility. Cassini's arrival at Saturn is timed for optimum viewing of the rings, during a period when they will be well illuminated by sunlight. Upon Cassini's arrival at Saturn in 2004 when launched in October 1997, the tilt of the ring plane and resulting illumination angle would allow Cassini's instruments an unsurpassed view of the ring disk.

  Cassini would allow detailed studies of ring structure and composition, dynamic processes, dust and micrometeoroid environments, and interactions among the ring systems, magnetosphere, and satellites.

- Saturn's magnetosphere is the region of space under the dominant influence of the planet's magnetic field. Cassini would carry instruments to study the configuration and dynamics of the magnetosphere; the nature, source, and fate of its trapped particles; and its interactions with the solar wind and Saturn's satellites and rings. A particular phenomenon of interest is the Saturn Kilometric Radiation—a poorly understood, very low frequency, electromagnetic radiation—which scientists believe is emitted by the auroral regions in Saturn's high latitudes.

Implementation of the proposed action would also ensure that the spacecraft would complete its orbital tour before 2010, when Saturn's rings would present themselves nearly edge-on to the Earth and Sun, severely limiting the ability for detailed observations.

The Cassini spacecraft incorporates three (3) Radioisotope Thermoelectric Generators (RTGs) to provide onboard electric power for spacecraft operation and scientific instruments. The RTGs generate electric power by utilizing the heat from decay of radioactive material. The material is an isotopic mixture of plutonium in the form of dioxide, along with small amounts of long-lived actinides and other impurities. About 71 percent of the oxide mixture (by weight) is plutonium-238 (Pu-238). The three RTGs onboard the Cassini spacecraft contain a total of 32.7 kg (about 72 lb) of PuO2, amounting to 1.49x1016 Bq (402,000 Ci). In addition, 129 Radioisotope Heater Units (RHUs) will be employed to regulate the temperature inside the spacecraft and for several instruments. Each RHU contains about 2.7 gm (0.006 lb) of mostly plutonium-238 dioxide, amounting to a collective total of about 0.35 kg (0.77 lb), or about 1.48x1014 Bq (4,000 Ci) of radioactive material in the 129 RHUs.

The 1995 Cassini EIS was made available to Federal, state and local agencies, the public and other interested parties on July 21, 1995. In addition to the No-Action Alternative, the 1995 Cassini EIS addressed three alternatives for completing preparations for and operating the Cassini mission to Saturn and its moons. On October 20, 1995, utilizing the impact analyses in the EIS, along with other important considerations such as [vi] programmatic, economic, and international relations, the ROD selecting the Proposed Action was rendered.

ALTERNATIVES EVALUATED

The Proposed Action and preferred alternative consists of completing preparations for and operating the Cassini mission to Saturn and its moons, with a launch of the Cassini spacecraft onboard a Titan IV(SRMU)/Centaur. The launch would take place at Cape Canaveral Air Station (CCAS) during the primary launch opportunity of October 6 through November 15, 1997. A secondary launch opportunity occurs from late November 1997 through early January 1998, with a backup opportunity from mid-March to early April 1999, both using the Titan IV(SRMU)/Centaur. The primary launch opportunity would employ a Venus-Venus-Earth-Jupiter-Gravity-Assist (VVEJGA) trajectory to Saturn; the secondary and backup opportunities would both employ a Venus-Earth-Earth-Gravity-Assist (VEEGA) trajectory. The Proposed Action would allow the Cassini spacecraft to gather the full science return desired to accomplish mission objectives.

Along with the No-Action Alternative, the 1995 Cassini EIS evaluated two other mission alternatives. The March 1999 Alternative would have used two Shuttle flights launched from Kennedy Space Center (KSC), with on-orbit integration of the spacecraft and upper stage, followed by injection of the spacecraft into a VEEGA trajectory to Saturn. The March 1999 Alternative is no longer considered reasonable at this time due to the long lead-time in developing and certifying the new upper stage that would be needed to implement this mission alternative. When combined with the significant additional costs associated with this alternative, the 1999 dual Shuttle alternative is no longer considered reasonable.

The other mission alternative evaluated in the 1995 Cassini EIS was the 2001 Alternative, which would use a Titan IV(SRMU)/Centaur to launch the spacecraft from CCAS in March 2001 using a Venus-Venus-Venus-Gravity-Assist (VVVGA) trajectory. A backup opportunity in May 2002 would use a VEEGA trajectory. The 2001 Alternative would require completing the development and testing of a new high-performance rhenium engine for the spacecraft, as well as adding about 20 percent more propellant to the spacecraft. Science returns from this alternative would meet the minimum acceptable level for the mission.

RADIOLOGICAL IMPACTS OF ACCIDENTS

Evaluation of the recently available safety analyses has indicated that the only parts of the previous Cassini EIS potentially affected are the analyses of the radiological consequences of accidents involving a potential release of plutonium dioxide (source term) from the RTGs and/or the RHUs onboard the spacecraft. The environmental impacts of completing preparations for the mission are unaffected by the updated analyses, and [vii] remain as presented in the 1995 Cassini EIS. In addition, the analyses of the environmental impacts of both an incident-free launch and incident-free interplanetary gravity-assist trajectory are also unaffected and remain as presented in the 1995 Cassini EIS.

The EIS's and recently available analyses overall assessments of the Cassini mission's risk are similar. The updated assessment of individual mission segment accidents has identified higher risks for launch segment accidents and lower risks for the Earth gravity assist (EGA) swingby segment. Both the EIS and the updated analyses indicate that only a fraction of conceivable launch accidents are calculated to result in releases of PuO_2.

The ongoing safety analysis process is similar to the process used for the earlier Galileo and Ulysses missions and has resulted in incremental improvements in the modeling and analysis techniques. The potential source terms are determined by using simulations to evaluate the response of the RTGs, RTG components, and RHUs to the defined accident environments. The ongoing analyses utilize probabilistic risk assessment techniques with computer simulation and modeling of RTG responses to accident environments, and are based upon safety test and analysis studies performed by and on behalf of DOE. The safety test and analysis studies have been performed over the past 12 years on General Purpose Heat Source (GPHS) RTGs and materials, and RHUs. These tests provide a database of the performance response of the RTGs and RHUs to simulated accident conditions such as high-velocity impacts on hard surfaces, impacts from high-velocity fragments, and exposure to thermal and mechanical stresses such as would be encountered in a reentry from Earth orbit or exposure to burning solid rocket motor propellant. It must be emphasized that for a release of plutonium dioxide (PuO_2) to occur, the initiating accident must be followed by other events to create an accident environment that threatens the integrity of the RTGs and RHUs.

Since the issuance of the 1995 Cassini EIS, the refinements in the evaluation of accidents and estimation of their potential consequences have resulted in revised estimates. Comparison between the 1995 Cassini EIS results and the updated results are presented in this SEIS. The 1995 Cassini EIS reported point estimates of the "expectation" and "maximum" cases. The expectation case utilized source terms for each accident scenario that were probability-weighted, and was based upon a range of release conditions considered in the analysis. The maximum case utilized source terms that corresponded to either the upper limit deemed credible for the scenario, based on consideration of supporting analyses and safety test data, or to a total probability greater than or equal to a probability cutoff of $1x10^{-7}$ (1 in 10 million). The updated analyses used probabilistic risk assessment techniques similar to those used for the Galileo and Ulysses missions to generate updated estimates of consequences and risk.

The 1995 Cassini EIS utilized the concept of risk as one of the key measures in the accident analyses. Risk, for the purpose of the 1995 Cassini EIS and for this supplement, is defined as the total probability of an event occurring (i.e., a release from an RTG or RHU), multiplied by the mean consequence of the event (i.e., health effects described as latent [viii] cancer fatalities over a 50-year period within the population potentially exposed by an accident). With respect to the Cassini accident analyses, the total probability of a release occurring is determined by multiplying the probability of the initiating accident that could threaten the RTGs and RHUs, times the conditional probability that the accident will result in a release. Risk estimates for the Cassini mission (expressed as health effects) have been developed for each mission phase/accident scenario and for the average exposed individual. The updated analyses report the best estimate of consequences and risks. While the overall probability of an accident that could threaten the RTGs or RHUs during the Cassini mission is 2.8×10^{-2}, or 1 in 36, the probability of an accident predicted to release PuO_2 is 2.8×10^{-3}, or less than 1 in 357. Such an accident could result in 0.089 mean health effects. This results in an overall mission risk of 2.5×10^{-4}, or 0.00025, health effects worldwide. This risk level is lower than the overall risk reported in the 1995 Cassini EIS (expected value of 1.7×10^{-3}, or 0.0017, health effects).

The total mission risk is distributed over four major mission segments—i.e., pre-launch (Phase 0), early launch (Phases 1 and 2), late launch (Phases 3-8) and Earth Gravity Assist (EGA). The pre-launch segment runs from 48 hours (T-48 hrs) prior to launch to T-0 seconds (s). The early launch segment starts with ignition of the SRMUs at T-0 s and extends through T+143 s when the SRMUs are jettisoned. The time period from T+143 s to T+206 s is not considered because there are no accidents that could result in a release of PuO_2 during this time period of the mission. The late launch segment starts at T+206 s and extends to the point where the spacecraft has escaped from Earth orbit. The EGA segment encompasses the period from Earth escape to completion of the Earth swingby.

Pre-launch accidents were not covered in the 1995 Cassini EIS because, at that time, none were postulated that could result in a release of PuO_2. However, information recently made available from the updated mission safety analyses indicates the total probability of a pre-launch accident that results in a release of PuO_2 is 5.2×10^{-5}, or about 1 in 19,200, and could result in 0.11 mean health effects and could contaminate 1.5 km2 (0.58 mi^2) of land above 7.4×10^3 Bq/m^2 (0.2 μCi/m^2) (the Environmental Protection Agency's [EPA's] guideline level for considering the need for further action).

The total probability of an early launch accident that results in a release of plutonium is 6.7×10^{-4}, or about 1 in 1,490, and could result in 0.082 mean health effects and could contaminate 1.6 km^2 (0.62 mi^2) of land above the EPA guideline level. In comparison to the 1995 Cassini EIS, this segment's mean mission risk is 0.000055 health effects, which exceeds the 1995 Cassini EIS estimate of 0.00000046.

The total probability of a late launch accident that results in a release of plutonium is 2.1×10^{-3}, or 1 in 476, and could result in 0.044 mean health effects and could contaminate 0.057 km^2 (0.02 mi^2) of land above the EPA guideline level. In comparison to the 1995 Cassini EIS, this segment's mean mission risk is 0.000092 health effects, which exceeds the 1995 Cassini EIS estimate of 0.00000037.

[ix] The total probability of an EGA accident that results in a release of plutonium is 8.0×10^{-7}, or less than 1 in 1 million, and could result in 120 mean health effects and could contaminate 15 km^2 (5.8 mi^2) of land above the EPA guideline level. In comparison to the

1995 Cassini EIS, this segment's mean mission risk is 0.000098 health effects, which is less than the 1995 Cassini EIS estimate of 0.0017.

In addition to these new best estimate analyses, DOE has conducted a study of the uncertainty in the underlying test data and models used to estimate accident risks and consequences. This information is presented in Chapter 4 of this SEIS.

Document II-43

Document title: "Statement from Daniel S. Goldin, NASA Administrator," NASA Press Release 96–159, August 6, 1996.

Source: Office of Public Affairs, NASA Headquarters, Washington, D.C.

Document II-44

Document title: "Meteorite Yields Evidence of Primitive Life on Early Mars," NASA Press Release 96–160, August 7, 1996.

Source: Office of Public Affairs, NASA Headquarters, Washington, D.C.

Document II-45

Document title: "President Clinton Remarks," August 7, 1996.

Source: Office of the Press Secretary, The White House, Washington, D.C.

Document II-46

Document title: Mars Expeditions Strategy Group, "The Search for Evidence of Life on Mars," September 26, 1996.

Source: Office of Space Science, NASA, Washington, D.C.

Mars captivated the attention of NASA and the public worldwide in 1996. That summer, a NASA-funded research team announced in the journal Science that it had identified evidence of organic molecules in a Martian meteorite found in Antarctica, which suggested that primitive life may have existed on early Mars. As these documents illustrate, the discovery stimulated enthusiasm at NASA and across the nation for Mars exploration. Prior to the meteorite study, NASA had initiated a program of Mars exploration in which the agency planned to send two spacecraft, an orbiter and a lander, to Mars about every two years over a decade's time. Determined to make the search for life its overarching goal for Mars exploration after the announcement, NASA formed a multidisciplinary group of scientists to develop strategies for searching for signs of life on future missions. The last document is an excerpt from the group's report.

Document II-43

[no pagination]

Laurie Boeder
Headquarters, Washington, DC August 6, 1996
(Phone: 202/358-1898)

RELEASE: 96-159

STATEMENT FROM DANIEL S. GOLDIN, NASA ADMINISTRATOR

"NASA has made a startling discovery that points to the possibility that a primitive form of microscopic life may have existed on Mars more than three billion years ago. The research is based on a sophisticated examination of an ancient Martian meteorite that landed on Earth some 13,000 years ago.

The evidence is exciting, even compelling, but not conclusive. It is a discovery that demands further scientific investigation. NASA is ready to assist the process of rigorous scientific investigation and lively scientific debate that will follow this discovery.

I want everyone to understand that we are not talking about 'little green men.' These are extremely small, single-cell structures that somewhat resemble bacteria on Earth. There is no evidence or suggestion that any higher life form ever existed on Mars.

The NASA scientists and researchers who made this discovery will be available at a news conference tomorrow to discuss their findings. They will outline the step-by-step 'detective story' that explains how the meteorite arrived here from Mars, and how they set about looking for evidence of long-ago life in this ancient rock. They will also release some fascinating images documenting their research."

Document II-44

[no pagination]

Donald L. Savage
Headquarters, Washington, DC August 7, 1996
(Phone: 202/358-1727)

James Hartsfield
Johnson Space Center, Houston, TX
(Phone: 713/483-5111)

David Salisbury
Stanford University, Palo Alto, CA
(Phone: 415/723-2558)

RELEASE: 96-160

METEORITE YIELDS EVIDENCE OF PRIMITIVE LIFE ON EARLY MARS

A NASA research team of scientists at the Johnson Space Center (JSC), Houston, TX, and at Stanford University, Palo Alto, CA, has found evidence that strongly suggests primitive life may have existed on Mars more than 3.6 billion years ago.

The NASA-funded team found the first organic molecules thought to be of Martian origin; several mineral features characteristic of biological activity; and possible microscopic fossils of primitive, bacteria-like organisms inside of an ancient Martian rock that fell to Earth as a meteorite. This array of indirect evidence of past life will be reported in the August 16 issue of the journal *Science*, presenting the investigation to the scientific community at large for further study.

The two-year investigation was co-led by JSC planetary scientists Dr. David McKay, Dr. Everett Gibson and Kathie Thomas-Keprta of Lockheed-Martin, with the major collaboration of a Stanford team headed by Professor of Chemistry Dr. Richard Zare, as well as six other NASA and university research partners.

"There is not any one finding that leads us to believe that this is evidence of past life on Mars. Rather, it is a combination of many things that we have found," McKay said. "They include Stanford's detection of an apparently unique pattern of organic molecules, carbon compounds that are the basis of life. We also found several unusual mineral phases that are known products of primitive microscopic organisms on Earth. Structures that could be microsopic [sic] fossils seem to support all of this. The relationship of all of these things in terms of location—within a few hundred thousandths of an inch of one another—is the most compelling evidence."

"It is very difficult to prove life existed 3.6 billion years ago on Earth, let alone on Mars," Zare said. "The existing standard of proof, which we think we have met, includes having an accurately dated sample that contains native microfossils, mineralogical features characteristic of life, and evidence of complex organic chemistry."

"For two years, we have applied state-of-the-art technology to perform these analyses, and we believe we have found quite reasonable evidence of past life on Mars," Gibson added. "We don't claim that we have conclusively proven it. We are putting this evidence out to the scientific community for other investigators to verify, enhance, attack—disprove if they can—as part of the scientific process. Then, within a year or two, we hope to resolve the question one way or the other."

"What we have found to be the most reasonable interpretation is of such radical nature that it will only be accepted or rejected after other groups either confirm our findings or overturn them," McKay added.

The igneous rock in the 4.2-pound, potato-sized meteorite has been age-dated to about 4.5 billion years, the period when the planet Mars formed. The rock is believed to have originated underneath the Martian surface and to have been extensively fractured by impacts as meteorites bombarded the planets in the early inner solar system. Between 3.6 billion and 4 billion years ago, a time when it is generally thought that the planet was warmer and wetter, water is believed to have penetrated fractures in the subsurface rock, possibly forming an underground water system.

Since the water was saturated with carbon dioxide from the Martian atmosphere, carbonate minerals were deposited in the fractures. The team's findings indicate living organisms also may have assisted in the formation of the carbonate, and some remains of the microscopic organisms may have become fossilized, in a fashion similar to the formation of fossils in limestone on Earth. Then, 16 million years ago, a huge comet or asteroid struck Mars, ejecting a piece of the rock from its subsurface location with enough force to escape the planet. For millions of years, the chunk of rock floated through space. It encountered Earth's atmosphere 13,000 years ago and fell in Antarctica as a meteorite.

It is in the tiny globs of carbonate that the researchers found a number of features that can be interpreted as suggesting past life. Stanford researchers found easily detectable amounts of organic molecules called polycyclic aromatic hydrocarbons (PAHs) concentrated in the vicinity of the carbonate. Researchers at JSC found mineral compounds commonly associated with microscopic organisms and the possible microscopic fossil structures.

The largest of the possible fossils are less than 1/100 the diameter of a human hair, and most are about 1/1000 the diameter of a human hair—small enough that it would take about a thousand laid end-to-end to span the dot at the end of this sentence. Some are egg-shaped while others are tubular. In appearance and size, the structures are strikingly-similar to microscopic fossils of the tiniest bacteria found on Earth.

The meteorite, called ALH84001, was found in 1984 in Allan Hills ice field, Antarctica, by an annual expedition of the National Science Foundation's Antarctic Meteorite Program. It was preserved for study in JSC's Meteorite Processing Laboratory and its possible Martian origin was not recognized until 1993. It is one of only 12 meteorites identified so far that match the unique Martian chemistry measured by the Viking spacecraft that landed on Mars in 1976. ALH84001 is by far the oldest of the 12 Martian meteorites, more than three times as old as any other.

Many of the team's findings were made possible only because of very recent technological advances in high-resolution scanning electron microscopy and laser mass spectrometry. Only a few years ago, many of the features that they report were undetectable. Although past studies of this meteorite and others of Martian origin failed to detect evidence of past life, they were generally performed using lower levels of magnification, without the benefit of the technology used in this research. The recent discovery of extremely small bacteria on Earth, called nanobacteria, prompted the team to perform this work at a much finer scale than past efforts.

The nine authors of the *Science* report include McKay, Gibson and Thomas-Keprta of JSC; Christopher Romanek, formerly a National Research Council post-doctoral fellow at JSC who is now a staff scientist at the Savannah River Ecology Laboratory at the University of Georgia; Hojatollah Vali, a National Research Council post-doctoral fellow at JSC and a staff scientist at McGill University, Montreal, Quebec, Canada; and Zare, graduate students Simon J. Clemett and Claude R. Maechling and post-doctoral student Xavier Chillier of the Stanford University Department of Chemistry.

The team of researchers includes a wide variety of expertise, including microbiology, mineralogy, analytical techniques, geochemistry and organic chemistry, and the analysis crossed all of these disciplines. Further details on the findings presented in the *Science* article include:

* Researchers at Stanford University used a dual laser mass spectrometer—the most sensitive instrument of its type in the world—to look for the presence of the common family of organic molecules called PAHs. When microorganisms die, the complex organic molecules that they contain frequently degrade into PAHs. PAHs are often associated with ancient sedimentary rocks, coals and petroleum on Earth and can be common air pollutants. Not only did the scientists find PAHs in easily detectable amounts in ALH84001, but they found that these molecules were concentrated in the vicinity of the carbonate globules. This finding appears consistent with the proposition that they are a result of the fossilization process. In addition, the unique composition of the meteorite's PAHs is consistent with what the scientists expect from the fossilization of very primitive microorganisms. On Earth, PAHs virtually always occur in thousands of forms, but, in the meteorite, they are dominated by only about a half-dozen different compounds. The simplicity of this mixture, combined with the lack of lightweight PAHs like napthalene, also differs substantially from that of PAHs previously measured in non-Martian meteorites.

* The team found unusual compounds—iron sulfides and magnetite—that can be produced by anaerobic bacteria and other microscopic organisms on Earth. The compounds were found in locations directly associated with the fossil-like structures and carbonate globules in the meteorite. Extreme conditions—conditions very unlikely to have been encountered by the meteorite—would have been required to produce these compounds in close proximity to one another if life were not involved. The carbonate also contained tiny grains of magnetite that are almost identical to magnetic fossil remnants often left by certain bacteria found on Earth. Other minerals commonly associated with biological activity on Earth were found in the carbonate as well.

* The formation of the carbonate or fossils by living organisms while the meteorite was in the Antarctic was deemed unlikely for several reasons. The carbonate was age dated using a parent-daughter isotope method and found to be 3.6 billion years old, and the organic molecules were first detected well within the ancient carbonate. In addition, the team analyzed representative samples of other meteorites from Antarctica and found no evidence of fossil-like structures, organic molecules or possible biologically produced compounds and minerals similar to those in the ALH84001 meteorite. The composition and location of PAHs organic molecules found in the meteorite also appeared to confirm that the possible evidence of life was extraterrestrial. No PAHs were found in the meteorite's exterior crust, but the concentration of PAHs increased in the meteorite's interior to levels higher than ever found in Antarctica. Higher concentrations of PAHs would have likely been found on the exterior of the meteorite, decreasing toward the interior, if the organic molecules are the result of contamination of the meteorite on Earth.

Additional information may be obtained at 1 p.m. EDT via the Internet at http://www.jsc.nasa.gov/pao/flash/

Document II-45

[no pagination]

Subject: President Clinton remarks
Date: Thursday, August 08, 1996 3:19PM

THE WHITE HOUSE

Office of the Press Secretary

For Immediate Release August 7, 1996

REMARKS BY THE PRESIDENT
UPON DEPARTURE

The South Lawn

1:15 P.M. EDT

THE PRESIDENT: Good afternoon. I'm glad to be joined by my science and technology adviser, Dr. Jack Gibbons, to make a few comments about today's announcement by NASA.

This is the product of years of exploration and months of intensive study by some of the world's most distinguished scientists. Like all discoveries, this one will and should continue to be reviewed, examined and scrutinized. It must be confirmed by other scientists. But clearly, the fact that something of this magnitude is being explored is another vindication of America's space program and our continuing support for it, even in these tough financial times. I am determined that the American space program will put its full intellectual power and technological prowess behind the search for further evidence of life on Mars.

First, I have asked Administrator Goldin to ensure that this finding is subject to a methodical process of further peer review and validation. Second, I have asked the Vice President to convene at the White House before the end of the year a bipartisan space summit on the future of America's space program. A significant purpose of this summit will be to discuss how America should pursue answers to the scientific questions raised by this finding. Third, we are committed to the aggressive plan we have put in place for robotic exploration of Mars. America's next unmanned mission to Mars is scheduled to lift off from the Kennedy Space Center in November. It will be followed by a second mission in December. I should tell you that the first mission is scheduled to land on Mars on July the 4th, 1997—Independence Day.

It is well worth contemplating how we reached this moment of discovery. More than 4 billion years ago this piece of rock was formed as a part of the original crust of Mars. After billions of years it broke from the surface and began a 16 million year journey through space that would end here on Earth. It arrived in a meteor shower 13,000 years ago. And in 1984 an American scientist on an annual U.S. government mission to search for meteors on Antarctica picked it up and took it to be studied. Appropriately, it was the first rock to be picked up that year—rock number 84001.

Today, rock 84001 speaks to us across all those billions of years and millions of miles. It speaks of the possibility of life. If this discovery is confirmed, it will surely be one of the most stunning insights into our universe that science has ever uncovered. Its implications are as far-reaching and awe-inspiring as can be imagined. Even as it promises answers to some of our oldest questions, it poses still others even more fundamental.

We will continue to listen closely to what it has to say as we continue the search for answers and for knowledge that is as old as humanity itself but essential to our people's future.

Thank you.

Document II-46

[no pagination]

The Search for Evidence of Life on Mars

(Excerpt from report of Mars Expeditions Strategy Group
Dan McCleese, JPL, Chairman)

26 September 1996

FOCUS

Did life ever exist on Mars? A multi-disciplinary group of scientists brought together by the National Aeronautics and Space Administration [NASA] is currently developing a strategy to seek the answer to that question. When complete, this strategy will form the basis for NASA's future program of Mars exploration. This report is a statement of work-in-progress by the group to identify a systematic approach, using robotic space missions and laboratory analyses of samples returned to Earth, to understand the possible origin and evolution of life on Mars.

NASA is today conducting a series of robotic missions to Mars with the goal of understanding its climate, resources and potential for harboring past or present life. The measurements to be made have in common the study of water and its history on the planet. The first mission to return to the surface of Mars since the Viking spacecraft in 1976 will be launched in December of 1996. Also this year, an orbiter will begin regional and global mapping of the surface, searching for sites potentially hospitable to life some time in the planet's past.

HYPOTHESES

The fundamental requirements for life as we know it are liquid water, an inventory of organic compounds, and an energy source for synthesizing complex organic molecules. Beyond these basics, we have yet to achieve consensus regarding either the environmental requirements or the processes of chemical evolution that lead to the origin of life. Comparisons of genetic sequences in living organisms suggest that the last common ancestor of life on Earth may have been a sulfur-utilizing bacterium that lived at high temperatures. This implies that hydrothermal environments were important in the early evolution of the biosphere. Given that hydrothermal systems have also been shown to be energetically favorable places for organic synthesis, some scientists believe that it was in such location that life actually originated. However, others argue quite convincingly for a low temperature origin of life.

Unfortunately for attempts to resolve this controversy, plate tectonics and extensive recycling of the crust have obliterated any record of pre-biotic chemical evolution on Earth. The story is, however, quite different for Mars. The absence of plate tectonics suggests that the Martian crustal record is much better preserved than that on Earth. The cratering record on Mars implies that vast areas of the Martian southern highlands are older than 3.8 billion years. Analysis of meteorites from Mars indicates that some highland terranes date back to the very earliest period of planetary evolution (~4.5 billion years), overlapping the period on Earth when pre-biotic chemical evolution first gave rise to life. Thus, even if life never developed on Mars, any inventory of biogenic elements and organic compounds that may be preserved in the rocks of the ancient cratered highlands will yield crucial information about the pre-biotic chemistry that led to living systems on Earth.

ENVIRONMENTS

The members of the Mars strategy group recommend that the search for life on Mars should be directed at locating and investigating, in detail, those environments on the planet which were potentially most favorable to the emergence (and persistence) of life. Three in particular can be cited for concentrated study:

(a) Ancient ground water environments: early in the planet's history liquid water, regarded as prerequisite for life, appears to have been widespread beneath the surface and may have provided a clement environment for the origin of life. Intense energy was dissipated by impacts associated with the final stages of planetary accretion and, along with volcanism, could have created warm ground water circulation systems favorable for the origin of life. In this scenario, evidence for ancient habitats may be found in the heavily cratered terranes of the Martian highlands.

(b) Ancient surface water environments: also during early Martian history, liquid water was apparently released from subsurface aquifers, flowed across the surface, and pooled in low-lying regions. Solar irradiance would have provided biologically useful energy. During this period habitats may have been formed, with evidence of life preserved in water-lain sediments in the valley systems and basins found in the highlands.

(c) Modern ground water environments: life may have formed at any time, including recently, in habitats where subsurface water or ice is geothermally heated to create warm ground water circulation systems. In addition, life may have survived from an early epoch in places beneath the surface where liquid water is present.

Given our present uncertainty about the environmental conditions necessary for the origin of life, and our limited knowledge of the geologic history of Mars, we urge strongly that the investigation strategy emphasize sampling at diverse sites. It is specifically recommended that the implementation of the program of exploration of Mars be aimed at the study of a range of ancient and modern aqueous environments. These environments may be accessed by exploring the ejecta of young impact craters, by investigating material accumulated in outflow channels, and by coring.

NEEDED INVESTIGATIONS

In-situ studies conducted on the surface of Mars are essential to our learning more about Martian environments and for selecting the best samples for collection. However, for the next 10 years or more, the essential analyses of selected samples must be done in laboratories on Earth. It is evident from studies of meteorites that it is difficult to predict the full suite of analytic techniques which will be needed to complete the analysis of returned samples. Further, based upon the results of the Viking landers and analyses of Martian meteorites, markers of life are thought to be at low concentrations and fossils, if present, are likely to be very small. Therefore, "high precision" (i.e., sophisticated, state-of-the-art) analytical techniques must be used, such as those found in only the most advanced laboratories here on Earth.

We also believe that to achieve widely accepted confirmation of Martian life, all three of the following must be clearly identified and shown to be spatially and temporally correlated within rock samples: 1) organic chemical signatures that are indicative of life, 2) morphological fossils (or living organisms), 3) supporting geochemical and/or mineralogical evidence (e.g., clearly biogenic isotopic fractionation patterns, or the presence of unequivocal biominerals). These characteristics can not be properly evaluated without the return of a variety of Martian samples to Earth for interdisciplinary study in appropriate laboratories.

Precursor orbital information must be obtained, as well, to select the best sites for surface studies. We can already say with reasonable certainty, however, that the ancient highlands represent a region of great potential, and that at least the initial focused studies should be performed there. Maps of surface mineralogy will be needed to enhance investigations within the highlands and enable searches elsewhere. This work begins with the launch of the Mars Global Surveyor (MGS) later this year. Additional measurements from orbit at higher spatial resolution are essential to identify productive sites (e.g., regions containing carbonates) at scales accessible by surface rovers. In addition, instruments capable of identifying near-surface water, water bound in rocks, and subsurface ice, would greatly accelerate and make more efficient our search for environments suitable for life.

We have found it useful to consider the factors that lead to the fossilization and long-term preservation of microorganisms and key compositional indicators in rocks. Based on studies of the microbial fossil record on Earth, the long-term preservation of organic signatures is most favored within sedimentary environments where aqueous minerals precipitate rapidly from solution, entrapping organic materials within an impermeable mineral matrix.

The best host minerals are those that have long crustal residence times by virtue of being chemically stable. In ancient rock sequences on Earth, organic materials tend to be found in association with a fairly restricted number of sedimentary precipitates, which include silica, phosphate, and carbonate. Preservation of polycyclic aromatic hydrocarbons within the carbonates of the Martian meteorite ALH 84001 indicates that such mineralization processes were an effective means for capturing organic materials in the early Martian crustal environment and, importantly, for preserving them for billions of years.

From these factors we judge that an implementation strategy for the initial phases of Mars exploration can already be affirmed:

(1) For ancient ground water environments, a sample return mission can occur relatively soon, since the necessary precursor information for site selection is already available from existing orbital photogeologic data, including Mariner 9 and Viking imagery, or will be provided by Mars Surveyor orbiters in '96, '98 and '01.

(2) For ancient surface water environments, orbital and surface exploration/characterization should precede sample return because identification of extensive areas of carbonates and evaporites is highly desirable. This implies the use of advanced orbital and in situ instruments for mineral characterization. Technologies which enable long-range surface exploration are also needed.

(3) For modern ground water environments, additional means for the identification of thermally active regions will be needed. Techniques for location of subsurface water (i.e., liquid and ice) are also needed.

Sample return missions will retrieve the most productive samples if they are supported by extensive searches, analyses and collections performed by sophisticated rovers. These should be capable of ranges of 10s of kilometers in order to explore geologically diverse sites. The specific samples to be returned to Earth would be selected using criteria that increase the probability of finding direct evidence of life as well as the geological context, age and climatic environment in which the materials were formed.

In order to retrieve scientifically meaningful samples, significant constraints must be placed on the way samples are handled during collection and return to Earth. We anticipate retrieving dry rocks and minerals for which mechanical preservation is a major factor; self-abrasion or shake-induced disintegration of the samples must be minimized. Almost certainly, the rocks will have been exposed already at the Mars surface so that packing can be accomplished using local Mars soils; individual containerization of different rocks might not be a strict requirement. For subsurface environments, where ices or brines are possible, sample materials must be handled in such a way that melting or evaporation of volatiles within the samples can be controlled. For volatile-rich samples, temperature control, individual containerization, and hermetic sealing to prevent mass loss or mass exchange are likely to be requirements. If extant life is found, even more stringent environmental controls may be required. For samples from all environments, preservation protocols must address the sensitivity of biogeochemical materials (organic compounds plus minerals containing the chemical elements H, C, N, O, S, and P) to material contamination and to thermal degradation.

Unmodified samples of the Martian atmosphere must also be brought back to Earth where they can be examined in our laboratories. The possibility of the origin and evolution of life on Mars must be fundamentally linked to the evolution of the atmosphere, through its contribution of biogenic elements and compounds (including water), through chemical reactions taking place at the atmosphere-surface interface and through regulation of the planetary climate.

Although precise requirements for sizes or masses of samples require further evaluation, our preliminary recommendation is that individual rock samples should be on the order of at least 10-20 grams. Experience with planetary samples, including Martian meteorites, has amply demonstrated that a representative 10-20-gram rock sample can be divided effectively and distributed to state-of-the-art laboratories to accomplish all of the important measurements. Even though larger samples are desirable for certain types of studies, the Apollo lunar program taught us that a limited sample payload mass is more profitably expended on numerous small samples than on a few large ones.

To summarize, our science strategy is predicated on the execution of several (at least three) mission sequences comprising precursor orbital and roving elements together with selected retrieval of samples for detailed analysis in Earth laboratories. To achieve efficiencies of time and cost, sample selection and caching may occur at more sites than sample return. An endeavor of this nature involves a number of uncertainties and should be expected to encounter occasional setbacks. The overall structure and implementation of the program must be sufficiently flexible to accommodate these perturbations and to adjust to discoveries as it progresses.

SAMPLE QUARANTINE

By long-standing international agreement, spacefaring nations take measures to protect planetary environments against biological cross-contamination during space exploration missions. We assume that some level of sample quarantine will be included in mission requirements. We recommend that any sample quarantine and sterilization protocols be closely coordinated with plans for analysis of returned samples and we urge that care be taken throughout the planning process to assure that tradeoffs among quarantine, sterilization, and science goals are clearly understood before implementation plans are adopted. Even though sample quarantine probably will be conducted in a restricted-access facility, and some preliminary characterization of the samples will occur behind the quarantine barriers, we believe that the maximum value of the samples can be extracted only if the samples are made available to scientists in their individual, specialized laboratories. Therefore, we recommend that, if sample quarantine and sterilization become operational requirements, some provision be made so that sterilized samples can be released to outside research laboratories, with suitable controls, and at the earliest possible opportunity in the execution of the program.

TECHNOLOGY REQUIREMENTS

Although this group of scientists has only recently begun to develop a road map for enabling technologies, we can already see several technology needs emerging:

1) Long-range rovers capable of surviving from months to years on the Martian sur-

face. Rovers must be capable of carrying a sophisticated battery of tools and instruments over distances of 10s of km.

II) Low-mass propulsion, power and communications systems for landed elements (e.g., Mars ascent vehicles and rovers).

III) High spatial resolution (orbital) remote sensing instruments. Spectrometers and radiometers are needed for mineralogy and detection of thermally active regions.

IV) In situ instruments, supported by sample preparation tools, able to identify aqueous minerals in rocks and relative ages of samples. A report by a NASA ad hoc working group on instruments for exopaleontology includes descriptions of promising techniques (Point Clear Exobiology Instrumentation Workshop, 13-17 May 1996; T. J. Wdowiak, D. G. Agresti, J. Chang, Eds.).

V) Tools are needed for shallow excavation, coring to depth, rock and soil manipulation, and sample preparation. Tools must be lightweight and low power.

VI) Development of advanced terrestrial laboratory instrumentation.

These requirements for technology will be refined and additional technologies identified in the near future as the exploratory strategy unfolds. It is clear today, however, that development should proceed apace with long-lead technologies (e.g. instruments, rovers, propulsion systems).

OPPORTUNITIES FOR INTERNATIONAL COLLABORATION

We view the exploration of Mars to be inherently an international undertaking. The strategy outlined above is well suited to, and likely to be dependent upon, foreign involvement. Participation by non-U.S. scientists and agencies could range from participation in individual instruments to entire missions being sponsored abroad.

HUMAN EXPLORATION

The science strategy described above requires a series of robotic sample return missions. This series may continue until either:

(a) it has been conclusively shown that life existed on Mars at some time in the past; or

(b) the evidence for Mars life is ambiguous, but little progress is being made, or expected, through additional robotic sample returns. (We note that it is impossible to prove that life never arose on Mars.)

In the former case, the questions of life's beginning, evolution and possible survival to the present become prominent scientifically. In the latter case, we will inevitably have learned much more about the environments that existed throughout Mars' history, but we will be hindered by lack of technology, lack of new ideas, or lack of resources. At present, we are encouraged in (a) above by the discoveries in Antarctic Meteorite ALH 84001. In either case, a re-examination of the strategy will be necessary after analysis of the initial returned samples.

Exploration involving humans may be required at this decision point. If past life were to be demonstrated, the questions then asked would be more complex, requiring substantially larger amounts of data, a reconnaissance mode of exploration would no longer be sufficient, and the observational and analytical capabilities that could be provided by humans could be the more effective approach. If the data were still ambiguous, but promising, the need for human in situ capabilities could prove compelling. For example, if the search turns to locating and drilling for extant subsurface warm aqueous systems, the observational and manipulative skills of humans could be important. Thus, the perceived difficulties of making further progress could form the basis for a decision to conduct human scientific exploration of Mars. The questions raised by the discovery of evidence for past or present life on Mars could become so important that they provide much of the rationale for human exploration.

Whether human missions become practical and desirable either from the scientific perspective, or from other rationales, the robotic orbital, surface and sample return program will provide important information to support human missions, through (a) characterization of the surface environment in which humans must establish their presence, such as the toxicity of dust, the availability of water, the radiation environment, and resolution of the forward/back-contamination issues; and (b) development and/or demonstration of technologies that would be used in human missions, such as Mars resource extraction systems, surface mobility, deep coring and analytical instrumentation, among others.

MARS METEORITE RESEARCH

In addition to pursuing an exploration program focused on missions to the planet, we strongly endorse NASA's efforts aimed at increasing the number of Martian samples available for laboratory study through expanded support of the NSF/NASA/Smithsonian-sponsored Antarctic Search for Meteorites (ANSMET) program. Five Martian meteorites have been discovered through the US Antarctic program since 1977, and an additional sample has been documented (but not yet extensively studied) in the similar effort by Japanese Antarctic teams. For the US program alone, this corresponds to approximately one Martian meteorite per 1000 Antarctic meteorites collected, or one Martian rock per four seasons of meteorite collection. Mars Expeditions encourages investigation of ways in which the productivity of ANSMET—measured in terms of the area searched each season—can be increased to allow the rate of discovery of Martian meteorites to be accelerated. Re-examination of the methodologies used to locate, document and collect samples might allow such an increase in productivity without calling for an increase in the number of participants involved in the field collection effort. In addition, NASA should expand the resources applied to the laboratory processing, cataloging, and organically clean handling of Martian meteorites so that research relevant to the search for Martian life can be supported at a faster pace.

Methodologies used in the handling and study of meteorites from Mars are similar to those that will be applied to samples retrieved from Mars by spacecraft. Continued support of ANSMET and Martian meteorite research will assist directly in preparation for eventual Mars sample analysis. It is our view moreover, that strong ties should be forged with other nations participating in meteorite searches (such as Japan) to further expand the effort. While we do not suggest that study of more meteorite samples will unequivocally answer the question of whether life ever existed on Mars, we have no doubt that analysis of a larger set of Martian meteoritic materials will enhance our understanding of the geological and possible biological history of the planet.

Mars Expeditions Strategy Group
26 September 1996

Chapter Three:

Exploring the Universe: Space-Based Astronomy and Astrophysics

Nancy Grace Roman

Astronomy before 1958[1]

For millennia until the Second World War, astronomical observations were limited to visible light, the type of electromagnetic radiation sensed by the human eye.[2] When people look at the sky with the naked eye, they see only stars and patches of dark against dense star backgrounds, as in the southern Milky Way. With a telescope, one can see nebulae, or clouds of gas, shining either by fluorescence or by reflected light. Large collections of stars that form distant galaxies much like the Milky Way galaxy can also be seen through telescopes.[3]

Although it had been known for several centuries that some stars vary in brightness, only a few such stars were known. It was not until 1718 that the English astronomer Edmund Halley noticed that three bright stars had changed their positions in the two millennia since they had been cataloged by Ptolemy, thus recognizing the tiny motions (the proper motions) of stars across the sky. Only with the use of spectroscopy in the early twentieth century could astronomers measure the motion of stars toward and away from Earth (the radial velocities). In 1939, physicist Hans Bethe proposed that the light observed from most stars results from the conversion of hydrogen into helium in the stellar cores and delineated a probable chain of reactions to accomplish this conversion.[4] As helium is slightly lighter than four hydrogen atoms, this reaction changes a bit of matter into energy. Therefore, most stars are changing with time, but this change is so slow that the Sun has remained essentially unchanged for about five billion years and will remain nearly the same for another five billion years. The heavens were considered the epitome of calm and

1. In this essay, astronomical observations are defined as those focused on objects and phenomena existing beyond the solar system. A short section on general relativity also is included.

2. Astronomers call light and similar radiation "electromagnetic radiation." They describe particular portions of this electromagnetic radiation by wavelength, which increases toward the red, and by frequency and energy, which increase toward the blue. The "rainbow" formed by the spread of the colors is called the spectrum. Wavelength and frequency consider electromagnetic radiation as a wave. The wavelength is the distance between the same portion of the successive cycles; frequency is the number of passages in one second of the same portion of the successive cycles past the same point. Thus, frequency is the velocity of the radiation divided by the wavelength. The wave number, a unit frequently used in the infrared, is the inverse of the wavelength in centimeters. Energy measurements consider the radiation as a stream of particles. The energy is proportional to the frequency.

3. Three galaxies are visible to the naked eye from dark viewing points: the Andromeda galaxy, a close relative of the Milky Way galaxy, and the two Magellanic Clouds, smaller systems that are much nearer to the Milky Way. The latter are visible only from the Southern Hemisphere.

4. Hans Bethe, "Energy Production in Stars," *Physical Review* 55 (1939): 434-56.

lack of change. Observations in other wavelengths were to show how misleading the observations in the visible region had been.

In the 1930s, astronomer Karl Jansky first detected radio emissions from the center of the Milky Way.[5] The first attempt to study celestial objects in wavelengths other

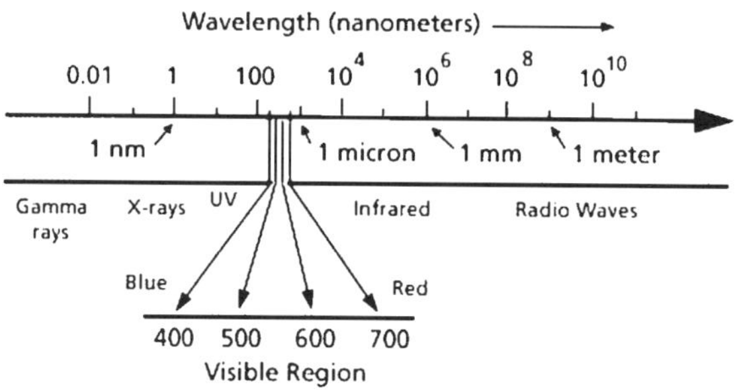

Electromagnetic Spectrum

than the visible was made as the result of the development of radar in the 1940s. Grote Reber, an amateur astronomer, observed strong emission from the constellation Sagittarius and weaker maxima in the constellations Cygnus, Cassiopeia, Canis Major, and Puppis.[6] These emissions at long wavelengths were puzzling. They did not show the variation of intensity with wavelength that would be expected for a thermal source. Eventually, I. S. Skhlovsky realized that some continuum radiation (that is, radiation not restricted to a narrow region of the spectrum), such as that from the Crab Nebula, resulted from electrons moving with nearly the speed of light in a strong magnetic field.[7] Other radio emissions appeared to come from regions in which particles slammed at high speed into material already present. Also during the 1940s, Hendrik van de Hulst, a young Dutch astronomer, recognized that neutral hydrogen had a very weak transition that radiated and absorbed in a narrow portion of the observable radio region.[8] In spite of the weakness of the transition, scientists soon observed a great abundance of hydrogen between the stars.[9] More recently, astronomers have detected many molecules in the radio region of the spectrum.

Since the invention of the telescope, astronomers have been frustrated by the multiple problems presented by Earth's atmosphere. First and foremost, the continual density fluctuations in the atmosphere have blurred astronomical images, preventing, until recently, even the largest telescope from observing details on planetary surfaces or in

5. Karl Jansky, "Electrical Disturbances Apparently of Extraterrestrial Origin," *Proceedings of the Institute of Radio Engineers* 21 (1933): 1387-98.

6. Grote Reber, "Cosmic Static," *Astrophysical Journal* 100 (1944): 279-87.

7. I. S. Skhlovsky, "On the Nature of the Radiation from the Crab Nebula," *Doklady. Akademii. Nauk SSSR* 90 (1983): 983-86.

8. Hendrik Christoffel van de Hulst, "Radio Waves from Space" (in Dutch), *Nederlandische Tijdschrift Natuurkunde* 11 (1945): 201, 210.

9. H. I. Ewen and E. M. Purcell, "Radiation from the Galactic Hydrogen at 1420 Mc/Sec," *Nature* 168 (1951): 356.

dense star fields any finer than those that can be seen easily with a good amateur telescope. Second, and almost as important, the constituents of the atmosphere block most of the electromagnetic spectrum, and electrons in the ionosphere block access from the ground to long-wave radio waves. Although the latter makes long distance radio reception possible, it also cuts out an important region of the astronomical spectrum. The atmosphere also scatters light, making it impossible to see a faint star near a bright one. Finally, the atoms and molecules in the atmosphere emit light, ensuring that the sky as seen from the surface of Earth is never completely dark.

For these reasons, some astronomers became interested in the possibility of observations from above the atmosphere.[10] In 1946, Princeton University astronomer Lyman Spitzer wrote a short paper in which he explained the advantages of a space-based telescope; the origins of planning for the Hubble Space Telescope can be traced to this paper.[11] [III-1] In 1952, Fred Whipple, a Harvard astronomer, discussed briefly some of the technical aspects of an ultraviolet (UV) telescope in space. He assumed that it would be operated in conjunction with a human-occupied space station, but not attached to the station.[12]

Astronomers soon had an opportunity to make observations from above the disturbing atmosphere. At the end of World War II, the United States had captured a number of German V-2 rockets and the Army was anxious to test them. The military solicited scientific experiments to serve as functioning payloads for these tests. (See Chapter 1 of this volume for more information on these experiments.) The first celestial photograph taken from a scientific payload flown on a V-2 was a spectrum of the Sun, obtained by Richard Tousey and his colleagues at the Naval Research Laboratory (NRL) in 1946.[13] Researchers from around the country flew a variety of instruments aimed at answering questions in solar and atmospheric physics. In the early 1950s, astronomer Jesse Greenstein, then at the University of Chicago, built a spectrograph for stellar observations. The rocket on which the experiment rode failed, as many others did in these early years.[14] In November 1955, researchers in the Rocket Branch at NRL succeeded in flying the first UV stellar photometers.[15] Since hot stars emitted much of their radiation in the UV that was not accessible from the ground, it made sense that the first astronomical observations of the night sky were directed to observations of this region; the earliest results, with very low angular resolution, proved to be unreliable. By then, the smaller, more reliable Aerobee rocket had replaced the V-2 and became the launch vehicle that dominated the sounding rocket astronomy program for several decades.[16]

---

10. Others, however, were skeptical of the usefulness of observing the heavens from space. See the section later in this chapter on the Great Observatories for more information on this subject.

11. Lyman Spitzer, *Astronomical Advantages of an Extra-Terrestrial Observatory* (Santa Monica, CA: Project RAND, 1946). For additional background on Spitzer's vision of a space telescope, see Lyman Spitzer and Jeremiah P. Ostriker, eds., *Dreams, Stars, and Electrons* (Princeton, NJ: Princeton University Press, 1996).

12. Fred L. Whipple, Lecture at Second Symposium on Space Travel at the Hayden Planetarium, American Museum of Natural History, New York, NY, 1952.

13. W. A. Baum et al., "Solar Ultraviolet Spectrum to 88 Kilometers," *Physical Review* 70 (November 1946): 781-82.

14. After his experiment's failure, Greenstein promised to have nothing more to do with trying to conduct experiments in space. Although he was never responsible for another instrument, and at first was very negative toward the space program, he remained interested in the possibilities of observing the ultraviolet spectra of stars and served as both a formal and informal advisor to the NASA astronomy program.

15. Byram et al., *The Threshold of Space*, ed. M. Zelikoff, (London, England: Pergamon Press, 1957).

16. For more information, see David H. DeVorkin, *Science with a Vengeance: How the Military Created the U.S. Space Sciences after World War II*, (New York: Springer-Verlag, 1993).

NASA Starts an Astronomy Program

When it began operations in October 1958, NASA was composed primarily of two groups of people: those who had been part of the National Advisory Committee for Aeronautics (NACA) and a large contingent from NRL. The latter included most of the NRL Rocket Branch and of those working on Project Vanguard at NRL.

The first astronomical activity at NASA was the continuation of the sounding rocket program already underway at NRL. James Kupperian, formerly of NRL, led a group at NASA's Goddard Space Flight Center (originally the Beltsville Space Center) that also included several others from the NRL rocket program. At the same time, some astronomers remained at NRL, including Herbert Friedman, who continued to lead a strong program there, particularly in x-ray astronomy.

Although astronomers originally wanted to explore the entire spectrum not accessible from the ground, many astronomers were particularly interested in the UV region. Molecular ozone restricts ground-based observations to the near UV.[17] It was known from studies in the visible that the maximum emission from hot stars is at wavelengths shorter than this ozone limit. Also, the resonance lines of many important light elements and ions such as those of oxygen, aluminum, silicon, carbon, nitrogen, and, particularly, hydrogen are located in the inaccessible region.[18] Although both x-rays and UV emission had been observed from the Sun years before the start of NASA, instruments launched on sounding rockets had observed other objects only in the UV. Hence, the early sounding rocket program at NASA concentrated on the UV.

Gerhardt Schilling, who had been an assistant to astronomer Fred Whipple at the Smithsonian Astrophysical Observatory (SAO), joined NASA as head of the astronomy program. John O'Keefe, who had recently joined the Theoretical Division at Goddard, assisted Schilling on a part-time basis. The first job of Schilling and O'Keefe was to start the development of several experiments and spacecraft that would become part of NASA's first astronomy satellites, known as the Orbiting Astronomical Observatories (OAOs). [III-4]

In February 1959, the author of this essay joined NASA from the Radio Astronomy Branch at NRL to become Head of the Optical Astronomy Program, which included the UV. Schilling left less than a year later, and the author took over the entire astronomy program. At that time, the program included all wavelengths—from high-energy gamma rays to long-wave radio waves—for all celestial objects observed from the vicinity of Earth, as well as geodesy.[19]

A primary activity in the first few years was alerting the astronomical community to the opportunities offered by the NASA program and, at the same time, learning what possibilities were of interest to various astronomers. [III-3] The latter, somewhat modified by the author's understanding of both astronomical questions and technical capabilities, was

---

17. Specifically, it restricts ground-based observations to wavelengths longer than 300 nanometers. A nanometer is 1 x 10^{-9} meter.

18. A resonance line is the line absorbed or emitted when an electron moves between the lowest (ground) level and the next higher level. The absorption continuum arises when an electron from the ground level is lost from the atom. The region between the resonance line of hydrogen and the hydrogen continuum is the far UV.

19. NASA's attempts to establish a geodetic satellite program were strongly opposed by the Air Force and traversed a rocky road until the program was finally established a few years later.

the basis of the planned program. Astronomers, practitioners of a very old science, deal with long-lived objects and thus tend to be conservative. Hence, it is not surprising that there were social as well as technical problems to be met in the development of the new NASA astronomy program.

Technical and Social Challenges of a NASA-Supported, Space-Based Astronomy Program

Technical Challenges

The early attempts to observe the sky in the ultraviolet used spinning rockets. Astronomical objects beyond the solar system, however, are faint, and except for studies of the very brightest objects, relatively long observations of a single target are required. Obtaining lengthy observations with the spinning rockets proved impossible because of the short exposure time for each part of the sky.

The development of satisfactory pointing controls was essential both for payloads on sounding rockets and for satellites. NASA's first orbiting missions designed to study the Sun, the Orbiting Solar Observatories (OSOs), were able to provide reasonable three-axis pointing in a particular direction by locking onto the Sun, but could not point to any other region of the sky. The first satellite to provide versatile three-axis pointing was the first of NASA's major astronomy missions, Orbiting Astronomical Observatory (OAO)-1.[20] The OAOs provided a breakthrough with even slightly better pointing than the sounding rockets of that era.[21]

Obtaining fine detail from astronomical sources requires good imaging. Astronomers also want to observe a long stretch of a spectrum at the same time. Hence from its astronomy program's inception, NASA has constantly needed to develop sensitive imaging detectors. In the 1960s, ground-based astronomers used photographic plates for the visible regions, but this procedure was too complex and expensive for astronomical observations from satellites.[22] Photographic film was used successfully in rockets, but film sensitive to the UV tended to scratch easily and was difficult to handle. Early on, researchers also used proportional counters, UV versions of Geiger counters, and various similar electronic detectors for the different spectral regions. Astronomers also used the photomultiplier, which had an extensive history in ground-based astronomy. Neither the proportional counter nor the photomultiplier had imaging capabilities. On OAO-3, a photomultiplier that measured each point individually was scanned across the spectrum. Intensified vidicons (a space variant of a television camera) were used in several satellites, including OAO-2 and the International Ultraviolet Explorer (IUE), but these were diffi-

20. The OAO program is discussed further in the section of this essay on optical astronomy.

21. It is interesting to note that both systems came to fruition in 1965. Both provided pointing that was accurate and stable to within one arcminute, a distance smaller than the apparent size of a half dollar placed at one end of a football field and viewed from the other. By contrast, the Hubble Space Telescope can point and hold its position to within 0.01 arcseconds. If an airplane taking off from New York could be guided with this accuracy, it could land on a dime in Los Angeles. As small as this distance seems, it is large compared with many details in astronomical objects.

22. The national security community had used films in photo-intelligence satellites and had recovered them.

cult to calibrate, had a distorted field, and were not particularly sensitive. By 1980, solid state detectors called digicons became available for one-dimensional imaging; they are still used for low-resolution spectra. Vidicons were finally replaced by charge-coupled devices (CCDs), which were developed by the national security community and, in the 1970s, for television.[23] The first one used in a satellite was flown in the Wide Field Planetary Camera on the Hubble Space Telescope (HST) to produce most of the familiar pictures from the telescope. CCDs are now being used generally for optical and high-energy space astronomy as well as for ground-based studies.

Social Challenges

Throughout the space astronomy program, NASA has had to address a number of "social" issues. An early challenge was arousing the interest of members of an astronomical community that was comfortable with the instruments they had used for decades. There was a clear division of interest between the astronomers in the West, who had extensive access to large, ground-based telescopes and those in the East, who lacked such facilities. Astronomers at Princeton, Harvard, and the University of Wisconsin were among those anxious to get involved in the space astronomy program. In contrast, those at the California Institute of Technology and the various campuses of the University of California, in general, thought that the space program was a waste of time and money.[24] Also, many astronomers in 1960 had relatively little background in developing sophisticated instruments. The influx of observers trained as experimental physicists solved this problem. With the availability of the IUE in the late 1970s, a satellite telescope became available that could be used much like a ground-based telescope. This mission allowed the majority of academic astronomers to become comfortable with space instruments as a natural addition to their repertoire, a comfort factor that later increased with the HST.[25]

Before World War II, most astronomy in the United States had been supported privately. The major involvement of scientists in the war effort led to substantial government funding of many sciences, including some support of astronomy by the Air Force and by the Office of Naval Research. After the establishment of the National Science Foundation (NSF) in 1950, that agency became the major supporter of American astronomical research. With the founding of NASA, it was obvious that making observations from sounding rockets and satellites was going to include astronomical observations. NSF Director Alan Waterman feared that the space-based research, which was so much more expensive than comparable ground-based astronomy, would overwhelm the latter activity, which still had a strong role to play in astronomical research. In an effort to ensure that both types of astronomy remained viable, Waterman and NASA Administrator T. Keith Glennan signed a memorandum of understanding in 1959 agreeing that NSF should be

23. Although a bare charge-coupled device is sensitive only to the red and near infrared, it can be coated with a phosphor sensitive to other wavelengths or used with another imaging device. Either of these acts as a wavelength converter for the CCD.

24. The issue of geographical differences of opinion is discussed further in Robert W. Smith, *The Space Telescope: A Study of NASA, Science, Technology, and Politics* (New York, NY: Cambridge University Press, 1989), pp. 47-48.

25. IUE and HST are discussed in the Optical Astronomy and Great Observatories sections of this essay, respectively.

responsible for ground-based astronomy and NASA only for space astronomy. Despite the agreement, the border of responsibilities between the agencies remained fuzzy. Although the division was clear for satellites and rockets, it was less clear for balloon observations. Moreover, NASA flight programs relied strongly on ground-based data to understand the space results. The problem was solved by close cooperation and information exchanges between the agencies at the program level. [III-9]

NASA's interest meant not only access to new techniques in astronomy and the avoidance of the many problems presented to astronomy from the ground, but also a new source of funding for instrumentation, observations, and theory. Moreover, the interest in space generated by Sputnik and the formation of NASA attracted many new people into astronomy. The membership of the American Astronomical Society, which includes almost all professional astronomers in the United States, tripled between 1960 and 1970.

The creation of an astronomy program operated by NASA also presented scientists with a new approach to managing government-provided funds. NSF used a hands-off approach, checking only that a scientist was making satisfactory progress in NSF-funded research. Because of the necessity to meet flight schedules and because of the higher cost overrun potential of space efforts, NASA has practiced more detailed management for most of its flight programs and the ground-based efforts on which they depend.[26] Most astronomers were not only unused to such detailed management, but in fact tried to rebel against it. Although astronomers and physicists involved with the design and development of satellites recognized the complexity of the undertaking and the valuable assistance of NASA engineers, submitting to paperwork requirements, scheduling constraints, and constantly changing budget restrictions continued to rankle. Most investigators also would have preferred a freer hand to do things their own way, going to NASA only for needed help.

Throughout the program, university-based astronomers have questioned the competence of civil service astronomers working for NASA.[27] On the whole, university astronomers felt from the early days of the space agency that NASA was overly bureaucratic and treated astronomers at NASA Centers preferentially.[28] Part of the problem was that the astronomical community generally had no appreciation of the complexity of satellite projects. This issue became particularly evident in 1966, when NASA Administrator James Webb asked Harvard professor Norman Ramsey to chair a committee to advise NASA on the execution of a National Space Astronomy Observatory, among other projects.[29] [III-11] The Ramsey Committee's final report suggested that the astronomy program be transferred to a consortium of universities.[30] [III-12] NASA did not accept the suggestion that the astronomy program be run entirely by an outside consortium, but attempted to curb the academic scientists' unhappiness with the degree of

26. Research not tied to launch deadlines and comparable in cost to that funded by NSF has been managed in much the same way as most NSF efforts, allowing the investigator substantial freedom with little detailed oversight.

27. This was somewhat less of a problem in the geophysics discipline, in which the scientists who were later part of NASA had played an active role in the International Geophysical Year.

28. NASA Headquarters made a serious attempt not to give preference to Center astronomers but to some extent it was unavoidable, as the lead project scientist was always from a NASA Center.

29. James Webb to Norman Ramsey, January 14, 1966.

30. NASA Ad Hoc Science Advisory Committee, "Report to the Administrator," August 15, 1966.

their involvement in program planning by establishing an Astronomy Missions Board, made up of external astronomers, that would advise NASA routinely.[31] [III-14, III-15] Since then, NASA has routinely received advice on its astronomy and astrophysics programs from both committees of the National Research Council/National Academy of Sciences and from external advisory committees reporting directly to NASA. [III-31, III-33] Although over the years there have been occasional tensions in the relationship between NASA and external scientists, in general, the relationship has been mutually productive. [III-35, III-36]

The issue of the position of university astronomers arose again in the decision of where to situate the organization that would manage the selection of scientific observers using HST. In this case, NASA maintained control and oversight of spacecraft operations, but resolved to locate a Space Telescope Science Institute outside of NASA, thus stimulating the anger of astronomers at NASA's Goddard Space Flight Center who had wanted that responsibility. [III-27] In the case of the Chandra X-ray Observatory, launched in 1999, operations were contracted to an academic institution, but the selection of observers still remained with NASA.

Adding to the discomfort of the academic astronomers has been the bureaucracy inherent in a government agency, which must assure Congress and the public that funds are being well spent and, as mentioned above, to meet flight schedules and keep costs under control. A part of the problem is that NASA has operated chiefly as an engineering organization, responsible for the solution of technical—as opposed to scientific—problems, and for the management of complex flight programs.

Scientists and engineers have very different cultures and approaches to problems. The scientist wants to know *why* things happen or have come to be. There are many approaches to the solution of such a question, and usually a number of approaches must be combined to find the answer. Moreover, along the scientist's way, new questions develop, often pursued instead of completing the original quest. The path to solution is rarely direct and sometimes not even in the original direction. In contrast, the engineer wants to know *how* things operate. He or she tries to solve a specific problem, usually under both time and money constraints. While the engineer may experiment with different approaches, he or she must remain on a direct path. Moreover, the final product from an engineer must work properly the first time; both property and lives depend on it. These differences in approach and objective give rise to different ways of looking at problems and cause difficulties when the two groups try to communicate. As a scientist who worked with engineers before joining NASA, the author has often said that in her first year at NASA one of her major jobs was to act as an interpreter between scientists and engineers.

Yet another issue debated by astronomers inside and outside NASA was the extent to which the same basic spacecraft, with minor modifications, should be used for several missions, as opposed to developing a unique satellite for each mission. The result has been a compromise. OSOs, OAOs, the Small Astronomical Satellites (SASs), and the High Energy Astronomical Observatories (HEAOs) used the same basic design for

31. NASA Management Instruction 1156.16, "NASA Astronomy Missions Advisory Board," September 25, 1967; ed. Robert Doyle, *A Long-Range Program in Space Astronomy: Position Paper of the Astronomy Missions Board* (Washington, DC: NASA SP-213, July 1969).

each member of the series, but allowed for improvements and modifications to suit each mission. This tactic was generally effective for the early period in which failures were not uncommon, money was plentiful, and the time between launches was brief. Nevertheless, mission-to-mission modifications increased costs, and thus it has never been clear whether individual spacecraft or a series of similar spacecraft have been more cost effective. In contrast, the Great Observatories have each been individually designed, as was the IUE, although the latter was based to some extent on the SAS design. The Extreme Ultraviolet Explorer (EUVE) was to be the first of a series of Explorers using a planned standardized platform, but so far it is the only one to have used that platform.

An additional issue with which astronomers have had to deal since NASA's space astronomy program's inception has been the question of access to the results of observations. In the beginning, the individual investigators responsible for each instrument tended to consider the data proprietary. Moreover, early instrumentation was sufficiently difficult to use and that the data were hard to interpret by anyone not intimately involved in the design. Yet, restrictions on use of the data were inconsistent with the fact that the data were paid for by the American public and hence were public property. Gradually, NASA developed a policy that gave individual investigators priority in the use of their mission's data for a finite period of time, often one year. After this time, the investigator would be responsible for depositing the data promptly in the NASA Space Science Data Center in a generally usable form together with full documentation. Although it took many years for some of the early data to be deposited, this responsibility has been well recognized, and scientists are submitting the data to the Center more quickly now. This data archive has been the source for numerous scientific papers, often in areas not envisioned by the original instrument designers.[32]

Modern satellite instruments are frequently general-purpose systems. Astronomers not on the development team are often anxious to address different scientific questions than those initially proposed. As space astronomy has become more routine and instruments have been designed that are easier to use, it has become customary to conduct a guest observer program on each major satellite. Thus, the selection of the data to be obtained is no longer restricted to the instrument developers. Although the fraction of time dedicated to the guest observer program varies with the satellite, it usually starts only after a period in which the designers have full use of the instrument. This practice insures that the instrument is working as expected and that its operation is well understood, and it rewards the developers with forefront data in return for the years they have spent on the project. After this period, the guest observer program is normally scheduled for an increasing portion of the time as the satellite ages. The guest observer program requires that the calibration and reduction of observations be standardized and made available quickly by the overseeing institution.

32. Modern software now makes it possible to find what observations have been made of an object or region of the sky by any space instrument, and then to request the appropriate data electronically. Many sources of ground-based data can also be accessed. For the new major observatories, it has become customary to release some data as soon as a reasonable calibration has become possible. For HST, data are archived quickly in raw form and calibrated "on the fly" when they are requested, although a specified proprietary period may still apply.

International Cooperation

International cooperation has always been an important component of NASA's astronomy program. Not only do scientists tend to pay less attention to national boundaries than politicians usually do, but NASA also wished to encourage space activity in the major European and other allied countries when the program started. Many cooperative sounding-rocket flights have taken place over the years with a variety of countries. While the Department of Defense's Transit satellite made the first low, single frequency radio astronomy measurements, the first such studies in which NASA was involved were made by Alouette I, a satellite designed and built by Canada to study the ionosphere.[33] UK-5, also known as Ariel 5, was designed and built by the United Kingdom and flown in 1974 on a NASA launch vehicle. It carried long-wave radio and x-ray astronomy experiments, including one developed by American astronomers.[34] The same British group involved in this mission had flown a similar x-ray instrument on OAO-3. In another cooperative program, the Netherlands Astronomical Satellite was built by the Dutch, but both the United States and the Netherlands participated in its design, and it carried instrumentation from both countries. The Infrared Astronomical Satellite (IRAS) entailed a similar division of responsibility between the Netherlands and the United States.

NASA also has taken advantage from time to time of an Italian launch facility, San Marco, off the coast of Kenya. Because this site is near the equator, satellites launched from San Marco can reach a sufficiently high altitude to minimize air drag and still stay below the Van Allen radiation belts. The particles in the Van Allen belts not only present problems for satellite electronics but also, perhaps more importantly, confuse many scientific instruments, particularly those designed to measure high-energy radiation. In addition, American astronomers have made use of the Woomera rocket launch site in Australia to launch sounding rockets to observe the southern sky, which cannot be observed by rockets launched from the United States.

Both the competition for guest observer time and access to the data from all instruments always have been open to all competent users, whatever their nationality. In addition, NASA has routinely selected the best scientific instruments for flight regardless of the nationality of the proposer. The only restriction is that NASA transfers no funds to a foreign country to support participation in a NASA mission; hence, investigators from other countries must find their own support.

Today, few major astronomy satellite missions are restricted to a single country. Much of the future activity in NASA's x-ray astronomy program is being planned in conjunction with Japan. A particularly successful radio astronomy effort has been the Very-Long Baseline Interferometry Space Observatory Program (VSOP), which was built and launched by the Japanese in 1997 as one component of a worldwide Very-Long Baseline Interferometer (VLBI) network.[35] Astronomers from the Massachusetts Institute of

---

33. For an interesting account of the early history and development of Alouette, see "Alouette/ISIS: How It All Began," *http://www.lark.ieee.ca/library/milestone/keynote.htm*.

34. Memorandum of Understanding between the United Kingdom and the United States National Aeronautics and Space Administration, November 2, 1970.

35. The VLBI technique links telescopes throughout the world to obtain a resolution equivalent to a telescope more than 11,200 kilometers in diameter. As the angular resolution of a telescope is proportional to the wavelength of the radiation divided by the diameter, this long baseline provides images in the radio region comparable to those possible in the optical with a large single mirror. The VSOP satellite extended this baseline by several times to provide correspondingly better resolution.

Technology, NASA's Jet Propulsion Laboratory, and the National Radio Astronomy Observatory as well as those from many other countries have participated in ground-based observations in conjunction with this satellite. Together, these measurements of radio sources provided the finest detail obtained in any part of the spectrum.

As satellites have become more complex, extensive efforts have been made to involve other countries in providing instruments and other spacecraft components. For very expensive missions such as the HST and those currently planned for coming years, sharing the costs among two or more countries makes the mission more affordable for all. Congress in the early 1970s required NASA to cooperate with other nations on what became HST. Europe provided the solar panels and a high-resolution camera on the spacecraft.[36] [III-29] The European Space Agency (ESA) has been included in two predominantly American astronomy missions, the Next Generation Space Telescope (NGST) and the Laser Interferometer Space Antenna (LISA), in its planning for the future.

Relations with the Human Space Program

Within the first few years of NASA's existence, it became clear that human endeavors in space would dominate the Agency's agenda. The question of the relation of the space science program—including astronomy—to the human spaceflight program arose once the Apollo program got underway.[37] [III-13] The earliest planning for the Large Space Telescope (later to become the Hubble Space Telescope) by the aerospace industry and by NASA's Langley Research Center, which also did early planning for the human flight program, envisaged active observing with a human riding with the instrument and perhaps looking through the telescope.[38] Astronomers were finally able to convince engineers that this was not practical. Not only did astronomers not normally observe visually through ground-based instruments, but also the human eye is not sensitive to many of the wavelengths to be observed from space. In addition, a human moves and thus would disturb the pointing of the instrument; humans also need the very air-filled environment that astronomers wanted to leave behind through the use of satellites.

During the Apollo program, enthusiasm for human participation was high among those astronomers interested in the space program. In 1965, the National Academy of Sciences' Space Science Board (SSB) conducted a summer study that discussed the possibilities of human maintenance, instrument exchange, and recovery for a space-based telescope.[39] [III-10] Astronomers understood that these functions could be carried out in low-Earth orbit, in geosynchronous orbit, or on the Moon. The question of putting an observatory on the Moon, however, became moot for some time when NASA decided not to return to the moon after the 1972 Apollo 17 mission. The planning for the Hubble

36. Memorandum of Understanding between the European Space Agency and the United States National Aeronautics and Space Administration, October 7, 1977.

37. A number of documents from the 1960s show some of the thinking of the time about human involvement in scientific projects. A document that provides great insight into some of this thinking is G. C. Augason, "Manned Space Astronomy," November 1966.

38. One of the leading studies on such a project was The Boeing Company Aerospace Group, "A System Study of a Manned Orbital Telescope," prepared for NASA Langley Research Center under contract NAS1-3968, (Seattle, WA: D2-84042-1, 1965).

39. Space Science Board, *Space Research: Directions for the Future* (Washington, DC: National Academy of Sciences, 1966).

Space Telescope took full advantage of these possibilities, at least in low-Earth orbit, and the program execution, which has included several human servicing efforts, has fully substantiated the value of human interactions with robotic facilities.

Various small astronomical experiments were flown on Gemini and Apollo missions. Gemini astronauts photographed the spectra of celestial objects using hand-held cameras. Early human flights provided a way in which instruments could be pointed at individual targets for times longer than sounding rocket flights. Later, during Apollo 16, astronauts successfully placed on the Moon's surface a far-UV camera and spectrograph developed by a team led by NRL astronomer George Carruthers. [III-16] This instrument provided a large number of photographic spectra, primarily of hot stars.

During the mid-1970s, NASA made a decision to tie its science program tightly to the human spaceflight program, arguing that the developing Space Shuttle would provide relatively inexpensive, frequent access to space. Because the Shuttle needed payloads and because projections were that Shuttle launches would cost less than expendable launch vehicle launches, all astronomy missions were planned for the Shuttle in that period.[40] [III-19] The *Challenger* accident changed planning precipitously. As a result of the extensive delays after the accident, the slower launch schedule, and the escalating costs of Shuttle launches, most scientific missions, including those devoted to astronomy, were dropped from the Shuttle manifest.[41] This change caused significant redesign problems for missions well along in planning at the time of the accident, greatly increasing the costs of these missions.

The planning for the Shuttle included an extensive study of the features the Shuttle would require in order to support scientific experiments and observations.[42] The European Space Research Organization (ESRO) decided in 1973 to provide a facility on the Shuttle in which to conduct experiments in a wide variety of scientific disciplines.[43] [III-20] This facility, Spacelab, flew several times, although perhaps not frequently enough to have justified its cost. It was comprised of several components that could be flown together or separately. A pressurized cabin provided facilities to accommodate numerous small experiments that benefited from human interaction or used the crew as experimental subjects. When flown, it occupied only part of the Shuttle payload bay. In the additional space in the bay, there were pallets on which experiments could be mounted and facilities to permit crew communication with the instruments on these pallets. This permitted astronaut manipulation of the experiments if desired. Another Spacelab component, an instrument pointing system, also could be flown in the unpressurized portion of the Shuttle bay. This could accommodate several sets of instruments pointing at the same object at the same time. Although this system was particularly suitable for solar observations, it also was used successfully for non-solar observations in the UV and in x-rays.

Spacelab 2, the third Spacelab mission, was flown in 1984, and was primarily dedicated to astronomy. The pointing system carried four solar telescopes, and the payload bay

40. NASA, *Final Report of the Space Shuttle Payload Planning Working Groups: Astronomy* (Washington, DC: Government Printing Office, May 1973).

41. A few missions, including the Great Observatories, remained on the Shuttle.

42. A NASA/European Space Research Organization (ESRO) Joint Users' Working Group made a study of the resources required on the Shuttle for a variety of science experiments.

43. NASA Astronomy Spacelab Payloads Project, *Interim Report of the Astronomy Spacelab Payloads Study: Executive Volume* (Washington, DC: Government Printing Office, July 1975).

carried a large, hard x-ray telescope on a pallet and a helium-cooled infrared (IR) telescope on its own mount. The largest experiment in this payload was a 2300-kilogram cosmic-ray detector.

In both 1990 and 1995, Astro flew on the Instrument Pointing System and a Broad Band X-ray Telescope (BBXRT) flew on its own pointing system. Astro included three instruments: a UV photopolarimeter,[44] a UV imaging telescope, and a 90-centimeter telescope feeding a UV spectrometer. Although optimized for the far UV, this spectrometer could be used to provide coverage of portions of the UV and the nearer portion of the extreme UV, including wavelengths shorter than the resonance line of hydrogen.[45] This instrument proved that some sources were observable in the extreme UV. The imaging telescope used an image intensifier with film. The ability to use and recover film allowed the astronomers to obtain numerous photographs in the UV of galaxies, clusters, and hot stars covering much more of the sky than the HST images. The BBXRT demonstrated the usefulness of a nest of many thin grazing incidence x-ray mirrors for imaging in the soft x-ray region.[46] Because they are very thin, many mirrors can be nested to provide a large collecting area with limited weight. This type of system is now being used on the European X-ray Multi-Mirror (XMM) satellite.

Another important way in which the Shuttle has accommodated scientific experiments is through the use of Spartan payloads. These are smaller satellites set free from the Shuttle with their own instruments, guidance, and tracking to operate for days rather than for the minutes provided by a sounding rocket. The satellites are then recovered by the Shuttle crew and can be flown again on later missions. Spartan payloads have revealed their value in reacting to unexpected circumstances: an instrument to observe Comet Halley in the UV was prepared in fourteen months to fly on a Spartan when NASA realized HST would not be ready in time for the observations. Unfortunately, this Spartan was lost in the *Challenger* accident. An American-German UV spectrograph, available as a guest-investigator instrument, flew aboard a Spartan payload for fourteen days in 1996 and observed more than two hundred targets for astronomers in a number of countries. Smaller experiments have been flown on a Hitchhiker bridge and still smaller experiments in Get Away Special cans.

There are thus both advantages and disadvantages to the use of humans to support astronomical instruments. The ability to compensate for the mirror problems on HST and to upgrade both the spacecraft and the instruments every few years has certainly not only rescued a major mission but also enhanced its capability immensely. On the other hand, the design, testing, and paperwork requirements inherent to human launches make instruments flown on such missions extremely costly, at least the first time they fly. In addition, the use of the Shuttle either confines an astronomy experiment to low-Earth orbit or requires an additional stage. Most astronomical observations benefit from being farther from Earth to provide longer, uninterrupted periods of observation and to avoid the thermal, radiation, and atomic environment of near-Earth space. At present, this fact makes

---

44. A photopolarimeter measures the brightness of an object as a function of the direction of the vibration of the light waves.

45. Only the Copernicus satellite had previously explored the region for which the instrument was optimized, and the shorter wavelength region had not been explored at all at that time.

46. The energy contained in each photon ranges from 0.3 to 12 kiloelectron volts (keV).

revisits impossible, although some in NASA are considering the possibility of servicing spacecraft at the Lagrangian 2 (L2) point.[47] NASA is planning to send the Space Infrared Telescope Facility (SIRTF), as well as several other next-generation astronomical instruments, to this location.

Exploring the Spectrum

This essay now turns to a review of space astronomy and astrophysics in various regions of the electro-magnetic spectrum.

Gamma-ray Astronomy

Gamma rays have the advantage of being able to traverse the entire universe to the top of Earth's atmosphere with little absorption and, unlike cosmic rays, retain information on the direction of their sources. Partly on the basis of an overly optimistic prediction of the intensity of cosmic gamma rays,[48] there were early, unsuccessful attempts to observe this radiation. Aside from their low intensity, a major problem with detecting gamma rays is that high-energy particles, both from cosmic rays and from the interactions of cosmic rays with the atmosphere, behave in the detectors much like gamma rays. Spacecraft themselves also contain small quantities of radioactive impurities that produce both gamma rays and high-energy particles. These background sources of radiation are much stronger than the gamma rays to be measured. Thus, in addition to good instrument sensitivity, it is essential to have excellent shielding and a way to determine the direction of arrival of the radiation.

The earliest attempts to observe cosmic gamma rays were with balloons.[49] Although these early flights were unsuccessful, the development of larger balloons capable of lifting heavier payloads to higher altitude led to many successful flights. Balloon studies have both made important discoveries and tested new approaches to instrumentation. For example, the electron-positron annihilation line at 0.511 million electron volts (MeV)[50] was first observed from a balloon.[51] However, the energy determination from these measurements was sufficiently uncertain that confirmation of the line position awaited the results from another balloon flight in 1981.

Cosmic ray researchers at the Massachusetts Institute of Technology (MIT) began in the mid-1950s to study the directional intensity of cosmic gamma rays using detectors flown to high altitudes on balloons. Soon they realized that only with a satellite would they be able to conduct gamma-ray experiments that surveyed the entire celestial sphere and avoided the interference of atmospherically produced background radiation. In 1958, the

---

47. The L2 point is a point on the Sun-Earth line, beyond Earth, at which a spacecraft orbits the Sun within the same period as Earth and, hence, remains in essentially the same position with respect to Earth.

48. See, for example, Malcomb P. Savedoff, "The Crab and Cygnus A as Gamma-Ray Sources," *Il Nuovo Cimento* 10 (1959): 12-18.

49. T. L. Cline, "Search for High-Energy Cosmic Gamma Rays," *Physical Review Letters* 7 (1961): 3.

50. This spectral line results when an electron and a positron (positive electron) merge and are both destroyed in a burst of energy corresponding to their total rest mass.

51. M. Leventhal et al., "Gamma-Ray Lines and Continuum Radiation from the Galactic Center," *Astrophysical Journal* 240 (1980): 338-343.

MIT group, led by William Kraushaar, made a proposal first to the National Science Foundation and then to the Space Science Board of the National Academy of Sciences for a satellite-borne gamma-ray experiment.[52] [III-2] On April 27, 1961, Kraushaar's experiment was launched aboard Explorer 11, the first astronomical satellite. Explorer 11 may have detected several galaxies and strong radio sources, but the data were marginal—only one or two gamma rays were observed from each.[53]

During the 1960s, NASA initiated a scientific spacecraft series, the Orbiting Solar Observatories (OSOs), designed to be the first major space program to study the Sun. The OSO satellites were essentially large gyroscopes. A heavy wheel stabilized the satellite, and two compartments rotated against the wheel to point at the Sun continuously. The wheel not only contained the necessary spacecraft components, but also had room for non-solar experiments.[54] The first reliable detection of high-energy cosmic gamma rays was from OSO-3, on which Kraushaar flew an improved version of the Explorer 11 instrument.[55] This experiment showed diffuse radiation to be concentrated in the plane of the Milky Way, with a peak intensity in the direction of the center of the galaxy.[56] Although later satellites improved the details of the distribution, the basic results from this observation have not changed. The gamma rays detected in this experiment, for the most part, result from the interaction of cosmic rays with interstellar material. Later OSOs also provided important gamma-ray data.

An interesting and exciting cosmic gamma-ray discovery was made with Department of Defense satellites in 1969. The Vela series of satellites had been launched to monitor worldwide compliance with the treaty outlawing nuclear testing in the atmosphere or above ground. These satellites detected various brief bursts in soft gamma rays.[57] These bursts often lasted for a number of seconds, with the intensity varying rapidly and chaotically in a fraction of a second.[58] There were also a number of x-ray bursts observed with these satellites, but only two were coincident with gamma-ray bursts. These measurements had a major effect on the later NASA program, which included various spacecraft entirely devoted to the study of these events as well as specialized instruments on other spacecraft. For example, observations with the Burst and Transient Source Experiment (BATSE) on the Compton Gamma Ray Observatory (CGRO) have shown that the gamma-ray bursts are evenly distributed over the sky. The spatial coincidence of a

52. William L. Kraushaar to J. Howard McMillen, May 20, 1958; William L. Kraushaar, "Research and Budget Proposal to the Space Science Board of the National Academy of Sciences for the Support of a High-energy Gamma-ray Satellite-borne Experiment to be Performed by the Cosmic Ray Group of the Massachusetts Institute of Technology Laboratory for Nuclear Science," October 10, 1958.

53. William Kraushaar et al., "Explorer XI Experiment on Cosmic Gamma Rays," *Astrophysical Journal* 141 (1965): 845. Interestingly, each source detected by Explorer 11 has since been observed in gamma rays.

54. Eight OSOs, with increasing capability, were eventually flown. Skylab followed in 1973-1974. This human mission produced spectacular results in the x-ray region, the UV, and the visible.

55. William Kraushaar, "Proposal to the National Aeronautics and Space Administration for the Support of the Development and Construction of an Instrument for Gamma Ray Astronomy to be Flown to the Orbiting Solar Observatory," November 8, 1962.

56. Carl E. Fichtel and Jacob I. Trombka, *Gamma-ray Astrophysics: A New Insight into the Universe*, 2nd ed. (Washington, DC: National Aeronautics and Space Administration, 1997). Some other information in this section of the essay has also been taken from this book.

57. The bursts had energies in the range 150 keV to 1.5 MeV.

58. J. Terrell et al., "Observation of Two Gamma-ray Bursts by Vela X-ray Detectors," *Astrophysical Journal* 254 (1982): 279-286.

gamma-ray burst, observed with the Italian-Dutch satellite, Beppo-Sax, with a following optical image permitted the identification of the source. A spectrum of this source proved that it was at a large, cosmological, distance from the Milky Way. However, even after thirty years, there is still no understanding of the nature of these bursts. A completely different type of gamma-ray burst was discovered later. These bursts appear to originate within the Milky Way and repeat irregularly. They probably arise from highly magnetic neutron stars.

In the 1970s, NASA launched a series of scientific satellites called Small Astronomical Satellites (SASs). The second Small Astronomical Satellite (SAS-2), launched in 1972, carried a gamma-ray spark chamber that had about twelve times the sensitivity of the OSO-3 gamma-ray experiment and an angular resolution of a few degrees.[59,60] SAS-2 gave a detailed picture of the diffuse background, which astronomers determined was correlated with known structural features in the galaxy. SAS-2 also provided observations of a number of types of discrete sources, including pulsars.

NASA followed these satellites with the much larger High Energy Astrophysical Observatories (HEAOs).[61] [III-21] HEAO-1, launched in 1977, was primarily devoted to x-rays, but also carried a soft gamma-ray detector. Its primary result was a nearly complete survey of the sky. HEAO-3 carried a hard x-ray, soft gamma-ray experiment. This was a large germanium spectrometer designed to detect gamma-ray lines from various sources. These include the excitation and de-excitation of interstellar nuclei and the decay of nuclei created in excited levels in supernovae.[62] Thus, these observations provide information on both the composition of sources and their physical natures. A team led by W. A. Mahoney observed aluminum in the galaxy.[63] Although it has roughly the same spatial distribution as the continuum radiation, the radiation from this long-lived isotope is more clumped. With the possible exception of the Vela supernova, the source of the clumps is unknown. This observation is providing information on the distribution of matter in interstellar space, although we do not yet understand the significance of the clumping.

In the 1970s NASA began planning for its next gamma-ray astronomy satellite. The result, the Compton Gamma Ray Observatory, was launched in 1991 as the second of NASA's Great Observatories. This mission is discussed in greater detail in the subsequent section on the Great Observatories. Thus, gamma-ray astronomy experienced a twelve-year gap between launches of missions; some balloon investigations, however, continued during the interim.

---

59. Resolution is extremely important for locating a source. It also helps distinguish a source from the background, which makes it possible to detect fainter sources. High angular resolution is the primary advantage of HST.

60. COS-B, a European satellite launched in 1975, carried an instrument with approximately the same sensitivity and angular resolution.

61. ASA, "HEAO Project Plan," June 13, 1973.

62. The maximum mass for a white dwarf is about three times that of the Sun. If a star is much heavier than that when it uses the last of its nuclear fuel, it condenses so rapidly that the material essentially bounces and most is ejected into space. Because this material had been near the core of the star it is very hot. Thus, the star becomes very large and bright, rivaling the brightness of an entire galaxy for a short time. This outburst is called a supernova because it looks like a nova but is much brighter.

63. W. A. Mahoney, "HEAO-3 Discovery of Al-26 in the Interstellar Medium," *Astrophysical Journal* 286 (1984): 578-85. Specifically, the team observed the Al-26 line at 1.809 MeV.

X-ray Astronomy

Encouraged by the observations of x-rays from the Sun by Herbert Freidman and his associates at NRL, astronomers made early attempts with sounding rockets to detect non-solar x-rays.[64] Not surprisingly, since even Alpha Centauri (Capella), the nearest star (and a solar twin), would have been too faint to be observed, it was not until 1962 that cosmic x-rays were detected by Riccardo Giacconi and his colleagues at American Science and Engineering.[65] Giacconi had been urged to search for celestial x-ray sources by MIT physicist Bruno Rossi, who believed that searching the universe in the x-ray region would enable astronomers to peer further into the universe than they had been able to see in other wavelengths. Using a spinning rocket and Geiger counters, Giacconi's team observed a strong x-ray source near, but probably not coincident with, the galactic center, and a second source in the vicinity of Cassiopeia-A and Cygnus-A, two strong radio sources.[66] The poor angular resolution of the detectors and the uncertainties in the direction of the sources precluded a closer identification. In addition, the team observed a diffuse x-ray background. The following year, Giacconi's group made a proposal to NASA to pursue a program of extra-solar x-ray astronomy studies.[67] [III-8] Later rocket observations located these sources more accurately.[68]

Subsequent NASA and non-NASA x-ray studies built on the work of Giacconi's 1962 experiment. Harder, or higher energy, x-rays were too weak to be observed in the short time available with sounding rockets, but could be observed from balloons; high-energy x-rays from the Crab Nebula, for example, were detected using balloons.[69] OSO-3 observed the hard x-ray diffuse background, and later OSOs also carried x-ray experiments that produced important results, including OSO-8's measurement of iron-line emission. Later in the 1960s, scientists detected x-rays from galaxy M87, proving that x-ray astronomy could allow astronomers to study objects beyond this galaxy.

The first satellite exclusively devoted to x-ray astronomy was the SAS-1. This spacecraft was launched from an Italian platform off the coast of Kenya to minimize problems with Earth's radiation belts. It was named *Uhuru*, the Swahili word for "freedom," in honor of its launch on Kenya's Independence Day, December 12, 1970. [III-17] It carried several proportional counters.[70] The satellite rotated slowly, thus monitoring the entire sky and having enough time in a given pointing direction to detect sources up to a thousand times

---

64. H. V. D. Bradt, "X-ray Astronomy Missions," *Annual Review of Astrophysics* 30 (1992): 391-427. Many portions of this section are based on this source.

65. Riccardo Giacconi et al., "Evidence for X-rays from Sources outside the Solar System," *Physical Review Letters* 9 (1962): 439-443.

66. The first radio sources to be discovered were given the names of the constellations in which they occur, followed by a letter, with A for the first source. Thus, the Crab Nebula is Taurus A. The constellation name is usually abbreviated to three letters. Sources of x-ray emissions discovered early followed a similar naming scheme. Thus, the first x-ray source discovered was Sco (Scorpio) X-1.

67. American Science and Engineering, "An Experimental Program of Extra-Solar X-ray Astronomy," September 25, 1963.

68. In 1963, NRL studies confirmed the detection of celestial x-ray sources and pinpointed the source near the galactic center source, which became known as Sco X-1.

69. Balloons are still used both to observe hard x-rays and to test new instrumentation for detecting both hard x-rays and gamma rays.

70. The proportional counters were sensitive to the energy range 2 to 20 keV and had angular resolution of one by ten degrees.

fainter than the Crab Nebula.[71] The final *Uhuru* catalog contained 339 objects, representing most of the common types of x-ray sources. Particularly interesting were the many binary sources in which x-rays were produced by *bremsstrallung*, or braking radiation, with material from one source impacting a compact companion.[72] Such sources play a major role in high-energy astronomy. One *Uhuru* source, Cyg (Cygnus) X-1, later detected optically, was found to be heavy enough that the compact object must be a black hole, thus providing convincing, if indirect, proof that black holes exist.

Observational x-ray astronomy was quite active between *Uhuru* and the launch of the first HEAO in 1977. Many observations were made from both sounding rockets and satellites. Launched in 1972, OAO-3, also called *Copernicus*, carried small grazing incidence mirrors that fed an x-ray proportional counter. The Astronomical Netherlands Satellite (ANS) carried both x-ray and ultraviolet experiments. OSO-7 and OSO-8 also carried several x-ray experiments. Among other things, these experiments showed that the intensity of Cen (Centaurus) A, an active galaxy, had changed by a factor of four in less than two years, confirmed that x-ray bursts displayed a black body spectrum,[73] and detected iron-line emission from several clusters of galaxies. ANS showed that bursting x-ray sources do not pulsate and that pulsating x-ray sources do not burst. A rocket instrument showed that radiation from the Crab Nebula is polarized, thus confirming its synchrotron source.[74] An image of the Cygnus loop, a supernova remnant, clearly showed shock waves. Emission from the corona, the hot, outermost region of a star, was observed from Capella, and soft x-rays were observed from a white dwarf star. Oxygen that had lost six electrons was detected in the diffuse background, thus confirming the thermal origin of the soft x-ray background and the ultraviolet result from *Copernicus*.

In 1974, Ariel 5, built by the British, carried a NASA pinhole x-ray camera. Both long-period pulsars and bright transient sources were discovered with this satellite. NASA's SAS-3, launched in 1975, could be spun slowly or pointed for up to thirty minutes. The first highly magnetic white dwarf binary was discovered with this satellite. It also provided precise locations for about sixty x-ray sources and a survey of the soft x-ray background.[75] These examples represent only a few of the many exciting discoveries made during this time.

The HEAO program in 1977 opened the era of large, high-energy instruments. These spacecraft were 2.5 by 5.8 meters in size, weighed about 3,000 kilograms, and had a high telemetry rate. The first had a limited pointing capability, used in its last year of operation, but was intended primarily for surveys. A proportional counter array with about the same sensitivity as *Uhuru* produced a catalog of 842 sources. The large area of the detector permitted searches for rapid brightness variations. One result was the discovery of irregular variation in Cyg X-1, with time scales down to three thousandths of a second. A smaller proportional counter array covered a broad higher energy region.[76] A catalog of 85 high-

71. Intensities in x-ray astronomy are often given in units of the intensity of the Crab Nebula. This unusually stable object is usually the brightest x-ray source in the sky.
72. This braking radiation results from the conversion of kinetic energy to thermal energy when rapidly moving material is stopped suddenly.
73. A black body is an object that is a perfect absorber of radiation.
74. That is, the radiation came from rapidly moving electrons in a magnetic field.
75. The survey was conducted between 0.1 and 0.28 keV.
76. This was the region between 0.2 and 60 eV.

latitude sources yielded improved x-ray brightness for active galactic nuclei and clusters of galaxies. This experiment showed that all quasars[77] emit x-rays. Particularly surprising was the detection of 100-second variability in a Seyfert[78] galaxy. A catalog of 114 soft x-ray sources also was produced. Positions were determined to about one arcminute, leading to several hundred optical identifications. The fourth experiment on this satellite was a high-energy experiment that produced a catalog of about 40 high-energy sources.[79]

The second pointed x-ray experiment, and the first to use moderately large grazing incidence optics, was carried on the second of the HEAOs, later named *Einstein*. Such optics produce true images like those in common photographs, but can only focus on moderately soft x-rays. Any one of four instruments could be rotated into the focal plane of the telescope.[80] The good resolution and imaging capability provided high sensitivity to weak point sources as well as to extended images, such as nebulae. The sensitivity and resolution of *Einstein* made observations in the x-ray region comparable in power to those in other wavelength regions. Much new information resulted. This was the first satellite to have a major guest-observer program.

Although other countries launched small x-ray astronomy satellites during the period, NASA launched no x-ray missions in the 1980s.[81] During that time, international cooperation in x-ray astronomy played a more major role and extensive guest-observer use of the instruments became common. In 1982, NASA agreed to work with Germany and the United Kingdom on the Roentgen Satellite (ROSAT), an x-ray observatory. The SAO provided the High Resolution Imager. This mission emphasized softer (less energetic) radiation.[82] In six months of scanning, ROSAT observed more than 150,000 discrete sources at higher energies and 479 in the soft band. The latter were primarily late-type, or cool, stars and white dwarfs (comparatively near the Sun).

NASA continued to participate in international missions throughout the 1990s. The fourth Japanese satellite, the Advanced Satellite for Cosmology and Astrophysics (ASCA), concentrated on the 0.4 to 10 keV range, using four nests of thin grazing incidence mirrors feeding two cameras and two spectrometers. Astronomers at NASA's Goddard Space Flight Center and MIT contributed instruments. As one of numerous examples of the sensitive spectroscopy from this satellite, it has produced much new knowledge of supernova remnants. Among other things, it has also located many previously unknown neutron stars associated within supernova remnants, thus solving the mystery of the apparent scarcity of these stars after supernova explosions. It has found synchrotron radiation in the outer

77. A quasar is the extremely bright nucleus of an active galaxy. It may outshine the remainder of the galaxy in the optical region and is bright in all other wavelengths as well. It may be evidence of material interacting with a black hole many millions of times more massive than the Sun compressed into the volume whose radius is about 1/10 times the distance of Earth from the Sun.

78. A Seyfert galaxy is an active galaxy with a bright nucleus but is the least luminous of active galaxies. The rapid variability indicates that the radiation comes from a region that light can traverse in 100 seconds, that is, less than 3,000 kilometers.

79. The sources had energies between 0.025 and 10 MeV.

80. An imaging proportional counter with high sensitivity and resolution near one arcminute, an imager with four-arcsecond resolution, a solid state spectrometer with appreciably higher spectral resolution than a proportional counter, and a Bragg crystal spectrometer with high spectral resolution.

81. During this time, however, NASA continued to carry out work begun in 1976 on a large x-ray spacecraft, the Advanced X-ray Astrophysics Facility, or Chandra, which was launched in 1999.

82. A wide-field camera on this mission was sensitive from 62 to 206 eV; a higher resolution camera was sensitive from 0.1 to 2.5 keV.

regions of these remnants, apparently resulting from electrons accelerated strongly in shocks. This indicates that these may be the sites of cosmic ray acceleration.

The European X-ray Multi-Mirror (XMM) telescope was launched in December 1999. It carries a dense nest of thin grazing incidence telescopes that provide an unusually large collecting area for its diameter. It is not competitive with Chandra (discussed below) for imaging, but complements Chandra by providing excellent spectroscopic capability. It also can image sources in the x-ray, UV, and visible wavelengths simultaneously. The visible limiting magnitude can be appreciably deeper than from the ground. Scientists from Columbia University and the University of California at Santa Barbara provided parts of the instruments.

By the 1990s, when NASA became active in launching its own x-ray satellites again, the initial surveys had been essentially completed, except in the extreme UV, and missions were more specialized in their purposes.[83] In spite of its name, extreme UV research was more closely related to the x-ray region than to the UV, and normally uses x-ray techniques. Shortward of the edge of the hydrogen continuum,[84] the absorption decreases slowly but hydrogen is so abundant that over the large distances between stars, most of the region between 30 and 90 nanometers was expected to be opaque. However, the distribution of neutral hydrogen is not uniform. As a result, in many regions, stars can be seen at moderate distances in some of this wavelength region. Several telescopes on the Shuttle and low-resolution spectrometers on planetary probes proved this.[85] The fields of view on the planetary probes were comparatively large so the instruments were good for observing extended objects such as globular clusters and nebulae.[86] The sky background as a function of wavelength in this region also was measured. Although the "telescopes" were small, the long exposure times available on the interplanetary trajectories compensated.[87]

NASA's first satellite dedicated to the extreme UV was the Extreme Ultraviolet Explorer (EUVE). This satellite carried three grazing incidence telescopes.[88] Surprisingly, more than twenty extragalactic sources were observed in directions with low hydrogen absorption. All of these sources have active galactic nuclei; at least one is a quasar. In fall 2000, NASA decided to de-orbit EUVE, not due to its inability to continue returning excellent science but because of budget constraints.

The Rossi X-ray Timing Explorer (RXTE), launched in 1995, is currently measuring the variability over time, in scales from milliseconds to years, in the emission of x-ray sources in a wide energy range.[89] Most x-ray sources vary in brightness. The variation in

83. The extreme UV is the region between 0.1 keV and the Lyman break (13.6 eV).

84. The edge of the absorption continuum of hydrogen starts at 90 nanometers, although the crowding of upper level lines causes a pseudo-continuum at somewhat longer wavelengths whose location depends on the spectral resolution.

85. Instruments on Voyager covered the region 50 to 170 nanometers with a resolution near 1.8 nanometers.

86. The fields of view were 0.10 by 0.87 degrees.

87. Much of this paragraph is based on Holberg, "Extreme and Far-Ultraviolet Astronomy from Voyagers 1 and 2," in Yoji Kondo, ed., *Observations in Earth Orbit and Beyond* (Boston, MA: Kluwer Academic Press, 1990): 49-57.

88. Each of the survey telescopes carried two band pass filters; together they surveyed the sky at 100, 200, 400, and 600 angstroms. Three spectrometers provide spectra from roughly 70 to 760 angstroms with a resolution $\lambda/\Delta\lambda \sim 300$ (λ stands for wavelength). Of course, this equation also works for frequency and energy.

89. This included the energy range from two to 250 keV.

brightness can tell a great deal about the nature of each source. The RXTE can also point to a chosen source rapidly to observe short-lived phenomena. This satellite has discovered kilohertz quasi-periodic objects (QPOs),[90] and, from a detailed study of a bursting pulsar, provided a stringent test of the way material falls onto a compact object.

In July 1999, NASA launched its most sophisticated x-ray spacecraft ever. Originally called the Advanced X-ray Astrophysics Facility (AXAF), this satellite was renamed the Chandra X-ray Observatory in honor of astronomer Subrahmanyan Chandrasekhar. One of the Agency's Great Observatories, this spacecraft is discussed in greater detail below.

Optical Astronomy[91]

Observations in the visible wavelengths from space offer two advantages over similar observations from the ground: freedom from atmospheric turbulence and lack of the air glow background.[92] Taking advantage of either of these improvements required longer exposures with better pointing than could be obtained with rockets; balloons, however, offered the possibility of observations from above the atmospheric turbulence that blurs the images. Princeton University astronomers developed two programs to exploit this capability. First, under Office of Naval Research sponsorship, Princeton scientists flew a 30-centimeter telescope to observe the Sun. The results were spectacular and proved the advantage of observations above the atmosphere. This success led to the development of a NASA-supported, balloon-borne, 91-centimeter telescope for other celestial observations called Stratoscope II. [III-7] Led by Martin Schwarzschild, the Princeton team obtained excellent images both of planets and nuclei of galaxies. However, while these flights proved the possibilities of the technique, they were much more complex and expensive than had been expected, and the effort was dropped after several flights of the 91-centimeter telescope.

In the 1960s and 1970s, NASA commenced a very active rocket program focusing on the study of stars and galactic nebulae in the UV. Sounding rockets also were used to test new instrument techniques before they were used on satellites. According to NASA Goddard Space Flight Center astronomer Theodore Stecher:

> The first flights were ultraviolet photometry where only the spin of the Aerobee rocket was controlled. These photometers covered a large fraction of the sky as the rocket spun and precessed in free fall. The rigid body problem was solved after the flight in order to ascertain which stars had been observed.[93] This technique was then extended to spectra with objective grating spectrometers where the controlled spin of the rocket did the spectral scans. These early UV observations provided information on the stellar energy distributions and also the nature of the interstellar extinction. The astronomers and other technical staff learned how to build

---

90. QPOs are objects that vary in brightness nearly, but not exactly, regularly.

91. In this essay, "optical" includes the far UV, UV, and visible. That is, it includes the region between the hydrogen continuum and the red part of the spectrum in which atmospheric molecules begin to cause serious absorption.

92. Background sources beyond Earth's vicinity do remain, however.

93. That is, the standard rules governing the behavior of an inflexible body were used to understand the motion of the rocket.

> experiments and how to make them work. An attitude control system was developed in stages with Goddard programs serving as the trial flights in many cases. First it was a stable platform. Then [it] could point an instrument at bright stars. And finally, a stable offset pointing system enabled the astronomer to observe anything that his instrument could detect.[94]

With the availability of the International Ultraviolet Explorer and, particularly, the Hubble Space Telescope, the UV sounding rocket program decreased in importance. A few are still used in this spectral region, particularly for solar system objects and targets of opportunity, but the cream provided by bright sources has been skimmed and longer exposure times than those available from rocket flights are required to investigate most modern problems in astronomy.

Balloons do not float high enough to make observations in the UV region, but it appeared that NASA's high-altitude experimental airplane, the X-15, could. Arthur Code, an astronomer from the University of Wisconsin, replaced one of the cameras normally carried on the plane with a two-channel UV photometer. Code explained:

> I was traveling [in the late 1950s] to one of many committee meetings when I noticed a sliver of sunlight on the back bulkhead of the plane. I went back and measured the motion of the light and of the distance from the window to the bulkhead and concluded that the autopilot was holding the aircraft steady to within a minute of arc. I looked out the window and the sky was a clear dark blue; certainly you could observe from such a platform. If only the plane could get above the ozone layer we could check on the UV flux of stars in a conventional way, we could get images using UV sensitive photographic emulsion. We approached NASA about utilizing the X-15 rocket plane. With the help of Ernest Ott at NASA Headquarters, this project was approved and we started by replacing one of the on-board movie cameras located in a bubble on the fuselage with a two-channel photometer providing a visual and a UV band pass. This photometer provided measurements of the sky brightness below and above the ozone layer. Martin Burkhead's Ph.D. thesis utilized this data to map the UV sky brightness. During this time we contracted with Astronautics Corporation of America to develop a pointing system for the aircraft. The gyro-stabilized pointing system replaced the instrument elevator located behind the pilot compartment on the X-15. As the plane moved into ballistic flight the hatches were opened and the cockpit flyball was biased so that if the pilot centered the needles, the line of sight was directed to the desired star position. A star tracker then took command of the platform position. We had mounted both UV cameras and a spectrograph on the platform. Observations from the X-15 showed no halos.[95] We also obtained the first UV photometry of a late-type star, Antares.[96]

---

94. Theodore Stecher, personal communication.

95. Based on early rocket observations, astronomers had announced that they observed halos around the few bright stars that they could measure. J. E. Kupperian et al., "Observational Astrophysics from Rockets I: Nebular Photometry at 1300 Angstroms," *Astrophysical Journal* 128 (1958): 453.

96. Arthur Code, personal communication.

Unfortunately, the modified X-15 crashed on its third flight; when it was rebuilt, NASA designed it for speed rather than altitude. It no longer appeared to be worth continuing the program.

When the United States was formulating plans for the International Geophysical Year (IGY) in 1954 and 1955, the National Academy of Sciences asked scientists to propose instrumentation for scientific investigations that they would like to conduct from a satellite. Four astronomers responded. Code proposed a UV photometer; Fred Whipple, from the SAO, proposed a television map of the sky in the UV; Leo Goldberg, from Harvard, proposed a UV telescope for studying the Sun; and Lyman Spitzer, from Princeton University, proposed a high-resolution UV spectrometer. Although they were scientifically interesting proposals, each of these instruments was too large for the small satellite the United States was developing for the IGY.

Almost immediately after the establishment of NASA, these proposals were revived. It was clear that the four experiments shared major characteristics. They were comparatively large (although the experiments from Code and Whipple were somewhat smaller than that from Spitzer) and each, except for Goldberg's, required the ability to aim the instruments accurately at any point in the sky and to hold that aim for a significant period of time. Of course, they also shared the requirements common to all space experiments, such as a way to collect the data and transmit it to the ground, a power supply, and a capability to command the spacecraft and the experiment. Because of the common pointing requirements, it was decided early that a standard spacecraft design would serve each experiment with very minor modifications. Moreover, the Code and Whipple experiments were sufficiently compact that they could share the same spacecraft, by pointing out opposite ends. Soon it was realized that the thermal characteristics of an experiment pointing to the Sun were so different from those of the other experiments that Goldberg's experiment was incompatible with the same spacecraft design, and thus this experiment was postponed to the Advanced Orbiting Solar Observatory (AOSO), then under discussion.[97] In its place, NASA substituted a low-resolution spectrograph fed by a 91-centimeter mirror, proposed by James Kupperian from Goddard. Thus three missions were definitely planned and NASA expected that there would be a continuing series following these, with minor modifications leading up to a larger primary mirror, possibly 1.5 meters in diameter. The resulting satellites were the OAOs, discussed earlier in this essay.[98] [III-5, III-6]

As was often the case, particularly early in the program, the technological problems proved more difficult than had been expected. All, except the problems with the vidicons (television tubes), were solved with a three-year slip of the originally planned first OAO launch from 1963 to 1966.[99] Television tubes for the visible region were common and it was not expected that the change to an UV-sensitive cathode would be difficult. This change of cathode indeed did not present a problem, but it was necessary for the tube to be evacuated. Because glass does not transmit the UV, the UV radiation from stars had to

97. AOSO was never developed. It was not until the 1990s that any other major solar satellites, produced with international cooperation, were launched. Goldberg never did fly an experiment although he remained interested in the space program. Solar research is discussed in Volume VI of this series.

98. Homer Newell to Abe Silverstein, "Proposed NASA Project—Orbiting Astronomical Observatories," March 16, 1959, with attachment, March 12, 1959.

99. Many technological innovations from OAO were the bases of future developments. For example, IBM later used the magnetic core memory data storage system it developed for the OAOs for a series mainframe computers.

pass through a quartz or lithium fluoride window. The entire tube could not be built of these materials, and the problem of sealing such a window to a metal tube proved to be nearly intractable. Although this problem was finally solved in time for the first successful OAO launch in 1968, the tubes never did perform as well as had been hoped.[100]

The first OAO mission was to carry the experiments of Code and Whipple. In spite of the delay in the Whipple experiment, NASA decided to go ahead with the launch. That meant a hasty substitution for the SAO experiment. Phillip Fisher of Lockheed Missiles and Space Systems had developed an x-ray experiment that proved to be suitable; a prototype of the Explorer 11 gamma-ray detector also could be used. Thus an x-ray and a gamma-ray instrument substituted for the SAO instrument in 1966. Despite a satisfactory launch, a problem in the power supply system of the spacecraft prevented the acquisition of any useful data from this mission.

A prototype of the Code experiment, along with Whipple's experiment, was flown on another OAO spacecraft in 1968; this was the first successful OAO mission. The SAO experiment produced a catalog of UV fluxes from more than 100,000 stars. The Wisconsin experiment made several important discoveries. Perhaps the most interesting was the confirmation and more detailed study of the peak in the interstellar opacity near 220 nanometers. The presence of graphite (carbon) is probably the primary cause of this opacity, but other elements may be present. The results also showed that spiral galaxies are appreciably brighter in the UV than had been expected, indicating the presence of numerous faint blue stars.

The Goddard experiment was launched in 1970, but, unfortunately, a technician had tightened a bolt on the shroud of the Goddard payload too much. The shroud did not come off as it was supposed to, and the satellite did not achieve orbit. Spitzer's experiment flew on an OAO mission launched in 1972 that became known as *Copernicus*. Until the launch of NASA's Far-Ultraviolet Spectroscopic Explorer (FUSE) in 1999, the Princeton spectrometer was the only free-flying satellite that could observe the far UV, and the only instrument that could obtain good spectral resolution. From the observation in this spectral region of oxygen that has lost five electrons, Spitzer and his colleagues determined that much of interstellar space is filled with a hot, ionized medium at about 300,000 Kelvin (K).[101] This is not only hotter than many regions of interstellar space, where temperatures are lower than 100 K, but also hotter than the ionized gas near hot stars, whose temperatures reach 10,000 K.

Early in the planning for a European space science program, the European Space Research Organisation (ESRO) had proposed an astronomical satellite similar to the OAO and had awarded a contract to United Kingdom astronomer Robert Wilson to design the satellite. Budgetary limitations, however, prevented the development of such a satellite by Europe. The failure of the 1970 OAO mission left UV astronomy with no low-resolution UV spectrometer or any spectrometer that could observe moderately faint stars. Wilson and Albert Boggess, the Goddard scientist who had replaced Kupperian on the OAO experiment, realized that if the United Kingdom and the United States pooled their planning,

---

100. For a discussion of the problems OAO encountered, see "The Orbiting Astronomical Observatory," *Spacecraft Technology*, Vol. VII, ESRO SP-15, October 1966.

101. Kelvin (K) indicates that the temperature is measured on the Centigrade scale from absolute zero (-460 degrees Fahrenheit). Human body temperature is about 310 K. Kelvin temperature is 273 degrees greater than the Centigrade temperature.

they might be able to amass the funds necessary to build an ultraviolet spectrometer.[102] Moreover, they could take advantage of technological developments since the planning of the OAOs. They estimated that with a low-resolution spectrometer, they could obtain spectra of the brightest quasar, 3C273. A vidicon would be used to detect the spectra.

A major innovation of the project was to place the satellite in a synchronous orbit. Since this orbit permits continuous communication with the satellite, astronomers could work with the satellite in the same way they were used to working with telescopes on the ground, changing the conditions of the exposure in response to the data and even changing the order of the program. A second advantage was that in the higher orbit Earth blocked less of the sky. Moreover, whereas a spacecraft in low orbit could only yield thirty- or forty-minute exposures at a time, in synchronous orbit it could observe a source for as long as eighteen hours without needing to re-point to the object.

This proposal resulted in the International Ultraviolet Explorer (IUE).[103] Funding came from not only the United States and the United Kingdom, but also from the European Space Agency (ESA), which replaced ESRO in 1975.[104] ESA established a tracking station in Spain that controlled the satellite eight hours a day while it was closer to Europe than to the United States, and also contributed to the calibration and reduction of the data. Launched in January 1978, IUE was almost immediately available for use by any astronomer with a satisfactory proposal. There were no restrictions based on country of origin, and even while the Cold War was still in progress, observers from the Soviet Union and China participated. About half of the world's astronomers used this telescope during its twenty-year life.[105] The possibility of obtaining observations, in much the same way as ground-based astronomers were used to working, largely overcame astronomers' earlier reluctance to get involved in space astronomy.

The sensitivity of IUE's spectrometers was surprisingly high. Not only was it possible to reach the brightest quasars, but a number of fainter ones were also accessible.[106] The results from IUE touched almost every field of astronomy. The satellite measured water on Mars, aurorae on Jupiter, spectra of hot stars and of stars with peculiar spectra, the chromospheres[107] of cooler stars like the Sun, many types of variable stars, and the nuclei of active galaxies.[108] [III-30] In all, as of August 2000, 3,600 scientific papers had resulted from observations made with this satellite.[109] Because of budget constraints, IUE was turned off after twenty years of operation, still working well; active use of the data continues.

---

102. They proposed a spectrometer with two resolutions, a low resolution of about 0.7 nanometers and a high resolution near 0.1 to 0.3 nanometers.

103. NASA had originally referred to the satellite as SAS-D.

104. Memorandum of Understanding between the European Space Research Organisation and the United States National Aeronautics and Space Administration, March 12, 1974.

105. Yoji Kondo, "The Ultraviolet International Explorer (IUE)," ed. *Yoji Kondo, Observations in Earth Orbit and Beyond* (Boston, MA: Kluwer Academic Press, 1990), pp. 35–40.

106. *Ibid.* The faintest source observed was seven magnitudes fainter than 3C273, i.e., more than 600 times fainter.

107. The chromosphere is the region of a stellar atmosphere just outside the apparent surface (as seen in the visible region). It is the coolest region of the stellar atmosphere, but also contains very hot active regions.

108. Thomas A. Mutch to NASA Administrator, "IUE Post Launch Report #2," August 16, 1979.

109. Yoji Kondo, personal communication.

NASA's FUSE mission, launched in 1999, investigated the far-UV region.[110] A key question in this region is the ratio of deuterium[111] to common hydrogen. This ratio is determined cosmically by the mass-density of the universe. However, as deuterium and common hydrogen are both destroyed in stars, with deuterium being destroyed faster than common hydrogen, only an upper limit to the original value can be determined. As might have been expected, observations with FUSE have shown that the ratio in the interstellar medium, as seen against hot stars, varies from star to star; it is surprising that the ratio varies by about fifty percent over scales possibly as small as thirty light years.[112] Several decades ago, radio astronomers discovered clouds of neutral hydrogen high above the galactic plane which were falling into the plane at high velocities. Surprisingly, FUSE observed that many of these clouds also contain oxygen that has lost three electrons, indicating that they also contain highly ionized gas.[113] The explanation for this combination of neutral hydrogen and highly ionized oxygen is unclear.

The most powerful satellite devoted to optical observations is HST. Politically and possibly technically the most complex scientific satellite to date, this spacecraft is one of NASA's Great Observatories and is discussed in detail below.

Infrared Astronomy

Parts of the near-IR region and longer wavelengths are observable from the ground, but the atmosphere is opaque in much of the region.[114] This region of the spectrum was the last to be explored from space. The lack of sensitive detectors was a major constraint. Largely as a result of research sponsored by the national security community, good infrared detectors gradually became available. As in the gamma-ray region, background noise is a major problem in the infrared, although the source of the background is very different. All material above the temperature of absolute zero[115] emits all wavelengths in an amount that depends on the material's temperature. Although hotter bodies emit more at every wavelength than cooler ones, the highest relative emission for bodies between 1500 and 3 K is in the IR. Thus the telescope used to collect celestial IR radiation also radiates, providing an unavoidable background. This background can be lessened by cryogenically cooling the telescope. The detectors must also be cooled both to increase their sensitivity and to decrease the background. The atmosphere above the telescope also provides an inescapable background at airplane and balloon altitudes.

---

110. In the 91.2-to-120-nanometer region, the resolution, $\lambda/\Delta\lambda$, is about 30,000; it is more moderate in the remainder of the range. W. Moos, "Lyman and the Far-Ultraviolet Spectroscopic Explorer," ed. Yoji Kondo, *Observations in Earth Orbit and Beyond* (Boston, MA: Kluwer Academic Press, 1990), pp. 171-176.

111. The nucleus of common hydrogen is a proton; the nucleus of deuterium contains a neutron also and thus is twice as heavy as hydrogen. It is often known as heavy hydrogen.

112. M. Lemoine et al., "Deuterium Abundances," *New Astronomy Letters* 4 (1999): 231-43.

113. W. Moos, "Overview of the Far-Ultraviolet Violet Spectroscopic Explorer," *Astrophysical Journal Letters* 538 (1999): 1-6.

114. Water vapor and other molecules cause problems in the IR, particularly for wavelengths longer than one micrometer (1×10^{-6} meter). The atmosphere is opaque in most of the region between 25 and 1000 micrometers.

115. Absolute zero is the temperature at which all atomic motion ceases.

A great deal of the preliminary information in this spectral region has been obtained from aircraft and balloons, both of which are still used extensively.[116] The first IR observations of objects other than the Sun were made from a business jet airplane flying at an altitude of fifteen kilometers. Most of the absorption of the atmosphere in the IR is by water vapor. Although there is still some water above the altitude at which the plane flew, most is below; the average transmission is of the order of sixty to eighty percent in the mid-IR.[117] The plane carried a gyro-stabilized, thirty-centimeter telescope mounted in the aft escape hatch, without a window. Early flights showed that the IR emission from the Orion Nebula was from dust, and that both the center of the Milky Way galaxy and that of a Seyfert galaxy were very bright in the IR. NASA conducted eighty-five flights with this system between October 1968 and January 1971.[118] Among many other results, observations confirmed that the cosmic background is a black body source at a temperature less than 3 K.[119] The success of the airplane program led to the construction of a 91-centimeter telescope that was mounted in a modified C-141. With its first flight in 1974, this Kuiper Astronomical Observatory (KAO) was used extensively until it was decommissioned in 1995. Results covered a broad range of areas including detailed studies of dust clouds, emission nebulae, cool stars, and galaxies.[120] Also, as for its predecessor, it played a major role in the development of instruments and techniques.

Advantages of airborne instrumentation compared to experiments carried by other space platforms include mobility, almost no restriction on weight and support resources, and access to the instrument during flight. The KAO also provided frequent flight opportunities, typically about seventy research flights per year, each 7.5 hours in duration.[121] The success of this program led to the development of the Stratospheric Observatory for Infrared Astronomy (SOFIA), a three-meter telescope on a Boeing 747SP, being constructed jointly by Germany and the United States.

Airborne instruments are good for studying point and angularly small sources and for quickly responding to targets of opportunity. Nevertheless, they can only study small regions in which they can rapidly switch between the source and a neighboring area unaffected by the source in order to determine what fraction of the brightness observed from the source region results from the background. Since the background varies from one area to another, the comparison must be done very near the source. Theory predicted that it should be possible to observe the result of the "big bang" at the time electrons and atomic nuclei started to combine. Because of the expansion of the universe, this originally very hot radiation should now appear to be only a few degrees above absolute zero. Although the black body nature of this cosmic

---

116. Balloons are used, particularly, in Antarctica where the air is very cold and dry.

117. H. H-G. Aumann, *Airborne Infrared Astronomy* (Rice University Ph.D. thesis, 1970), (Ann Arbor, MI: University Microfilms, 1973).

118. F. J. Low, "Airborne Infrared Astronomy: The Early Years," Airborne Astronomy Symposium, NASA Ames Research Center, NASA Conference Publication 2353 (1984): 1-8.

119. P. E. Boynton and R. A. Stokes, "Airborne Measurements of the Temperature of the Cosmic Microwave Background at 3.3 mm," *Nature* 247 (1974): 528-530.

120. As of 1990, the NASA Airborne Observatory Publication list included 789 referenced publications resulting from airborne observations. H. P. Larson, "The NASA Airborne Astronomy Program: A Perspective on its Contribution to Science, Technology, and Education," *ASP Conference Series* 73 (1995): 591-607.

121. H. P. Larson, "The NASA Airborne Astronomy Program: A Perspective on its Contribution to Science, Technology, and Education," *ASP Conference Series* 73 (1995): 591-607.

microwave background (CMB) was approximately established from aircraft measurements, a detailed study of this background could not be conducted at airplane altitudes. Balloons reach altitudes more than twice as high with a corresponding decrease in atmospheric background. Thus, balloon observations have complemented aircraft observations. They have been particularly useful in studies of CMB. Although there were still problems with the result, Weiss and Muehlner published their observation in the *Physical Review* in 1973.[122]

Sounding rockets have played a smaller role in IR astronomy than in the UV and x-ray regions, although a number were flown. The Air Force Geophysical Laboratory produced a catalog of 2,000 sources using data from rocket flights but this was somewhat a *tour-de-force*. Time at high altitude for a rocket is too short to allow adequate outgassing of instruments. Residual water vapor was a major problem and most of the rocket flights produced little useful data.

The first satellite to study the infrared was not launched until 1983. This satellite, the Infrared Astronomy Satellite (IRAS), was a joint effort among the United States, the Netherlands, and the United Kingdom. The Netherlands built the satellite and two small instruments, the United States built the major instrument and provided the launch, and the United Kingdom assisted with the data. The primary mission of the satellite was to provide a photometric survey of the sky in four wavelength regions.[123] Care was taken to eliminate signals from charged particles and nearby dust by requiring that a source be seen twice within seconds. Extraneous objects at medium distances were eliminated by duplicate observations within hours, and asteroids were identified by repeats six months later. The telescope and detectors were in a well-shielded dewar (a container that keeps things hot or cold like a thermos bottle) filled with liquid helium at a temperature of 1.8 K.[124] The IRAS catalog contained 250,000 sources, including both point sources and extended sources. IRAS also obtained spectra for the brighter of these sources. Thus, after a long wait, astronomers had an excellent map of the IR sky. It remains for the fourth Great Observatory, the Space Infrared Telescope Facility (SIRTF), still under construction, to both observe fainter sources and obtain more spatial and spectral detail of interesting objects.

IRAS was unsuited to studying CMB. The Cosmic Background Explorer (COBE), launched in 1989, was a major advance toward addressing this problem. [III-22] It carried three instruments to make different, complementary observations of the background. One instrument, the Far-Infrared Absolute Spectrometer (FIRAS), compared the CMB to an accurate black body.[125] This experiment demonstrated that the background radiation is extremely close to that of a black body over a broad range of wavelengths.[126] The Differential Microwave Radiometer (DMR) was designed to search for primeval fluctua-

122. Weiss and Muehlner completed their work before Boynton and Stokes had published their measurement.

123. The wavelength regions were near 12, 25, 60, and 100 micrometers.

124. One Dutch instrument provided low-resolution spectra in the region 11 to 22.6 micrometers: the other Dutch instrument provided high spatial resolution (1 arcsecond) in a nine-by-nine-arcsecond field at 50 micrometers and 100 micrometers.

125. FIRAS has two spectrometers with about 5 percent resolution covering the wavelengths 0.1 to 10 millimeters. The instrument was cooled to 1.5 K.

126. Specifically, the temperature is 2.726 K +/- 0.010 K.

tions in the brightness of the CMB radiation.[127] The Diffuse Infrared Background Experiment (DIRBE) was designed to study the cosmic IR background.[128] While DIRBE put only upper limits on this background, it mapped the entire sky in ten IR wavelengths. The plane of the Milky Way galaxy was particularly obvious. The observations confirmed that this plane is slightly warped, as had been suggested earlier from radio observations, and indicated that the Milky Way is a barred spiral in shape. It also provided important information on the distribution of interplanetary dust.

The United States participated in the development of two IR satellites built by other nations and launched in 1995. One from Japan, the Infrared Telescope in Space, which had a small mirror, was optimized for studies of low surface-brightness objects. It carried two spectrometers for the near IR, a spectrometer for the mid-IR, and a photometer for the far IR. A European satellite, the Infrared Space Observatory, which had a larger, cooled mirror, performed spectroscopy, imaging, photometry, and polarimetry at a broad range of IR wavelengths.[129] This satellite was used primarily by guest observers and produced interesting results in many areas.

Two small NASA IR satellites followed. The Submillimeter Wave Astronomy Satellite (SWAS), launched in 1998, uses radio techniques to observe molecules of astrophysical interest in the submillimeter region. The Wide-field Infrared Explorer (WIRE) was launched in 1999 to study the evolution of starburst galaxies—that is, galaxies forming new stars in large numbers—and to search for ultra-luminous galaxies and protogalaxies. However, a control problem that occurred just after launch prevented the acquisition of useful scientific data.

The program of relatively small satellites will be followed by SIRTF, the fourth Great Observatory, which is discussed below.

Radio Astronomy

Much of the radio region is easily observable from the ground, but the two ends of the region must be observed from space. The submillimeter and millimeter regions were discussed with the infrared region, to which they are an extension. At the other end of the window, the long-wave end, the ionosphere is opaque. At even longer wavelengths, interplanetary space is also opaque, but there is a region from about thirty to near 500 meters that can be observed from the vicinity of Earth but not satisfactorily from the ground. A very difficult observation made from Tasmania, where the ionosphere tends to be thinner, and observations from several sounding rocket flights gave contradictory measurements of the spectral distribution of the radio background in this region.

In 1968 and 1973, NASA launched two essentially identical satellites to measure the spectrum more accurately. Called Radio Astronomy Explorers, the satellites each carried

127. The DMR had three channels in each of three wavelength regions: 31.5, 53, and 90 Gigahertz that compare 7-degree beams 60 degrees apart. Very small variations were observed that probably indicate the density variations that led to the development of galaxies early in the history of the universe.

128. The DIRBE measured radiation at 1.25, 2.2, 3.5, 4.9, 12, 25, 60, 100, 140, and 240 micrometers. The Cosmic Infrared Background is at shorter wavelengths than the CMB and results both from the cosmic red shift and reprocessing of radiation by dust. It comes from a younger region of the universe than the CMB.

129. This range extended from 2.5 to 240 micrometers.

two, oppositely directed "rabbit-ear" antennas, each 225 meters from base to tip, in order to obtain at least modest angular resolution. The primary astronomical receiver covered the range from thirty three to 667 meters. Other receivers covered the range from thirty eight to 1500 meters. The longer wavelengths were primarily of interest for studying the ionosphere. The first flight successfully observed the terrestrial ionosphere and the major planets, but terrestrial radiation interfered with observations of the galaxy. Therefore, the second instrument was placed in orbit around the Moon, thus shielding the spacecraft from terrestrial radiation during the lunar occultation of Earth. Although these missions clarified the wavelength distribution of radio radiation from beyond the solar system, the results essentially agreed with predictions and otherwise provided little new information about this region. Obtaining more useful information will require higher angular resolution.[130] NASA is discussing in its long-range space science plans flying a low-frequency interferometer with a very long baseline.

As discussed above, Japan was responsible for launching a very productive radio mission, the Very Long Baseline Interferometry Space Observatory Program (VSOP). This spacecraft provided one element of a VLBI network. The various ground-based radio observatories that normally participate in VLBI measurements, including some in the United States, provided other elements. Since the separation of the satellite from the other observing sites was not limited by the diameter of Earth, astronomers were able to obtain higher resolution images of sources such as of the nuclei of active galaxies than those previously available.

General Relativity

Albert Einstein's General Theory of Relativity has proved successful for predicting the behavior of light and material bodies at scales ranging from those of atomic nuclei to galaxies but the differences between the predictions of the gravitational theories of Einstein and Isaac Newton are subtle. There are other theories of gravity that agree with Einstein's within the accuracy with which the effects can be measured currently. Experimental relativity is difficult on Earth because the large gravitational field of Earth masks the small effects predicted by Einstein's and newer theories. The possibility of moving away from Earth into a different gravitation environment has interested physicists in several experiments.

The first test in space of the current theory arose as an operational rather than as a basic science problem. In order to predict the orbits of both the planets and of space probes sufficiently accurate to target the probes properly, relativistic corrections must be applied to the trajectories of both the probes and the solar system objects. The accuracy with which space probes can now be aimed continually confirms this aspect of Einstein's theory. Additional tests of Einstein's theory were provided by lunar laser measurements and planetary radar, as well as by dual frequency measurements of the delay of telemetry signals. Nevertheless, the General Theory of Relativity makes predictions that are not confirmed by these measurements.

---

130. The maximum angular resolution of a telescope is inversely proportional to the wavelength of the radiation being collected. Specifically, the resolution in degrees is 70 times the wavelength divided by the diameter of the collector. Thus, even at 33 meters the resolution of each rabbit ear was only ten degrees. This meant that little could be learned of the detailed distribution of the radiation.

Einstein predicted that a rapidly moving clock should run more slowly than a stationary clock. The flight of an atomic clock around the world in an airplane confirmed that a clock runs more slowly when moving at high velocity.[131] Einstein also predicted that a clock runs faster in a strong gravitational field than in a weak field. The gravitational field at 10.5 kilometers altitude is still too strong compared to that on the ground to accurately confirm the predicted gravitational effect on clock rate. The desire to confirm the prediction more accurately led to Gravity Probe A, the first space experiment specifically designed to test the General Theory of Relativity. In 1976, Robert Vessot of SAO flew a hydrogen maser in a Scout rocket on a suborbital trajectory. The frequency of the clock at an altitude of 10,000 kilometers was compared accurately with the frequency of a similar clock on the ground. The frequency of the clock downlink was set so that the effects of the ionosphere on the different telemetry uplink and downlink frequencies could be removed.[132] The sum of the delays of both the uplinked and downlinked signals canceled the large correction for the relative velocity of the probe and the ground. A correction also had to be made for the second-order Doppler effect, which depends on the square of the difference in the vector velocities of the two clocks. The experiment required very accurate tracking of the probe trajectory. When all necessary corrections were applied, the frequency change agreed with that predicted by the General Theory of Relativity within an accuracy of seventy parts per million. The second-order red shift also matched the prediction of the General Theory of Relativity. These results meaningfully constrain the degree to which competing theories can differ from Einstein's.

According to the General Theory of Relativity, a gyroscope in a high-altitude satellite will change its pointing very slowly (by seven arcseconds per year) because it is moving in the curved space-time around Earth. In addition, there is a small effect on the pointing of the gyroscope (0.05 arcseconds per year) because Earth is rotating and, hence, drags its gravitational field with it. To measure these effects, William Fairbanks in 1964 proposed Gravity Probe B (GP-B). Although work was started nearly forty years ago, GP-B still had not flown at the time of this writing.[133] [III-18] This experiment contains two pairs of cryogenically cooled quartz gyroscopes, with the members of each pair pointing in orthogonal directions. The pointing of each gyroscope with respect to a star must be measured to within approximately one milli-arcsecond, equivalent to the angle subtended by a human hair at a distance of 16 kilometers. The absolute drift rate resulting from the relativity effects is ten million times smaller than that of the best Earth-bound gyroscopes. A small telescope accurately pointed to a bright star is to be tightly held relative to these gyroscopes. The gyroscopes and the telescope are cooled in an enclosure filled with liquid helium. These gyroscopes and the telescope are to be well shielded by an outer shell. The entire satellite will be stabilized to 0.1 arcseconds and flown in a polar orbit at 800 kilometers. A comparison of the readout of the two gyroscopes with the direction of the star can measure the frame dragging and curved field effect. After Fairbanks' death, his colleague, Francis Everitt, took over the development of the experiment.

---

131. This phenomenon has also been confirmed by the fact that radioactive particles in cosmic rays decay more slowly than they do in a laboratory.

132. Robert Vessot, personal communication.

133. Along the way, there have been a number of technological advancements. One of particular importance to astronomy was the development of the porous plug. This allows the escape of helium gas, formed as liquid helium slowly warms but not the escape of the liquid helium itself. This type of plug has been used on all infrared astronomy satellites and probably made such satellites successful.

The Great Observatories

By the early 1980s, NASA had four large astronomical spacecraft in various stages of development. Between them, they covered the wavelength regions from high-energy gamma rays to the short radio region. In order of increasing wavelength, they were: the Gamma Ray Observatory (GRO, now the Compton Gamma Ray Observatory, CGRO), the Advanced X-ray Astrophysics Facility (AXAF, now Chandra), the Hubble Space Telescope, and the Space Infrared Telescope Facility (SIRTF, originally the Shuttle Infrared Telescope Facility).[134] NASA's Director of Astrophysics, Charles Pellerin, came up with idea of calling these spacecraft the "Great Observatories." The labeling was quite effective as a way of identifying the set of missions as a unique combination, and has been used since. [III-34]

The four Great Observatories shared various problems in their development. Each, except CGRO, took more than twenty years from the beginning of development until launch. Each was squeezed by financial restraints that both lengthened the program (and thus increased the total cost) and, except for SIRTF, caused descoping of the project. As each was planned for a Shuttle launch, each was affected, although in different ways, by the *Challenger* accident.

Hubble Space Telescope (HST)

The first of the Great Observatories to be launched was HST[135]. Even before NASA was created, astronomers had dreamed enthusiastically of orbiting a large space telescope (LST). [III-1] As early as 1962, a Space Studies Board (SSB) summer study suggested that it was time to start planning of such an instrument.[136] This was an exciting possibility, and not only for the astronomers. NASA's Langley Research Center started a study of the project, with a human along as an observer. Several aerospace companies, partly funded by NASA, began studies of how such a telescope might be launched and controlled.[137] Aden Meinel, an early proponent of a large space telescope, started a Space Division at the Kitt Peak National Observatory even before the start of the Apollo program. He was a major proponent of the telescope at both the 1962 and 1965 SSB meetings.

Not all astronomers were enthusiastic about the project. To quote Meinel, "Ira Bowen [the director of the Mount Wilson and Palomar Observatories] said at one meeting that one could never stabilize a space telescope enough to yield high resolution. He said that simply pulling out the dark slide would disturb it. He also remarked that higher [angular] resolution wouldn't be of much importance to astrophysics."[138]

In spite of the strong division of opinion about a large space telescope, by the 1965 SSB summer study, momentum behind the project had grown to the point that NASA

---

134. SIRTF will measure wavelengths almost ten billion times longer than those CGRO measured.

135. For an outstanding history of HST, with special emphasis on the political complications the project had to navigate, see Robert W. Smith, *The Space Telescope: A Study of NASA, Science, Technology, and Politics* (New York, NY: Cambridge University Press, 1989).

136. Space Science Board, *A Review of Space Research* (Washington, DC: National Academy of Sciences, 1962).

137. The Boeing Company, "A System Study of a Manned Orbital Telescope."

138. Aden Meinel, personal communication.

Headquarters decided that it was important to start planning for the mission. Various additional studies were funded to prove the feasibility of the idea and to investigate the areas thought most likely to require extensive development. A committee of the SSB, under the chairmanship of Lyman Spitzer, began a four-year activity to define the scientific uses of a large space telescope.[139] The Astronomy Program at NASA Headquarters and astronomers on the Astronomy Working Group (an advisory committee that was composed of astronomers from both NASA Centers and the non-NASA astronomy community) began to develop the arguments for such an instrument.

In 1970, NASA established two committees: an LST[140] Task Group to map out the engineering requirements of the project, and a Scientific Advisory Committee to define the scientific requirements. NASA Headquarters officials chaired both committees. The Task Group was primarily an in-house committee from NASA Centers; the Advisory Group had a primarily, but not exclusively, non-NASA membership.

In 1971 and early 1972, Goddard Space Flight Center and Marshall Space Flight Center conducted competitive Phase A (preliminary) studies of the LST. However, when it came to deciding how to partition work between the Centers, the decision was based primarily on the fact that Goddard already was fully involved with other science projects, while Marshall, whose work was declining after the push for Apollo, was anxious for a new responsibility. Hence, the overall management of the project was assigned to Marshall in 1972. Nevertheless, Goddard, with its experience in astronomy, retained the management of the scientific instruments. At the urging of the scientific community, C. Robert O'Dell was brought to Marshall as the project scientist. Because Marshall would be managing the project, the Science Advisory Group was transferred to Marshall under O'Dell's leadership. Typical instruments were defined, and various groups were selected to work with the project to ensure that the spacecraft could accommodate such instruments. At about the same time, it was decided that the project should be divided into three sections—the Support Systems Module, the Optical Telescope Assembly, and the Scientific Instruments—each to be contracted for separately. This made the management of the project particularly complex.

In early 1973, politically astute NASA managers realized that the cost of the LST would limit their ability to sell it to either the Administration or Congress. Hence, Marshall was given a cost target well below its estimate of the cost of the telescope concept then under examination. Various cuts were made in the plans to reduce the cost; these reductions often had to be reinstated later in the program. The flight of a precursor 1.5-meter telescope to test the many complicated systems on the LST was dropped at this time.

In 1974, Congress appeared unenthusiastic about the LST. The House cut all funds for the project. At this point a few astronomers, primarily in Princeton, rallied their colleagues nationwide to lobby for the LST. A major argument made by skeptical Congressmen was that the National Academy of Science's study of astronomy in the 1970s barely mentioned the LST. This was partly the case because the study's chairman, Jesse Greenstein—perhaps

139. Space Science Board, *Scientific Uses of the Large Space Telescope* (Washington, DC: National Academy of Sciences, 1969).

140. Although LST stood for Large Space Telescope, in the minds of many astronomers it also stood for the Lyman Spitzer Telescope, given Spitzer's seminal role in proposing the concept.

because he had been burned almost three decades earlier by his V-2 experience and also because of his West Coast connections—was unenthusiastic about the large space telescope idea. More importantly, the study committee doubted that the telescope could be launched before 1980, thus falling outside the range of the committee's responsibility. By this time, the Academy had embarked on a new study that was to elevate the LST to top priority, but this study had not yet been completed. To counteract the impact of the Greenstein report, the study committee was again polled for its views on the LST. This time, after additional lobbying within the astronomical community, the Academy committee unanimously gave the LST top priority. Influenced by this result and extensive lobbying, the Senate was convinced to include the requested funding. As often happens, the House-Senate conference committee split the difference; NASA received half of the amount that had been requested.

Congress agreed to supply additional funds for the project only if significant foreign involvement in the LST was included; this would decrease the cost of the project to the United States. After extensive negotiations between NASA and the ESRO (later succeeded by ESA), Europe agreed to supply a major scientific instrument and the solar arrays. In return, European astronomers were guaranteed 15 percent of the observing time. [III-29] Although both the decision to accept a European instrument without competition and the guarantee of observing time upset some U.S. members of the study teams, it was likely that the Europeans could have successfully bid for fifteen percent of the observing time in any open competition. Moreover, it was unlikely that NASA would have been able to fund an additional instrument, or even get Congressional approval for the LST overall without the European contribution.

In October 1975, President Gerald Ford cut the federal budget by $28 billion in order to try to balance the budget in three years. NASA's response to its share of the cut was to drop the new start for the LST in the Fiscal Year (FY) 1977 budget request. The Office of Management and Budget also felt that because of a slip in the Shuttle schedule, FY 1977 was too early to start the LST, and James Fletcher, the Administrator of NASA, believed that the new start was politically unfeasible. Instead, NASA requested a new start for the Solar Maximum Mission in FY 1977, and no funds specifically for the LST. Again the astronomical community launched a major lobbying effort, both in Congress and with NASA. The NASA Administrator then argued for support of the LST with President Ford. The result was that a new start for the project slipped to FY 1978. The "L" was dropped in references to the project—making it just "ST"—so as not to advertise its cost, although some astronomers were concerned that the name change was an indication that the project's scope might be cut further. [III-24, III-25]

At about this time, Senator Proxmire asked NASA why the average American taxpayer should want to pay for such an expensive project. NASA's answer was that for the price of a night at the movies, the average American could enjoy fifteen years of exciting discoveries. Although it is unlikely that this response made any difference, it is interesting that as both the ST and movies have increased in cost, the statement is still approximately true.

NASA Headquarters directed the Marshall Space Flight Center to find ways to cut the cost of the project in preparation for a FY 1978 new start. Marshall suggested various ways, of which the most draconian was to decrease the size of the telescope's mirror. The original plan called for a three-meter mirror. Both contractors and scientists were asked to look at the impact of including a mirror in each of three sizes: 3, 2.4, and 1.8 meters.

A major objective of the ST was to improve knowledge of the Hubble constant. This is the ratio between the speed of recession of a galaxy and its distance. The Milky Way is a member of a group of thirty to fifty galaxies that interact gravitationally. Thus their motions are affected by this gravitational interaction in addition to the expansion of the universe. To measure the Hubble constant, it is necessary to determine the distances of galaxies outside this Local Group. The most significant collection of the nearest such galaxies lie in the Virgo cluster. Thus, it had been assumed from the beginning that the LST must be able to observe Cepheid variable stars in the Virgo cluster. It had been known for most of a century that the period of the variation of a Cepheid is closely correlated with its intrinsic brightness. Hence, to measure its distance, it is only necessary to measure the period of the variation and the mean or maximum brightness. The astronomers determined that a 2.4-meter telescope could still obtain these measurements; a 1.8-meter telescope could not. Therefore the astronomers on the Science Advisory Group agreed that they could accept a 2.4-meter objective, but that they would recommend that the project be ended rather than settle for a 1.8-meter mirror. [III-23]

Also, facilities existed for the manufacture of a precise 2.4-meter mirror, while new facilities would have to be built for a three-meter mirror. This would greatly increase the cost of the Optical Telescope Assembly. Reducing the mirror size to 2.4 meters would also relax the pointing requirements and simplify the pointing and control system. Moreover, using a 2.4-meter mirror would simplify the control design even more by allowing the designers to wrap the heavy Support Systems Module around the telescope.

By the time the FY 1978 budget was ready to go to Congress, NASA had gotten both the President and the Office of Management and Budget enthusiastic about the project. Moreover, after several years of experience, the astronomers had become more skillful and sophisticated lobbyists. There was also quieter lobbying behind the scenes. Although there were no astronomers in a high position at NASA, there were several good scientists who understood the objectives of the project. Thus, the first task was to transmit the enthusiasm and wishes of the astronomers with whom NASA was working to NASA managers and to get them equally enthusiastic about the project. Next, when they had become enthusiastic, NASA Administrator James Webb, an astute politician, set about relaying that enthusiasm to various groups of politically influential individuals. In the late 1960s, he held a series of dinners for small groups of these people. After each dinner, representatives of the Physics and Astronomy Program Office presented the concept of the LST, the design features, its feasibility to the extent that these had been determined, and the scientific arguments for the mission. These "dog and pony shows" proved to be very successful in ultimately gaining political support for the project. Finally, potential contractors began an extensive lobbying campaign well before the astronomers became involved. They also provided significant political guidance to the astronomers as the latter started their campaigns.

A new start for the ST was approved at last in the President's FY 1978 budget proposal. [III-28] Technical problems now came to the fore. Because of stringent restrictions on overall NASA personnel as well as on the project's budget, and because Marshall had a reputation of excessively enlarging project personnel, Marshall was given a very stringent personnel cap for the project. With far too few capable people, Marshall had to manage two associate contractors, an international partner, and another Center, each of which was in turn dealing with a number of subcontractors. Partly for this reason and probably because of the reluctance of the national security community to allow "outsiders" full access to those portions of the project with a national security heritage, NASA was unable to monitor its contractors closely. Also, relations between Marshall and Goddard were severely strained for the first few years of the project.

Almost immediately after the Phase C/D (development, construction, and preparation for launch) contracts were awarded, each of the contractors increased their cost estimates substantially. Yet, Marshall was not allowed to budget for any additional funds. These factors led to a continuing series of severe problems until NASA Headquarters intervened in a major way in 1983. Project managers were replaced at both Marshall and Goddard. The new managers made a determined effort to work together, thus solving one problem. Also, NASA Headquarters, after careful review of the project, agreed that substantially more money and manpower should be allotted. Although, as in any complex technological project, there were many problems after this, they were under more control. There were also schedule slips, but a launch in late 1986 still seemed possible. The 1986 *Challenger* accident eased the schedule problem, but also substantially increased the cost of the program as the spacecraft remained in storage in a clean room in Palo Alto, California, for three years, while the project team had to be kept together until the launch.

As the Ramsey Committee had stated in the 1960s, university astronomers wanted a non-NASA institute to manage the science of the project. In contrast, astronomers at Goddard were anxious to have scientific control of the project. This led to a major fight, which the university-based astronomers won. [III-27] In addition to granting the wish of the scientific community, NASA Headquarters recognized that the size of the necessary institute would overwhelm Goddard, and particularly its small astronomical staff. The Space Telescope Science Institute (STScI) got off to a rocky start in its relations with NASA. Riccardo Giacconi, the director selected, had ambitious plans for STScI, and immediately indicated that the staff had to grow significantly above that described in the proposal. Just as NASA Headquarters officials had failed to respond to the sometimes desperate requests for funds from Marshall, they also tried to squelch the staffing and budget growth demanded by STScI. Finally, after a careful look at the functions for which NASA believed STScI should be responsible, some of which had not been included in the original specifications, NASA agreed to a major increase in personnel and space. Over time, the relations between Giacconi and NASA became smoother, with each developing a better understanding of the other's problems.

STScI maintains an archive not only of HST observations but also of UV observations from other satellites, particularly the IUE. Rather than depending on the observer to produce reduced data from HST, STScI archives the raw data and calibrates these "on the fly" when they are requested from the archive. This procedure removes any delay (beyond an agreed proprietary period) in making the data available to other astronomers. This archive has been quite successful, attracting many users and resulting in a number of scientific papers.

There was great delight among astronomers in April 1990 when the space telescope was finally launched. By then it had been named the Hubble Space Telescope after Edwin Hubble, the astronomer who first demonstrated that the more distant a galaxy, the higher is its velocity of recession. A little later, the joy turned to dismay when it was discovered that the images were not of the expected quality. Analysis showed that the telescope was suffering from spherical aberration. Even if a backup mirror had been completed (work on it was stopped to save money), it would have been impossible to exchange mirrors in orbit. Return of the telescope to the ground had been ruled out earlier because of the cost, the danger of contamination, and the possibility of damage to the telescope from re-entry and landing. Therefore, an intensive period of study ensued, led by STScI but including NASA and other optics experts, to determine the most effective remedy. [III-37] The individual instruments could have been redesigned to correct the problem but, because of the financial problems, no backup instruments were available except for the Wide Field/Planetary Camera (WF/PC).

Finally, it was realized that the backup WF/PC could be easily corrected and that a single system could be designed to correct the image for each of the other instruments. The problem was how to install such a system with stringent alignment requirements in a tight space. While taking a shower in Germany, Jim Crocker, a HST engineer, was inspired by the showerhead to create a mechanical design that could meet the restrictions.[141] To add the correction system, called the Corrective Optics Space Telescope Axial Replacement (COSTAR), it was necessary to remove one of the original instruments. The High Speed Photometer was selected for removal. As this instrument's principal investigator remarked to the author some years later, "What wonderful results we could have obtained with the improved image quality!" Three years passed before the new instruments could be completed and a Shuttle repair mission could be launched. [III-38] In the meantime, mathematical methods were developed to get reasonable images from HST, but they did not work well for extended sources or crowded regions. Also, the poor light concentration in the image limited the faintness that could be reached.

The remarkable images obtained after the corrective optics were installed vindicated the hopes of astronomers who had worked so hard for large, diffraction-limited optics in a satellite that they could point with sufficient accuracy to avoid degrading the image. The problem of improving the determination of the Hubble constant started as soon as possible after the correction of the optics problem. The results to date are still somewhat controversial, but most astronomers believe that that the constant is now known within ten percent, in contrast to the fifty percent uncertainty before HST observations. An impressive and surprisingly fruitful observation entailed keeping the telescope pointed continuously to the same "uninteresting" place for ten days. In the resulting image, sources were detected which are as faint as 1/10,000,000,000 of the brightness of the faintest star normally visible to the human eye in a clear, dark sky. Some of the galaxies (there were very few individual stars in this tiny field) are so far away that their light left them when the universe was only a few percent of its present age. These images not only show that galaxies formed very early in the history of the universe, but that most are somewhat different from the modern galaxies near the Milky Way. The ability to resolve small details near the centers of active galaxies has established almost beyond any doubt that these centers contain black holes. Images and spectra of objects ranging from comets and planets to very distant galaxies have impacted modern astronomy (and the public's perception of the cosmos) as much as Galileo's telescope did more than three centuries earlier.

---

141. David Leckrone, personal communication.

Compton Gamma Ray Observatory (CGRO)

The second Great Observatory was CGRO, launched in 1991. It was named to honor physicist Arthur Holly Compton, who had studied the behavior of gamma rays. This spacecraft also had a somewhat tortuous history.[142] Originally, a somewhat smaller version of CGRO's Energetic Gamma Ray Experiment (EGRET) was proposed for the HEAO program, but as a result of cost overruns on the Mars Viking project, three large experiments, including EGRET, were removed from the HEAO program. EGRET was then studied as an independent Explorer mission, with the spacecraft to be built by the Johns Hopkins Applied Physics Laboratory (that had built the SASs). A year later, NASA Headquarters decreed that the Multi-Mission Spacecraft (MMS) should be used, but that proved to be so expensive that the mission was cancelled. By this time, 1976, it was realized that other gamma-ray experiments were also important, and the concept of a multi-experiment gamma-ray mission, designated the Gamma Ray Observatory (GRO), was developed. After some study and an announcement of opportunity, five experiments were selected in 1978.

By 1981, it appeared that a spacecraft with these five experiments would be too large and too heavy. The Gamma-Ray Line Experiment was, therefore, dropped. [III-32] This was one of the same experiments that had previously been dropped from the HEAO. As all programs were significantly delayed by the *Challenger* accident, the GRO launch date was reset for around 1990. There were, of course, additional costs due to the launch delay. The final launch date was slipped again, this time to 1991. An attempt made to develop an optimum technical and budgetary schedule led to the GRO being ready about nine months before it was actually possible to launch it. (Probably the last year of the delay resulted from the desire to launch the HST first.)

Four instruments were carried on the final spacecraft.[143] The Burst and Transient Source Experiment (BATSE) was composed of eight gamma-ray modules placed on the spacecraft to provide all-sky coverage.[144] Not long after launch, the tape recorder on CGRO failed, thus necessitating real-time data transmission. This proved to be a great advantage, as it allowed the information about a burst detection to reach the ground within seconds rather than in the two hours that had been planned.[145] The Oscillating Scintillation Spectrometer (OSSE) covered the low-energy range.[146] The Compton

142. Aside from the advantage of not being the first, CGRO benefited from involving only a single center in the management (although instruments came from other institutions). In addition, it did not have to deal with national security problems.

143. Together, the instruments covered the energy range from below 0.1 to about 3×10^4 MeV.

144. Each module contains two detectors, one designed for high sensitivity and the other for higher energy resolution. They can measure gamma-ray temporal variations on time scales down to several microseconds and energy spectra in the range 30 keV to 1.9 MeV.

145. The decision not to depend much on Shuttle servicing turned out to be a blessing. Both tape recorders started to give trouble after about six months and failed completely after the first year. In order to get real-time data from the satellite, NASA added a Tracking and Data Relay Satellite System (TDRSS) receiving station in Australia, thus closing the previous gap in satellite coverage. This continual real-time receipt of data from the satellite permitted prompt alerts to gamma-ray bursts.

146. The range of OSSE was 0.1 to 10 MeV. A phoswitch system was used with cesium iodide crystals behind sodium iodide crystals. The field of view was limited to 3.8 by 11.4 degrees by a tungsten alloy shield.

Telescope (COMPTEL) was based on Compton scattering.[147] This instrument detected both the energy and the direction of the gamma ray. EGRET covered the high-energy range.[148] This was a much larger version of the SAS-2 spark chamber with the addition of good energy measurement. The accuracy to which a point source could be located varied from five arcminutes for strong sources to forty-five arcminutes for the weakest sources. Originally, a major guest-investigator program was planned for CGRO, but it was not approved due to budget constraints. It was reintroduced when CGRO became part of the Great Observatory program.

The CGRO was originally designed to be serviced by the Shuttle and returned to the ground for repair. The changes in the Shuttle program after the *Challenger* accident increased the cost of launches sufficiently that this was no longer cost effective. The degree to which the spacecraft could be refurbished in orbit also was reduced to save money. By 2000, several of CGRO's gyros had failed. NASA was concerned that if another failed, the spacecraft would be uncontrollable and could reenter Earth's atmosphere and drop heavy pieces in a populated area, causing damage and, possibly, loss of life. The gyros could not be serviced individually inorbit, but the entire unit could have been replaced. This was considered to be too expensive, and recapture was considered dangerous as well. Therefore, though it was still producing excellent science, the spacecraft was commanded in 2000 to reenter the atmosphere. It burned up over the Pacific Ocean.

CGRO was exceedingly productive in areas of study ranging from the solar system to distant regions of the universe. Fichtel and Trombka list the following accomplishments:

- the finding of new objects including high-energy, gamma-ray blazars (a kind of active galaxy);
- a very clear separation of the gamma-ray properties of blazars and Seyferts;
- a major increase in knowledge of gamma-ray bursts;
- the observation of an increased fraction of pulsar electromagnetic radiation being emitted as gamma rays as pulsars age up to one million years, and the detailed knowledge of their spectra;
- the determination with high certainty that cosmic rays are galactic;
- the detailed mapping of the galactic diffuse radiation, including the aluminum line and the measurement of the *pi meson* bump in the high-energy gamma-ray spectrum;
- the detection of gamma-ray lines from SN1987A[149] and Cas (Cassiopeia) A;
- the absence of microsecond bursts and its implication for certain unification theories;
- the existence of energetic particles near the Sun for over ten hours following a flare and the associated implication for the shock acceleration theory; and

---

147. COMPTEL detected gamma rays by the occurrence of two successive interactions: first a Compton scatter collision occurred in a detector of material with low atomic number; then a second interaction took place in a lower plane of material of high atomic number in which, ideally, the scattered gamma ray was totally absorbed. Gamma rays below about 2 MeV cannot be detected; the upper limit to the energy for which neutrons can be discriminated from gamma rays is about 100 MeV.

148. EGRET covers the region above 20 MeV.

149. SN1987A is the supernova that occurred in 1987 in the Large Magellanic Cloud, a nearby galaxy.

- the measurement of the spectrum of the diffuse, presumably extragalactic, gamma radiation with a flat spectrum in the high-energy region consistent with a blazar origin.[150]

Advanced X-ray Astrophysics Facility/Chandra X-ray Observatory

The third of the Great Observatories, Chandra, was a follow-on to HEAO-2, *Einstein.* Like *Einstein,* but much larger, it carries grazing incidence mirrors with excellent image quality. With a focal length of ten meters, the spacecraft can detect point sources more than twenty times fainter than previous x-ray telescopes and provides eight times better angular resolution.

AXAF started in 1976 with a proposal from Giacconi and SAO's Harvey Tannenbaum.[151] [III-26] After a competition among NASA Centers, the project was assigned to Marshall in 1977. There were originally two spectrometers on AXAF. A Bragg crystal spectrometer from MIT's Claude Canizares was at the focal plane of the telescope. A calorimeter from Stephen Holt of Goddard was also included. The Bragg instrument was dropped in 1989 to save money. Originally plans were to launch the spacecraft into a low orbit from which the Shuttle could service it. Because of the severe increase in Shuttle launch costs after the *Challenger* explosion, this no longer seemed feasible. Eliminating this possibility saved substantial money, including both servicing costs and additions in spacecraft construction. Instead, project officials decided to launch AXAF into a high orbit where the spacecraft would be less affected by Earth's radiation belts and in which there would be no temptation to service the mission. The combined weight of the spacecraft and the additional rocket stage needed to reach the desired high orbit from Shuttle altitude turned out to be too heavy for a Shuttle launch. Two significant changes were made to the spacecraft to reduce the weight: the calorimeter was dropped and the number of mirrors was decreased from six to four. The higher observing efficiency in the new orbit compensated for the decrease in the total mirror area. Plans were to fly the calorimeter on a separate spacecraft; that spacecraft was cancelled in 1993, again because of funding constraints. Instead, the calorimeter was put on the Japanese satellite Astro-E, which failed. [III-39]

AXAF, like the GRO, had to wait for the HST launch, which was delayed by the *Challenger* accident. Spacecraft integration proved to be more difficult than anticipated and there were some problems with components. These technical problems benefited from the launch delay.

Launched in 1999 (and renamed Chandra after astronomer Subrahmanyan Chandrasekhar), AXAF/Chandra had a productive first year observing objects from comets to quasars. It discovered that the x-rays that had been observed previously from comets were a result of the collision of the solar wind with ions in the comet. A flare was

150. Carl E. Fichtel and Jacob I. Trombka, *Gamma-ray Astrophysics: A New Insight into the Universe,* 2nd ed. (Washington, DC: National Aeronautics and Space Administration, 1997). The information on the CGRO instruments is also from this book.

151. Smithsonian Astrophysical Observatory, "Proposal to NASA for the Study of the 1.2-Meter X-ray Telescope National Space Observatory," April 1976.

observed from a brown dwarf, a star-like body that is too light to fuse hydrogen for energy. The observatory has observed two galaxies merging.[152] Many galaxies are extremely bright in the x-ray region but optically faint. There are many low-luminosity black holes that are not understood. As Chandra Project Scientist Martin Weisskopf remarked, "Every image leads to a discovery."[153]

Space Infrared Telescope Facility (SIRTF)

The fourth, not yet launched, Great Observatory is the SIRTF. SIRTF will carry an 85-centimeter telescope that will be cooled to 1.6 K. To cover the broad wavelength range and provide both imaging and spectroscopy, SIRTF will carry three focal-plane instruments.[154] The Infrared Array Camera will use large-area, two-dimensional IR array detectors to provide diffraction-limited angular resolution in the nearer IR.[155] The IR Spectrometer will cover the entire range of wavelengths in which SIRTF will be used, with a variety of resolutions and modes.[156] The Multi-band Imaging Photometer will provide both imaging and low-resolution spectrometry in the mid- and far IR.

SIRTF was originally called the Shuttle Infrared Satellite Facility. The plans were to keep the spacecraft attached to the Shuttle or at least in the Shuttle's vicinity and to return it to Earth at the end of the Shuttle's mission. By 1983, IRAS had shown that a long-lived IR satellite was feasible. Also, there was some concern that material around the Shuttle might cause problems. The name of the mission was therefore changed to the Space Infrared Telescope Facility, and it was decided to fly the spacecraft in a 900-kilometer orbit, above the strongest radiation belts. In 1989, the planned orbit was raised to a 100,000-kilometer orbit and later to a heliocentric, Earth-trailing orbit. This change will improve both scientific performance, because of the lower background in the far IR, and observing efficiency, as Earth becomes a small target. The move to a heliocentric orbit was accompanied and somewhat enabled by decreases in payload complexity.

Both the SIRTF schedule and the spacecraft, instrument, and mission design were severely delayed by funding constraints. However, as Project Scientist Michael Werner noted: "The long delay allowed us to invest in enabling technology—detector arrays, cryogenic technology, and lightweight optics—and the tough funding encouraged very creative thinking on the part of the scientists and the engineers. As a result, the $500 million

152. NASA Marshall Space Flight Center press release, August 22, 2000.

153. Martin Weisskopf, personal communication. Much of the history of the project is also based on this conversation.

154. The Multiband Imaging Photometer (MIPS) for SIRTF will provide background-limited imaging and photometry in the range from 30 to 200 micrometers and a low resolution spectrometer for spectral energy distributions. It will also use an array detector to provide broad band photometry and mapping from 200 to 700 micrometers with a possible extension to 1.2 millimeters. The Infrared Spectrograph (IRS) consists of several long-slit and echelle-mode spectrographs covering the interval from 2.5 to 200 micrometers. Resolving power will vary from 100 to 2,000. Its large collecting area and sensitive array detectors will provide sufficient capability to observe many different types of sources. Finally, the Infrared Array Camera (IRAC) will map large fields using a step-and-stare method, at 3.6, 4.5, 5.8, and 8.0 micrometers.

155. The telescope will provide diffraction-limited images from 2 to 27 micrometers.

156. The instrument will cover the energy range between 2.5 and 200 micrometers.

SIRTF we now have has almost the same mirror size, the same lifetime, and the same basic instrument functionality as did the $2 billion-plus version talked about in 1990."[157] The project got back on track after a long launch delay by a combination of ingenuity and technology advances, plus the fact that it became an example of NASA's 1990s "faster, better, cheaper" approach to mission development and operations.

The Future

With the launch of SIRTF, planned for late 2001, every region of the electromagnetic spectrum not observable from the ground, with the exception of long-wave radio radiation, will have been surveyed and observed with good sensitivity and angular resolution. It is probable that most types of celestial sources will have been identified, although there will certainly be surprises. Indeed, many cosmological phenomena are not yet completely understood. A test of Einstein's General Theory of Relativity will have been conducted successfully and another will be far along in development.

Plans for the next decade are ambitious. [III-40] They include small missions dedicated to answering specific questions, and very complex missions aimed at increasing angular resolution, always a major *desiderata* in astronomy. The increase in resolution will permit detailed study of crowded sources, such as the vicinities of black holes in galactic centers. Improved resolution also will allow for the comparison of galaxies as they existed early in the life of the universe with those near the Sun that we see now, some thirteen billion years later.

The smaller missions are an extension of the Explorer program—a program of small scientific satellites started early in the NASA program—with several important changes. The most critical is that the new program includes three mission classes (mid-sized, small, and university class), each with a strict funding cap. In addition, there is a fourth class for participation in non-NASA missions, also with a strict funding cap.

FUSE was the first mission within this new scheme (although it started at least twenty years ago as a much more ambitious project). At least four missions per year, with a total funding cap of $226 million are planned. Included in the cap are the costs of project definition, development, launch service, mission operations, and data analysis. A major problem in the past has been that when a mission was accepted, no detailed design study had been conducted. Hence, the proposed costs were highly uncertain and were often greatly exceeded by the final cost. A new approach is to select missions tentatively, with final selection after a period of design study sufficient to provide a meaningful estimate of costs. If the costs, including contingencies, exceed the cap, the mission will be stopped or descoped. A third change is that the proposing institution will be given more responsibility for many of these missions. An example of the largest new Explorer missions is Swift, which will monitor the sky for gamma-ray bursts. When one is discovered, it can start x-ray and optical observations of the site within fifty seconds and send initial coordinates of the burst to the ground within fifteen seconds. In this way, scientists should get much important information on the nature and origin of such bursts.

---

157. Michael Werner, personal communication. Much of the discussion of SIRTF is based on this communication.

The complex missions are ambitious indeed. They are a new generation of "Great Observatories," going beyond the capabilities of the earlier ones with high sensitivity as well as high angular resolution. Again, they have a number of characteristics in common. All are much larger and have greater collecting area than the preceding generation of instruments. Because of their size, most must be launched in a collapsed configuration and assembled automatically in orbit. Most are based on interferometry in order to combine information from independent instruments. Interferometry has been used on the ground by radio astronomers for many years but has been used successfully in the optical region only in the past decade. Although interferometry will be far from trivial even in the IR region, it will be exceedingly difficult at high energies, as the relative positions of the component telescopes must be known to a small fraction of a wavelength. All of these missions will be expensive enough, as well as capable enough, so that international cooperation is imperative. Finally, most if not all of the observing time will be open to all astronomers in a guest observer mode. That is, each will be an international facility.

In addition to the technical challenges presented by the hardware, data handling from these large missions will be a major problem. Data handling involves not just collecting and transmitting the data, but also producing well calibrated data in a form that can be used by someone familiar with astronomical observation generally but not familiar with the quirks of a particular instrument. Interferometry involves much more data and more complicated data processing than do single telescope techniques. Finally, many of these instruments will be placed near the L2 point to avoid both the occultation of a large portion of the sky by Earth and its radiation environment.

An example of one of these missions is the Terrestrial Planet Finder. For this mission, two or more medium-sized near-IR telescopes will be linked interferometrically to provide sufficient angular resolution to separate a medium-sized planet from its parent star and to observe it spectroscopically. At present, only much larger planets can be detected by their gravitational influence on their parent stars or, in special orientations, by planetary eclipses. In the portion of the radio region that can be observed from the ground, a satellite in orbit will be linked with ground-based instruments to provide baselines several times longer than the diameter of Earth. In the longer wavelengths, antennas and receivers very widely spaced in orbit will provide significant angular resolution for the first time. To detect gravity waves longer than those observable from the ground, a pair of satellites whose separations are accurately measured will look for tiny changes in the separation as a result of the passage of the wave.

The possible future of space-based astronomy and astrophysics is thus both exciting and daunting.

Acknowledgements

Thanks are due to many former colleagues who helped with this essay, particularly Arthur Code, L. Kaluzienski, Aden Meinel, Theodore Stecher, Michael Werner, and Martin Weisskoff, who provided material cited in the essay. Others provided background information. Thanks also is due to Amy Paige Snyder who collected most of the documents and provided much helpful advice and editing.

Document III-1

Document title: Lyman Spitzer, Jr., "Astronomical Advantages of an Extra-terrestrial Observatory," Project RAND, July 30, 1946.

Source: The RAND Corporation, reprinted with permission.

Prior to World War II, Earth-orbiting telescopes only existed in science fiction stories. The advent of guided missiles by Germany during the war, however, made a few astronomers optimistic that this new rocket technology would soon be able to loft telescopes and other astronomical instruments into space. Among the believers, Princeton University's Lyman Spitzer authored a paper for the Douglas Aircraft Company's Project RAND (the think tank established by the Army Air Corps after World War II) on the scientific benefits of a space-based telescope. The paper became part of a larger 1946 RAND report on the feasibility of developing and launching a scientific spacecraft. Originally classified, the Spitzer study was unknown to other astronomers for several years. When his ideas became known, many astronomers remained skeptical of the worth of space-based instruments. Over time, however, astronomers began to embrace the astronomical studies Spitzer described in his paper and eventually attributed the Hubble Space Telescope's development to Spitzer's efforts.

[no page number]

YALE UNIVERSITY OBSERVATORY
PROSPECT AND CANNER STREETS
NEW HAVEN 11, CONNECTICUT

July 30, 1946.

ASTRONOMICAL ADVANTAGES
OF AN
EXTRA-TERRESTRIAL OBSERVATORY

It has been proposed that rockets be used to accelerate a small mass, containing scientific equipment, up to a speed of 5 miles a second, at which speed the mass could revolve around the earth indefinitely, forming a small satellite. Such a development is certainly not out of the question within the next few decades, in view of the rapid strides already made in rocket research, and the emphasis now being placed on research in this field. The present memorandum points out, in a very preliminary way, the results that might be expected from astronomical measurements made with such a satellite. The discussion is divided into three parts, corresponding to three different assumptions concerning the amount of instrumentation provided. In the first section it is assumed that no telescope is provided; in the second a 10-inch reflector is assumed; in the third section some of the results obtainable with a large reflecting telescope, many feet in diameter, and revolving about the earth above the terrestrial atmosphere, are briefly sketched.

It should be emphasized that this is only a preliminary survey of the scientific advantages that astronomy might gain from such a development. The many practical problems, which of course require a detailed solution before such a satellite might become possible, are not considered, although some partial mention is made of certain problems of purely astronomical instrumentation. The discussion of the astronomical results is not intended to be complete, and covers only certain salient features. While a more exhaustive analysis would alter some of the details of the present study, it would probably not change the chief conclusion that such a scientific tool, if practically feasible, could revolutionize astronomical techniques and open up completely new vistas of astronomical research.

I. Solar Spectroscopy with a Small Ultra-Violet Spectroscope

The simplest astronomical instrumentation for a satellite would be a small spectroscope, analyzing the ultraviolet radiation which it receives from any portion of the sky; in practice, this would be the solar spectrum whenever the sun was visible. Such a spectroscope could analyze, either the light incident on a diffuse reflector or the light passing through a small LiF sphere, or bead. Such a system has the advantage that it would not need to be accurately oriented m any particular direction. The intensity in the spectrum could presumably be radioed down to earth. An instrument of this sort would have the following uses:

[2] 1. Continuous Recording of the Solar Ultra-Violet Spectrum

The scientific and military importance of information on the sun's ultraviolet spectrum has already been pointed out.[i]

Occasional spectra of the sun in the far ultra-violet can presumably be obtained with high altitude rockets which subsequently fall to earth. However, for an adequate picture of the sun's probably large variability in ultra-violet radiation, more frequent measurements may be necessary. For a complete examination of the effect which solar disturbances produce on terrestrial phenomena, especially on conditions in the ionosphere, a relatively continuous portrayal of the sun's output of ultraviolet energy may be required. For example, if a radio fade-out occurs at some particular time, only a record of the solar spectrum during the time immediately preceding can show what the relationship between sun and earth was for that particular fade-out. More important still, for detailed predictions of ionosphere conditions, and thus for practical advance information on radio transmission conditions, daily measurements of the sun's ultra-violet spectrum are believed to be essential. These can probably be obtained most simply by a satellite observatory.

2. Detailed Analysis of the Earth's Upper Atmosphere

As seen from the satellite, the sun will rise and set at frequent intervals. On each such occasion, the sun's ultra-violet light will change markedly as the sun's rays shine through atmospheric layers of changing height. By observing changes of the spectrum with time it would be possible to obtain a detailed picture of how the densities of different types of atoms in the earth's upper atmosphere change with changing height. While essentially similar information could be obtained from a rocket which rose out of the earth's atmosphere and then fell back to earth, the observations from a satellite could be obtained much more frequently. In view of the probable variability of the ionosphere, resulting from the variability of the sun's ultra-violet radiation, rather frequent spectrographic observations of the structure of the ionosphere, as well as of the sun's ultra-violet spectrum, are probably required to indicate exactly what is happening. It may well be the case that this information can be obtained at less cost with such a satellite than with a series of rockets of lower velocity.

II. Spectroscopy of the Sun and Stars with a 10-inch Reflecting Telescope

To obtain information about the ultra-violet spectrum of [3] the stars, or to analyze in detail the sun's surface as seen in ultra-violet light, a telescope is required, together with means for orienting the instrument in any desired direction. Orientation might be accomplished in principle by reducing the angular momentum of the satellite to zero by means of external jets; thereafter the satellite could be rotated by internal means to any particular direction, and would point in that direction indefinitely unless hit by a meteorite. Since the telescope would be designed for spectroscopic purposes only, the shape of the mirror would not need to be highly accurate.

A 10-inch reflecting satellite telescope, equipped with one or more spectroscopes, would be a powerful astronomical tool. While it would intercept less light than the large reflecting telescopes on earth, it would have the advantage that the background light from the night sky would be much reduced, provided that the satellite was above the atmospheric layers responsible for this night illumination; 500 miles should be adequate for this purpose. Thus the faintest star which could be reached with such a telescope might be as faint as that which can just be photographed with the 100-inch telescope, provided that photocell techniques can reach the point where they are as effective as the photographic plate. A photon counting technique, with the use of long "exposures" or, more appropriately, "counting intervals" would probably serve this purpose. Such a telescope-spectroscope combination could measure the spectra of stars, planets, etc., down to at least 1000 A and also out to the infra-red, without the absorption of the earth's atmosphere, which blots out all the ultra-violet and obscures many regions in the infra-red. Listed below are some of the astronomical uses of such an instrument.

It may be noted that practical uses of this instrument would not be immediate; this would be an instrument which might be expect to increase very basically our understanding of what goes on in the stars and in the spaces between them. Since in this way we obtain information on the behavior of matter under conditions not attainable in the laboratory, knowledge of fundamental physics would thereby be enhanced.

1. Detailed Information on Solar Meteorology

With a reflecting telescope and accessory equipment, sunspots, prominences, and other types of storms on the sun could be examined in ultra-violet light of different wavelengths. In particular, the behavior of the resonance line of hydrogen (Lyman) at 1216 A would give basic information on the nature of these puzzling and complicated disturbances, which are related to the variability in the output of ultra-violet radiation from the sun.

2. Composition of Planetary Atmospheres

The small amount of O_2 and H_2O present in the atmosphere of Mars and Venus cannot be detected spectroscopically because of the absorption produced by these same molecules in our own atmosphere. A spectroscopic satellite telescope could observe the spectra of planetary atmospheres without any such interferences, and could supplement observations in the infra-red with equally useful ultra-violet data.

[4] 3. Structure of Stellar Atmospheres

Among the most abundant elements in typical stars are hydrogen, helium, carbon, nitrogen and oxygen. The absorption lines produced by these atoms in their lowest states (called "resonance lines") all lie in the ultra-violet; the absorption lines of these atoms in the visible spectrum all arise from states whose excitation potential is at least seven volts;

since few atoms are so highly excited, the visible absorption lines produced by these atoms are all very weak, except for hydrogen, whose great abundance makes up for its high excitation potential. Thus practically no direct evidence is available on the behavior of helium, carbon, nitrogen, or oxygen in most stars. While the resonance lines of helium lie in the far ultra-violet at about 500 A, those of carbon, nitrogen and oxygen all lie between 1000 and 2000 A; the resonance lines of these three elements are unquestionably very strong in the spectra of most stars, and should be readily observable with a satellite spectroscopic telescope. Such observations should indicate any differences in composition between different stars – these differences are important in stellar evolution and stellar energy generation. In addition, the nature of unusual stellar atmospheres – expanding, rapidly rotating, etc. – would be more clearly indicated by information on the behavior of such abundant elements as carbon, nitrogen and oxygen as well as by the behavior of the resonance lines of hydrogen.

4. Color Temperatures of Hot Stars

For stars hotter than about 15,000oC, the color of the star, as measured in visible radiation, is independent of temperature. Measurements in the ultraviolet would help to determine the surface temperatures of hot stars, a basic item in astrophysical research.

5. Bolometric Magnitudes

The determination of the total energy radiated by a star depends on the measurement of the total heat energy reaching the earth from the star; i.e., on the "bolometric magnitude". For stars whose surface temperature is similar to that of the sun, corrections for infra-red and ultraviolet absorption in the earth's atmosphere are not too serious, but for very cool or very hot stars the result depends heavily on the assumed corrections. Bolometric measurements made on a satellite observatory would give bolometric magnitudes directly for stars nearby, unobscured by interstellar dust.

6. Analysis of Eclipsing Binaries

Much of our present information about the masses, radii and structure of stars has been derived from eclipsing binaries. Measurements in the ultraviolet would be a powerful new tool in such research. For example, to determine stellar masses it is necessary to observe the Doppler shifts in the lines produced by each of the two stars, and in this way to measure the velocity of each. When the stars are of unequal luminosity this is difficult. However, the less luminous star is frequently smaller and hotter. In ultraviolet radiation the smaller star will frequently be more luminous, and from a satellite observatory its ultraviolet spectrum could be observed, and its velocity thus determined. Changes in the shape of the light curve during eclipse with changing frequency would also give important information on the structure of the atmosphere and on the nature of the opacity of matter in the stars.

[5] 7. Absolute Magnitudes and Stellar Distances

If the surface temperature of a star is approximately known from its spectrum, its absolute magnitude can be found if its radius can be estimated. Since the surface gravity and resulting pressure decrease together with increasing radius, a measurement of pressure suffices to give the absolute magnitude, which in turn gives the distance of the star. Observations of visible stellar spectra have given extremely important results along this line by determining the relative numbers of neutral and ionized atoms, which depend on the pressure. Measurements in the ultraviolet would yield data on the presence of highly ionized atoms, not detectable in visible radiation, and would greatly increase the sensitivity of this method for determining stellar brightnesses and distances.

8. Composition of Interstellar Gas

Interstellar atoms and molecules are known to be present between the stars, and to have a total aggregate mass about equal to that of the stars. Such particles are all in their ground state; hence observations of stellar spectra in the visible give no information on the presence of many of the atoms and molecules that may be expected. Measurements in the ultraviolet would give information on the density of interstellar hydrogen in space near the sun, and would indicate how much if any of this material was in the form of molecules. Such measurements would also indicate how much carbon, nitrogen and oxygen was present. Detailed information on the nature of interstellar gas may be important in understanding the origin of stars and of cosmic rays, which may both be produced from interstellar matter.

9. Properties of Interstellar Absorbing Grains

In addition to atoms and molecules, small grains of matter, about 10^{-5} cm. in diameter, absorb starlight in space. This absorption, generally important only for distant stars, is greater for shorter wavelengths. The distribution of these grains is known to be very uneven. Measurement of stellar spectra in the ultraviolet should therefore provide a very sensitive indication of the presence of these obscuring particles; comparison of this absorption with that in the visible region of the spectrum should yield information about the composition of these particles, which is an important item in the evolution of interstellar matter and in related cosmogonic problems.

10. Nature of Supernovae

These exploding stars must be the result of some gigantic cataclysm, possibly a chain reaction involving the entire star. The spectrum of the brighter supernovae is a complete puzzle. Measurements in the ultraviolet would be difficult to obtain with a 10-inch reflector, owing to the great distance and resultant faintness of these objects, but if obtainable might yield an important clue to the nature of the processes involved.

[6] III. Astronomical Research with a Large Reflecting Telescope

The ultimate objective in the instrumentation of an astronomical satellite would be the provision of a large reflecting telescope, equipped with the various measuring devices necessary for different phases of astronomical research. Telescopes on earth have already reached the limit imposed by the irregular fluctuations in atmospheric refraction, giving rise to "bad seeing". It is doubtful whether a telescope larger than 200 inches would offer any appreciable advantage over the 200-inch instrument. Moreover, problems of flexure become very serious in mounting so large an instrument. Both of these limitations disappear in a satellite observatory, and the only limitations on size seem to be the practical ones associated with sending the equipment aloft.

While a large reflecting satellite telescope (possibly 200 to 600 inches in diameter) is some years in the future, it is of interest to explore the possibilities of such an instrument. It would in the first place always have the same resolving power, undisturbed by the terrestrial atmosphere. If the figuring of the mirror could be sufficiently accurate, its resolving power would be enormous, and would make it possible to separate two objects only .01" of arc apart (for a mirror 450 inches in diameter); an object on Mars a mile in radius could be clearly recorded at closest opposition while on the moon an object 50 feet across could be detected with visible radiation. This is at least ten times better than the typical performance of the best terrestrial telescopes. Moreover, in ultraviolet light the theoretical resolving power would of course be considerably greater; ideally an object 10 feet across could be distinguished on the moon [139] with light of 100 A wavelength. In addition, with such a large light-gathering surface and such low background light, the positions and spectra of stars and galaxies could be analyzed out to much greater distances than is now possible. If the shape of mirror could not be figured so accurately without excessive effort, a large spectroscopic satellite telescope would still have many important uses.

The practical problems of operating such a large installation would of course be enormous. Telemetering back to earth the two-dimensional picture obtainable with such an instrument would involve many problems. With such high angular resolutions, some guiding of the telescope might be necessary to correct for changes in the aberration of light during the satellite's orbit. Absorption and radiation of the light received from both sun and earth would require careful consideration to ensure a constant temperature in the mirror and its mounting (to reduce distortion of the mirror's shape by thermal expansion and contraction) and to give a very low temperature in the photo-electric measuring equipment (to reduce the background of thermal emission from the photo-sensitive surface). To provide for a leisurely orbit and thus for relatively constant conditions, such an observatory should preferably be some distance away from the earth, probably as far as telemetering techniques and celestial mechanics might allow. [7] Most astronomical problems could be investigated more rapidly and effectively with such a hypothetical instrument than with present equipment. However, there are many problems which could be investigated only with such a large telescope of very high resolving power. A few of these problems are given in the following partial and tentative list. It should be emphasized, however, that the chief contribution of such a radically new and more powerful instrument would be, not to supplement our present ideas of the universe we live in, but rather to uncover new phenomena not yet imagined, and perhaps to modify profoundly our basic concepts of space and time.

1. Extent of the Universe

The 200-inch telescope is designed to push back the frontiers [of] explored space. It is not likely that this instrument will reach to the greatest distance possible. Further measurements with the more powerful instrument envisaged here would help answer the questions whether space is curved, whether the universe is finite or infinite. This instrument would help in particular to resolve individual stars in a distant galaxy and to analyze their spectra, thus identifying particular stars of known absolute magnitude and in this way determining accurately the distance to the galaxy. At present the distances of most galaxies are known only very approximately.

2. Structure of Galaxies

With such great resolving power, such an instrument could explore the details of the structure of galaxies, individual stars could be resolved and the nature of the as yet enigmatic spiral arms could be investigated. Measurement of radial velocities by spectral analysis would yield velocities of rotation in a number of galaxies and thus provide direct information about their masses – information now available for only a few galaxies.

3. Structure of Globular Clusters

These objects contain so many stars that resolution of individual stars has been possible only for the brighter members. With such great resolving power a much greater percentage of the individual stars could be resolved, some spectra and radial velocities obtained, and a serious attempt made to explore the structure of these stellar aggregations.

4. Nature of Other Planets

The controversy as to the presence of intelligent life on Mars could perhaps be settled by measurements with such a giant telescope. Similarly the type of surface detail present on the other planets could be accurately explored with such high resolving power and invariably perfect seeing.

Signed Lyman Spitzer, Jr.

---

i. "The Importance High-Altitude Spectroscopy," by L. Goldberg and L. Spitzer, Jr., July 15, 1946.

Document III-2

Document title: William L. Kraushaar, "Research and Budget Proposal to the Space Science Board of the National Academy of Sciences for the Support of a High-energy Gamma-ray Satellite-borne Experiment to be Performed by the Cosmic Ray Group of the Massachusetts Institute of Technology Laboratory for Nuclear Science," October 10, 1958.

Source: William Kraushaar, personal collection, reprinted with permission.

In the earliest days of NASA's space science program, the National Academy of Science's Space Science Board, and not NASA, attempted to assume responsibility for reviewing and recommending space-based scientific experiments proposed by the scientific community for the new space agency to pursue. One such proposal came in 1958 from a team at the Massachusetts Institute of Technology that believed that satellite-based studies of cosmic gamma rays would yield far more precise results than those obtained from balloon-borne experiments, which endured background radiation produced by the atmosphere. Accepted by NASA, the group's gamma-ray experiment flew aboard Explorer 11, launched on April 27, 1961, as the first U.S. satellite devoted to astronomy.

[cover sheet]

RESEARCH AND BUDGET PROPOSAL TO THE
SPACE SCIENCE BOARD OF THE
NATIONAL ACADEMY OF SCIENCES
FOR THE SUPPORT OF A HIGH-ENERGY GAMMA-RAY
SATELLITE-BORNE EXPERIMENT
TO BE PERFORMED BY THE
COSMIC RAY GROUP OF THE
MASSACHUSETTS INSTITUTE OF TECHNOLOGY
LABORATORY FOR NUCLEAR SCIENCE

Proposal Submitted:
October 10, 1958

[no page number]

RESEARCH AND BUDGET PROPOSAL FOR THE SUPPORT OF
A HIGH-ENERGY GAMMA-RAY SATELLITE EXPERIMENT

INTRODUCTION

The following is a proposal for research in cosmic rays involving gamma-ray detecting apparatus similar to that which has already been balloon-flown to a very high altitude by the principal investigators of this proposal. The apparatus is clearly adaptable to modifications which will make it suitable for inclusion in a satellite.

Funds requested for this work are for a period of eighteen months and in the total amount of $152,000. The need for these funds arises mainly for the purchase of components and construction materials for four such apparatuses; for funds to be used to cover travel in connection with the experiment; and for an estimated fifty hours of electronic computer time. A comparatively small fraction of the total funds will be used for the salaries of personnel, since it it [sic] intended to carry out the program largely with personnel now engaged in cosmic ray research as members of the laboratory's existing Cosmic Ray Group.

[1]

PART I
SCIENTIFIC PROGRAM
High Energy Gamma-Ray Satellite Experiment

I. Introduction

Somewhat less than a year ago, we at M.I.T. initiated a program to survey the directional intensity of cosmic gamma-rays. Gamma-rays, unlike the proton and heavy nuclei components of the cosmic radiation, are undeflected by terrestrial, solar and galactic magnetic fields and so should arrive from the direction of their sources. In this sense, gamma-ray astronomy (if the subject ever develops enough to warrant that name) is similar to optical and radio astronomy. Cosmic electromagnetic radiation having frequencies in the optical and radio frequency regions results from atomic phenomena and phenomena which involve the relatively large-scale motion of charged particles. The radiation in the gamma-ray region, on the other hand, should involve distinctly nuclear phenomena, and this fact together with the property of straight-line propagation is what makes the investigation seem an attractive one.

We shall not discuss here in any detail the various possible sources of cosmic gamma-rays. Generally speaking, two energy regions seem most promising. Gamma-rays in the first region, 0.2 to 5 Mev, should result from the radio-active decay of excited nuclei, fusion of light elements and possibly electron-positron annihilation. Gamma-rays in the second region, 50 to 200 Mev should result from the decay of neutral pi mesons produced in either high energy nuclear interactions or possibly in the annihilation of matter and anti-matter. Some sources in these categories have been discussed by Philip Morrison in a recent issue of *Nuovo Cimento.*[i] In addition to the mechanisms discussed by Morrison, [2] there is an almost certainly present intensity of gamma-rays in the second energy region (50 to 240 Mev) with a very high information content. To the best of our knowledge cosmic rays exist throughout our galaxy, and should occasionally collide with the nuclei of galactic gas atoms. These nuclear collisions should give rise to gamma-rays through the decay of the produced neutral pi mesons, and the directional intensity of these gamma-rays should yield important information about the cosmic ray and (non stellar) matter distribution in our galaxy.

There has been some previous study of gamma-rays in this energy region.[ii, iii, iv] Scintillation counters and Geiger Counter detectors of gamma-rays have been carried in ballons [sic], and Geiger counter detectors have been attached to rockets. Even the most significant of these experiments, however, have surveyed but a small portion of the sky with very unspecific solid-angle definition and have suffered from a large atmospherically-produced background. Consequently, only upper limits to the intensity of possible cosmic gamma-ray intensities are presently available and these upper limits are quite large (a few Mev cm^{-2} sec^{-1}).

With these factors in mind, and encouraged by the recent possibility of sending large pay loads aloft in balloons, we have designed two gamma-ray experiments both of which are scheduled for flight this summer. The [3] first is sensitive to gamma-rays in the 0.2 to 5 Mev region and includes a scintillation detector with good energy-resolution surrounded by enough lead to insure reasonable solid-angle definition. The second is sensitive to gamma-rays in the 50 to 200 Mev region and has an angular resolution of about 0.01 steradian. This equipment is described in somewhat more detail in the next section.

The U.S.S.R. delegates to the October, 1957, IGY meeting in Washington, released a number of papers, among which is one by S. N. Vernov, Yu. I. Logachev, A. Ye, Chudakov, Yu. G. Shafer. In this paper they too point out the attractive possibilities of gamma-ray astronomy, and describe a satellite-borne nuclear emulsion experiment. Interestingly, the public press has carried the story that Sputnik III carries gamma ray detection equipment of special significance.

II. <u>The High Energy Balloon-Borne Gamma Ray Experiment</u>

As mentioned previously, both high and low energy gamma-ray detectors are severely handicapped by atmospherically produced background and to a lesser extent by atmospheric absorption even when balloon-borne to heights of 100,000 feet. It is difficult to evaluate without the results of our balloon borne experiments in hand which type of experiment would benefit most by being operated above the atmosphere in an earth satellite. Our best guess at present is that the high energy experiment will be the most severely handicapped by the residual atmosphere. Further, there exists in the high-energy region the galactic flux of gamma-rays with a predictable intensity, and while this intensity is probably too small for significant study in a first satellite experiment, the data will be most important in planning future experiments.

[4] The accompanying sketch shows the high-energy gamma-ray detector that will be flown late this summer or early this fall. The mercury and lead at the top of the apparatus serve as a collimator. Only those gamma-rays with directions more or less along the axis of the apparatus can pass unimpeded between the mercury columns through

the plastic anti-coincidence counter and produce electron pairs in the mercury radiator. The electron pairs are detected first by the CsI scintillation crystal and then by the lucite Cerenkov detector. The light flashes from these two sources, CsI and Lucite, are detected by a single photomultiplier and distinguished for coincidence purposes by the difference in the characteristic time during which the light is emitted. The Cerenkov detector discriminates against particles incident from beneath, the CsI pulses are biased for the passage of two minimum ionizing particles, and the large encompassing anti-coincidence detector insures that incident charged particles will not be recorded. We are certain that there will be present a quite large more-or-less isotropic flux of gamma-rays from neutral mesons produced by cosmic rays in interactions with the nuclei of the residual air above the apparatus, and the heavy shielding is designed to prevent their being recorded when incident from off-axial directions. The solid angle of the collimator is 0.01 steradian when the mercury is in the collimator and is 0. 3 steradian when the mercury is removed. The small solid angle is designed for a study of possible cosmic point sources and the larger solid angle is designed for a general survey and study of the atmospherically-produced intensity. The solid angle will be changed in flight. The mercury radiator can be removed and reinserted in flight for measurements with and without the radiator providing a convincing test as to whether gamma-rays are really being detected. The apparatus is mounted [5] on a horizontal axis and is programmed to observe the zenith angles appropriate to several possible point sources (Cygnus A, the Sun, the Crab). The azimuth angle is changed continuously, one rotation every three minutes. Pertinent data will be recorded photographically and examined following recovery of the apparatus. The total weight of the apparatus including batteries for 20 hours of operation is 500 pounds.

III. The Satellite Experiment

The balloon-borne high energy gamma-ray apparatus has been described in some detail because the results of this experiment bear directly on the design of the proposed satellite experiment and because it itself is a possible prototype for satellite borne equipment.

The advantages of satellite over balloon borne gamma-ray experiments are several. The most important is the background question. To be above the atmosphere will not completely eliminate background, for there will still be albedo gamma-rays produced in the earth's atmosphere but travelling up. These will not, however, be coming from the direction of possible cosmic sources. The galaxy, for example, has a maximum thickness of about 0.1 g cm^{-2} looking across the local spiral arm towards the galactic center. A balloon experiment, if the detection of galactic gamma-rays were attempted, would have to distinguish between an intensity proportional to this 0.1 g cm^{-2} in a background proportional to the 10 g cm^{-2} of residual atmosphere. It is safe to say that if the galaxy is at all as we picture it, galactic gamma-rays cannot be studied with balloon borne experiments. The same sort of argument holds true, of course, for other possible low intensity sources. Another important advantage is that the entire celestial sphere can be surveyed from a satellite, while a single balloon-borne experiment can survey at most 10 per cent of the celestial sphere, [6] and an entire survey would require several flights both night and day (or day only if extended over 6 months) every 30 or 40 degrees in latitude. This is because atmospheric background increases as the zenith angle increases and only a band of 25 or 30 degrees in declination can be surveyed from any one latitude.

At present we know of no point cosmic gamma-ray sources and even if one or more should be uncovered by the balloon experiment, we wish to propose that a first satellite experiment scan the celestial sphere uniformly. To know where the apparatus is pointed is probably a technical problem far less severe than to actually keep the apparatus pointed in a given direction. Once source positions are known or suspected, controlled observations will be very important.

It should be pointed out that random scanning with an apparatus of solid angle Ω results in an observation time for any particular direction of only $\Omega\ T/4\pi$ where T is the total available scanning time. For this reason, in the presence of an isotropic noise backgrounds the effective ratio signal-to-noise is independent of the solid angle. This does not mean, however, that the solid angle can be made arbitrarily small, for the detection is inherently a counting process, and statistical significance comes only with large numbers. The optimum solid angle, therefore, depends upon the intensity of the isotropic background, and our best guess at present favors a half-angle of about 10 degrees. The solid angle as well as certain other design features can best be decided when the results of our balloon borne experiments are in hand. In any event, the basic scheme of the balloon experiment seems well suited to a satellite experiment.

[7] The following are specific points regarding the feasibility of the experiment.

1. Weight The balloon experiment weights [sic] 500 lbs. including batteries for 20 hours operation and the pressurized gondola. Most of the weight is lead, and in view of the anticipated lower background the weight of the lead can be reduced from 400 to perhaps 200 lbs.

With an effective detector area equal to that of the balloon experiment, the weakest point source resolvable will have an intensity of about $5 \times 10^{-4}\ cm^{-2}\ sec^{-1}$, 10 times smaller than the similar intensity for the balloon experiment. This weakest intensity varies as $\sqrt{(J/\Omega AT)}$, where J is the isotropic background, Ω is the solid angle, A is the area and T is the available running time. In the above estimate we have used T=10 days. Since the weight of lead necessary increases somewhat faster than the area, very little is to be gained by say, doubling the weight of lead, and for the first experiment we propose that the effective area be kept near 120 cm^2.

For reasons of expedience, and the relatively small additional weight involved, some vacuum tubes have been used in the balloon experiment. These can certainly be replaced by transistors, and the total weight of electronic instruments exclusive of telemetering equipment but including power supplies, should not be over 30 pounds. This assumes 40 days of operation and 3 watts dissipated power.

Miscellaneous hardware, the scintillators and the radiator will add another 30 lbs, and the total is thus 260 lbs. This is not needless to say, a very precise estimate.

2. Size The balloon borne experiment is within a 37" diameter gondola. A satellite experiment can be made smaller than this because no internal rotations are contemplated.

[8] 3. Telemetering With the apparatus as described, coincidence counting rates up to one per second may be expected. Each count must be correlated with the direction (to within a degree or so) of the axis of the detector. This problem, while formidable,

does not seem impossible and has probably been considered in connection with other satellite experiments. For proper evaluation of the data it would be convenient to know two or three auxiliary counting rates. These need not be recorded simultaneously with the principle data, but could if necessary, be telemetered say once every ten earth traversals.

4. Satellite Aspect If left to itself a satellite probably rotates for long periods about a single axis. We, on the other hand, wish to scan the entire sky and, therefore, must provide some method for changing the moments of inertia. This too, it seems likely has been considered in connection with other experiments. Related is the problem of removal. And reinsertion of the mercury radiators for a redistribution of mass is involved. Possibly these two requirements and problems can be arranged to aid each other.

5. Time At least 10 days of actual data seem necessary. Since the axis of the detector will point towards the earth half the time, and since it seems questionable that satellite aspect can be extrapolated around the night side of the earth (it has been assumed that aspect will be obtained from the sun) we have allowed 40 days of satellite time for 10 days of useful data.

6. Orbits The orbit requirements are now [sic] severe, except that the apparatus should not enter the region of intense X-radiation reported by Van Allen and his co-workers.

W. Kraushaar, MIT

i. P. Morrison,. *Nuovo Cimento*, 6, 858 (1958)
ii. T. Bergstralh and C. Schroeder, Phys. Rev. 81, 244 (1951);
iii. L. Reiffel and G. U. Burgwald, Phys. Rev. 95, 1294 (1954);
iv. Rocket Exploration of the Upper Atmosphere, R. Boyd and M. Seaton, editors, London, 1954, p. 306, by C. Johnson, L. Davies, and J. Siry.

Document III-3

Document title: Notes from meeting of Dr. J. E. Kupperian and members of the Smithsonian and Harvard College Observatories, January 7, 1959.

Source: Fred L. Whipple papers, Archives, Smithsonian Institution, Washington, D.C.

Document III-4

Document title: Memorandum to Homer E. Newell, Jr., from Abe Silverstein, "Proposed NASA Project – Orbiting Astronomical Observatories," March 16, 1959.

Source: NASA Historical Reference Collection, History Office, NASA Headquarters, Washington, D.C.

Document III-5

Document title: Homer Newell, Deputy Director, Space Flight Programs, NASA, "Memorandum For The Files: Conference Report – A Review of the Orbiting Astronomical Observatory Project," February 18, 1960.

Source: Fred L. Whipple papers, Archives, Smithsonian Institution, Washington, D.C.

Document III-6

Document title: Memorandum to Dr. Nancy Roman from A. D. Code, L. Goldberg, L. Spitzer, and F. L. Whipple, June 15, 1960.

Source: Fred L. Whipple papers, Archives, Smithsonian Institution, Washington, D.C.

Prior to 1959, no one had ever before attempted to put a telescope into space; a number of leading American astronomers thus found themselves struggling in NASA's early years with precedent-setting scientific, technical, and managerial decisions as they attempted to build up NASA's astronomy program. Document III-3 reveals that despite the numerous issues that needed to be resolved to make a space-based astronomy mission possible, NASA did not want to ease into space exploration with small, conservative missions, but instead desired to embark on a program of large space observatories. Within the next few months, NASA began planning for a series of astronomical satellites called the Orbiting Astronomical Observatories (OAOs) [Documents III-4 and III-5]. Although, as shown in Document III-6, some of the nation's senior astronomers criticized NASA for its neglect of smaller missions as well as for its management of certain elements of the OAO program, these very astronomers went on to conduct pioneering experiments on the OAOs in a variety of wavelengths.

Document III-3

[no page number]

The following notes outline the high points of a meeting on 7 January 1959 between Dr. J. E. Kupperian, of NASA, and the Directors and several members of the staffs of Smithsonian and Harvard College Observatories. The notes refer primarily to the meeting, in Dr. Whipple's office, attended by Dr. F. L. Whipple, Dr. G. F. Schilling, Dr. C. A. Whitney, Dr. R. E. McCrosky, and Mr. R. J. Davis.

NASA has been considering a 1,000-pound "space observatory" project. They are pushing for early 1961 for this. They want to be able to offer the experimenter a "standard" stable platform, perhaps attached to the last stage rocket shell, so that they need only let one contract for the development of the platform and launch vehicle. There is a possibility that this platform would be spinning during launch; the spin to be removed later. They hope to provide for six experiments from such platforms; these experiments will be placed in orbit, no matter how many launching attempts are required, but additional experiments will not be added at the end to take care of unused rockets should there be any. It is not entirely certain at present what launching system will be used. However, they plan to have the scientists do science; it will be up to the engineers to satisfy them, and the scientist will not have to worry explicitly about launch systems, telemetry, etc. There will, of course, be much liaison between scientists and engineers.

Presumably, an ordinary-looking telescope could be placed on this platform, although, of course, this will probably not be the configuration chosen.

The sun is a different problem from the rest of the celestial sphere; the earth is also a specialized problem. It will probably be best not to try to combine solar, terrestrial, and non-solar celestial telescopes. NASA is going to try to avoid a hodgepodge system.

There are 2 possible types of fine pointing: a servo loop in the satellite itself, or one going back to the ground.

Current thinking on requirements for pointing accuracy and stabilization have the telescope people in mind, since their requirements will probably be most exacting. Other types of experiments will thus have higher stability than minimum requirements.

There are three reasons for getting above the atmosphere: elimination of airglow for observations of faint objects; improvement of seeing; getting rid of an absorbing agent that is variable in time, position, and spectral selectivity, and that blocks out some types of information entirely. It is in the third area that NASA expects the most rapid advances, and they therefore are placing the most urgency on developments aimed at this area. They are interested in increasing resolution, for the time being, via balloon experiments; this facet was discussed more thoroughly at lunch. They also want to support a large rocket program, since no experiment should be [2] operated from a satellite if it can be done from a rocket, and since rockets will be necessary to provide experimental data on which to base satellite design.

NASA wants to start work as soon as possible on actual implementation of these ideas. They are definitely thinking of a quantum jump here, rather than slow development. Dr. Kupperian has already visited Dr. Goldberg at Michigan; Dr. Spitzer, of Princeton, is now in Washington talking to NASA about this same problem. What they want is a meeting of working scientists and engineers in Washington the first week of February to thrash the whole thing out in both large and small groups, ending up with definite decisions concerning where we should go from the standpoint of experiments, hardware and organization. Present thinking is that the experimenter will be Smithsonian-Harvard for one experiment, Princeton for another, Michigan for another, possibly etc. up to 6, but only Wisconsin was mentioned as an additional party to the February meetings; the experimenter will set the specifications for the experiment; NASA will provide engineering and logistic support, including ground stations, perhaps through contractors; the experimenter will have complete control over the operation of his experiment. All of this is still fluid, subject to thrashing out at the February meetings.

One of the outcomes of the February meetings is hoped to be the instigation of a working group on astrophysical space experiments which will include members of the participating organizations. NASA wants to have all the problems thoroughly considered before drawing up the actual proposals, so that sending through the proposal will be almost a formality. They want to be able to tell the systems people what is required, and let them take two years to develop it.

In this scheme of things, we would have to separate the ground station part of our proposal from the rest of it, and submit it separately to the systems people at NASA. The experiments must be kept separate from the systems to work.

Glennan himself agrees that the experiment comes first. Experiments must have scientific value, not merely publicity value.

They want to get the working group together to get an experiment lined up. Thus we would have the plan developed together rather than by the staff at NASA.

NASA encourages preliminary experiments in rockets. They hope to provide rocket service for people. They see a service organization to run the ground station network.

Mr. Dunkelman, of NASA, has had experience in ultraviolet sensing equipment and systems work (as well as infrared) since 1942. He might be able to help us with industrial liaison.

[3] NASA will support projects, but will not support "institutions." They can put out "3-year" money, but cannot promise continuing support (or any support at all, for that matter) to a scientific organization, except in regard to <u>specific</u> projects.

Drs. La Gow and Meredith are handling meteor matters.

****

In addition to those at the morning meeting, the following Harvard-Smithsonian scientists were at the luncheon meeting: Dr. D. H. Menzel, Dr. J. A. Hynek, Dr. K. C. Henize, Dr. G. de Vancouleurs, Dr. T. E. Sterne, Mr. H. Ingrao, and Mr. G. Nielson. The major new topics of discussion were rocket and balloon astronomy.

Notes taken by Mr. Robert J . Davis.

Document III-4

16 March 1959

Abe Silverstein

Homer E. Newell, Jr.

[no page number]

The Office of Space Sciences has completed the initial planning phase of a project for Orbiting Astronomical Observatories, to be launched as part of the National Space Program by the end of 1961. The present planning status of the project is described in the attachment to this memorandum. It is based on staff review, involving the selective adoption and synthesis of concepts suggested in preliminary proposals received from interested institutions, and the results of a Space Sciences Discussion Group meeting held on 9-10 February 1959.

The preliminary budget estimate totals $57 million, $600,000 of which is to be funded from FY 59 allocations. A breakdown of these figures by fiscal years and major categories is given on page 8 of the attachment. It appears that the projected Vega vehicle system will have adequate capability to perform the mission of this project.

Supplementary information:

1. According to agreements reached between Drs. Glennan, Dryden, and Waterman on March 11, 1959, the National Science Foundation will fund and activate, within a few days, a proposal by the Association of Universities for Research in Astronomy, Inc. for preliminary studies on the long-range problem of placing a large astronomical telescope in orbit about 1965.
2. Specific proposals for initial studies as part of the research phase of the NASA project described in the attachment have been received from the Smithsonian Institution and Harvard University, and complementary proposals are expected within the next four weeks from the University of Michigan, the Space Sciences Division of the NASA Beltsville Space Center, the University of Wisconsin, and the University of Rochester. These are detailed research project proposals based on the results of the above mentioned Discussion Group meeting.

[2]

Recommendations

It is recommended that the Director of Space Flight Development

a. Approve in principle a project of Orbiting Astronomical Observatories as part of the national space program;

b. Authorize the Assistant Director for Space Sciences to submit research project proposals which constitute initial phases of the Project for early funding of FY 59 allocations, up to a limit of $600,000;

c. Authorize the Assistant Director for Space Sciences to proceed with detailed planning, specifications, and improvement of the project.

Attachment

Bcc: Wyatt
Stoller
Clark
Kupperian
Roman
Fuhrman
Schilling

[no page number] [attachment]
12 March 1959

Proposed National Aeronautics And Space Administration Project

Project Title: Orbiting Astronomical Observatories

Project Objective: The objective of this project is to establish and operate astronomical observatories orbiting above the absorbing atmosphere of the earth. Precision telescopic observations, with ground control, will be made of the emission and absorption features of the sun, stars, and nebulae in the unexplored ultraviolet, infrared, and X-ray regions of the electromagnetic spectrum.

Specific Experimental Objectives: The observing techniques developed for ground based telescopes form the basis of measurements proposed to extend our knowledge into short wave regions of the spectra. The difficulties that will be encountered in detecting and analyzing radiation in the far ultraviolet and X-ray as well as infrared portions of the spectrum dictate quite unique optical and detection systems. This will result in optical and detection systems that are optimized for specific experimental objectives both as to observational technique and wave length region. Some of the proposed observing techniques can be listed:

1. Objective prism spectra
   a. Stellar spectra
   b. Nebular spectra and monochromatic images
2. High resolution stellar spectra
3. Broad band stellar photometry

[2] 4. Broad band nebular photometry
   a. Total energy
   b. High resolution images
5. Monochromatic solar images
   a. Total solar disk
   b. High resolution investigations of specific areas of the disk or limb of the sun.

6. Solar spectra
    a. Center-limb variations
    b. Specific areas

While some overlap of these techniques can be expected, a single compromise system would have marginal potential. It has developed, however, that a stabilized observing "platform" can be constructed which with the addition of the suitable optical and detection systems could support various experimental objectives employing different observing techniques on different flights.

Potential Contractors: To date, the following institutions have shown interest in participating and in instrumenting such an observing "platform":

Smithsonian Astrophysical Observatory
University of Michigan Observatory
University of Wisconsin Observatory
Beltsville Space Center (NASA) Space Sciences Division
Jet Propulsion Laboratory (NASA)
[3] Princeton University Observatory
Associated Universities for Research in Astronomy
University of Rochester
University of California
Harvard University

Project Scope: The present project schedule includes the orbiting of six astronomical observatories over a period of at least two years. Each would consist of a common vehicle-stabilized platform system instrumented with an optical and detection system optimized for a specific observational objective. This concept is based on considerations of:

a) wavelengths region to be covered and observational techniques to be used.
b) initial development cost of basic stabilized platform.
c) the need for day-to-day programming of telescope.
d) capacity of the astronomical community to instrument the platforms, and to recover and analyze the data.
e) desirability of exploiting discoveries obtained in the earlier phases of the project.

SUPPORT REQUIREMENTS:

Research phase: Two areas of endeavor need strong support as a preliminary phase to the orbiting observing programs. First, basic research and product development are needed in optical materials and detection systems for use in the ultraviolet, infrared, and X-ray spectral [4] regions. A continued and expanded effort is needed in the study of the interaction of ultraviolet and X-ray quanta with surfaces, solids and gases. Studies of the photoelectric phenomena are needed in all wavelengths since the detectors will encounter large background light levels in the visible and near ultraviolet. Product development will be needed to devise image devices sensitive to selective bands of radiation, and to improve the range and stability of gas ionization detectors currently in use. Additional studies will be needed on the design of the optical equipment to withstand high accelerations and the remote readjustment of the components.

Second, rocket borne sky surveys and associated theoretical studies are needed to define the radiation intensity range and the wavelength regions of most promise. While the sky surveys are dependent upon future detector development, the present state of the detector art is such that both areas can be vigorously pursued at this time. Past experience would indicate that from 3 to 5 small rocket flights would be desirable in support of each satellite payload.

Technological Phase: Of prime importance is the engineering and development of the directable and stabilized mount for the optical system. The basic design criteria necessary to support the scientific objectives are:

Weight:

a. Structure, stabilization and coarse orientation control: 1300 lbs
b. Observing equipment (Optics, detection, apparatus, and fine orientation) 600 lbs
c. Power supply, telemetering and command functions: 600 lbs

[5] Stabilization in orbit:

a. Angular drift rates: Drift rates should be less than the libration produced by tidal effects. These should be less than 0.5 degree per orbit. There should be no moving parts after stabilization is accomplished.
b. Course aspect: 3 axis control, 0.25 degree steps at 0.5 degree/second.

Telemetering:

a. Bandwidth – 200 KC band width, suitable for TV type scanning detectors (raster sync pulses could be supplied from ground). Total orbit telemetering is not necessary.

Command Systems:

a. Coarse aspect (3 axis)
   Fine aspect (3 axis)
b. 20 channels of three position controls (up-down-stop)

Physical Configuration:

a. Capable of accommodating 36" dia. optical element.
b. Spherical mass distribution. Departure will cause libration due to the tidal effect of the earth. (See drift rates)

Orbit:

Perigee: 500 miles desirable – 300 miles minimum
Apogee: 500 miles desirable – 600 miles maximum
Inclination: 34°

Power Requirements and Life:

a. Power requirements for experimental equipment: 5 watts average, 50 watts peak for 5 minute intervals during telemetering period.
b. Life: 2 years.

[6] Vehicle Phase:

The projected Vega vehicle system appears to have adequate capability to perform the mission of this project. A nearly circular orbit is desirable. The choice of apogee is such as to be below the Van Allen radiation belt. The choice of perigee depends both upon telemetering considerations and the unstabilizing effect of atmospheric drag. Below about 250 miles, atmospheric absorption would also have undesirable effects.

Six Vega vehicles will be required to support the observational phase of the project.

About 30 small sounding rocket flights will be required to complete the sky surveys and to check detector and guidance systems.

Tracking and Telemetering Phase:

Ground stations will serve the dual function of command control and programming and data recovery. Eighty min. telemetering recovery per day should be allowed.

The various experiments will employ somewhat different data acquisition techniques. There will be a common requirement for command contact during the satellite pass to maintain the telescope under semi-active control during the telemetering period. Active on-board guidance and data storage can be employed in some cases. The ability to reduce the data from successive transmissions and to program the telescope to new regions of the sky will undoubtedly be the criterion in determining ground station coverage. These considerations would seem to limit the requires installations to two or three receiving and control stations. It has been suggested that [7] the installation be limited to one master station with slave stations to extend coverage.

Management :

Project management and technical direction will be provided through the Office of Space Sciences.

SCHEDULE

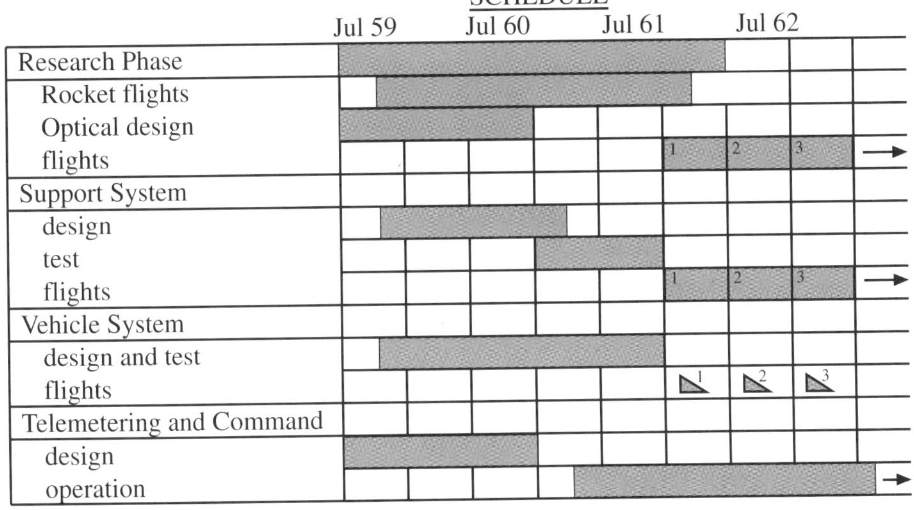

It is important to note that only after the analysis of the first data obtained from an orbiting observatory is available will its full potential as an astronomical tool be apparent. It can be expected that this analysis will suggest modifications to those systems yet to be flown. Thus, the actual flight schedule will depend some what on this uncertain scientific factor.

[8] PRELIMINARY BUDGET ESTIMATE FOR SIX ORBITING OBSERVATORIES:

	Total	FY 1959	FY 1960	FY 1961	FY 1962
(1) Research Phase	$ 6 M	0.6	2	3	.4
(2) Support Systems	$ 8 M		2	6	
(3) Vehicle Systems	$36 M		6	18	12
(4) Data reduction and Ground Support	7 M		2	2	3
	$57 M	0.6	12	29	15.4

REMARKS:

(1) Based on participation by 10 research institutions.

(2) Assuming use of certain techniques for stabilized platforms developed under NASA Meteorological Satellites and Project Mercury.

(3) Based on production costs for 6 vehicles.

(4) Assuming use of basic facilities at existing telemetering and tracking stations and full data recovery.

Astronomy and Astrophysics Programs
Office of Space Sciences
March 12, 1959

Document III-5

MEMORANDUM FOR THE FILES

SUBJECT: CONFERENCE REPORT – A Review of the Orbiting Astronomical Observatory Project – Held 18 February 1960

PARTICIPANTS: Dr. Malcolm Hebbs, General Electric Research Laboratory, Schenectady, New York
Mr. Donald F. Ling, Bell Telephone Laboratory
Mr. Jesse Mitchell, White House
Mr. Homer E. Newell, NASA
Dr. Edward M. Purcell, Harvard University
Dr. George Rathjens, White House
Dr. Nancy Roman, NASA

At Dr. Purcell's request, Dr. Newell reviewed broadly the steps leading up to the present time in connection with the Orbiting Astronomical Observatory, and its current status. He pointed out that the responsibility for this project has been assigned to the Goddard Space Flight Center, and Mr. Ziemer of GSFC has been named project manager. A vehicle has been assigned for the first such observatory, and it is planned to have some 3 to 6 firings in the course of a two year period following the launching of the first.

Dr. Roman then described the details of the planning for the Orbiting Astronomical Observatory. She reviewed the various meetings of the Working Groups, and passes out copies of the preliminary specifications drawn up by the Ames Research Center. Also given out were copies of the information on general specifications handed out to industry, as well as a copy of the minutes of the December 1959 Working Group meeting.

The responsibilities of the manager were described. The manner of working with the experimenters was also described, as well as the steps taken to insure that the proper interests of the experimenters were protected.

The members of the Purcell Committee were scheduled to meet with Dr. Kupperian at the Goddard Space Flight Center for a presentation on some of the laboratory work to date that has been put in on the Orbiting Astronomical Observatory project.

The review of the Orbiting Astronomy Observatory given to Dr. Purcell and his colleagues was essentially that contained in the attached writeup.

Homer E. Newell
Deputy Director
Space Flight Programs

Attachment

****

[no page number]

18 February 1960

Memorandum for: Hugh L. Dryden
Subject: Orbiting Astronomical Observatory

Jesse Mitchell, Secretary of the Purcell Committee on Space, showed Nancy Roman and me a writeup of the Committee's conclusions concerning the Orbiting Astronomical Observatory based on their deliberations and review of the subject over the last several months. The writeup indicates strong interest of the group in having the Orbiting Astronomical Observatory pushed with vigor and as rapidly as possible. In addition, the writeup indicates dissatisfaction with having only one Atlas-Agena scheduled at the present time for use on this project. The Purcell Committee would like to see perhaps two Atlas-Agenas scheduled for each of three successive years for this specific project. Finally, the Committee indicated some dissatisfaction with having a NASA group in a position of having to judge between out-house and in-house scientific experiments. This point is introduced by stating that there is considerable disagreement on the approach that NASA is taking to the Orbiting Astronomical Observatory project, specifically that the experimenters disagree with having a universally useful and adaptable stabilized platform. In exploring into this question, it was brought out that there is indeed a Working Group on Astronomical Observatories in which all of the experimenters are members, that these points have been discussed and considered in the Working Group meetings, and that NASA has carefully taken them all into consideration in arriving at its approach. It was further pointed out that a number of the individual experimenters are obviously interested in controlling the project and directing it to meet primarily their own needs. In the early days of getting going on this activity, Professor Whipple had in a proposal to undertake the management and conduct of the entire job. With this arrangement the proposed satellite would do very well for Professor Whipple's experimental needs, but how it would suit the requirements of other experimenters is open to question. Likewise, Professor Spitzer recently, in a private conversation with Dr. Roman, suggested that the management of the whole [2] Orbiting Astronomical Observatory project be given to him. He said that in this way the needs of Code of Wisconsin and Spitzer of Princeton could be integrated into a single satellite. Dr. Roman inquired as to whether or not this would take care of other interests like the University of Michigan and the Smithsonian Astrophysical Observatory. The answer indicated that this really would not. In fact, what develops is that each astronomer would apparently like to run a specific project on his own cover his needs. This is really not very efficient way of managing the project and is certainly a costly way.

The gist of the discussion appears to be that NASA is attempting to manage the project in such a way as to take care of the requirements of many experimenters in an efficient and reasonable fashion, while at the same time not delaying things unreasonably. The experimenters on the working group appear to have been going along with this, but at the present time seem to be seizing upon the opportunity afforded by the new activity of the Purcell Committee to reopen the question. My suspicions are that by so doing, each individual hopes that he himself be given the entire job.

We suggested that Drs. Purcell, Roman, and Newell get together to talk over the Orbiting Astronomical Observatories activity and try to come to an understanding of the real problems involved and of the proper way to approach the solution of those problems. Mitchell agreed to arrange such a meeting, which has been done.

Homer E. Newell
Deputy Director
Space Flight Programs

cc: Dr. Silverstein

Document III-6

[no page number]

June 15, 1960

To: Dr. Nancy G. Roman
From: A. D. Code, L. Goldberg, L. Spitzer, and F. L. Whipple

We are writing this memorandum to put on the record our views on two major items in the NASA astronomy program: first, the vehicles to be used for astronomical observations in space, and second, the organization of the OAO (Orbiting Astronomical Observatories) program. In summary, we are relatively well satisfied with the proposed organization of the OAO program, but we believe that in addition a strong effort should be made to put smaller astronomical equipment into orbit, at least during the next few years.

1. Vehicles

The OAO spacecraft, launched by an Atlas Agena rocket, will make possible a wide variety of astronomical programs in space, and should open new vistas of fascinating and important research. In fact, this equipment might well revolutionize our knowledge of astronomy. To make the fullest use of this powerful equipment, however, it would be very helpful to have preliminary experience with smaller astronomical payloads. Until guided telescopes and their accessory equipment have functioned in space for appreciable periods it will be difficult to design the most efficient and reliable equipment to function in orbit, unmanned, for many months. In addition, some preliminary data on the intensity of stellar radiation in ultraviolet wave length will be of very great importance in designing detection systems to measure this radiation with the large OAO system. The first OAO satellites launched will certainly be much more reliable and useful if some engineering and astronomical data on stellar space astronomy can be obtained beforehand.

To some extent such data can be obtained from satellites already launched or scheduled for the near future, and from rockets. In particular, rockets can obtain limited astronomical data on stellar ultraviolet radiation. A guided stellar telescope can be sent up with an Aerobee rocket, and several programs of this sort are now underway. However, a rocket is above the absorbing layers of the atmosphere for only a few minutes, and the amount of information obtained is strictly limited. According to present plans no stellar telescope will be placed in orbit until the first OAO launching, scheduled for 1963. This would require a very large jump indeed from simple rocket experiments to the massive and sophisticated OAO system. We recommend, therefore, that attempts be made to place smaller, simpler astronomical telescopes in orbit during the next few years.

[2] Two possibilities would appear along this line. The first would be to include small astronomical telescopes as secondary equipment in satellites launched for other purposes. Wide-band photometry of the brighter stars should be possible [handwritten underline] in this way. [handwritten note added: if they can be stabilized.] Such equipment would give vastly more information than can be obtained with rockets, and would give important data on the lifetime and reliability of the components that will be used in the OAO program.

A second possibility would be to launch small astronomical payloads separately. Development of a separate guided satellite system would appear too costly for such an interim program. However, the S-16 satellite, designed for solar research, could be used for stellar observations. If additional rockets of the Thor-Delta type can be made available for this purpose, it seems likely that a small astronomical satellite could be launched in 1962, well before the first OAO launching, and in time to influence materially the engineering and scientific plans for this important program. The early availability of scientific results from the smaller vehicles would also stimulate programs of analysis and interpretation in advance of the launchings of the OAO program.

We recommend that the possibility of implementing these two proposals be explored vigorously.

2. Organization of OAO Program

In the preliminary document of June 10 a detailed plan for the organization of the OAO program is set forward. We have examined this plan carefully and believe it is reasonable and workable. We have been concerned in the past by the absence of such a specific plan, but our concerns are now largely dissipated by the present document.

One area which will doubtless present a continuing problem is the relationship between the experiments and the engineering effort for the OAO. Success of this difficult and important enterprise seems more likely if there is close integration between these two aspects of the program. It is clear from the present plans that the experimenters have no official responsibility for the engineering aspects of the OAO. On the other hand, it is obviously highly desirable that the experimenters have a chance to comment on engineering plans when important decisions are being made. Continuing attention to this requirement for integrating engineering and scientific aspects should be given by the various groups concerned, as procedures for carrying out the program gradually develop.

[no page number]

ADDENDUM

It should be pointed out that serious problems of morale and scientific efficiency may arise during a long delay following failure in a major satellite launching. Related programs of lesser scope will provide valuable material to "tide" research groups over such periods.

Furthermore, extrapolating from previous astronomical experience, we believe that using smaller and relatively inexpensive satellites will have continuing long-term value for specialized researches suggested by the major program.

While the use of smaller vehicles prior to the OAO and possibly on a continuing basis can prove of great benefit, we do feel that the development of such programs should not divert energy or funds in such a manner as to jeopardize either the performance or the time schedule of the Orbiting Astronomical Observatory.

Document III-7

Document title: "Memorandum on the Stratoscope II Balloon Project to Assistant Director for Lunar and Planetary Programs," from Morton J. Stoller, Assistant Director for Satellite and Sounding Rocket Programs, November 4, 1960.

Source: NASA Historical Reference Collection, History Office, NASA Headquarters, Washington, D.C.

In 1946, the Office of Naval Research (ONR) initiated a program of upper atmosphere cosmic radiation research from high-altitude plastic balloons. A few years thereafter, Princeton University astronomer Martin Schwarzschild made a proposal to ONR that a telescope be carried in a balloon to photograph the Sun's surface. The success of this project, Stratoscope, led to the sponsorship by ONR, the National Science Foundation, and NASA of a larger, more sophisticated balloon-borne telescope called Stratoscope II. Between 1963 and 1971, Stratoscope II instrumentation observed the atmospheres of red giant stars and planets and also obtained high-resolution images of the nuclei of several bright galaxies.

[no page number]

4 November 1960
In reply refer to: DG(NGR:sds)

MEMORANDUM for Assistant Director for Lunar and Planetary Programs

Attn: Mr. R. C. Moore

Subject: Support for the Stratoscope II Balloon Project being conducted by Princeton University

1. In mid October Dr. Keller of the National Science Foundation telephoned to inquire whether NASA could provide funding for the Stratoscope II Balloon Project. Princeton University had requested $1,070,000 for this project of which the Office of Naval Research was able to supply $100K and the National Science Foundation $470K. Dr. Keller explained that if NASA could supply $250K, the project could proceed without severe delay.

2. This project is the first serious attempt to obtain high resolution photographs of astronomical objects other than the Sun without the complication of atmospheric seeing. The project is highly thought of in the scientific community and is considered to have the highest priority of any of the projects funded by the National Science Foundation Mathematics, Physics, and Engineering Division.

3. One portion of the astronomy program planned for the Orbiting Astronomical Observatories and their successors is the collection and analysis of high resolution images of astronomical bodies in the visual region of the spectrum. Questions have been raised as to whether it is preferable to do this work from a satellite free from the influences of wind and gravity or from a balloon. Balloons fly at sufficiently high altitudes to permit observations free of most atmospheric turbulence as well as much of the infrared atmospheric absorption. Since balloon payloads are recoverable and usually somewhat less expensive than satellite payloads of comparable complexity, appreciable savings can be made in the space program if balloons can [2] provide satisfactory long term observations of this nature. Therefore, the Stratoscope II Project is of direct interest to the NASA astronomy programs. First, it will provide the first attempt to remotely control and oper-

ate a high altitude telescope for long period observations of faint celestial bodies and, secondly, it should provide significant answers to the questions as to whether satellites or balloons are most efficient for astronomical observations in those regions of the spectrum which can be studied with high altitude balloons.

4. One hundred and twenty five thousand dollars of the money listed in the FY 1961 budget for Satellite and Sounding Rocket Programs under high altitude recoverable astronomical experiments can be used for this project.

[stamped "Morton J. Stoller"]
Morton J. Stoller
Assistant Director for
Satellite & Sounding Rocket
Programs

Cc: D/Silverstein
DD/Howell
NGRoman/sds
3 Nov 1960

Document III-8

Document title: American Science and Engineering, Inc, "Proposal for an Experimental Program of Extra-solar X-ray Astronomy, Prepared for NASA," September 25, 1963.

Source: Riccardo Giacconi, personal collection, reprinted with permission.

Researchers at American Science and Engineering were the first to detect cosmic x-rays. Encouraged by their discovery, Riccardo Giacconi and Herbert Gursky authored and submitted to NASA in September 1963 a paper describing their vision of a possible x-ray astronomy mission program. Beginning with simple rocket experiments and ending with the launch of a 1.2-meter x-ray observatory, the program outlined the ideas that NASA ultimately turned into reality. Their work provided the foundation for missions including the Uhuru satellite launched in 1970 and the Chandra observatory launched in 1999.

[cover page]

A Proposal for

AN EXPERIMENTAL PROGRAM
OF EXTRA-SOLAR X-RAY
ASTRONOMY

Prepared for
National Aeronautics and Space Administration
Washington 25, D. C.

Prepared by .
American Science and Engineering, Inc..
11 Carleton Street
Cambridge 42, Massachusetts
25 September 1963

Approved:
[signature]
Riccardo Giacconi
Vice President
Space Research and Systems Division

****

[12] [section] II. REQUIREMENTS FOR FUTURE X-RAY OBSERVATIONS

The discovery of galactic X-rays, together with the various hypothesis [sic] that have been put forward to explain them raise two obvious questions which must be answered in future observations:

1. What are the precise position, distances and dimensions of the discrete sources?
2. Do all X-rays come from discrete sources or is there a general diffuse background?

A great elaboration of the observational techniques will clearly be required in order to answer these questions and the new ones which will arise in the course of the development of X-ray astronomy. In common with technical developments for astronomical observations in other regions of the spectrum, those needed for X-ray astronomy will be directed toward:

1. All-sky surveys with increased angular resolution and increased sensitivity to distinguish discrete sources and the diffuse background;
2. Higher resolution studies of the structure of individual sources;
3. Increased spectral resolution both for discrete and diffuse sources:
4. Study of the detailed properties of X-ray emissions such as secular changes and polarization.

[13] In view of recent developments in X-ray optics and methods of detection it is now possible to plan along range program of X-ray observations with the assurance that the technical means exist for carrying it out.

[14] III. INSTRUMENTATION

The instrumentation which is utilized in most of the experiments here proposed is based on two new instruments recently developed at American Science and Engineering, Inc. a photoelectric X-ray detector and an X-ray telescope.

These instruments furnish orders of magnitude greater sensitivity and finer angular resolution than the conventional instrumentation.

1. Photoelectric X-ray Detector

The photoelectric X-ray detector utilizes the alkali halides as a photoelectron emitting surface for X-rays in the region between 2 and 100 Å. (Lukirskii 1960) .

A prototype of the detector has been constructed and tested at American Science and Engineering, Inc. The laboratory tests have confirmed the results of Lukirskii, et al., for SrF2 and have been so promising that this detector was chosen for the OSO-D wheel experiment presently being carried out under NASA Contract NAS 5-3569. The principal advantages of this type of detector with respect to conventional Geiger counters are:

1. The use of extremely thin windows which extends the range of observable soft X-ray wavelengths to the 20 - 100 Å region;
2. The ease of obtaining extremely large detecting areas;

[15] 3. The relatively low background noise;

4. The extremely long useful life of the device.

The detector is described in detail in the ASE proposal ASE-334, "Experiment to Measure Extra-Solar X-Radiation from the Rotating Wheel Section of the OSO-D Satellite".

2. The X-Ray Telescope

The X-ray telescope which is proposed for use in phases VI and VII of the proposed research program utilizes the principle of total external reflection of X-radiation at grazing incidence to form X-ray images in the focal place of the device. This instrument was first proposed by Giacconi and Rossi in 1960 (Giacconi 1960). It has been developed under NASA sponsorship (NAS 5-660) and is presently utilized on OSO-D for a pointed wheel solar X-ray experiment being carried out by ASE (NAS 5-3569). It is also being used on a pointed rocket experiment to obtain an X-ray picture of the Sun on recoverable photographic emulsion which is being carried out as a joint ASE-GSFC program (NAS 5-3401).

A description of the principle of operation is given in the NASA document X-614-63-112, "High Resolution (5 arc sec) X-Ray Telescope for Advanced Orbiting Solar Observatory".

The principle advantages of using an X-ray telescope are:

1. Large areas of collection;
2. Extremely fine angular resolution (the theoretical limit is a few seconds of arc) coupled to ease of alignment.

[16] 3. Orders of magnitude improved signal to noise ratio due to the focussing;

4. Ease of construction by use of traditional optical instruments manufacturing techniques.

[17] IV. THE PROPOSED RESEARCH PROGRAM

This program consists of three major steps. First, an all-sky survey is proposed from rockets (Phase I), OSO-D (Phase II) and a scanning X-ray satellite (Phase III) which culminates in the detection of sources 100 to 1000 times weaker than presently detected with a resolution of 1 degree and with preliminary spectrum information in the 0.1 to 60 Å region.

The second step consists of use of pointing systems to study in detail some of the observed sources or pre-selected potentially interesting objects. Phase N is an experiment to be performed by a crew member in the Gemini capsule. Phase V is a non-prime experiment to be performed from OAO. The possibility of pointing permits us to devote much longer time to the detailed study of single sources than is possible in a scanning experiment.

The third step is the introduction of imaging techniques in galactic and extragalactic X-ray observations. Phase VI is a prime experiment from OAO with a 10 foot telescope which will furnish a collecting area of about 38 cm^2. An improved version of this experiment is Phase VII, where a 30 foot telescope with an area about 400 cm^2 is proposed. The angular resolutions which now become possible are of the order of seconds of arc for detailed study of the structure of galactic and extragalactic sources.

Even though of necessity the latter phases of the program become less specific, it is believed that the execution of the entire program is well within the state of the art. A preliminary time schedule is shown in Figure 2.

[18]

PROPOSED TIME SCHEDULE

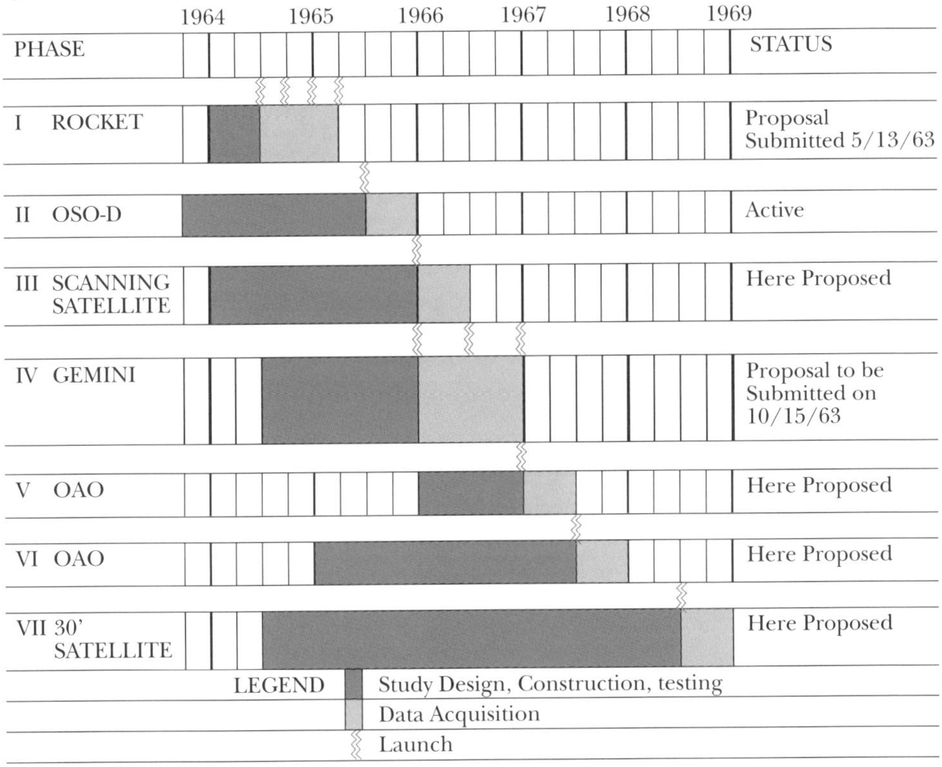

Document III-9

Document title: Nancy Roman to NASA Associate Administrator for OSSA, "NASA Support of Ground-based Astronomy," March 16, 1965.

Source: NASA Historical Reference Collection, History Office, NASA Headquarters, Washington, D.C.

In March of 1959, NASA Administrator T. Keith Glennan and Deputy Administrator Hugh Dryden had reached an agreement with National Science Foundation (NSF) Director Alan Waterman regarding a division of responsibility between the two agencies for funding of astronomy research. Under the memorandum of understanding, NSF received authority over ground-based astronomical research, while NASA would fund space-based astronomical studies. Officials from both agencies soon realized the difficulty of making such a strict division work when research efforts often relied on the results of studies conducted both in space and on the ground; NASA thus adopted a policy of funding ground-based research that strongly supported its space program. In 1965, a team led by NASA Astronomy Chief Nancy Grace Roman reviewed this policy in response to a request by the White House Office of Science and Technology that NASA and other Government agencies evaluate a National Academy of Sciences commission's assessment of the needs of ground-based astronomy. Roman's group did not advocate any changes to the policy, but recommended that NASA increase funding to support ground-based astronomy.

---

[1]
[stamped "MAR 16 1965"]

S/Associate Administrator for
Office of Space Science and Applications

SG/Chief of Astronomy
Physics and Astronomy Programs

NASA Support of Ground-based Astronomy

The excellent review of the needs of astronomy and astronomers prepared by the National Academy of Sciences committee chaired by Dr. A. E. Whitford has stimulated a detailed re-examination of government support of astronomy. As part of this re-examination, I have discussed the NASA policy on support of ground-based astronomy with the Astronomy Subcommittee of the Space Science Steering Committee and with Drs. Liddel, Brunk, Holloway, Smith, and Mr. Scott. For reference the policy which has been followed within OSSA may be summarized as follows:

"General support of astronomy is the province of the National Science Foundation. NASA cooperates with this and other government agencies in an attempt to insure support of deserving programs and to discourage any possible attempt to stimulate competition between agencies. However, it is obvious that the exploration of space must intimately involve both astronomers and astronomical research. It is difficult to conceive of any area of astronomical research which has no contact with space exploration. Moreover, in many areas of its program, NASA urgently needs additional astronomical research both to plan its programs intelligently and to interpret the results obtained. Therefore, NASA has undertaken the support of both astronomical research and astronomical facilities in those areas of astronomy which most directly affect the space program. In addition it has used the flexibility of its SRT program, its field center research, its training grants, and its facilities grants to support, at a lower level, an important broader a of astronomical research from which more closely mission-oriented research must arise."

Both groups consulted endorsed the present policy and urged that NASA support of astronomy be enlarged somewhat within the broad guidelines of this policy. They further agreed that, on the whole, the Whitford committee report is well thought out and outlines a program which is well conceived and, at least in optical astronomy, conservative. However, it is our recommendation that the major responsibility for the [2] implementation of this report lies more appropriately with the National Science Foundation. Within NASA, support of astronomy should remain as it is now, a program activity in OSSA. Program funds should be in-creased to cover a relatively modest increase in support of ground-based astronomy, including research, support of existing observatories, and support of new instruments. A line item in the NASA budget is neither necessary nor desirable.

In this context, we have reviewed NASA support of various areas of astronomy, with particular emphasis on the instruments recommended by the Whitford report. We agreed that it is probably not desirable for NASA to take an active part in the design and construction of either the largest feasible optical reflector or the largest feasible steerable paraboloid for radio astronomy. These general instruments fall within the traditional province of the National Science Foundation which should continue to retain responsibility for them. At the present time NASA has no foreseeable requirement for large optical telescopes in the 150 - 200 inch class. Funds for the three such instruments recommended by the Whitford paper appear to be available elsewhere. In addition to the telescopes now being built at the University of Arizona and the University of Texas, we foresee the need for major NASA access to two or three additional telescopes of the 60- to 120-inch class in the next few years. We also foresee the desirability of NASA support of at least one or two optical telescopes per year in the 24- to 50-inch class.

For tracking satellites and space probes, NASA requires instruments similar to those used for astronomical observations. The instruments should be made available for astronomical research on a non-interference basis to the maximum feasible extent as the Baker-Nunn telescopes are now used for flare star photometry and the 85-foot antenna at Rosman is being adapted to radio astronomy. Particularly in the case of the larger installations, such as the 210-foot dishes of the deep space net, the possibility of radio astronomical use should be considered in this design stages. NASA is exploring large phased arrays for tracking purposes. The results of researches on such arrays should be made freely available to the astronomical community to direct and support research of mutual benefit to both NASA tracking requirements and radio astronomy.

At the present time we do not foresee an obvious programmatic need for NASA to support the other major radio astronomical facilities recommended in the Whitford report. Hence, the responsibility for these [3] facilities should remain with the National Science Foundation. We should undoubtedly continue to support smaller special purpose instruments of particular interest to our programs, as we have in the past supported the low frequency radio astronomy arrays at the University of Maryland and the National Bureau of Standards and the accurate millimeter radio telescope at the University of Texas. It may also be desirable for NASA to support special purpose instruments for the near, intermediate, and far infrared regions of the spectrum.

Two areas of ground-based astronomy are conspicuously lacking from the Whitford recommendations. These are radar astronomy and solar astronomy. NASA should continue to partially support the use of existing large radars for special astronomical problems, including both instrumentation directly controlled by NASA and those radars built by other agencies. We do not, at present, foresee NASA support for the construction of new radar facilities, although further experience with radar exploration of the solar system may modify this conclusion. In solar astronomy, specialized, modest scale instruments like coronographs and cinemonochromators play a vital role in supporting satellite observations of the sun; two or three additional installations will be required to adequately back up the solar observatories flight program. We also estimate that NASA will want to support about five additional flare patrol photoheliographs and at least one intermediate-sized solar tower (telescope and spectrograph). We may wish to provide partial support of the Sacramento Peak vacuum telescope, if necessary, to guarantee its erection. Some support of other existing facilities may also be desirable.

In addition to support of telescope construction, several other areas of ground-based astronomy support merit consideration. NASA should plan to provide a major share of the continuing operational support for those instruments the construction of which we have funded as long as the research conducted is of interest to NASA. We should also increase support of other telescope operations to insure optimum productivity in those areas particularly relevant to the NASA flight missions. This support should include purchase or construction of auxiliary instrumentation, as well as of specific research projects.

NASA is in an unusually good position to cooperate in the development of automated instrumentation for ground-based astronomy. NASA has developed many automated techniques for both unmanned observations in space and ground-based reduction of large quantities of data which [4] could be adapted to astronomy. In turn, the further development of such techniques in ground-based astronomy may contribute significantly to NASA's increasing need for automated techniques in data acquisition and reduction

As long as NASA is charged with the scientific exploration of space, we shall continue to rely as astronomers both as full time participants in our program or as consultants on special problems. Therefore, we should continue our present policy of supporting astronomical education in the following ways:

1. Encouragement of the award of NASA training grants to pre-doctoral astronomy students.
2. Continued use of pre- and postgraduate student assistants on research projects of interest to NASA.
3. Continued support and encouragement of summer institutes for astronomy and related areas.

In summary, we recommend no major change in current NASA policy, but recommend an increase in the amount of program funds available to support astronomical research and facilities.

[signature]
Nancy G. Roman

cc: SG/Naugle, Smith
SL/Brunk, Liddel
SC/Holloway, Scott
SS/Clark
S/Newell

SG [signature] John E. Naugle
[stamped "16 MAR 1965"]

Document III-10

Document title: "Space Research: Directions for the Future – Report of a Study by the Space Science Board," Woods Hole, Massachusetts, 1965.

Source: Space Studies Board, National Research Council, National Academy of Sciences, Washington, D.C.

In 1962, the Space Science Board brought scientists together to conduct a major review of recent and future space research. One of the study's most significant outcomes came from the Working Group on Astronomy, which envisioned a large, space-based telescope as the next logical step after the Orbiting Astronomical Observatories. Three years later, when the Space Science Board convened another major review of space research, the Working Group on Optical Astronomy was tasked with assessing the utility of large aperture telescopes for optical astronomy. As its members were enthusiastic supporters of space astronomy, the group strongly recommended in the report of the study that NASA develop what they referred to as a Large Orbital Telescope. Soon thereafter, NASA began research on the pointing system for the facility that eventually became the Hubble Space Telescope.

---

[147] [Chapter] II: Optical Astronomy
1. SUMMARY AND RECOMMENDATIONS

The Working Group on Optical Astronomy was organized "to examine the future needs of optical astronomy for large-aperture orbiting telescopes of a generation beyond the orbiting astronomical instruments which are now being readied for launching." The Group interpreted this charge to include the space program for optical astronomy generally, since consideration of large instruments requires study of the scientific data as well as engineering experience gained with small instruments. As applied to the Working Group's area of concern, optical astronomy in space was defined to include all astronomical research carried out with reflecting telescopes in space at wavelengths from 800 Å to 1 mm, excluding solar studies. In terms of the instruments used, this definition is logical, since a conventional optical telescope with near-normal-incidence reflecting optics can be used for a wide variety of observational studies in this wavelength range. At the lower wavelength limit, somewhat shorter than 912 Å, mirror reflectivities tend to be low, and stellar radiation is probably completely cut off by the interstellar hydrogen absorption. Above the upper limit of 1 mm, the atmosphere becomes transparent and larger radio telescopes on the Earth's surface are more effective. (Solar research, with different prob-lems of thermal control and guidance requirements, needs different types of telescopes from those used for observing stars, stellar systems, nebulae, and planets, and was therefore the subject of study by a different Working Group.)

The space astronomy discussions at Woods Hole in 1965 were in some ways a continuation of earlier discussions by the Astronomy Working Group at the Iowa Summer Study Group in 1962 ("A Review of Space Research," Publication No. 1079 of the National Academy of Sciences-National Research Council). During the three-year interval since that earlier study, great strides in space technology have been made. Large rocket boosters have placed tons of equipment in orbit, and the Gemini flights in the spring of [148] 1965 have shown that man can operate effectively in space, even outside the spacecraft. The progress of optical space astronomy in the study of objects other than the Sun has been impeded by the difficult pointing requirements, but the accumulating data on ultraviolet stellar spectra obtained with sounding rockets (including a recent spectrogram with 1 Å resolution), and the progress made in fabricating and testing Orbiting Astronomical Observatories, suggest that rapid progress in this field can now be expected. The Woods Hole discussions naturally reflect the confidence resulting from these developments.

The present report is designed primarily to present the recommendations made by the Working Group, together with enough background material to explain the chief reasons underlying each specific recommendation. Many of the auxiliary points discussed by the Group are not mentioned here. To provide general background information, Section 2 presents a brief discussion of some of the most important and striking research objectives of astronomy in general and of optical astronomy in space in particular. Section 3 [sections 3-5 not included] discusses the short-range program in optical space astronomy, including flights planned during the next ten years, and related programs in astronomical instrumentation, optical design, and ground-based research generally. Section 4 is devoted to the longer-range goal of a large space telescope. Section 5 comprises three appendixes – the working papers of the Group.

RECOMMENDATIONS

The Working Group on Optical Astronomy has considered the possibilities for studying stars, star systems, nebulae, and planets by means of telescopes in space sensitive to electromagnetic radiation at wavelengths between 800 Å and 1 mm. For the short-range program (1965-1975), the following recommendations (all summarized here) have been made:

(1) The number of coarse-pointing sounding rockets available each year for optical space astronomy should be increased to twice the present level.
(2) Two or more telescopes having apertures of 40 inches or larger should be included in the Apollo Extension Systems (AES) program. The Orbiting Astronomical Observatory (OAO) program should be continued until AES launchings are definitely scheduled.
(3) Development of various detectors required in space telescopes should be supported by NASA.
(4) Development of improved gratings would be of central importance in the space astronomy program.
(5) Development of optical interferometers should be pressed, with probable initial operation on the ground.
(6) Research and development concerned with problems of space [149] telescope optics, especially with the primary mirror, should be supported by NASA.
(7) Support of ground-based astronomy should be increased, as such support is urgently needed for the continuing healthy growth of astronomy in general and of space astronomy in particular.

With regard to the long-range program (after 1975), the Working Group has concluded that the focus of the national effort in optical space astronomy generally should be toward, and in the context of, a very large orbital telescope to be used with a wide variety of astronomical instrumentation. To help pursue this objective the following recommend-ation (given in full here) was adopted:

(8) We conclude that a space telescope of very large diameter, with a resolution corresponding to an aperture of at least 120 inches, detecting radiation between 800 Å and 1 mm, and requiring the capability of man in space, is becoming technically feasible and will be uniquely important to the solution of the central astronomical problems of our era. We recommend that the Space Science Board of the National Academy of Sciences appoint an ad hoc panel to work toward this Large Orbital Telescope and to encourage studies of those critical areas where particular research and development is required in the near future to further this program.

****

[162] <u>Infrared telescope</u>. A telescope designed to be diffraction-limited at a wavelength somewhere between 10 and 100 microns might conceivably be made very much larger than an instrument designed for ideal optical performance at 0.5 micron. Until this field of research has been explored more fully from the ground and from space, the value of such a specialized instrument cannot be assessed.
<u>Interferometer</u>. The beam interferometer, designed to achieve very high resolution on particular objects, would be a useful instrument in optical astronomy. Current efforts to use this technique from the ground have been discussed above, and further information is required before the need for interferometric equipment in space can be evaluated.

After study of these various points the Working Group concluded that at present the long-range program in optical astronomy should be concentrated on a single general-purpose telescope, though special-purpose instruments might be included at a later date, when and if a clear demonstration of their value can be made.

Following considerable discussion the Working Group adopted the following recommendation:

Recommendation 8

We conclude that a space telescope of very large diameter, with a resolution corresponding to an aperture of at least 120 inches, detecting radiation between 800 Å and 1 mm, and requiring the capability of man in space, is becoming technically feasible and will be uniquely important to the solution of the central astronomical problems of our era. We recommend that the Space Science Board of the National Academy of Sciences appoint an ad hoc panel to work toward this Large Orbital Telescope and to encourage studies of those critical areas where particular research and development is required in the near future to further this program. (See p. 2-21 for considerations leading to the last part of this Recommendation.)

Confidence in the technical feasibility of a diffraction-limited 120-inch space telescope was based on the various technical studies carried out for NASA directly or indirectly by various groups (Boeing, American Optical, Perkin-Elmer); the engineering problems of such a large instrument were discussed only briefly by the Group. The design goal of a 120-inch aperture was adopted in the belief that a long-range instrument should be a very significant advance over the insturements [sic] used in the Stratoscope and OAO programs, whose apertures are in the 30 to 40 inch category. The aperture could well be greater than 120 inches, if that proves technologically feasible (see page 17).

[163] It was the conviction of the Group that this large instrument could provide a dramatic central focus for the optical space astronomy program, and that it would be an appropriate major space program for the nation. It was to help emphasize the central character of this instrument in the national space effort that the name "Large Orbital Telescope" (LOT) was proposed. While the term "orbital" was used for this large-span telescope, the possibility of a lunar location was not strongly excluded.

Clearly, adoption of the LOT program would have a significant impact on the short-range program in optical space astronomy. While the short-range program discussed in Section 3 is designed primarily to obtain significant scientific results, the data obtained and experience gained would be absolutely essential for the LOT effort. In particular, the AES effort could be an important forerunner of the manned high-resolution LOT. In general, considerable expansion of much of the short-range program might be required if the LOT were to be effectively used within the time scale outlined below.

The subsequent sections discuss the possible design parameters for the Large Orbital Telescope, a time schedule that may be visualized for its construction, and some administrative problems that might be associated with this enterprise.

DESIGN PARAMETERS

The general characteristics of a large space telescope, discussed in earlier sections of this report, apply to the LOT as well. Thus, this large telescope would be a general-purpose instrument, focusing electromagnetic radiation in the wavelength range from about 800 Å to 1 mm. The Group discussed briefly the engineering problems of this telescope and the design parameters that might be chosen in view both of these problems and of the scientific objectives. While no recommendations were adopted on most of these items, the conclusions are summarized here for reference.

Aperture

For reasons already outlined, the goal of designing a diffraction-limited 120-inch telescope was adopted by the Group. The actual diameter of the instrument would depend, of course, on the technical situation at the time the instrument was designed. One possibility discussed by the Group was that the actual diameter might substantially exceed 120 inches, but with the image size corresponding to a diffraction-limited 120-inch mirror. Such an increase in light-gathering power would be desirable for many researches and might be technically feasible if a corresponding decrease [164] in angular image diameter were not required. (If the Saturn V were used to place the LOT in orbit, and the primary were a single mirror, the diameter could not exceed 250 inches; without doubt, other engineering considerations would limit the diameter to a substantially smaller figure.)

Role of Man

It was generally agreed that the LOT should be usable for many decades, with occasional changes and improvements in the instrumentation provided at the focal plane. This requirement can presumably not be met unless a man is intimately involved in maintaining and repairing the equipment, and presumably a man will also be required for the initial adjustment and operation. The design of the LOT should provide for ease in trouble shooting, for access to all parts of the telescope, and for replacement of defective modules. The <u>extent</u> to which a man should actually operate the telescope is a matter of debate, and it is not excluded that the entire system should be completely automated. Guidance on stars will presumably be automatic, and, during this time, man should probably not be coupled to the instrument. However, guidance by man might prove useful for observations of a rotating planet, for which automatic guidance would be difficult. Similarly, in a crowded star field, acquisition of the desired object by a man might be useful, though this could be done through use of a television camera rather than by looking through the telescope. There was agreement that the instrument should be completely controllable at will, either by equipment on the ground or by a man nearby. There was some discussion of the likelihood of failures resulting from human error.

Location

After reviewing the recommendations of the Report on Lunar Exploration Systems after Apollo (LESA, North American, 1965), the Group discussed the relative advantages of the following three different locations for a large space telescope: low orbit (below the Van Allen belts), at 400 km altitude or less; high orbit (above the Van Allen belts), at 30,000 km altitude or more; and on the Moon. Most of the considerations examined would appear to favor the high orbit. As compared with location on the lunar surface, the advantages of a high orbit include no gravitational flexure, no secondary micrometeorites, and lower cost. A possible major disadvantage of the high orbit is greater risk of exposure of equipment and men to high-energy radiation from solar flares, though evidence presented to the Group suggests that adequate shielding is no problem. Objects close to the Sun, however, might be more difficult to observe from a high orbit than from the lunar surface. As compared with a low orbit, the advantages of the high orbit are: negligible occultation of objects by the Earth [165] (in a low orbit, occultations complicate the programming and are likely to reduce the net observing time by about one half); nearly constant thermal environment, which much simplifies the maintenance of the mirror figure; reduction of external torques due to gravity gradients, magnetic fields, and air drag by at least two orders of magnitude, with resultant simplification of the guidance problem; darker sky than in low orbit, where airglow may contribute light; and virtual absence of oxygen atoms striking the telescope and oxidizing the aluminum. From a high orbit, communication with the ground might be simplified by continuous radio contact, but, as compared with a low orbit, communication would be complicated by the increased distance. The greater exposure to solar flare radiation may be an important disadvantage of the high orbit, especially in view of the longer time (at least 10 hours) required to return a human operator to Earth from the high orbit. A very clear disadvantage of a high orbit is that it requires a Saturn V for launching instead of a Saturn IB; since this additional cost would be required for each visit by men, this could be a conclusive argument for the low orbit. The Working Group unanimously came to the conclusion that, on technical grounds, the high orbit appears at the moment to be the optimum location for the LOT.

Optical Design

A conventional parabola-hyperbola or a Ritchey-Chrétien system seems indicated. The primary should have a relatively low focal ratio to minimize the over-all length of the instrument. Use of the prime or Newtonian focus would not seem to offer any particular advantages, and all of the instrumentation would presumably be at the Cassegrain focus, possibly with tiltable mirrors to direct the light toward the desired instrument or sensor. Careful baffling would be required to keep earthlight as well as sunlight out of the optical path, and the secondary supports should presumably be apodized (with Couder strips). Automatic focusing would presumably be required and, probably, automatic collimation as well.

[167] ADMINISTRATIVE PROBLEMS

Three different phases of the program were considered: (a) preliminary phase, (b) design and construction, and (c) post-launch operation. As entirely different administrative problems would be encountered in each of these phases, they are discussed separately here.

Preliminary phase

Such a major astronomical effort as the LOT should not be undertaken until a majority of the astronomical community supports the program with enthusiasm. It appears to the Working Group that progress in space research generally, and in space astronomy particularly, combined with increasing awareness of the close interdependence of space astronomy and ground-based astronomy, may help in generating enthusiasm for the LOT among U.S. astronomers.

To help in explaining LOT plans to their colleagues, and in pressing for the program generally, the Working Group concluded in effect that the Group as a whole, or a representative fraction of it, should continue in existence, as an ad hoc panel, and requested the National Academy of [168] Sciences to endorse a proposal to this end as contained in Recommendation 8, page 3. The purpose of the panel would be:

(i) To attempt to broaden the base of support, for (a) the space astronomy program in general, and for (b) an eventual launching of a large astronomical instrument in particular. By discussion with their colleagues, they would hope to clarify the issues involved and to stimulate the interest of astronomers who are at present unfamiliar with the aims of the space program.

(ii) To begin an orderly examination of some of the technical problems that will arise in the design of a large orbiting telescope, anticipating that more permanent arrangements will be made later.

(iii) To implement these two aims by holding fairly frequent informal meetings, preparing discussions of specific subjects, inviting the participation of other astronomers, and generally to keep alive the idea of working toward a large orbital telescope.

Design and construction phase

In the initial organization of the program and during all successive stages until launch, there must be close and effective contact between NASA and its engineering contractors, on the one hand, and the astronomical community on the other. How this contact can best be maintained and integrated into the vast administrative structure required for such a large program is a question that deserves careful study. Perhaps a group of astronomers might be organized to carry out detailed design studies, with advice from engineers and optical experts; such a group might then serve in an advisory capacity during the engineering design phase that would follow. Perhaps a committee under the National Academy of Sciences, with representatives from various interested groups, might serve a useful function in this context, and might help to provide a bridge between the NASA organization for the LOT and the. scientific community. Further exploration of these and other possibilities is desirable.

Operations phase

Clearly, the LOT would be a truly national facility, and should be administered as one. The plan should be workable from the standpoint of NASA's internal administration, since the situation would be complicated by the fact that flights would be involved. The Working Group visualizes that the detailed program for operating the LOT (allotment of observing time, expeditious recovery of data, proposals to place auxiliary instruments of newer design on board, etc.) would need to be managed in a way analogous to present ground-based national facilities. Responsibility for detailed scheduling must be defined, as it would depend not only on the scientific [169] program but also on such factors as the relative position of the telescope, the Earth, the Sun, the object to be observed, communications, etc. Experience with the OAO-D program, in which two-thirds of the observing time will be allotted to guest investigators (i.e., investigators other than the principal investigator, who is responsible for the experiment), may help to reveal some of the administrative problems in these areas.

Document III-11

Document title: Letter to Dr. Norman F. Ramsey, Harvard University, from James Webb, Administrator, NASA, January 14, 1966.

Source: NASA Historical Reference Collection, History Office, NASA Headquarters, Washington, D.C.

Document III-12

Document title: NASA Ad Hoc Science Advisory Committee, "Report to the Administrator," August 15, 1966.

Source: NASA Historical Reference Collection, History Office, NASA Headquarters, Washington, D.C.

In January 1966, NASA Administrator James Webb sent a letter to Norman Ramsey of Harvard University requesting that he head an ad hoc committee to recommend how NASA ought to proceed on a variety of space science activities, including the ways in which NASA could involve the academic community. Among its tasks, the committee was asked to provide advice on how NASA should develop and manage a major space observatory project that the National Academy of Sciences' Space Science Board had proposed the prior summer. The Ramsey committee's response to NASA's request regarding the space observatory focused largely on the management of such a facility. In particular, the committee strongly endorsed the concept that a consortium of universities, which it called the Space Telescopes for Astronomical Research, Inc., or STAR, be established and charged by NASA to select and manage the scientific investigations that would be conducted on the space observatory. Although NASA was reluctant to give so much authority to an external group so quickly, the space agency did establish an astronomy advisory group made up of non-NASA scientists, and ultimately awarded management of the Hubble Space Telescope's scientific operations to a non-NASA entity.

Document III-11

[1]

NATIONAL AERONAUTICS AND SPACE ADMINISTRATION
WASHINGTON. D.C. 20546
January 14, 1966

OFFICE OF THE ADMINISTRATOR

Dr. Norman F. Ramsey
Harvard University
Cambridge, Massachusetts

Dear Dr. Ramsey:

The purpose of this letter is to request you to assemble and chair an Ad Hoc Science Advisory Committee to advise NASA in the execution or [to ascertain] the major new projects which are either underway or under serious consideration within the agency. These projects are the Voyager/Automated Biological Laboratory Program, the post-Apollo lunar exploration program, and the National Space Astronomy Observatories recommended by the Woods Hole Summer Study.

As you are well aware, the objectives of those large complex programs have been developed through an intensive interaction with the scientific community and, I believe, are well accepted by that community. The objectives and goals of Voyager have been reviewed by the President's Science Advisory Committee, so you are well aware of the aims of that Program.

In view of the interaction which has taken place with the scientific community, you may question the need for further consultation and work on the part of people such as yourself and the others we would like to see on the advisory group.

The need for such a group arises from the problems involving the scientific community due to the size, [2] complexity, and long lead time of these projects. In the past, scientists at universities and Government laboratories have been able to participate directly in the program by conceiving, designing, and building their own experimental apparatus, which was then integrated into a spacecraft and flown. After the flight, the scientist received the data and was able to analyze the data and publish the results in the established tradition of academic research.

We in NASA think it is essential that competent scientists at academic institutions participate fully in the next generation of space projects, and we believe that we will need new policies and procedures and perhaps new organizational arrangements in order to enable them to participate.

The next generation of space astronomy projects is an example of some of the problems we face. The astronomers at Woods Hole recommended that NASA develop 120-inch telescopes for space astronomy. We envision such an instrument as a National Astronomical Observatory in Space – an astronomical facility in space for astronomers to use, rather than the separate instruments developed by individual scientists such as we have on the Orbiting Astronomical Observatory. We estimate that it will require at least a decade to bring such a facility into being. During that period NASA will have to work closely with many of the most competent astronomers in the country. We have a nucleus of competent astronomers and engineers at the Goddard Space Flight Center; however, we expect that nucleus will have to be strengthened. It is exceedingly important that we have highly competent scientists and engineers at the Center responsible for a major astronomical facility and that we have the proper kinds of people in positions of leadership at the Center.

[3] As another example, it would appear that the Automated Biological Laboratory (ABL) will have to be a "laboratory" to be used by scientists, rather than a "collection" of

individual experiments designed and built at several institutions. The requirements for sterilization and the severe constraints on weight imposed by the low surface pressure on Mars dictates that the ABL must be a tightly integrated package, designed and built under the most exacting conditions in complex facilities which will be too expensive to duplicate. However, the most competent biologists and physicists in the country must be deeply involved in the conception, design, fabrication, and testing of ABL to assure the success of the mission. Furthermore, this must be a continuing involvement over a period of six to eight years prior to launch. Obviously, great care must be taken in the assignment of responsibility for ABL and the selection of key personnel to design and build it, so that competent scientists will want to participate in the development and proper relationships can be established with scientists at academic institutions.

Clearly, one of the major tasks of this advisory group would be to review the resources at our NASA field centers, and such other institutions as would be appropriate, against the requirements of the next generation of space projects and advise NASA on a number of key problems, such as:

(1) How can we organize these major projects so that the most competent scientists and engineers can participate?

(2) How can academic personnel participate and at the same time continue in strong academic roles?

(3) What mechanism should be used to determine the scientific investigations which should be conducted?

[4] (4) How does a scientist continue his career development during the six to eight years it requires to develop an ABL or a large astronomical facility?

(5) Should we change the orientation of some of our NASA Centers?

(6) What steps should be taken in scientific staffing, both inside and outside NASA, over the next few years to assure that we have the proper people at the proper places to do the job?

(7) How can we obtain the competent scientists to take the key roles in these major projects?

If you will undertake this job, we at NASA will work closely with you. Dr. Newell and the staff of the Office of Space Science and Applications will arrange any briefings which you desire, and will arrange for tours of the NASA Centers.

We envision a group of astronomers, biologists, physicists, and geologists from both departments and administration in the universities. NASA would provide per diem and would pay the usual consulting fee.

I will be glad to meet with you and discuss this further at your convenience.

Sincerely yours,

[signature]
James E. Webb
Administrator

Document III-12

[i]
[cover letter]

HARVARD UNIVERSITY
DEPARTMENT OF PHYSICS
LYMAN LABORATORY OF PHYSICS
CAMBRIDGE, MASSACHUSETTS 02138

August 26, 1966

Mr. James Webb, Administrator
National Aeronautics and Space Administration
Washington, D. C. 20546

Dear Mr. Webb:

The Ad Hoc Science Advisory Committee, which you asked me to assemble and Chair in your letter of January 14, 1966, has convened a series of meetings since that time and has prepared its recommendations. The Committee in its deliberations has been greatly assisted by members of the NASA staff, by participants in the NASA program, and by other individuals.

The Committee herewith forwards its report. We believe that the changing nature of the National Space Program will create a multitude of new and deep questions of science, administration, and management. We have, therefore, made a number of recommendations to help you in meeting these anticipated new problems.

Please feel free to call on us if we can be of any assistance to you in the interpretation of these recommendations.

Sincerely yours,

[signature]
Norman F. Ramsey, Chairman
Ad Hoc Science Advisory Committee
National Aeronautics and Space Administration

****

[17] G. Large Astronomical Facilities in Space

Clearly, if NASA is to engage in a scientifically orderly development of a large earth-orbiting astronomical facility, which is a long-term undertaking of great cost, NASA needs the assurance of a continuing, responsible commitment from the scientific community towards this endeavor. Reciprocally, in view of the long-term character of this commitment, the scientific community needs the assurance of a reasonably strong role in the scientific direction of the program. The following plan suggests a mechanism to fulfill these two needs.

Let us assume that the general plan of building a large multi-purpose space astronomical facility has preliminary approval, and then attempt to envision the various steps toward an orderly organization of the task. A legally constituted organization of universities active or interested in the fields of optical, radio, X-ray, gamma-ray astronomy and other terrestrial radiations can be used. The Board of Trustees of this organization can appoint consultants who are representative of other interested persons and related research fields.

The organization should be primarily designed to achieve the following:

1. A long-term commitment of the astronomical community, solidly based on institutional commitments of a representative group of universities, to assist to the fullest extent required in the planning, design, construction, and use of an orbiting astronomical facility.
2. A means whereby leading astronomers can provide the necessary scientific direction in the creation and use of the facility, while still preserving their university roles in research and education.

[18] 3. A means for securing international participation in the creation and use of the facility.

The facility will take many years to build, and it should operate for a very long time—ground-based telescopes 60 years old still produce useful research. We conceive of the facility as a number of scientific instruments, most of them under common operational control, relatively closely grouped around a space station in an approximately 200-mile orbit, so their scientific re-supply and engineering access is possible and economical. Earlier astronomical telescopes, including OAO, OSO, X-ray and other newer high-resolution telescopes, will provide essential, preliminary and continuing experience.

The corporate entity would provide continuity of scientific planning, instrumentation, research and development, and programming. The function of this organization can be fulfilled either by an existing not-for-profit interuniversity corporation, or by a new one. In either case we shall, for convenience here, call the organization Space Telescopes for Astronomical Research, Inc., or STAR for brevity. STAR would obtain the best and most authoritative advice from the scientific community in defining the goals of the scientific program, and the details of the instruments that would become part of the scientific facility in space. STAR would develop preliminary specifications that would be transmitted to NASA Headquarters, which in turn would direct the appropriate field Centers to prepare alternative and competing designs and engineering approaches. Rival simultaneous and separated approaches in this phase seem desirable, considering the complexity of the problem; any apparent waste of manpower and time in this preparatory phase would be small compared to that in constructing and launching inefficient devices, because of the high costs of orbiting large payloads. A first-phase report prepared by STAR should be transmitted to NASA, and it should give the major scientific goals, conclusions on [19] feasibility, and studies to indicate costs. STAR would work closely with NASA Headquarters, the NASA Centers, and the industrial designers to evaluate the designs and conduct trade-off studies.

In phase two, let us assume that either one or more of the NASA Centers is assigned the management of the large astronomical facility. Its major instruments may include a large optical telescope, special -purpose solar telescopes, high-frequency radio telescopes, X-ray and gamma-ray telescopes, other radiation detectors, and possibly an associated space station.

STAR would have an administrative, planning and small scientific staff at its headquarters. The Advisory Board of STAR would report its operations both to the member universities and to NASA at a high level. In return, the STAR Board would be kept informed by NASA as to the direction the program is taking. STAR would also provide to the chosen NASA Centers continued scientific review. Because of the enormous complexity of the scientific and engineering problems involved, a close liaison must be maintained between NASA Headquarters, NASA Centers, and the engineering groups, contractors, and scientists in STAR. Although a NASA Center would be responsible for the construction of the major elements of the facility, certain sub-systems and scientific auxiliaries could be designed and built most efficiently by STAR, or by individual universities or scientists.

The relative permanence of the principal elements of the facility places a heavy burden on its planning because further developments of astrophysical knowledge will affect the scientific goals. Maintaining continued close relations between the management Center and the university community will help to provide the most fruitful multi-purpose large astronomical facility. Above all, the design should be kept responsive to the scientific goals, not only during the planning phases, [20] but also as late as possible during construction; also great care must be taken to ensure that the demands of engineering do not unduly compromise these goals.

The sequence of devices to be built for X-ray astronomy will depend largely upon the present rapid advances in this field, and they will be better defined as observation and technology advances. Gamma ray astronomy, now in an early stage of growth, can also be integrated into the orbiting facility. A submillimeter or millimeter wave length radio antenna may be in the same orbit if scientifically justified, but a much higher orbit is required for a very low frequency, large radio telescope. Although physically separate, the latter, organizationally, belongs within the facility and STAR organization. Likewise, the STAR organization may be involved in providing the smaller orbiting telescopes that precede the large one.

The large orbiting astronomical facility will involve manned attendance with much hardware already available, possibly from the Apollo program, for re-supply, updating of sensors, repairs of defective modular subsystems, and recovery of photographic material. This is an essential operational feature because of the great cost of the large telescopes. The involvement of man would permit flexibility for future changes in the scientific program, would provide a closer link for cooperation between scientists and those interested in the continued development of our national capability for man-in-space, and would familiarize man with the problems of operating elaborate scientific devices under less difficult conditions than are normal in the smaller space vehicles of two or three man capacity.

For the operational phase STAR would develop policy for allocating telescope observing time to scientists from STAR, other universities, Government laboratories, NASA Centers, ground-based observatories, and to other interested scientists. The scientific program and definition of the operation policy of the telescopes, [21] including changes of subsystems, would be under the direction of the Board of STAR. Operational control of the facility, including engineering, re-supply, and safety, would be the responsibility of the NASA Centers.

Astronomy has traditionally been a field that excites the imagination and interest of a wide variety of inquiring minds, and one in which there is a long history of fruitful international cooperation. The nature of the orbiting observatory is such that participation in the scientific use and program of observations could form an interesting and natural basis for international cooperation and for providing a powerful and constant reminder of the genuine interest of the United States in learning how to engage in helpful international cooperation.

It would hence be very valuable to make clear from the beginning that this unique facility, the orbiting observatory, is intended to serve astronomers on a world-wide basis, and to work out meaningful ways in which the world's most creative astronomers could participate and contribute to the scientific program. The extent or mechanisms through which international representatives should be involved in the management and planning of the observatory have not been examined. However, we believe that STAR could facilitate the working out and operation of appropriate arrangements.

Document III-13

Document title: G. C. Augason, "Manned Space Astronomy," NASA Headquarters, November 1966.

Source: Space Telescope History Project, Archives, Smithsonian Institution, Washington, D.C.

NASA's emphasis on human space flight in the 1960s affected nearly all of its activities, including the astronomy program. For example, Boeing's 1965 feasibility study for NASA on the Large Space Telescope—the first extensive NASA-funded study on what became the Hubble Space Telescope—assumed that the observatory would be supported in various ways by human inhabitants of a near-by space station. Many other studies likewise considered the possibility of using humans in space to service or operate astronomical facilities. This NASA overview highlights many of these studies.

[cover sheet]

MANNED SPACE ASTRONOMY
G.C. AUGASON
NASA HEADQUARTERS
NOVEMBER 1966

[no page number]

MANNED SPACE ASTRONOMY

SECTION I – EARLY EFFORTS

Astronomy Subcommittee

The purpose of this review is to recount the efforts made by the National Aeronautics and Space Administration (NASA) to use man in space as an aid in astronomical research. The Astronomy Subcommittee of the Space Science Steering Committee (SSSC) early anticipated the importance to astronomy of man in space.

At the second meeting of the Astronomy Subcommittee, May 23-24, 1960, the role of man was discussed. J. A. O'Keefe was given the assignment of determining the effect of motion by the astronaut on observations. He reported in a later meeting that it was possible for the astronaut to guide on objects which were several minutes of arc in angular subtense. At the fourth meeting, October 24-25, 1960, the fact that an astronaut could make useful observations was once again emphasized and it was stated that astronomical training of the astronaut would be worthwhile. At the October 30-31, 1961, meeting an Ad Hoc Committee for Scientific Tasks for Man-in-Space was formed under the chairmanship of Dr. Jocelyn Gill. One of the tasks of the new committee was to develop a training program that would help the astronaut to make meaningful astronomical observations.

[2] On February 20, 1962, the United States accomplished its first orbital launch when John Glenn made a three-orbit flight in the Mercury capsule, Friendship 7 (MA-6). In addition to his assigned tasks, Glenn made several observations which were of astronomical interest. At the subcommittee meeting on February 25-27, 1962, it was proposed that an astronomer be present during the astronaut's debriefing to note and interpret any observation the astronaut had made. The Ad Hoc Committee for Scientific Tasks for Man-in-Space coordinated and reviewed experiments for the Mercury flights. Because of the space limitations, the lack of a suitable telescope, the lack of a large viewing port, and because of other demands on the astronaut's time, none of the experiments performed were of an astronomical nature.

In subsequent meetings, one of the main areas of concern of the Astronomy Subcommittee was the role of man, his limitations, and his training and preparation. Attention was shifted from the Mercury flights to the Gemini-Apollo series in an attempt to have meaningful experiments ready for these flights. A careful set of ground rules was chosen for these experiments. In addition to fulfilling the space, weight and power limitations of the space capsule, and being of great value scientifically, the experiments had to be such that their performance was greatly enhanced by the presence of a man and/or some [3] portion of the experiment required reliable recovery. One of the main concerns at this time was that the observing port be of adequate size and have good transmission in the ultraviolet and visible. The February 4-5, 1963, subcommittee meeting was held at Houston at the Manned Spacecraft Center. The facilities were inspected and, in a detailed presentation, D. Slayton of MSC explained the training background and special problems of the astronaut.

Manned Flight Experiments Working Groups

Toward the end of the Mercury program, the Ad Hoc Committee for Scientific Tasks for Man-in-Space was disbanded and a Panel on In-Flight Scientific Experiments (POISE) was formed which was chaired by Drs. Gill and John O'Keefe. This panel was originally established to provide coordination between the Manned Spacecraft Center and in-flight experimenters. This panel reviewed over 100 experiments, 12 of which were then sent to the Manned Space Flight Experiments Board (MSFEB) and the Space Science Steering Committee (SSSC) for approval. Early in 1963 after the initial experiments had been selected for the Gemini flight program, POISE was discontinued. Subsequent manned space experiment review was performed by the regular subcommittees of the SSSC. A new group was formed consisting of flight experimenters, Manned Spacecraft Center personnel, and NASA Headquarters personnel which is called the In-Flight Experimenters [4] group. Its purpose is to continue the coordination activities of POISE. This group was chaired by Dr. Gill and has been responsible for bringing about some improvements in the experimenters conditions on board Gemini and Apollo. A specific example is that of the Gemini observations port which is now protected during launch to prevent scratches and contamination from obscuring the pane.

In order to provide coordination between the various subcommittees of the SSSC and manned flight experiments program, a committee was formed called the Manned Space Science Working Group. This committee was composed of a member of each of the subcommittees and was chaired by Mr. Willis B. Foster, the Director of Manned Flight Experiments in the Office of Space Science and Applications (OSSA). This group met twice, January 30 and March 26, 1964. The meetings were largely informative, i.e., for the purpose of explaining the manned space science programs. The main concern of the members of the group was with experiment opportunities on the various spacecraft and the methods which should be used to publicize these opportunities in the scientific community. The group members as a whole were quite critical of NASA's apparent delay in doing this.

Concurrent with the Manned Space Science Working group, separate disciplinary panels were meeting to determine the need and possible [5] experiments which might be performed using an Orbiting Research Laboratory (ORL). The Astronomy Panel met twice on October 26, 1963, and June 24, 1964. Dr. Peter C. Badgley of Manned Flight Experiments of OSSA presided and 18 astronomers and physicists attended who had demonstrated a previous interest in space astronomy. This panel set as its goal an 120-inch diffraction-limited telescope in orbit around the earth. The panel then formulated a general list of experiments to be performed with such a telescope. The panel noted in its recommendation that a growing group of astronomers felt an urgent need for an immediate study of the detailed problems of manned space astronomy. They then delineated some of the specific problems expected. They felt that solutions to these problems should involve a large part of the astronomical community and that a consensus be obtained, although the panel did not know how to initiate such a program.

Another group also met at this time to formulate experiments for the early Apollo missions. This was known as the Ad Hoc Working Group on Apollo Experiments and Training on the Scientific Aspects of the Apollo Program. This group, under the direction of Dr. Charles P. Sonett, had as its prime interest the study of selenography, planetology and aeronomy. They published one report December 15, 1963. They considered the possibility of the astronaut's doing astronomical experiments but concluded that "the initial Apollo missions should not [6] be burdened by astronomical activities."

Apollo and Manned Space Station Studies

At the May 23, 1963, Astronomy Subcommittee meeting, the members were asked their views on the scientific value of a manned space station. Most of the members felt that it was not justifiable at that time on a scientific basis, but that this might change once the limitations of unmanned observatories were determined.

At the following meeting, August 8-9, 1963, Dr. L. Roberts of the Langley Research Center presented a report to the Astronomy Subcommittee on a feasibility study for a large manned orbiting astronomical telescope. The study had been carried out at Langley. Although the astronomical uses of the telescope generated by the study were felt by the subcommittee to be unrealistic, they felt it was a valuable study because the engineering aspects that were developed. It is necessary to know the engineering design parameters because these determine the constraints which must be placed on a related telescope. At the same meeting a representative from the Manned Spacecraft Center presented their long range plan for a Manned Space Station.

At the November 7-8, 1963, Astronomy Subcommittee meeting, several proposals were reviewed for flight on Gemini spacecraft. The proposals [7] that were recommended by the subcommittee were for a general purpose telescope by the University of Arizona; an X-ray astronomy proposal by the American Science and Engineering; and a proposal for the operation of a small fully stabilized telescope by Lowell Observatory. The advantages and disadvantages of a lunar-based observatory were compared with those of an orbiting platform at the May 6, 1964, Astronomy Subcommittee meeting. Also, at the same meeting, Dr. Harlan Smith made a report on the activities of the Manned Space Working Group. He described the method of selecting experiments for the Apollo program. This was done by approaching certain scientists directly. Then, at a meeting held at the MSC on June 15, 1964, the potential experimenters were to present their ideas for experiments. On August 11, 1964, a letter was sent out to many astronomers at various institutions, informing them of the opportunity to participate in the design of flight instruments and telescope systems which were larger than those currently scheduled and which would use a man in conjunction with these instruments. Those responding to this letter were to define the research objectives, to estimate the technical requirements to reach those objectives, and to identify engineering problems which would the demand particular attention. Essentially, the only response to this invitation was by Princeton University, with Dr. Lyman Spitzer as the principal investigator. Their proposal was for the design of a [8] diffraction-limited orbital telescope for direct imagery and ultraviolet spectrophotometry.

At the time of the May 6-7, 1964, joint subcommittee meeting of the Astronomy Subcommittee and the Solar Physics Subcommittee, the following recommendation was jointly adopted: Both subcommittees resolved that it was their belief "that the vital part of both stellar and solar astronomy lies in the establishment of large astronomical observing equipment in space. This equipment should be a logical outgrowth of the currently planned OAO and AOSO programs and should utilize the best technology available at that time." At the October 28-30, 1964, meeting of the Astronomy Subcommittee, it was decided that the subcommittee would review all proposals which had been submitted for a manned orbiting telescope in answer to a request for proposals originating with Dr. Leonard Roberts of the Langley Research Center. As a result of this invitation, the University of Virginia developed the astronomical objectives and requirements of a large orbiting telescope, the Fecker Division of the American Optical Company studied the optical feasibility of such a telescope and the Boeing Aircraft Company investigated the mission and engineering requirements.

At this time considerable debate was occurring about whether the future needs of manned astronomy and science in general could be served [9] by adaptation and modification of the existing Apollo spacecraft system or if an entire new generation of spacecraft would be required. This problem was further complicated by the military requirements of the Department of Defense. In order to answer these questions, several industrial companies were funded to generate lists of manned scientific experiments which could be flown on Apollo. They were aided in this by several government panels.

On February 1, 1965, an ad hoc panel was chaired by Dr. Henry Smith. The purpose of this panel, known as the Astronomy and Astrophysics Panel for Experiment Recommendations for the Earth Orbital Apollo Missions, was to review the astronomical experiments which had been proposed for Apollo. Dr. Smith's panel reviewed over two dozen experiments, fifteen of which were found worthy of flight consideration. These experiments were used as a planning basis for a post-lunar Apollo program which was to become known as the Apollo Extension Program (AEP). These experiments were later reviewed in detail by other reviewing groups. The AEP program, later to become known as the Advanced Apollo Program (AAP), will be discussed in greater detail under the section entitled Flight Experiments.

[10] SECTION II - PROGRAM PLANNING

Several alternatives were considered as possible manned astronomical programs to follow the Gemini and Apollo Lunar Missions. These alternatives required that a decision be made as to whether the Apollo Hardware could be modified to allow the accomplishment of future scientific requirements or would a new spacecraft have to be developed. This decision was further complicated by the requirement that NASA and the Department of Defense (DOD) might have to share the same vehicle.

This choice was simplified when the President of the United States, on August 25, 1965, authorized DOD to develop a Manned Orbiting Laboratory for military use. This laboratory was to have a two-man crew and unmanned launches were to begin in 1968. The Air Force selected Douglas Aircraft Company, Inc., to build the spacecraft, with the General Electric Company being responsible for the experiments. The launch vehicle was to be a Titan 3C booster with a NASA Gemini capsule used for the return to the earth. With this decision, military requirements no longer had to be considered when developing a civilian manned space program.

NASA chose to plan their future programs around the use of the modified Apollo spacecraft and this new series of missions was to be known as the Apollo-Extension Program (AEP), which later became the Apollo Applications Program (AAP). Following the AAP, it was assumed [11] that experiments would probably be done from a space station orbiting the earth and for planning purposes this space station was named the Manned Orbiting Research Laboratory (MORL).

The decision to use Apollo for the advanced manned scientific missions was based on the following reasons: first, the Apollo capsule was able to meet many mission requirements as a result of the capability which was built into the Apollo system when the Lunar Orbit rendezvous mission was selected as the mode to be used for Lunar exploration; secondly, the Apollo spacecraft would be a fully developed and tested vehicle by the time the AAP began; and third, the Apollo/Saturn system represented a national capability which might be lost if a program of this type were not developed for it.

Once the decision had been made to pursue a manned astronomical observatory program based on the use of the Apollo hardware, the definition of particular missions was possible. In defining these missions the work of the previous panels, committees and working groups was incorporated. Experiments which were previously planned for unmanned flight were reevaluated to see if they could gain by being flown in a manned environment. If the presence of a man would increase the lifetime or improve the reliability or reduce the signal bandwidth requirements, some of these experiments were reconsidered [12] for manned flight assignment. New meetings were held to help advise NASA in the formulation and priority assignment of different experiments. Many of these meetings were necessary because of the rapid advances which had been made in several areas of astronomy because of the use of unmanned vehicles. X-ray and gamma-ray astronomy are examples of such rapidly progressing subdisciplines. In addition, new invitations went out to publicize the new opportunities for experiment. An invitation describing the advanced Apollo missions and the opportunities for experiment was sent out to many potential experimenters March 11, 1966.

Studies by the National Academy of Sciences

Space Science Board Meeting, Woods Hole, Massachusetts. An important meeting in the NASA planning process was held under the auspices of the Space Sciences Board of the National Academy of Sciences at Woods Hole, Massachusetts, on June 20 to July 16, 1965. A report of this meeting entitled "Space Research, Directions for the Future," defined the scientific objectives which should be achieved and made recommendations on how they could be accomplished. The broad research objectives for astronomy were defined as:

"(1) Is the Universe finite or infinite, and if it is finite, what is its size?

(2) Is the Universe in a steady state, and if not, how did it begin and how will it end?

[13] (3) Do the laws of physics as deduced on the Earth apply without change for all times and overall distances? Alternatively, are there fundamental physical laws or phenomena still undiscovered in terrestrial laboratories, that are observable only on an astronomical scale?

(4) Were the chemical elements that form all matter built up out of hydrogen, and if so, how?

(5) How are stellar systems, stars, and planets, formed?"

Their general conclusions were that: 1) Successful study of the planets will require the presence of scientists, either on the planet's surface or orbiting the planet because of communication time delay and/or power bandwidth requirements. 2) Man may be successfully employed in space as an observer; for assembly, placement, repair and operation of scientific instruments; for analysis, collection, storage and retrieval of data. 3) The cost of manned space flight can not be justified on the basis of its "scientific value" alone.

The working group visualized four types of programs: 1) The current Gemini and Apollo programs; 2) More advanced laboratories such as the Apollo Applications Program (AAP); and 3) Large, special-purpose space laboratories such as optical, radio, X-ray or gamma-ray observatories. Both the optical and solar astronomy panels made recommendations that the AAP be utilized as an interim program to gain [14] experience with telescopes in the 60- to 80-inch category with a goal of a manned orbiting telescope (MOT) with a 120-inch diameter mirror and usable in the 500 A to 1 mm spectral region.

Space Science Board - Large Space Telescope Panel. A large national facility was recommended strongly by the Woods Hole Summer Study and other panels and committees. Although such a facility has been given various titles and it may be lunar rather than orbital, NASA has come to call it MOT (for manned orbiting telescope). As recommended at the Woods Hole meeting, a panel under the chairmanship -of Dr. Lyman Spitzer, Jr., of Princeton University has continued to hold meetings to discuss the problems involved in developing a large space telescope (April 29-30, 1966, and July 17-19, 1966). In their meetings they have considered the scientific program, technical problems and managerial problems. The last meeting mentioned was devoted to developing the scientific objectives in detail. The scientific recommendations of the Woods Hole meeting have been used as guidelines. The committee has accepted the recommendations of NASA as regards the largest diameter mirror which can be placed in orbit with the present generation of launch vehicles. They believe it is reasonable to assume that such a mirror (120 inches in diameter) may be made diffraction-limited. If this is not possible, they would prefer to sacrifice the diffraction limitation to total aperture.

****

[22]

NASA Headquarters OMSF. The Office of Manned Space Flight supported a major study by the International Business Machines Corporation entitled, "ORL Experiment Program," under Contract NRSw-1215. The ORL (Orbiting Research Laboratory) concept is a general one which included many disciplines. This study had as its very commendable goal the determination of experiments to be performed by man in space. This goal was not to be accomplished from the "bottom up" by selecting a list of experiments compiled from submissions by interested experimenters but rather experiments would be chosen on the basis of their ability to answer fundamental and important questions in a particular discipline. IBM pointed out the following shortcomings in the "bottom up" approach:

"a. It results in a collection of individual tests, rather than in a cohesive program; the interrelationships of the individual experiments and the extent of their overlap are obscured.

b. It lacks a rationale to determine whether the most important experiments have been identified and are being pursued.

c. Few of the suggested experiments are explicitly tied to requirements or ultimate benefits; as a consequence, the resulting [23] programs frequently fail to demonstrate the value of the space station vis-a-vis its cost."

Under the Astronomy/Astrophysics discipline IBM defined four main objectives, many subobjectives and 132 knowledge requirements. This list of experiments has been used as a guide on occasion for various studies, although it is incomplete and sometimes trivial.

North American Aviation, Inc.: Under Contract NAS 2-1047, North American Aviation did a study of "The Lunar Exploration System for Apollo" (LESA). The purpose of this investigation was to develop a description of scientific operations that can be carried out on the moon with LESA. The LESA system includes a family of prefabricated modules that can be assembled on or below the lunar surface in a variety of arrays. Recommended scientific investigations were obtained from each of nine -panels. Three of the panels which considered astronomical experiments were: Optical Astronomy, Radio Astronomy and Wave Propagation, and Radiation which treated Ultraviolet, X- and Gamma-Ray, and Neutrino Astronomy. The various experiments were divided up to be performed on early and on late missions. In case of Optical Astronomy the early mission would use a 40-inch lunar–based telescope for multihour exposures. The later missions would employ a 120 inch instrument.

Langley Research Center. Langley Research Center, an Office of Advanced Research and Technology (OART) center took an early interest [24] in the problems associated with manned astronomical experiments in space. Most of their studies were done by contract under Associate Director, C. J. Donlan. This work was carried out by P. Hill who headed the Manned Orbital Research Laboratory (MORL) Steering Committee, Dr. L. Roberts who was in charge of the Manned Orbital Telescope (MOT) Steering Committee, and W. Gardner who headed the MORL Study Group.

The Langley people came to the conclusion (based on their early studies) that a telescope employing a 120-inch mirror which had the structural integrity to be diffraction limited was the largest telescope which could be launched on the Saturn series of vehicles. This determination has been generally accepted in all other studies.

Douglas Aircraft Co.: One of the earliest studies was that by Douglas Aircraft Co. with International Business Machines (IBM) as a subcontractor. This study was initiated by the MORL Study Group and was completed January 1966. Its purpose was to define and optimize typical missions using MORL. This study was accomplished in three phases: Phase I (June-September 1963) System Comparison and Selection Study of a Manned Orbital Research Laboratory; Phase I A[sic; should be IIA] (December 1963 - November 1964) Optimization of the MORL System Concept and Phase II B (December 1964 - January 1966) Development of the MORL System Utilization Potential.

Phase I demonstrated the feasibility of launching, operating, and maintaining a manned research laboratory. Phase II A defined the MORL concept [25] to include a (1) 260 inch dia. laboratory launched by the Saturn IB in a 200 nautical mile orbit inclined at 28.72° to the equator; (2) a Saturn IB launched Apollo logistics spacecraft and; (3) supporting ground systems. Phase II B examined experiments from NASA, and Department of Defense (DOD) sources redefined them and collated them, and commonality was searched for. This was done to determine the utilization potential of the MORL system concept.

University of Virginia: The MOT Steering Committee supported an investigation by Dr. Laurence W. Fredrick of the University of Virginia entitled, "Applications in Astronomy Suitable for Study by Means of Manned Orbiting Observatories and Related Instrumentation and Operational Requirements." The resulting report considered the following areas appropriate for research with a MOT: Stellar Mass Determination: The ability of an orbiting telescope to determine mass depends on (a) the theoretical resolution limit and (b) the accuracy of guiding during exposure. With a guiding accuracy of only 0.3 seconds of arc, masses of at least 30 new binaries may be determined. Cosmological Tests: The concept of uniformity in general relativity may be tested by an intercomparison between brightness distances and size distances. High Resolution Photography and Spectroscopy: Photography with the resolution possible in space is important to all fields of astronomy. Increased spatial resolution will allow increased spectral resolution which is needed in all spectral regions. Infrared Observatories: Because of the longer observing times available and low sky temperatures, infrared observations from a MOT should be two orders of magnitude better than observations from a comparable telescope on earth even in the [26] windows and of course infinitely better in the opaque portions of the atmosphere. Measurement of stellar diameters: Interferometric and image orthicontechniques on a MOT should allow stellar diameters to be measured down to 0.0005 seconds of arc. The Study of Regions of Polarization: Techniques which have been used to study polarization in the visual spectral region could be extended to other spectral regions. The Search for Very Faint Stellar Companions: An apodizing device could be constructed for a MOT which would not increase the definition of a star but would enable close faint objects to be detected. Integrated Studies of Comets: The large aperture of the MOT would allow studies of comets to be made as they become fainter after leaving the sun and long intergarion times would not be needed. Also comets could be observed in all spectral regions and their behavior monitored.

American Optical Company: In 1964 the MOT Steering Committee funded the J. W. Fecker Division of the American Optical Company at Pittsburgh, Pennsylvania, to do a "Feasibility Study of a 120-inch Orbiting Astronomical Telescope" report AE-1148. They looked at the optical problems involved. They did ray tracing and developed some optical configurations. They chose a cassegrain configuration with an f/2 parabola for the primary mirror which could be used with different optical components to give a range of focal ratios from f/2 to f/100. Although their design was not optimized they felt that they had demonstrated feasibility. One of the most severe problems they encountered was maintaining the alignment of the secondary mirror, but they felt this could be accomplished by the use of an active system [27] with a laser and interferometric techniques.

Boeing Aircraft Company: The University of Virginia and American Optical Company studies served as a starting point for a rather thorough study by the Boeing Company of Seattle entitled "A System Study of a Manned Orbital Telescope" D2-84042-1, October 1965 and a continuation, "Synchronous Orbit Study" D2-84042-2, April 1966. General Electric was retained as attitude stability and control subcontractor and Drs. Zdenek Kopal of the University of Manchester, and James G. Baker of Harvard served as consultants. This report did much to establish the 120 inch telescope as the largest size which could be launched. The limiting stellar magnitudes presented may be too bright by as much as five magnitudes.

The principal results of the Boeing Study were: Astronomy and Optics—The astronomical objectives of The Virginia Study were accepted almost in total. A Ritchey-Chretein modification of the pure Cassegrainian System was selected with an f/4 primary and two secondary mirrors which permit operation of the telescope at f/15 and f/30. Necessary instrumentation was designed conceptually. Operations Analysis—The role and contributions of man were carefully delineated. The study of the operation of the MOT was narrowed down to three basic modes which used the MOT in conjunction with the MORL as defined by Douglas in their study.

The preferred modes were a detached mode capable of docking with the MORL but employing a shuttle for normal operations and a soft [28] gimballed mode. Configuration—Various configurations were examined. Launch and orbital configurations for the two selected modes of operation were developed. Scientific instrumentation and cabin arrangement for the MOT were also defined. Structures—The primary structural design was based on the boost condition. Special attention was given to the primary mirror stresses during the boost and docking, the dynamics of the soft gimbal made, thermal distortions of the primary mirror during operation and meteoroids and radiation. Attitude Stability and Control—The observational requirements were used as the basis from which to synthesize an attitude control system. The study showed that it would be feasible to stabilize the telescope to within 0.01 seconds of arc. A pointing error for the soft gimbal concept of about 0.003 seconds greater than for the detached mode was indicated. Thermal Analysis—Computer studies were performed to determine the thermal gradients in the primary and secondary mirrors and the telescope structure. The use of an earthshade and doors that closed when the optics were pointed towards the earth greatly aleviated [sic] many thermal problems.

The general conclusions were that the MOT was a feasible system and that the soft gimbal mode would be preferable. A synchronous orbit is feasible and it would allow longer observation times with reduced thermal problems. In addition to the above, Boeing outlined several special problems including optics manufacture, film handling an [sic] attitude stability control.

Marshall Space Flight Center. The Marshall Space Flight Center (MSFC) has supported studies to develop programs to follow the early Apollo lunar missions and which would utilize the talent and hardware which were developed for Apollo.

****

[43]

The Manned Orbiting Telescope (MOT)

The term "manned orbiting telescope" has been used rather ambiguously in NASA. It originally referred to any program which employed a telescope and utilized a man to set it in operation and to maintain and operate it. It was usually assumed that this telescope was in an earth orbit but telescopes operating from the moon have also been considered under this title. More recently the designation MOT has been reserved for telescopes not connected with AAP. Earth orbiting telescopes connected with AAP are now called ATM and MOT is reserved for another generation of missions. In recent usage, MOT also refers more specifically to a major national astronomical facility in space.

In addition to the MOT studies referred to in Section II, NASA Headquarters has made two additional studies. The first was "The National Astronomical Space

Observatories Working Group Report," August 16, 1966. This report was generated for the Planning Coordination Steering Committee of NASA Headquarters. This document describes the major astronomical problems that exist and then describes how a 120–inch diffraction-limited telescope would help find the answers to these problems.

It has been thought by many that, to be most effective, at least one and perhaps several astronomers will have to spend their full time with MOT. In order to support these astronomers some sort of a space station will be required. A study has been made of space stations and [44] their associated problems which is entitled "The Needs and Requirements for a Manned Space Station," September 28, 1966. This report was prepared by the Space Station Requirements Steering Committee, and discusses in depth the justifications for a space station; an MOT is one [sic] the more important justifications.

Document III-14

Document title: NASA Management Instruction 1156.16—"NASA Astronomy Missions Advisory Board," September 25, 1967.

Source: NASA Historical Reference Collection, History Office, NASA Headquarters, Washington, D.C.

Document III-15

Document title: NASA, "A Long-range Program in Space Astronomy, Position Paper of the Astronomy Missions Board," NASA, edited by Robert O. Doyle, Harvard College Observatory, July 1969.

Source: NASA Historical Reference Collection, History Office, NASA Headquarters, Washington, D.C.

While it did not immediately embrace the idea of establishing an external organization to manage the scientific operations of a space-based astronomical facility, NASA's head of space science, Homer Newell, in reaction to the Ramsey report, suggested creating a standing group of academic astronomers to advise NASA on astronomy program objectives and strategies. The new group, the Astronomy Missions Board (AMB), was chartered in 1967 to address astronomy needs, just as the Lunar and Planetary Missions Board supported the solar system exploration program. In July 1969, the AMB issued a long-range program in space astronomy, outlining the most pressing problems and the types of observations and instruments necessary to solve these problems. After more than two years of meeting regularly, after preparing this report the AMB ceased functioning because of board members' frustration that NASA never acted on its recommendations due to lack of funds to conduct suggested programs.

Document III-14

[1]

NMI 1156.16
September 25, 1967
Effective date

Management Instruction

SUBJECT: NASA ASTRONOMY MISSIONS ADVISORY BOARD

1. PURPOSE

This Instruction establishes the NASA Astronomy Missions Advisory Board and sets forth its functions and scope of interests.

2. AUTHORITY

a. Section 203 (b) (7) of the National Aeronautics and Space Act of 1958 (42 U.S.C. 2473 (b) (7)).
b. Executive Order 11007 (February 26, 1962).
c. NMI 1150.2.

3. ESTABLISHMENT

a. (1) The NASA Astronomy Missions Advisory Board (hereafter referred to as the "Board") is hereby established to assist NASA in the planning and conduct of all NASA missions to create and operate astronomical telescopes in space.
(2) The long-range continuing objectives of astronomy are to learn by remote observation the structure and behavior, the origin, growth, and demise of all types of celestial bodies, ranging from the smallest (meteorites, comets, and planets) to the largest, including stars, star systems, the matter in space, and the entire cosmos. The telescopes utilized by space astronomy work over the whole range of the electromagnetic spectrum, from gamma rays, through X-rays, the ultraviolet, visible light, infrared light, and radio frequency radiation.
(3) Excluded from the Board's area of responsibility are missions to study the moon and planets from a close vantage point, or in earth orbit for study of the earth.

b. Pursuant to Executive Order 11007 and paragraph 6 of NMI 1150.2 it has been determined that the formation and use of the Board is in the public interest. The Board is not an "industry advisory committee" within the meaning of paragraph 5b of NMI 1150.2.

[2]4. FUNCTIONS

The Board will serve in an advisory capacity only. Organizationally, the Board will be responsible to the Associate Administrator. Determination as to any action to be taken which is based in whole or in part on proposals or recommendations of such Board shall be made solely by appropriate full-time salaried officials or employees of NASA. The scope of the Board's activities include, but are not limited to:

a. The development and review of the scientific objectives and general strategy for space astronomy and associated ground-based astronomy.
b. The review of support of ground-based astronomy and observations from sounding rockets, balloons and aircraft as well as satellites.
c. The formulation of guidelines and specific recommendations for the design of space astronomy missions, and for the various telescopes and auxiliary equipment to be developed and used on these missions.
d. The continuing review of the way these missions are meeting the needs of current scientific objectives and strategy.
e. The continuing examination of policies relating to the operation of these telescopes in the space observatory once they have been made operational and are available for observations by the scientific community.
f. The development and improvement of mechanisms by which the NASA space astronomy program can get the best assistance from, and give the most help to, the entire community of astronomers and space physicists.

5. MEMBERSHIP

a. The Board shall be composed of about 12 scientists and engineers, all of whom are broadly experienced in the technology and scientific discipline of astronomy. To be proportionately representative of the astronomical community at large, the membership majority will normally not be made up of full-time NASA employees. The Board Chairman will be designated from among the non-NASA members of the Board. The Board will also have an Executive Director who shall be a full-time salaried employee of NASA, but who will not be a member of the Board.
b. The Board and its Chairman will be appointed by the Administrator, and will serve at his pleasure. The Executive Director will be appointed by the Associate Administrator. The Administrator will consult with the Board on any appointments to the Board.

6. BOARD RELATIONSHIPS

a. The Board will have a close working relationship, through the Associate Administrator, with senior NASA officials and organizational elements involved. Requests for advice and recommendations will be made to the Board through the Associate Administrator. Similarly, the Board will request studies or other assistance required from other groups in or out of NASA through the Associate Administrator.

Document III-15

[cover sheet]
NASA SP-213

A
LONG-RANGE
PROGRAM
IN
SPACE
ASTRONOMY

Position Paper
of the
Astronomy Missions Board

July 1969

Edited by
ROBERT O. DOYLE
Harvard College Observatory
Cambridge, Mass.

[iii] **PREFACE**

The Astronomy Missions Board was established by the National Aeronautics and Space Administration by charter in September 1967 to assist in an advisory capacity in the planning and conduct of all NASA missions to create and operate astronomical experiments in space. The scope of the Board's activities includes development and review of the scientific objectives and general strategy for space astronomy and associated ground-based astronomy; the formulation of guidelines and specific recommendations for the design of space astronomy missions, and for the various experiments and auxiliary equipment to be developed and used on these missions; the continuing examination of policies relating to the operation of these space observatories once they have been made operational and are available for observations by the scientific community. The work of the Board encompasses the many aspects of space astronomy including direct observations of electromagnetic radiation from astronomical sources, cosmic-ray particles and the supporting research that is necessary, but its scope does not include the study of the Moon and planets from close vantage point or study of the Earth.

The Astronomy Missions Board is presently composed of 18 members of the scientific community with a wide diversity of interests and experience. They are drawn largely from universi-ties, but include members from national laboratories (see appendix for a list of members of the Board and its panels) [appendix not included]. The Board's activities are supported and supplemented by seven panels and two ad hoc working groups to whom specific areas of responsibility are assigned. The panel compositions are similar to that of the Board itself and involve an additional 31 scientists. This wide membership provides a broad representation of current thought in space astronomy both directly through its membership and from the wider astronomical community by means of letters and discussions.

The activity of the Board has been intensive. With few exceptions, it has met monthly for 2 days at locations appropriate to its current activities. In addition to extensive deliberations and [iv] discussions, the meetings have included reports and resumes from NASA personnel about matters such as the current status of projects then underway, present NASA plans for the future, technical reports on areas of special relevance, and budgetary aspects of current and planned programs. The panels have met several times during the past year and have taken the opportunities for obtaining firsthand information about the activities in space astronomy at various NASA centers relevant to their particular fields of interest. Again, briefings as to technical capabilities and current planning were obtained and the panels prepared detailed programs and recommendations for activities in their areas.

An important continuing activity of the Board is the presentation of specific recommendations to the Associate Administrator of NASA. Many of these recommendations have been ad hoc answers to questions raised by NASA, while others have been of a more general nature and have, in most cases, been incorporated into the body of this report. Many of these ad hoc recommendations were for the purpose of assisting NASA to optimize a low-level program, and should not be construed as approval of such a program by the Board or the scientific community.

The Board has created a long-range national program for space astronomy-including discussions of the major problems of astronomy and astrophysics, an observing program describing the next important measurements from space, and examples of the instruments, spacecraft, and missions needed to make those measurements. Specific mission descriptions are not intended as concrete definitions of future missions, but as part of an exemplary program which is used to establish the best current balance between the subdisciplines. The plan contains sufficient mission priorities and interdependences on which to base AMB advice to NASA at various foreseeable levels of effort, and should enable NASA management to assess the impact on scientific progress of the various future options available to them. The purpose of this position paper is to describe the long-range plan as it appears in July 1969.

Past experience has shown that astronomy is a field full of surprises and the unexpected, and it would be extremely shortsighted to expect this report to remain up to date for very long. This report is not intended to be a static document. It is, rather, a working paper to be updated and altered continuously by the Board as technical capabilities change and scientific opportunities and priorities evolve. Nevertheless, it seems appropriate to publish [v] this version of the position paper, just as it was submitted to NASA as part of the fiscal year 1971 budget planning cycle, in order to acquaint a wide community of astronomers, astrophysicists, physicists, and other interested scientists with the workings of the Astronomy Missions Board, as well as with the national space astronomy program. NASA and the Astronomy Missions Board hope in this way to continue to improve the mechanisms by which the NASA space astronomy program can get the best assistance from, and give the most help to, the entire community of astronomers and space physicists. From time to time, as the extent of the revisions makes a major part of this work obsolete, the Board will again publish an updated position paper.

The detailed reports on the subdisciplines of space astronomy, authored by the panels and endorsed in substance by the Board, will be found in Part II[not included]. Part III describes how the panels' programs were evaluated, and how parts of them were combined into long-range plans at two levels of effort-a minimum balanced program and an optimum program-both of which do not attempt simply to do everything suggested by the subdisciplines, but rather emphasize research on those problems judged astrophysically most important by the greatest consensus of the Board.

A summary of the position paper and key features of the long-range plan will be found in Part VII [not included].

****

[223] III

AMB LONG-RANGE PLAN FOR SPACE ASTRONOMY

[227] THE TWO AMB LONG-RANGE PLANS

Before presenting the final mission schedules, we should briefly define each of the long-range plans—describing the characteristics that the AMB hoped to impart to these composite programs.

The Minimum Balanced Program

The definition of a minimum program, especially the concept of a sharp break below which progress becomes substantially more difficult if not impossible, is often an exercise fraught with the possibility of misunderstanding. We therefore emphasize the significance of the AMB minimum balanced program as that level below which one or more of the subdisciplines of astronomy must be dropped to maintain the others above their minimum thresholds of efficient and scientifically profitable operation.

Such a negative step would then seriously undermine a central assumption of the Board's planning; namely, that the agreement between the subdisciplines on the most important astrophysical problems requires an orchestration of the multiwavelength observing programs. This assumption is that for many problems a few relatively unsophisticated, and sometimes less expensive, measurements in different wavelength regions might lead to a deeper [228] understanding of the physics of a process—than a most beautiful detailed picture achieved at great cost at a single wavelength. This is not to say that in some cases the narrow, highly specialized approach might not produce the essential, even indispensable, key. It is simply a judgment of the Board about the requirements for greatest progress in most problems before us at the present time. We should also note in passing that the Board found it was often these problems requiring a multiwavelength, multidisciplinary approach, which are the problems attracting so many physicists and scientists from other disciplines to come to work in modern astrophysics.

A timely example of the lengths to which scientists will go to achieve the completeness of the multiwavelength approach, and an illustration of the uses to which the future capability in space astronomy will be put, is the current standby alert trying to catch the brilliant flash of a flare in the X-ray star, Sco XR-1. Starting in May, astronomers at Cerro Tololo Observatory in Chile have been continuously measuring the visible radiation, watching for the onset of a flare. At the first sign of activity, they will radio other astronomers at Caltech where the 200-inch Mount Palomar telescope will make infrared observations, in Hawaii where a rocket will be launched to record the X-ray spectrum, and at Goddard Space Flight Center where University of Wisconsin experimenters will turn the Orbiting Astronomical Observatory's ultraviolet telescopes toward the X-ray star. Since the Australians recently discovered Sco XR-1 to be a strong variable at radio wavelengths, a large steerable radio telescope may also be used to complete the wavelength coverage. The great hope is to achieve several measurements whose combined value in terms of scientific understanding might greatly exceed the combined cost relative to that of any one of the measurements standing alone.

The Optimum Program

Although the concept of an optimum program is usually less controversial than a minimum program, the Board wants to take pains to stress that this is not simply a program in which the subdisciplines all are encouraged to do their maximum unconstrained programs. First, it was composed from a set of subdiscipline maximum programs where the principal constraint, or upper limit, was the projected availability of excellent people-scientists and supporting teams of specialists-to carry out the recommended missions. Second, the level of the resulting optimum Board program was about 20 percent below the sum of the maximum subdiscipline [229] Finally, the Board's priority assignments again raised and lowered the levels of the individual programs to accomplish a unity and balance with prospects for greatly multiplied combined benefits.

FLIGHT SCHEDULES

On the following pages we present the schedules of space astronomy missions which implicitly contain the best judgment of the Astronomy Missions Board concerning the present optimum balance of effort between the various subdiscipline programs, as those programs were described in part II. *These schedules cannot be adequately interpreted without recourse to the subdiscipline reports where the observational programs and mission objectives are developed.* We present here a brief description of the structure of the schedules and a glossary of terms and abbreviations which will assist the reader in referring back to a particular subdiscipline report for further information. The minimum balanced program flight schedules are shown in table 1 (Astronomy Missions) and table 2 (Space Physics and Interplanetary Missions). The optimum programs are shown in tables 3 and 4. The general plan of the Astronomy Tables 1 and 3 shows increasingly expensive types of missions arranged vertically in successive blocks. Each block contains spacecraft which are approximately a factor of 2 more expensive than the preceding block. Within each block one row

is given to each of the subdisciplines, with one exception (X—X and g—ray, O—Optical UV and IR, R—Radio, S—Solar, P—Planetary).
[230]

TABLE 1.—*Minimum Balanced Program—Astronomy*

Fiscal year	1971	72	73	74	75	76	77	78	79	80	81	82	83	84	85
IR airplane	Continuous—Fly best available platforms—$3M/yr effort														
Sounding rocket flights:															
X	15	16	17	18	22	24	26	28	29	31	33	35	37	39	40
O	12	12	12	12	12	12	12	12	12	12	12	12	12	12	12
R	1		1		1		1		1		1		1		1
S	8	8	8	8	8	9	9	9	9	9	10	10	10	10	10
P	3	3	3	3	3	3	3	3	3	3	3	3	3	3	3
Explorers:															
SAS X	1	1	1	2	2	2	2	2	2	2	2	2	2	2	2
SAS O		1			1			1			1			1	
RAE R			1	2					1	1	2				
S															
SAS P			1		1		1		1		1		1		1
"Heavy explorer":															
X				A					B					C	
O															
R															
S		OSO–I		OSO–J	OSO–K		OSO–L	OSO–M		OSO–N	OSO–O		OSO–P	OSO–Q	
P															
10-km radio 5-arcsec solar:															
X				OAO–E[a]		OXO–A						OXO–B			
O		OAO–D		OAO–E			OAO–F			OAO–G					
R							ORO–A							ORO–B	
S		ATM–A			5"–B			5"–C			5"–D	([b])		5"–E	([b])
P		OAO–D[a]													
ASTRA 1-arcsec solar															
X															
O								ASTRA–A					ASTRA–B		
R															
S															
P															
5–10-meter LST 0.1-arcsec solar															
X															LST→
O															
R															0.1"→
S															
P															

[a] Partial use.
[b] Evolves to effect 1" observations in 1980's.

TABLE 2.—*Minimum Program for Fields and Particle Astronomy (Astronomy Portion)*

	1971	1972	1973	1974	1975	1976	1977	1978	1979	1980	1981	1982	1983	1984	1985	1986	1987
Explorers: Interplanetary	[H]	[J]	K	L, M	N, O	P, Q	R, S	T, U	V, W	X, Y							
Scout (Sun Blazer)	[B, C]	[D, E]	F, G	H, I													
Pioneers (Jupiter/galactic probe)		F	G	H ↓	I ↓												
Outer Planets Explorer * (Jupiter/galactic probe or "multiplanet tours)":				or A	or B or B		C	D	E	F (End B?)			(End A?)				(End D?)
(a) In ecliptic plane																	
(b) Out of ecliptic plane																	
(c) Mariner-Mars (or Viking)	[A, B]		C, D														
Solar Probes:																	
(a) Venus/Mercury Flyby			A		B												
(b) German/U.S. solar probe				A													
(c) "New" solar probe to 0.2 AU						A	B	C	D	E							
Large automated experiments (or Apollo Applications with unmanned instrument standards)			A	B, C	D		E	F	G	H							
Planetary Explorers:																	
(a) Venus			A		B												
(b) Mars				6	C		D										
Rockets	6	6	6	15	6	6	6	6	6	6							
Balloons, large	20	18	15		15	15	15	15	15	15							
SR&T																	

* This program based on flyby-type missions without capsule landers.
☐ Letters in boxes indicate approved missions as of March 1969.

****

[234]

X-Ray and γ-Ray Missions

Balloons.—Continued pointed flights—γ-ray continuum studies—reach 10^{-3} CM^{-2} sec^{-1} level at 10^5 eV for selected X-ray discrete sources.

Rockets.—Broad participation–quick turnaround.

Explorers.—SAS-A: sky survey sensitivity 10^{-4} Sco X-1; 1-8 keV, 0.5° resolution, broad-band spectral resolution; SAS-B: high-energy γ-ray survey 10^{-6} CM^{-2} sec^{-1}; SAS-C and beyond: extend energy response 200 eV to 1 MeV, larger spark chamber devices, improve pointing to study sources, time variations.

OSO.—Continued use of wheel section for surveys, monitoring.

OAO or equivalent.—First stellar X-ray imaging telescope positions to ~1 arcsec, $\Delta\lambda/\lambda$ to ~ 0.01 for sources to 10^{-4} Sco X-1. Instrumentation state of art.

OWSE.—High-sensitivity X-ray and nuclear – g-ray surveys, nuclear lines and continuum studies; large Cerenkov telescopes, 10^{-8} CM^{-2} sec^{-1} above 500 MeV, crude energy resolution.

Heavy Explorers.—A: high-sensitivity X-ray survey 10^{-6} Sco X-1, 0.1° resolution, nuclear γ-ray survey chamber 10^{-7} cm^{-2} sec^{-1}, Cerenkov telescope; B: extend sensitivity, broad energy resolution, increase angular resolution–study continuum γ-rays from known extragalactic X-ray sources.

OXO.—Stellar X-ray imaging telescopes, design goal ~1-m aperture, interchangeable instruments at focus to accommodate image detectors, polarimeters, spectrometers.

5-10-m X-ray telescope.—A permanent National Space Observatory.

Optical Highlights

Structure and processes in the outer atmospheres of stars—especially extremes such as very hot, very cold, and very unstable objects—hold clues to the history and fate of stars. Stellar spectrophotometry in the UV can give information on such stellar chromosphere and coronas, adding to our knowledge of similar solar activity which controls solar-interplanetary-terrestrial relationships.

Absorption line measurements in the UV are three orders of magnitude [235] more sensitive than visible lines in detecting the interstellar gas, helping to determine chemical composition, physical state, and energy balance of the interstellar medium.

Continuous spectrophotometric measures of planetary atmospheres, comets, and the interplanetary medium will help us to understand the origin and present nature of our planetary system.

Spectrophotometric UV and IR observations of gas ejection from galactic nuclei, together with high-resolution images of these objects, will help unravel these explosive events whose extreme dynamic conditions play a fascinating role in the evolution of stellar systems and which may lead to new knowledge of fundamental physics.

Optical Missions

Airplanes and balloons.—IR telescopes, from 36 inches in minimum program to a possible 120 inches in maximum; balance flights between large platforms such as Convair 990 and smaller single experiment flights such as Lear Jets. High-resolution visible telescopes (Stratoscope).

Rockets.—UV spectrographs, 1-X resolution, for studies of stellar atmospheres and interstellar absorption lines. Possible standby for a bright comet.

Far IR broad-band scan of sky for emission from interstellar dust grains and sources with peak intensities at wavelengths greater than 20 m.

SAS.—Broadband UV photometer and polarimeter, selective extinction and polarization by grains, variable stars and galaxies, solar-system objects.

UV sky survey.—Interstellar gas emission at several wavelengths.

IR telescope.—Probably refrigerated-10-100 m for the study of planets, stars, gas, grains.

OAO-B (GSFC).—UV scanning spectrometer 2-X resolution, stars and nebulae; OAO-C (Princeton) : UV scanning spectrometer 0.1- and 0.4-Å widths, interstellar absorption in stars to 6th magnitude, stellar spectra; OAO-D (national facility) : UV scanning spectrometer with offset guidance, 0.3-0.5 Å width, spectra to 8th magnitude, 40-Å resolution to 13th magnitude; OAO-E (National) : wideband UV spectrophotometer and polarimeter, offset guidance, galaxies, variable stars, interstellar grains; OAO-F (National) UV echelle spectrometer with integrating TV tube, width 0.1 Å to 9th magnitude, 100 Å to 18th magnitude; OAO-G (National) : IR interferometric spectrometer plus broadband IR photometer, lightweight cryogenics system, later flights include improved versions of E, F, and G.

Astra.—A: UV echelle spectrometer, high-resolution imagery with filters to magnitude 26 in visible; B: more flexible instrumentation; C: include IR capability.

Large space telescope.—Aperture 120 inches or more with resolution corresponding to 120 inches, indefinite life. See LST report of the National Academy of Sciences.

Infrared Highlights

Infrared detector technology and infrared astronomical discovery are together undergoing revolutionary developments in which there are increasing advantages in making observations outside the atmosphere. Infrared space observations will permit—

(1) Observations of extended faint objects.

(2) Observation in the five octaves of spectrum from 25-700 m.

[237] (3) Broadband observations of extremely faint objects with detectors limited only by celestial radiation.

Fields of study already known to be able to profit from these capabilities are—

(1) The stellar and dust structure of our galaxy.

(2) High-energy processes that occur in some galactic nuclei and quasars.

(3) The role of dust envelopes in the evolution of stars.

(4) The role of dust envelopes in the formation of, planetary systems around young stars.

(5) The thermal mission and heat balance of planets and the Moon.

(6) Infrared background radiation. High-temperature phenomena at remote epochs will have their radiation shifted into the infrared. Thus these studies have potential cosmological significance.

Infrared Missions

Because of the rapid development possible in detector technology, and the high rate of discovery of new classes of astronomical infrared phenomena from the ground, a major effort must be made in these areas to insure that full possible benefit is obtained from the space observations.

Significant developments are expected to occur from the use of small high-altitude aircraft, small balloon-borne equipment, and some rocket flights.

Major installations appropriate for the next few years include a 36-inch telescope to be used in a stratospheric airplane primarily for point source observations, and a Small Astronomy Satellite to be used for studies of extended objects and for surveys for new objects. Eventually, technological advances and astronomical discoveries may slow down, and it will be necessary to use larger platforms such as an OAO or a Large Space Telescope. Should detector improvement be difficult, this phase would come earlier.

Radio Highlights

Measure the flux densities of 50 to 100 extragalactic and galactic sources at a number of frequencies around 1 MHz.

Map the cosmic background noise level of the full sky at a number of frequencies from 0.5 MHz to 10 MHz.

Measure dynamic radio astronomical phenomena and, in particular, record variations of radio emission from the Sun, Jupiter, and other variable radio sources.

Measure the brightness distribution across a few individual radio sources which are occulted by the Moon.

Obtain data on the statistical parameters of cosmic background noise fluctuation at a few frequencies near 1 MHz.

Study variable interplanetary absorption and interplanetary scintillation effects.

Solar-system observations will concentrate on understanding physical processes in the solar corona and in the magnetospheres of the planets, especially Jupiter and the Earth. The region in the corona from 1 to 50 solar radii is particularly difficult to reach by optical observations or space probes.

Radio Missions

Rockets.—High apogee (1000 km) experiments; e.g., absolute calibration, 1-5, 5-10 MHz, 100-150 lb.

[237] *Explorers.*—RAE-C ionospheric focusing and magnetospheric noise; RAE D&E: two or more element interferometer–supersynthesis test, location experiments; RAE-F: cosmic radio noise background; RAE-G: solar-system radio monitor.

Orbiting radio observatory.—Ten-km filled-aperture antenna, circular polar synchronous orbit–2000-4000 lb.

Solar Highlights

Improved angular resolution XUV spectra and spectroheliograms may lead to understanding the mechanism of nonthermal energy production-plasma and magnetic-field interactions; steep density, temperature gradients; shock and magnetohydrodynamic waves; particle transmission and ejection (perhaps the cosmic accelerator); a flow of energy that controls the state of interplanetary space and planetary ionospheres.

Absolute photometry of XUV resonance lines of atoms and ions will lead to improved abundance determinations and perhaps settle the question of different abundances at different levels in the solar atmosphere.

Absolute photometry of the UV continuum will provide direct observation of the temperature inversion.

White-light coronagraphs may reveal outward-moving disturbances from flares and other active regions.

Visible spectrum observations with very high angular resolution exceeding that possible from below the atmosphere will reveal details of sunspots, flares, prominences, plages, spicules, and the fine network structure.

Solar Missions

OSO-1: K coronagraph; OSO-J: spectrograph-absolute photometry 300-3000 Å; OSO-L: spectrograph absolute photometry _ < 30 Å; OSO-M: scanning spectroheliograph _ < 300 Å; OSO-N: K coronagraph; OSO-O: scanning spectroheliograph 300-1300 Å; OSO-P: spectrograph line profiles 300-1600 Å; OSO-Q: spectrograph line profiles _ < 300 Å.

5-Arcsec Spacecraft-ATM-A:

1. Scanning spectrometer and spectroheliometer (300-1300 Å).
2. Slitless spectroheliograph, photographic (300-650 Å).
3. Spectrograph, photographic, high X/AX (900-4000 Å).
4. Small-field, large-scale X-ray telescope, and slitless spectroheliograph, photographic (2-60 Å).
5. Large-field X-ray telescope with filters, photographic (2-60 Å).
6. White-light coronagraph, photographic.

5" Spacecraft No. 2:

1. High resolution (_ ~ 0.2") internally pointed telescope, with filters and spectrograph (1100-30 000 A).
2. Spectrometer for line profiles and spectrum mapping (300-1000 Å). (This is probably a full load.)

5" Spacecraft No. 3:

1, 2. Spectrometers for X-ray spectrum mapping and absolute photometry (2-300 Å). (Probably at least two instruments.)

3, 4. X-ray line profile spectrometers to measure at least the strongest lines (2-300 Å).

5. X-ray imaging instrument (pinhole? Fourier shadowgraph? something else?) (A < 3 Å).

[238] **Planetary Highlights**

Small pointed satellites will allow spatial scans of the planets in different spectral regions, which will allow us to deduce the vertical structure of hazy atmospheres (e.g., Venus, Jupiter).

Ultraviolet photometry, as well as infrared observations from high-altitude aircraft, will provide critical knowledge of the planetary albedos, necessary to an understanding of the planetary heat budgets.

A cloud model is fundamental to interpretation of spectroscopic measurements, which in turn can yield compositions, temperatures, and pressures.

Measurements of the planetary hydrogen corona in Lyman-a radiation with high angular resolution (< 1 arcsec) can yield the escape temperature, an essential quantity to studies of evolution of the atmosphere and ionospheric structure.

Essential features of a planet's meteorology could be obtained by long-period monitoring of the atmospheric fine structure with large-aperture instruments in Earth orbit.

Planetary Missions

Rockets.—Survey spectrophotometry at several-Å resolution, 1800-3300 Å. Photometry in far-UV resonance lines, especially Lyman-a (1215 Å). Possibly high-resolution scans of narrow spectral regions of special interest.

Explorers.—Extension of sounding rocket objectives but to fainter objects and improved spectral resolution, and with the important addition of spatial resolution over disk.

OA O-A2 (WE P).—Broadband photometry, Lyman-a photometry; OAO-B: spectrophotometry with 10 Å resolution; OAO-C: resolution of 0.1 and 0.4 Å over narrow spectral regions of special interest, possibly with spatial scans.

Particles and Fields Highlights

Observations within the solar system of magnetic fields and particles with energies from 0.5 keV to many GeV yield information on such diverse astrophysical problems as the 3° blackbody radiation, supernovae, the interstellar medium, the dynamical behavior of the galactic disk, nucleosynthesis and the origin of the elements, and stellar abundances—information not available through any other kind of observations or experiments.

Cosmic-ray particle studies are related to radio, infrared, ultraviolet, X-ray, and g-ray astronomy measurements across the spectrum from the microwave background to megavolt photons to form an overall picture which would be inaccessible from optical and radio studies alone. Through couplings between high- and low-energy processes such as the inverse Compton effect, all of the data related to a given object are related to each other and eventually all inputs are needed for a full understanding of the environment.

High-energy astrophysics not only adds new windows to the cosmic electromagnetic spectrum by providing X-ray, g-ray, and particle and field astronomy, but also represents the first scientific unification of those disciplines and all other experimental studies of cosmic processes.

NEW DIRECTIONS FOR THE SPACE ASTRONOMY PROGRAM

Comparisons with the current NASA space astronomy program reveal some of the new directions which will be required to implement [239] the AMB plan. Perhaps the most significant change is an increased effort in X-ray and gamma-ray astronomy. Less than 10 percent of the current NASA effort, X- and γ-ray astronomy amounts to about a quarter of the AMB program, which assigns approximately equal levels of effort to optical, solar, and high-energy astronomy. The increase needed in the minimum balanced program is a major start in fiscal year 1971 on a new spacecraft with the pointing, telemetry, and general sophistication of an Explorer-class spacecraft, but with a payload size capable of carrying large area X-ray detectors, spark chambers, and Cerenkov telescopes, as well as particle and field experiments in the 1- to 5-ton range. Also included is adaptation of a future OAO spacecraft or an equivalent vehicle to carry a state-of-the-art stellar X-ray imaging instrument comparable to existing solar instrumentation. Later, stellar imaging X-ray telescopes of about 1-m aperture, 10-m focal length will be required.

The optical ultraviolet astronomy program has a mid-1970's goal of a 1- to 1.5-m telescope with diffraction-limited performance, as an essential intermediate scientific and technological step toward the Large Space Telescope of the 1980's. This could be achieved either through a new spacecraft design or by upgrading an evolutionary OAO program.

The infrared astronomy program has a most pressing need for research and development of detectors and small cooling systems which will permit infrared observations with the much greater efficiency that is commonplace at both shorter and longer wavelengths. Such advances could continue the present high rate of discovery of new classes of astrophysical phenomena from the ground and from airplane observatories.

Observations of astrophysical objects in the longwave radio portion of the spectrum with the minimum angular resolution required to distinguish sources may require an antenna made of wires surrounding an area 10 km in diameter. However, a remote possibility of making similar observations by "supersynthesis" interferometric techniques must be studied before this large electronically filled aperture is initiated.

The continuing need for observations of the solar surface with an effective angular resolution of 5 arcsec will require the development of a ground-controlled solar spacecraft with the instrumental sophistication of the ATM-A.

Observations of the planets from Earth orbit will be accomplished with the instruments of the planned OAO's and a Small Astronomical Satellite optimized for planetary observations.

[240] The acquisition of data on cosmic-ray particles and fields in the interplanetary medium requires a careful programing [sic] of small fractions of the missions to the planets, and the joint use of the "heavy Explorer" spacecraft for high-energy astronomy.

An important element in the balanced acquisition of essential astrophysical data in the AMB plan is the continuing requirement for the smaller space experiments: the aircraft, balloons, rockets, and small Explorer-class satellites. Though less dramatic and unimposing by their nature, they have a great potential for economic and timely measurements of important data that can complement the other space-based and ground-based multiwavelength observations.

An essential part of the AMB exercise to project the level of space astronomical research as far as possible into the future was an assessment of the availability and enthusiastic interest of excellent people-scientists and supporting specialists, including several engineering and technical groups skilled in the measurement of astronomical radiation. Continuity, breadth, and active competition for flight opportunities must be maintained by a strong NASA program in Supporting Research and Technology (SR&T).

Both SR&T and NASA's Advanced Research and Technology program must press forward to develop essential instrumentation such as lightweight optical mirrors, improved X-ray reflectors and detectors, X-ray photometric standards, electronic imaging systems, improved grating technology, infrared sensors and small cryogenic systems, devices which will be useful in ground based observatories of the future as well as space experiments. Support is also essential for the experimental and theoretical research in related areas of atomic and nuclear physics that will insure progress in analyzing the new observations resulting from these technological advances.

In a properly integrated program of federally supported astronomy, NASA should have a responsibility to support particular ground-based instruments, especially those which are most closely and directly related to NASA's mission. Specific instruments, which are of comparable expense to some spacecraft and might be defended as separate line items in the NASA budget, should include special-purpose monitoring telescopes of intermediate (60- to 100-inch) aperture, large optical telescopes in both hemispheres, and a large steerable paraboloid radio telescope.

Document III-16

Document title: NASA Experiment Proposal for Manned Space Flight, Lunar-Surface Ultraviolet Camera, April 1970.

Source: George Carruthers, personal collection, reprinted with permission.

After initial test flights to the Moon, NASA attempted to maximize the scientific returns of the remainder of the Apollo lunar landing program, when possible inviting scientists to propose experiments to be conducted in the lunar environs. The Naval Research Laboratory (NRL) was successful in receiving NASA approval to fly an astronomy mission to the Moon's surface on the Apollo 16 mission, launched in April 1972. This is an excerpt from the NRL's proposal: a far-ultraviolet camera and spectrograph to be placed on the lunar surface. Under Principal Investigator George Carruthers, an NRL astronomer, the instrument recorded the spectra of many hot stars.

[cover sheet]

NATIONAL AERONAUTICS AND SPACE ADMINISTRATION
EXPERIMENT PROPOSAL
FOR
MANNED SPACE FLIGHT

TITLE LUNAR-SURFACE ULTRAVIOLET CAMERA

EXPERIMENT NUMBER

(A combination of MSC Proposal "Lyman-alpha Ultraviolet Camera," 27 Aug 1969 with NRL Proposal P-11-69 "Far-ultraviolet spectroscopy of Diffuse Background Radiation, Diffuse Nebulae, and Stars, from the Lunar Surface" by George R. Carruthers and Herbert Friedman, dated 14 October 1969.)

PRINCIPAL INVESTIGATOR

(Signature) Thornton L. Page (Date)

PRINCIPAL ADMINISTRATOR

(Signature) Anthony J. Calio (Date)

****

[9] SECTION II. TECHNICAL INFORMATION

1. Objectives:

From the lunar surface, outside the Geocorona, to obtain:

a. Imagery in the band 1050 - 1230A (mostly Lyman-alpha 1216A) by differencing unfiltered (1050-1600A) and filtered (1230-1600A) photographs, all exposures accurately timed, of:
   (1) The Geocorona and Earth's Atmosphere, half sunlit
   (2) Possible clouds of Solar Wind and sky background near the band of the Zodiac at 45° and 135° from the sun, and changes after 5 or 10 hr.
   (3) Milky-Way star clouds, preferably in Sagittarius
   (4) Two or three nearby Galaxies, preferably M31, M32, and M33
   (5) One or two Clusters of Galaxies, preferably the Coma Cluster

b. Spectra in the range 300 to 1600A of:
   (1) Quasars and Seyfert-type galaxies
   (2) Small, bright galactic nebulae and stars
   (3) Interstellar extended HII regions
   (4) Interplanetary gas, zodiacal light, and solar wind near ecliptic
   (5) Background sky light in Coma Cluster and near pole of ecliptic

c. A valid operational test of optical instruments in the lunar environment.

2. Significance

The electronographic Schmidt camera will detect extended surface brightness as small as 0.3 Rayleighs (0.004 erg/sec.m^2 sterad at Lyman alpha). Although absorption in the line of sight will complicate imagery and spectra of distant sources, the Lyman-alpha resonance line of hydrogen is an extremely sensitive detector of hydrogen, the most abundant element in the universe. Lyman-alpha emission or absorption indicates the presence of hydrogen gas clouds, and the ratio of its intensity to that of Balmer-alpha 6563A, and of the 21-cm emission line, is related to the density and electron temperature. In normal stellar spectra, there will probably be Lyman-alpha absorption, partly due to cold interstellar hydrogen along the line of sight. Because of the Geocorona Lyman-alpha emission, observations from OAO and from other law Earth-orbiters are limited to relatively bright and compact objects.

It is possible that the Solar-Wind clouds emit Lyman-alpha, and that other interplanetary hydrogen will be detected near the plane of the ecliptic. Away from the ecliptic, interstellar hydrogen should be detected where it is excited by electron collisions or UV starlight. The spatial distribution of Lyman-alpha emission in the Milky-Way is thus related to the locations of interstellar gas and hot blue stars (Population I).

[10] It is possible that very hot hydrogen exists in intergalactic space, and may be detected between the galaxies in clusters. The sensitivity and wide angular field of the electronographic Schmidt camera will show concentrations of intergalactic hydrogen if they exist in clusters, and may thus explain the "Mass discrepancy" in clusters of galaxies. Note that the Coma-Cluster redshift avoids the absorption of its Lyman-alpha in nearby interstellar clouds.

Spectra will probably include absorption or emission lines at 1165A (CIII), 1206A (SiIII), 123-43A (NV), 1302-06A (OI), 1400A (SiIV) 1550A (CIV), 1610A (FeII), 1134A

(NI), 1200A (NI), and possibly lines of CI, CII, and as well as Lyman-alpha 1216A, and can eventually be used to determine chemical abundances in stellar atmospheres, nebulae, the interstellar medium, and the solar wind. Quantitative measurements in this wavelength range will therefore have significance in studies of all these objects, as well as Seyfert-type galaxies and quasars. A slat collimator provides spectra of background light with 30A resolution; with the LiF corrector plate removed from the camera, these should show Lyman-beta (1026A), HeI (584A) and HeII (340A). The latter two, and Lyman-alpha are strongly absorbed by the interstellar medium, but should appear in spectra of the Solar Wind.

The operation of a camera-spectrograph on the lunar surface is the first step in tests of how effective an astronomical instrument may be on the lunar surface. This manned operation for 20 or 30 hours should be followed by use of a larger, remote-controlled telescope for geophysical and astrophysical purposes for a year or more. If the geo-astrolunar telescope (GALT) proves effective, a larger telescope may be appropriate at a lunar base.

3. <u>Disciplinary Relationship</u>

The Geocorona and Lyman-alpha background are related to a decade of research on the Earth's magnetosphere and the Solar Wind, most of it done under NASA auspices. The most recent data were obtained by WEP and Celescope on OAO-A2.

The Lyman-alpha emission from nearby galaxies is related to WEP measures from OAO-A2 of far-UV spectra described by Code (1969), and to 21-cm radio measures such as those reported by M. S. Roberts (1967). In connection with the mass discrepancy in clusters of galaxies, there has been speculation by several astrophysicists such as Rood (1969) about intergalactic hydrogen. In each of these areas, there is indirect evidence, but no direct confirmation of the amount and excitation of hydrogen.

(It should be noted that Lyman-alpha imagery will provide dramatic photographs of the Earth's environment, and of distant regions of the universe. These are bound to stimulate public interest in the geophysics and astrophysics that can be done from the lunar surface.)

Two measurements of Lyman-alpha sky background have been made at large distances from the Earth; one by Barth (196) from MARINER V, and one by Kurt and Syunyaev (196) from the Soviet VENURA space probe. Both showed a concentration toward the galactic plane, which indicates that nearby interstellar [11] clouds are not completely black in Lyman-alpha. In fact, Barth found particularly strong emission from the region of the Gum Nebula (RA 8hr. 2 min., dec.-39°55'). Far-UV stellar spectra obtained by Bless and Code (196) from OAO-A2, and by Carruthers (196) from Aerobee Rocket flights, seem to show ten times less Lyman-alpha than predicted from 21-cm radio measurements. Kurt and Syunyaev consider that the Solar-Wind protons interact with the magnetic field of the Galaxy to produce a ring of recombining hydrogen in the plane of the Galaxy far from the sun.

The observations proposed here will help sort out Lyman-alpha emission from these several sources, and will determine the changing ionization of the Solar Wind at different distances from the sun.

4. <u>Experiment Approach</u> (See Summary, page 1.) [not included]

A small electronographic Schmidt camera, shown in Figures 1 and 2 [not included], has been designed and built at the Naval Research Laboratory, and similar units have been flown by Carruthers on three Aerobee Rocket flights. Light is focused by the Schmidt camera with LiF correcting plate onto a curved KBr photocathode shaped to fit the focal surface. This photocathode, and its opaque backing, are kept at about 20 kV negative with respect to the film cassette, mirror, and rest of the instrument (which is at ground potential, and completely shields the high voltage from the astronaut observer). A lontirudinal magnetic field is provided by bar magnets in the cylinder surrounding the camera, so that the photoelectrons are accurately focused on the film in the cassette (NTB-3 nuclear-track emulsion). This results in a detection efficiency about 20 times higher than the best UV-sensitive photographic emulsions (such as Eastman SC-5).

The field of the camera is a 20°-diameter circle; a 27-mm circle on the film at scale 45 arc-min/mm. For direct photography, resolution will be 2 arc-min or better, and the density of the developed film is accurately proportional to the integrated photon flux over a wide range of densities.

A sunshade is desirable, even though the camera is to be deployed in the shade of the Apollo LM, to reduce scattered light from the nearby sunlit surface of the moon. A CaF_2 filter (or separate corrector plate) is provided to cut out the wavelength band 1050-1230A. The KBr photocathode is sensitive to the band 1050-1600A, and two photographs must be taken to obtain a Lyman-alpha image.

Spectra are to be obtained by placing a plane (reflecting) grating in front of the camera as shown in Figure 2. Using a 1200-line/mm replica blazed for 1300A first order, a field of stars or other small sources will produce spectra displaced about 10° from the zero-order (specularly reflected) positions on the focal surface. The film will then record spectra with fairly high efficiency over most of the 20° circular field, centered exactly 90° off the camera axis by proper adjustment of the 3 x 4.3-inch grating. The dispersion is about 75 A/mm and resolution in stellar spectra about 2A. For extended nebulae and sky background, a "venetian-blind" (slat) collimator is [12] necessary to provide spectral resolution. Slats about 40 mm wide at 0.2 mm spacing along the grating dispersion limit the field to 20°x 0°.25 and provide spectral resolution of 30A. The camera must be pointed so that this narrow field crosses the target nebula (usually 2° or more in extent), and stars that happen to be in the field are easily distinguished from the nebular spectrum by the limited width of their spectra. The 30-A resolution is adequate for nebulae and background since their spectra are expected to consist of widely spaced emission lines. The lower spectral resolution, set by the slat collimator, makes it unnecessary to have a corrector plate. If this is removed, spectra will extend down to 300A in the extreme UV.

We therefore propose to modify the present NRL design to allow 3 choices for corrector plate (CaF_2 or LiF, or none) and 3 choices for the front end (straight-ahead view through a sunshade, 90° reflection off the grating without collimator, and 90° reflection off the grating with slat collimator). The first selection can best be made by rotating a 6.5-inch wheel with two corrector plates and an empty 3-inch hole in it. The second selection can be made by sliding the sunshade-grating assembly fully up or fully down, and by folding the slat collimator in or out, as shown in Figure 4.

Because the astronauts will have difficulty seeing stars through their space suit visors, it is necessary to point the camera by means of setting circles, one in declination (N-S) and one in hour angle (E-W). For objectives a(1-5) and most of b(1-2), pointing accuracy of 5° to 10° is adequate, since the camera field is 20°. For the background-light spectra, objectives b(4-5), the slat collimator requires pointing accuracy of 3° to 5°. A few of the targets for objectives b(1-3) are of about 1° extent in the sky, and the E-W strip field about 0°.5 wide must be pointed within 0°.5 in declination to get their spectra. Moreover, the faint objects require long exposures (up to 4 hours) during which the camera must be turned ("driven") westward at 0°.5 per hour. Hence a drive motor is required on the polar axis shown in Figure 3, and the axis must be parallel to the Moon's axis within 7°. The Moon's axis is inclined to the horizontal by an angle equal to the latitude of the landing site, and the polar axis is so inclined to the "table" in Figure 3. This "table" is to be supported by a tripod that can be levelled using screws on the feet, which may be 10-inch threaded spikes tapped firmly into the soil.

After levelling the table, the camera mounting must be rotated until the polar axis points north. This can most easily be accomplished relative to the sun line at the landing site. Because the camera will be set up in the shade of the LM, the second astronaut can walk down-sun out of the IM shadow about 100 feet, where he holds a staff or other tool at arm's length toward the sun with its shadow on his space suit. The first astronaut sets the camera circles at two predetermined settings (both about 90°) and adjusts the mounting so that the sighting bar on top of the camera points at the second astronaut's staff when the shadow is directly behind the staff. He then tightens the two lever-screws. fastening the mounting to the table with its polar axis within 1° or 2° of the correct direction north. A check on [13] the circle readings is then desirable. They can be set to predetermined settings (about 0° on each) to point the camera toward the Earth. A small mirror at the lower end of the almost vertical sighting bar will, show whether the camera is centered on the 2° Earth. If necessary, the settings should be adjusted to center the earth, and the differences in circle settings should be added or subtracted from all later settings in the program of camera targets.

After turning on the power supply (from a 28-volt 2-amp-hour battery pack in the camera mount, the astronaut selects the proper corrector plate and front end by gently tapping lever switches to operate solenoids. He sets the circle settings to predetermined settings and initiates a series of three to six exposures on each target. Care must be taken not to jar the camera tripod. Each exposure sequence starts with moving a new frame of film into position, exposing it 15 seconds, then moving another frame in for a 1-minute exposure, then 4-, 16-, 64-, and 256-minute exposures. After the 4-min. exposure is completed, a small blue light is switched on, after the 16-minute exposure a green light, after the 64-minute one a white light, and after the 256-minute exposure no film transport takes place, and a red light is switched on. Depending on the expected faintness of the target, 3, 4, 5, or 6 exposures will be taken. Then a new corrector plate will be selected by the astronaut, possibly a new front end, and new circle settings. Then he initiates a new exposure sequence (which overrides any exposure in process).

The three or more exposures ensure the maximum amount of information from each target, since each part of the image or spectrum will reach a density on one of the exposures which is on the linear part of the characteristic curve of photographic film. In general the imagery exposures can be short (0.25, 1, 4, and 16 minutes), while some of the spectroscopic targets will require long exposures. The program of targets will be planned to fit in with other astronaut activities; it is desirable that the camera be deployed, adjusted, and started exposing early in the first EPA. A long exposure should be started at the end of the first EPA since it can continue for 5 hours while the astronauts are resting.

Before boarding the LM for the ascent stage, the astronaut must remove the film cassette from the camera-spectrograph and carry it aboard for careful stowage (probably in a sealed can) during the return to MSC Houston.

The only difficulty anticipated in the procedure outlined above is the astronaut's accurate sighting along the sighting bar attached to the camera. This bar will be designed and tested with astronaut help at MSC. If the desired accuracy cannot be achieved, one or two of the nebular spectra must be omitted.

Of course, the astronaut is essential to accomplishing this experiment–in deploying the camera-spectrograph on its tripod, in pointing it with the aid of the setting-circles, in starting the exposure sequence, and in recovering the film for return to MSC Houston.

[14]

[15] 5. <u>Baseline or Control Data</u>:

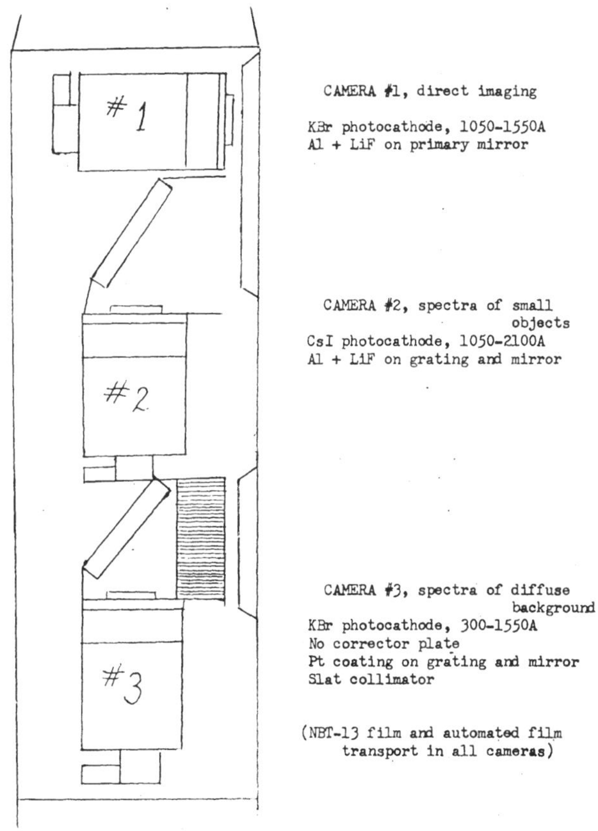

Figure 5. Cameras for Aerobee-150 Trial Flight

It would be desirable, though not required, to have simultaneous observations of the Geocorona (inside view) and Solar Wind made from OAO or OGO. The best landing site will be near the lunar equator at the time of first-quarter moon. The best time of year for observing the Coma Cluster and pole of the Galaxy is between April and July. In order to get Solar Wind near time of maximum solar activity, it is desirable to fly this experiment soon (1971).

It would be highly desirable to fly three of the cameras with different corrector plates and front ends (as shown in Figure 5) on an Aerobee 150 rocket before the Apollo flight. The purpose of this test flight is not to test instrumental design (which will be fixed by astronaut handling of mockups at MSC) but to check exposure times on various targets, and to learn which classes of targets are most interesting in their far UV spectra. A good test field would be the Constelletion [sic] Orion which has a wide variety of early-type stars, a compact emission nebula (the Orion Nebula), and two extended emission nebulae (λ Orionis and Hothead Nebula region). The Geocorona Lyman-alpha emission would provide widespread foreground. The NRL can arrange this flight if it is approved, by early 1971 when Orion is visible from the dark site of the Earth. Of course, the data obtained from this first flight would be of scientific value in themselves.

Document III-17

Document title: John E. Naugle, Associate Administrator for Space Science and Applications to NASA Administrator, Memorandum and Post Launch Mission Operation Report, May 17, 1971.

Source: NASA Historical Reference Collection, History Office, NASA Headquarters, Washington, D.C.

Seven years after the team from American Science and Engineering proposed to NASA a long-range plan for the study of x-ray astronomy from space (Document III-8), NASA launched the first satellite dedicated to x-ray astronomy. One of the space agency's many scientific Explorer satellites and the first of its Small Astronomy Satellites, the spacecraft was renamed Uhuru, the Swahili word for "freedom," because it was launched off the coast of Kenya on that country's Independence Day. Uhuru was a pioneering force in x-ray astronomy, helping astronomers to identify and catalog more than 160 x-ray sources within and beyond our galaxy.

[no page number]

Post Launch
Mission Operation Report
No. S-878-70-01
17 May 1971

MEMORANDUM
To: A/Administrator
From: S/Associate Administrator for
Space Science and Applications
Subject: Explorer 42 (Small Astronomy Satellite) Post Launch Report #2

Explorer 42 has been adjudged a success, based upon the results of the mission with respect to the approved prelaunch objectives.

As of 12 April 1971, 4 months after launch on 12 December 1970, the SAS-A has completed a full systematic scanning of the galactic plane and a substantially complete scan (95 percent) of the entire celestial sphere. The satellite remains operative with data being acquired in real time at a number of ground stations located around the earth on or near the equator. This planned backup real time mode of operation was initiated after failure of the tape recorder. It permits 60 percent acquisition of each full orbit of data to be obtained and allowed us to achieve the mission objectives.

Experiment instrumentation continues to function in an outstanding manner, exceeding many design objectives. The spacecraft control section performance, with the exception of the tape recorder and some decrease in telemetry modulation which has not affected data quality, is excellent.

To date, the acquired scientific data has touched on every aspect of observational X-ray astronomy. In addition to the expectation that many more X-ray sources would be discovered, significant unexpected phenomena have also been observed. In particular, the discovery of three new X-ray pulsars, identified as Cygnus X-1, Centaurus X-3, and Lupus X-1, has revealed a completely different class of pulsating X-ray source which differs in many respects from the previously known X-ray pulsar in the Crab nebula. The Centaurus source exhibits an even more startling characteristic in that its pulsation frequency appears to be extremely variable while all known pulsars, optical, radio, or X-ray exhibit no more than a slight variability while consistantly [sic] showing decreases in frequency over time.

Some of the initial results have already been submitted to the Astrophysical Journal (letters) and presented to the American Astronomical Society at the annual meeting in Baton Rouge, Louisiana, on 31 March 1971 by the Principal Investigator, Dr. Riccardo Giacconi of American Science and Engineering, Incorporated. The results with emphasis on the Centaurus pulsar has also been presented to the annual meeting of the American Physical Society in Washington, D.C. on 27-30 April 1971.

[no page number] The significant results already achieved have been derived solely from quick look data which is only a small fraction of the total amount of data actually acquired. It is expected that the analysis of the production tapes will develop and expand the catalog of known X-ray sources by many more sources than the 20 or so discovered to date. Based upon the results so far, it is probable that further surprises in X-ray source characteristics will appear in the data already acquired and throughout the continued lifetime of the satellite.

Dr. Giacconi, in a letter to the Director of Physics and Astronomy Programs discussing experiment performance and scientific results, stated his belief that Explorer 42 "has given us an unqualified scientific success and it has fulfilled and surpassed every expectation we had before launch."
[signature]
John E. Naugle

Document III-18

Document title: Memorandum to Director of Physics and Astronomy Programs, Office of Space Science, from Program Manager, Astronomical and Solar Observatories, Physics and Astronomy Programs, "Status of the Gravitational Physics Program," April 12, 1972.

Source: NASA Historical Reference Collection, History Office, NASA Headquarters, Washington, D.C.

A space-based test of Einstein's theory of relativity intrigued NASA from its inception. With access to the reduced gravity environment of space, scientists might finally be able to test the validity of Einstein's prediction that the passage of time would slow down in a lower gravitational field relative to the passage of time on Earth. NASA's Office of Space Science held discussions with astronomers and clock manufacturers just months after the space agency opened regarding the feasibility of flying a clock aboard a satellite in Earth orbit to test the theory. By the early 1970s three space experiments designed to test relativity seemed possible. This memo describes the three experiments, called Gravity Probes, with which NASA became involved. While Gravity Probe A flew on a sounding rocket in 1976, Gravity Probe C, proposed by the European Space Research Organization, was ultimately cancelled. Gravity Probe B, designed to fly on a satellite, is still in development by NASA.

[stamped] APR 12 1972

MEMORANDUM

TO: SG/Director of Physics and Astronomy Programs
Office of Space Science

FROM: SG/Program Manager, Astronomical and Solar Observatories, Physics and Astronomy Programs

SUBJECT: Status of the Gravitational Physics Program

NASA is actively engaged in three Gravitational Physics Projects:

1. GP-A, The Smithsonian Maser Clock Experiment
2. GP-B, The Stanford Gyroscope Relativity Experiment
3. GP-C, The ESRO Relativity Experiment

These three experiments are described in the attached summary document for Gravity Probes that was prepared for "Proposed FY 73 New Starts" and the individual descriptions of the GP-A, GP-B, [a]nd GP-C Projects.

The Gravitational Physics Program was reviewed by a Panel on 12 and 13 October 1971. The Smithsonian experiment was voted Category I unanimously and recommended for flight with the highest priority. It was the consensus of the Panel that both the Stanford and the ESRO experiments were excellent to meet their scientific objectives, and that NASA should continue to support both groups to more clearly define the technical feasibility of the mission.

The present status of the Projects is as follows:

GP-A: The experiment was selected for flight in late 1971 by Dr. Naugle. The flight project has been approved and is now under active development by the Marshall Space Flight Center (MSFC). The planned launch date is early in 1975.
[2] The only uncertainty at the present time, and this is minor, is whether the experiment instrument and the associated spacecraft subsystems can be constrained to about 150 pounds. This weight limitation is imposed by the ability of the Scout vehicle to fly a 10,000nm high trajectory.

I expect an affirmative answer to the weight question by the end of April.

There are no other problem areas of which I am aware. In fact, Smithsonian is very encouraged by a break through that permits the clock cavity to be cast rather than requiring lengthy machining. Further, the ERTS Project will loan MSFC an X-band transponder for initial tests, thus minimizing costs and compacting schedules.

GP-B: As you are aware, Stanford has been supported with SRT funds for 8 to 10 years. During this time, they have determined the feasibility of using a highly precise gyroscope (accuracy of 0.001 arc seconds) to determine relativistic effects in low earth orbit (500 nm attitude inclined 90°); they have developed a ground-based dewar; and worked extensively on the gyroscope and the spacecraft control system.

In July 1972, BBRC was awarded a Mission Definition Study contract of four months duration. The study, which is now being reviewed at MSFC, indicates the feasibility of the GP-8 mission, provided that the following items are proven: gyroscope, dewar operation in zero gravity, dewar-gyroscope operation in zero gravity, and optical contacting for assembling the telescope and gyroscopes.

The gyroscope is being developed at both Stanford (M-H contract) and MSFC. Recently, Stanford has demonstrated the levitation of the gyro rotor, and by June 1 expect that complete gyro operation in a 1-g field will be demonstrated at either or both Stanford and MSFC.

A Cryogenic Workshop was held at MSFC in March to determine the state of the art in cryogenic systems at liquid helium and liquid hydrogen temperatures, and to attempt to permit a coordinated attack on the requirements of the planned and potential uses of cryogenic systems. I expect that recommendations resulting from the Workshop will be available in the next one to two months. This should provide a mechanism for developing a dewar for the Stanford experiment.

[3] MSFC will take over the responsibility for the Stanford SRT effort in FY 1973. Present plans call for a $625K GP-B SRT budget at MSFC in FY 1973, of which $325 is earmarked for Stanford. The SRT support will continue until a flight project is approved in either FY 1974 or FY 1975. The earliest launch date would be the late 1970's .

GP-C: ESRO has been constrained in their Solar Probe Experiment to funding for studies. In the past two years, they have studied the mission feasibility, the laser clock, the X- and S-band transponders, and a number of related items. Although they would like to do laboratory work on the drag-free system required by the experiment, no funding has been available.

As you know, NASA is committed to aid ESRO in the Mission Definition. This commitment has resulted in reviews of the studies by MSFC, JPL, and Headquarters personnel, and participation in their Mission Definition Group meetings in Europe.

A year ago ESRO wanted NASA's cooperation for a Titan vehicle and in-orbit operations support. Now, in addition, I understand they plan to discuss with Dr. Naugle on 18 April 1972 the possibility of further cooperation (maser clock, X- and S-band transponders, and perhaps lasers).

The additional support is brought about by the level of their scientific budget ($25 to 30M per year) and the present estimate of the GP-C mission of $50 to 100M which they probably cannot fund alone. They are also seeking cooperation from Germany with the use of the HELIOS spacecraft system.

The future of this program is, in my mind, less firm than the GP-B mission, both from the technical and the funding standpoint. Technically, they will have to begin demonstrating the feasibility of systems, particularly the drag-free system, as we have done with the gyro experiment. In the funding area, they need help from outside ESRO.

ESRO has been shooting for a luanch [sic] in the late 1970s.

In summary, we are moving out with the GP-A mission. Both the GP-B and the GP-C missions are uncertain and do have technical areas that require feasibility demonstration.

[4] You should also keep in mind that MSFC, under Dr. Decher, is undertaking a comparative study to determine the relative merits of the various methods now being used or under consideration (ground-based efforts by MIT and JPL, GP-B, and GP-C) to determine the relativity parameters β [Beta] and γ [Gamma] and the solar quadruple moment J_2. This study will probably not be completed for one year. It will involve all or the majority of the groups working in this area.

C. Dixon Ashworth

Enclosure [not included]

Document III-19

Document title: Space Shuttle Payload Planning Working Groups, Goddard Space Flight Center, Final Report, Volume 1: Astronomy, May 1973.

Source: Nancy Grace Roman, personal collection, reprinted with permission.

Document III-20

Document title: Interim Report of the Astronomy Spacelab Payload Study, Astronomy Spacelab Payloads Project, NASA, July 1975.

Source: NASA Historical Reference Collection, History Office, NASA Headquarters, Washington, D.C.

With President Nixon's January 1972 decision to pursue the development of the Space Shuttle as NASA's next centerpiece project after Apollo came the need to define the purposes it would serve. Like their counterparts focusing on other disciplines, the working group examining astronomical uses of the Shuttle determined that the new vehicle could serve a number of functions. Such purposes ranged from flying astronomical instruments for the duration of a Shuttle mission, to enabling astronauts to launch and service the Large (later Hubble) Space Telescope.

The contribution to the Space Shuttle program by the European Space Research Organization (succeeded by the European Space Agency in 1975) of Spacelab, a scientific facility that could be flown in the Shuttle's payload bay, provided yet another way in which astronomers could take advantage of the Shuttle. Many astronomers believed the frequency of Shuttle flights anticipated (two per month) combined with Spacelab's versatility would offer them a cost-effective and readily available means to fly their instruments in space. The 1975 report enumerates the astronomical uses and required subsystems for instrument integration identified by a NASA working group created to assess Spacelab's value to astronomy. Astronomers used the Spacelab facility on several occasions in the 1980s and 1990s with great success.

Document III-19

[cover sheet]

FINAL REPORT
OF THE
PAYLOAD PLANNING WORKING GROUPS

Volume 1
Astronomy

MAY 1973
NATIONAL AERONAUTICS & SPACE ADMINISTRATION
GODDARD SPACE FLIGHT CENTER
Greenbelt, Maryland 20771

[xv] ASTRONOMY WORKING GROUP

EXECUTIVE SUMMARY

The principal advantages of space astronomy over ground-based observations reside in the greatly increased spectral coverage and angular resolution attainable from above the earth's atmosphere. For the first time celestial objects can be studied over virtually the entire electromagnetic spectrum from radio to gamma-ray frequencies. Even at the present early stage, this ability has produced a number of major surprises–for example, the overwhelming infrared emission from a variety of objects including planetary nebulae and galactic nuclei. Higher angular resolution will not only permit more detailed study of the structure of individual objects but, because of night sky suppression, will also allow observation of substantially fainter and hence more distant sources. By exploiting these advantages during the coming decades we will be able to solve, or at least to greatly increase, our understanding of such major scientific problems as the evolution of the early universe, the nature of quasars, galactic nuclei and radio sources, the formation of galaxies and of the stars within them, the origin of the chemical elements, and the origin of the solar system and of life itself. Solutions to these problems will impact all branches of human endeavor that have been seriously hampered in the past by the limited view of the universe available from the ground.

The immense potential of space astronomy has been amply demonstrated during the last decade with comparatively small, exploratory instruments, limited to the observation of relatively bright sources. The time is now appropriate to establish in space the full range of observing facilities required to solve longstanding astronomical problems. The advent of the Space Shuttle renders this not only technically feasible but even moderately inexpensive as compared to earlier ventures in space science.

The cornerstone of our recommendations for the 1980s is the Large Space Telescope (LST), a three meter aperture, diffraction-limited telescope optimized for the ultraviolet and visible regions of the spectrum but usable also in the infrared. It will be operated as an automated satellite and will be periodically serviced by the Shuttle. The LST will extend significantly the distance to which we are able to probe the universe and offers, for example, a prospective solution to the cosmological problem, which has not proved possible from the ground. A balanced program requires that this major instrument be supplemented by other more specialized instruments, as indeed are also required in ground-based observatories.

[xvi] Because the LST is not planned primarily for the infrared, early emphasis in the Shuttle Sortie program is placed on this spectral region. Two infrared telescopes are proposed.

- A 1.5-meter aperture telescope, cryogenically cooled to about 20°K specifically for the 10-50 mm wavelength region.

- A very large uncooled telescope for the far-infrared and microwave region, and for planetary studies and narrow-band spectroscopy over the whole infrared range.

Although both telescopes could operate as automated free-flyers based on the same spacecraft Support System Module (SSM) developed for the LST, both would gain by operation on the Shuttle. For the uncooled telescope the Shuttle allows the accommodation of larger optics than would be possible with the Titan-compatible SSM, as well as the possibility of interchanging instruments at the focal plane during flight. The cryogenic system for the cooled telescope would be much simpler and less expensive on the Shuttle. These telescopes will be powerful tools in the exploration of such diverse phenomena as the immense infrared energy output of galactic nuclei, the conditions in the interstellar medium leading to star formation, and the physical properties and composition of planetary atmospheres and surfaces.

In the ultraviolet, there is a definite need for a wide angle telescope to provide a UV survey in one broad wavelength band if the LST is to be used for many years to maximum effect. Subsequent use for studies at different wavelengths or for an ultraviolet spectral survey would be valuable but less urgent. A one meter diffraction-limited telescope for the ultraviolet and visible will provide high angular resolution imaging over relatively wide fields of view (0.5°). Such a capability is required, for example, for photometric studies of stellar evolution in globular and open clusters and to supply observations of nearby galaxies as the basis for LST studies of faint ($>21^m$) extragalactic sources. Unless or until the LST makes possible the frequent monitoring of solar system bodies, the 1-meter telescope can provide the needed synoptic coverage. The major advantage of the Shuttle for both these instruments is that it will allow use of photographic and electronographic detectors with their very large information storage capability. The 1-meter telescope will also provide an important test bed for auxiliary instrumentation for LST, allow specialized observations of a "one-of-a-kind" nature and relieve LST of observations of relatively bright sources.

In addition to these five instruments, which the panel considered in detail, several other instruments which were considered briefly are typical of those which the Shuttle program should include. Examples are a very wide angle ultraviolet camera for the study of large scale, low surface brightness nebulae and star [xvii] clouds, a grazing incidence telescope for the extreme ultraviolet between the normal X-ray region and the Lyman limit of hydrogen, Explorer-class free flyers (to measure the cosmic microwave background for example), and rocket-class instruments which can fly frequently on a variety of missions.

Except for the LST, each of the major astronomy instruments requires approximately half of the space, weight, and other support of a Sortie flight. While each could be operated remotely from the ground, our present impression is that in most cases it would be preferable to have the support of a four man Shuttle crew, in addition to the pilot and copilot, and a small laboratory to provide workspace, data storage, communications and access to the focal plane of at least one telescope. Although the individual instruments could share a Sortie mission with another discipline, compatibility requirements are severe. Astronomy requires stabilization of the Shuttle to near one arc minute (by means of control moment gyros), control of the pallet pointing direction throughout operation as dictated by the astronomical program, and a contamination-free environment. We therefore believe that we would be our own best companion. Most scientific direction must be from the ground, making it necessary to have excellent communication, including picture transmission, on both up and down links. A data relay satellite would be very helpful, although astronomy can use the intermittent communication provided by a ground network of tracking stations if adequate capacity compensates for limited time and if real-time communications are possible from the receiving station to a central control station at the same rate.

Document III-20

[cover sheet]

INTERIM REPORT OF THE
ASTRONOMY SPACELAB PAYLOADS
STUDY

Executive Volume

Prepared by the
Astronomy Spacelab Payloads Project

JULY 1975

NATIONAL AERONAUTICS & SPACE ADMINISTRATION
GODDARD SPACE FLIGHT CENTER
Greenbelt, Maryland

****

[51] **CONCLUSIONS**

At this stage of the Astronomy Spacelab Payload Study, several conclusions about astronomical investigations and the effective methods of using the Spacelab for research in astronomy during the early 1980s have been identified. In each of the scientific areas of the study, solar physics, UV and optical astronomy, and high energy astronomy, a substantial and valuable scientific program has been identified based on the experience of our past research in space, from recommendations from available studies and from consultations with scientists in the fields. The direct utilization of instruments operated from balloons, sounding rockets and satellites appears readily feasible and, in many cases, desirable in the pallet mode of the Spacelab missions. Furthermore, the huge volume and weight available with the Space Shuttle affords the opportunity of incorporating large instruments and, in fact, facilities in the Spacelab program. The costs for the design and construction of such instruments for use with Spacelab appear very reasonable; significantly less than instrument costs used in satellite payloads as the development of these instruments in many ways appears to parallel the techniques used with sounding rockets, balloons and aircraft.

The methods of carrying out experiments with Spacelab are of a special nature with many similarities and disparities with the past techniques. Although the Spacelab missions represent full scale satellite-of-the-Earth operations, the missions are relatively shortlived, they may be amended by the crew of specialists on hand, and the return to Earth of the scientific equipment for maintenance and modifications is a guaranteed aspect of this mission mode. In addition, the flight-into-orbit schedule is like the streetcar approach of the old Orbiting Geophysical Observatory, with an expected launching schedule of two space shuttles a month and probably about ten launches a year which may be available for some astronomical research. In effect, in about five years from now, the capacity for carrying instruments into Earth orbit will be increased by more than an order of magnitude and certainly more than the increase in the number of scientists, funds and other resources for carrying out research. It is essential that the methods for utilizing Spacelab match and adjust to such constraints.

The Astronomy Spacelab Payloads Study has, from the engineering and mission analysis investigations, found several requirements to effectively use the Spacelab for astronomy. These requirements include a set of pointing platforms for a variety of instruments, special instrument containers for rapid and easy integration of scientific instruments, some standardization of power, telemetry and operational functions, and modular overall integration into pallets at the integration center for the scientific program. The conclusions so far derived from this study are listed below. They are divided into groups defined by scientific areas and by required subsystems to integrate the scientific instruments and by the cost of such integration and schedule procedures.

[52] Scientific Program

1. Astronomers may have available simple and regular access to extended wave lengths into ultraviolet, superb image quality and a dark sky with a one meter class Spacelab UV Optical Telescope (SUOT). This ultraviolet facility can provide regular opportunities for a great number of astronomers, and with the wide field and regular access to focal plane instruments it would complement the Large Space Telescope. The SUOT should be developed for early Spacelab operation in 1981.
2. A solar telescope of large aperture for diffraction-limited observations extending over near UV and visible wavelengths can be of great value in studies of the heating of the solar chromosphere, for studying mass transport, magnetic field configurations, fine scale phenomena in sunspots and abundance distributions of elements in solar structure. Such a spectroheliograph or One-Meter Telescope Facility should be developed for the 1980-1981 Spacelab program in solar physics.
3. A Solar EUV-XUV Soft X-ray Facility covering the solar spectral region from 2000Å to 2Å and a Hard X-ray Imaging Facility consisting of instruments to study X-ray, gamma ray and neutron emissions from the flaring and nonflaring sun, should be constructed for the early 1980 period of Spacelab operations. These facilities will be used for observations and studies of processes in the tenuous transition region and the corona, and studies of the physics of flares.

4. The field of high energy astrophysics encompassing X-ray, gamma ray and cosmic ray astronomy includes an outstanding group of scientists with the developed technologies, instrumentations and experiments that can fully utilize the expanded capability of the early Spacelab modes. One of the first Spacelab missions should be devoted to high energy astrophysics and regular opportunities for about two dedicated missions a year should be planned.
5. A wide variety of experiments derived from experiments using sounding rockets, balloons and satellites have been identified in each of the astronomy disciplines. Considerable flexibility exists in combining experiments and integrating instruments on pallets and segments of pallets and these experiments are compatible with many Spacelab missions. An organized instrument preparation, integration and scheduling system for effectively and fully using each Spacelab mission would give scientists a powerful, productive and continuing means for carrying out research in astronomy and astrophysics.

[53] <u>Experiment Integration and Mission Management Operations</u>

1. Three classes of pointing systems have been identified to fulfill the scientific requirements for astronomical observation with Spacelab.
    (1) For facilities and large high energy instruments, the Instrument Pointing System (IPS) using an inside-outside gimbal, is under development by the European Space Agency. A pointing accuracy and stability in the one arc second range with limited roll is required for solar and astronomical observations. For several of the X-ray experiments more modest, near one arc minute pointing, and instrument capacities of close to three tons are needed. Based on the preliminary projected scheduling of this pointing system for astronomy and applications, a total of three (3) IPSs are required.
    (2) For pointing instruments of moderate weight a double-mount Small Instrument Pointing System (SIPS) has been under study. The SIPS can accommodate the moderate weight ATM class of solar instruments and the great majority of solar and astronomical instruments with a pointing accuracy and stability approaching the one to two arc second range. Four SIPS units are required for astronomy.
    (3) A low-cost, one arc minute accuracy and 10 arc second stability system is needed for the many rocket-class instruments. This system may readily be developed in-house by personnel of the Sounding Rocket Division of GSFC. Six of these units are needed.
2. <u>Instrument canisters</u> are required for thermal control and ease of integration of the wide variety of instruments considered for Spacelab astronomy flights. Canister configurations for compatibility with the SIPS and various instrument and mounting requirements can be developed. <u>Contamination</u> control is available with the <u>instrument canister</u>. The flexibility of the <u>instrument canister</u> is substantial, as it not only is used to control the environment of the instrument, but it also may afford a means of remote integration and becomes a shipping container for the instrument on Earth and in space.
3. Astronomical research with Spacelab involves mission planning and scheduling, instrument integration and mission operations, and requires Payload Operations

Control Center (POCC) at the GSFC. The experimenters would use the POCC during the installation and check out of instruments on pallets and later during the operation of the instruments in orbit. The POCC would incorporate in-flight experiment operations, Spacelab communications, and data reduction operations. Investigator Stations would be incorporated into POCC for the operation and control of individual and sets of experiments during the mission.

[54] 4. For Spacelab mission planning, the assignment of prime mission goals to a particular astronomical discipline, a "dedicated mission," is scientifically and operationally efficient because the orbit, orientation, and mission sequences may be optimized. Solar physics, UV/optical astronomy and High Energy Astrophysics are generally mission compatible and combinations of experiments in these fields also would be scientifically productive. The interrelationships among mission parameters are complex and necessitate iterative and continuing mission analyses studies and operations.

The Astronomy Spacelab Payload Study has identified the mode for astronomical research using scientific facilities and instruments evolved through research using sounding rockets, balloons, aircraft, and satellites and the large instruments and instrument evolution making use of the Space Shuttle capacity and instrument return capability. The use of the pressurized module, the interface with free-flyers and space stations, and the general effects of working with the Spacelab mode requires further study. Of special concern is the ordering of the developments of facilities, the focal plain instruments and the support for experiments for the early missions.

Although the actual selection of experiments will be made from proposals submitted according to the NASA Announcements of Opportunity, early guidance in the relative value and comparison factors for the scientific and technological program is required. This is the initial year for Astronomy Spacelab Payload Study—in the next year the start and the ordering of the facilities will be made, the critical engineering subsystems for pointing, environment, power and data handling will be under development and the evaluation of experiment proposals and the selection of early experiments will be initiated.

The newly evolving capabilities of the Space Shuttle will not only permit a new approach to scientific investigations, but can influence lowering the costs of scientific instruments and their supporting subsystems. The availability of the shuttle as an Engineering test bed, the substantial payload carrying capacity, the presence of man in the operation and the capability to return the instruments should permit the development of ASP payloads in an evolutionary manner and enable the scientist and engineer to take risks. Cost savings should be expected. In addition the capability to refurbish and fly payloads should further increase the cost effectiveness of the ASP payloads. To take full advantage of this new potential, cost consciousness and constantly look[ing] for the "cost drivers" will continue to be a prime concern.

Document III-21

Document title: High Energy Astronomy Observatory Project Plan, NASA, June 13, 1973.

Source: NASA Historical Reference Collection, History Office, NASA Headquarters, Washington, D.C.

While ultraviolet astronomy dominated NASA's interest in the early days of the space program, the results of the gamma-ray and x-ray experiments carried aboard spacecraft during the 1960s and conducted from the ground expanded the popularity of high-energy wavelength studies among astronomers. Just as NASA had developed a program of large spacecraft, the Orbiting Astronomical Observatories, to explore the optical wavelengths, in the early 1970s the space agency initiated a series of multi-ton High Energy Astronomical Observatories (HEAOs) dedicated to gamma-ray and x-ray astronomy. After providing initial funding for the first two of three planned HEAOs in 1971, NASA suspended the program in early 1973 due to money shortages in the space program. This document portrays the state of the program as it underwent restructuring to accommodate the funding changes. In the end, NASA managed to fly three HEAO spacecraft, albeit less ambitious in scope than those originally planned; none of the HEAO missions planned for the Space Shuttle were flown.

HMA- 1001
June 13, 1973

Revision C
HMA-1001
HIGH ENERGY ASTRONOMY OBSERVATORY
PROJECT PLAN

APPROVED:
[signature]
Rocco A. Petrone
Director,
Marshall Space Flight Center

APPROVED:
[signature]
J. Naugle
Associate Administrator for
Space Science

[iii]

FOREWORD

This document is the agency plan for conducting the activities associated with NASA Flight Project 832, High Energy Astronomy Observatory (HEAO). This plan is prepared and maintained in accordance with NMI 7121. 1B, "Planning and Approval of Major Research and Development Projects," (dated July 1, 1972), and the proposed draft of NMI 7120. 1A, "Authorization and Control of Space Science (OSS) Research and Development Projects."

This document describes the overall plan and approach, and is the basic agreement between the Associate Administrator, Space Science, and the Director, Marshall Space

Flight Center for proceeding with the HEAO project. This plan is the basis for all project and lower level detail planning necessary for project operations, and is to be followed by all involved NASA organizations. It has been revised to reflect the agency decision to restructure the HEAO program to achieve most of the scientific objectives through a lower cost approach over a longer period of time. The restructured program consists of two groups of missions, designated Block I and Block II. Block I missions utilize conventional medium class launch vehicles while Block II concepts utilize Space Shuttle capabilities and approved experiments not assigned to Block I missions. Detailed planning as reflected herein addresses only Block I missions due to the preliminary nature of the second block. The planning will be further expanded at significant points in the life cycle to incorporate current project experience.

This plan supersedes prior revisions in their entirety. The document is prepared and maintained for OSS by the MSFC HEAO Office in coordination with the OSS Program Office. The document will be updated when the degree of content change is sufficient to justify a new issuance. Proposed changes to this plan shall be submitted to the MSFC HEAO Office for coordination and staffing for management approval.

****

[1-1] SECTION 1 - INTRODUCTION

A. GENERAL

This document contains the overall plan for proceeding with the first group of missions (Block I) established for the NASA project identified as:

NASA Flight Project 832, entitled "High Energy Astronomy Observatory (HEAO)." The NASA code number is 85-850-832.

B. PROJECT AUTHORIZATION

The HEAO Project was initially established and authorized as one of the projects of the Physics and Astronomy Program by Project Approval Document (PAD) 71-85-001 dated July 2, 1971. The HEAO project was further delineated in Enclosure 6 of that FY 71 PAD. The PADS concerning the HEAO Project which had been issued as of the date of this plan are listed in Table 1-A in chronological order:

TABLE 1-A

CHRONOLOGY OF AUTHORIZATIONS

1. PAD 71-85-001, "FY 1971 Project Approval Document, Research and Development" and its Enclosure 6, both dated July 2, 1971.
2. PAD-85-850-P&A, "Project Approval Document, Research and Development," and its enclosure 6, both dated November 2, 1972.
3. PAD- 85-850-P&A, "Project Approval Document, Research and Development," and its enclosure 6 both dated (TBD).

****

[1-3] C. PROJECT DESCRIPTION

The Physics and Astronomy Program is primarily directed to extend the present knowledge of the earth's space environment, the sun, the stars, and the more distant celestial bodies. This research is being conducted through a combination of various tasks such as Supporting Research and Technology and flight projects such as HEAO.

The HEAO Project will search for and obtain high resolution data concerning high energy radiation from space (i.e., celestial X rays, gamma rays, and cosmic ray flux) by means of large unmanned earth orbiting observatories which will be built, launched, and operated as independent missions with complementary mission objectives and scientific experiments. The HEAO Project as presently contemplated is divided into groups of missions called "blocks." Block I observatories use the Atlas/Centaur as the launch vehicle. Block II would use the Space Shuttle capabilities with the observatories containing experiment hardware which could not be accommodated in Block I.

Block I consists of three missions, designated HEAO-A, HEAO-B and HEAO-C. The basic elements of Block I are shown in Figure 1-1 [no figures included]. HEAO-A is a scanning mission which will conduct a total sky survey for X-ray sources. HEAO-B is a pointing mission which logically continues the program by using an X-ray telescope to accurately locate, define, and determine the properties of the major X-ray sources.. HEAO-C is a scanning mission surveying the sky for gamma ray and cosmic ray. Figure 1-2 and 1-3 illustrate the current observatory configurations for HEAO-A and HEAO-B. The HEAO-C configuration is being defined using candidate experiments to formulate the payload, hence an illustration is not shown.

****

[1-6] D. PROJECT STATUS

The HEAO program as initially approved was suspended by an agency decision on January 5, 1973. The suspension is expected to remain in effect for approximately one year, during which time the previously planned program will be restructured and rebaselined as described herein. A recommended approach for restructuring the program was presented to the Associate Administrator for Space Science, on February 13, 1973, and then to the Administrator of NASA on February 20, 1973. Authorization to proceed with the redefinition and to establish firm cost and schedule plans was issued by the Office of Space Science (OSS) on March 5, 1973. OSS assigned experiments to missions HEAO-A and HEAO-B, and determined the candidate experiments to be considered in defining the HEAO-C observatory. Redefinition of the selected Block I missions has been initiated, and revised proposals are being prepared. Preliminary and conceptual design is underway. Definition of Block II payloads is proceeding at a low level of effort, pending clarification of the overall definition approach and development of funding plans.

[2-1] SECTION 2 - PROJECT PLAN SUMMARY

A. GENERAL

This section summarizes the approaches and planning to be used in accomplishing the HEAO Block I missions. Each of the sections following this summary contains the detail necessary to provide a complete agency plan for the project.

B. PROJECT OBJECTIVES

The objective of the HEAO program is to extend the present knowledge of celestial X- rays, gamma rays, and cosmic-ray flux through studies facilitated by means of large earth-orbiting observatories.

The objective of the X-ray studies is to survey the entire sky for X-ray sources of about one-millionth of the intensity of the brightest known source, SCO X-1, and to investigate the shape and structure of these sources. Many sources show flares and flickering; the observatories will monitor these intensity variations. The initial survey mission will be followed by missions capable of performing studies on the spectra, structure, and location of these sources.

The gamma-ray studies will be directed at the measurements of the gamma-ray flux and at determining source locations. Line spectra will be obtained and analyzed from the sources discovered. These studies, along with the X-ray studies, should contribute to our understanding of such phenomena as pulsars and quasars.

The high energy particle experiments will examine the composition and synthesis of cosmic-ray nuclei. The isotopic composition of cosmic rays will provide information on the age and nuclear interactions producing the rays. The existence of very heavy nuclei will provide [2-2] an opportunity to probe into their origin, age and propagation through the interstellar medium. These studies will form the basis for future orbiting cosmic-ray laboratories, utilizing the capabilities afforded by the Space Shuttle.

The objective will be achieved incrementally on an integrated basis by the HEAO missions. Each mission is planned to achieve certain goals and to provide information which can be utilized in subsequent activities to achieve the project objectives. Section 3 [not included] describes the project and mission objectives in detail.

Document III-22

Document title: J. Mather, Goddard Institute for Space Studies et al. "Cosmological Background Radiation Satellite," October 1974.

Source: John Mather, Goddard Space Flight Center, NASA.

One of the most important astrophysical measurements is that of the Cosmological Microwave Background (CMB), the remnant of the radiation resulting from the "big bang" at the origins of the universe. Because of the expansion of the universe in the 13-15 billion years since its origin, this originally very hot radiation was thought to now be only a few degrees above absolute zero. The Cosmological Background Explorer (COBE), first proposed in 1974 and launched in 1989, gathered data of fundamental scientific importance on the CMB.

[cover sheet]

COSMOLOGICAL BACKGROUND RADIATION SATELLITE

J. Mather
P. Thaddeus
Goddard Institute for Space Studies

R. Weiss
D. Muehlner
Massachusetts Institute of Technology

D. T. Wilkinson
Princeton University

M. G. Hauser
R. F. Silverberg
Goddard Space Flight Center

OCTOBER 1974

[i]

COSMOLOGICAL BACKGROUND RADIATION SATELLITE INVESTIGATORS AND INSTITUTIONS

John C. Mather 212-678-5609
Patrick Thaddeus 212-678-5621
Goddard Institute for Space Studies
2880 Broadway
New York, New York 10025

Rainer Weiss 617-253-3527
Dirk Muehlner 617-253-4824
Massachusetts Institute of Technology
20F006
Cambridge, Massachusetts 02139

David T. Wilkinson 609-452-4406
Princeton University
Department of Physics, Jadwin Hall
Princeton, New Jersey 08540

Michael G. Hauser 301-982-2468
Robert F. Silverberg 301-982-2468

Infrared Astronomy Research Program Laboratory for High Energy Astrophysics
Goddard Space Flight Center
Code 661
Greenbelt, Maryland 20771

[ii]

TABLE OF CONTENTS

[no pagination]

A. Summary

A group of four instruments to measure the cosmological background radiation is proposed for a Delta-class Explorer. The experiments address fundamental issues in observational cosmology. Three of the experiments deal with the 2.7° K cosmic background radiation. The first is a definitive measurement of the spectrum of this radiation between 0.1 and 3 mm using Fourier transform spectrometry. The experiment will map the spectrum at shorter wavelengths than have been possible from balloons and will measure the spectrum in the region around the blackbody peak with a precision of 10-4. It will also look for the emission from cold dust clouds and from infrared galaxies. The second and third experiments are devoted to measuring the large scale isotropy of the background radiation at a number of wavelengths to a precision of 10-5. It should be possible to measure the motion of the Earth relative to the co-moving frame defined by the expansion of the Universe. Measurements at several wavelengths are required in order to distinguish anisotropy in the background radiation itself from anisotropy due to discrete sources. Definite observation of the Earth's motion relative to this radiation will be further confirmation of the primeval fireball interpretation of the 2.7° K radiation.

The fourth proposed experiment searches for diffuse radiation in the 5-30 micron wavelength range, expected to arise from interplanetary dust, interstellar dust, and, in particular, from the integrated luminosity of very early galaxies. The experiment is designed to separate these contributions by their spectral and directional properties.

These four experiments have similar spacecraft requirements. Three require liquid helium, and two require slow rotation. The required scan of the sky can be provided by the orbital motion combined with precession of the orbital plane. The spacecraft could be shared with other experiments requiring near vertical pointing and the same simple scanning mode. The scientific importance of these experiments, their need for a space platform, and the relatively modest spacecraft requirements they impose all recommend this mission as an attractive first application of a liquid helium cryostat in space.

B. Objectives and Significant Aspects

An Explorer spacecraft equipped with cryogenically cooled instrumentation will provide a uniquely sensitive system for study of diffuse cosmic radiation. It is proposed to develop a mission in which such a system is used to make definitive measurements on the radiative relics of the earliest stages of the universe. Four experiments are proposed, characterized by their common cosmological motivation and by compatible and relatively modest demands upon the spacecraft. The experiments proposed here include:

1. Spectrum of the 2.7 K Cosmic Background from 0.1 to 3 mm
2. Isotropy of the 2.7 K Cosmic Background between 0.5 and 3 mm Wavelength
3. Isotropy of the 2.7 K Cosmic Background at 3, 5, 9, and 16 mm Wavelength
4. Search for Diffuse Cosmic Radiation at 5-30 micron Wavelength.

The personnel responsible for each experiment and principal requirements for each are summarized in Table 1. It should be noted that experiment (3) does not require cryogenic cooling, but it is intimately related to the first two experiments, and does require a satellite platform for high quality results.

C. Plan of the Report

In order to facilitate understanding of each experiment, the remainder of Section I discusses the objectives, concept, and spacecraft requirements for each experiment separately. It should be emphasized, however, that this separation is only for convenience: we strongly believe that this mission should be considered as a whole, since it represents a scientifically exciting and technologically modest first application of a cryogenic satellite. To this end, a technical plan providing for a suitable spacecraft and orbit for the complete mission is also presented in Section I. Management and cost plans for the mission are given in Section II. Appendix I contains a summary of the current status of cryostat development prepared by Ball Brothers Research Corporation. The operating principles of the Michelson interferometer used in experiment (1) are discussed in Appendix II. Biographies of the proposers are contained in Appendix III.

Table 1 COSMOLOGICAL BACKGROUND RADIATION EXPERIMENTS

Title	Spectrum of the 2.7°K Cosmic Background from 0.1 to 3mm Wavelength	Isotropy of the 2.7°K Cosmic Background Wavelength	Isotropy of the 2.7°K Cosmic Background at 3, 5, 9, and 16mm Wavelength	Search for Diffuse Cosmic Radiation at 5-30μ Wavelength
Description	Polarizing Michelson interferometer, 0.1 to 3 mm range, liquid helium cooled	Broad-band radiometer 3mm to 0.5mm, liquid helium cooled detector	Microwave radiometers, 3, 5, 9, and 16mm Radiative cooling only	High sensitivity Spectrophotometry 5m-30m, liquid helium cooled
Experimenters *Institution	John Mather, P.I., GISS Patrick Thaddeus, Co-I, GISS Rainer Weiss, Co-I, MIT Dirk Muehlner, Co-I, MIT Michael Hauser, Co-I, GSFC Robert Silverberg, Co-I, GSFC	Rainer Weiss, P.I., MIT Dirk Muehlner, Co-I, MIT	David Wilkinson, P.I., Princeton	Michael Hauser, GSFC Robert Silvery, CO-I, GSFC David Wilkinson, Co-I, Princeton John Mather, Co-I, GISS
Instrument Cost	\$500,000 - 1,000,000	\$100,000 - 150,000	\$1,000,000 - 1,500,000	\$500,000 - 1,000,000

Size	30cm diam x 1m long inside cryostat which is 1m diam x 1.5m long	80 cm x 30cm x 60cm on outside of cryostat (1/4 of circumference)	on outside of cryostat 3/4 of circumference, 30 cm thick	20 cm diam x 30cm long inside cryostat
Mass	20 kg	20 kg	40 kg	20 kg
Pointing	near vertical, away from sun	same, but spinning about near vertical	spins about near vertical axis at 1 rpm	near vertical, away from sun
Telemetry	200 bits/sec	200 bits/sec	200 bits/sec	200 bits/sec
Power	2W	5W	15W	2W
Required Experiment Lifetime	6 months	1 year	1 year	1 year

Document III-23

Document title: Nancy G. Roman, Program Scientist, and C. R. O'Dell, Project Scientist, to Members of the LST Operations and Management Working Group, with attached Minimum Performance Specifications of the LST, February 12, 1975.

Source: Space Telescope History Project, National Air and Space Museum, Smithsonian Institution, Washington, D.C.

Document III-24

Document title: George B. Field, Director, Center for Astrophysics, Harvard College Observatory, to Dr. James Fletcher, NASA Administrator, February 12, 1976.

Source: NASA Historical Reference Collection, History Office, NASA Headquarters, Washington, D.C.

Document III-25

Document title: James C. Fletcher, NASA Administrator, to James L. Mitchell, Associate Director for Natural Resources, Energy and Science, Office of Management and Budget, April 12, 1976.

Source: NASA Historical Reference Collection, History Office, NASA Headquarters, Washington, D.C.

Before providing NASA with funds to plan for the Large Space Telescope (LST), the White House, as part of its pressure to reduce the NASA budget in the post-Apollo period, stipulated that the space agency had to find a way to reduce the telescope's cost from the price of $325 million that NASA was estimating in 1974. NASA thus undertook the task of developing an LST that would satisfy the financial concerns of the White House and Congress as well as the scientific objectives of the astronomy community. As even LST "father" Lyman Spitzer realized and advocated, reducing the size of the telescope's 3-meter primary mirror and thus scaling down the entire spacecraft represented one of the most obvious ways to cut the mission's costs. After reviewing the results of studies comparing telescopes with 3-meter, 2.4-meter, and 1.8-meter mirrors, the LST Science Working Group informed NASA that it would be unwilling to support a telescope with a mirror less than 2.4 meters in diameter. In this letter, LST Program Scientist Nancy Roman and LST Project Scientist C. R. O'Dell asked the working group whether they in fact drew the line at 2.4 meters or were willing to negotiate on this number with NASA Headquarters, which was inclined to pursue an even smaller telescope. Astronomers replied that a 2.4-meter mirror was their minimum. Among other factors, the strong opinions of these astronomers ensured that NASA indeed would proceed with a 2.4-meter LST.

NASA was unsuccessful in convincing the White House to approve a new start for the LST in the FY 1977 budget, which President Ford transmitted to the Congress in February 1976. This was an outcome that was deeply disappointing to the leaders of the astronomical community, as suggested by the letter from Harvard astrophysicist George Field to NASA Administrator James Fletcher. NASA and the astronomy community redoubled their efforts to get an early start on LST approval and then have the project included in the FY 1978 budget. NASA requested such an approach in Fletcher's April 1976 letter to the Office of Management and Budget. The campaign in support of the LST was successful, and the budget sent to Congress by the outgoing Ford administration in January 1977 contained funds to begin project development. (The FY 1978 budget also contained a new start for another major science project, the Jupiter Orbiter-Probe mission that became Galileo.)

Document III-23

[no page number]

February 12, 1975

TO: Members of the LST Operations and Management Working Group

Dear Colleague:

We appear to be approaching a very pivotal time for LST. As you know, NASA will submit its budget for FY77 (starts 1 October 1976) by May 15 of this year. If LST does not appear as a new start in this budget there is a very real risk of the program foundering. Should this occur, we expect that it might be several years before LST could be restarted and that precursor to LST missions might begin to dominate NASA planning.

The issue that will ultimately decide the question is program cost, i.e., if NASA can afford it, we'll have an LST and starting in FY77. As you saw at the December Working Group meeting, the cost figures for all elements except the SSM are by now very well defined and the program is austere. Within a few weeks we will have the SSM contractor's estimates of their costs and the picture will be complete.

The program that we are now costing is very significantly reduced from that which we started Phase B with in the summer of 1973. We have de facto reduced the aperture to 2.4-m, allowed the angular resolution to degrade, reduced the number of Scientific Instruments, etc. All of us were party to these decisions and we hope that you feel the Program Scientist and Project Scientist have not made these concessions without your full knowledge and consent. We were requested to prepare a document describing the "Minimum Performance Specifications for the LST", which was done in December. (The current version of this is enclosed herewith, slightly revised according to Al Schardt's insistence that the resolution number agreed to at the October meeting was 0.10" and the fact that a theoretically perfect system with a secondary mirror observation of 32% will not give more than about 64% within the first dark ring.)

[2] The question for us to address is "Do we defend this as the true minimum LST against efforts to bring the program down in cost by sacrificing performance?" We feel that the costs being identified are well studied and justified for this program and any significant cost savings would be at the expense of performance. Therefore, we are in a position of standing-by [sic] this definition of the minimum LST and running the risk of foundering or renegotiating the cost at a significant cost and performance reduction.

Both of us feel that the program has already been so drastically reduced that we are at the point that the line must be drawn and that the agency must either allow us to move ahead with at least this minimum LST, or that we are confronted by an agency money problem and put LST into hold until the future. Although we feel that the derivation of the definition of a minimum LST was done by all of us, we feel that you should have the opportunity to individually and collectively state your opinion at this crucial time, as the risks are quite real.

A further complication is the proposal by Al Schardt that we try to reduce program costs by constructing an evolving complexity LST, i.e., one that at initial launch would not meet the minimum performance specifications; but, one that would possess the potential for upgrading later upon successive ground returns to a full LST capability. Although detailed calculations have not been made, it appears that such a plan is plausible and at some reduction in cost to initial launch. The overall cost to eventually reach LST performance with this plan would certainly be higher than the direct approach. Although Dr. Schardt considers this plan very seriously, he will not support a plan that is not supported by you, the ultimate users and justifiers of LST.

It now appears that a decision on FY77 may be made in March, before our next scheduled Working Group meeting. Therefore, we'd like to ask each of you to write us your opinion on "Do we bargain or do we draw the line?" by February 25. Please write to either of us with a copy to the other. We will then set up a telephone conference to summarize the situation. Preferably this conference would be on the afternoon of March 4. Please let us know where you will be on that date, and what are "impossible" times for you.

Nancy G. Roman
Program Scientist

C. R. O'Dell
Project Scientist

[no page number] [attachment]
Revised 27 Dec 1974

Minimum Performance Specifications of the LST

The performance specifications imposed by the scientific users of the LST define that set of conditions necessary to have broad scientific support, the ultimate justification for its construction. The present specifications for a minimum LST were prepared by the Project Scientist after consultation with the Program Scientist, the LST Operations and Management Working Group and many other individual astronomers. Individual opinions on specific points vary; however, the following does represent the composite view:

1. LST is a versatile, long life-time observatory; i.e., it must have the capability to accommodate a variety of scientific instruments and vary the complement of instruments with time.
2. The optical image should satisfy the following requirements in the visual and near-vacuum ultraviolet wavelengths: Resolution using the Rayleigh criterion of 0. 08 arc-seconds; A full width half intensity diameter of 0.08 arc-seconds; 80% of the total energy of a stellar image must be contained within a diameter of 0.15 arc-seconds.
3. The overall LST system must work efficiently down to wavelengths permitting the study of the Ly (alpha) line at 1216Å, requiring reaching to about 1150Å. Likewise, it must allow efficient observations at infrared wavelengths longer than those readily accessible from the ground.
4. The system should accommodate at least four scientific instruments.
5. It must be capable of measuring objects appreciably fainter than those accessible from the ground. At the present this means going to about magnitude 27 with a signal to noise ratio of 10 in 4 hours of observing time.

6. It must be capable of measuring extended sources of surface brightness 23.0 m/[illegible] with a signal to noise ratio of 10 in 15 hours.
7. The LST must have the capability of using Scientific Instrument entrance apertures comparable in size to the image.

Document III-24

[no page number]
Center for Astrophysics
60 Garden Street
Cambridge, Massachusetts 02138

Harvard College Observatory
Smithsonian Astrophysical Observatory

February 12, 1976

Dr. James Fletcher, Administrator
Code A
NASA Headquarters
Washington, D .C. 20546

Dear Jim:

I am writing to you in my capacity as a member of the ST Working Group, not as Chairman of PSC, to share my thoughts with you on ST. I hope my comments will be helpful; they are meant to be.

As I am sure you have been, I was surprised by the depth and breadth of the reaction among astronomers to the decision not to have a 77 new start. I felt that the explanation you gave in the authorization hearings was intelligible; certainly the major budget cut thrust upon NASA at the last minute would have made it very difficult to get a new start for ST in 77. Why was the reaction so powerful, then?

To me, it simply testifies to the very deep and widespread excitement ST has generated in the astronomical community. I have watched the ST concept grow since I first heard about it from Lyman Spitzer in 1952. The response of the community, at first hesitant, has now reached the point of virtual unanimity among professional astronomers. The reason for this is that the ST, with its diffraction-limited images and its UV and IR capability, promises significant forward steps in virtually all branches of astronomy, and major steps forward in several of them. As a result, astronomers find, when they think about ST, that there are problems they are personally excited about which can be solved by it. Perhaps the ST is unique among space instruments in this respect. ST is not a specialized spacecraft of interest to only a small subdiscipline, but a true observatory with all the broad capabilities one finds at a facility such as Kitt Peak. Such national observatories on the ground are used by large numbers of astronomers for a great diversity of purposes, and for that reason, they enjoy strong nationwide support. The same is true of ST, but even more so because of its enormous capabilities.

I dwell on the enthusiastic support among astronomers to help you understand the depth of feeling you are now [2] encountering. Astronomers who are normally quite restrained have become passionate because they feel so frustrated. In this connection, Jim, I was disappointed that you could not meet with the ST Working Group (although I understood why you could not), because that would have been a good opportunity to defuse their feelings and allay their fears. I feel particularly disappointed because strong statements of support like those you made in Williamsburg would have made it very apparent that you mean to start ST in 78. As things are now, there are serious doubts among the senior astronomers as to NASA's true intentions. I think you will candidly admit that even with the best of intentions, it will be very difficult, in view of a possible change of administrations, and the usual problems with the budget, to get a new start in 78. The critical issue, as I see it, Jim (and here I speak completely frankly) is how to transform your statement in Congress into something tangible that will give us something to work with. I am concerned that if this is not done, there will be a really serious loss of morale among the top astronomers who are supporting the project - and I am sure that you are aware that Burbidge, Spitzer, and the others are at the absolute top of their profession. This would be a disaster which we must somehow avoid. If at all possible, Jim, I urge you to commit NASA to major activity for the ST in 77, if possible via a direct authorization by Congress.

I want to speak to you candidly about the public attitude toward the ST. Certainly very few of even the educated public are aware of the ST program, and the scientific results that would flow from it. The astronomers, of whom I am one, must take major responsibility for not publicizing the concept even better than has been done. But I am absolutely sure that with the proper approach, we can tap the wonder of distant and beautiful things that ordinary people feel when given a chance.

I would estimate that there are several million people in the U.S. who are fascinated by astronomy, cosmology, deep space, and the universe. Among them are many of our leaders, executives, and publicists. There is a great and enduring fascination with the depths of space and time out of which the earth, life, and finally humanity, arose. If we can show [3] how ST will probe the depths of space and time, we will have opened the door to public support.

How can we do this, Jim? An increased program of public education is needed. There are many astronomers who will participate willingly in such work in the year ahead. I am going to try to contact leading publications to offer articles on space astronomy and the ST. Others have offered to do the same. Much can be done through the American Astronomical Society's Task Group on Education in Astronomy, to get materials into newspapers, classroom resource materials, and other educational literature.

In all of this, we desperately need the assistance of NASA offices at all levels, including your own, the program offices, and the public information office. We need contacts with media people, we need guidance, and we need written and visual materials. For some reason, NASA has not been as effective in presenting its space astronomy program as it has its planetary exploration program, and NASA should strive for parity in this area. With proper collaboration between NASA and the scientific community, we can get our message across.

Most of all, Jim, we the astronomers need to meet with you face to face, so that you can allay the fears which cripple effective action, and so that your own interest and excitement about ST can be communicated to all. I suggest that you consider addressing a meeting of the American Astronomical Society, much as you do the societies associated with the national aerospace effort, for a frank discussion of your hopes for a meaningful space astronomy program. I would be happy to discuss these and any other step we can take toward the ST at any time.

Sincerely,
[signature]
George B. Field
Director

Document III-25

[stamped] APR 12 1976

Mr. James L. Mitchell
Associate Director for Natural Resources, Energy and Science
Office of Management and Budget
Washington, DC 20503

Dear Jim:

Following up John Naugle's recent discussion with Hugh Loweth on the Space Telescope, I would like to summarize our current approach to this project. As I am sure you recognize, a new start for the Space Telescope will be a very high priority item in NASA's FY 1978 budget recommendation.

The history of the Space Telescope extends back to studies and scientific discussions in the early 1960s; by 1965, there was a small but growing cadre of astronomers that recognized the unique contributions to many disciplines that could flow from a large telescope operating at the diffraction limit above the interference of the atmosphere.

With the advent of the shuttle in 1971, it became clear that some of the major problems of maintaining an astronomical facility in space could now be overcome at reasonable costs, and more specific project planning began. The astronomical community eventually settled on a set of technical parameters for a large space telescope: to warrant the considerable investment required, the instrument should be some 10 times more precise than any ground based telescope and should be able to resolve stars of at least the 27th magnitude.

These requirements translated into an instrument with a 3-meter aperture operating at or near the diffraction [2] limit. Our early feasibility and design concept studies indicated that it was quite possible to build such a telescope, but that its costs would be higher than either we or the scientific community felt desirable. We then began to examine alternate aperture sizes in relation to probable cost and to technical performance. We found that with an aperture of some 2.4 meters we could meet the scientific requirements and at the same time significantly reduce overall system development complexities and therefore cost.

The astronomical community, after some initial concerns that were dissipated by thorough engineering analyses, enthusiastically accepted the revised specifications: a 2.4-meter, diffraction-limit aperture and a pointing accuracy of .1 arcsecond. In FY 1975 and 1976, therefore, we funded competitive industry studies for the Space Telescope with all elements designed to come together to support a new start in FY 1977.

As it turned out, the Space Telescope was deferred for consideration until the FY 1978 Presidential decision cycle.

It appears we had underestimated the commitment to the Space Telescope project that had grown over the years among scientists, industrial organizations, and members of Congress. We have been inundated with severe criticism from virtually every academic institution associated with astronomy. We have been urged forcefully by the aerospace and optical contractors to do something to alleviate the high costs of their holding together effective engineering teams in order to be able to bid on the telescope project if and when it were authorized and funded. (Here we are dealing with three aerospace firms in competition for the spacecraft and two or more optical houses competing for the telescope assembly; all were geared up to propose in FY 1977 and now are faced with the economics of retaining their technical teams for another year.) The House has chosen to authorize an FY 1977 new start on the Space Telescope at the level of $3 million and our [3] Subcommittee, supported by the urging of the industrial and scientific community, has required us to provide them a plan that would permit early selection of the winning spacecraft and optical contractors (thereby relieving the losers of the considerable financial liability inherent in trying to retain a competitive posture well into next year). Our response to this request from Chairman Fuqua is due early this month.

The Senate did not follow the House lead. The Senate bill does not authorize a new start for the telescope in FY 1977, but the report language accompanying the bill is very strong in urging NASA to proceed with the project as "the item of highest priority" in FY 1978 and requesting that NASA sustain the Space Telescope pre-contract activities to assure an orderly and efficient transition into development.

The House-Senate conference to accommodate the differences between the two bills is expected to be completed by the Easter recess. We believe we can take certain actions now that would preserve the momentum of the program without committing the Administration and that would permit the House to recede gracefully over the question of the authorization for the Space Telescope.

What we propose to do is the following:

a. In August of this year, we would issue a formal Request for Proposal (RFP) for the telescope assembly; this RFP would clearly state that: no contract would ensue until and unless there were authorization, appropriation, and apportionment actions appropriately taken in the future. The responses to this RFP would be available in mid-October which would help in the formulation of the President's FY 1978 budget. The NASA source evaluation process would proceed during the winter and be completed in the March-April period of 1977 – well after the decisions on the President's [4] budget and even after the completion of the authorization and appropriation hearings. Depending on the situation at that time, we could then terminate the process (if the project were not approved) or proceed with detailed negotiations aimed at a contract effective on October 1, 1977.

b. We would follow essentially the same pattern with the spacecraft RFP, except it would be released in October of 1976, responded to in December of 1976, and the responses evaluated by May or June of 1977.

c. We would solicit Principal Investigator responses for the Space Telescope's scientific instrumentation beginning around September of this year and extending well into 1977 before any decision would be made.

We feel that this limited action, not committing the Administration in advance of the normal budget process, will place us in a sound position to proceed with the project if it is approved during the coming cycle. Conversely, we feel that to take no steps until mid-1977 would result in serious program discontinuities and even some inequities-:

— Some of the contractors have indicated that they might be unable to maintain their teams and competitive postures until mid-1977 and would therefore be unable to bid.

— All the contractors have indicated a strong desire for an early selection process even in the absence of funding.

— All the contractors have stated that their technical positions would have seriously eroded by mid-1977, leading to weaker technical proposals and greater cost uncertainties (some have said they would require an additional preliminary design phase, extending the program and increasing its costs).

[5] — The Congressional supporters of the Space Telescope might be moved to force unnecessary and undesirable confrontations with NASA and the Administration on the question of responsiveness to their guidance.

— The scientific supporters of the Space Telescope would continue to exert what pressures they could for an overt action by the Congress and the Administration, perhaps thereby reducing some of the positive impact of the President's recent science and technology message.

In summary, we feel that the most prudent course to take is to inform the interested communities on a low-key basis that we will proceed with an early selection process beginning in late summer. We believe that this approach will not commit the Administration in advance of the normal budget process, that it will retain the necessary flexibility on the part of the Government to adjust its plans to fit future situations as they become real, and that it will capitalize on current support for the Administration's posture on science and technology.

Sincerely,
[Original signed by]
James C. Fletcher
Administrator

bcc:
A/Dr. Fletcher
AD/Dr. Low
AA/Dr. Naugle
ADA/Gen. Crow
B/Mr. Lilly
X/Mr. Williamson
SD/Dr. Calio
AEM-3/Ms. LeCompte

Document III-26

Document title: Smithsonian Institution Astrophysical Observatory, "Proposal to National Aeronautics and Space Administration for the Study of the 1.2 Meter X-ray Telescope National Space Observatory," April 1976.

Source: Harvey Tannenbaum, Smithsonian Astrophysical Observatory, Cambridge, Massachusetts, reprinted with permission.

By the mid-1970s, NASA had taken a strong interest in high-energy astronomy, as evidenced by the launch of Uhuru and the development of the High Energy Astronomy Observatories. Encouraged by NASA's commitment to studying high-energy wavelengths, astronomers at the Smithsonian Astrophysical Observatory presented to NASA in 1976 an unsolicited proposal for the development of a major space-based x-ray observatory. The proposed spacecraft would enable astronomers to see farther than ever before possible into a universe of exploding stars, colliding galaxies, and enigmatic black holes—phenomena evident only by the x-rays they emit. Embracing the idea of this mission, which became known as the Advanced X-ray Astrophysics Facility, NASA began conceptual studies and assembled a working group to define scientific objectives for the mission in 1977. The satellite was finally launched, and named Chandra, in 1995.

[cover page]

PROPOSAL TO
NATIONAL AERONAUTICS AND SPACE ADMINISTRATION
FOR THE
STUDY OF THE 1.2 METER X-RAY TELESCOPE
NATIONAL SPACE OBSERVATORY
(Volume I - Technical Proposal)
P605-4-76
For the period 1 July 1976 to 30 September 1978
April 1976

Smithsonian Institution
Astrophysical Observatory
Cambridge, Massachusetts 02138

The Smithsonian Astrophysical Observatory
and the Harvard College Observatory
are members of the
Center for Astrophysics

****

[1-1] 1. 0 INTRODUCTION

A 1. 2 meter X-ray Telescope National Space Observatory has been included in recent program studies such as the National Academy of Sciences' "Scientific Uses of the Space Shuttle" and the Ad Hoc Planning Group of the High Energy Astrophysics Management Operations Working Group's "A Program for High Energy Astrophysics (1977-1988)."

The scientific objectives of the 1.2 meter X-ray Telescope Mission can be broadly stated to include study of stellar structure and evolution, study of large-scale galactic phenomena, study of the nature of active galaxies, and study of rich clusters of galaxies and cosmology. The discoveries of the past few years have clearly established that X-ray observations are an essential tool in the study of many of the objects of greatest current astrophysical interest such as pulsars, quasars, Seyfert galaxies, clusters of galaxies, and the intergalactic medium. The study of compact X-ray emitting objects in binary systems permits investigations of the properties of stars near the end point of stellar evolution and of the physics of matter at extreme pressures, densities, and magnetic fields. With the 1. 2 meter telescope, X-ray observations can be extended to the coronas of main sequence and giant, late type stars, as well as to peculiar stars, such as flare stars. It will also be possible to detect and resolve clusters of galaxies at extreme distances ($Z > 3$) and study their evolution over times comparable with the age of the universe. X-ray emission from clusters of galaxies is likely to originate in the heretofore unobserved intergalactic medium which may contain a large fraction of the total observable mass of the universe. These studies will profoundly influence our understanding of the dynamics and evolution of the cosmos.

[1-2] If X-ray astronomy is to move from its current exploratory state to take its place among the established branches of observational astronomy, then continuity in the observations is a

primary requisite. The present HEAO X-ray missions represent significant increases in sensitivity beyond previous X-ray experiments. In the case of the HEAO-B X-ray telescope the nominal design life is only one year. Even if we are successful in extending this lifetime by being judicious in the use of the RCS gas supply, and by being fortunate enough to escape catastrophic component failures in orbit, there will come a time shortly after 1980 at which this mission will end for some reason or other. It is clear that while such short-lived missions can accomplish a great deal in elucidating the nature of the most interesting X-ray objects known, and in extending the level of sensitivity of the previous surveys by a few orders of magnitude, they do not provide a permanent, continued capability to carry out X-ray measurements. By analogy, it is very much as if each new visible light telescope were to be placed in operation, run for a year, and then dismantled. The unexpected and fortunately long operation of the UHURU mission has given us a clear indication of the need for long-term observations even of sources in which dramatic changes occur in times as short as milliseconds. By extending the observations of Her X-1, Cyg X-1 and other binary X-ray sources over a period of years, we have been able to detect previously unsuspected changes of the basic energetics and kinematics of the stellar systems to which these sources belong. Moreover, it is clear that only through a long period of continued and correlated observations carried out in visible light, radio, and X-rays, can some of the most fundamental problems in stellar evolution and cosmology be resolved. [1-3] The 1.2 meter X-ray Telescope we are proposing here will represent a major improvement with respect to HEAO-B for the following reasons:

1. Long Life (~ 10 years). This not only permits the detection of a great number of targets of interest, but will also allow astronomers to receive data, fully comprehend their significance and import, carry out related optical or radio observations, and then again seek new data to clarify the nature of phenomena of interest. In addition, it is clear that the time scales of interest in X-ray astronomy extend to years, as for instance, in the study of changes in the orbital periods of binary X-ray sources. In order to insure a long-term utilization of the facility, we assume that a program of refurbishment can be carried out for focal plane instruments every three years. This would serve both for maintenance and for replacement of older instruments with newly developed ones.

2. Higher Angular Resolution (~ 0.5 arcseconds). A number of extremely important astrophysical problems cannot be attacked at the HEAO-B 2-arc second resolution level. Many of the arguments which have been advanced to support this view in the optical wavelength range axe applicable in X-ray astronomy, as well.

In particular, in extragalactic research the study of the nuclei of active galaxies demands that the resolution achieved be comparable to the characteristic sizes of interest. A resolution as fine as 0. 51, would permit us to study the nucleus of our own galaxy with a resolution of 2×10^{-2} pc, of a nearby galaxy with 20 pc resolution.

For clusters of galaxies at 100 Mpc, a resolution of 200 pc would allow us to resolve 1/50 of the linear size of a galaxy in a cluster, to locate the source [1-4] of the X-ray bright spot in the Perseus Cluster, and to study its relation to the structure of filaments and to the location of the nucleus of NGC1275.

For the study of clusters at very great distances ($Z > 3$), a high resolution X-ray telescope would prove particularly valuable since the radiation tends to be red-shifted to the range of greatest sensitivity and the fine angular resolution permits the study of the angular structure and size.

3. Greater Sensitivity. For the 1.2 meter X-ray telescope, we expect the sensitivity to increase by a factor of 20 or more over HEAO-B. This increase in sensitivity arises from a factor of 4 increase in mirror area and a factor of at least 5 increase in high resolution imaging detector efficiency that will be provided by the negative electron affinity detectors that NASA has recently approved for development by us. This increase in sensitivity is extremely important not only because of the improved ability to explore further in the Universe and much expand our catalog of X-ray sources, but also because the increased sensitivity will make it possible to study in detail known objects within reasonable observing time. High resolution spectroscopic measurements, which in HEAO-B are only marginally possible for a few targets due to the long observation times required (weeks), will suddenly become entirely routine for a very large number of targets.

We firmly believe that NASA and the astronomical community have recognized the value of the 1. 2 meter mission and that there is a strong commitment to carry out this mission by all concerned. In order that a launch take place in the early 1980's so as to minimize the hiatus in X-ray telescope observations between HEAO-B and the 1. 2 meter telescope, it is imperative that this proposed study be undertaken as soon as possible. We therefore have divided [1-5] the period of performance into two phases: (1) 1 July 1976 - 30 September 1977 and (2) 1 October 1977 - 30 September 1978.

During the first phase we will carry out studies of the mission requirements and spacecraft interfaces, of the 1. 2 meter mirror, of the overall experiment structure, of the thermal design philosophy, of the electrical interface, of the experiment aspect system, and of the experiment integration and test planning. We will also carry out studies of imaging detectors and of objective grating spectrometers. The purpose of these studies will be to establish a sufficient understanding of the mission to be able to identify and assess all major cost and schedule drivers. This in turn will allow us to conduct certain tradeoff studies and to establish valid cost estimates for the C/D phase of the program. These data will be provided to NASA in June 1977.

During the second phase of the study we will continue to work in those areas described above but in much greater detail. Conceptual designs will be generated for all areas of the mission, critical technical areas (such as the mirror) will be designed in detail and actual fabrication techniques will be tested and selected, initial CEI specifications will be drafted, full-up working brassboard imaging detectors will be fabricated and tested, interfaces will be established for other focal plane and non-focal plane experiments, and interfaces with the spacecraft will be coordinated. As a result of the above activity we will submit a detailed C/D phase proposal including the necessary technical and management plans for the program. If we assume that the C/D phase is approved for an Fy 79 start and if we allow 4 years for the C/D phase activity we can anticipate a launch sometime late in 1982.

[2-1]

2.0 TECHNICAL APPROACH

Based upon our HEAO-B experience we have generated a preliminary conceptual description of the 1. 2 meter X-ray telescope in support of an MSFC feasibility study. Since this conceptual description is the starting point for the technical effort proposed here for study, we summarize the major features below.

Figure 2-1 is a schematic drawing (provided by MSFC) showing a strawman payload with the major elements of the payload indicated. The 1.2 meter mirror consists of a number of nested paraboloid-hyperboloid pairs combined by structural elements to constitute a single assembly. The focal plane of the mirror assembly contains a number of instruments which can interchangeably be brought to the focus of the telescope. These instruments are mounted on a major structure with a carousel or equivalent interchange mechanism and are intended for carrying out imaging, spectroscopic and polarimetric observations. The strawman payload also shows two non-focal plane candidate instruments - a monitor proportional counter for measuring source continuum fluxes during spectroscopic and polarimetric observations and an all-sky monitor for detecting transient phenomena over the entire sky as candidates for telescope observations. Also indicated in the figure are the experiment aspect sensors and the optical bench structure required between the mirror and the focal plane instruments. Not shown in the figure but certainly important candidate elements are an objective grating spectrometer and a filter spectrometer which in HEAO-B are located directly behind the mirror.

In support of the feasibility study, we have ray-traced candidate mirror configurations. Figure 2-2 shows the effective area versus wavelength (energy) for a representative mirror configuration for a source on-axis. In this configuration

[2-2]

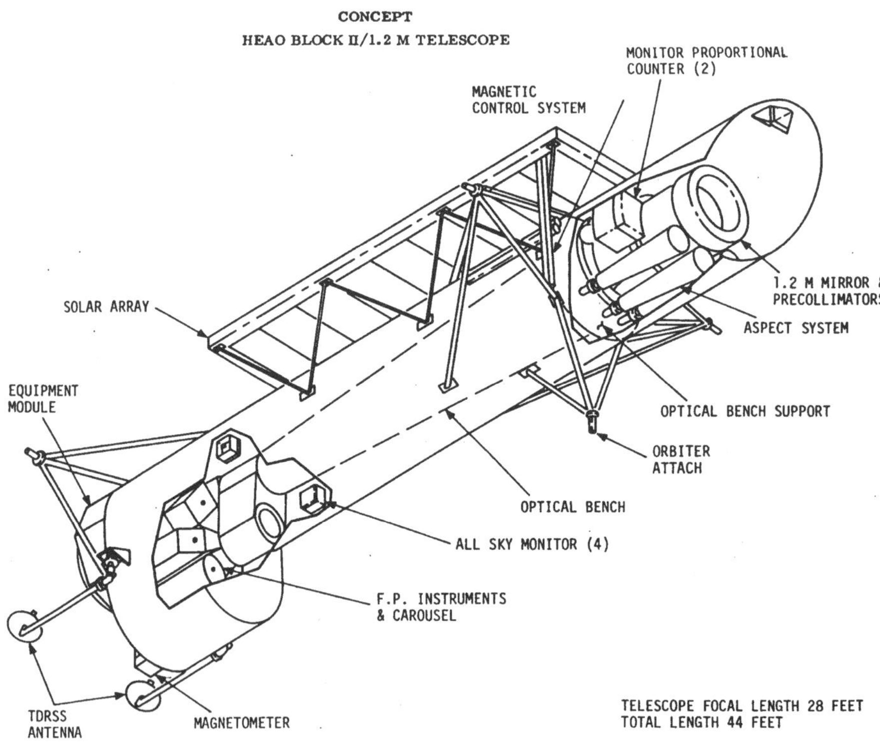

Figure 2-1.

[2-3]

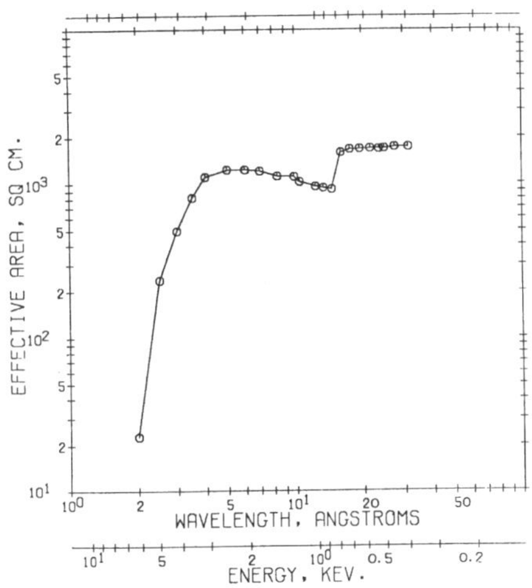

1.2 METER, NO APERTURE LOSSES

Figure 2-2
2-3

[2-4] the mirror would consist of 6 nested paraboloid-hyperboloid pairs with the diameter of the outermost mirror being 1. 2 meters and the diameter of the innermost mirror being ~ 0.6 meters. The mirror segments would be 32 inches in length and the focal length 28 feet (we have also ray-traced 25-ft. focal lengths in support of the MSFC study). The mirror resolution would exceed 0.5 arcseconds for sources within 2.5 arcminutes of the optical axis and would decrease approximately as the square of the angle off-axis. The effective area would be approximately 1700 cm2 at wavelengths longer than 20Å, between 1000 and 1700 cm^2 for wavelengths between 4 and 20Å, 500 cm^2 at 3Å, and 100 cm^2 at 3Å which can be considered the short wavelength cutoff. As a result of the longer focal length, the focal plane scale would be increased by a factor of 2.34 over HEAO-B with 0.5 arcseconds corresponding to ~ 20 microns (~ 3/4 mils) at the focus of the 1.2 meter telescope. The field of view would be similar to HEAO-B with the effective area decreasing by almost a factor of 2 at approximately 1/2° off-axis. At the same time the resolution at 1/2° off-axis will have decreased to approximately 1 arc minute.

These parameters have been used below as guidelines for proposed instrument developments. For example, an imaging detector covering an effective 1/2° (half-cone angle) field of view would require a radius of approximately 75 mm. A high resolution imaging detector covering a 12.5 arcminute (half angle) field of view - over which the telescope resolution exceeds 10 arcseconds - would require a radius of approximately 30 mm.

In the subsections that follow we propose the studies that should be carried on each of these various hardware elements as well as on the overall mission requirements, on the experiment system design, integration, and testing, [2-5] and on the spacecraft requirements. These technical studies are summarized in section 3.0 as a Statement of Work for this proposal.

One final point concerns the rationale for the development of three types of imaging detectors proposed below. One detector would be based on the current HEAO-B microchannel plate (MCP) high resolution imaging (HRI) detector where we have already achieved a 10 micron spatial resolution over a 25 mm diameter field of view. The primary development work for this detector would involve using larger microchannel plates and an appropriately scaled up version of the HEAO-B multiwire crossed grid charge detector. A second detector would be based on a reflection mode negative electron affinity (NEA) photocathode to achieve quantum efficiency close to unity and moderate spectral resolution as well. This detector depends on a successful development program which we have already proposed to NASA to carry out during FY 76 and FY 77 and which has recently been approved. The third imaging detector would be an improved version of the HEAO-B Imaging Proportional Counter (IPC). If we qualitatively compare these three imaging detectors to understand the rationale for having all three, we have the following matrix:

	Detection Efficiency	Spatial Resolution	Spectral Resolution	Field of View	Development Required for 1.2 Meter
MCP	Moderate	High	None	Moderate	Moderate
NEA	High	High	Moderate	Moderate	Substantial
IPC	High	Moderate	Moderate	Large	Moderate

[2-6] 2.1 Mission Requirements Definition

The mission requirements definition is primarily an SAO responsibility which will be performed with the assistance of the overall experiment systems design and integration study contractor(s). The objective of this activity is to define an initial set of requirements which will form the basis of the remainder of the study. Typical items which will be included are

- Mission Constraints
  - Weight, size, power, telemetry, and command constraints
  - Initial system design loads (NASA provided)
  - Strawman spacecraft interface constraints (NASA provided)
  - Orbital parameters
- Experiment Design Requirements
  - Mirror size and overall configuration
  - Area and resolution required and desired
  - Telescope focal length and overall length
  - Error budget including initial allocation to various hardware elements during manufacturing, initial alignment, and orbital performance

- Strawman Payload Definition
    - Number of instruments to be considered in design and possible options on this number
    - Initial weight, size, volume, power, and telemetry allocations
    - Definition of typical instrument peculiar requirements which could be design drivers
    - Typical allocation for consumables
    - Number and typical characteristics of non-focal plane instruments if any

[2-7] Initial Interface Definition
- preliminary allocation of system parameters such as weight, power, volume, data rate, command capability to various elements of experiment system including instruments, mirror, star trackers, integration hardware
- Identification of initial available contingencies
- Definition of lifetime requirements
- Cleanliness requirements and associated restrictions on choice of materials, parts, etc.

The output of this phase will be used for a number of purposes:
- Inputs for an AFO for additional focal plane and non-focal plane instruments
- Basis for design parameters to be used by designers of mirror, star trackers, instruments, and integrating hardware
- Guidelines for selection of experiment configuration to be pursued in more detail
- Basis for selection of tradeoff studies to be included in remainder of study.

The initial requirements definition should be reviewed and updated periodically during the study by all parties involved. SAO will retain responsibility for overall control of this activity. During the second phase of this proposed study the mission requirements definition activities will be expanded to include preparation of initial drafts of the experiment CEI specification, mechanical and electrical interface control documents for the instruments, star trackers, and mirror. In addition, initial inputs will be provided as appropriate for the [2-8] spacecraft/experiment interface control documents. These documents will be prepared by the study contractors) under the direction of SAO.

Document III-27

Document title: Institutional Arrangements for the Space Telescope: Report of a Study at Woods Hole, Massachusetts, July 19–30, 1976.

Source: Space Studies Board, National Research Council, National Academy of Sciences, Washington, D.C.

A decade earlier, the academic astronomy community had recommended to NASA that a university-based institution be set up to manage the utilization of the proposed Large Space Telescope. (See Document III-12.) As the possibility that such a telescope would actually be constructed became more real, the scientific community returned to this theme. Even though NASA's Goddard Space Flight Center had hoped to act as the link between the astronomy community and the Large Space Telescope, NASA decided to establish a Space Telescope Science Institute along the lines recommended in this report.

[cover sheet]
Institutional
Arrangements for the
Space Telescope

Report of a Study
at Woods Hole, Massachusetts
July 19-30, 1976

sponsored by the Space Science Board
for the Assembly of Mathematical and Physical Sciences
The National Research Council

NATIONAL ACADEMY OF SCIENCES
1976

[iii] PREFACE

Since the early days of the space program, it has been recognized that one of its major scientific contributions would come through telescopes placed in orbit outside of the degrading effects of the earth's atmosphere. In the succeeding years, NASA has carried through a number of highly successful programs, including the Orbiting Astronomical Observatories and the Small Astronomy Satellites, which have taken increasing advantage of the possibilities made available through space astronomy.

All of these programs, as well as the program of planetary observation and exploration, have posed the problem of achieving an interface between NASA, which is itself a complex organization involving close cooperation among headquarters, various centers and industrial contractors, and appropriate segments of the scientific community. Since the missions to date have had reasonably specific goals, to be achieved within limited periods of time, this interface has generally been achieved by *ad hoc* arrangements appropriate to the specific mission. By and large this has involved the designation of a principal investigator (PI) who is responsible for the scientific conduct of the mission and whose precise mode of operation is spelled out in a contract between NASA and the PI's institution. It is then the responsibility of the principal investigator to involve other scientists – through a guest investigator program, for example.

When the Space Telescope (ST) is considered, the questions involved in integrating the efforts of NASA with those of the worldwide astronomical community are much more complex. Within NASA the ST program involves the Marshall Space Flight Center, the Goddard Space Flight Center, NASA Headquarters, and the NASA worldwide communications and tracking network. The satellite will be a general-purpose facility, an observatory from which observations can be carried out simultaneously by at least four different investigators. It is hoped that the satellite will have a long life (10 to 20 years), so that its equipment can be upgraded from time to time and occasionally replaced by a new generation of instruments, utilizing the Space Shuttle either to effect the repairs or to bring the observatory back to earth for more extensive refurbishing. Moreover, it is expected that the investigations conducted on the ST will be at the frontier of astronomical research, with respect to both planetary and galactic astronomy, so that it will have an important part in shaping the evolution of astronomical knowledge. [iv] Finally, the ST digital data will be returned at a higher rate than were the data from previous telescopes; ST observations will involve an intimate interaction between scientific investigators and the NASA control and communications organization.

Recognizing both the initial cost and the continuing operating costs of the ST, NASA is anxious to achieve the greatest possible scientific return. With this in mind, the Office of Space Science requested the National Academy of Sciences to undertake a study of possible institutional arrangements for the scientific use of the ST. The study was undertaken by the Space Science Board under the aegis of the Assembly of Mathematical and Physical Sciences. The charge given to the study group by the Assembly is presented as Appendix I.[not included] All of the items in the charge were considered by the study group, but with respect to some points no conclusions were reached.

The basic question posed concerned "the general principles applicable to those ground-based facilities that may be needed to provide the interface between large space observatories and the scientific user community internal to and external to NASA, including relevant communities outside the U.S." More specific questions related to the need for and functions of a Space Telescope Science Institute, the auspices under which the Institute should function and the structure of its policy-making mechanism, and the considerations that govern the location of the Institute.

The study group included astronomers who have previously been involved in space experiments and other astronomers who have done all of their work from ground-based observatories. Some of the group's members have been connected with national facilities in other fields of science, and others have extensive experience in the transmission and processing of digital data.

The work of the study group was carried out in two sessions. At a three day meeting in Washington, including a full day at the Goddard Space Flight Center, the group was briefed regarding the plans for the ST program, the consideration of its organization by members of the NASA staff, experiences with related satellite missions, and considerations arising from the experience at existing scientific institutions.* Subsequently, the group met for two weeks of discussion at Woods Hole, Massachusetts. Despite the variety of backgrounds of the members, the group arrived at a consensus regarding the conclusions of this report. The report has been reviewed and concurred in as a whole by all members of the study group, but no effort has been made to secure unanimity regarding every detail of the presentation. For the details the Chairman assumes responsibility.

DONALD F. HORNIG

*Kitt Peak National Observatory, National Radio Astronomy Observatory, National Accelerator Laboratory, and National Center for Atmospheric Research.

[1] **CONCLUSIONS AND RECOMMENDATIONS**

The following conclusions and recommendations are offered as a summary of this report; amplifying and supporting statements are contained in the main body of the report.

1. Because of the increase in resolving power, greater ability to detect faint objects, and ability to make observations in spectral regions in which the earth's atmosphere is opaque, the probability is high that the Space Telescope (ST) will lead to important discoveries.

2. The productive use of the ST depends upon the safe, reliable operation and maintenance of the spacecraft and its associated communications and data-processing systems, and upon the quality of the astronomical research which is conducted with it.

3. Whereas the operation of the ST and its associated systems is best carried out by NASA, optimum scientific use of the ST requires the participation of the astronomical community.

4. An institutional arrangement, which we call the Space Telescope Science Institute (STScI), is needed to provide the long-term guidance and support for the scientific effort, to provide a mechanism for engaging the participation of astronomers throughout the world, and to provide a means for the dissemination and utilization of the data derived from the ST.

5. We recommend that the STScI be operated by a broad-based consortium of universities and non-profit institutions. The consortium could be developed from an existing consortium or a combination of existing consortia, or a new one could be created for this purpose. The consortium would operate the Institute under a contract with NASA.

6. We recommend that the policies of the STScI be set by a policy board of about ten people representing the public interest, as well as the astronomical community and the broader scientific community. The quality and independence of the policy board is essential to the success of this enterprise.

7. The Institute should have a director and staff of the highest professional stature.

8. The Institute should be of sufficient size, in facilities and staff, to carry out its functions, but should not become so large as to absorb an inordinate fraction of the resources devoted to astronomical research. The institute we [2] envision would be comparable in budget and manpower to other national astronomical facilities.

9. A productive institute requires suitable facilities and a first-rate central staff which, through its own involvement in research with the ST, will ensure the optimum use of the telescope by monitoring and improving the technical performance of the ST and by assisting visiting scientists in making observations and in processing data.

10. We believe the Institute should develop its own engineering and instrumentation facilities at a level which permits the staff to participate, along with NASA centers and oth-

ers, in generating new capabilities for the ST. Unless it is thus involved in advancing the state of the art, it will be difficult for the Institute to recruit and keep a professional staff of the requisite quality.

11. By means of an extended staff and decentralization of some of its functions, the Institute should ensure broad, intimate, and responsible involvement of the astronomical community.

12. The Institute must be responsible for the scientific direction of the ST and should be involved in the provision of modified and second-generation instrumentation.

13. In order to make the best use of scarce and valuable observing time, the astronomical community should be involved through the Institute in the evaluation of experiments and the allocation of time on the telescope.

14. We suggest that initially a portion of the observing time on the ST be allocated to the Institute for two reasons: to assist in recruiting a Director and staff of the highest quality, and to permit the staff to become acquainted rapidly with the possibilities and limitations of the ST and its associated systems.

15. The Institute does not need to duplicate the full image-correction capability at the Goddard Space Flight Center (GSFC), which requires a very large computer, nor need it have access to a high-capacity land line.

16. For scientific interpretation of data, the Institute needs moderate scale computational facilities which can be linked to remote minicomputers at perhaps ten centers in the country.

17. The Institute should coordinate the development of software that can be used at remote sites to reduce and interpret data derived from the ST.

18. We recommend that arrangements be made for close liaison with appropriate NASA organizations. In particular, we recommend that the principal responsibility for liaison with the NASA Project Manager rest with a senior staff member of the Institute and that representatives of the Institute be resident in the Mission Operations Center at GSFC.

19. We recommend that arrangements be made for international participation in the Institute, including its policy-making bodies.

20. It would be advantageous to the Institute if its basic funding could be [3] supplemented with private funds to provide discretionary resources. This might be achieved through an initiation fee and annual dues from members of the consortium. However, to provide long-term stability for the Institute, we believe it important eventually to raise an endowment from foundations and individuals.

21. We recommend that the Institute be favorably located for recruiting a high-caliber staff. We believe this requires proximity to a first-rate scientific center, availability of good schools and housing, and a stimulating environment.

22. We recommend that the Institute be located so that land, buildings, shops, and engineering facilities are available on a scale that will meet the initial needs of the Institute.

23. We recommend that the Institute be located where it has easy access to a major international airport.

24. We have not found any compelling data-handling, managerial, or cost reasons for locating the Institute at an existing NASA center.

25. Special arrangements should be developed to ensure ready access by the Institute

to large, ground-based telescopes that may be needed to support the operation of the Space Telescope.

26. Although the model of an institute we have proposed may be applicable to other space-based astronomy projects, we have not addressed the question of their possible inclusion in the STScI.

27. The selection of a consortium and the search for a site should be initiated in the near future.

Document III-28

Document title: "Announcement of Opportunity for Space Telescope," March 1977.

Source: Space Telescope History Project, National Air and Space Museum, Smithsonian Institution, Washington, D.C.

With the Space Telescope approved as a new start by the White House, NASA in early 1977 felt free to announce to the scientific community the opportunities for its use. This initial Announcement of Opportunity spelled out the conditions for scientists' access to the facility.

---

[1] [Stamped "18 Mar 1977"]
AO No. OSS-1-77

ANNOUNCEMENT OF OPPORTUNITY
FOR
SPACE TELESCOPE

I. DESCRIPTION OF OPPORTUNITY

The National Aeronautics and Space Administration (NASA) announces the solicitation of proposals for scientific investigations and related participation in the Space Telescope (ST). This mission is under consideration for launch by the Space Shuttle during the fourth quarter of Calendar Year 1983. While the implementation of the ST program has not yet been approved by Congress, an early selection of scientific participants will permit a prompt start on scientific planning and related hardware definition, if and when such approval is received. This solicitation does not constitute an obligation on the part of the U.S. Government to carry the proposed effort to completion.

The ST Program is conceived as a long-term program in space astronomy that will provide mankind with an astronomical capability achievable by any current or foreseeable ground-based telescope. A high-resolution 2.4 meter telescope will be placed in a circular Earth orbit at an altitude of approximately 500 km with an inclination of 28.8∞ to the equator. The telescope will be an F/24 Ritchey-Chretien design (34% central obscuration diameter ratio) with a focal plane data field ≥ 0.30 m in diameter. It will provide point-source images with 70% energy within a diameter of £ 0.2 arc seconds at 633 nm and will have useful sensitivity over the wavelength range 120 nm to 1 mm. Up to five Scientific Instruments (SI's) will be accommodated at the focal plane.

[2] The ST will differ from existing automated satellites in that it will be designed to permit on-orbit maintenance and repair by a space-suited astronaut and be retrievable by the Space Shuttle for return to Earth for refurbishment and subsequent relaunch. A feature of the design will be the provision for replacement of any of the focal-plane SI's at the time of on-orbit visits or during ground refurbishment. This will allow updating of the instrumentation and the use of the ST to fulfill a broad range of scientific requirements over its lifetime, which is expected to exceed a decade. The on-orbit visits and/or refurbishments are nominally scheduled for 30-month intervals; however, the exact timing will depend on the operating efficiency and scientific program of the ST.

The overall ST Project is being managed by Marshall Space Flight Center (MSFC). Goddard Space Flight Center (GSFC) is responsible for managing the development of the SI's and for post-launch operation of the observatory.

This Announcement of Opportunity (AO) is a solicitation of proposals for the scientific investigations and the definition and development of focal-plane SI's required to carry out the investigations and for individual "Observatory Scientists" who will be selected to perform scientific investigations using the initial complement of focal-plane SI's and to assist the ST Project in working out scientific mission parameters, as delineated later in this AO. One of the initial focal-plane SI's, a Faint Object Camera (FOC), will be provided by the European Space Agency (ESA), subject to satisfactory negotiation of a formal agreement between NASA and ESA.

Scientists selected through this AO and who participate substantially in the development of the ST and its initial focal-plane SI's will be allocated observing time in the early months of operation.

Specifically, proposals are now solicited from:

A. Investigation Definition Teams (IDT's), each consisting of a Principal Investigation (PI) and appropriate Co-Investigators (Co-I's) proposing a scientific investigation and the definition and development of focal-plane instrumentation required to carry out that scientific investigation;

[3] B. Individual investigators who desire consideration as a Co-I on an IDT (including U.S. representation on the ESA FOC team); and,

C. Individual investigators who desire to participate as Observatory Scientists.

For a more detailed description of all categories of participation, see Section V.A. [not included]

The PI's, Observatory Scientists, and appropriate NASA and ESA scientists will constitute an ST Science Working Group, chaired by the NASA Project Scientist. This Working Group will assist the ST Project in working out scientific mission parameters.

II. MISSION OBJECTIVES

Scientific Objectives

The scientific objectives of the ST are to determine:

A. The constitution, physical characteristics, and dynamics of celestial entities;

B. The nature of processes which occur in the extreme physical conditions existing in and between astronomical objects;
C. The history and evolution of the universe; and,
D. Whether the laws of nature are universal in the space-time continuum.

The ST will increase the sensitivity and resolving power and extend the spectral range of astronomical observations decisively beyond those achievable from ground-based observatories. While it is likely that the ST will reveal unimagined phenomena and, hence, will open new areas of scientific inquiry, the following are examples of the type of specific scientific objectives, within the broader objectives above, which have been suggested. (These specific objectives are not intended to limit the scope of scientific programs which may be proposed by respondents to this AO. However, to be considered for selection, proposals must be within the scope of the scientific objectives of the ST, as stated above.) [4]

- Precise determination of distances to galaxies out to expansion velocities $\sim 10^4$km s^{-1} and calibration of distance criteria applicable at cosmologically significant distances
- Determination of the rate of the deceleration of the Hubble expansion of the universe, its uniformity in different directions, and possibly its constancy with time
- Testing of the basic reality of the universal expansion by determination of the surface brightness versus red shift relation for distant galaxies
- Establishment of the history of star formation and nuclear processing of matter as a function of position in nearby galaxies and determination of the variations from galaxy to galaxy
- Determination of the nature of stellar populations in the early stages of galactic evolution, based on "lookback" observations of distant galaxies
- Estimation of the He/H ratio in quasars by observation of red-shifted He I and He II resonance lines
- Search for multiple-red-shift absorption line groups in the ultraviolet spectra of low-red-shift quasars
- Intercomparison of total spectra of high-red-shift quasars, low-red-shift quasars, and active galactic nuclei
- Resolution of densely-packed nuclei of globular star clusters in search of massive black holes
- Identification and flux measurement in ultraviolet and optical wavelengths of faint x-ray sources and radio pulsars
- Resolution of the complex internal structure of Herbig-Haro objects to investigate their possible links to star formation

[5]
- High spatial resolution, infrared observations of proto-stars
- Direct imaging and astrometric search for planetary companions of nearby stars
- Determination of bolometric luminosities of faint, hot stars for studies of stellar evolution
- Determination of composition, temperature, density, and ionization structure of the gas in the galactic halo, in high-velocity clouds, and in the intergalactic medium
- Precise mapping of the 100 mm flux sources in compact H II regions

- Determination of composition of clouds in the atmospheres of Jupiter, Saturn, Uranus, and Neptune
- Surface mapping of the Galilean satellites of Jupiter and of asteroids
- Synoptic mapping of atmospheric features on Venus, Jupiter, Saturn, and Uranus
- Intensity measurements of atomic and molecular ultraviolet emission lines important to understanding the chemistry of comets

III. SPECIFIC REQUIREMENTS AND CONSTRAINTS ON INVESTIGATION DEFINITION TEAM (IDT) PROPOSALS

The ST and its SI's will be designed and developed within the framework of a tightly cost-controlled program, leading to the proposed launch in the fourth quarter of 1983. The SI's will be delivered to GSFC 28 months prior to launch for NASA acceptance testing and verification. The number and amount of associated development of instruments required for investigations selected for the first mission will be limited by the total funding available in the program budget. It is imperative that proposers establish a credible low-cost approach and schedule for the development of flight-quality instruments and associated hardware.

[6] Initial IDT selections will be tentative. Following tentative selection, each PI will be funded to carry out a preliminary design study, including breadboarding or other activities necessary to verify the detailed approach and costs for the final instrument development phase. At the end of this initial activity, the PI (if responsible for instrument development) will submit an updated implementation plan and schedule for the instrument development. NASA will then reevaluate the scientific, technological, and cost aspects of each instrument. Based on this reevaluation, the NASA Associate Administrator for Space Science will confirm the participation as originally determined, direct modifications to meet Project schedule and funding limitations, or terminate the proposed participation. At that time, firm commitments to performance specifications, costs, schedule, and scope will be established for each instrument development to be completed. Contracts for participation in the operations phase of the Project and for the timely deposit of the resulting data in the National Space Science Data Center will be negotiated later.

Preliminary studies have been carried out for the types of instruments listed below. Final reports on these studies, as well as on separate studies of various detectors appropriate for use in the instruments, are available by use of the attached order form (Attachment M) [not included].

A. Ultraviolet Spectrographs
   1. Faint Object Spectrograph (FOS)
   2. High Resolution Spectrograph (HRS)
   3. Combined FOS-HRS
B. High Resolution Cameras
   1. Wide Field Camera (WFC)
   2. Faint Object Camera (FOC) (to be provided by ESA)
   3. Planetary Camera (PC)
C. Photometers
   1. Infrared Photometer (IRP)
   2. High-Speed Photometer (HSP)

D. Guidance System (Astrometry)

[7] NASA has determined that, for reasons of cost-effectiveness, astrometry will be carried out with the spacecraft guidance system and no separate SI will be developed for astrometry. NASA will, however, select a scientific investigation in this area involving a PI and possibly several team members.

Also, NASA (with the advice of the astronomical community) has designated the WFC and the FOS as particularly important. If qualified proposals for investigations involving these instruments are received, a WFC and an FOS will be selected for inclusion in the initial instrument complement. Qualified proposals are those for investigations which are considered to be well conceived, scientifically and technically sound, and pertinent to the goals and objectives of the ST Program. In addition, they must be offered by a competent investigator from an institution capable of supplying the necessary technical and management capabilities to ensure that flight hardware or other support can be delivered on time and within budget and that data can be properly reduced, analyzed, interpreted, and published in a reasonable time. Finally, qualified proposals are those recommended, with high priority, for tentative selection.

Proposals may be submitted for investigations responsive to the stated objectives of this AO and entailing the design, development, and scientific application of any suitable instrument (excepting an FOC or astrometry instrument), whether or not it was previously studied. The proposer must establish the scientific merit, technical feasibility, interface compatibility, consistency of the effort with the proposed budget and schedule, and the existence of wide interest within the scientific community.

A working description of the FOC to be provided by ESA is contained in the attached summary of scientific performance goals (Attachment H) [not included].

IV. ALLOCATION OF OBSERVING TIME

Those scientists selected for participation in the ST Program as a result of this solicitation will be involved in the development of the ST and its initial complement of instruments. After launch, they will be involved in the on-orbit checkout expected to last about one month, after which the ST will be declared operational.

[8] As has been the custom for ground-based telescopes, observing time will be assigned primarily on the basis of the scientific merit of the competitive observational proposals. During the first 30 months after the ST is declared operational, a portion of the observing time will be allocated to the IDT's, the Astronomy Team, and the Observatory Scientists selected as a result of this solicitation. The remainder of the observing time until the first on-orbit maintenance will be available for the general ST Observer Program which will be the subject of subsequent AO's, the first of which will be issued about nine months before the initial launch. Scientists selected now will be free to compete for this additional time.

The monthly percentages of the total observing time, which will be allocated to the IDT's, the Astronomy Team, and Observatory Scientists, are as follows:

Period	% Allocated
Months 1-2 (after checkout period)	100%
Months 3-8	50%
Months 9-20	25%
Months 21-30	10%

In dividing the above time allocations among the various participants, the Observatory Scientists, as a group, will be treated in the same manner as a single IDT or Astronomy team. The U.S. member(s) chosen for the ESA FOC Team will be allocated time comparable to that allocated to Co-I's on U.S. IDT's. While the IDT's will have priority on the use of their own instrument during this allocated time, they may negotiate observing time on other SI's which may be required to complement their own investigations.

NASA is considering the establishment of an ST science operations facility which would manage the scientific use of the ST during its operational life. Observing time for all observers, including those selected by this and subsequent AO's, would be scheduled by this operations facility. PI's will have an ongoing responsibility during the period after post-launch checkout to assist in the operation, calibration…

[9] V. EVALUATION AND SELECTION PROCESS

A. Categories of Participation

1. Investigation Definition Teams (IDT's)

Proposals are solicited for scientific investigations and for definition and development of scientific focal-plane instruments for the ST required to carry out the investigations. In addition to a complete description of the scientific investigation, proposals should cover the development of the entire required instrument, including detectors. The proposing Team should consist of a PI and the Co-I's necessary to complete the proposed investigation. NASA may elect to delete members of the proposed Team and/or augment the membership of Teams with selected scientists who have proposed individually or as part of other Teams.

2. Individual Investigators

Proposals are invited from individual scientists not allied with a proposing Team who wish to propose scientific investigations which may impact the design of a particular instrument or who are interested in the astrometric use of the fine guidance system. NASA may select such individual scientists as PI's or as additional Co-I's on IDT's. The U.S. scientists interested in performing investigations with the FOC and participating on the ESA FOC Team should also propose as individuals. It should be noted that the one or more U.S. members of the FOC Team will need to travel extensively. Individual instrumentalists, e.g., detector specialists, are encouraged to affiliate with more than one IDT proposal. Separate detector proposals should not be submitted to NASA in response to this AO.

[10] 3. Observatory Scientists

Proposals are invited from scientists wishing to serve in one of the following capacities: Telescope Scientists, Interdisciplinary Scientists, and Data and Operations Team Leader.

Telescope Scientists will carry out scientific investigations and will assist the ST Project to assure that the spacecraft is compatible with scientific requirements. In addition to proposing a qualified scientific investigation for the ST, individuals proposing as Telescope Scientists should have a broad knowledge of imaging optical systems, with a working knowledge of the relation of optical and thermal mechanical systems and their effects on the performance of the observatory.

Interdisciplinary Scientists are generalists or theoreticians who, in addition to carrying out scientific investigations using the ST, will serve the Project by maintaining a broad and critical scientific overview of the ST development. Proposers as Interdisciplinary Scientists should not only propose qualified specific scientific investigations using the ST but should also have a thorough knowledge of the forefront problems of modern astronomy and astrophysics and a clear interest in applying the capabilities of the ST to those problems. Moreover, these scientists must be able to relate the potential attainment of general scientific goals to specific characteristics of the ST and SI performance.

The Data and Operations Teams Leader will carry out scientific investigations and will assist the Project to assure that the SI Control and Data Handling System is consistent with scientific requirements. He will head a Team consisting of a representative from each of the IDT's, which will assist in developing the detailed requirements for the SI Control and Data Handling System and in providing integrate SI operations and data management requirements. Proposers for the Data and Operations Team Leader, in addition to proposing a qualified scientific investigation for the ST, should have a good understanding of instrument and control systems, flight operation, and ground data handling systems.

Document III-29

Document title: "Memorandum of Understanding Between The European Space Agency and The United States National Aeronautics and Space Administration," October 7, 1977.

Source: NASA Historical Reference Collection, History Office, NASA Headquarters, Washington, D.C.

In order to gain Congressional approval of the Space Telescope, NASA had agreed that it would seek significant international participation in the effort. In practice, given the status of other space programs in the 1970s, this meant inviting European participation in the project. Europe agreed to supply a major scientific instrument, the Faint Object Camera, and solar panels for the telescope in return for a guarantee of fifteen percent of the observing time. While NASA agreed to this arrangement, it was controversial among some members of the U.S. scientific community, both because they could not compete to supply the scientific instrument and the arrangement might decrease the observing time available to them.

[no pagination]

MEMORANDUM OF UNDERSTANDING
BETWEEN
THE EUROPEAN SPACE AGENCY
AND
THE UNITED STATES NATIONAL AERONAUTICS AND SPACE ADMINISTRATION

ARTICLE 1
Purpose

The European Space Agency (ESA) and the United States National Aeronautics and Space Administration (NASA), desiring to extend the fruitful cooperation developed in previous space projects, agree that ESA will participate in the NASA 2.4 Meter Space Telescope (ST) Project, as described below.

ARTICLE 2
Mission

The mission of the ST Project is to provide a space observatory for use by the international astronomy community to extend the sensitivity, resolving power and spectral range of astronomical observations decisively beyond those achievable from earth observatories.

ARTICLE 3
Participation

1. To carry out this project, NASA plans to launch the ST by means of the Space Shuttle in1983/1984 and to operate it for a period of 10 to 15 years. ESA agrees to assist in the provision of scientific instruments and subsystems for the ST, in the operation of the ST and related facilities, in the in-orbit maintenance, major refurbishments and reflights, at anticipated intervals of two or three years, and to arrange for participation of ESA-sponsored astronomers in the observation programs.

2. ESA will carry out its participation in accordance with the plans, specifications and schedules contained in the NASA/ESA ST Project Plan annexed to this Memorandum of Understanding. [not included] This plan may be subject to change as defined in Article 7 (d) below.

ARTICLE 4
ESA responsibilities

To implement the cooperation in this project ESA will:

(a) Provide the Faint Object Camera (FOC), a scientific instrument of high sensitivity for high resolution imagery in the ultraviolet, visual, and near infrared portions of the spectrum. The performance specification for this instrument is included in the NASA/ESA ST Project Plan annexed to this Memorandum of Understanding. With respect to the provision of the FOC, ESA will:

(i) design, fabricate, test, calibrate and deliver for integration into the ST the FOC, comprising the camera optics and an Imaging Photon Counting System (IPCS). The FOC models and associated hardware together with the schedule for their delivery are defined in the NASA/ESA ST Project Plan;

(ii) provide personnel and equipment to support NASA activities relating to testing, integration of the FOC with the ST, and launch site operations, as defined in the NASA/ESA ST Project Plan;

(iii) set up an ESA Instrument Science Team (IST) to ensure the scientific integrity of the definition and design of the FOC, and its compatibility with the telescope, and to provide scientific advice for the ESA project management, which will manage FOC development. It is agreed that at least one NASA-appointed scientist will be a member of the IST;

(iv) provide participation in NASA reviews in accordance with Article 8 (b) (iii) below;

(v) develop the ground and flight software packages uniquely required for the FOC, as defined in the NASA/ESA ST Project Plan;

(vi) provide a team of two to four instrument and data scientists to support the operations of the FOC after launch, as mutually agreed; and,

(vii) refurbish the FOC for reflight, and provide necessary support for maintenance or modification of the FOC in-orbit, when mutually agreed. In the event NASA wishes to refurbish or modify the FOC and ESA does not desire to do either, ESA will provide the available documentation and other support as agreed between the Project Managers.

(b) Provide the ST Solar Array. In this respect, ESA will:

(i) design, fabricate, test and deliver to the NASA Support Systems Module (SSM) contractor for integration with the Space Telescope a complete solar array and associated hardware, the performance specification, interface requirements, number of models and hardware parts and their delivery schedule being defined in the NASA/ESA ST Project Plan;

(ii) provide personnel and equipment to support the handling and testing of the solar arrays while they an not attached to the ST spacecraft, as defined in the NASA/ESA Project Plan;

(iii) provide appropriate representation in NASA reviews in accordance with Article 8 (b) (iii) below;

(iv) maintain continuing engineering liaison with NASA ST project elements which interface with the ESA solar array;

(v) provide personnel and equipment to support NASA activities relating to testing, integration of the solar array with the ST, and launch site operations, as defined in the NASA/ESA ST Project Plan; and

(vi) refurbish, repair, and/or replace the solar array and provide support to NASA for its maintenance or modification in-orbit, as mutually agreed and reflected in the NASA/ESA ST Project Plan.

(c) Participate in the activities of the science operations facility as described in Article 7 (g) below. In this respect ESA will, in particular:
   (i) provide a team of approximately six persons to support one shift of the scientific planning and operation tasks of the science operations facility;
   (ii) provide a team of approximately seven persons to support the ST scientific data reduction shift of the science operations facility;
   (iii) provide membership to this facility, as mutually agreed, in order to participate in the definition and implementation of the ST science activities; and
   (iv) provide a representative to NASA to serve as a focal point for NASA/ESA consultations regarding the establishment and operation of the science operations facility; it being understood that the members of the ESA teams mentioned in (i) and (ii) above will be considered functionally as members of the science operations facility.

ARTICLE 5
NASA responsibilities

To carry out the cooperation in this project, NASA will:

(a) Design, fabricate, test, integrate and prepare for launching the complete ST assembly.
(b) Define jointly with ESA the solar array interfaces, design requirements, and test and handling specification, and provide this information to ESA in accordance with the schedule defined in the NASA/ESA ST Project Plan.
(c) Test the solar array when it is integrated on the ST and supply ESA with relevant engineering data. NASA is responsible for handling of the solar array after its acceptance at the integration site.
(d) Provide appropriate representation at ESA reviews in accordance with Article 8 (b) (iv) below.
(e) Provide and transport to the launch site all necessary ST ground support equipment, including items delivered by ESA to the test and integration sites to support the FOC and the solar array.
(f) Transport to the launch site the flight qualified ST and perform necessary tests and checkout prior to launch.
(g) Provide initial launching of the ST on a Space Shuttle, release the ST into the designated orbit, and conduct, with ESA participation, checkout of the ST as required for a period of approximately 30 days following launch.
(h) Provide all tracking and data acquisition services during the lifetime of the project.
(i) Exercise all in-orbit control functions. To accomplish this, NASA will organize, equip, staff, train and operate an operations center.
(j) Include the FOC in the payload of the first flight of the ST and operate it in accordance with the provisions of Article 6 below.
(k) Establish a science operations facility to conduct scientific operations as outlined in Article 7 (g) below.

(l) Provide in-orbit maintenance, and retrieve and relaunch the ST when necessary according to operations plans, and conduct refurbishment and in-orbit modification operations assisted by ESA as provided in Articles 4 (a) (vii) and 4 (b) (vi) above.

ARTICLE 6
Flight arid Operation of the ESA-supplied FOC

I. The following principles shall apply to the flights of the ESA-supplied FOC:
   (a) Assuming normal functioning, it will be operated by NASA for a nominal initial period of thirty months.
   (b) Following this period, it will be flown as long as it is decided to be a component of the scientifically optimum payload.
   (c) Following the initial or subsequent operating periods, both parties may nevertheless agree that NASA will continue to fly it in modified form, in application of Article 4 (a) (vii), or replace it for a given period by another scientific instrument.
   (d) If the FOC fails to meat the minimum scientific requirements laid down in the performance specification as defined in the NASA/ESA ST Project Plan, or if the delivery by ESA would substantially delay the first flight, NASA may decide, after consultation with ESA, to launch the ST without the FOC. In this event, provided that, at the relevant time, the FOC does meet the minimum scientific requirements referred to above, NASA will take its availability into account when funding the development of ST instruments for subsequent flights and accept it as a candidate instrument for those flights.

II. Decisions concerning the implementation of the provision of the foregoing paragraph shall be made by the NASA Associate Administrator for Space Science and the appropriate Director at ESA Headquarters, subject to the application of the provisions of Article 18 of this Memorandum of Understanding.

ARTICLE 7
Management and Organization

Unless otherwise provided in this Memorandum of Understanding, the management arrangements are understood by NASA and ESA to be as follows:

(a) NASA will establish a ST Project Office to provide for project planning and management with the following responsibilities:
   (i) Overall responsibility for the design, fabrication, test, integration, launch, in-orbit verification, and operation of the ST;
   (ii) flight performance evaluation, and planning for and carrying out in-orbit maintenance and modification; and
   (iii) planning and carrying out ST refurbishments.
(b) The Project Office will be headed by a NASA Project Manager. To carry out the ST Project, the Project Manager will be supported by a full-time staff of managers and engineers. Full responsibility for management of the ST Project resides with the NASA Project Manager. The Project Office provides the principal means for carrying out these management responsibilities.

(c) ESA will:
  (i) Designate an ESA Project Manager for overall coordination, planning and execution of the ESA tasks described in this Memorandum of Understanding, and will assign appropriate representation to the ST Project Office, as mutually agreed; and
  (ii) Appoint an ESA Project Scientist.

(d) Management and technical decisions which have a bearing on the execution of ESA tasks as described under this Memorandum of Understanding or affect the contents of the annexed NASA/ESA ST Project Plan, and items with respect to which mutual agreement is necessary in accordance with provisions of this Memorandum, shall be taken in common by the NASA and the ESA Project Managers. If they are unable to come to an agreement on a particular issue, it shall be referred to the NASA Associate Administrator for Space Science for resolution in consultation with the appropriate Director at ESA Headquarters, subject to the application of the provisions of Article 18 of this Memorandum of Understanding.

(e) NASA will establish a Space Telescope Science Working Group (ST-SWG) which will be the principal mechanism for scientific input to the Project Office during the development of the ST. The ST-SWG will be chaired by the ST Project Scientist designated by NASA. The ESA membership on the ST-SWG will consist of the ESA Project Scientist referred to in paragraph (c) of this Article and of the Chairman of the ESA IST referred to in Article 4 (a) (iii), who will be assisted by such members of his team as he desires to be present for specific ST-SWG meetings.

(f) NASA will establish a mission operations center, as defined to the NASA/ESA Project Plan to:
  (i) Carry out mission operations planning;
  (ii) execute ST command and control;
  (iii) acquire data on and evaluate ST engineering performance;
  (iv) report ST anomalies;
  (v) reduce ST engineering data; and
  (vi) acquire and process ST scientific data.

(g) NASA will establish a science operations facility to carry out scientific management of the observatory in orbit. Operational support of the investigators by this facility will include, but is not limited to:
  (i) Planning long-term scientific operations;
  (ii) scheduling daily scientific observations;
  (iii) conducting real time scientific observations; and
  (iv) performing ST scientific data management, including making available ST scientific data to investigators in a form suitable for analysis.

ESA will be represented and participate in this facility in accordance with the provisions of Article 4 (c).

ESA participation in the daily activities of the science operations facility will include, but will not be limited to the provision of the support provided under Article 4 (a) (vi) and 4 (c) above.

ARTICLE 8
Technical interfaces

The management of the technical interfaces on the ST will be carried out with a minimum of documentation and formal reviews. It is understood between ESA and NASA that the following principles and procedures will apply:

(a) General Responsibilities for Technical Interfaces
(i) The NASA Project Manager, in accordance with Article 7 (b) above, is responsible for the management of the interfaces, including documentation, general control of the use of such documentation, and the conduct of technical reviews on all systems for which NASA is responsible:
(ii) The ESA Project Manager is responsible to work to such requirements and interfaces as he and the NASA Project Manager have mutually agreed to, and for the conduct of technical reviews to insure that all systems and hardware for which ESA is responsible comply with ST interface requirements.
(b) Specific Responsibilities and Procedures
(i) The NASA Project Manager will review and mutually agree with the ESA Project Manager, as to which standards and specifications will be considered to constitute the requirements for control purposes in the ST Project. The agreed standards and specifications, if any, will be referenced as part of the NASA/ESA ST Project Plan.
(ii) ESA will supply data to NASA for the generation of appropriate Interface Requirements Documentation (IRDs) Interface Control Drawings (ICDs) and Contract End Item (CEI) Specifications Part I for the Solar array and the FOC. The CEI Specifications Part II for the Solar array and the FOC will be generated by ESA. The Project Managers will mutually agree on these documents, and the NASA Project Manager will approve them in accordance with a schedule in the NASA/ESA ST Project Plan. Subsequent modifications to either the IRDs, the ICDs, or the CEI Specification will be approved by appropriate change control procedures identified in the NASA/ESA ST Project Plan. Should such modifications seem unacceptable to either Project Manager for financial or schedule reasons, the provisions of Article IS will apply.
(iii) NASA will make final determination of the overall readiness of the ST for launching. This determination will be based on periodic reviews chaired by NASA to address the concept, design, and readiness for flight of the ST. ESA will have appropriate representation at selected reviews and will furnish engineering data as agreed by the Project Managers.
(iv) Determination of the readiness for integration of the solar array and the FOC will be based on periodic reviews, chaired by ESA, of the concept, design and readiness for flight of the hardware. NASA will have appropriate representation at these reviews as agreed between the Project Managers and will furnish engineering data as agreed by the Project managers. Final determination of the readiness for integration of the solar array and the FOC will be the

responsibility of the NASA Project Manager, based on recommendations from the ESA Project Manager and review committees.

(v) NASA and ESA shall have full access to, and the right to use and disclose, non-proprietary data necessary to discharge their respective responsibilities under this Memorandum of Understanding. In principle, there will be no transfer of documents bearing proprietary or other restrictive markings. To achieve this, optimal use will be made of mathematical models, mock ups and simulators, as appropriate to assure hardware interface and operations compatibility. Should either party consider that the above limitation prevents it from carrying out a particular responsibility under this Memorandum of Understanding, the Project Managers will determine a mutually agreeable solution on a case-by-case basis. In the event that the solution must involve the transfer of proprietary data, the furnishing party, with the consent of the proprietor, shall furnish the data and shall merit them with a notice limiting the use and disclosure of the information for ST Project purposes only, and the receiving party will use its best efforts to comply with such limitations.

(vi) Detailed arrangements for working level technical interfaces, including NASA and ESA contractors, are defined in the NASA/ESA ST Project Plan.

ARTICLE 9
Apportionment of Observing Time

1. The term "observing time" as used in this Article is understood to mean that time during which the ST instruments are in operation, less idle time and time necessary for calibration, testing and maintenance.
2. NASA and ESA agree that ST observing time will be made available to investigators from the international community of astronomers on the basis of the scientific merit of proposals made.
3. Subject to the application of this principle, and in consideration of ESA's participation as defined in this Memorandum of Understanding, ESA will obtain, for use by ESA-sponsored astronomers, a portion of the observing time on the total complement of scientific instruments of the ST. It is expected and intended that this portion will be not less than 15 % of the observing time on the average over the lifetime of the ST Project.

ARTICLE 10
Selection of Observing Programs

1. All proposals for observing programs from astronomers in ESA member States will be submitted in accordance with ESA procedures to the ST Proposal Review Committee (STPRC), which will be the primary body for the review and evaluation of all proposals for observing programs to be carried out on the ST.
2. The STPRC will have an appropriate European membership (minimum of two), the number to be agreed between ESA and NASA Headquarters. It will make recommendations for observing programs to the NASA Associate Administrator for Space

Science, who, after consultation with ESA as necessary, will make a final determination of the observing programs.

3. Should ESA consider that the observing programs so determined are inconsistent with the provisions of Article 9, or should there be a major alteration in the ESA participation in the ST Project, either party may request a joint review of the evaluation and selection process.

ARTICLE 11
Deposit Accounts

Should ESA desire that NASA procure goods and services on ESA's behalf to assist ESA in carrying out an ESA responsibility under this Memorandum of Understanding, NASA is prepared to consider such requests on a case-by-case basis under the provisions of a Deposit Account Agreement to be negotiated separately. Similar requests from NASA will be treated by ESA accordingly.

ARTICLE 12
Funding Arrangements

Each Agency will arrange to meet the cost of discharging its responsibilities, including travel and subsistence for its own authorized personnel and transportation charges on all equipment and flight hardware for which it is responsible. Other than deposit account transactions referred to in Article 11, there will be no exchange of funds between ESA and NASA.

ARTICLE 13
Custom and Visas

ESA and NASA will use their best efforts to arrange free customs clearance for equipment required in the ST project. NASA will use its best efforts to facilitate the issuance of visas to European astronomers and ESA contractors collaborating in the ST Project.

ARTICLE 14
Data rights

Use of ST scientific data for scientific analysis will be reserved to investigators for a twelve month period, beginning with the receipt of data and any associated spacecraft data in a form suitable for analysts. Investigators may occasionally be requested to share data to enhance efficient utilization of the observatory and of ground observing operations. Immediately after the period reserved to the investigator ,reduced data will be deposited with the National Space Science Data Center (NSSDC) and with the science operations facility. In addition, European investigators will deposit their data in the Data Library of the European Space Operations Center (ESOC). Such records will then be available to the international scientific community through the World Data Center for Rockets and Satellites. It is agreed that a listing of all observations will be published at least every six months in sources readily available to astronomers.

ARTICLE 15
Publication of Results

Subject to the provisions of Article 14 above, results of the experiments will be made available to the scientific community in general through publication in appropriate journals or other established channels as soon as possible and consistent with good scientific practice. Reprints of scientific and technical reports and publications resulting from this project will be exchanged between ESA and NASA. In the event that such reports or publications are copyrighted, ESA and NASA shall have a royalty free right under the copyright to reproduce and use such copyrighted work for their purposes. Final reports and publications will be placed in the Data Library of ESOC and in the science operations facility.

ARTICLE 16
Public Information

Each Agency may release information to the public regarding its own activities covered by this Memorandum of Understanding. Each Agency undertakes to coordinate with the other in advance those public information activities which relate to the other Agency's responsibilities or performance in the ST project. Implementing arrangements for these public information activities will be agreed separately.

ARTICLE 17
Limits of Obligation

It is understood that the ability of ESA and NASA to carry out their obligations under this Memorandum of Understanding is subject to the availability of appropriate funds.

ARTICLE 18
Disputes

1. Any dispute as to the interpretation or implementation of the terms of this Memorandum of Understanding shall be referred to the NASA Administrator and the Director General of ESA for settlement.
2. Should the NASA Administrator and the Director General of ESA be unable to resolve such disputes, they will be submitted to such other form of resolution or arbitration as they may agree.

ARTICLE 19
Liability

1. NASA shall bear responsibility for damage to US nationals in the course of this cooperative project, unless such nationals are employees of ESA.
2. ESA shall bear full responsibility for such damage to ESA employees.
3. NASA shall be liable for damage to those items delivered to it by ESA in accordance with Article 4, after the accomplishment of the relevant receiving inspections defined

in the NASA/ESA ST Project Plan, but shall not be liable for damage occurring to such items in connection with the Space Shuttle launch, flight or descent.

4. In the event of damage to other persons or property, for which damage there is liability under international law or the principles of the Convention on International Liability for Damage caused by Space Objects, NASA and ESA shall consult promptly on an equitable sharing of any payments that have been or may be agreed in settlement. If agreement is not reached within 180 days, the two Agencies will act promptly to arrange for early arbitration to settle the sharing of such claims following the 1958 model rules on arbitral procedure of the International Law Commission.

ARTICLE 20
Patent use – Authorization, Consent and Indemnification

1. In order to avoid any possible interruption to the conduct of this cooperative project which might arise from patent infringement litigation in U.S. Courts, NASA hereby gives authorization and consent (without prejudice to any rights of indemnification) for all use or manufacture by ESA of any invention described in and covered by a patent of the United States in the performance of any obligations under this Memorandum of Understanding, including the performance of any such obligations by any contractor or subcontractor, providing such use and manufacture is confined entirely to the discharge of the obligations of this Memorandum of Understanding.
2. In the event any liability is incurred by the US Government for the practice of inventions covered by privately owned U.S. patents, either as royalties owed under an existing patent license inuring to the benefit of NASA or as judgement and litigation costs resulting from a suit for patent infringement in the U.S. Court of Claims, and such liability is incurred as a result of ESA's and/or any of its contractors' or subcontractors' performance of obligations under this Memorandum of Understanding, or as a result of NASA's use under this Memorandum of Understanding of the items furnished by ESA under this Memorandum of Understanding, ESA agrees to indemnify NASA or any other U.S. Agency against, and make reimbursement for such royalties and/or costs. ESA shall provide such information and assistance as it has available in the defense of any such. patent infringement suit brought in the U.S. Court of Claims.

ARTICLE 21
Amendments

Each party may propose to the other amendments to this Memorandum of Understanding in writing. Agreements on such amendments shall be established by the parties in the form of riders to this Memorandum of Understanding.

ARTICLE 22
Termination

This Memorandum of Understanding shall enter into force when both the NASA Administrator and Director General of ESA have signed it, and it shall remain in effect for a first period of eleven years from the first launching of the ST. It is anticipated that this period wilt include at least ten years of ST space operations. At least one year, before the expiration of the eleven year period, the parties agree to consult as to continuation or termination of this Memorandum of Understanding. The Memorandum of Understanding will continue in its present form unless it is terminated or amended.

For the National Aeronautics and Space Administration	For the European Space Agency
[signature] R. Frosch	[signature] R. Gibson
7 October 1977	7 October 1977

Document III-30

Document title: Post Launch Mission Operation Report to NASA Administrator from Thomas A. Mutch, NASA Associate Administrator for Space Science, August 16, 1979.

Source: NASA Historical Reference Collection, History Office, NASA Headquarters, Washington, D.C.

After NASA's fourth Small Astronomy Satellite, an ultraviolet mission, was postponed due to budget constraints, the space agency teamed with the European Space Research Organization (later the European Space Agency [ESA]) and the United Kingdom's Science Research Council to redefine the project, which became known as the International Ultraviolet Explorer (IUE) mission. Launched from the United States in 1978, IUE succeeded in meeting all of its scientific objectives. IUE became one of the world's most successful astronomy missions ever, generating more than 3,000 scientific papers among astronomers worldwide. Although the satellite was still operating well twenty years after launch, it was turned off so funds for its operation could be used for other space science purposes.

[no page number]

Post Launch
Mission Operation Report
No. S-868-78-03
August 16, 1979

TO: A/Administrator

FROM: S/Associate Administrator for Space Science

SUBJECT: International Ultraviolet Explorer (IUE)
Post Launch Report # 2

The International Ultraviolet Explorer (IUE) is adjudged successful based upon the results of the mission with respect to the approved prelaunch objectives.

The IUE, an Explorer-class ultraviolet astronomy mission, is an international cooperative program between the United States, the United Kingdom (UK), and the European Space Agency (ESA) which provides for a single launch into a geosynchronous orbit to conduct spectral distribution studies of celestial and solar system ultraviolet sources. The spacecraft and scientific instrument were [sic] designed and fabricated at the Goddard Space Flight Center. The spectrograph camera system was provided by the UK; ESA provided the Solar Array as well as the European Ground Station.

The available observing time is shared roughly equal to the respective contributions, with the US having two-thirds and the UK and ESA sharing equally in the remaining one third.

The IUE observatory system was designed to functionally resemble a ground-based optical observatory at which guest observers could execute observing programs in real time. Observations are made from ground stations at GSFC and Madrid, Spain.

In the 15 months since IUE commenced routine guest observer operations on April 3, 1978, Observatory performance has substantially exceeded design and mission objectives. At high resolution, spectra of stellar sources has been obtained as faint as 12th magnitude while at lower resolution, observations have been made of extragalactic sources fainter than 17th magnitude. The latter observations required 3-axis stabilized pointings in excess of 14 hours continuously.

The secondary mission objectives have also been met. The IUE gyros have been selected for Space Telescope (ST) use; the IUE Spectrograph is a forerunner of the ST High Resolution Spectrograph; the IUE cameras have influenced the design of detectors for the ST Faint Object camera; and the IUE operational software and guest observer operations will provide an experience base for ST.

In addition to the high quality of the output, the data productivity is also great. NASA guest observers have obtained over 6000 images supporting more than 100 different research programs. The UK and ESA guest observers have produced almost 3000 images in support of 150 or so research programs. As a result, scientific results are [2]being widely reported. Well over 100 papers have been presented at various meetings and symposia both in this country and abroad. By the time of the first anniversary in orbit, January 1979, 15 publications had already appeared in NATURE and the Astrophysics Journal Letters (Attachment). [not included] Many more have been published or are in preparation (Attachment 2 is a partial summary of results from NASA observers). [not included]

A summary of a selected number of the most important results obtained to date follows:

a. Discovery of mass loss in hot subdwarfs and of "cool" stellar winds in G and K supergiants.

b. Delineation of the region in the HR diagram exhibiting chromospheric phenomena.
c. Discovery of short-term variability in line profiles of OB supergiants.
d. Discovery of gold in A peculiar stars.
e. Detection of CR II in the interstellar medium.
f. Discovery of hot circumstellar shells around stellar X-ray sources.
g. Discovery of bright UV sources at the centers of some globular clusters.
h. The first ultraviolet observations of a recurrent nova indicating that the ejected mass is an order of magnitude less than for classical novae.
i. The first ultraviolet spectra of a supernova.
j. The first ultraviolet spectra of supernova remnants.
k. The first direct observational evidence of a high temperature corona about our galaxy indicating Te ~ 105 and Ne ~ 4 x 10-4.
l. Observations of UV line intensities for several low and intermediate redshift QSO's; detection of continuum radiation in two high redshift QSO's down to rest wavelengths below 400 A.
m. Discovery of acetylene in the atmosphere of Saturn.
n. Discovery of ultraviolet limb brightening on the Jovian disc, requiring the existence of an extensive pure Rayleigh atmosphere.

IUE performance continues to be excellent. The only expendable limitation to IUE lifetime is the onboard hydrazine for momentum wheel unloading and station keeping. At the present usage rate, IUE could last for 30 years. All the essential spacecraft subsystems are redundant; the only failure that has occurred is in a redundant Panoramic Attitude Sensor (PAS), but the PAS is not required for in-orbit operations. Some anomalies have occurred with the onboard computer (OBC) but they have been corrected through internal reprogramming. It should be noted that during the course of the anomalies, backup and survival modes were implemented successfully.

[3] Scientific Instrument performance has also been excellent. The only problem is with a redundant Short Wavelength Spectrograph Camera which operates intermittently. Both Long Wavelength Spectrograph cameras are operational as are both Fine Error Sensors.

In summary, the IUE is working very well and shows every expectation of continuing. The great productivity and large number of exciting and even unexpected results constitute a substantial scientific and technical achievement and give promise to future substantial scientific results.

[signature]
Thomas A. Mutch

Document III-31

Document title: The Management Operations Working Group for Space Astronomy, NASA, "Space Astronomy Program Plan for the 1980s and 1990s," July 1981.

Source: NASA Historical Reference Collection, History Office, NASA Headquarters, Washington, D.C.

As NASA entered the 1980s, there was a perceived need to define a new astronomy program for the rest of the twentieth century. Astronomers felt ready to design space missions to address some of the fundamental scientific questions about the origin and evolution of the universe.

[cover sheet]

SPACE ASTRONOMY PROGRAM PLAN
FOR THE
1980s AND 1990s

The Management Operations Working Group
for Space Astronomy
July, 1981

[i] PREFACE

Space astronomy is on the threshold of a new era of exploration. Over much of the electromagnetic spectrum, surveys during the past two decades have systematically mapped out the wonders of the Universe. Future missions must confront the challenge of understanding the complexity and physical nature of the diverse astrophysical objects that astronomers have already begun to catalogue. Only by firmly grasping the unique perspectives provided by space astronomy can we hope to solve the great mysteries and unfold the grand designs that determine the evolution of stars and galaxies. Exciting problems that may be resolvable over the next two decades include the nature of the underlying energy source in quasars and the nuclei of active galaxies, the origin and eventual fate of the Universe, and the mechanism of stellar birth and planetary system formation.

The great questions which express our curiosity about our cosmic environment often persist unanswered for generations. What factors determine the long-term climactic variation on Earth? Has intelligent life evolved elsewhere? Will the Universe end in a cataclysmic collapse, will all life eventually succumb to entropy, or will the expansion of the Universe offer infinite scope for suitably evolving intelligences? Are the laws of physics, which underpin our attempts to understand Nature, truly eternal and universal?

The means by which we seek to answer such questions, and the myriad subsidiary questions, change almost every decade. In this report, we present a plan for the U. S. space astronomy program, with specific projects for the near future, development programs for subsequent projects, and projects which await a more extensive presence in space.

****

[63] V. RECOMMENDATIONS

This chapter describes recommendations made by the MOWGSA [Management Operations Working Group for Space Astronomy] regarding the implementation of missions and programs discussed in the proceeding [sic] chapters, and summarizes those recommendations (Section V.A). In a later section (V. B), a brief review of the recommended missions and their impact on the science goals outlined in Chapter 1 is presented, showing which of those goals will be accomplished by the recommended program.

A. Summary of Recommendations

The MOWGSA has discussed programs and missions that fall into several general categories: current flight programs (Section II.A); pending flight programs (II.B); supporting programs (II.C); future missions (Chapter III); and technology development (Chapter IV). These categories are summarized separately in the following paragraphs.

1. Current Flight Programs

Programs considered current are those that are already funded and under development. These include the Space Telescope (ST), the Infrared Astronomical Satellite (IRAS), Spacelab and Shuttle experiments, and the Explorer program. The MOWGSA considers all of these programs to be important and endorses each.

The Space Telescope will be the mainstay of the U. S. space astronomy program for the rest of this century, and the MOWGSA heartily endorses it and urges that every effort be made to provide opportunities for refurbishment and upgrading of the focal-plane instruments.

The Infrared Astronomy Satellite will provide the first comprehensive survey of the heavens at far-infrared wavelengths, and will therefore set the stage for a new era of infrared astrophysics by locating and determining the properties of some 10 sources.

In addition to endorsing ST and IRAS, the MOWGSA has agreed upon the following specific recommendations regarding the Explorer and Spacelab programs: that the Explorer program should have enhanced funding, at least sufficient to recover losses to inflation over the past decade; and that the Spacelab program should be augmented, allowing more numerous flight opportunities.

2. Pending Flight Programs

Pending flight projects are those currently in the NASA five-year plan, but not yet under development. These include the Cosmic Background Explorer (COBE), the Extreme Ultraviolet Explorer (EUVE), and the Shuttle Infrared Telescope Facility (SIRTF). Again, the MOWGSA endorses all of these programs, and makes one specific recommendation: that funding for SIRTF be significantly enhanced, [64] providing for many more refurbishments and flights than currently planned. The SIRTF, with its broad potential wavelength coverage and the versatility to support a great variety of focal-plane instruments, will open new areas of astrophysics for infrared astronomy, and could form the backbone of future infrared observational programs if the community is given sufficient opportunities to take full advantage.

3. Supporting Programs

These are programs that provide various kinds of support services, often with direct scientific benefits, and in other cases with significant indirect effects. Among the former are the suborbital programs (sounding rockets, balloon and airborne astronomy) and theoretical astrophysics, while the latter includes laboratory astrophysics, some aspects of data analysis, and the funding of scientific personnel.

The MOWGSA endorses continued support for all of these programs, but particularly emphasizes a few. One of these is the continued provision for the development of new instrumentation, through the suborbital programs, or eventually the Experiments of Opportunity Payloads (EOP) program. It is vital for the entire space astronomy program that support be provided to PI-class investigators for the development of new instruments, because this is the most efficient way to test new concepts without commitment to a major mission. The MOWGSA has made no attempt to recommend a specific balance among the programs that fall into this area, but simply emphasizes the importance of the overall concept.

Among the other supporting programs, the MOWGSA specifically makes recommendations for the augmentation of support for: theoretical astrophysics, because of its important, but sometimes overlooked, rate both in planning new missions and in the interpretation of data; laboratory astrophysics, because there is an increasing need by astrophysicists for atomic data relevant to the interpretations of new observational areas (such as the extreme ultraviolet), while at the same time there are increased difficulties (due in part to lack of support) in successfully encouraging atomic and molecular physicists to provide the needed data; and data-analysis, because of the growing complexity and quantity of scientific data being returned by space missions, particularly with the advent of two-dimensional electronic detectors.

4. Future Missions

Generally speaking, a mission or mission concept was assigned to one of three recommended classes: (1) highly recommended, meaning that the MOWGSA feels strongly that the program should be implemented (or at least formally studied) as soon as possible; (2) recommended, meaning that the program is desirable and should become a part of NASA's planning, so that opportunities for implementation may be sought and taken when possible; and (3) concept to be studied, meaning that the idea has sufficient merit to justify some expenditure of resources in developing the concept to a point where its feasibility and importance may be adequately assessed.

[65] Table V.I. Future Missions

Mission	Status	Cost Class*	Readiness Category
Far-UV Spectroscopic Explorer (FUSE)	Highly recommended	A	II
Large Deployable Reflector (LDR)	Highly recommended	C	III
Orbiting VLBI Mission	Highly recommended	B	I
Relativity Explorer (REX)	Highly recommended	A	II
Simultaneous Astrophysics Mission (SAM)	Recommended	A	I
Planetary Spectroscopy Telescope	Recommended	A	I

UV Wide-Field Imaging Mission	Recommended	A	I
Extreme UV Spectroscopic Explorer (EUSE)	Concept for study	A-B	IV
Very Large Space Telescope (VLST)	Concept for study	C	IV
Optical Interferometry Mission	Concept for study	A-B	III
Molecular Astrophysics Pencil-Beam Explorer (MAPPER)	Concept for study	A-B	III
Solar Sail Telescope	Concept for study	A	IV
Laser Gravitation Wave Antenna	Concept for study	B	IV
Orbiting Eotvos Experiment	Concept for study	A	III
Large Airborne Telescope (LAT)	Concept for study	A	III

*Cost Classes:
A: < $100 M
B: $100 - 500 M
C: > $500 M

+Technological Readiness Categories: I. Technology already demonstrated in flight.
II. Technology known; not yet used in flight.
III. Concepts well developed; not yet demonstrated.
IV. Concepts not yet fully developed.

[66] The missions that have been discussed in Chapter 3 are summarized in Table V.1, where the MOWGSA recommendation category, as well as a technological readiness category and a broad cost-class estimate are also presented. The definition of both the readiness and cost classes are given in footnotes to this table.

The MOWGSA has identified four future missions that are given "highly recommended" status. The committee has not prioritized among these four, so no specific recommendation regarding their sequence can be given. Most likely, the technological readiness and the method of funding (e. g. whether or not a mission can be developed as an Explorer, or whether there is partial support from abroad) will ultimately dictate the schedule for implementation of these missions.

The Far-Ultraviolet Spectroscopic Explorer (FUSE) mission is a natural and necessary-adjunct to such past and present programs as Copernicus, IUE, and Space Telescope, providing as its highest priority high-resolution spectroscopy between 912 A and 1200 A, a region rife with important features for both stellar and interstellar problems but not covered by IUE or ST, and only weakly covered by Copernicus. The Large Deployable Reflector (LDR) will provide infrared and millimeter-wave astronomers with a powerful tool for exploring important new wavelength regions, attaining high spatial resolution. The Very Long Baseline Interferometry (VLBI) Explorer will allow a baseline for high spatial-resolution radio mapping with coverage of much of the sky, something not possible with ground-based VLBI measurements. The Relativity Explorer (REX) will measure two new gravitational effects (geodetic effect and motional effect) in earth orbit to test gravitational theory.

An additional three missions are listed as "recommended". Again, no prioritization within this group has been made.

A Simultaneous Astrophysics Mission (SAM) would provide simultaneous observations of variable stars and other objects in the soft x-ray, ultraviolet, visible, and possibly the near-infrared regimes, so that different physical regions in these objects can be monitored at the same time. The Planetary Spectroscopy Telescope would be optimized to provide spectroscopic data on planetary atmospheres, allowing synoptic studies and those requiring special pointing that are not feasible with other instruments. After lengthy discussion concerning the UV imaging missions that were considered, it was decided to treat as distinct missions two alternatives: a wide-field survey intended to eventually cover the entire sky; and a moderate-field imager with superior sensitivity and spatial resolution, for detailed studies of specific fields. A consensus was reached that the former, the wide-field survey, should be given higher priority, but that the latter should also be encouraged. A full-sky survey in ultraviolet wavelengths would provide a natural extension towards short wavelengths of the kind of information available in the Palomar Sky Survey, and would become a fundamental storehouse of information that would be especially important in the era of Space Telescope. The moderately wide-angle imaging experiment was also generally favored, but is probably not of sufficient importance to be considered for support unless NASA obtains substantial participation from abroad.

In addition, it is recommended that every opportunity be utilized to perform add-on gravitational physics experiments in conjunction with interplanetary missions including search for gravitational radiation, precision radio tracking of planetary orbiters and landers, and relativity experiments on the proposed Star Probe.

[67] Finally ten more possible future missions were chosen to be listed as "concepts for study," in this case with some rough prioritization. Five of these ten were placed in a higher priority category: these five are the Extreme Ultraviolet Spectroscopy Explorer (EUSE), the Very Large Space Telescope (VLST), gravity wave experiments using laser interferometry, the Large Airborne Telescope (LAT), and an optical interferometry mission. The MOWGSA recommends particularly highly the study of these five possible missions, and urges that funding for such studies be provided as soon as possible.

5. Technology Needs

In the course of its study of possible future programs, the MOWGSA identified several general areas of technology development, which will be required by one or more of the recommended missions. No prioritizations among these areas were attempted, and here they will only briefly be summarized. Most are already under development at some stage; the MOWGSA's endorsement is meant to emphasize the needs and, in specific cases, calls for enhanced support.

The MOWGSA urges the development of space platforms, to serve as support systems for long-term (e.g. 6-month) orbital experiments. This is a natural follow-up to the Spacelab program which will support short-duration flight, and may also be a step in the direction of permanent manned space stations. The space platform should provide some sort of pointing capability as well as power.

Another area of interest to the MOWGSA is the development of suborbital platforms, such as the semi-buoyant aircraft described in the discussion of the Large Airborne Telescope (Appendix) [Appendix not included]. Telescopes placed in the atmosphere above 40,000 feet would have significant capabilities, both in wavelength coverage and in spatial resolution, and would be useful tools throughout the spectrum from the near-ultraviolet far into the infrared. The potential for maintaining heavy payloads at these altitudes for long periods with minimal fuel expenditure is very attractive.

The MOWGSA also encourages the development of generalized spacecraft pointing and control systems, which could be applicable in many situations, thereby reducing the cost of development of such systems as needed for individual missions. This would have application to a wide variety of situations, including the space platform mentioned above.

The general area of detector development, which already receives significant support, is an important one which the MOWGSA feels should be augmented. At present NASA is unable to support competitive parallel development of detector concepts, generally being forced instead to limit funding to one development program for each type of detector. This does not necessarily ensure that the best possible detector is produced in the end, nor does it allow NASA the flexibility to pursue new concepts when they arise. To support this program at a higher level will be of paramount importance, until the time when photon-counting area detectors with the highest theoretically-possible quantum efficiency have been developed for each wavelength region of interest.

Another area of concern to the MOWGSA is the development of optical systems for space astronomy, including optical surfaces and coatings, the development of [68] filters of various kinds, and the design of optimized optical system needed for the various missions recommended for future implementation.

Data management, already mentioned under supporting programs, will require some technology development. This includes not only hardware devices such as image display systems, fast data lines, and super-compact storage device, but also the development of data management systems. The need for increased support in these areas grows with the launch of each new instrument, and will make a quantum jump when ST is in operation.

The MOWGSA also recognizes the need for technology developments in support of various gravitational physics experiments, many of which are recommended in this report. The needs include the development of cryogenic technology, stable clocks, and radio ranging and doppler tracking systems for experiments flying aboard interplanetary probes.

Finally, the MOWGSA encourages support for the study of technology for interstellar flight. This refers to unmanned probes to nearby stars, which could sample and report; on conditions and materials in interstellar space, as well as obtaining close-up data on other stars. While such a mission is obviously far in the future, it is appropriate to begin related studies soon. Of particular importance is the need to study possible propulsion systems.

B. Impact on the Science Program

Having made its recommendations, the MOWGSA felt that it would be useful to summarize here how the recommended program fulfills the science goals outlined in Chapter I. The following paragraphs briefly do so, taking things in the sequence established in that chapter.

1. Cosmology

The principal missions discussed in this report that will have an impact on cosmology are: COBE, which will make accurate measurements of the spectrum and isotropy of the cosmic background radiation; LDR, which will attempt to measure the anisotropy of the cosmic background; FUSE, which will be capable of directly measuring the D/H ratio towards distant and/or reddened stars; and ST, which will make enormous strides towards measuring intergalactic gas and the study of primordial galaxies. The latter goal will not only have direct bearing on the properties of the Universe as a whole, but will also be important for studying galactic evolution, as mentioned below. SIRTF, IRAS, COBE, and LDR will yield information on the properties of very young galaxies whose emission is red-shifted all the way into the infrared or millimeter-wave portions of the spectrum.

All of these missions, except for LDR and FUSE, are already either under development or pending, LDR and FUSE are among the most highly recommended future missions.

[69] 2. Gravitation

The Objective of gravitational physics experiments and missions is the investigation of relativistic gravity effects to test gravitational theories. Relativistic gravitation plays a major role in many astrophysical phenomena and in cosmology. The highly recommended REX mission will measure the frame dragging or mass current effect which is the gravitational counterpart to electromagnetism. This phenomenon, which is important for fast rotating and dense astrophysical objects, has not been measured yet. The proposed Star Probe mission would provide an opportunity to measure relativistic gravitational effects in the strong gravity field close to the Sun and to determine the solar quadrupole moment to high accuracy. The Mercury Orbiter mission would provide an alternate way for accurate determination of the solar quadrupole moment and of the relativistic perihelion advance. Add-on gravitational experiments on planetary lander missions would provide unique opportunities for long term observations of relativistic effects in the motion of planets and to test the constancy of the gravitational constant.

The search for gravitational radiation is a very important scientific objective. Search for low frequency gravitational radiation can be done by precision radio doppler tracking of interplanetary space probes. A much more sensitive detector would be a laser gravitational wave antenna in space, a proposed future mission requiring rather advanced technology. Other proposed future missions would perform extremely high precision measurements of relativistic effects to probe for the limit of validity (possible breakdown) of gravitational theories: a test of the relativistic deflection of light to second order could be accomplished with the POINTS mission and an extremely sensitive test of the Weak Equivalence Principle with the Orbital Cryogenic Eötvös Experiment.

3. Galaxies and Galactic Evolution

The strongest tool proposed for studies of galaxies and galactic evolution is ST (under development) which will provide spectroscopy and high-resolution imagery of galaxies and galactic nuclei, with potential for studying all the related problems of chemical enrichment, stellar and galactic winds, stellar populations, and the diffuse interstellar medium. In addition, a VLBI mission (highly recommended) would yield a wealth of information on morphology of galaxies; and in the infrared, IRAS, SIRTF, COBE, and the LDR (highly recommended) would all contribute to studies of the distribution of primordial galaxies and their spectral energy distributions, as well as the distribution of dust and star-formation regions within nearby galaxies. Variability in galactic nuclei would be usefully observed with the SAM, (recommended), while both the wide-field and moderate-field UV-imaging missions (both recommended) would contribute enormously to studies of the morphology and chemical evolution and gradients of galaxies. FUSE (highly recommended) would provide data on the far-UV spectra of galactic nuclei, as well as on the interstellar gas and dust in nearby galaxies.

The heavily-obscured regions in galaxies could be probed with HAS (under development); SIRTF (pending), and LDR (highly recommended) which will penetrate these regions, providing photometry on objects inside as well as spectroscopy of both the embedded objects and the intervening interstellar material. A large airborne observatory with infrared capabilities (concept for study) would allow [70] significant work to be done in these areas from a suborbital platform, and the MAPPER mission (concept for study) would trace the distribution of molecular clouds in nearby galaxies, as well as our own. Finally, FUSE (highly recommended) would significantly extend coverage of the diffuse interstellar medium, by probing far-UV wavelengths where a number of important transitions lie, inaccessible to ST or IUE. The EUSE (concept for study) would extend this even further towards short wavelengths, providing coverage of more highly-ionized species in the coronal gas.

4. Stellar Research

A number of the missions mentioned in the galactic astronomy section will. also be important tools in stellar astrophysics. Stellar flux distributions will be determined by IRAS (under development) and EUVE (pending), while spectroscopy will be carried out from the infrared (SIRTF; pending) to the ultraviolet (ST, under development; and FUSE, highly recommended) and possibly even the extreme ultraviolet (ELISE, concept for study). The development of the VLST (concept for study) and the LDR (highly recommended) will augment both areas by allowing an extension to fainter objects.

Stellar chromospheres and coronae will be especially well studied by FUSE (highly recommended) and ELISE (concept for study, covering a wide range of highly-ionized species, and extended atmospheres in cool stars will be observed efficiently with SIRTF (pending) and LDR (highly recommended). Stellar winds will best be observed in the UV (with ST; under development) and the far-UV (FUSE; highly recommended), and infrared data from HAS (under development), SIRTF (pending), and the LDR (highly recommended) will also be important in probing the interactions between the interstellar medium and winds from embedded stars.

In all manner of variable star research, SAM (recommended) will play an important role by providing simultaneous observations of activity in different wavelength regions, hence different layers in stellar atmospheres and envelopes.

Finally, the extension to fainter stellar objects by many of the recommended instruments, particularly ST, will allow studies of individual stars in nearby galaxies, opening the way for analyses of the effects on stellar properties of galaxy type and evolution.

5. Planetary Astronomy

Many of the instruments and missions discussed in this report will have important applications to studies of solar system objects. ST (under development) will provide high-resolution images of planets, extending studies of atmospheric motion begun with the Pioneer and Voyager probes, and will be capable of detecting certain classes of planets orbiting nearby stars. FUSE (highly recommended) could provide important ultraviolet spectroscopic data on planetary atmospheres, as could SIRTF (pending) and LDR (highly recommended) in the infrared. Great contributions to the study of the planets, particularly synaptic studies of their atmospheres, would be made by the Planetary Spectroscopy Telescope (PST; recommended), [71] which would have pointing and scheduling characteristics optimized for such studies.

Finally, spectroscopy of comets could be accomplished by a variety of missions such as ST (under development), FUSE (highly recommended), SIRTF (pending), and LDR (highly recommended) all of which will be sufficiently sensitive not only for emission-line measurements, but also for absorption-line observations, using background stars as continuum sources.

6. SUMMARY

Nearly every major research goal outlined in Chapter I can be accomplished, at least in part, by missions described in this report. A large fraction will be carried out by those listed as "under development", "pending", "highly recommended", or "recommended", so that prospects are strong for accomplishing much of what the MOWGSA sees as desirable before the end of this century. The success of this program depends not only on the specific missions mentioned in this section, but also on the supporting programs and technological developments outlined earlier.

The MOWGSA hopes that this planning document will prove to be useful in the coming years, as NASA seeks to carry its functions in space astronomy.

Document III-32

Document title: Gamma-Ray Observatory Science Working Team, The Gamma-Ray Observatory Science Plan, September 1981.

Source: Alan Bunner, Office of Space Science, NASA Headquarters, Washington, D.C.

The Space Science Board endorsed the development of a major space-based facility devoted to gamma-ray astronomy in 1976. A year later, NASA released an announcement of opportunity inviting scientists to propose instruments for the spacecraft, which became known as the Gamma-Ray Observatory (GRO). While five instruments were tentatively selected for definition studies, that list was narrowed to four when one of the experiments could not meet cost and programmatic constraints. President Jimmy Carter in 1979 approved the GRO for development in preference to a U.S. mission to comet Halley, because he was convinced that it would produce more important scientific data than would a comet mission. In September 1981, the GRO Science Working Team developed this science plan in light of the four experiments selected and the goal to keep total mission costs below $100 million (FY 1981 dollars). These four instruments made up the payload of spacecraft, which took the name

Compton Gamma Ray Observatory, after physicist Arthur Holly Compton, when it was launched aboard the Space Shuttle in 1991. The Compton GRO spacecraft was purposely deorbited in 2000 because its control gyroscopes were failing.

[cover sheet]

THE GAMMA-RAY OBSERVATORY
SCIENCE PLAN
SEPTEMBER 1981
Prepared by: Gamma-Ray Observatory Science Working Team

[1] GAMMA-RAY OBSERVATORY SCIENCE PLAN

I. INTRODUCTION

Gamma-ray astronomy, the study of the highest energy electromagnetic radiation from the cosmos, occupies a unique position in the search for understanding the Universe. This high energy radiation is produced in a wide variety of astrophysical processes which would otherwise remain unobservable. These processes include nuclear reactions, matter-antimatter annihilation, elementary particle decays, and some general relativistic effects. The great penetrating power of gamma rays allows them to reach the top of the atmopshere [sic] from almost anywhere in the Universe. On the other hand, the atmopshere [sic] is opaque to gamma rays, and, hence, the observations must be made from space. The astrophysical sites where gamma-ray emission is a major source of energy release are some of the most energetic objects in the Universe–e.g., supernovae, neutron stars, black holes, cores of galaxies, and quasars. Among the problems addressed by gamma-ray astronomy are the formation of the elements in the Universe, the structure and dynamics of the Galaxy, the nature of pulsars, the possible existence of large amounts of antimatter in the Universe, phenomena occurring in the nuclei of galaxies–especially explosive galaxies–and the origin and evolution of the Universe itself. For many such problems, gamma rays are the only source of information about the high energy reactions taking place.

Because gamma-ray astronomy requires complex detectors operating outside the Earth's atmosphere, it is only in recent years that this field has begun to develop. The discoveries in gamma-ray astronomy parallel those in other new branches of astronomy in that the unexpected results have been as significant as those which had been predicted in providing new insight into a number of astrophysical problems.

[2] For example, it has been found that some pulsars emit several orders of magnitude more energy in the form of gamma rays than in the form of radio waves and that the quarsar [sic] 3C273 appears to radiate as much energy in gamma rays as in any other form of electromagnetic radiation. Also, many energetic gamma-ray sources have been found which at present have not been correlated with objects observed at other wavelenghts [sic]. These observations suggest the possibility of a class of celestial objects not previously known. Further, intense bursts of low energy gamma rays have been detected; the ori-

gin of these events remains a mystery. In all these cases, these objects cannot be fully understood without a thorough knowledge of their gamma-ray emission, because this emission represents such a significant fraction of the total radiated energy. The understanding of gamma-ray-luminous sources is one of the most important open problems for all astronomy.

Other important astronomical questions for which gamma-ray astronomy can provide decisive answers include nucleosynthesis, via the study of gamma-ray line emission; Galactic structure, as revealed by the gamma rays produced in the interactions of cosmic rays with interstellar matter; and the origin and evolution of the Universe, through observations of the isotropic gamma radiation. Beyond these known returns lies the anticipation of further unexpected results in gamma-ray astronomy as the sensitivity of the observations improves, particularly because much of the gamma-ray energy range is just now being explored and much of the gamma-ray sky has not been observed.

The Gamma-Ray Observatory (GRO), which will provide the first comprehensive, coordinated observations covering the entire spectrum of gamma-ray astronomy, with much better sensitivity than any previous mission.[sic]This approach requires four separate detector systems with quite different characteristics, each emphasizing a particular aspect of the observations.

[3] In this Science Plan for the GRO, Section II [not included] discusses in depth the scientific rationale for gamma-ray astronomy. Section III presents the specific scientific objectives for the GRO and describes how the four selected instruments have a combined capability to achieve these objectives. Section IV [not included] contains a summary of each of the four investigations chosen for the mission.

****

[21] III. GAMMA-RAY OBSERVATORY

A. Scientific Objectives

Based on the foregoing scientific rationale and the recommendation of the Committee on Space Astronomy and Astrophysics of the National Academy of Science's Space Science Board, GRO has adopted the following scientific objectives:

- A study of discrete objects such as black holes, neutron stars, and objects emitting only at gamma-ray energies.
- A search for evidence of nucleosynthesis - the fundamental process in nature for building up the heavy elements in nature and other gamma-ray lines emitted in astrophysical processes.
- The exploration of the Galaxy in gamma rays in order to study the origin and dynamic pressure effects of the cosmic-ray gas and the structural features revealed through the interaction of the cosmic rays with the interstellar medium.
- A study of the nature of other galaxies as seen at gamma-ray wavelengths, with special emphasis on radio galaxies, Seyfert galaxies and QSO's.
- A search for cosmological effects, through observations of the diffuse gamma radiation, and for possible primordial black hole emission.
- Observations of gamma-ray bursts, their luminosity distribution, the spectral and temporal characteristics and their spatial distribution.

[22] In the section that follows, a brief description of the observatory requirements necessary to achieve these objectives, the specific spacecraft parameters needed to support these requirements and a brief description of the instruments to be used in these observations will be presented.

B. Observatory Requirements

To achieve these scientific objectives, the Gamma-Ray Observatory must be capable of conducting a comprehensive survey of the gamma-ray sky over an energy range extending from the upper end of existing x-ray observations up to the highest practical energy. The GRO sensitivity for discrete sources, diffuse radiation, and gamma-ray lines should be significantly greater than any previous instruments.

No single scientific instrument is capable of meeting all the requirements. The band of wavelengths encompassed by gamma-ray astronomy is more than 100 times as broad as that of x-ray astronomy, and more than 104 times broader than the visible region. Different detection methods are needed in different parts of the gamma-ray spectrum. Further, even within a part of the energy range, energy and angular resolution can usually be improved only at the expense of sensitivity. A complementary set of experiments is required, therefore, in order to meet the scientific objectives. The spacecraft supporting these instruments must be capable of pointing them accurately and with stability to any part of the sky for a period of two weeks, provide adequate power and thermal control, supply attitude and timing data as precise as needed by the instruments, and handle the data from all these instruments efficiently.

[23] C. Spacecraft Summary

The Gamma-Ray Observatory will be a shuttle-launched, free-flyer satellite. The nominal circular orbit will be about 400 kilometers with an inclination of 28.5°. The radius should remain below 450 kilometers to prevent excessively high trapped particle dosages during passage through the South Atalantic Anomally [sic]. An orbital radius below about 350 kilometers causes excessive aerodynamic drag on the Observatory. The spacecraft must be capable of accommodating 5500 kilograms of instruments and must supply 600 watts of experiment power. The 17 kilobits per second of experiment data will be supported via NASA's Tracking and Data Relay Satellite system. Celestial pointing to any point on the sky (excluding the Sun) will be maintained to an accuracy of ±0.5°. This is determined by the precision to which exposure to a given region of the sky must be known in order to determine the sensitivity of an observation. Knowledge of the pointing direction will be determined to an accuracy of 2 arc minutes so that this error contributes negligibly to the over all determination of the direction of gamma-ray source. Absolute time will be accurate to 0.1 milliseconds to allow precise comparisons of pulsars and other time varying sources with observations at other wavelengths from ground observations and other satellites. The attitude and timing data together with orbital position will be encoded into the telemetry data. These spacecraft support requirements are summarized in Table I.

[24]

Table I

SPACECRAFT SUPPORT REQUIREMENTS

Scientific Payload Weight	5500 kilograms
Instrument Power	600 watts
Experiment Data Rate	17 kilobits
Pointing Accuracy	±0.5°
Attitude Determination	2 arc minutes
Absolute Timing Accuracy	0.1 milliseconds

Brief capsule descriptions of each experiment are given as follows: More detalied [sic] descriptions can be found in Section IV.

1. Gamma-Ray Observatory Scintillation Spectrometer (OSSE):

This experiment utilizes four large actively-shielded and passively-collimated-Sodium Iodide (NaI) Scintillation detectors, with a 5° x 11° FWHM field of view. The large area detectors provide excellent sensivity [sic] for both gamma-ray line and continuum emissions. An offset pointing system modulates the celestial source contributions to allow background substraction. It also permits observations of off-axis sources such as transient phenomena and solar flares without impacting the planned Observatory viewing program.

[25] 2. Imaging Compton Telescope (COMPTEL):

This instrument is based on a newly established concept of gamma-ray detection in the 1-30 MeV range. It employs the unique signature of a two-step absorption of the gamma-ray, i.e., a Compton collision in the first detector followed by total absorption in a second detector element. This method, in combination with effective charged particle shield detectors, results in a more efficient suppression of the otherwise inherent instrumental background. Spatial resolution in the two detectors together with the well defined geometry of the Compton interaction permits the reconstruction of the sky image over a wide field of view (~1 steradian) with a resolution of a few degrees. In addition, the instrument has the capability of searching for polarization of the radiation. The instrument has good capabilities for the search for weak sources, weak galactic features and for the search for spectral and spatial features in the extragalactic diffuse radiation.

3. Energetic Gamma-Ray Telescope (EGRET):

The High Energy Gamma-Ray Telescope is designed to cover the energy range from 20 MeV to 30 x 10^3 MeV. The instrument uses a multi-thin-plate spark chamber to detect gamma rays by the electronpositron pair process. A total energy counter using NaI(Tl) is placed beneath the instrument to provide good energy resolution over a wide dynamic range. The instrument is covered by a plastic scintillator anticoincidence dome to prevent readout on events not associated with gamma rays. The combination of high energies and good spatial resolution in this instrument provides the best source positions of any GRO instrument.

[26] 4. Burst and Transient Source Experiment (BATSE):

The Burst and Transient Source Experiment for the GRO is designed to continuously monitor a large fraction of the sky for a wide range of types of transient gamma-ray events. The monitor consists of eight wide field detector modules. Four have the same viewing path as the other telescopes on GRO and four are on the bottom side of the instrument module viewing the opposite hemisphere. This arrangement provides maximum continuous exposure to the unobstructed sky. The capability provides for 0.1 msec time resolution, a burst location accuracy of about a degree and a sensitivity of 6 x 10^{-8} erg/cm^2 for a 10 sec burst.

The salient features of the four experiments are summarized in Table II. As mentioned above, each instrument represents a significant step forward over its predecessors. For example, the sensitivity for line gamma-ray detection has been improved by more than an order of magnitude over the HEAO-A4 and HEAO C-1 instruments. The continuum sensitivity in the MeV range is typically improved by a factor of twenty or more. Improvements of about an order of magnitude in source location capability are also expected due to the improved instruments and the greatly increased exposure factors. The addition of a massive NaI calorimeter crystal has markedly improved the energy resolution (a factor of ≥ 2 better than SAS-2) in the > 100 MeV range and extended the range to 20 GeV. Also in this range the total effective area (i.e., area X geometry factor) is 25 times larger than that of COS-B.

[27]

Table II
SUMMARY OF GRO DETECTOR CHARACTERISTICS

	OSSE	COMPTEL	EGRET	BATSE
Energy Range (MeV)	0.10 to 10.0	1.0 to 30.0	20 to 3x10^4	0.05 to 0.60
Energy Resolution	8.0% at 0.66 MeV	5 - 8%	15%	35% at 0.1 MeV
Maximum Effective Area (cm^2 efficiency)	2310	50	2000	5500
Position Resolution (strong source)	10 arc min square error box (special mode)	7.5 arc min (1σ radius)	5 arc min. (1σ radius)	1°
Maximum Effective Geometric Factor (cm^2sr effeciency) [sic]	12	30	1000	15000
Estimated Threshold (source sensitivity) Line	2x$10^{-5}$$cm^{-2}s^{-1}$	3x10^{-5} to 3x10^{-6}		0.1 Crab-transient
Continuum	~ 3x$10^{-5}$$cm^{-2}s^{-1}$	5x$10^{-5}$$cro^{-2}s^{-1}$	5x$10^{-8}$$cm^{-2}s^{-1}$	6x10^{-8}erG cm^{-2}-burst
Weight (Kg)	1730	1477	1708	570
Average Power (watts)	140	195	170	100
Height (m) x Width (m)	1.5x(1.5x2.3)	2.8x1.7	2.25x1.65	0.7x0.6x0.7
Bit Rate (kbps)	6.0	4.5	5.0	1.5

Document III-33

Document title: Astronomy Survey Committee, National Research Council, "Astronomy and Astrophysics for the 1980s," January 1982.

Source: National Academy of Sciences, Washington, D.C.

The U.S. astronomical community, under the auspices of the National Research Council of the National Academy of Sciences, each decade prepared a blueprint for what it hoped would happen in astronomy and astrophysics in the coming ten years. This is the third in the series of these "decadal" reports. It was prepared by a committee chaired by Harvard astronomer George Field, and became known as the Field report. Of special note is the high priority the report assigns to space-based astronomical investigations.

[cover]
Astronomy and Astrophysics for the 1980s
VOLUME 1: Report of the Astronomy Survey Committee

Astronomy Survey Committee
Assembly of Mathematical
and Physical Sciences
National Research Council

NATIONAL ACADEMY PRESS
Washington, D.C. 1982

****

[ix]
Astronomy Survey Committee

GEORGE B. FIELD, Harvard-Smithsonian Center for Astrophysics, Chairman
MICHAEL J. S. BELTON, Kitt Peak National Observatory
E. MARGARET BURBIDGE, University of California, San Diego
GEORGE W. CLARK, Massachusetts Institute of Technology
S. M. FABER, University of California, Santa Cruz
CARL E. FICHTEL, NASA Goddard Space Flight Center
ROBERT D. GEHRZ, University of Wyoming
EDWARD J. GROTH, Princeton University
JAMES E. GUNN, Princeton University
DAVID HEESCHEN, National Radio Astronomy Observatory
RICHARD C. HENRY, The Johns Hopkins University
RICHARD A. McCRAY, Joint Institute for Laboratory Astrophysics and the University of Colorado

JEREMIAH OSTRIKER, Princeton University
EUGENE N. PARKER, University of Chicago
MAARTEN SCHMIDT, California Institute of Technology
HARLAN J. SMITH, University of Texas, Austin
STEPHEN E. STROM, Kitt Peak National Observatory (*ex officio*)
PATRICK THADDEUS, NASA Goddard Institute for Space Studies and Columbia University
CHARLES H. TOWNES, University of California, Berkeley
ARTHUR B. C. WALKER, Stanford University
E. JOSEPH WAMPLER, University of California, Santa Cruz

PAUL BLANCHARD, *Executive Secretary*
DALE Z. RINKEL, *Administrative Secretary*

****

[13] 2- RECOMMENDED PRIORITIES FOR ASTRONOMY AND ASTROPHYSICS IN THE 1980s

The Astronomy Survey Committee takes note at the outset of the support provided to U.S. astronomy and astrophysics over the past decades through the scientific programs of the National Science Foundation (NSF), the National Aeronautics and Space Administration (NASA), and other federal agencies. This support has enabled U.S. astronomical research to maintain an overall position of world leadership and has vastly widened our horizons for exploration of the Universe.

The programs recommended in this report have been selected from research activities that were, at the beginning of the Survey, candidates for implementation in fiscal year 1983 and beyond. Before presenting a summary of its recommendations, however, the Committee wishes to emphasize the importance of approved, continuing, and previously recommended programs to the progress of astronomical research during the remainder of the decade. The present Committee's recommendations take explicit account of such programs and build upon them.

The Committee calls particular attention to the need for support of the following approved and continuing programs, for which the order of listing carries no implication of priority: Space Telescope and the associated Space Telescope Science Institute; second-generation Space Telescope instrumentation; the Gamma Ray Observatory; NASA level-of-effort observational programs, including research with balloons, aircraft, and sounding rockets, together with the [14] Explorer and Spacelab programs; the Solar Optical Telescope and the Shuttle Infrared Telescope Facility for Spacelab; facilities for the detection of neutrinos from the solar interior; federal grants in support of basic astronomical research at U.S. universities; and programs at the National Astronomy Centers. The 25-Meter Millimeter-Wave Radio Telescope, which was recommended in an earlier form in the Greenstein report, has not yet been implemented. The present status of these approved, continuing, and previously recommended programs is described later in this chapter; their importance for the health of U.S. astronomy in the 1980s is discussed in Chapter 4. [not included]

SUMMARY OF THE RECOMMENDED PROGRAM

The Astronomy Survey Committee recommendations for a program in astronomy and astrophysics for the 1980s fall into three general categories:

- Prerequisites for new research initiatives;
- New programs; and
- Programs for study and development.

As noted in the Preface, the observational components of these recommendations are restricted to remote sensing from the Earth or its vicinity. A background and overview of the recommendations follows later in this chapter.

Prerequisites for New Research Initiatives

In order to be effective, the recommended new research initiatives for the coming decade must be supported by a set of Prerequisites that apply to both the gathering and the analysis of the data produced. These Prerequisites are essential for the success of major programs but are inexpensive by comparison. Although significant support already exists for each, the Committee strongly recommends substantial augmentations in the following areas, in which the order of listing carries no implication of priority:

A. *Instrumentation and detectors*, to utilize the latest technology to enhance the efficiency of both new and existing telescopes in the most cost-effective manner;
B. *Theory and data analysis*, to facilitate the rapid analysis and understanding of observational data; [15]
C. *Computational facilities*, to promote data reduction, image processing, and theoretical calculations;
D. *Laboratory astrophysics*, to furnish the atomic, molecular, and nuclear data essential to the interpretation of nearly all astronomical observations; and
E. *Technical support at ground-based observatories*, to ensure that modern astronomical instrumentation is maintained in the best condition permitted by the state of the art.

A detailed consideration and justification of these Research Prerequisites appears in Chapter 5. [not included]

New Programs

The Astronomy Survey Committee recommends the approval and funding of new programs in astronomy and astrophysics for the 1980s. These have been arranged into three categories according to the scale of resources required.

A. *Major New Programs* The Committee believes that four major programs are critically important for the rapid and effective progress of astronomical research in the 1980s and is unanimous in recommending the following order of priority:

1. *An Advanced X-Ray Astrophysics Facility* (AXAF) operated as a permanent national observatory in space, to provide x-ray pictures of the Universe comparable in depth and detail with those of the most advanced optical and radio telescopes. Continuing the remarkable development of x-ray technology applied to astronomy during the 1970s, this facility will combine greatly improved angular and spec-

tral resolution with a sensitivity up to one hundred times greater than that of any previous x-ray mission.

2. *A Very-Long-Baseline (VLB) Array* of radio telescopes designed to produce radio images with an angular resolution of 0.3 milliarcsecond. Among many potential applications of profound importance, this instrument will probe the small-scale structure surrounding the enigmatic energy sources in the cores of quasars and active galactic nuclei and will directly determine the distance scale within our Galaxy with unprecedented accuracy.
3. *A New Technology Telescope* (NTT) of the 15-m class operating from the ground at wavelengths of 0.3 to 20 μm, to provide a tenfold increase in light-gathering capacity at visual wavelengths [16] and a hundredfold increase in speed for spectroscopy at infrared wavelengths, with application to a very wide range of scientific problems. The Committee finds the scientific merit of this instrument to be as high as that of any other facility considered and emphasizes that its priority ranking does not reflect its scientific importance but rather its state of technological readiness. The design studies needed before NTT can be constructed are of the highest priority and should be undertaken immediately.
4. *A Large Deployable Reflector* in space, to carry out spectroscopic and imaging observations in the far-infrared and submillimeter wavelength regions of the spectrum that are inaccessible to study from the ground, thus extending the powerful capabilities of NTT to these longer wavelengths. Such an instrument, in the 10-m class, will present unprecedented opportunities for studying molecular and atomic processes that accompany the formation of stars and planetary systems.

B. *Moderate New Programs* In rough order of priority, these are:

1. *An augmentation to the NASA Explorer program,* which remains a flexible and highly cost-effective means to pursue important new space-science opportunities covering a wide range of objects and nearly every region of the electromagnetic spectrum.
2. *A far-ultraviolet spectrograph in space,* to carry out a thorough study of the 900-1200-Å region of the spectrum, important for studies of stellar evolution, the interstellar medium, and planetary atmospheres.
3. *A space VLB interferometry antenna in low-Earth orbit,* to extend the powerful VLBI technique into space in parallel with the rapid completion of a ground-based VLB Array, in order to provide more detailed radio maps of complex sources, greater sky coverage, and higher time resolution than the Array can provide alone.
4. *The construction of optical/infrared telescopes in the 2-5-m class,* to observe transient phenomena, conduct long-term survey and surveillance programs, provide crucially needed ground-based support to space astronomy, and permit the development of instrumentation under realistic observing conditions. The Committee particularly encourages federal assistance for those projects that will also receive significant nonfederal funding for construction and operation. [17]
5. *An Advanced Solar Observatory in space,* to provide observations of our Sun—the nearest star—simultaneously at optical, extreme ultraviolet, gamma-ray, and x-ray wavelengths, to carry out long-term studies of large-scale circulation, internal dynamics, high-energy transient phenomena, and coronal evolution.

6. *A series of cosmic-ray experiments in space*, to promote the study of solar and stellar activity, the interstellar medium, the origin of the elements, and violent solar and cosmic processes.
7. *An astronomical Search for Extraterrestrial Intelligence (SETI)*, supported at a modest level, undertaken as a long-term effort rather than as a short-term project, and open to the participation of the general scientific community.

C. *Small New Programs* The program of highest priority is:

- An antenna approximately 10 m in diameter for submillimeter-wave observations, at an excellent ground-based site.

Other programs of outstanding scientific merit, in which the order of listing carries no implication of priority, are as follows:

- A spatial interferometer for observations of high angular resolution in the mid-infrared region of the spectrum;
- A program of high-precision optical astrometry; and
- A temporary program to maintain scientific expertise at U.S. universities during the 1980s through a series of competitive awards to young astronomers.

Detailed discussion and justification of the New Programs appears in Chapter 6. [not included]

Programs for Study and Development

Planning and development are often time-consuming, especially for large projects. It is therefore important during the coming decade to begin study and development of programs that appear to have exceptional promise for the 1990s and beyond. Projects and study areas recommended by the Committee in this category include the following, in which the order of listing carries no implication of priority:

A. Future x-ray observatories in space;
B. Instruments for the detection of gravitational waves from astronomical objects;
[18] C. Long-duration spaceflights of infrared telescopes cooled to cryogenic temperatures;
D. A very large telescope in space for optical, ultraviolet, and near-infrared observations;
E. A program of advanced interferometry in the radio, infrared, and optical spectral regions;
F. Advanced gamma-ray experiments; and
G. Astronomical observatories on the Moon.

Detailed discussion of the Programs for Study and Development appears in Chapter 7. [not included]

Document III-34

Document title: "The Great Observatories for Space Astrophysics," NASA Astrophysics Division, 1984.

Source: Charles Pellerin, personal collection.

By the mid-1980s, NASA had plans in place for four major space-based astronomical facilities to explore the infrared, optical, x-ray, and gamma-ray wavelengths: the Space Infrared Telescope Facility, the Hubble Space Telescope, the Advanced X-ray Astrophysics Facility, and the Gamma Ray Observatory. Constraints on NASA's space science budget during that time, however, led astronomers to debate which of these missions should have priority. Charles Pellerin, then Director of Astrophysics at NASA, believed that the astronomy community should take a different course, pushing not for one or two missions but for all four missions, because data in all of these wavelengths were necessary to provide the most information about the universe. Needing a way to explain clearly the importance of funding all four of these observatories to Congress and the public, Pellerin and other astronomers developed an illustrated brochure explaining the four missions and their objectives. Impressed with Pellerin's packaging of the astronomy missions, George Field, leader of the National Research Council's study of astronomy and astrophysics in the 1980s, dubbed the four spacecraft the "Great Observatories." As the title of the original "comic book" and this second printing of the document suggest, the name stuck. Taken to the Office of Management and Budget as well as to various congressional offices by astronomers, the booklet's message and easy-to-understand language and graphics were a major influence in securing funding for all four Great Observatories.

[cover sheet]

[redacted]

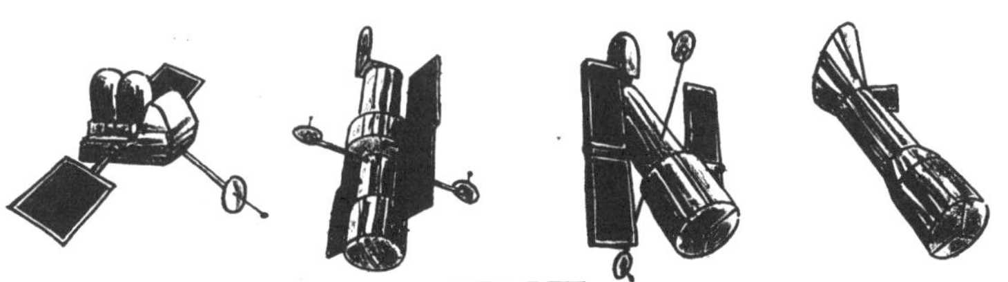

THE GREAT OBSERVATORIES

FOR SPACE ASTROPHYSICS

[1]

THE GREAT OBSERVATORIES
FOR SPACE ASTROPHYSICS
The origin of the Universe
The fundamental laws of physics
The birth of stars, planets and life

NASA
National Aeronautics and Space Administration
Astrophysics Division

[2] **An Astronomical Heritage**

Although astronomy is a science that has been practiced since ancient times, the Universe remains veiled in mystery. The ruins of Stonehenge and Chichen Itza, the clay tablets of Babylon, the cosmic models of Greek schools of thought, and the celestial mythologies of various cultures offer historical evidence for widespread astronomical observations. Until Galileo revolutionized astronomy with the telescope, however, our understanding of the Universe owed more to preconceptions than to precise observations and measurements. Since Galileo first peered into the heavens with a device more sensitive than the human eye, telescopes and observatories have proliferated, revealing a richly varied Universe.

Astronomers study the nature of the Universe by observing its contents and behavior, while astrophysicists seek to understand these observations in terms of consistent laws of physics. Limited for centuries to observations in the visible band of the electromagnetic spectrum, scientists now have access to the Universe at virtually all wavelengths. The tremendous advance that has occurred in our lifetimes became possible with spaceflight: Placing sensitive astronomical instruments above the filtering atmosphere opened new windows onto the cosmos and revealed intriguing objects and events there.

Today we are still motivated by the ancient urge to observe, measure, compute, and thereby come to greater understanding of the nature of the Universe. We have at our disposal the most advanced technology, and we have new opportunities to place entire observatories into space for investigations across the spectrum. Our astronomical heritage flourishes on the insights and discoveries of this new Era of Space Observatories.

[3]

[4] **The Popular Appeal of Astronomy**

Astronomy is in the midst of its most exciting period since Galileo probed the heavens with the first telescope. Widespread public interest is evident in the flowering of amateur astronomy societies, the popularity of space science and astronomy publications, and the attendance figures for astronomy lectures, films, museum exhibits, and planetarium shows.

In the United States today, there are at least a quarter million amateur astronomers, many of them children who will become the scientists and engineers of tomorrow. The National Air and Space Museum, the country's principal museum dedicated to space exhibits, has attracted up to 15 million visitors annually, and attendance continues to grow. More than 350,000 visitors a year pay admission to attend the planetarium shows there. A thousand other planetaria exist around the country, most of them in high schools, where their educational value is especially significant. Audiences for astronomy lectures and films are typically large and enthusiastic.

Television productions based on astronomy are extraordinarily popular, attracting millions of viewers. Many of the mass-circulation magazines (including Time, Newsweek, National Geographic, Smithsonian, Omni, Scientific American, Discover, and others) vividly report astronomical discoveries to millions of readers.

The popular appeal of astronomy, for education and entertainment, is enormous. Almost everyone is curious about the Universe.

[5]

[6] **Careers in Astronomy**

An education in astronomy and space science can lead to a variety of careers. Some graduates apply their skills to the design of new techniques for observing and interpreting cosmic processes at great distances across the Universe. Others study our more immediate environment within the solar system, in part to determine causes of climatic variations. Still others become teachers or put their talents to use in industry and government.

Research:

Scientists extend the frontiers of knowledge in the various disciplines of astronomy and astrophysics by observation, analysis, and theory. They find employment in universities, observatories, and government centers. Teaching: The study of the Universe is important, and popular, in the curriculum at all levels:

* General science in elementary and secondary schools
* Basic astronomy in colleges and universities
* Graduate and postgraduate courses at dozens of universities.

Industry:

Many people interested in astronomy join industry to conduct applied research in optics, electronics, and computer science. Others become involved in instrument design and fabrication. These scientists and engineers are responsible for the advanced technology that makes further discovery possible.

NASA:

Astronomers provide leadership for the nation's space program, managing the pioneering exploration of space and meeting the challenges of tomorrow.

[7]

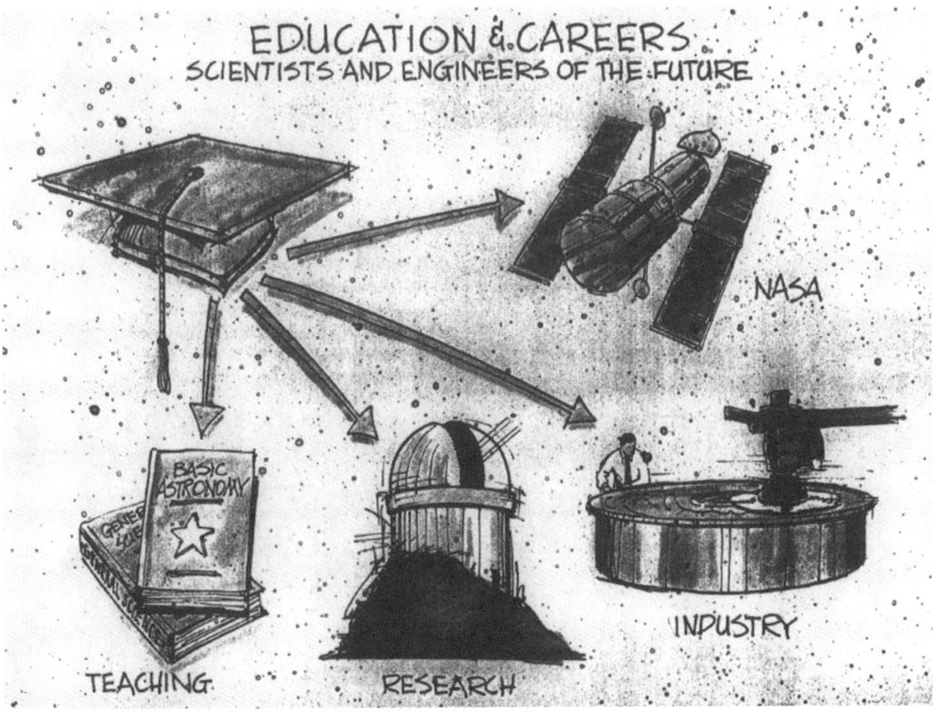

[8] **Astronomy and Technology: A Continuing Exchange of Novel Ideas**

Over the centuries, astronomy and technology have progressed hand in hand. The study of the Universe has benefited from improved observational devices and techniques. By the same token, developments in astronomy have led to practical applications in other disciplines.

1500-1600

* Increasingly accurate maps of the sky for navigation

1600-1700

* Christian Huygen's invention of the pendulum clock for navigational time keeping

* Newton's development of the calculus, the laws of motion and the law of universal gravitation as a means to explain the motions of planets and comets

1800-1900

* Increasingly sophisticated optical innovations by astronomers (William Herschel, Fraunhofer, Lord Rosse, Alvan Clark, and many others)
* Development of increasingly sensitive photographic techniques
* Lockyer's discovery of a new chemical element, helium, on the sun before it was known on Earth

1900-NOW

* Hans Bethe's theoretical prediction of hydrogen fusion at the center of the sun, a precursor for all modern fusion efforts
* Lyman Spitzer's development of astrophysical plasma theory, the basis of present devices for releasing energy from controlled fusion
* Very long baseline radio astronomy techniques used in high-precision geodesy to survey the structure of the Earth
* Techniques of celestial mechanics, precursors to the development of accurate spacecraft navigation.

The mutually beneficial interaction between astrophysics and technology continues today.

[9]

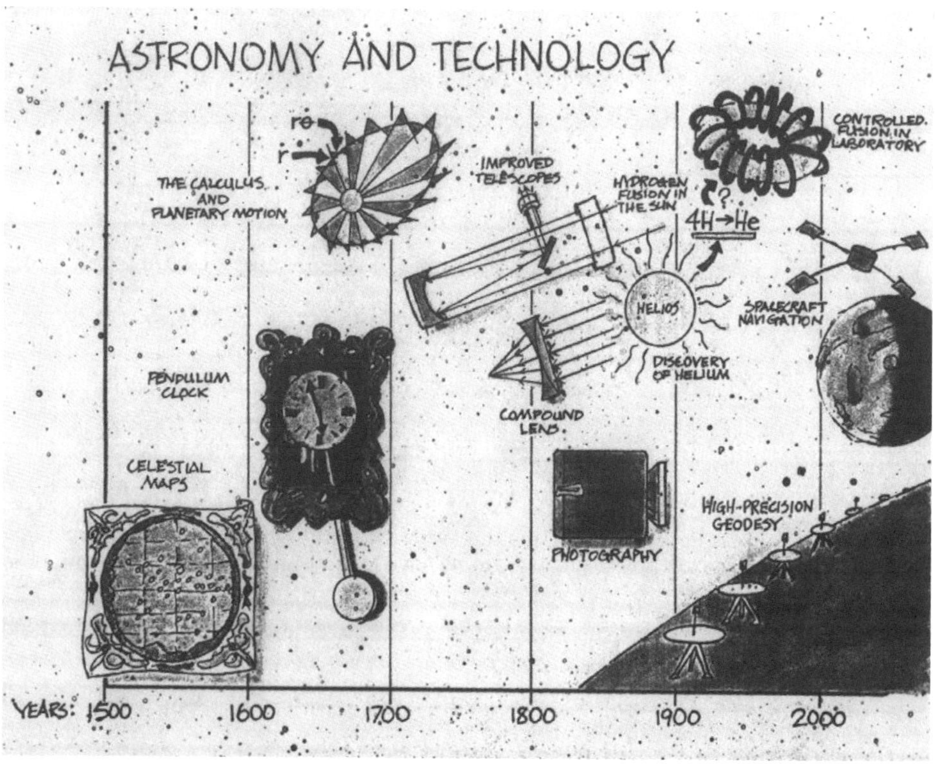

[10] **Observing the Universe with Improved Sensitivity**

Over the past two decades, NASA has introduced increasingly sensitive instruments into space. In astronomy, families of telescopes have been developed and placed in orbit for observations across the entire electromagnetic spectrum, especially those parts blocked by the atmosphere.

Each successive telescope has extended the limits of sensitivity and provided greater insight into the structure of stars, galaxies, and the cosmos. For these successes, new technologies had to be created and exploited.

Members of the new generation of space observatories offer significant new gains in sensitivity through state-of-the-art technology.

* The GAMMA RAY OBSERVATORY (GRO) will explore the most energetic part of the spectrum across a much greater wavelength range than its predecessors.
* The ADVANCED X-RAY ASTROPHYSICS FACILITY (AXAF) will cover the X-ray portion of the spectrum with a hundred-fold improvement in sensitivity.
* The HUBBLE SPACE TELESCOPE (HST) will penetrate deep into the Universe in visible and ultraviolet light, expanding the volume of observable space several hundred times.
* The SPACE INFRARED TELESCOPE FACILITY (SIRTF) will span the infrared part of the spectrum with a thousand-fold increase in sensitivity.

To complement these sensitive space telescopes, the astronomical community is currently considering two powerful new ground-based observatories: the Very Long Baseline Array (VLBA), an intercontinental network of radio telescopes, and the National New Technology Telescope (NNTT), a large optical telescope. In addition, NASA's Solar Optical Telescope (SOT) will provide detailed data on our nearest star, the sun, to augment our studies of distant stars and cosmic processes. With these new observatories, we will be able to open more of the Universe to scrutiny, to look back in time and space for order and meaning.

[11]

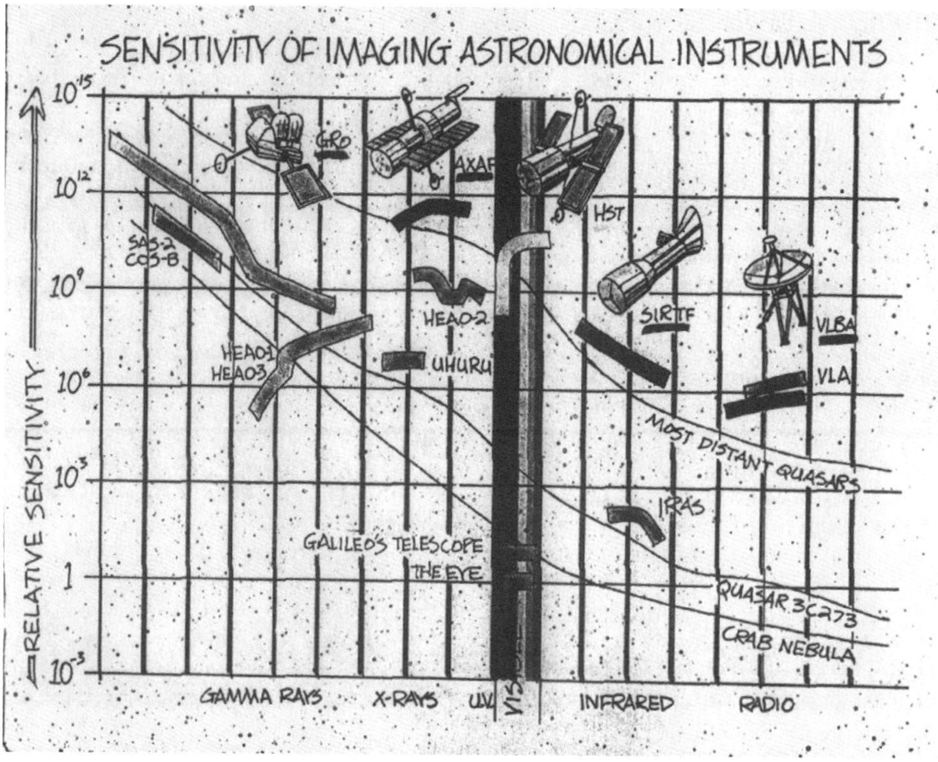

[12] **Technology for the Future**

Astronomical observations pose some of the greatest challenges to modern technology. Engineering difficulties overcome by astrophysicists often provide solutions to more general technical problems. We expect new technologies developed for the four space observatories to stimulate future applications in space and on Earth.

* GRO introduces a propulsion system that can be refueled in orbit to extend the lifetime of the observatory.

* AXAF's nested grazing-incidence mirrors will provide the most advanced X-ray focussing optics presently known. Previous X-ray astronomical advances have already led to commercial applications in low-dosage imaging systems.

* HST has already pioneered the construction of higher precision optics than any ever built. Its spacecraft pointing control will cross new thresholds of accuracy.

* SIRTF will provide longer endurance for ultra-low-temperature apparatus in space, a requirement for many other high-precision technologies. Previous infrared astronomical missions pioneered the handling of liquid helium, the ultimate refrigerant, in space. However, SIRTF introduces the new capability of replenishing liquid helium coolant in orbit.

These advanced technologies will be of benefit not only for scientific research but also for practical down-to-Earth uses. We can only guess what new applications will result from the spread of these technologies into everyday life.
[13]

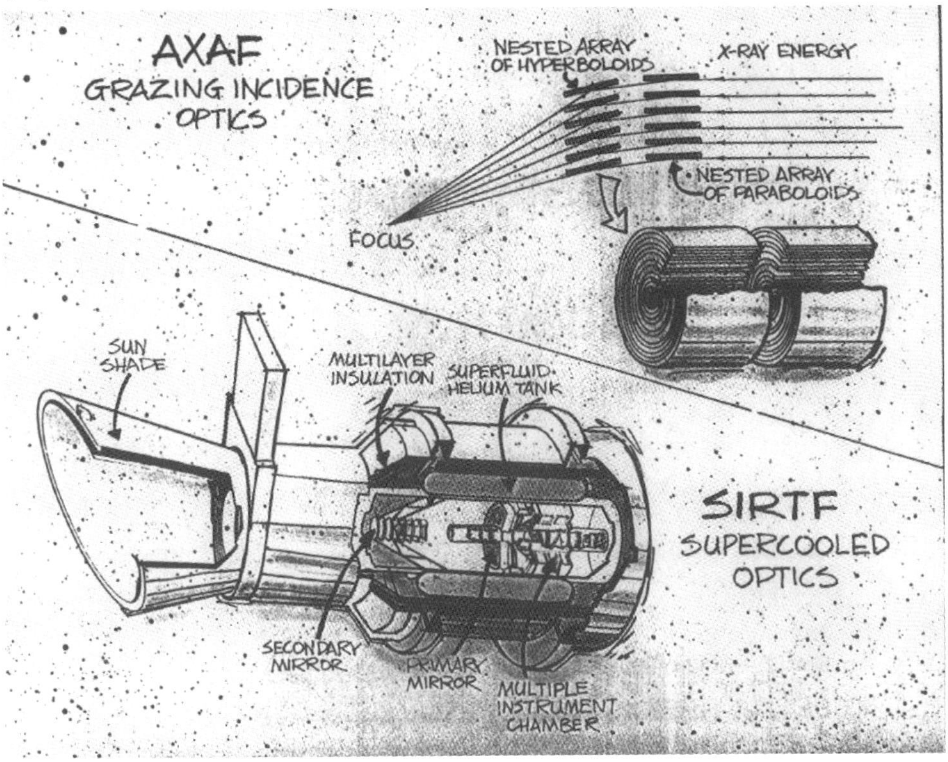

[14] **Discovering the Nature of the Universe**

Astronomical discoveries have been occurring at a quickening pace since the development of the first telescope, and particularly in the past few decades. These discoveries are closely linked to the introduction of new technologies into the field.

The discovery of X-ray stars and X-ray galaxies in the 1960's was made possible by the flight of first-generation X-ray telescopes on some of the early rockets. Infrared stars and galaxies were discovered about the same time with novel detectors that had just become available.

The listed discoveries, though impressive, are only a fraction of those remaining to be made. New technologies in the era of space observatories will certainly lead to further discoveries just as striking as those of the past.

What might these be? Possibly black holes at the distances of the nearest stars, having masses similar to those of ordinary stars and detected through the X-ray emission produced as the black hole gravitationally accretes interstellar matter. Or perhaps an infrared planet orbiting a nearby star with a period identical to that of some unusual radio signals from the same part of the sky, suggesting the existence of an intelligent, technically advanced civilization. Or else bizarre "shadow galaxies" or networks of massive cosmic "strings," predicted by some of our Grand Unified Theories of elementary particle physics but never seen in the laboratory. These particles, produced only at the enormous energies prevalent in the early exploding Universe, would be revealed through the X-ray emission from hot gas gravitationally attracted to them.

Discoveries:

1. *Stars*
2. *Planets*
3. *Novae*
4. *Comets*
5. *Moons*
6. *Rings*
7. *Galactic Clusters*
8. *Clusters of Galaxies*
9. *Interplanetary Matter*
10. *Asteroids*
11. *Multiple Stars*
12. *Variable Stars with Nebulosity*
13. *Planetary Nebulae*
14. *Globular Clusters*
15. *Ionized Gas Clouds*
16. *Cold Interstellar Gas*
17. *Giants/Main Sequence Stars*
18. *Cosmic Rays*
19. *Pulsating Variables*
20. *White Dwarfs*
21. *Galaxies*
22. *Cosmic Expansion*
23. *Interstellar Dust*
24. *Novae/Supernovae*
25. *Galaxies With/Without Gas*
26. *Supernova Remnants*
27. *Radio Galaxies*
28. *Magnetic Variables*
29. *Flare Stars*
30. *Interstellar Magnetic Fields*
31. *X-Ray Stars*
32. *X-Ray Background*
33. *Quasars*
34. *Microwave Background*
35. *Masers*
36. *Infrared Stars*
37. *X-Ray Galaxies*
38. *Pulsars*
39. *Gamma-Ray Background*
40. *Infrared Galaxies*
41. *Superluminal Sources*
42. *Gamma-Ray Bursts*
43. *Unidentified Radio Sources*

[15]

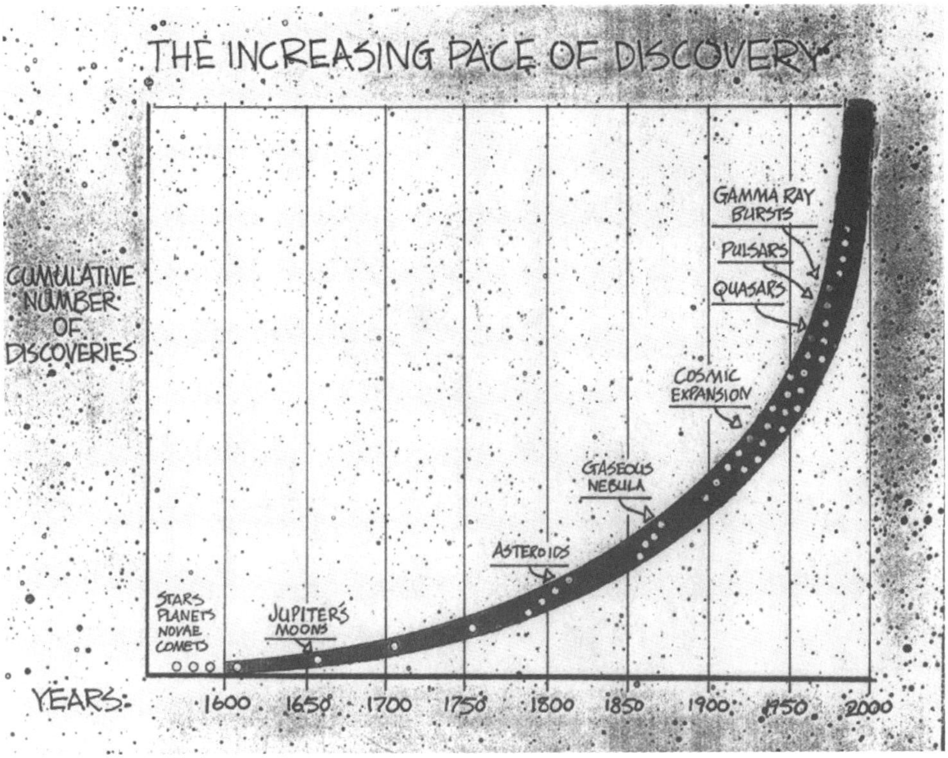

[16]

NASA's major contribution to modern astrophysics has been the agency's ability to place powerful new telescopes into orbit. From their vantage point in space, these observatories can sense gamma rays, X–rays, and ultraviolet, optical, and infrared radiation undisturbed by the distorting, absorbing atmosphere.

Discoveries of ultra-hot gas in clusters of galaxies, gamma background-radiation from the Universe, and galaxies which emit virtually all their energy in the infrared have all resulted from this capability. Observations in each wavelength band reveal a new Universe.

[17]

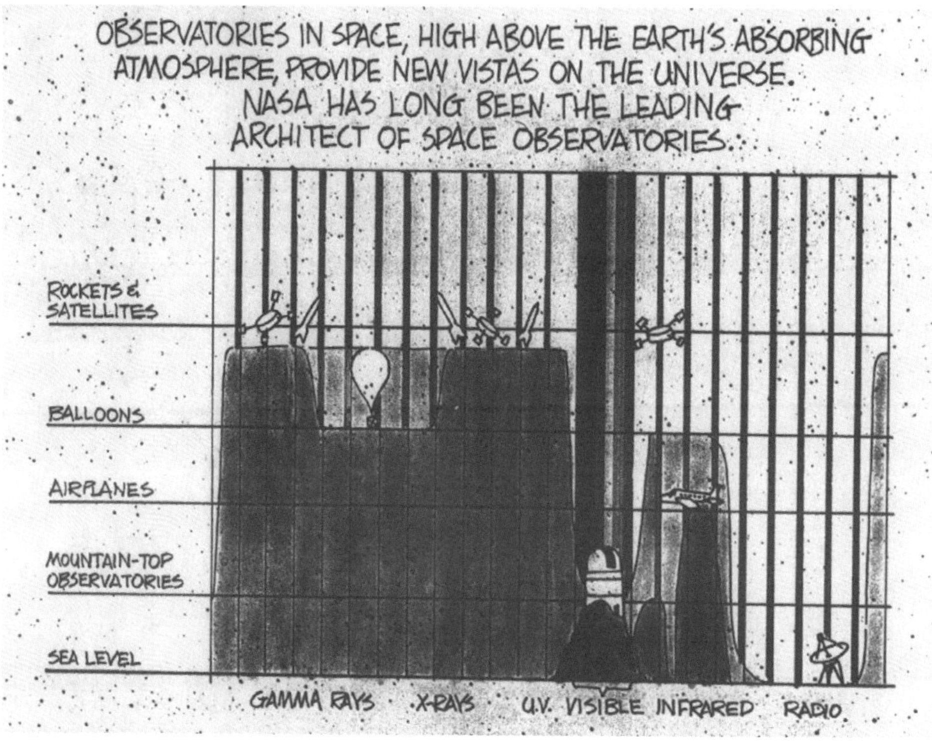

[18] **The Milky Way**

Our galaxy, the Milky Way, is populated by star clusters, dusty clouds of turbulent gas, exploding or collapsing stellar masses, and gradually evolving systems of stars - phenomena that are revealed by observations at widely differing wavelengths. Through the use of powerful telescopes, we hope to observe and comprehend these processes that reflect the birth of stars, their eventful lives, and their ultimate death.

* At radio wavelengths we detect cool clouds in space; some are destined to contract to form new stars, while others are ejected at high speeds, emitting radio waves characteristic of water vapor masers.

* At infrared wavelengths we probe clouds warmed by stars that have formed within them; we also register dying stars throwing off shells of matter.

* At visible wavelengths we see millions of stars like our sun, and we can study their evolution as they consume their nuclear energy.

* At ultraviolet wavelengths we detect the hottest stars. Some are still actively consuming nuclear energy; others, like white dwarfs, are dying remnants of small stars which once were active.

* At X-ray wavelengths we see matter at ultra-high temperatures falling on neutron stars—the remains of more massive dead stars.

* At gamma-ray wavelengths we detect sudden bursts of intense emission from sources not yet understood.

By combining these different pictures of our galaxy, we gain greater understanding, while any one of these observations alone would leave us puzzled.

[19]

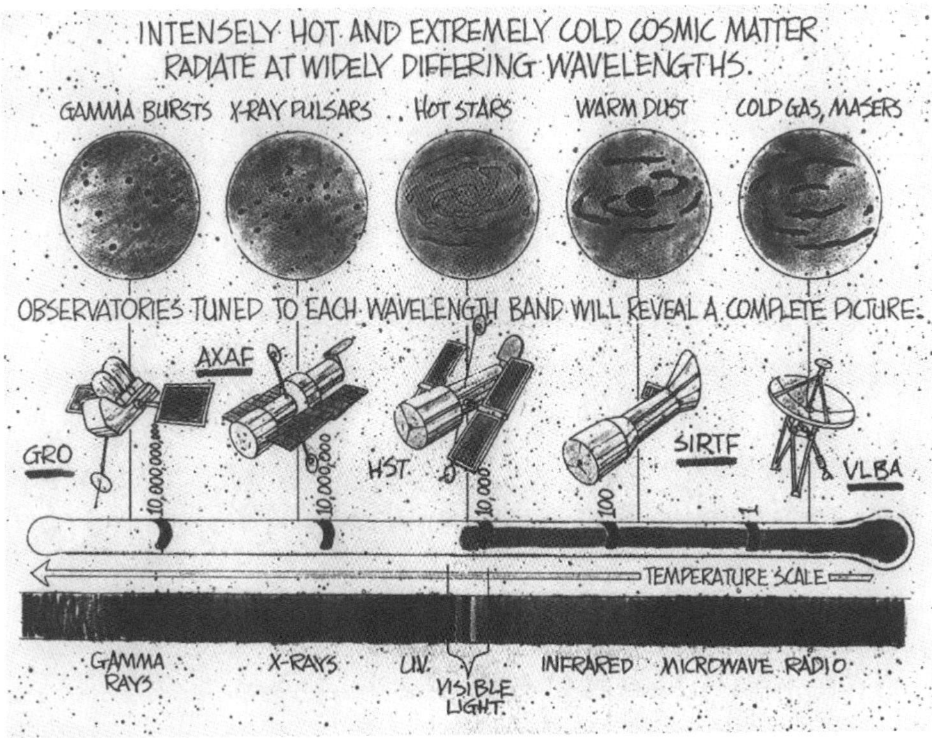

[20] **A Typical Cluster of Galaxies**

The gamma ray emission from clusters of galaxies is expected to emanate primarily from quasars and from the nuclei of active galaxies. With the next generation observatory, GRO, we will be able to determine whether most of the known gamma radiation arriving from the Universe originates in quasars, or whether there are other powerful, but presently unknown, sources of gamma ray emission.

The X-ray map of a cluster frequently is dominated by a hot, diffuse plasma permeating intergalactic space. Quasars also show up on such a map, while individual galaxies appear much fainter.

Pictures recorded at visible wavelengths show starlight from all the members of a cluster of galaxies.

Infrared radiation predominantly comes from dusty galaxies in which dust grains absorb virtually all the starlight and re-emit this energy at longer wavelengths.

The radio view is dominated by the luminous core of a massive central galaxy, from which magnetically channeled jets of electrons and protons are ejected at nearly the speed of light.

[21]

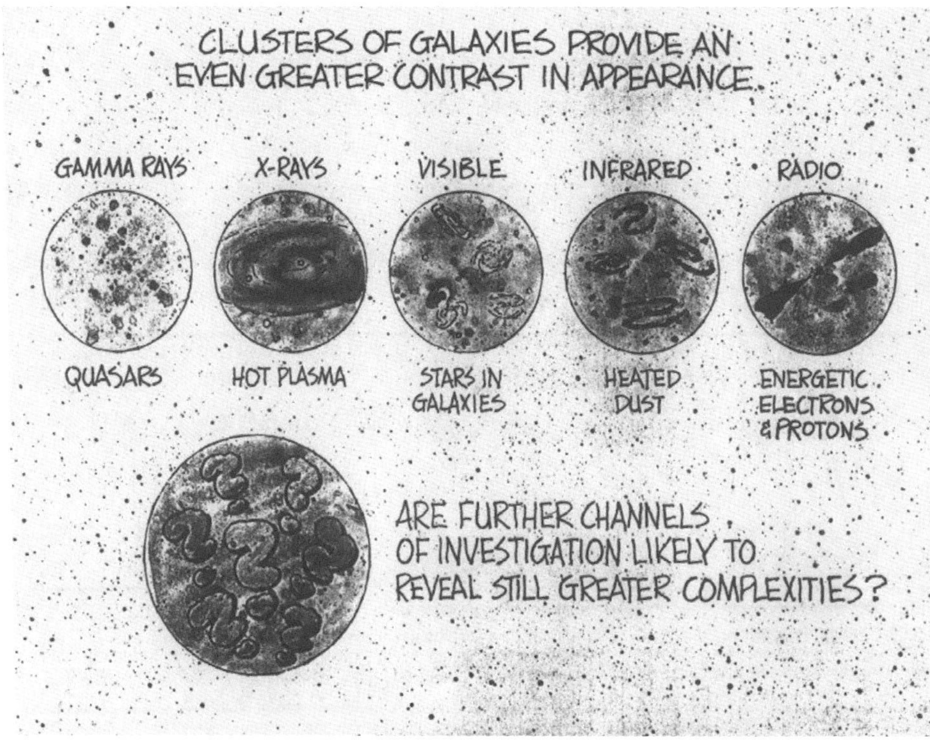

[22] **What If We Could Observe in Just One or Two Wavelength Bands?**

Many discoveries become apparent only through a combination of observations.

For example, quasars, discovered in 1963 by virtue of their powerful radio emission, had been recorded on photographic plates for many decades. Nobody had noticed them because they looked so much like normal stars. Later, more extensive optical data showed quasars to lie far out in the Universe. Recent observations with NASA's HEAO-2 (Einstein) observatory have shown quasars to be even more powerful emitters of X-rays than of radio or light waves.

To understand the nature of quasars and many other celestial objects, we need to study them at all wavelengths.

[23]

[24] **Fundamental Questions in Astrophysics**

Our thoughts about the long-term future of the human race involve fundamental questions about the nature of the cosmos - its past and its future, its governing physical laws, its harsh explosions, and its potential for hospitable planetary systems. We ask:
* How did the Universe form and evolve in the first few seconds? Can we learn more about the basic laws of physics from the effects they have had on the structure of the Universe?
* How did galaxies and clusters of galaxies initially form and how have they evolved?
* Will we need new laws of physics to describe observed phenomena? Will the Law of Gravitation have to be modified or will new fundamental particles be found to play a central role?
* Can massive stars or galaxy-size aggregates collapse to form black holes, liberating enormous amounts of energy? Are such black holes the energy sources of quasars and active galactic nuclei? How do these powerful sources affect the galaxies in which they reside?
* How do stars and star clusters form and die, and how do they interact with interstellar matter? Do shock waves from stars dying in supernova explosions induce star formation? How do magnetic fields arise in interstellar matter and in stars?
* How are planetary systems formed? How many stars have planets and how many might

be habitable? Where and how did life start? Are there intelligent civilizations elsewhere in the Universe?
[25]

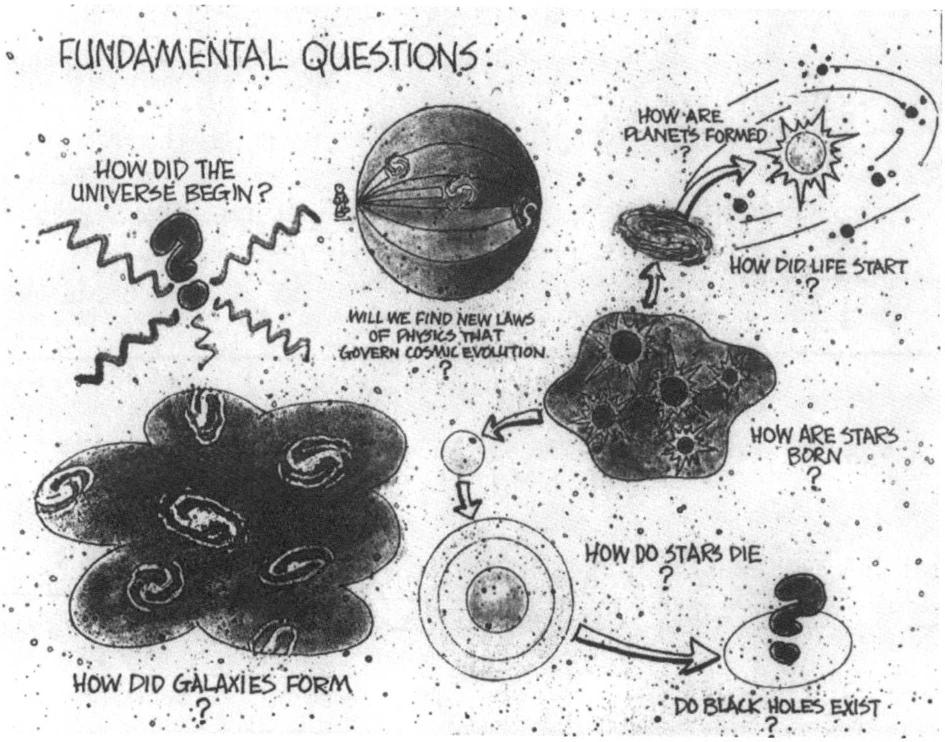

[26] **Birth Places of Stars**

We know that stars must be forming in our galaxy today. Although the Milky Way is more than ten billion years old, we see stars that are thousands of times younger. A star can continue to shine only as long as it has a supply of energy to radiate away into space. The most luminous stars quickly exhaust these limited supplies and must be young, no older than a few million years. Young stars are always found near dark, dusty gas clouds, the birth places of stars.

The gaseous central portions of a cloud contract, becoming ever more compact until a star is born. The early collapse of a contracting core can be detected only with infrared and radio observations that penetrate the dust–shrouded regions. SIRTF and radio telescopes have this capability. Once a young star is formed and its cocoon of dust is blown away by powerful stellar winds, eruptive magnetic processes that mark the final stages of star formation can be studied with sensitive optical telescopes such as HST.

Similar magnetic phenomena occur, on a smaller scale, on the surface of the sun, our best laboratory for studying these violent outbursts. The Solar Optical Telescope (SOT) will be a powerful tool not only for investigations of the sun but also for insight into the storage of magnetic energy at the surfaces of young stars. This energy is later unleashed sporadically in enormous flares.

[27]

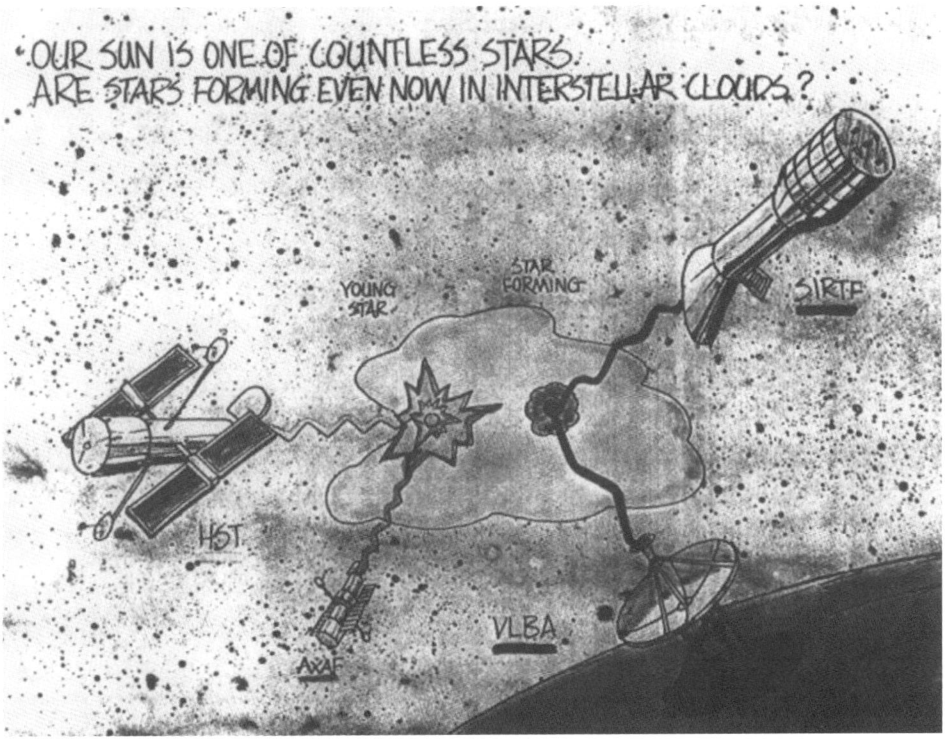

[28] **Life Cycles of Stars**

The birth of stars may well be triggered by the explosion of a supernova that compresses a nearby dusty cloud of gas, which then collapses to form a new group of stars. Some of these are more massive than others and begin to shine thousands of times more brightly than the sun. Such stars consume their supply of nuclear fuel in a few million years and collapse to form a neutron star, or possibly a black hole. In this collapse, enormous amounts of energy are suddenly liberated, and the outer shell of the star is hurled into space in another supernova explosion. X-ray observations of the remains of such explosions can tell us much about the original star as well as the exploding shell.

Less massive stars, like our sun, never explode as supernovae. Instead, they shine steadily, at a far more subdued rate, for ten billion years before continuing their lives, briefly as red giant stars and finally as faint white dwarfs. Some of these stars may originally be enveloped by a disk from which a system of planets settles out.

Currently we have no way of knowing how many stars are orbited by planets or how many stars are encircled by disks. We hope to answer these questions by making optical observations with HST and infrared observations with SIRTF.

[29]

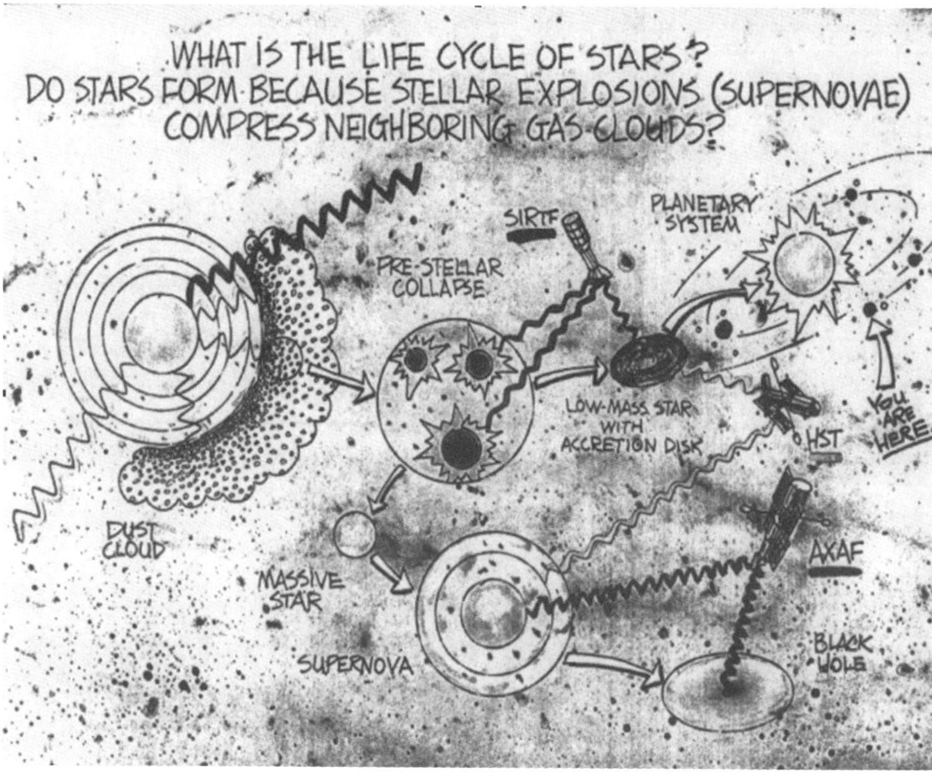

[30] **Quasars**

Quasars are distant, massive bodies so luminous that they outshine surrounding galaxies a hundred times. We do not know how to explain this immense power. One possible model of a quasar consists of an intensely hot central source emitting gamma rays and embedded in X-ray-emitting plasma. Enveloping dust clouds absorb much of the emitted energy, re-radiating it at far-infrared wavelengths. An outermost, unobscured layer also radiates at ultraviolet and visible wavelengths. Plasma beams ejected from the central source at nearly the speed of light power distant radio lobes.

How can so much energy be radiated from so compact a source? How can we account for the rapid variations in luminosity, from one month to the next, sometimes even from one hour to the next? Are quasars powered by a rapid succession of supernova outbursts in a central core, or is it more likely that matter falling onto a single, central black hole supplies all the energy?

To clear up many of these questions, we need the full complement of our most powerful observatories, often working together to trace outbursts as they evolve -sometimes emitting successively in different wavelength bands, sometimes simultaneously varying across the entire spectrum.

These detailed observations should clarify the nature of the central engines powering quasars and explain the structure of ambient regions.
[30]

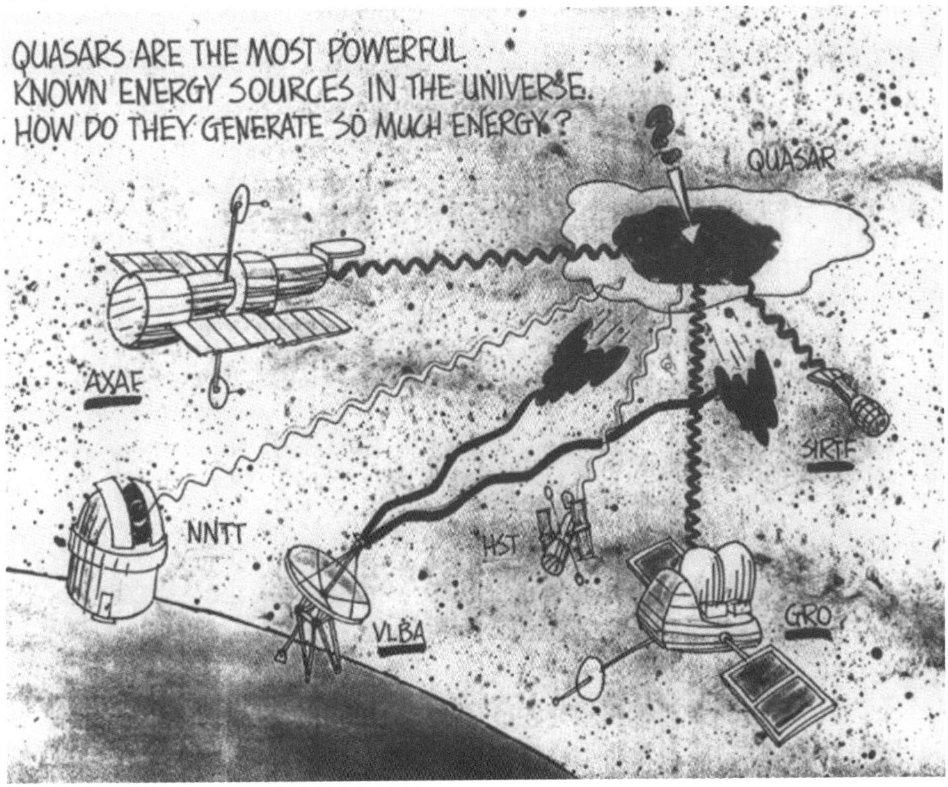

[32] **Black Holes**

Black holes are enormously compact bodies, so dense that matter falls into them under an irresistible gravitational pull. So far we are not sure whether nature produces such holes. If they do exist, black holes could be very massive, or quite small, or just about as massive as a star.

The largest black holes might be the power sources for quasars; each could have a mass comparable to that of an entire galaxy of a hundred billion stars.

The smallest black holes could have masses of only a billion tons - roughly the mass of the Rock of Gibraltar - and could be capable of exploding at any time, annihilating themselves totally in an enormous flash of gamma radiation lasting no more than a few seconds.

A stellar-sized black hole could have a mass five or ten times greater than the sun's. Such a black hole in a binary system with a giant star could syphon [sic] matter off the giant's surface, giving rise to X-ray emission as this matter crashed down onto an accretion disk encircling the black hole.

A different stellar-sized black hole, also a member of a binary system but sufficiently distant from its companion star to leave it intact, might be detected through careful observations of the companion's orbital motion. The companion would appear to be circling a massive center, but there would be no radiation coming from that point; it would seem as though nothing were there except a strong gravitational pull, a black hole.

We are not yet sure that we have observed any one of these, but we know of likely candidates that need to be studied with the most powerful observatories we can build.

[33]

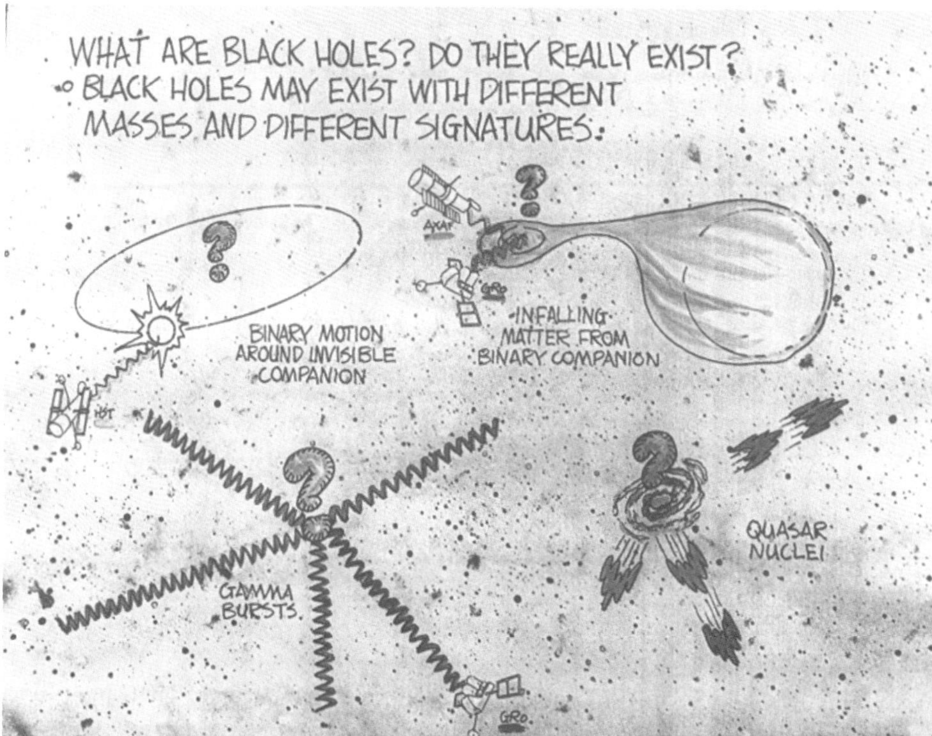

[34] **Magnetic Energy Storage**

Magnetic fields - on the surface of the sun, in inter-planetary space, in far-reaching stellar jets, in interstellar clouds, in the spiral arms of galaxies, and in the giant intergalactic jets spanning an entire cluster of galaxies - are able to store enormous amounts of energy. Sometimes the stored energy is released in an explosive flare, through processes we do not understand at all. In fact, we have no convincing theories to explain the generation and existence of such strong magnetic fields.

To gain greater understanding, we must not only look beyond the solar system but also observe more carefully within it, looking at magnetic processes occurring in the interplanetary medium and magnetic events taking place on the sun's surface. The Solar Optical Telescope (SOT) will help us to understand solar magnetic events, interplanetary probes will help us to understand transformations in the magnetized interplanetary plasma, and our other observatories should enable us to relate these local effects to phenomena taking place on galactic and intergalactic scales.

[35]

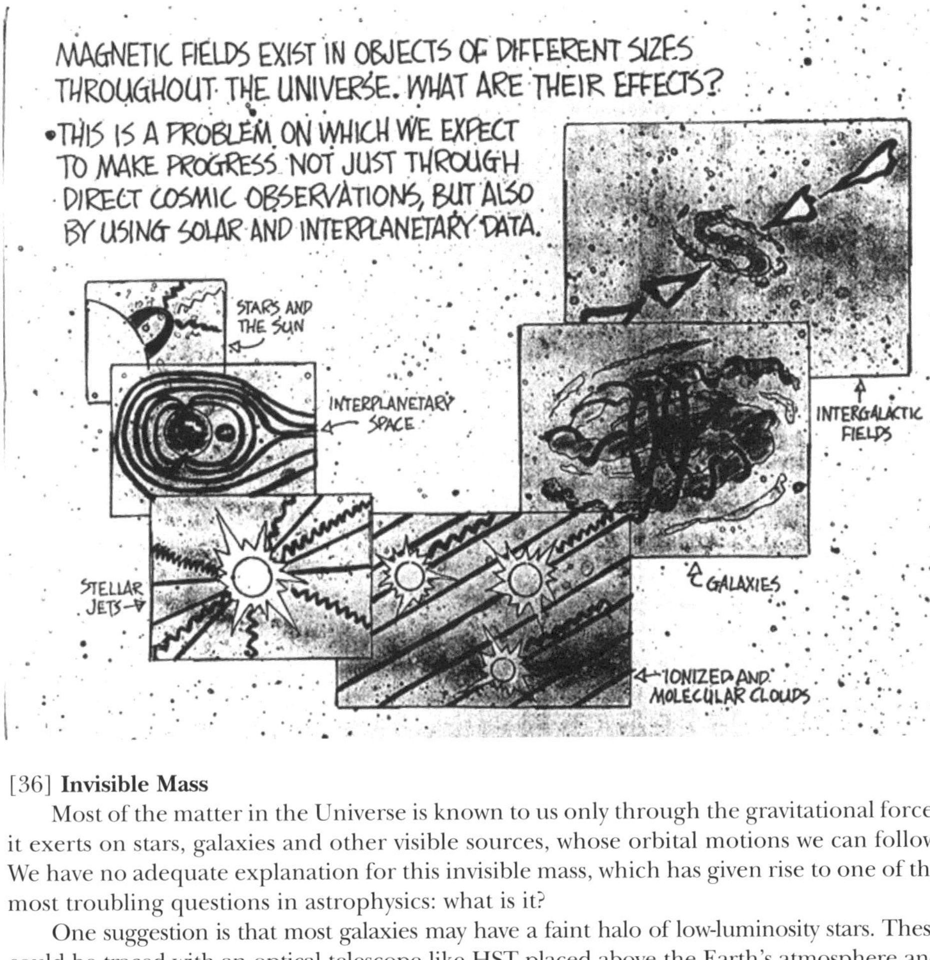

[36] **Invisible Mass**

Most of the matter in the Universe is known to us only through the gravitational forces it exerts on stars, galaxies and other visible sources, whose orbital motions we can follow. We have no adequate explanation for this invisible mass, which has given rise to one of the most troubling questions in astrophysics: what is it?

One suggestion is that most galaxies may have a faint halo of low-luminosity stars. These could be traced with an optical telescope like HST placed above the Earth's atmosphere and therefore capable of seeing fainter diffuse distributions of stars. The matter might also be distributed in galactic halos in the form of brown dwarfs, bodies intermediate in mass, between Jupiter-sized objects and the least massive stars known to emit visible light. Brown dwarfs would emit primarily at infrared wavelengths and be observed with SIRTF A further possibility is a halo of black holes or of low-mass stars. In either case, a faint diffuse glow of X-rays would emanate from the halos of galaxies, a glow that AXAF would permit us to detect.

An entirely different tracer of invisible mass in clusters of galaxies is intensely hot intergalactic plasma. X-ray emission from this plasma is brightest in the innermost portions of the cluster where most of the mass is concentrated. The distribution of X-ray brightness across the cluster provides us with a measure of total mass. Using this measure, AXAF would permit us to search for invisible mass in clusters at extreme distances across the Universe.

Finally, families of new, exotic particles, like axions or gravitinos, or else networks of massive cosmic strings required by some elementary particle theories, could be responsible for this invisible mass. Further study may permit us to distinguish among these different kinds of particles and provide insight into fundamental forces that govern their interactions.

[37]

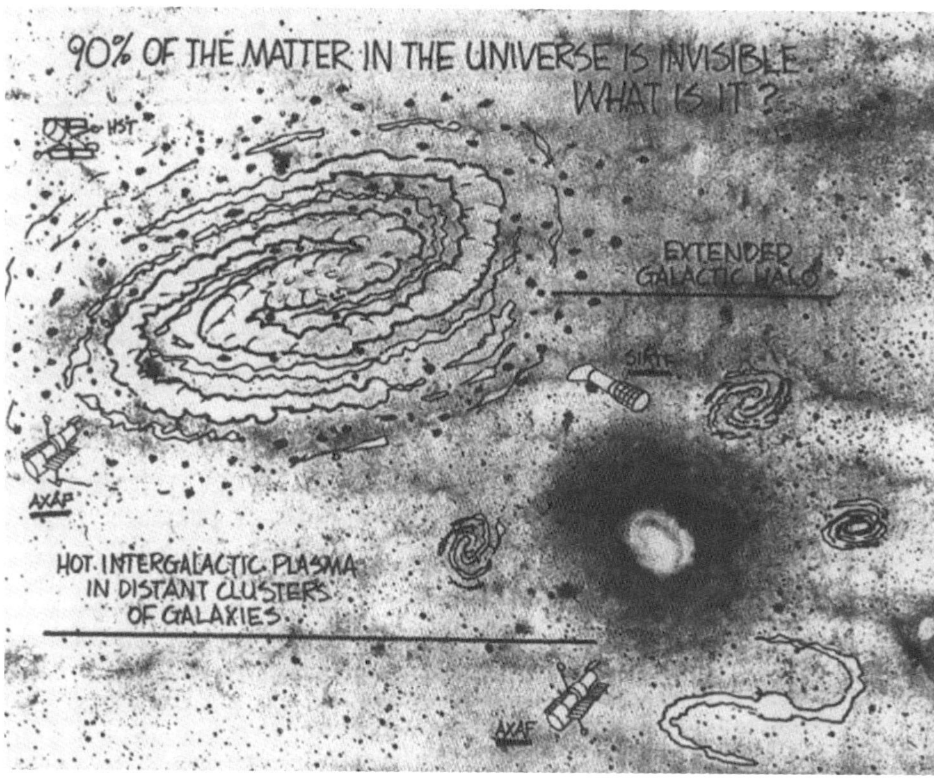

[38] **Looking Out into the Universe Means Looking Back in Time**

The Universe is so large that even signals travelling at the greatest speed that can be attained - the speed of light - require billions of years to cross major portions of the tracts we can survey. This long delay in the arrival of radiation can work to our advantage.

To understand how galaxies or quasars originated in a rapidly expanding Universe, we can look back in time to observe the contraction of protogalactic clouds, expected to emit far-infrared radiation, and young galaxies emitting radiation at visible and infrared wavelengths. Young quasars should be powerful sources of X-rays as well as radio waves.

These sources are beyond the range of present instruments, but those limits will be surpassed with the next generation of space observatories capable of surveying the sky out toward the moment when galaxies began to form, and beyond, to the impenetrable barrier that lies at a distance and time when electrons and protons were combining to form atoms of hydrogen. Currently, only radio telescopes can look back at that barrier from which the cosmic microwave background radiation emanates. Some day we may devise ways of looking even further back, but that may have to await the construction of gravitational wave detectors or neutrino observatories.

[39]

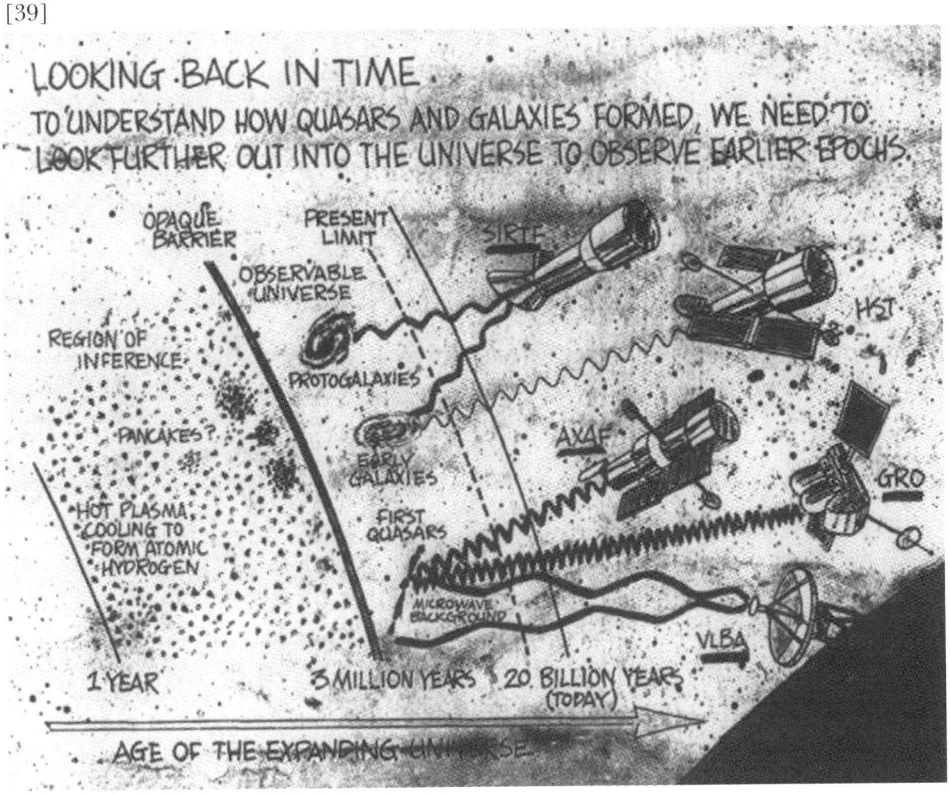

[40] **What is the Geometry of the Universe?**

On Earth, distant objects appear small; their angular diameters diminish as they recede. In a curved, expanding Universe all that is changed.

Distances across the Universe can be gauged by the extent to which radiation reaching us is shifted toward longer wavelengths -the extent to which it is red shifted. The more distant the emitting source, the greater is the red shift.

In a closed Universe, the angular diameter of a galaxy or quasar observed at ever-increasing red shift - distance -first shrinks but then expands.

In an extreme open Universe, the angular diameter at first also declines but then slowly approaches a constant value at increasing red shift.

With AXAF we will be able to locate the most distant quasars in the Universe, and with HST we will determine red shifts and diameters of the most distant galaxies and quasars to investigate whether our Universe is open and expanding forever, or closed and bound to collapse on itself billions of years from now.

[41]

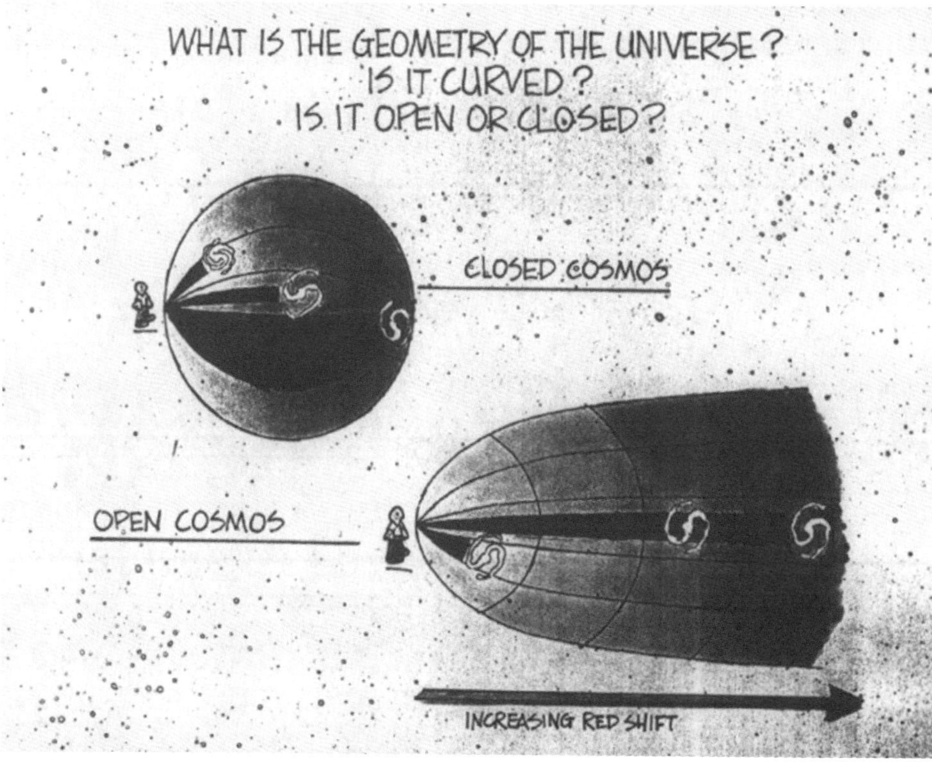

[42] **Matter and Antimatter in the Universe**

Much of the Universe we observe consists of hydrogen with an admixture of helium and heavier elements. However, everything we know about the Universe suggests that an equal amount of matter and antimatter - antihydrogen, antihelium and heavier antielements - should have existed at one time.

We can search for traces of antimatter, because we know that matter and antimatter annihilate on contact. If there existed distant galaxies composed entirely of antimatter, we should be able to detect the gamma radiation emitted when gas ejected from such a galaxy encountered and annihilated ordinary matter from a galaxy like ours.

If substantial amounts of matter and antimatter existed at earlier epochs, before galaxies ever formed, remnants of this annihilation radiation might still persist, red shifted but observable at gamma-ray and X-ray wavelengths.
[43]

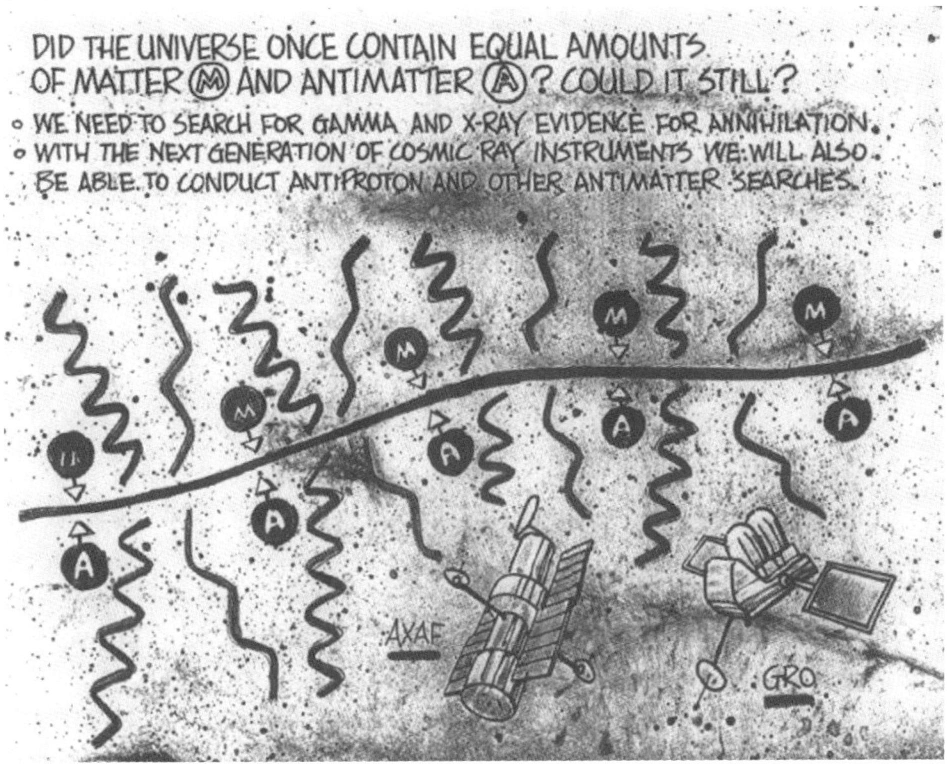

[44] **The Search for Other Planetary Systems**

Analysis of data from the Infrared Astronomical Satellite (IRAS) has shown disks of warm rocks and pebbles orbiting several stars. Such a protoplanetary disk might be a precursor of a planetary system or might co-exist with a system of planets like ours. By studying the planets of our own solar system, we should be able to gain increasing insight into how planets elsewhere might be formed and how we might best search for planets around other stars.

Distant planets will be detected most readily through infrared radiation, since planets are too cool to emit visible light, and stars are often less bright at infrared than at visible wavelengths. A visible spectrum of a planetary system mainly will register stellar emission and reflect the chemical composition of the star. An infrared spectrum will show planetary contributions to the system's emission and could provide evidence for molecules, like methane, found on planets in our solar system but destroyed on the hot surface of a star like the sun.

SIRTF will be able to search for planets around the nearest stars. Spectra for any planets detected could tell us the chemical composition of the atmosphere and help us determine whether it might sustain life similar to that on Earth. Once planets are detected, a Search for Extraterrestrial Intelligence (SETI) would become more focused.

[45]

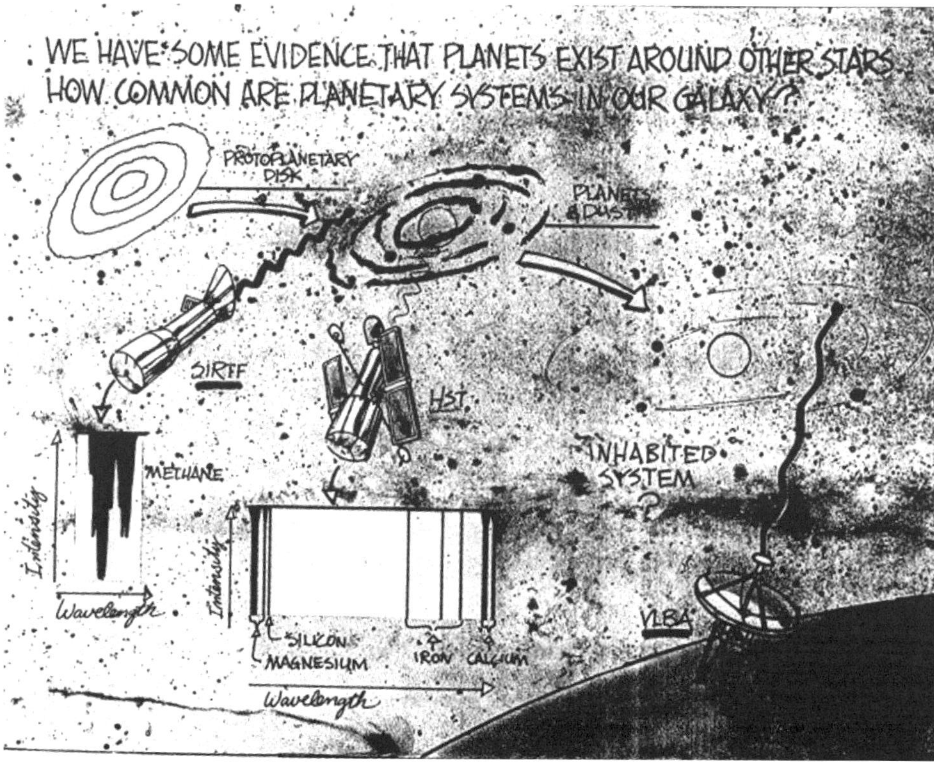

[46] **Is Life on Earth Unique?**

The Earth is an insignificant companion of our sun, an unremarkable star: there are a thousand billion, billion stars just like the sun all over the Universe. Can we reasonably expect life to be unique here on Earth?

There is no scientific basis on which that question can be answered today. However, the search for other planets may help us locate other solar systems in which we could pursue our quest for extraterrestrial life. Primitive life forms are likely to remain undetectable for a long time to come; but technologically advanced civilizations could be identified by artificial signals they generate.

We know that stray television and FM broadcast signals radiated into space from Earth could be picked up by powerful radio observatories if they existed in the vicinity of nearby stars. Similarly, highly sensitive receivers on Earth might detect comparable signs of technological expertise around other stars in nearby parts of the Milky Way. The only

question that would then remain is how we could be sure that such signals were artificial rather than generated by some previously unidentified natural phenomenon.
[48]

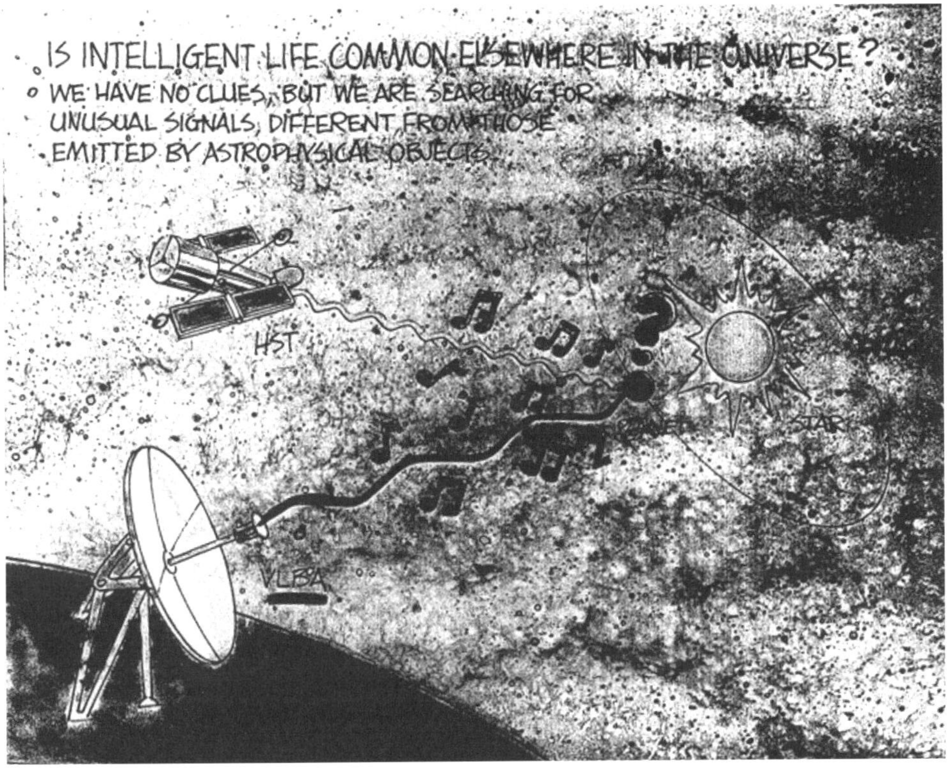

[48] **Epilogue**

Astronomical searches have occupied human thought for millennia. Over the generations, we have succeeded in gaining ever greater insight into the underlying forces at work in the cosmos. In the Space Station era, the family of permanent observatories in space will open the way to new, comprehensive studies of key remaining problems in astrophysics, helping us understand:

* The birth of the Universe, its large-scale structure, and the formation of galaxies and clusters of galaxies;

* The fundamental laws of physics governing cosmic processes and events;

* The origin and evolution of stars, planetary systems, life and intelligence.

If we succeed, we will leave a legacy to rank us with the great civilizations of the past.

[49]

[50] Prepared under the auspices of The NASA Astrophysics Division, Dr. Charles J. Pellerin, Jr., Director by Dr. Martin Harwit, Cornell University and Dr. Valerie Neal, Essex Corporation in consultation with the Astrophysics Management Operations Working Group
Graphic design and illustration by Brien O'Brien

Document III-35

Document title: Letter to Dr. L. J. Lanzerotti, Chairman, Space & Earth Science Advisory Committee, Bell Telephone Labs, from Glenn Mason, Associate Professor of Physics, University of Maryland, March 26, 1987.

Source: Alan Bunner, Office of Space Science, NASA Headquarters, Washington, D.C. Reprinted with permission.

Document III-36

Document title: Letter to G. M. Mason, Associate Professor of Physics, University of Maryland, from Dr. Martin Weisskopf, Chief, X-ray Astronomy, NASA, May 1, 1987.

Source: Alan Bunner, Office of Space Science, NASA Headquarters, Washington, D.C.

In 1982, a National Research Council committee headed by astronomer George Field recommended that the Advanced X-ray Astrophysics Facility (AXAF) should be the nation's top priority in astronomy because the mission was essential to answering many pressing astronomical questions. By the mid-1980s, NASA's Space and Earth Science Advisory Committee (SESAC) had likewise expressed its support for major, observatory-class missions such as AXAF. Not all astronomers agreed with NASA's trend of pursuing larger, more ambitious missions. Warning that investing resources only in major, space-based observatories was highly risky, such scientists advocated that NASA apply the lesson learned from the 1986 Challenger disaster to depend on a "mixed fleet" of launch vehicles for its space science program and employ a mix of mission sizes. These letters, exchanged among a university professor, the SESAC chairman, and a NASA astronomer, illustrate the differences of opinion among astronomers regarding the merit and risks of major space-based observatories.

Document III-35

[no pagination]

THE UNIVERSITY OF MARYLAND
COLLEGE PARK CAMPUS
Department of Physics and Astronomy
College Park, Maryland 20742
(301) 454-3401

Space Physics
Tel: (301) 454-3135

26 March 1987

Dr. L.J. Lanzerotti, Chairman
Space & Earth Science Advisory Committee
Bell Telephone Labs
600 Mountain Ave.
Murray Hill, NJ 07974

RE: The ultra-high risk NASA space science programs

Dear Lou,

In the post-Challenger examination of NASA science programs, many promising possibilities are being explored including the role of ELVs, moderate missions and possibly even an Explorer augmentation. These healthy discussions of future directions and strategies leave out consideration of the fact that the OSSA community has inherited what can now be seen to be a program with ultrahigh risks in which the fates of major subdisciplines are tied to single pieces of hardware. There are several examples, but perhaps the best are the "Great Observatories": HST, GRO, AXAF and SIRTF. Since HST and GRO are basically finished, consider the case of AXAF. The United States X-ray astrophysics community, which has had no mission since the Einstein observatory in the late 70's, is tying its entire future to a single device that might be launched in the mid-90's. Last year's Challenger explosion, followed shortly thereafter by the immensely costly Titan launched reconaissance [sic] satellite explosion, should remind us that failures will occur. Prudent science policy should ensure that when failures occur the results are not fatal, but indeed we now have several of our science subdisciplines thus exposed to single point failures.

It seems to me that we got into this situation in the late 70's for two reasons. First, the credulous (including me) rated the chance of a Shuttle launch catastrophe to be negligibly low, and we believed also that the shuttle would provide frequent access to space. Given this premise, there is nothing silly about putting all your eggs in one basket, particularly since any difficulties encountered in orbit could be fixed up on the next visit. I think it must now be admitted that this premise is bankrupt. Secondly, in the Field report a mission such as AXAF was promoted in the context of a vital, diversified Space Science program. The Field report not only recommended AXAF, it also recommended (pp 141-143) three other new X-ray Explorers for the 80's. So, the program conceived by the Field committee was reasonable and balanced, given the premises accepted at the time. But we now have the reality which is: risky launch vehicles, no Explorers, and a 10-year stretchout of AXAF. What is now left of the original program is a tremendous high risk venture in which a major subdiscipline has unwittingly put its neck on the block.

What can be done? I've occasionally asked NASA managers this, and get a response "but that's what the scientists want." Such responses are not adequate. The rationale for AXAF and other big missions was developed when NASA was advertising a super-reliable and inexpensive shuttle: take that away, along with the Explorers and moderate missions, and you have a whole new ball game. The X-ray astrophysics community is, unhappily, trapped in this situation. They have their AO and selction [sic], they've been blessed by all the appropriate committees, and they don't dare rock the boat for fear of being sent back to the starting gate. Step-by-step they've been led into a tremendously exposed and risky situation from which there is no obvious escape.

I think that what must be done here is somewhat analogous to the mixed-fleet study: it must be recognized at the outset that the post-Challenger era is a new ball-game that requires rethinking of the old programs. Just as the mixed-fleet study recommends changing implementation plans for launching certain missions, a reexamination of the super-risky OSSA programs may lead to revised implementations. Without this, we will have a situation wherein a large part of the 1990's will be spent with OSSA carrying out expensive and risky programs planned under seriously flawed premises. To allow this would be a serious mistake.

OSSA cannot carryout such a sweeping reexamination of its programs by itself. However, I believe it should start out with the adoption of a management policy that avoids programs which expose the existence of entire subdisciplines to single point failures. Advisory committees such as SESAC need to play a key role in any such reassessment, and this role in fairness should recognize the priority established for some programs even if they are significantly restructured. Since it is obvious that there will not be funds to build multiple copies of huge observatories, a less risky strategy would be to scale down any new ones to a size where it would be possible to carry on in parallel other Explorer-type missions in the same subdiscipline. Such a mix would help restore a healthy balance to the OSSA spectrum of program sizes, and is, I believe, better in keeping with the spirit of the Field report than the present single high risk missions.

In my own area of low energy solar and cosmic ray research, we are uncoupled from these big, risky observatories. I have discussed the problem in terms of AXAF, but other big programs are in the same situation. I urge you and your committee to look into this very difficult problem, and seek to find ways to reduce the undue risks to which large segments of our space science community are now exposed.

Sincerely,
[signature]
Glenn M. Mason
Associate Professor of Physics

cc: Dr. B.I. Edelson
Dr. L. A. Fisk
Mr: S.W. Keller
Dr. J.D. Rosendhal
Dr. C. Pellerin
Dr. L. Peterson
Mr. T. Perry

Document III-36

[no pagination]
ES65 May 1, 1987
Prof. G. M. Mason
Associate Prof. of Physics
Department of Physics and Astronomy
University of Maryland
College Park, Maryland 20742

Re: The "ultra-high risk" NASA space science programs

Dear Prof. Mason:

I have read with interest your letter to Dr. Lanzerotti and you certainly raise some interesting and thought provoking questions. I most certainly agree with your statements that balanced programs are a necessary ingredient for success in the Space Sciences. On the other hand, I totally disagree with your assessment of the situation concerning the discipline of X-ray astronomy in general and your discussion of the AXAF in particular. To me your letter makes the following assumptions which I comment on as follows:

1. There is no balanced program in X-ray Astronomy.

Admittedly if one looks back at the last decade, X-ray astronomy would certainly appear to have come on hard times. Indeed this is true for all of high energy astrophysics which had its last "new start" in 1979 with the commencement of the GRO program. The fantastic results of the HEAO-2/Einstein which placed X-ray astronomy on equal footing with the more traditional astronomical disciplines makes this all the more surprising. On the other hand, if we look at current activities we see that a) the homework for AXAF is done and this major program is about to commence; b) there is a U. S. instrument and guest investigator program in collaboration with West Germany on the ROSAT satellite; c) the X-ray Timing Explorer is being studied and is definitely in the Explorer queue and d) there are a wide variety of sounding rocket, balloon, and small (Scout Class) programs underway. Thus, although not necessarily ideal, it is certainly incorrect, today, to draw the conclusion that there is not a balanced NASA program in this discipline. The conclusion was true 10 years ago.

2.) The X-ray Astrophysics Community (with AXAF) is unhappily trapped.

We are trapped, but not because we support the AXAF. As the SESAC itself has recommended, "Major facility-class missions have become essential for answering fundamental scientific questions in each of the Space and Earth Science disciplines and must be provided in turn on an appropriate schedule." The AXAF emerged as the number one priority from the Field Committee because it is essential for the science it can accomplish, and not simply to fill a menu of small, medium and large. The trap that we are in is the difficulty we have faced in getting the AXAF program started.

3.) AXAF is ultra high risk

For many not intimately connected with the space program the Challanger [sic] accident came as a rude awakening that things can go wrong. But, a more careful examination of space launches from sounding rockets on up would show, in fact, that, if one considers all the types of failures that can and have taken place to prevent scientific accomplishment, the risk is more or less the same independent of the size of the venture. Thus, I feel, the more relevant issue is what is the pay off for the risk? In the case of X-ray astronomy, even a slew of smaller less ambitions missions of limited duration and scope have extremely limited return. They are incapable of addressing the fundamental questions and of serving the larger astronomical community. They do of course serve to keep the smaller X-ray community alive, but for what purpose? Admittedly if they fail, "few" dollars are lost (although the typical Explorer may cost well over 200 million in the 1990's) but how much is gained? If AXAF is successful, the life and vitality of the discipline, as whole, are guaranteed for more than a decade and astronomy and astrophysics will take a gigantic step forward.

A second point, also worth emphasizing, is that, in a macabre but very real, way, the Challanger [sic]accident has lessoned [sic] not enhanced, the risk for AXAF because of the great visibility and emphasis placed on the program to put the shuttle in operational status again and the implications for the future of the Agency, should another failure occur.

4.) Smaller is better

There is always a vague feeling, amongst many scientists, that smaller is better, more productive, and of higher scientific yield. I think the real truth is that, for the working experimentalist, smaller is more fun. The really crucial factor, as Martin Harwitt has noted, is improvement in sensitivity and expansion of parameter space. This may be accomplished either in the large or in the small. In astronomy, size (aperture) and angulation of the telescope are the vital factors and X-ray astronomy is no exception. The AXAF telescope's properties are required to answer the questions we now pose and the sensitivity enhancements justify the claims of potential unforeseen exciting discoveries and breakthroughs. To me, your suggestion of descoping AXAF and spreading the resources to a number of smaller, "less risky" missions is equivalent to the idea of legislating that one should only build a number of 16" telescopes in as opposed to one large diameter telescope because of a perceived higher risk of failure of the latter.

Sincerely yours,
[signature]
Dr. Martin C. Weisskopf
Chief, X-ray Astronomy

Document III-37

Document title: "Report of the HST Strategy Panel: A Strategy for Recovery," August-October, 1990.

Source: NASA Historical Reference Collection, History Office, NASA Headquarters, Washington, D.C.

Document III-38

Document title: NASA, "Report of the Task Force on the Hubble Space Telescope Servicing Mission," May 21, 1993.

Source: NASA Historical Reference Collection, History Office, NASA Headquarters, Washington, D.C.

NASA and the international astronomical community were shocked when they learned in June 1990 that the Hubble Space Telescope had been launched with an incorrectly shaped primary mirror. The nature and cause of the problem with the mirror was quickly identified as being a "spherical aberra-

tion" not detected during the testing of the mirror before launch. (Volume I, Document IV-19). The issue was how to correct for the problem. This report represents the response of a special committee set up to address this question.

The recommendations of the HST Strategy Panel were accepted by NASA, and used as the basis for planning the first Hubble Space Telescope Mission, which took place in December 1993. Given the difficulty and high stakes involved in this mission, NASA wanted to make sure that all elements of the mission were well understood, and that the hardware and mission crew were well prepared. An external review committee headed by Apollo-era NASA manager Joseph Shea was constituted as part of the process of providing such mission assurance.

Document III-37

[title sheet]

REPORT OF THE
HST STRATEGY PANEL:
A STRATEGY FOR RECOVERY
THE RESULTS OF A SPECIAL STUDY
AUGUST-OCTOBER 1990

EDITED BY R. A. BROWN AND H. C. FORD

[1] PREFACE

Astronomers and engineers realized that there was a problem with the images of Hubble Space Telescope (HST) shortly after it was launched in April 1990. The quality of the images failed to improve despite attempts to adjust the alignment of the optics. NASA concluded in June 1990 that the HST primary mirror had been manufactured with the wrong shape. Compared with the desired profile, the mirror surface is too low by an amount that from the center to the edge grows from zero to 0.002 mm or four wavelengths of optical light. NASA convened an investigatory board in July 1990 under Dr. Lew Allen, which reported in November 1990 how the error probably occurred. In late 1980 or early 1981, a technician had improperly assembled a measuring device used to figure the primary mirror. Though tests at the time indicated a problem, the warning was not heeded, and the HST was assembled and launched with the flawed mirror.

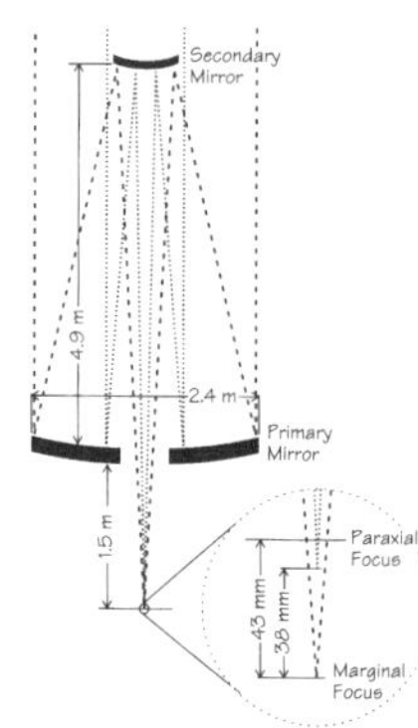

Figure 1. Spherical aberration means that light rays from different radii on the primary mirror come to focus at different distances. The marginal focus is 38 mm below the focus of the innermost rays, which graze the secondary mirror. The paraxial focus is obscured. Currently, the adopted focus (not shown) is 12 mm below the paraxial focus.

The deformity of the HST mirror causes spherical aberration in the images. This means light rays come to a focus at different distances depending on the radius at which the rays strike the mirror, as shown in Figure 1. Light from the edge of the primary mirror comes to a focus about 38 mm beyond where the innermost rays converge.

No positions, orientations, or other adjustments of the primary and secondary mirror can produce the diffraction-limited images required by much of the HST science program. The center of a star image in visible light has a core of radius 0.1 arcsec containing about 15% of the light; 70% was expected. The rest is spread about in a complex halo of radius 3 arcsec. Since aperture diffraction sets the size of the image core, the size is smaller at shorter wavelengths. The size of the halo, on the other hand, is set by geometrical optics and is constant. (The pattern of the halo varies with wavelength because it is an interference pattern.)

Spherical aberration degrades the science capacity of HST. Good science is being accomplished with HST as it is, but many crucial investigations-including many of the original justifications for HST-are on hold until the problem is solved.

[2] When the optical problem was announced, NASA began to seek solutions and develop a recovery plan. In the first phase, NASA focused on how to modify the scientific instruments already under development. These instruments are the Space Telescope Imaging Spectrograph (STIS), the Near-Infrared Camera and Multi-Object Spectrometer (NICMOS), and the second Wide Field and Planetary Camera (WFPC II), which NASA began to build in 1985 as a "clone" of the WFPC now in HST. NASA found it is feasible to correct these future instruments to compensate for spherical aberration. Based on this finding, NASA adopted an initial baseline plan to install the corrected WFPC II in place of WFPC on the first servicing mission in 1993, and later, on a second mission in 1996, to install STIS or NICMOS either to recover spectroscopic capabilities (in the case of STIS) or to add new infrared capabilities (with NICMOS).

This initial recovery plan of NASA restored faint source detection, one of the most critical capabilities crippled by spherical aberration. However, the plan delayed improving spectroscopy until the second half of the HST mission, and did not address full-resolution imaging at all. For these reasons, the HST Strategy Panel was formed in mid-August 1990 with a charter to search briskly for additional or alternative solutions.

In this second phase of NASA study, the HST Strategy Panel sought the best overall strategy to recover all primary HST science capabilities at an early time. The Panel did not adopt the WFPC II fix as a groundrule, but started "with a clean sheet of paper;" and tried to identify and review all potential options to alleviate the negative effects of spherical aberration on the HST science program. However, the Panel's recommendations and deliberations were firmly rooted in the assumption that the schedule for the two second generation instruments, STIS and NICMOS, would be adhered to by NASA.

The HST Strategy Panel's findings and recommendations were presented to Dr. Riccardo Giacconi, Director of the Space Telescope Science Institute, on October 18,1990. The Panel proposed a new program component as part of an augmented recovery strategy. The new component is the Corrective Optics Space Telescope Axial Replacement (COSTAR), a device to deploy corrective optics in front of the Faint Object Camera (FOC), High Resolution Spectrograph (HRS), and Faint Object Spectrograph (FOS). The strategy is to install both COSTAR and WFPC II into HST on the first servicing mission in 1993, and to fix the HST pointing problems. This strategy recovers essentially all the science capabilities expected at launch.

Dr. Giacconi endorsed the oral recommendations of the Panel and took the findings to NASA management. The Panel made a presentation at NASA Headquarters on October 26, 1990. In the following weeks, NASA conducted an intensive study of the feasibility and costs of COSTAR. In December 1990, NASA Headquarters authorized the implementation of the COSTAR program to proceed.

[3] SYNOPSIS

The HST strategy panel held four meetings between mid-August and mid-October 1990.

At these meetings, a wide variety of options for correcting spherical aberration were identified and debated. This report, as outlined below, presents the Panel's findings and recommendations.

The OPTICAL PROBLEM is now understood well enough to design and install a highly effective optical correction.

The OPTICAL SOLUTION is a pair of mirrors for each Science Instrument (SI) field of view. The first corrective mirror forms an image of the HST primary mirror on the second corrective mirror; the second corrective mirror has spherical aberration in precisely the same amount as the primary mirror-but with the opposite mathematical sign, thus cancelling [sic] the effect.

The COSTAR is the proposed device to carry and deploy the corrective optics for three scientific instruments, the FOC, HRS, and FOS. COSTAR would replace the High Speed Photometer (HSP).

The POINTING of HST must be improved to gain full value from the restored HST optical performance. The solar array "snap" that causes HST to lose pointing lock at day/night transitions must be fixed. The Panel further recommends that the operational parameters of the guidance system be adjusted to reduce jitter in the coarse tracking mode.

The WFPC II is being corrected with the same optical solution used in COSTAR. The Panel found that the alignment of the corrective optics is critical, which COSTAR can assure by special mechanisms. No comparable mechanisms exist in the original design for WFPC, and because WFPC II is a close copy of the original, the Panel recommends that the issue of WFPC II alignment be addressed with critical attention.

The 1993 SERVICING Mission can install the WFPC II and COSTAR. This currently planned mission can solve the spherical aberration problem for the SIs, fix the solar array disturbances, and replace other subsystems, as necessary.

The RECOMMENDED STRATEGY is to develop COSTAR on an urgent basis, continue WFPC II development with special attention to the alignment concerns, and improve the coarse track pointing performance by operational measures. Then, the 1993 HST servicing mission restores the scientific functionality expected at launch.

The FRESH REASONS to commit new resources to fix HST are abundant in the science program awaiting sharp images and precise pointing. This science program is the culmination of decades, even centuries, of maturing questions about the universe. It is also a program proposed largely by young astronomers, who need a restored HST to make the discoveries that will propel astronomical exploration into the twenty-first century.

The APPENDICES document the approach, options, background findings, and analyses of the HST Strategy Panel. [pp. 5-119 not included]

Document III-38

[cover page]

REPORT OF THE
TASK FORCE
ON THE
HUBBLE SPACE TELESCOPE
SERVICING MISSION

21 May 1993
Washington, DC

[1] 1. INTRODUCTION

The Task Force was chartered by the Administrator of NASA, Daniel S. Goldin, to review all aspects of the first Hubble Space Telescope (HST) servicing mission scheduled for December 1993. Mr. Goldin, in a letter dated 26 January 1993 to Dr. Joseph F. Shea, Chairman of the Task Force, stressed the "importance, complexity, and visibility of this mission." The review was conducted over a period of five weeks (4 February 1993 through 9 March 1993) and centered on a series of informational briefings conducted at the Goddard Space Flight Center (GSFC), the Johnson Space Center (JSC), Lockheed Missile & Space Company Headquarters, and NASA Headquarters. The members of the Task Force are listed in Appendix 1 [appendices not included].

HST is one of the great orbital observatories planned by NASA. Launched 24 April 1990, HST can produce exceptional image detail for bright, high-contrast objects such as nearby star clusters, the cores of galaxies, and solar system planets. The HST has proved to be an extremely valuable scientific resource and has already enabled several major discoveries.

Despite its successes, HST has experienced a series of problems and anomalies. To correct for the problems and restore redundancy, a Space Shuttle crew will conduct the first servicing mission in December 1993. During the course of the review, it became clear to the Task Force that this mission is the most comprehensive and challenging on-orbit servicing mission NASA has ever attempted. It will require more extravehicular activity (EVA) than ever before planned for a single Space Shuttle mission. The mission has three basic goals:

- Correct for spherical aberration of the primary mirror.
- Reduce the observatory's pointing fitter.
- Restore redundancy in key systems.

Many of the servicing tasks are complex and all must be performed in a limited amount of time. During this time, astronauts will be asked to perform the following tasks:

- Install an optics package called the Corrective Optics Space Telescope Axial Replacement (COSTAR).
- Install a replacement for the current Wide Field/Planetary Camera (WF/PC I), called WF/PC II.
- Install two or more sets of gyroscopes and associated electronics.
- Replace the solar arrays.
- Install a computer co-processor.
- Install a repair kit for the Goddard High Resolution Spectrometer (GHRS).
- Replace the Magnetic Sensing System #1 (MSS-1)

[2] In evaluating the effort required and the risks, it must be remembered that HST was designed for servicing and is therefore more compatible with crew servicing requirements than any other satellite produced to date.

[3] 2. EVOLUTION OF HST SERVICING PLANS

HST was designed and built for periodic on-orbit servicing. The current plan calls for servicing missions roughly every three years throughout the course of HST's 15 year life. Critical components are located in 50 unique types of orbital replacement units (ORUs). The ORUs were designed for easy access, removal, and replacement. In addition, 225 feet of crew translation hand holds and 31 portable foot restraint receptacles were provided on the surface of HST. These were placed to allow a servicing crew to move across the surface of the satellite, position themselves at any servicing location, and then reach the ORUs or related hardware.

At the start of HST development in 1977, all HST components except wiring, structure, thermal control surfaces and heaters, and the Optical Telescope Assembly optics were to be serviceable on-orbit. Early plans also included periodic return to Earth via the Shuttle for full refurbishment.

In 1980, however, as part of a cost reduction program, the requirements for on-orbit serviceability was eliminated for a number of components, such as the multiple access transponder, the data management unit, the data interface units, the power distribution units, and the solar arrays. Four years later, the decision was made to eliminate ground return for periodic refurbishment as an option. This decision was based on unanswered technical concerns regarding the impact of ground return and relaunch, many of which still exist today. As an example, HST has strict contamination control requirements which are necessary for successful ultraviolet performance and which would be difficult to maintain during ground return and servicing. In addition, as Shuttle-coupled loads increased, the predicted mechanical stresses on many HST primary structural elements increased to the point that fracture control restrictions limited the HST to a single launch. Another major consideration was the significant cost of ground return. Estimates indicated that manpower levels during refurbishment would approach those experienced during HST's initial development far exceeding the on-orbit servicing costs.

Following the decision to eliminate HST ground return, an effort was initiated to significantly upgrade HST on-orbit serviceability. By that time, budget constraints had left only 20 unique ORUs with designs allowing complete EVA serviceability. EVA-compatible aids, such as captive fasteners, wing-tab connectors, tether loops, alignment decals, and connector maps, were added to the design of an additional 24 ORUs. This set of redesigned ORUs is now referred to as Block II, while the original set of 20 ORUs is referred to as Block I. A third category, Block III, is composed of six ORU spares which were minimally enhanced to support EVA serviceability. These three blocks of ORUs have decreasing levels of EVA compatibility, but collectively enhance the basic serviceability of the HST design. Table 2.1 below, [4] provided by the HST Flight Systems and Servicing Project lists the candidates for repair or replacement during the first servicing mission and their block designation.

TABLE 2.1 PAYLOAD COMPLEMENT AND PRIORITIES

Item	Priority Class	Priority Sequence	Block
Solar Arrays (SA)	1	1	I
Wide Field/Planetary Camera II (WF/PC II)	1	2	I
Corrective Optics Space Telescope Axial Replacement (COSTAR)	1	2	I
Rate Sensor Unit-2 (RSU-2)	1	3	I
Coprocessor	1	4	I
RSU-3	1	5	I
Electronic Control Unit-3 (ECU-3)	1	6	I
Goddard High Resolution Spectrograph (GHRS) Repair Kit	2	7	I
Magnetic Sensing System-1 (MSS-1)	2	8	III
ECU-1	2	9	I
Fuse Plug	2	9	I
RSU-1	2	9	I
MSS-2 Blanket	3	10	N/A

Although design compromises were made which reduced the EVA serviceability of certain ORUs, the Task Force believes that in-orbit serviceability is adequate and lower risk than ground servicing. Ground return should be considered only in the event of equipment failures which threaten HST and cannot be remedied by on-orbit servicing.

****

[7] 4. REQUIREMENTS AND PLANS FOR THE FIRST SERVICING MISSION

The first HST servicing mission is designed to restore the full scientific capability to the spacecraft and to ensure the telescope systems continue to function and produce quality observations. The current priority of the various repairs has been divided into two categories - primary and secondary, as shown in the following table (Table 4.1) provided by the HST Project, Associate Director's Office. The primary repairs are those which are necessary for the mission to be regarded as successful. The secondary objectives will be met as time and resources allow. The actual mission plan does not directly follow this sequence because of Shuttle and HST limitations.

Table 4.1 HST First Servicing Mission Payload Overview

Failure/ Anomaly	Cause	Response	Comments	Category
Spherical Aberration	Primary mission ground to wrong prescription	WF/PC-III and COSTAR with corrective optics	Correction necessary to meet Level I requirements	Primary
Solar Array Jitter	Thermal deformations of array structure	Solar Array II with thermal shielding and improved mechanical design	Solar Array experiencing deformations outside of design envelope causing unacceptable attitude transient motion	Primary
Gyroscope Failures	Gyro 4 and 6: generic problem with Bendix pulse rebalance loop hybrid circuits	Retrofit of spare rate sensing units with Teledyne hybrids	3 of 6 gyros necessary to conduct science mission	Primary
		Changeout of RSU 2 and 3	5 gyros following first servicing mission increase failure tolerance	Primary
	Gyro 6 motor spindown: random failure of spin motor power phase–open circuit most likely originating in electronics	Changeout of ECU-3	Change out of RSU-3 already planned	Primary
	Gyro 1 shutdown: short circuit causing fuse to open, most likely fuse plug originating in ECU	Change out of RSU-1, ECU-1 and fuse plug	Investigation not yet complete	Secondary

DF-224 Memory Unit Failures	Memory Unit 3: open of plated through hole on 7 layer board	Augment memory with coprocessor/ shared memory unit	Problem believed to be generic board problem	Primary
	Memory Unit 4: failure of read/ write enable circuitry	Coprocessor/ shared memory augmentation	Stress cycling of spare boards proved tolerance	N/A
GHRS Low Voltage Power Supply Failure	Intermittent open of lug solder joint	Science date cross-strapping via relay box	Use of built in redundancy requires switch over of entire HST data management	Secondary
Magneto-meter Anomalies	Intermittent open on signal path of V2	Replace Magnet Sensing System #1	Attitude/thermally induced Not generic problem	Secondary
Gyro Fuse Derating	Appears to be improper fusing for gyro power circuits	Under review	3.0 Amp fuse must carry 2.0- 2.3 Amps for 27 sec.	N/A

[8] The requirements and order of priority are constamtly [sic] changing, given the extended mission preparation period and the changing status of the observatory. This instability impacts the mission preparation and mission operations training. It also places an additional load on astronaut and support team training. An additional concern is the escalating nature of the mission. Starting with a few objectives, the mission has grown to an unprecedented number of EVAs. While the rate of failures is not unusual for a spacecraft of this complexity, the current servicing plans include replacing some very large and sensitive ORUs.

The HST Project, the integration contractors, and the mission operations personnel have done a good job of pre-mission design and preparation for the periodic servicing of HST given the changing servicing philosophy throughout the development of the program. The development of EVA tasks, tools, and support hardware has progressed over the years of designing and building the HST. The prelaunch verification of the tools, including fit-checks to the flight unit and manned thermal vacuum testing and the planning for verification prior to the first servicing mission, has been methodical and adequate.

Although a great deal of planning and care has clearly gone into contamination control, the Task Force is concerned about the possibility for contamination of HST optics and other sensitive systems during the EVA. Possible sources of contamination include the EVA suits and tools. We recommend that orbital testing be conducted prior to the servicing mission to assess potential contamination using such techniques as witness plates.

As the launch nears, there is the potential for new failures which could change the relative priority of repairs and impact mission training and other preflight planning. New failures also have the potential to overload this single mission. This concern is heightened by the unknown causes of some of the existing failures. This mission already challenges the capability of the Shuttle and crew. In-depth planning for credible, but currently unknown, failures is required to prepare for this flight.

Past EVA history suggests that complex EVAs often encounter unexpected events which can significantly alter plans and time lines. This experience, coupled with the complexity of this servicing mission strongly suggest the need to plan for a second servicing mission. A second mission may also be required to compensate for a greater than anticipated failure rate following the initial mission. The schedule for the next planned mission may have to be accelerated or an additional mission may need to be inserted to maintain the telescope's health and enable the continuing science mission.

[9] 5. MISSION MANAGEMENT AND TRAINING

The HST servicing mission is the most complex Shuttle mission ever attempted. The accompanying mission planning and training also are quite complex. Mission preparation activities are managed by the Space Shuttle Program within the Office of Space Flight. Mission planning and training primarily are the responsibility of the Johnson Space Center with support from the Marshall Space Flight Center. The actual mission planning and training are accomplished by a large team composed of members from these two Centers and from the HST Program Office at the Goddard Space Flight Center. The Task Force commends the people involved in this mission for the tremendous amount of work they have already done in preparation for this mission. We recognize that more advanced planning and effort has gone into preparing for this mission than for almost any previous Shuttle mission.

The Task Force, however, found the interfaces between the organizations involved to be numerous and confusing. No single individual was charged with addressing all of the issues which might arise during mission preparation and who would be totally accountable for mission success. This problems appears to be under work, however.

In response to criticism by various review teams, a mission director has been assigned by NASA Headquarters. Effective cooperation between the Centers involved in this mission is absolutely essential to its success. A more complete distribution of these data could result in a better overall examination of the options. There should be no reluctance to hear proposals based on shared information; it is an accepted fact that, in the end, a decision must be made and that such a decision cannot be shared.

A chief concern of the Task Force is the amount of work that must be completed during a single Space Shuttle mission. The mission preparation for the first servicing mission follows the standard mission planning and training format for Shuttle missions and thus places much of the work in the last year prior to launch. The training and operations products are currently on schedule. However, the number of servicing tasks, the telescope's complexity, and the large size and numbers of ORUs to be repaired or replaced make the first servicing mission an EVA job that is unprecedented, and thus strongly indicate that pre-mission planning should have begun earlier than in past EVA missions. Given the mission's complexity, we also recommend that back-up crew members be assigned to train for the critical mission tasks.

Current EVA time line development activities indicate that four or possibly five days of EVA may be necessary to complete the objectives of this first servicing mission. In general, we find the current EVA time lines to have insufficient slack. The EVA time [10] required has increased over 25 percent during February alone. In addition, some of the hardware appears to be behind schedule (e.g., COSTAR). These factors raise concerns about the viability of the overall schedule and further support the need for accelerated mission operations preparation.

****

[14] 7. CONCLUSIONS AND RECOMMENDATIONS

Conclusions

1. The Hubble Space Telescope (HST) system is well designed to meet its continuing mission objectives. This design includes adequate redundancy within the spacecraft, a satisfactorily structured repair and refurbishment plan, and long-term science upgrades. We judge the first servicing mission to be feasible.
2. A major amount of in-depth effort has been put into this mission and significant progress has been made. The coordination and cooperation among the Office of Space Flight, the Office of Planetary Science and Astrophysics, the Goddard Space Flight Center, the Johnson Space Center, and the Marshall Space Flight Center is especially noteworthy.
3. The mission is very complex and contains the following risks:
   - The EVA time line is very tight and has grown about 25% during February.
   - The EVA planning, training and scheduling is late, considering the complexity of the time line.
   - The management structure has been diffuse. Recent changes are encouraging, but require further senior management attention.
4. The schedule of all activities between now and launch has inadequate margin. The principal replacement units are late and major testing still remains.
5. The component failure rate is worrisome:
   - The impact of failure rate on risk has not yet been adequately assessed. Some failures are not yet fully understood.
   - No redundancy remains in some key areas (e.g., gyro systems).
   - The recent fine guidance sensor anomaly is a matter of concern.
   - Additional failures before the servicing mission could impact the planning schedule.
6. The current flight plan is designed solely for Orbiter 105. This limits flexibility as other Orbiters are less capable.
7. The current plan does not include verifying the operability of all redundant HST subsystems.
8. A major effort has been applied to controlling contamination. Orbital testing is needed to assess potential contamination from EVA suits and tools.

[15] Recommendations

1. Focused management is essential to this mission. The recent management changes are positive and must be sustained. Close senior management attention, including formal reviews, must be continued to ensure that the new organization remains effective.

2. Development of the EVA time line and choreography, including contingencies, is necessary in order to properly plan the entire mission and should be accelerated. As soon as the time line is available, it should be reviewed by senior management.
3. Training for the mission must include joint integrated end-to-end simulations of the EVA portions.
4. HST recovery and return to Earth should not be considered unless repair is not achievable on-orbit. The overall risk and cost of the return and refurbish option is not otherwise justifiable.
5. Assure that the upgrades to the Neutral Buoyancy Simulator (e.g., Nitrox, video and RMS) are completed in time to support the joint simulations.
6. Designate and train a backup EVA crew member now.
7. Ensure that the current plan to exercise on-orbit HST repair procedures on an earlier Shuttle flight is implemented. This should include the use of witness plates to assess potential contamination issues.
8. Reassess priorities of the repair sequence. Give higher priority to replacing all gyroscope units.
9. Plan now for an early contingency mission to cover potentially incomplete tasks from the first servicing mission or additional failures on the spacecraft.
10. Conduct a quantitative assessment to determine the risk of verifying the status of all subsystems (both A– and B-sides) before the first servicing mission.
11. Conduct an integrated assessment of contamination control procedures and plans for the HST servicing missions.

****

[21]

APPENDIX B

TASK FORCE ON
THE HUBBLE SPACE TELESCOPE SERVICING MISSION
Membership List

Dr. Joseph F. Shea, Committee Chairman
Adjunct Professor of Aeronautics and Astronautics
Massachusetts Institute of Technology

Dr. Eugene E. Covert
Professor of Aeronautics and Astronautics
Massachusetts Institute of Technology

Dr. Maxime A. Faget
Chairman of the Board Space Industries, Inc.

Dr. Richard L. Garwin
IBM Fellow
IBM T. J. Watson Research Center

Dr. George J. Gleghorn

Maj. Gen. Ralph H. Jacobson (Ret.)
President and CEO
The Charles Stark Draper Laboratories

Dr. Herbert Kottler
Head, Aerospace Division
Massachusetts Institute of Technology
Lincoln Laboratories

Mr. Edmund Nowinski

Prof. Bradford Parkinson
Gravity Probe B
Hansen Experimental Physics Laboratories
Stanford University

Mr. Charles Roth

Mr. Harris Schurmeier

Lt. Gen. Thomas P. Stafford, USAF (Ret.)
Stafford, Burke and Hecker, Inc.

Document III-39

Document title: "AXAF Restructuring," Briefing for the NASA Administrator, August 15, 1992.

Source: Alan Bunner, Office of Space Science, NASA Headquarters, Washington, D.C.

More than a decade after the mission was proposed, the Advanced X-ray Astrophysics Facility (AXAF) received a new start in NASA's FY 1988 budget; limited space science funds, consumed in large part by the Hubble Space Telescope's development, accounted for most of this delay. Even after the new start, NASA had a tough battle to fight to keep AXAF alive. In addition to having to prove to Congress within three years that AXAF's mirrors could perform adequately (which NASA did), the agency had to find a way to reduce the mission's weight and complexity in order to reduce its cost. This was part of NASA's strategy of moving away from very large and costly missions. In 1992, NASA decided to restructure the mission to meet the new cost constraints. The most significant changes were to split the AXAF mission into two elements, with AXAF-I dedicated to imaging and AXAF-S devoted to spectroscopy, and to reduce the numbers of mirrors and instruments that would fly. The restructuring was *projected to save $290.7 million of the project's original cost of $2,021.7 million. AXAF-I became the AXAF spacecraft that was successfully launched in 1999; the satellite was renamed Chandra after launch. The AXAF-S instrument was eventually flown on Japan's Astro-E mission, also launched in 1999; that mission failed.*

[no pagination]

AXAF RESTRUCTURING
AUGUST 15, 1992

AXAF PROGRAM RESTRUCTURING

OBJECTIVES
- PRESERVE THE SCIENCE MISSION
- REDUCE BUDGET GROWTH FROM FY93 TO FY 94
- REDUCE PEAK YEAR FUNDING REQUIREMENTS
- REDUCE MISSION COMPLEXITY
- HOLD OR ADVANCE LAUNCH DATE
- PRODUCE ROBUST AND RESILIENT PROGRAM
- GAIN MAXIMUM BENEFIT FROM PREVIOUS EXPENDITURES

PROGRAMMATIC GUIDELINES
- ASSUME FY 93 BUDGET APPROVED
- ASSUME BUDGET CAPPED AT $250M PLUS INFLATION BEYOND FY93
- REDUCE OPERATIONS AND SERVICING BUDGET BY 30%
- REDUCE MISSION COMPLEXITY
- SEEK LOWEST COST LAUNCH OPTIONS
- HOLD OR ADVANCE THE BASELINE LAUNCH DATE OF 4/99
- MAXIMALLY EMPOWER PRIME CONTRACTOR FOR AXAF-I
- PERFORM AXAF-S IN-HOUSE AT MSFC

PROCESS
- AXAF TEAM BEGAN WORK IN JANUARY TO FIND OPTIONS WHICH WOULD MEET OBJECTIVES
- MANY OPTIONS (19) IDENTIFIED AND EVALUATED.
- AXAF TEAM RECOMMENDATIONS TO AA/OSSA ON APRIL 30,1992 – RESTRUCTURED AXAF INTO TWO COMPONENTS
- PROCESS CONTINUED TO DEVELOP MANAGEMENT STRATEGIES AND IMPLEMENTATION ASPECTS

DEVELOPMENT OF AXAF-S AT MSFC

- AXAF-I and AXAF-S
  - separate programs with distinct budgets

  but:
  - maximize common buys and shared expertise
- AXAF-S

- Risk is substantially reduced by:
  - Technical synergy between AXAF-I and AXAF-S at MSFC
  - Management flexibility between AXAF-I and AXAF-S at MSFC
- Centralized science and technical management of both missions
- Demonstrated enthusiastic response to in-house development at MSFC

<u>RESULTS</u>

- program architecture - two components
  - high spatial resolution imaging (AXAF-I)
    – 4 mirrors, 2 focal plane instruments, 2 gratings, Titan IV/centaur or STS/upper stage to high altitude elliptical orbit
  - high energy, high spectral resolution spectroscopy (AXAF-S)
    – foil or replicated mirrors, polar orbit, 1 focal plane instrument, Delta launch to polar orbit
- separate, lighter mission for imaging permits high elliptical orbit
  – high orbit mission is simpler and has better observing efficiency than baseline
  – fewer instruments on each mission element means less time-sharing.
  – offsets throughput and lifetime considerations
  – comparable to baseline mission for imaging science
- spectroscopy mission uses simpler, lighter and cheaper metal foil or replicated mirrors (decision 9/92).
- programmatic guidelines are met with 9/98 launch of AXAF-I and AXAF-S as soon as possible thereafter
- management strategy guidance
  – implement innovative management strategies to save substantial funding on imaging mission.
  – AXAF-S to be done in house with primarily govt [sic] labor.

[image not supplied]

AXAF RESTRUCTURING			
PARAMETER	OLD BASELINE	AXAF-I	AXAF-S
Orbit	600 Km Circular	10,000 X 100,000 Km (STS) 60,000 X 120,000 (T-IV)	600 Km/Polar Sun Sync.
On-Orbit Weight	32,800 Lbs.	< 11,500 Lbs.	< 6000 Lbs.
Launch System	Shuttle/ASRM	Shuttle/Upper Stage or Titan IV/Centaur	Delta II
Operational Life	15 Years W/Servicing	5 Years	3 Years
Launch Date	April, 1999	September, 1998	December, 1999
Mirror Configuration	6 Mirror Pair HRMA	4 Mirror Pair HRMA	Foil or Replicated Mirrors
Focal Length	10 Meters	10 Meters	3.5 to 5.0 Meters
Focal Plane SI's	HRC, ACIS, XRS	HRC, ACIS	XRS
Gratings	HETG and LETG	HETG and LETG	None
AXAF IMPLEMENTATION			
MSFC ROLE	OLD BASELINE	AXAF-I	AXAF-S
CONTRACT MONITORING	Maximum penetration of contractor	Maximal contractor empowerment with emphasis on incentive/award fees	Expect only S/C subsystems and SI's to be contracted for
IN-HOUSE EFFORT	Significant MSFC in-house redundancy of contractor effort	Selective MSFC task redundancy in high risk areas	Develop in-house at MSFC with flexibility to phase staff in and out of project

****

AXAF ISSUES

(1) HAVE ADEQUATE RESERVES BEEN IDENTIFIED AND ALLOCATED?

(2) WHAT IS THE APPROPRIATE CENTER ASSIGNMENT FOR AXAF-S?

RESERVE POSTURE

• Comparison of AXAF-I and AXAF-S reserves with the old baseline POP 92-1 reserves:

POP 92-1	20.8%
AXAF-I	20.5%
AXAF-S	17.7%

RESERVE LEVELS ARE ACCEPTABLE:

• AXAF-I
  - *Simplified* derivative of the old baseline
  - High earth orbit - simpler comm., thermal, power, ops, etc
  - *fixed* HRMA
  - No servicing

- P1 /H1 *success* - metrology debugged
- HDOS now ahead of schedule

• AXAF-S
- Maximum flexibility as an in-house development project.
- Instrument design mature, but will descope if cost growth appears
- Mirror technology demonstrated (Foil on BBXRT, Replicated on XMM)

Document III-40

Document title: Structure and Evolution of the Universe Subcommittee of the Space Science Advisory Committee, NASA, "Cosmic Journeys: To the Edge of Gravity, Space, and Time – Structure and Evolution of the Universe Roadmap 2003-2023," September 1999.

Source: Office of Space Science, NASA Headquarters, Washington, D.C.

For more than 40 years, NASA has crafted a program of space-based astronomy missions that have explored the universe in every major wavelength range between radio waves and gamma rays. The direction of that program has largely been influenced by the various groups of non-NASA astronomers to whom the Agency has looked for input throughout its existence. The situation is no different today. This document, created by NASA's current astronomy advisory group made up of scientists from outside the Agency, reflects the objectives of astronomers for exploring the universe at the end of one century and into the next.

[cover sheet]

Cosmic Journeys
To the Edge of Gravity, Space, and Time
Structure and Evolution of the Universe Roadmap: 2003-2023

prepared by
The Structure and Evolution of the Universe Subcommittee of the
Space Science Advisory Committee
National Aeronautics and Space Administration
September 1999

[no page number]
Executive Summary

The Roadmap for the Structure and Evolution of the Universe (SEU) Theme embraces three fundamental, scientific quests:

- To explain structure in the Universe and forecast our cosmic destiny
- To explore the cycles of matter and energy in the evolving Universe
- To examine the ultimate limits of gravity and energy in the Universe

We develop these quests into six focused research campaigns:

- Identify dark matter and learn how it shapes galaxies and systems of galaxies
- Explore where and when the chemical elements were made
- Understand the cycles in which matter, energy, and magnetic field are exchanged between stars and the gas between stars
- Discover how gas flows in disks and how cosmic jets are formed
- Identify the sources of gamma-ray bursts and high-energy cosmic rays
- Measure how strong gravity operates near black holes and how it affects the early Universe

These campaigns lead to a portfolio of future major missions of great scientific interest and popular appeal, strongly endorsed by the scientific community. Many have undergone significant initial study. Some are in a state of readiness that make them ideal candidates for the present Office of Space Science Strategic Plan; others may well feature in the next Plan. Each provides a golden scientific opportunity to advance our understanding of the Universe.

We have identified three top-priority near-term science objectives together with missions to accomplish these goals. The three problems span a diverse range of subdisciplines, of observational technique, of timescales, and of cost, and are thus complementary, forming a coherent core program for the SEU theme in the 2003-2007 timeframe.

- Obtain precise measures of the chemical composition and physical conditions in objects ranging from the closest stars to the most distant quasars via X-ray spectroscopy of unprecedented sensitivity.
- Utilize, for the first time, gravitational radiation as a probe of supermassive black holes throughout the Universe, compact binary sources within our Galaxy, and a possible gravitational wave background, using a 5 million kilometer arm-length laser interferometer in space.
- Determine the nature of the highest-energy cosmic rays, one of the most important questions in this fundamental field, via a measurement of the characteristics of individual elements over a wide range of mass and energy, utilizing the International Space Station as a platform.

We also describe a small number of exciting missions which are strong candidates for new start status in the midterm, 2008-2013, pending technology development. These missions tackle fundamental problems through the entire electromagnetic spectrum, from the radio through gamma rays, and in many cases develop fascinating technologies with applicability not only elsewhere in NASA but outside of space science as well. Finally, we describe a set of "vision missions," which stretch our scientific imagination and set technology challenges for our field.

A vigorous program of education and public outreach will bring the wonderful array of past and current scientific achievements in this theme to the public.

Biographical Appendix

B

James M. Beggs (1926–) was nominated by President Reagan on June 1, 1981, to become administrator of the National Aeronautics and Space Administration. He was the sixth individual to head the nation's civilian space agency. Beggs took his oath of office as head of the agency and entered the new post on July 10, 1981. Prior to his appointment as NASA administrator, Beggs had been executive vice president and a director of General Dynamics in St. Louis, Missouri. Beggs served with NASA from 1968 to 1969 as associate administrator for advanced research and technology. From 1969 to 1973, he was under secretary of the U.S. Department of Transportation. He went to Summa Corporation in Los Angeles, California, as managing director for operations, and then joined General Dynamics in January 1974. Prior to joining NASA, he had been with Westinghouse Electric Corporation in Sharon, Pennsylvania, and Baltimore, Maryland, for thirteen years. His resignation from NASA was effective on February 25, 1986. Since leaving NASA, Mr. Beggs has worked as a consultant from his offices in Bethesda, Maryland. See "Beggs, James M.," biographical file, NASA Historical Reference Collection, NASA History Office, NASA Headquarters, Washington, D.C.

Lloyd Berkner (1905–1967) was involved in most of the early spaceflight activities of the United States in some capacity. Trained as an electrical engineer, he was at first interested in atmospheric propagation of radio waves, but after World War II became a scientific entrepreneur of the first magnitude. He was heavily involved in the planning for and execution of the International Geophysical Year in 1957 and 1958, and served in a variety of positions in Washington, DC, where he could influence the course of science policy. See "Berkner, Lloyd V.," biographical file, NASA Historical Reference Collection.

Hans Bethe (1906–) was born in Strasbourg, Alsace-Lorraine. A mathematical prodigy, he received a Ph.D. in physics in 1928 from the University of Munich. After Hitler came to power in Germany, he left for England and then the United States, landing at Cornell University in 1935. A key figure in atomic physics, he was the head of the Los Alamos theoretical division from 1943 to 1945. After having helped develop the atomic bomb, he later became an outspoken advocate of nuclear arms reduction. He won a Nobel Prize in 1967 for his discovery of how stars nourish their nuclear fires. See "Bethe, Hans A.," biographical file, NASA Historical Reference Collection.

Albert Boggess served as the Hubble Space Telescope project scientist for operations at NASA's Goddard Space Flight Center in the 1980s. He also served as the project scientist for the International Ultraviolet Explorer spacecraft from its development phase through the first several years of operation, including its January 1978 launch. See "Boggess, Albert," biographical file, NASA Historical Reference Collection.

Detlev Bronk (1897–1975) was president of the National Academy of Sciences, 1950–1962, and a member of the National Aeronautics and Space Council. A scientist, he was president of the Johns Hopkins University from 1949 to 1953 and Rockefeller University from 1953 to 1968. See "Bronk, Detlev," biographical file, NASA Historical Reference Collection.

Percival Brundage (1892–1981) was the director of the Federal Bureau of the Budget during the Eisenhower administration. He earned an A.B. *cum laude* from Harvard University in 1914 and joined the New York staff of Price Waterhouse & Company accounting firm immediately after graduation. In 1930, he was made a partner of the firm and was a senior partner when he left to enter government service. He served as president of the American Institute of Accountants and chairperson of the executive committee of the New York Chamber of Commerce. See Marjorie Dent Candee, *Current Biography Yearbook: 1957* (New York, NY: The H.W. Wilson Company, 1958).

C

Claude Canizares (1945–) is the Bruno Rossi Professor of Experimental Physics at the Massachusetts Institute of Technology (MIT) and director of the Center for Space Research. He came to MIT as a postdoctoral fellow in 1971 and joined the faculty in 1974, progressing to professor of physics in 1984. He is a principal investigator on NASA's Chandra X-ray Observatory, leading the development of the High Resolution Transmission Grating

Spectrometer for this major space observatory, and is associate director of the Chandra X-ray Observatory Center. He also has worked on several other space astronomy missions, including as co-investigator on the Einstein Observatory (HEAO-2). His main research interests are high resolution spectroscopy and plasma diagnostics of supernova remnants and clusters of galaxies, cooling flows in galaxies and clusters, x-ray studies of dark matter, x-ray properties of quasars and active galactic nuclei, and gravitational lenses. Professor Canizares received B.A., A.M., and Ph.D. degrees in physics from Harvard University. He has authored or co-authored more than 135 scientific papers. He is a member of the NASA Advisory Council; chair of the Space Studies Board of the National Research Council; a member of the Board of Trustees of the Associated Universities, Inc.; formerly chaired NASA's Space Science Advisory Committee; a member of the National Academy of Sciences; a fellow of the American Physical Society; a corresponding member of the International Academy of Astronautics; and a fellow of the American Association for the Advancement of Science. See "Canizares, Claude R.," biographical file, NASA Historical Reference Collection.

George Carruthers (1939–) won NASA's Exceptional Scientific Achievement Medal in 1972 for development of the first lunar-based space observatory, which was carried to the surface of the Moon by the Apollo 16 crew. A leading African American astrophysicist, Dr. Carruthers worked at the Naval Research Laboratory's Space Science Division at the time. He received his Ph.D. in aeronautical and astronautical engineering from the University of Illinois in 1964, and won national recognition in 1970 when an instrument he developed found molecular hydrogen in interstellar space. In 1977, he went through screening to become a mission specialist astronaut. See "Carruthers, George R.," biographical file, NASA Historical Reference Collection.

Jimmy Carter (1924–) was the thirty-ninth president of the United States from 1977 to 1981. He served as a naval officer and businessman before entering politics in the Georgia State Legislature (1962–1966). He also served as the governor of Georgia from 1971 to 1975. See "Carter, Jimmy," biographical file, NASA Historical Reference Collection.

Subrahmanyan Chandrasekhar (1910–1995) was an astrophysicist who studied in India before moving to the United States in 1936. A specialist in the final stages of stellar evolution and white dwarf stars, he won a Nobel Prize in 1983. NASA named the Chandra X-Ray Observatory after him. See David Millar, Ian Millar, John Millar, and Margaret Millar, *The Cambridge Dictionary of Scientists* (Cambridge, England: Cambridge University Press, 1996) and *http://chandra.harvard.edu/about/chandra.html* on the Web.

Bill Clinton (1946–) was the forty-second president of the United States from 1993 to 2001. He earned a B.A. from Georgetown University and a law degree from Yale University, and also studied at Oxford University as a Rhodes Scholar. In 1976, he was elected attorney general of Arkansas, and in 1978 became the youngest governor of the United States. Elected to the presidency in 1992, he served two consecutive terms before leaving office. See "Clinton, William Jefferson," biographical file, NASA Historical Reference Collection.

Edgar B. Cortright (1923–) earned an M.S. in aeronautical engineering from Rensselaer Polytechnic Institute in 1949, the year after he joined the staff of Lewis Laboratory. He conducted research at Lewis on the aerodynamics of high-speed air induction systems and jet exit nozzles. In 1958, he joined a small task group to lay the foundation for a national space agency. When NASA was created, he became chief of advanced technology at NASA Headquarters directing the initial formulation of the agency's meteorological satellite program, including projects Tiros and Nimbus. Becoming assistant director for lunar and planetary programs in 1960, Cortright directed the planning and implementation of such projects as *Mariner, Ranger,* and *Surveyor.* He later became deputy director, then deputy associate administrator for space science and applications. In 1967, he became deputy associate administrator for manned spaceflight, and later director of the Langley Research Center in 1968, a position he held until 1975 when he went to work for private industry, becoming president of Lockheed-California in 1979. See "Cortright, Edward M.," biographical file, NASA Historical Reference Collection.

D

William Gould Dow (1895–1999) earned B.S., E.E., and M.S. degrees from the University of Minnesota. A faculty member of the University of Michigan for over 30 years, he served as chair of the department of electrical engineering. During World War II he led research and development for the continuous wave ultra-high-frequency high power transmitter at the Radio Research Laboratory at Harvard University. Dow published *Fundamentals of Engineering Electronics* (1937) and many other articles about electronics. He was a member of the American Society of Electrical Engineers, the Engineering Society of Detroit, the Cosmos Club (Washington, DC), and a fellow of the American Institute of Electrical Engineers and the Institute of Radio Engineers (now the Institute of Electrical and Electronic Engineers). See *Who's Who in Engineering: 1964* (New York, NY: Lewis Historical Publishing Co., Inc., 1964).

Hugh Latimer Dryden (1898–1965) was director of the National Advisory Committee for Aeronautics (NACA) from 1947 until the creation of the National Aeronautics and Space Administration (NASA) in 1958. He was named deputy administrator of the new aerospace agency, created in response to the *Sputnik* crisis. Before NASA, he was associate director of the National Bureau of Standards, where he had served since 1918 in scientific research. Influenced by Dr. Joseph S. Ames, who for many years was chairperson of NACA and was himself a pioneer in aerodynamics, Dryden undertook a study of fluid dynamics at the Bureau of Standards while continuing his courses at the Johns Hopkins University Graduate School. The university accepted his laboratory work and he received his Ph.D. in mathematics and physics in 1919. He served as the deputy administrator of NASA until his death on December 2, 1965. For further information on Hugh Dryden see Michael Gorn, "Hugh L. Dryden's Career in Aviation and Space" in *Monographs in Aerospace History*, No. 5 (Washington, DC: National Aeronautics and Space Administration, 1996), or Richard K. Smith, *The Hugh L. Dryden Papers, 1898–1965* (Baltimore, MD: The Johns Hopkins University Library, 1974).

E

Dwight D. Eisenhower (1890–1969) was the thirty-fourth president of the United States from 1953 to 1961. A Career U.S. Army officer, during World War II he was the supreme allied commander in Europe. As president, he was deeply interested in the use of space technology for national security purposes and directed that ballistic missiles and reconnaissance satellites be developed on a crash basis. For more information on Eisenhower's space efforts, see Rip Bulkeley, *The Sputniks Crisis and Early United States Space Policy* (Bloomington, IN: Indiana University Press, 1991); R. Cargill Hall, "The Eisenhower Administration and the Cold War: Framing American Astronautics to Serve National Security," *Prologue: Quarterly of the National Archives* 27 (Spring 1995): 59–72; Robert A. Divine, *The Sputnik Challenge: Eisenhower's Response to the Soviet Satellite* (New York, NY: Oxford University Press, 1993).

F

Michael Ference, Jr. (1911–1996) earned a Ph.D. in physics from the University of Chicago in 1936 and worked as a professor at the university from 1937 to 1946. Upon leaving the university, he worked for Signal-Corps Engineering Laboratories until 1953, serving as chief scientist from 1948 to 1951, and technical director from 1951 to 1953. He then became the chief scientist for Ford Motor Company's Scientific Laboratory, and was promoted to executive director in 1959. He was a member of the American Physics Society and chairperson of the Science-Engineering Activity Committee. His memberships also included the Geophysical Research Board, the National Academy of Sciences, and the Advisory Group on Weather Modification. He was the author of *Analytical and Experimental Physics* (1943). See *Who's Who in Engineering: 1964* (New York, NY: Lewis Historical Publishing Co., Inc., 1964).

James C. Fletcher (1919–1991) was the NASA administrator who gained the approval of the Nixon administration on January 5, 1972, to develop the Space Shuttle as the follow-on human spaceflight effort of the agency. He also served as NASA administrator a second time from 1986 to 1989, following the loss of the Space Shuttle *Challenger* on January 28, 1986. Fletcher received an undergraduate degree in physics from Columbia University and a doctorate in physics from the California Institute of Technology. After holding research and teaching positions at Harvard and Princeton Universities, he joined Hughes Aircraft in 1948, and later worked at the Guided Missile Division of the Ramo-Wooldridge Corporation. In 1958, Fletcher co-founded the Space Electronics Corporation in Glendale, California. He was later named systems vice president of the Aerojet General Corporation in Sacramento, California. In 1964, he became president of the University of Utah, a position he held until he was named NASA Administrator in 1971. Dr. Fletcher died at his home in suburban Washington on December 22, 1991. See "Fletcher, James C.," biographical file, NASA Historical Reference Collection.

Gerald Ford (1913–   ) (R-MI) was elected to the U.S. House of Representatives in 1948 and served there until he became vice president in 1973 following the resignation of Spiro Agnew. He was president of the United States from 1974 to 1977, following Richard M. Nixon's resignation. See "Ford, Gerald," biographical file, NASA Historical Reference Collection.

William Alfred Fowler (1911–1995) won the Nobel Prize for Physics in 1983 for his work with Subrahmanyan Chandrasekhar in the development of theories of element generation. He was awarded the Apollo Achievement Award in 1969, and the National Medal of Science, the nation's highest honor for scientific achievement, in 1974. He served on many science advisory boards, including NASA's Space Program Advisory Council from 1971 to 1973, and was a member of the Space Science Board of the National Academy of Sciences from 1970 to 1973 and from 1977 to 1980. Finally, he was the chairperson of the Office of Physical Scientists from 1981 to 1984. Fowler earned a bachelor's degree in engineering at Ohio State University and a Ph.D. from the California Institute of Technology in 1936. Upon earning his doctorate, he began research at the California Institute of Technology as an assistant professor of physics, and in 1982 he was named professor emeritus. See *Who's Who in the World 10th Edition, 1991–1992* (Wilmette, IL: Marquis Who's Who, 1990).

Herbert Friedman (1916–2000) earned his Ph.D. in physics from the Johns Hopkins University in 1940. He conducted his first experiments in rocket astronomy with a V-2 rocket in 1949. He performed hundreds of experiments including having traced the solar cycle variations of x-rays and ultraviolet radiations from the Sun and measured the ultraviolet fluxes of early-type stars. Dr. Friedman received the National Medal of Science, the nation's highest honor for scientific achievement, as well as numerous other awards and merits. His scientific and technical contributions included 50 patents and about 300 published papers. He served on many science advisory committees, including the President's Science Advisory Committee, the General Advisory Committee of the Atomic Energy Commission, and the Space Science Board of the National Academy of Sciences. See "Friedman, Herbert," biographical file, NASA Historical Reference Collection.

G

Riccardo Giacconi (1932–   ) became head of the Hubble Space Telescope Science Institute in 1981 and served through that spacecraft's launch. Previously he served as associate director of the High Energy Astrophysics Division of the Harvard-Smithsonian Center for Astrophysics. A pioneer in the field of x-ray astronomy, he led the team that sent up the first x-ray satellite, *UHURU*, in 1970. See "Giacconi, Riccardo," biographical file, NASA Historical Reference Collection.

T. Keith Glennan (1905–1995) was the first administrator of NASA, formally established on October 1, 1958, under the National Aeronautics and Space Act of 1958. Within a short time after NASA's formal organization, Glennan incorporated several organizations involved in space exploration projects from other federal agencies into NASA to ensure that a viable scientific program of space exploration could be reasonably conducted over the long term. A resident of Reston, Virginia, for twenty years after his retirement, he moved to Mitchellville, Maryland, in the late 1980s. He died in Mitchellville on April 11, 1995. See "Glennan, T. Keith," biographical file, NASA Historical Reference Collection.

M. J. E. Golay (1902–1989) was an accomplished physicist and inventor. The author of over fifty scientific and technological publications, he was the owner of at least fifteen U.S. patents. Dr. Golay received his Ph.D. in physics from the University of Chicago in 1931 and became the developing engineer and later the chief scientist of the Computer Division of Signal Corporation Laboratories. His inventions included the Golay infrared detector, Golay delay line, Golay coils, and Golay chromatographic columns. See *Who's Who in Engineering: 1964* (New York, NY: Lewis Historical Publishing Co., Inc., 1964).

Leo Goldberg (1913–1987) was the director of the Kitt Peak National Observatory from 1971 to 1977. Previously, he served as a professor of astronomy and observatory director at the University of Michigan and Harvard University from 1948 to 1971. A former president of the International Astronomical Union and American Astronomical Society, he received three degrees, including his Ph.D., from Harvard. See "Goldberg, Leo," biographical file, NASA Historical Reference Collection.

Daniel S. Goldin (1940–) initiated a revolution to transform America's aeronautics and space program during his tenure as NASA's longest continually serving administrator. Before coming to NASA, Goldin was vice president and general manager of the TRW Space and Technology Group in Redondo Beach, California. During a twenty-five-year career at TRW, Goldin led projects for America's defense and conceptualized and managed production of advanced communication spacecraft, space technologies, and scientific instruments. He began his career at NASA's Lewis Research Center in Cleveland, Ohio, in 1962, and worked on electric propulsion systems for human interplanetary travel. See "Goldin, Daniel," biographical file, NASA Historical Reference Collection.

Albert Gore (1948–) (D-TN) was the forty-fifth vice president of the United States. Prior to being elected vice president in 1992, he served in the U.S. House of Representatives from 1977 to 1985 and in the U.S. Senate from 1985 to 1993. He graduated with a degree in government, with honors, from Harvard University in 1969, and attended Vanderbilt Law School after serving in the Vietnam War. In 2000, he received the Democratic Party nomination for president of the United States. See "Gore, Albert," biographical file, NASA Historical Reference Collection.

Charles F. Green received his bachelor's and master's degrees from the University of Kansas and, in 1915, joined the University of Illinois as a graduate assistant in mathematics. World War I interrupted his work and he enlisted in the Air Corps, serving overseas as a test pilot. Upon his return, he received his Ph.D. from the University of Illinois and remained on the staff until joining General Electric in Schenectady, New York, in 1929. Dr. Green was among the group of experts sent to Europe early in 1945 to investigate engineering achievements of the Axis powers. When he returned he brought with him information on the Germans' progress in guided missiles and jet aircraft, which he obtained by visiting their military, industrial, and research centers. See "Green, Charles F.," biographical file, NASA Historical Reference Collection.

H

Edmund Halley (1656–1742) was an English astronomer and physicist. He made a number of significant astronomical discoveries, including the well-known comet that bears his name. He also cataloged the stars of the Southern Hemisphere. See David Millar, Ian Millar, John Millar, and Margaret Millar, *The Cambridge Dictionary of Scientists* (Cambridge, England: Cambridge University Press, 1996), and *The Encyclopedia Americana International Edition*, Volume 13 (Danbury, CT: Grolier, Inc., 1996).

Philip Handler (1910–1981) was chairperson of the Department of Biochemistry at the Duke University Medical Center. Dr. Handler served as president of the National Academy of Sciences from 1969 to 1981, where he was a leading spokesman for excellence in American scientific endeavors. In addition to his Academy presidency, Dr. Handler served as a member, and subsequently as vice chairperson and chairperson, of the National Science Board from 1962 to 1970. He was instrumental in the development of the National Science Foundation. See "Handler, Philip," biographical file, NASA Historical Reference Collection.

Harry H. Hess (1906–1969) was one of the ten members of the Lunar Sample Analysis Planning Team researching samples returned to Earth by the Apollo spacecraft. He was predominantly a geologist, serving as presidents of the Mineralogical Society of America and the Geological Society of America. Hess earned his doctorate at Princeton University and became the Blair Professor of Geology. During the Apollo era, he was chairperson of the Space Science Board of the National Academy of Sciences. See "Hess, Harry H.," biographical file, NASA Historical Reference Collection.

Richard A. Horner (1917–) has been associated with aerospace activities throughout his career. He served as a pilot in the U.S. Army Air Forces during World War II, and was director of flight test engineering at Wright Field, Ohio (1944–1945 and 1947–1949). He was promoted to colonel in 1948. Between 1950 and 1955, he was first technical director and then senior engineer for the Air Force Flight Test Center at Muroc, California. In May 1955, Horner became deputy for requirements in the office of the assistant secretary of the Air Force, and in 1957 he became assistant secretary of the Air Force for research and development. In June 1959, he left the Air Force to become NASA associate administrator. He resigned from NASA in July 1960 and became senior vice president of the Northrop Corporation. In 1970, he joined the E. F. Johnson Company as president and chief executive officer. See "Horner, Richard E.," biographical file, NASA Historical Reference Collection.

Edwin P. Hubble (1889–1953) was considered by many people to be the greatest astronomer of the twentieth century. Hubble made a number of key discoveries about the nature of galaxies, such as classifying them into spiral, elliptical, and irregular categories. Perhaps his most famous discovery became known as Hubble's Law and states that all galaxies except those closest to the Milky Way are receding from us and at speeds proportional to their distances from us. NASA's Hubble Space Telescope is named after him. See "Hubble, Edwin P.," biographical file, NASA Historical Reference Collection.

Josef Allen Hynek (1910–1986) contributed much to the world of astrophysics, but he is known best for his work in the study of UFO sightings. Hynek dedicated much of his life to the Air Force, working as a consultant in a special project assessing reports of UFO sightings, bringing a more scientific reputation to the field. In 1960, he became the chairperson of the department of astronomy at Northwestern University and was also the director of its Dearborn Observatory. He retired from that position in 1974 after founding The Center for UFO Studies in Evanston, Illinois, in 1972. Dr. Hynek is also credited for coining the phrase "Close encounters of the third kind," which was used in his 1972 book, *The UFO Experience*, and the movie of the same title. Dr. Hynek received his bachelor's degree and his doctorate from the University of Chicago. See "Hynek, Josef A.," biographical file, NASA Historical Reference Collection.

J

Karl G. Jansky (1905–1950) was a scientist at Bell Telephone Laboratories who discovered celestial radio waves in the early 1930s, founding the field of radio astronomy. See "Jansky, Karl G.," biographical file, NASA Historical Reference Collection.

Lyndon Johnson (1908–1973) (D-TX) was elected to the U.S. House of Representatives in 1937 and served until 1949. He was a U.S. senator from 1949 to 1961, U.S. vice president from 1960 to 1963, and then the thirty-sixth president of the United States from 1963 to 1969. Best known for the social legislation he passed during his presidency and for his escalation of the war in Vietnam, he also was highly instrumental in revising and passing the legislation that created NASA. He showed his support for the U.S. space program as chairperson of the Committee on Aeronautical and Space Sciences and as chairperson of the Preparedness Subcommittee of the Senate Armed Services Committee. He later served as chairperson of the National Aeronautics and Space Council when he was vice president. On his role in support of the space program, see Robert A. Divine, "Lyndon B. Johnson and the Politics of Space," in *The Johnson Years: Vietnam, the Environment, and Science*, Robert A. Divine, ed. (Lawrence, Kansas: University of Kansas Press, 1987): 217–53; and Robert Dallek, "Johnson, Project Apollo, and the Politics of Space Program Planning," unpublished paper delivered at a symposium on "Presidential Leadership, Congress, and the U.S. Space Program," sponsored by NASA and American University, March 25, 1993.

K

Joseph Kaplan (1902–1991) was born in Tapolcza, Hungary, and came to the United States in 1910. He trained as a physicist at the Johns Hopkins University and worked on the faculty of the University of California-Berkeley from 1928 until his retirement in 1970. He directed the university's Institute of Geophysics, later the Institute of Geophysics and Planetary Physics, from the time of its creation in 1944. Kaplan was heavily involved in efforts in the 1950s to launch the first artificial Earth satellite, serving as the chairperson of the U.S. National Committee for the International Geophysical Year, 1953–1963. See "Kaplan, Joseph," biographical file, NASA Historical Reference Collection; Joseph Kaplan, "The Aeronomy Story: A Memoir," in R. Cargill Hall, ed., *Essays on the History of Rocketry and Astronautics: Proceedings of the Third Through the Sixth History Symposia of the International Academy of Astronautics* (Washington, DC: NASA Conference Publication 2014, 1977), 2:423–27; Joseph Kaplan, "The IGY Program," *Proceedings of the IRE*, June 1956, 741–43.

W. W. Kellogg (1917–) was a meteorologist with the Rand Corporation between 1947 and 1959. He has held a senior position with the National Center for Atmospheric Research in Boulder, Colorado, since 1959. See *Who's Who in America, 2000 Edition* (New Providence, NJ: Marquis Who's Who, 1999).

Ernst Henry Krause (1913–) earned his bachelor's, master's, and doctorate degrees from the University of Wisconsin, then served as the associate director of research at the Naval Research Laboratory from 1938 to

1954. After leaving this post, he was the director of research laboratories, Missile Systems Division, at Lockheed Aircraft Corporation until 1955. He then became a member of the board of directors of Aeronutronic Systems, Inc. from 1956 to 1960. He was director of technical staff for the Aeronutronic Division of Ford Motor Company until 1962. After leaving Ford, he was the vice president of the Aerospace Corporation. Dr. Krause also served as a member of the Science Advisory Board in Redlands, California, and earned a Distinguished Civilian Service Award from the United States Navy. See *Who's Who in Science from Antiquity to Present* (Chicago, IL: Marquis Who's Who, Inc., 1968).

L

O. B. "Bill" Lloyd (1916–1990) graduated from Northwestern University's Medill School of Journalism in 1938, then became a staff member for U.S. Senator Lyndon B. Johnson (D-TX). He left this position in 1961 to join NASA as director of public services until retirement in 1979. He was awarded NASA's Exceptional Service Medal in 1969. See "Lloyd, O. B.," biographical file, NASA Historical Reference Collection.

George Harry Ludwig (1927–) is credited as one of three discoverers of the Van Allen radiation belts. He earned his Ph.D. in electrical engineering in 1960 from the University of Iowa and began working at Goddard Space Flight Center in the Fields and Particles Instrumentation Section. After 12 years at Goddard (including service as associate director for data operations), Dr. Ludwig changed careers and began working for the National Oceanic and Atmospheric Administration (NOAA). His employment at NOAA in the National Environmental Satellite Service lasted for eleven years (1972–1983), during which he became the director of Environmental Research Laboratories. He served as assistant to the chief scientist at NASA Headquarters from 1983 to 1984, and was involved in the designing of the Space Station from 1983 to 1992. Additionally, Dr. Ludwig was the principle designer of radiation detection instrumentation for several scientific spacecraft, including *Explorer 1*. He oversaw development and operation for the United States National Environmental Satellite System from 1972 to 1980, and was awarded the Program Administration and Management Award from NOAA in 1977. He was also a Van Allen scholar (1958), a research fellow for the U.S. Steel Foundation (1958–1960), a recipient of NASA's Exceptional Service medal (1969), and a recipient of NASA's Exceptional Science Achievement medal (1984). A life member of the Institute of Electrical and Electronics Engineers, he is also a member of the American Meteorological Society and the American Geophysical Union. See *Who's Who in America, 2000 Edition* (New Providence, NJ: Marquis Who's Who, 1999).

M

Neil H. McElroy (1904–1972) was U.S. secretary of defense from 1957 to 1959. He had previously been president of Procter & Gamble and returned there in December 1959 to become chairman of the board. He served in that position until October 1972, a month before his death. See "McElroy, Niel H.," biographical file, NASA Historical Reference Collection.

Carl Edwin McIlwain (1931–) was on President Johnson's Science Advisory Committee on the Fields and Particles and Anti-Submarine Warfare subcommittees from 1964 to 1967. After receiving his Ph.D. from the State University of Iowa in 1960, he became a member of NASA's Space Science Steering Committee from 1962 to 1966, and, in 1967, he was a recipient of a Guggenheim fellowship. Additionally, Dr. McIlwain is a member of the American Institute of Physics and the American Geophysics Union. He has published works on measurements of charged particles producing bright auroral displays and Van Allen radiation. See *Who's Who in Science from Antiquity to Present* (Chicago, IL: Marquis Who's Who, Inc., 1968).

Barbara Ann Mikulski (1936–) (D-MD) became the first Democratic woman elected to the U.S. Senate in 1987. She was immediately elected chairperson of the Senate Appropriations Subcommittee on Veteran Affairs, Housing and Urban Development, and Independent Agencies, with jurisdiction over NASA. Though her subcommittee covers a wide range of subjects, she is best known in the science community for her defense of NASA during intense budget cuts. Because many Goddard Space Flight Center employees are Maryland residents, Senator Mikulski has fought countless battles for increased funding and against downsizing. She earned the American Astronautical Society's John F. Kennedy Award in 1995. See "Mikulski, Barbara Ann," biographical file, NASA Historical Reference Collection.

Thomas A. (Tim) Mutch (1931–1980) was the NASA associate administrator for space science from 1979 to 1980, when he died in a climbing accident in the Himalayas. Previously a professor of geological sciences at Brown University, he led the *Viking* spacecraft's imaging science team from 1969 to 1977. He earned a Ph.D. in geology from Princeton University in 1960. See "Mutch, Thomas A.," biographical file, NASA Historical Reference Collection.

N

John E. Naugle (1923–) was trained as a physicist at the University of Minnesota and began his career studying cosmic rays by launching balloons to high altitudes. In 1959, he joined NASA's Goddard Space Flight Center in Greenbelt, Maryland, where he developed projects to study the magnetosphere. In 1960, he took charge of NASA's Fields and Particles Research program. He also served as NASA's associate administrator for the Office of Space Science and as the agency's chief scientist before his retirement in 1981. See John E. Naugle, *First Among Equals: The Selection of NASA Space Science Experiments* (Washington, DC: NASA SP-4215, 1991).

Homer Newell (1915–1983) earned his Ph.D. in mathematics at the University of Wisconsin in 1940 and served as a theoretical physicist and mathematician at the Naval Research Laboratory from 1944 to 1958. During part of that period, he was science program coordinator for Project Vanguard and was acting superintendent of the atmosphere and astrophysics division. In 1958, he transferred to NASA to assume responsibility for planning and development of the new agency's space science program. He soon became deputy director of spaceflight programs. In 1961, he assumed directorship of the Office of Space Sciences; in 1963, he became associate administrator for Space Science and Applications. Over the course of his career, he became an internationally known authority in the field of atmospheric and space sciences as well as the author of numerous scientific articles and seven books, including *Beyond the Atmosphere: Early Years of Space Science* (Washington, DC: NASA SP-4211, 1980). He retired from NASA at the end of 1973. See "Newell, Homer," biographical file, NASA Historical Reference Collection.

Oran Nicks (1925–1998) was the deputy director of NASA's Langley Research Center from 1970 to 1980. Prior to this position, he was the deputy associate administrator of the Office of Space Science and Applications from 1968 to 1970 and the associate administrator of the Office of Advanced Research and Technology in 1970. He was also the director of Lunar and Planetary Programs from 1961 to 1968. Upon retiring from NASA, Mr. Nicks was the director of the Space Research Center at Texas A&M University from 1985 until his death in 1998. See "Nicks, Oran," biographical file, NASA Historical Reference Collection.

O

Hugh Odishaw (1916–1984) was assistant to the director of the National Bureau of Standards (1946–1954), served as executive director of the U.S. National Committee for the International Geophysical Year (1954–1965), and then became the executive secretary of the Division of Physical Sciences in the National Academy of Sciences (1966–1972). See "Odishaw, Hugh," biographical file, NASA Historical Reference Collection.

John O'Keefe (1917–) is an astronomer who worked at the Goddard Space Flight Center from 1958 until 1995. Previously he worked for the Army Corps of Engineers doing geodesy for sixteen years. In 1992 he received NASA's Award of Merit. O'Keefe is an expert on tektites—small glassy meteorites. He received his Ph.D. in astronomy from the University of Chicago. See "O'Keefe, John A.," biographical file, NASA Historical Reference Collection.

William J. O'Sullivan (1915–1971) invented the world's first lightweight inflatable satellite, which was used for the first transcontinental telephone call via space. He was awarded a $5,000 NASA grant for his "significant contribution to space science and technology," and awarded the NASA Medal for Exceptional Scientific Achievement in 1961. In addition to being a NASA scientist, he also worked for the National Advisory Committee for Aeronautics. See "O'Sullivan, William J.," biographical file, NASA Historical Reference Collection.

P

Rocco Petrone (1926–) was an instrumental member of the Apollo team. After earning his bachelor's degree at West Point and a master's degree in mechanical engineering from the Massachusetts Institute of Technology, Petrone worked at the Missile Firing Laboratory of the U.S. Army's Guided Missile Development Division at the

Redstone Arsenal in Huntsville, Alabama. Here he developed some of the launch vehicle technology used later in the Apollo launches. In 1959, he joined the NASA team and became Saturn project operator in 1960. Four years later he transferred to the Kennedy Space Center, where he was the director of Plans, Programs and Resources. In 1966, he was promoted to Apollo program manager, and after the success of the lunar landing, he became director of the Apollo Program in 1969. In 1973, he succeeded Dr. Eberhard Rees as director of the Marshall Space Flight Center, and became the third highest-ranking NASA official. He left NASA in 1975 to become president and chief executive officer of the National Center for Resource Recovery. See "Petrone, Rocco," biographical file, NASA Historical Reference Collection.

William Pickering (1910–) obtained his bachelor's and master's degrees in electrical engineering, then a doctorate in physics from Caltech, before becoming a professor of electrical engineering in 1946. In 1944, he organized the electronics efforts at the Jet Propulsion Laboratory (JPL) to support guided missile research and development, and became project manager for Corporal, the first operational missile JPL developed. From 1954 to 1976, he was director of JPL, which developed the first U.S. satellite (*Explorer 1*), the first successful U.S. cislunar space probe (*Pioneer 4*), the *Mariner* flights to Venus and Mars in the early to mid-1960s, the *Ranger* photographic missions to the Moon in 1964–1965, and the *Surveyor* lunar landings of 1966–1967. See "Pickering, William H.," biographical files, NASA Historical Reference Collection.

Richard Porter, an electrical engineer, worked on missile programs with the General Electric Company before working on Earth sciences programs at the National Academy of Sciences. In 1964, he was the Academy's delegate to the Committee on Space Research (COSPAR). He also chaired the Technical Panel for the Earth Satellite Program. See "Assorted Government Officials" biographical file, NASA Historical Reference Collection.

Frank Press (1924–) served as President Carter's science advisor and director of the Office of Science and Technology Policy from 1977 to 1981. Upon leaving this post, he was elected nineteenth president of the National Academy of Sciences. Press earned his Ph.D. in geophysics from Columbia University, and has earned twenty-eight additional honorary doctorates. See "Press, Frank," biographical file, NASA Historical Reference Collection.

William Proxmire (1915–) (D-WI) was a U.S. Senator from Wisconsin who served from 1957 to 1989. He was well known for his "Golden Fleece Awards," which he presented to various federal government agencies for projects that he felt wasted taxpayers' money. See *Biographical Directory of the American Congress, 1774–1996* (Alexandria, VA: CQ Staff Directories, Inc., 1996), and "Proxmire, William," biographical file, NASA Historical Reference Collection.

Ptolemy (87–150 A.D.) was a Greek mathematician who lived in the second century. His conception of the universe as Earth-centered remained until Copernicus' theory was published in the sixteenth century. See "Ptolemy" biographical file, NASA Historical Reference Collection.

Q

Donald Quarles (1894–1959) was a deputy secretary of defense between 1957 and 1959. After World War II, he served as vice president for the Western Electric Company and later at Sandia National Laboratories, but in 1953 he accepted the position of assistant secretary of defense for research and development. He was also secretary of the Air Force between 1955 and 1957. See "Quarles, Donald," biographical file, NASA Historical Reference Collection.

R

Norman F. Ramsey (1915–) is a physicist who shared the Nobel Prize in 1989 for his work on a cesium atomic clock and the hydrogen maser. He received his Ph.D. from Columbia University in 1940, after also studying abroad at Cambridge University. During World War II, he worked on radar systems and on the atomic bomb project at Los Alamos. After working as a professor at Columbia and helping found the Brookhaven National Laboratory, he became a professor at Harvard University in 1947, where he has worked ever since. See Emily J. McMurrary, editor, *Notable Twentieth-Century Scientists*, Volume 3 L-R, (New York, NY: Gale Research, Inc., 1995), and "Ramsey, Norman F.," biographical file, NASA Historical Reference Collection.

Ernest Clark Ray (1930–) began work as an aerospace technologist at the Goddard Space Flight Center in 1965. He received his Ph.D. from the State University of Iowa in 1956, where he became an assistant professor of physics. He was a National Academy of Sciences fellow at Goddard Space Flight Center from 1962 to 1963, and is a member of the American Physics Society and the American Geophysics Union. He has researched and published works in theoretical studies of the motion of cosmic rays trapped in radiation in Van Allen radiation belts. See *Who's Who in Science from Antiquity to Present* (Chicago, IL: Marquis Who's Who, Inc., 1968).

Ronald Reagan (1911–) was elected as the fortieth U.S. President in 1980 and served two consecutive terms from 1981 to 1989. He was in office during the beginning of Space Station Freedom in 1984, and during the *Challenger* tragedy on January 28, 1986. A graduate of Eureka College, he was a radio announcer until 1937, when a screen test won him a contract with Hollywood. Over the following twenty years, he appeared in fifty-three films, was president of the Screen Actors Guild, and was a national spokesman for conservatism. In 1966 he was elected governor of California, paving the road to his nomination as the Republican Party candidate in the 1980 presidential election. See "Reagan, Ronald," biographical file, NASA Historical Reference Collection.

Eberhardt Rechtin (1926–) was one of three engineers to design the digital image transmission system technology that allowed us to receive pictures of Jupiter, Saturn, Uranus, and Neptune during the *Voyager* missions, and the radar technology that allowed mapping of the surface of Venus. He is the founder of the Deep Space Network (DSN), and worked at the Jet Propulsion Laboratory from 1949 to 1967. He was president and CEO of the Aerospace Corporation and assistant secretary of defense for telecommunications under President Nixon from 1972 to 1973. He earned both his B.S. and Ph.D. degrees from the California Institute of Technology. See "Rechtin, Eberhardt," biographical file, NASA Historical Reference Collection.

Bruno B. Rossi (1905–1993) was considered a pioneering figure in the study of high-energy astrophysics, x-ray astronomy, and interplanetary plasma (space physics). Born in Venice, he received a Ph.D. in physics from the University of Bologna in 1927. He left Italy in 1938 for Denmark and England, before coming to the United States and joining Cornell University's faculty in 1940. From 1943 to 1946, he worked at Los Alamos, where the atomic bomb was developed. Early in his career, he developed significant new techniques for observing cosmic rays. With his colleagues, he created a detector aboard the *Explorer 10* satellite, which in 1961 discovered the magnetopause, the edge of the Earth's magnetic field. After its launch in December 1995, NASA renamed its X-Ray Timing Explorer spacecraft in honor of Rossi; the spacecraft is now known as RXTE. See Bruno Rossi obituary, *The New York Times*, November 24, 1993, page D19, and "Rossi, Bruno B.," biographical file, NASA Historical Reference Collection.

S

Robert C. Seamans, Jr. (1918–) was born on October 30, 1918, in Salem, Massachusetts. He attended Lenox School in Lenox, Massachusetts; earned a B.S. degree in engineering at Harvard University in 1939; a M.S. degree in aeronautics at the Massachusetts Institute of Technology (MIT) in 1942; and a doctor of science degree in instrumentation from MIT in 1951. Dr. Seamans also received the following honorary degrees: doctor of science from Rollins College (1962) and from New York University (1967); and doctor of engineering from Norwich Academy (1971), from Notre Dame University (1974), and from Rensselaer Polytechnic Institute (RPI) in 1974. In 1960, Dr. Seamans joined NASA as associate administrator. In 1965, he became deputy administrator, retaining many of the general management-type responsibilities of the associate administrator and also serving as acting administrator. During his years at NASA, he worked closely with the U.S. department of defense in research and engineering programs, and served as co-chair of the Astronautics Coordinating Board. Through these associations, NASA was kept aware of military developments and technical needs of the department of defense and Dr. Seamans was able to advise that agency of NASA activities that had application to national security. For further information on Robert C. Seamans, Jr., see his autobiography, *Aiming at Targets* (NASA SP-4106, 1996).

Abe Silverstein (1908–), who earned a B.S. in mechanical engineering (1929) and an M.E. (1934) from Rose Polytechnic Institute, was a longtime NACA manager. He had worked as an engineer at the Langley Aeronautical Laboratory between 1929 and 1943 and at the Lewis Laboratory (later, Research Center) in a succession of management positions, the last (1961–1970) as director of the Center. When T. Keith Glennan arrived at NASA, Silverstein was on a rotational assignment to the Washington headquarters as director of the office of space flight development (later, space flight programs) from the position of associate director at Lewis, which he had held since 1952. During his first tour at Lewis, he had directed investigations leading to significant improvements in reciprocating and early turbojet engines. At NASA Headquarters, he helped create and direct the efforts leading to the spaceflights of Project Mercury and to establish the technical basis for the Apollo program. As Lewis's director, he oversaw a major expansion of the center and the development of the Centaur launch vehicle. He retired from NASA in 1970 to take a position with Republic Steel Corporation. On the career of Silverstein see, Virginia P. Dawson, *Engines and Innovation: Lewis Laboratory and American Propulsion Technology* (Washington, DC: NASA SP-4306, 1991), and "Silverstein, Abe," biographical file, NASA Historical Reference Collection.

John A. Simpson (1916–) is the founder of the Laboratory for Astrophysics and Space Research at Enrico Fermi Institute for Nuclear Studies. Additionally, he is the Arthur H. Compton Distinguished Service professor emeritus at the University of Chicago, and the Martin Marietta chair in space history at the Smithsonian Institute's National Air and Space Museum. He earned his Ph.D. from New York University in 1943. See "Simpson, John A.," biographical file, NASA Historical Reference Collection.

Tony Spear is a thirty-six-year veteran of the Jet Propulsion Laboratory (JPL) in Pasadena, California. As project manager for the *Mars Pathfinder* mission, he oversaw the mission from its conception to the successful landing in 1997. After the success of *Viking Lander 1* in 1976, he stepped down from his position as project manager and joined the Advanced Deep Space System Development Program (called X2000). Upon joining NASA in 1962, he was an engineer in several positions. In 1974, he was the advanced projects planning manager for the NASA/JPL Deep Space Communications and Spacecraft Tracking Network. Spear was manager of the 1989 *Magellan* mission to map the surface of Venus, manager of the synthetic aperture imaging radar instruments that flew aboard several Space Shuttle missions in the early 1990s, and was an engineer on the 1978 *Seasat* oceanographic satellite mission. Spear earned a B.S. in electrical engineering from Carnegie Mellon University, an M.S. in electrical engineering from the University of Southern California, and an M.S. in mechanical engineering from the University of California-Los Angeles. He retired from JPL in 1998. For more information on Tony Spear, see the Media Relations Office at JPL.

Athelstan Frederick Spilhaus (1911–1998) was born in Cape Town, South Africa, and earned a B.Sc. and D.Sc. from the University of Cape Town before coming to the United States in 1931. He then earned a S.M. from the Massachusetts Institute of Technology (MIT) in 1933, and a D.Sc. from Coe College in 1961. He was a research assistant at MIT from 1934 to 1935, and then became assistant director of technical services for the Union of South African Defense Forces until 1936. In 1947, he served as meteorological advisor for the Union of South African Government. Additionally, he was the U.S. commissioner for the Seattle World's Fair in 1961–1962, the chairperson of the National Fisheries Center and Aquarium Advisory Board for the U.S. department of the interior, and a member of the Advanced Commission for Armed Forces. He is credited with the research and development of meteorological equipment, radar and radio upper wind finding, spherics, and the development of meteorological instruments for measurements from aircraft in flight. His awards included a Decorated Legion of Merit Exceptional Civilian Service Medal from the U.S. Air Force, and a Patriotic Civilian Service Award from the U.S. Army. See *Who's Who in Science from Antiquity to Present* (Chicago, IL: Marquis Who's Who, Inc., 1968).

Lyman Spitzer, Jr. (1914–1997) earned his B.A. and D.Sc. from Yale University (1935, 1958); Ph.D. from Princeton University (1938); D.Sc. from Case Institute of Technology (1960); and his LL.D. from Toledo University (1963). He was an instructor of physics and astronomy and an associate professor of astrophysics at Yale from 1946 to 1947. As a Charles A. Young Professor of Astronomy, he taught at Princeton and became the chairperson of the astrophysical sciences department and director of the observatory in 1947. Dr. Spitzer was the director of Project Matterhorn (1953–1961) and chairperson of the executive committee of the Plasma Physics Laboratory (1961–1966). He was a member of the National Academy of Sciences, American Academy of Arts and Sciences, American Philosophical Society, International Academy of Astronautics, and was the president of the American Astronomical Society. Dr. Spitzer received the Rittenhouse Medal (1957), NASA Medal (1972), Bruce Medal (1973), and Draper Medal (1974). He was the author of *Physics of Fully Ionized Gasses* (New York, NY: Interscience Publishers, 1956), *Diffuse Matter in Space* (New York, NY: Interscience Publishers, 1968), and was the editor of *Physics of Sound in the Sea* (Washington, DC: U.S. Department of the Navy, 1969). Included in his research projects were research on interstellar matter, cosmogony, stellar atmospheres, and plasma physics. He also pioneered research on controlled thermonuclear fusion and in space astronomy. See *Who's Who 1976–77: An Annual Biographical Dictionary* (New York, NY: St. Martin's Press, 1976).

Ernst Stuhlinger (1913–) is a physicist who earned his Ph.D. at the University of Tubingen in 1936, and continued research into cosmic rays and nuclear physics until 1941 while serving as an assistant professor at the Berlin Institute of Technology. He then spent two years as an enlisted man in the German army on the Russian front before being assigned to the rocket development center at Peenemunde, Germany. There he worked principally on guidance and control of rockets. After World War II, he came to the United States as part of Project Paperclip and worked with Wernher von Braun at Fort Bliss, Texas, and then at the Redstone Arsenal in Huntsville, Alabama. Transferred to the Marshall Space Flight Center in 1960, he was director of its space science laboratory from 1960 to 1968 and then its associate director for science from 1968 to 1975, when he retired and became an adjunct professor and senior research scientist with the University of Alabama at Huntsville. He directed early planning for lunar exploration and the Apollo telescope mount, which flew on Skylab and produced a wealth of scientific information about the Sun. He also was responsible for the early planning on the high energy astronomy observatory and contributed to the initial phases of the space telescope project. His work included studies of electric propulsion and of scientific payloads for the Space Shuttle. See "Stuhlinger, Ernst," biographical file, NASA Historical Reference Collection.

T

Eldon Taylor (1929–) served as President Carter's inspector general for NASA from 1979 to 1981. He was the first director of administration for the Virginia Center for Innovative Technology, and assistant director of administration for the National Science Foundation from 1973 to 1979. Mr. Taylor was a Navy civilian from 1949 to 1959 (with time out for military service), and graduated from American University with a B.S. and M.S. in public affairs. He has earned several awards, including the William A. Jump Meritorious Award, NASA's Exceptional Service Award, the Environmental Protection Agency Special Achievement Award, and the National Science Foundation Distinguished Service Award. See "Taylor, Eldon," biographical file, NASA Historical Reference Collection.

Maj. Gen. Holger Toftoy (1903–1967) was a career U.S. Army officer, an expert in ordnance, and was responsible for bringing the German Rocket Team under the leadership of Wernher von Braun to the United States in 1945. He became commander of the Redstone Arsenal in Huntsville, Alabama, in 1954, and worked closely with von Braun's teams in the development of the Redstone and Jupiter missiles. In the aftermath of *Sputnik 1* in 1957, he persuaded the department of defense to allow the launch of the United States' first Earth-orbiting satellite aboard the Jupiter missile and the result was the orbiting of *Explorer 1* on January 31, 1958. He also held a number of other positions in the Army, such as head of the Rocket Research Branch of the Chief of Ordnance in Washington, DC, and commander of the Aberdeen Proving Ground in Maryland. He retired from the Army in 1960 with the rank of major general. See "Maj. Gen. Holger Toftoy Dies; Leader in U.S. Rocket Program," *New York Times*, April 20, 1967, p. 41.

Richard Tousey (1909–1997) received a Ph.D. in optical physics from Harvard University in 1933. He worked as a solar physicist at the Naval Research Laboratory for a number of years, leading the rocket spectroscopy branch of its space science division. Tousey specialized in vacuum ultraviolet spectroscopy and he designed a camera that the Skylab astronauts used to photograph the Sun. See Richard Tousey obituary, *The Washington Post*, April 16, 1997, p. B5, and "Tousey, Richard," biographical file, NASA Historical Reference Collection.

Charles Townes (1915–) was trained in physics at Duke University and specialized in the development of laser and maser technology. He first worked for the Bell Telephone Laboratories and, in 1948, joined the faculty of Columbia University, leaving there in 1961 to move to the Massachusetts Institute of Technology and on to the University of California. For his work on the maser, Townes received the Nobel Prize in 1964. See David E. Newton, "Charles H. Townes," in Emily J. McMurray, ed., *Notable Twentieth-Century Scientists* (New York, NY: Gale Research Inc., 1995):2042–44.

John W. Townsend, Jr. (1924–) was the deputy director of the Goddard Space Center (1965–1968) and director (1987–1990). Additionally, he worked at the Naval Research Laboratory from 1949 to 1958, serving as branch head from 1955 to 1958. He has held several positions in various scientific fields, including deputy administrator of the

Environmental Sciences Services Administration (1968–1970), associate administrator of the National Oceanic and Atmospheric Administration (1970–1977), president of the Fairchild Space and Electronics Company (1977–1982), and president of the Fairchild Space Company (1983–1987). He also has been involved in the International Academy of Astronautics, NASA Advisory Council, National Academy of Engineering, National Research Council Space Applications Board, and the Office of Technology Assessment Advisory Board Panel on International Cooperation and Competition in Civilian Space Activities. He earned his B.A., M.A., and Sc.D. from Williams College. See "Townsend, John W., Jr.," biographical file, NASA Historical Reference Collection.

V

James Van Allen (1914–) is a pathbreaking astrophysicist best known for his work in magnetospheric physics. Van Allen's January 1958 Explorer 1 experiment established the existence of radiation belts—later named for the scientist—that encircled the Earth, representing the opening of a broad research field. Extending outward in the direction of the Sun approximately 40,000 miles, as well as stretching out with a trail away from the Sun to approximately 370,000 miles, the magnetosphere is the area dominated by Earth's strong magnetic field. See James A. Van Allen, *Origins of Magnetospheric Physics* (Washington, DC: Smithsonian Institution Press, 1983); David E. Newton, "James A. Van Allen," in Emily J. McMurray, ed., *Notable Twentieth-Century Scientists* (New York, NY: Gale Research Inc., 1995):2070–72.

Wernher von Braun (1912–1977) was the leader of what has been called the "rocket team," which had developed the German V-2 ballistic missile in World War II. At the conclusion of the war, von Braun and some of his chief assistants—as part of a military operation called Project Paperclip—came to America and were installed at Fort Bliss in El Paso, Texas, to work on rocket development and use the V-2 for high altitude research. They used launch facilities at the nearby White Sands Proving Ground in New Mexico. Later, in 1950, von Braun's team moved to the Redstone Arsenal near Huntsville, Alabama, to concentrate on the development of a new missile for the Army. They built the Army's Jupiter ballistic missile, and before that the Redstone, used by NASA to launch the first Mercury capsules. The story of von Braun and the "rocket team" has been told many times. See, as examples, David H. DeVorkin, *Science With a Vengeance: How the Military Created the US Space Sciences After World War II* (New York, NY: Springer-Verlag, 1992); Frederick I. Ordway III and Mitchell R. Sharpe, *The Rocket Team* (New York, NY: Thomas Y. Crowell, 1979); Erik Bergaust, *Wernher von Braun* (Washington, DC: National Space Institute, 1976).

W

Gerry Wasserburg (1927–) earned his B.S., M.S., and Ph.D. from the University of Chicago, and has taught at the California Institute of Technology, University of Kiel, Harvard University, University of Bern, and the Swiss Federal Technical Institute. Primarily his research is in the fields of geology, geochemistry, and geophysics. His awards and recognitions include NASA's Group Achievement Award (Lunar Sample Analysis Planning Team, 1969), Arthur L. Day Medal (Geological Society of America, 1970), Medal for Distinguished Public Service (NASA, 1972), J. F. Kemp Medal for Distinguished Public Service (Columbia University, 1973), and Leonard Metal (Meteoritical Society, 1975). See "Wasserburg, Gerry," biographical file, NASA Historical Reference Collection.

Alan Waterman (1892–1967) was a prominent physicist who served as director of the National Science Foundation and president of the American Association for the Advancement of Science. He received his Ph.D. from Princeton University. He was the deputy chief and chief scientist in the Office of Naval Research from 1931 to 1948. In 1964, he was sworn in as a consultant to NASA. See "Waterman, Alan," biographical file, NASA Historical Reference Collection.

James Edwin Webb (1906–1992) was the second administrator of the National Aeronautics and Space Administration. Mr. Webb was educated at the University of North Carolina, where he received an A.B. in education 1928. He also studied law at George Washington University and was admitted to the Bar of the District of Columbia in 1936. President Harry S Truman asked Mr. Webb to serve as under secretary of state in the U.S. Department of State. When the Truman administration ended early in 1953, Mr. Webb left Washington for a position in the Kerr-McGee Oil Corporation in Oklahoma. James Webb returned to Washington on February 14, 1961, when he accepted the position of administrator of NASA. Under his direction, the agency undertook one of the most impressive projects in history, the goal of landing an American on the Moon before the end of the decade through the execution of Project Apollo. After retiring from NASA, Mr. Webb remained in Washington, DC, serving on several advisory boards, including as a regent of the Smithsonian Institution. He died on March 27, 1992. For more information about James E. Webb see the 1995 biography published by the Johns Hopkins University Press, issued in the "New Series in NASA History." Written by W. Henry Lambright of Syracuse University, *Powering Apollo: James E. Webb of NASA*, emphasizes the leadership style and method of management Webb brought to complex organizational issues.

Fred L. Whipple (1906–) received his Ph.D. in astronomy from the University of California, Berkeley, and served on the faculty of Harvard University. He was involved in efforts in the early 1950s to expand public interest in the possibility of spaceflight through a series of symposia at the Hayden Planetarium in New York City and articles in *Collier's* magazine. He also was heavily involved in planning for the International Geophysical Year, 1957–1958. As a pathbreaking astronomer he pioneered research on comets. See Raymond E. Bullock, "Fred Lawrence Whipple," in Emily J. McMurray, ed., *Notable Twentieth-Century Scientists* (New York, NY: Gale Research Inc., 1995):2167–70.

Z

Harold Adelbert Zahl (1904–1973) earned his B.A., M.S., and Ph.D. from North Central College in Naperville, Illinois. He was a physicist for the U.S. Army from 1931 to 1966, where he was the director of research of the Electronics Laboratory, director of the Atmospheric Sciences Laboratory, and worked for the U.S. Electronics Command. In addition to authoring *Electrons Away...or Tales of a Government Scientist* (New York, NY: Vantage Press, 1968), he researched and published works regarding verification of wave particle dualism of atoms, and propagation of sound through the ocean, radar, and electron tubes. He developed the infrared detecting cell, tubes used in radar tube (i.e., the Zahl tube), and radar switching tubes. His decorations included the Department of the Army Decoration for Exceptional Civilian Service, Scientific Achievement Award from the Service Clubs of Long Island, Federal Business Association of New York Outstanding Civilian Award, and the Distinguished Alumnus award of North Central College. See *Who's Who in Science from Antiquity to Present* (Chicago, IL: Marquis Who's Who, Inc., 1968).

ISBN 0-16-061774-X
90000
9 780160 617744

Index

A

B

C

D

E

F

G

H

I

J

K

L

M

N

O

P

Q

R

S

T

U

V

W

X

Y

Z

The NASA History Series

Reference Works, NASA SP-4000

Grimwood, James M. *Project Mercury: A Chronology* (NASA SP-4001, 1963).

Grimwood, James M., and Barton C. Hacker, with Peter J. Vorzimmer, *Project Gemini Technology and Operations: A Chronology* (NASA SP-4002, 1969).

Link, Mae Mills. *Space Medicine in Project Mercury* (NASA SP-4003, 1965).

Astronautics and Aeronautics, 1963: Chronology of Science, Technology, and Policy (NASA SP-4004, 1964).

Astronautics and Aeronautics, 1964: Chronology of Science, Technology, and Policy (NASA SP-4005, 1965).

Astronautics and Aeronautics, 1965: Chronology of Science, Technology, and Policy (NASA SP-4006, 1966).

Astronautics and Aeronautics, 1966: Chronology of Science, Technology, and Policy (NASA SP-4007, 1967).

Astronautics and Aeronautics, 1967: Chronology of Science, Technology, and Policy (NASA SP-4008, 1968).

Ertel, Ivan D., and Mary Louise Morse, *The Apollo Spacecraft: A Chronology, Volume I, Through November 7, 1962* (NASA SP-4009, 1969).

Morse, Mary Louise, and Jean Kernahan Bays, *The Apollo Spacecraft: A Chronology, Volume II, November 8, 1962–September 30, 1964* (NASA SP-4009, 1973).

Brooks, Courtney G., and Ivan D. Ertel, *The Apollo Spacecraft: A Chronology, Volume III, October 1, 1964–January 20, 1966* (NASA SP-4009, 1973).

Ertel, Ivan D., and Roland W. Newkirk, with Courtney G. Brooks, *The Apollo Spacecraft: A Chronology, Volume IV, January 21, 1966–July 13, 1974* (NASA SP-4009, 1978).

Astronautics and Aeronautics, 1968: Chronology of Science, Technology, and Policy (NASA SP-4010, 1969).

Newkirk, Roland W., and Ivan D. Ertel, with Courtney G. Brooks, *Skylab: A Chronology* (NASA SP-4011, 1977).

Van Nimmen, Jane, and Bruno, Leonard C., with Rosholt, Robert L. *NASA Historical Data Book, Volume I: NASA Resources, 1958–1968* (NASA SP-4012, 1976, rep. ed. 1988).

Ezell, Linda Neuman. *NASA Historical Data Book, Volume II: Programs and Projects, 1958–1968* (NASA SP-4012, 1988).

Ezell, Linda Neuman. *NASA Historical Data Book, Volume III: Programs and Projects, 1969–1978* (NASA SP-4012, 1988).

Gawdiak, Ihor Y., with Helen Fedor. Compilers. *NASA Historical Data Book, Volume IV: NASA Resources, 1969–1978* (NASA SP-4012, 1994).

Rumerman, Judy A. Compiler. *NASA Historical Data Book, Volume V: NASA Launch Systems, Space Transportation, Human Spaceflight, and Space Science, 1979–1988* (NASA SP-4012, 1999).

Rumerman, Judy A. Compiler. *NASA Historical Data Book, Volume VI: Space Applications, Aeronautics, and Other Topics, 1979–1988* (NASA SP-4012, 1999).

Astronautics and Aeronautics, 1969: Chronology of Science, Technology, and Policy (NASA SP-4014, 1970).

Astronautics and Aeronautics, 1970: Chronology of Science, Technology, and Policy (NASA SP-4015, 1972).

Astronautics and Aeronautics, 1971: Chronology of Science, Technology, and Policy (NASA SP-4016, 1972).

Astronautics and Aeronautics, 1972: Chronology of Science, Technology, and Policy (NASA SP-4017, 1974).

Astronautics and Aeronautics, 1973: Chronology of Science, Technology, and Policy (NASA SP-4018, 1975).

Astronautics and Aeronautics, 1974: Chronology of Science, Technology, and Policy (NASA SP-4019, 1977).

Astronautics and Aeronautics, 1975: Chronology of Science, Technology, and Policy (NASA SP-4020, 1979).

Astronautics and Aeronautics, 1976: Chronology of Science, Technology, and Policy (NASA SP-4021, 1984).

Astronautics and Aeronautics, 1977: Chronology of Science, Technology, and Policy (NASA SP-4022, 1986).

Astronautics and Aeronautics, 1978: Chronology of Science, Technology, and Policy (NASA SP-4023, 1986).

Astronautics and Aeronautics, 1979–1984: Chronology of Science, Technology, and Policy (NASA SP-4024, 1988).

Astronautics and Aeronautics, 1985: Chronology of Science, Technology, and Policy (NASA SP-4025, 1990).

Noordung, Hermann. *The Problem of Space Travel: The Rocket Motor.* Ernst Stuhlinger and J.D. Hunley, with Jennifer Garland. Editors (NASA SP-4026, 1995).

Astronautics and Aeronautics, 1986–1990: A Chronology (NASA SP-4027, 1997).

Gawdiak, Ihor Y. and Shetland, Charles. *Astronautics and Aeronautics, 1990–1995: A Chronology* (NASA SP-2000-4028).

Orloff, Richard W. *Apollo by the Numbers: A Statistical Reference* (NASA SP-2000-4029).

Management Histories, NASA SP-4100

Rosholt, Robert L. *An Administrative History of NASA, 1958–1963* (NASA SP-4101, 1966).

Levine, Arnold S. *Managing NASA in the Apollo Era* (NASA SP-4102, 1982).

Roland, Alex. *Model Research: The National Advisory Committee for Aeronautics, 1915–1958* (NASA SP-4103, 1985).

Fries, Sylvia D. *NASA Engineers and the Age of Apollo* (NASA SP-4104, 1992).

Glennan, T. Keith. *The Birth of NASA: The Diary of T. Keith Glennan.* Hunley, J.D. Editor (NASA SP-4105, 1993).

Seamans, Robert C., Jr. *Aiming at Targets: The Autobiography of Robert C. Seamans, Jr.* (NASA SP-4106, 1996).

Project Histories, NASA SP-4200

Swenson, Loyd S., Jr., James M. Grimwood, and Charles C. Alexander. *This New Ocean: A History of Project Mercury* (NASA SP-4201, 1966; rep. ed. 1998).

Green, Constance McL., and Milton Lomask. *Vanguard: A History* (NASA SP-4202, 1970; rep. ed. Smithsonian Institution Press, 1971).

Hacker, Barton C., and James M. Grimwood. *On Shoulders of Titans: A History of Project Gemini* (NASA SP-4203, 1977).

Benson, Charles D. and William Barnaby Faherty. *Moonport: A History of Apollo Launch Facilities and Operations* (NASA SP-4204, 1978).

Brooks, Courtney G., James M. Grimwood, and Loyd S. Swenson, Jr. *Chariots for Apollo: A History of Manned Lunar Spacecraft* (NASA SP-4205, 1979).

Bilstein, Roger E. *Stages to Saturn: A Technological History of the Apollo/Saturn Launch Vehicles* (NASA SP-4206, 1980, rep. ed. 1997).

SP-4207 not published.

Compton, W. David, and Charles D. Benson. *Living and Working in Space: A History of Skylab* (NASA SP-4208, 1983).

Ezell, Edward Clinton, and Linda Neuman Ezell, *The Partnership: A History of the Apollo-Soyuz Test Project* (NASA SP-4209, 1978).

Hall, R. Cargill. *Lunar Impact: A History of Project Ranger* (NASA SP-4210, 1977).

Newell, Homer E. *Beyond the Atmosphere: Early Years of Space Science* (NASA SP-4211, 1980).

Ezell, Edward Clinton, and Linda Neuman Ezell. *On Mars: Exploration of the Red Planet, 1958–1978* (NASA SP-4212, 1984).

Pitts, John A. *The Human Factor: Biomedicine in the Manned Space Program to 1980* (NASA SP-4213, 1985).

Compton, W. David. *Where No Man Has Gone Before: A History of Apollo Lunar Exploration Missions* (NASA SP-4214, 1989).

Naugle, John E. *First Among Equals: The Selection of NASA Space Science Experiments* (NASA SP-4215, 1991).

Wallace, Lane E. *Airborne Trailblazer: Two Decades with NASA Langley's Boeing 737 Flying Laboratory* (NASA SP-4216, 1994).

Butrica, Andrew J. Editor. *Beyond the Ionosphere: Fifty Years of Satellite Communication* (NASA SP-4217, 1997).

Butrica, Andrew J. *To See the Unseen: A History of Planetary Radar Astronomy* (NASA SP-4218, 1996).

Mack, Pamela E. Editor. *From Engineering Science to Big Science: The NACA and NASA Collier Trophy Research Project Winners* (NASA SP-4219, 1998).

Reed, R. Dale, with Darlene Lister. *Wingless Flight: The Lifting Body Story* (NASA SP-4220, 1997).

Heppenheimer, T.A. *The Space Shuttle Decision: NASA's Search for a Reusable Space Vehicle* (NASA SP-4221, 1999).

Hunley, J.D. Editor. *Toward Mach 2: The Douglas D-558 Program* (NASA SP-4222, 1999).

Swanson, Glen E. Editor. *"Before this Decade is Out..." Personal Reflections on the Apollo Program* (NASA SP-4223, 1999).

Tomayko, James E. *Computers Take Flight: A History of NASA's Pioneering Digital Fly-by-Wire Project* (NASA SP-2000-4224).

Center Histories, NASA SP-4300

Rosenthal, Alfred. *Venture into Space: Early Years of Goddard Space Flight Center* (NASA SP-4301, 1985).

Hartman, Edwin, P. *Adventures in Research: A History of Ames Research Center, 1940–1965* (NASA SP-4302, 1970).

Hallion, Richard P. *On the Frontier: Flight Research at Dryden, 1946–1981* (NASA SP- 4303, 1984).

Muenger, Elizabeth A. *Searching the Horizon: A History of Ames Research Center, 1940–1976* (NASA SP-4304, 1985).

Hansen, James R. *Engineer in Charge: A History of the Langley Aeronautical Laboratory, 1917–1958* (NASA SP-4305, 1987).

Dawson, Virginia P. *Engines and Innovation: Lewis Laboratory and American Propulsion Technology* (NASA SP-4306, 1991).

Dethloff, Henry C. *"Suddenly Tomorrow Came . . .": A History of the Johnson Space Center* (NASA SP-4307, 1993).

Hansen, James R. *Spaceflight Revolution: NASA Langley Research Center From Sputnik to Apollo* (NASA SP-4308, 1995).

Wallace, Lane E. *Flights of Discovery: 50 Years at the NASA Dryden Flight Research Center* (NASA SP-4309, 1996).

Herring, Mack R. *Way Station to Space: A History of the John C. Stennis Space Center* (NASA SP-4310, 1997).

Wallace, Harold D., Jr. *Wallops Station and the Creation of the American Space Program* (NASA SP-4311, 1997).

Wallace, Lane E. *Dreams, Hopes, Realities: NASA's Goddard Space Flight Center, The First Forty Years* (NASA SP-4312, 1999).

Dunar, Andrew J., and Stephen P. Waring. *Power to Explore* (NASA SP-4313, 1999).

Bugos, Glenn E. *Atmosphere of Freedom: Sixty Years at the NASA Ames Research Center* (NASA SP-4314, 2000).

General Histories, NASA SP-4400

Corliss, William R. *NASA Sounding Rockets, 1958–1968: A Historical Summary* (NASA SP-4401, 1971).

Wells, Helen T., Susan H. Whiteley, and Carrie Karegeannes. *Origins of NASA Names* (NASA SP-4402, 1976).

Anderson, Frank W., Jr. *Orders of Magnitude: A History of NACA and NASA, 1915–1980* (NASA SP-4403, 1981).

Sloop, John L. *Liquid Hydrogen as a Propulsion Fuel, 1945–1959* (NASA SP-4404, 1978).

Roland, Alex. *A Spacefaring People: Perspectives on Early Spaceflight* (NASA SP-4405, 1985).

Bilstein, Roger E. *Orders of Magnitude: A History of the NACA and NASA, 1915–1990* (NASA SP-4406, 1989).

Logsdon, John M. Editor. With Linda J. Lear, Jannelle Warren-Findley, Ray A. Williamson, and Dwayne A. Day. *Exploring the Unknown: Selected Documents in the History of the U.S. Civil Space Program, Volume I: Organizing for Exploration* (NASA SP-4407, 1995).

Logsdon, John M. Editor. With Dwayne A. Day and Roger D. Launius. *Exploring the Unknown: Selected Documents in the History of the U.S. Civil Space Program, Volume II: External Relationships* (NASA SP-4407, 1996).

Logsdon, John M. Editor. With Roger D. Launius, David H. Onkst, and Stephen J. Garber. *Exploring the Unknown: Selected Documents in the History of the U.S. Civil Space Program, Volume III: Using Space* (NASA SP-4407, 1998).

Logsdon, John M. Editor. With Ray A. Williamson, Roger D. Launius, Russell J. Acker, Stephen J. Garber, and Jonathan L. Friedman. *Exploring the Unknown: Selected Documents in the History of the U.S. Civil Space Program, Volume IV: Accessing Space* (NASA SP-4407, 1999).

Siddiqi, Asif A. *Challenge to Apollo: The Soviet Union and the Space Race, 1945–1974* (NASA SP-2000-4408).